Medical Image Analysis

THE ELSEVIER AND MICCAI SOCIETY BOOK SERIES

Chairman of the Series Board: Alejandro F. Frangi

Advisory Board

Titles

Balocco, A., et al., Computing and Visualization for Intravascular Imaging and Computer Assisted Stenting, 9780128110188.

Dalca, A.V., et al., Imaging Genetics, 9780128139684.

Depeursinge, A., et al., Biomedical Texture Analysis, 9780128121337.

Munsell, B., et al., Connectomics, 9780128138380.

Pennec, X., et al., Riemannian Geometric Statistics in Medical Image Analysis, 9780128147252.

Trucco, E., et al., Computational Retinal Image Analysis, 9780081028162.

Wu, G., and Sabuncu, M., Machine Learning and Medical Imaging, 9780128040768.

Zhou S.K., Medical Image Recognition, Segmentation and Parsing, 9780128025819.

Zhou, S.K., et al., Deep Learning for Medical Image Analysis, 9780128104088.

Zhou, S.K., et al., Deep Learning for Medical Image Analysis, 2nd Ed., 9780323851244.

Zhou, S.K., et al., Handbook of Medical Image Computing and Computer Assisted Intervention, 9780128161760.

Burgos, N. and Svoboda, D., Biomedical Image Synthesis and Simulation, 9780128243497.

Tian, J., et al., Radiomics and Its Clinical Application, 9780128181010.

Liao, H., et al., Deep Network Design for Medical Image Computing, 9780128243831.

Nguyen, H.V., et al., Meta Learning with Medical Imaging and Health Informatics Applications, 9780323998512.

MICCAI

Medical Image Analysis

Edited by

Alejandro F. Frangi
Department of Computer Science, School of Engineering
Faculty of Science and Engineering, and
Division of Informatics, Imaging and Data Sciences
School of Health Sciences, Faculty of Biology, Medicine,
and Health, The University of Manchester
Manchester, United Kingdom
Katholieke Universiteit Leuven
Departments of Electrical Engineering and
Cardiovascular Sciences
Leuven, Belgium

Jerry L. Prince
The Johns Hopkins University
Image Analysis and Communications Laboratory
Electrical and Computer Engineering Department
Baltimore, MD, United States

Milan Sonka
The University of Iowa
Iowa Institute for Biomedical Imaging
Electrical & Computer Engineering Department
Iowa City, IA, United States

Academic Press is an imprint of Elsevier
125 London Wall, London EC2Y 5AS, United Kingdom
525 B Street, Suite 1650, San Diego, CA 92101, United States
50 Hampshire Street, 5th Floor, Cambridge, MA 02139, United States
The Boulevard, Langford Lane, Kidlington, Oxford OX5 1GB, United Kingdom

ISBN: 978-0-12-813657-7

For information on all Academic Press publications
visit our website at https://www.elsevier.com/books-and-journals

Publisher: Mara Conner
Acquisitions Editor: Tim Pitts
Editorial Project Manager: Maria Elaine D. Desamero
Production Project Manager: Prem Kumar Kaliamoorthi
Cover Designer: Greg Harris

Typeset by VTeX

To my wife Silvia, our children Bernat, Núria, Pau, Inés, Aina, Alex,
Silvia Jr and Leo, and our parents, who continuously inspire me.
To all the teachers, students, and collaborators from whom I learned.

Alejandro F. Frangi

To my family and to all the students, postdocs, and collaborators who have made my career possible.

Jerry L. Prince

To my wife Jitka – for her love, support, and decades of stimulating cross-disciplinary discussions.

Milan Sonka

Section editors

Carlos Alberola-López
University of Valladolid, Valladolid, Spain

Bulat Ibragimov
University of Copenhagen, Copenhagen, Denmark

Ivana Išgum
University of Amsterdam, Amsterdam, The Netherlands

Andreas Maier
Friedrich-Alexander-Universität, Erlangen, Germany

Jon Sporring
University of Copenhagen, Copenhagen, Denmark

Gonzalo Vegas Sánchez-Ferrero
Harvard University, Boston, MA, United States

Contents

PART IV Medical image registration

For additional information on the topics covered in the book, visit the companion site: https://www.elsevier.com/books-and-journals/book-companion/9780128136577

Editors

Alejandro F. Frangi is the Bicentennial Turing Chair in Computational Medicine and Royal Academy of Engineering Chair in Emerging Technologies at The University of Manchester, Manchester, UK, with joint appointments at the Schools of Engineering (Department of Computer Science), Faculty of Science and Engineering, and the School of Health Sciences (Division of Informatics, Imaging and Data Science), Faculty of Biology, Medicine and Health. He is a Turing Fellow of the Alan Turing Institute. He holds an Honorary Chair at KU Leuven in the Departments of Electrical Engineering (ESAT) and Cardiovascular Sciences. He is IEEE Fellow (2014), EAMBES Fellow (2015), SPIE Fellow (2020), MICCAI Fellow (2021), and Royal Academy of Engineering Fellow (2023). The IEEE Engineering in Medicine and Biology Society awarded him the Early Career Award (2006) and Technical Achievement Award (2021). Professor Frangi's primary research interests are in medical image analysis and modeling, emphasizing machine learning (phenomenological models) and computational physiology (mechanistic models). He is an expert in statistical shape modeling, computational anatomy, and image-based computational physiology, delivering novel insights and impact across various imaging modalities and diseases, particularly on cardiovascular MRI, cerebrovascular MRI/CT/3DRA, and musculoskeletal CT/DXA. He is a co-founder of adsilico Ltd., and his work led to products commercialized by GalgoMedical SA. He has published over 285 peer-reviewed papers in scientific journals with over 34,000 citations and has an h-index of 75.

Jerry L. Prince is the William B. Kouwenhoven Professor in the Department of Electrical and Computer Engineering at Johns Hopkins University. He is Director of the Image Analysis and Communications Laboratory (IACL). He also holds joint appointments in the Departments of Radiology and Radiological Science, Biomedical Engineering, Computer Science, Applied Mathematics, and Statistics at Johns Hopkins University. He received a 1993 National Science Foundation Presidential Faculty Fellows Award, was Maryland's 1997 Outstanding Young Engineer, and was awarded the MICCAI Society Enduring Impact Award in 2012. He is an IEEE Fellow, MICCAI Fellow, and AIMBE Fellow. Previously he was an Associate Editor of IEEE Transactions on Image Processing and an Associate Editor of IEEE Transactions on Medical Imaging. He is currently a member of the Editorial Boards of Medical Image Analysis and the Proceedings of the IEEE. He is co-founder of Sonavex, Inc., a biotech company located in Baltimore, Maryland, USA. His current research interests include image processing, computer vision, and machine learning with primary medical imaging applications; he has published over 500 articles on these subjects.

Milan Sonka is Professor of Electrical & Computer Engineering, Biomedical Engineering, Ophthalmology & Visual Sciences, and Radiation Oncology, and Lowell

C. Battershell Chair in Biomedical Imaging, all at the University of Iowa. He served as Chair of the Department of Electrical and Computer Engineering (2008–2014) and as Associate Dean for Research and Graduate Studies (2014–2019). He is a Fellow of IEEE, Fellow of the American Institute of Medical and Biological Engineers (AIMBE), Fellow of the Medical Image Computing and Computer-Aided Intervention Society (MICCAI), and Fellow of the National Academy of Inventors. He is the Founding Co-director of an interdisciplinary Iowa Institute for Biomedical Imaging (2007) and Founding Director of the Iowa Initiative for Artificial Intelligence (2019). He is the author of four editions of an image processing textbook, Image Processing, Analysis, and Machine Vision (1993, 1998, 2008, 2014), editor of one of three volumes of the SPIE Handbook of Medical Imaging (2000), past Editor-in-Chief of "IEEE Transactions on Medical Imaging" (2009–2014), and past editorial board member of the "Medical Image Analysis" journal. His >700 publications were cited more than 42,000 times, and he has an h-index of 80. He cofounded Medical Imaging Applications LLC and VIDA Diagnostics Inc.

Contributors

Santiago Aja-Fernández
Image Processing Lab. (LPI), Universidad de Valladolid, Valladolid, Spain

Carlos Alberola-López
Laboratorio de Procesado de Imagen (LPI), Universidad de Valladolid, Valladolid, Spain

Andre Altmann
Inria Sophia Antipolis, Epione Research Group, UCA, Sophia Antipolis, France

Sophia Bano
Wellcome/EPSRC Centre for Interventional and Surgical Sciences (WEISS), Department of Computer Science, University College London, London, United Kingdom

Vincent Christlein
Pattern Recognition Lab, Friedrich-Alexander-Universität Erlangen-Nürnberg, Erlangen, Germany

Shan Cong
Department of Biostatistics, Epidemiology and Informatics, University of Pennsylvania Perelman School of Medicine, Philadelphia, PA, United States

Sailesh Conjeti
Siemens Healthineers, Erlangen, Germany

Tim Cootes
Division of Informatics, Imaging and Data Sciences, The University of Manchester, Manchester, United Kingdom

Ariel H. Curiale
Applied Chest Imaging Laboratory, Brigham and Women's Hospital - Harvard Medical School, Boston, MA, United States
Medical Physics Department, CONICET - Bariloche Atomic Center, Río Negro, Argentina

Marleen de Bruijne
Department of Computer Science, University of Copenhagen, Copenhagen, Denmark
Department of Radiology and Nuclear Medicine, Erasmus MC - University Medical Center Rotterdam, Rotterdam, the Netherlands

Stefanie Demirci
Computer Aided Medical Procedures, Technical University of Munich, Garching, Germany

Bob D. de Vos
Department of Biomedical Engineering and Physics, Amsterdam UMC, Amsterdam, the Netherlands

James S. Duncan
Division of Bioimaging Sciences, Departments of Radiology & Biomedical Imaging, Biomedical Engineering and Electrical Engineering, Yale University, New Haven, CT, United States
Department of Statistics & Data Science, Yale University, New Haven, CT, United States

Alejandro F. Frangi
Department of Computer Science, School of Engineering, Faculty of Science and Engineering, The University of Manchester, Manchester, United Kingdom
Division of Informatics, Imaging, and Data Sciences, School of Health Sciences, Faculty of Biology, Medicine, and Health, The University of Manchester, Manchester, United Kingdom
Department of Electrical Engineering (ESAT), KU Leuven, Leuven, Belgium
Department of Cardiovascular Sciences, KU Leuven, Leuven, Belgium

Simon Graham
Department of Computer Science, University of Warwick, Coventry, United Kingdom

Mattias Heinrich
Institute of Medical Informatics, Universität zu Lübeck, Lübeck, Germany

Ivana Išgum
Department of Biomedical Engineering and Physics, Amsterdam UMC, Amsterdam, the Netherlands
Informatics Institute, University of Amsterdam, Amsterdam, the Netherlands

Marco Lorenzi
CMIC, University College London, London, United Kingdom

Cheng Lu
Medical Research Institute, Guangdong Provincial People's Hospital (Guangdong Academy of Medical Sciences), Southern Medical University, Guangzhou, China

Anant Madabhushi
Department of Biomedical Engineering, Emory University and Georgia Institute of Technology, Atlanta, GA, United States
Atlanta Veterans Administration Medical Center, Atlanta, GA, United States

Andreas Maier
Pattern Recognition Lab, Friedrich-Alexander-University Erlangen-Nuremberg, Erlangen, Germany

Melissa Martin
Penn Statistics in Imaging and Visualization Center, Department of Biostatistics, Epidemiology, and Informatics, University of Pennsylvania, Philadelphia, PA, United States

Thomas Moreau
Inria, Saclay Île-de-France, Palaiseau, France

Sean Mullan
Iowa Institute for Biomedical Imaging, The University of Iowa, Iowa City, IA, United States

Ipek Oguz
Department of Electrical Engineering and Computer Science, Vanderbilt University, Nashville, TN, United States
Department of Computer Science, Vanderbilt University, Nashville, TN, United States

Nikos Paragios
Digital Vision Center (CVN), CentraleSupélec, Université Paris-Saclay, Paris, France
TheraPanacea, Paris, France

Jens Petersen
Department of Computer Science, University of Copenhagen, Copenhagen, Denmark
Department of Oncology, Rigshospitalet, Copenhagen, Denmark

Jerry L. Prince
Electrical and Computer Engineering, Johns Hopkins University, Baltimore, MD, United States

Nasir Rajpoot
Department of Computer Science, University of Warwick, Coventry, United Kingdom

Gabriel Ramos-Llordén
Athinoula A. Martinos Center for Biomedical Imaging, Department of Radiology, Massachusetts General Hospital, Charlestown, MA, United States
Harvard Medical School, Boston, MA, United States

Nishant Ravikumar
Centre for Computational Imaging and Simulation Technologies in Biomedicine (CISTIB), Schools of Computing and Medicine, University of Leeds, Leeds, United Kingdom

Julia Schnabel
School of Biomedical Engineering and Imaging Sciences, King's College London, London, United Kingdom

Li Shen
Department of Biostatistics, Epidemiology and Informatics, University of Pennsylvania Perelman School of Medicine, Philadelphia, PA, United States

Russell T. Shinohara
Penn Statistics in Imaging and Visualization Center, Department of Biostatistics, Epidemiology, and Informatics, University of Pennsylvania, Philadelphia, PA, United States

Hessam Sokooti
Department of Radiology, Leiden University Medical Center, Leiden, the Netherlands

Milan Sonka
Iowa Institute for Biomedical Imaging, The University of Iowa, Iowa City, IA, United States

Aristeidis Sotiras
Department of Radiology, and Institute for Informatics, School of Medicine, Washington University in St. Louis, St. Louis, MO, United States

Jon Sporring
Department of Computer Science, University of Copenhagen, Copenhagen, Denmark

Lawrence H. Staib
Division of Bioimaging Sciences, Departments of Radiology & Biomedical Imaging, Biomedical Engineering and Electrical Engineering, Yale University, New Haven, CT, United States

Marius Staring
Department of Radiology, Leiden University Medical Center, Leiden, the Netherlands

Danail Stoyanov
Wellcome/EPSRC Centre for Interventional and Surgical Sciences (WEISS), Department of Computer Science, University College London, London, United Kingdom

Mathias Unberath
Department of Computer Science, Johns Hopkins University, Baltimore, MD, United States

Gonzalo Vegas Sánchez-Ferrero
Department of Radiology, Harvard Medical School, Brigham and Women's Hospital, Boston, MA, United States

Demian Wassermann
Inria, Saclay Île-de-France, Palaiseau, France

Yan Xia
Division of Informatics, Imaging, and Data Sciences, School of Health Sciences, Faculty of Biology, Medicine, and Health, The University of Manchester, Manchester, United Kingdom

Paul A. Yushkevich
Penn Image Computing and Science Laboratory, University of Pennsylvania, Philadelphia, PA, United States
Department of Radiology, University of Pennsylvania, Philadelphia, PA, United States

Arezoo Zakeri
Division of Informatics, Imaging, and Data Sciences, School of Health Sciences, Faculty of Biology, Medicine, and Health, The University of Manchester, Manchester, United Kingdom

Honghai Zhang
Iowa Institute for Biomedical Imaging, The University of Iowa, Iowa City, IA, United States

Lichun Zhang
Iowa Institute for Biomedical Imaging, The University of Iowa, Iowa City, IA, United States

Miaomiao Zhang
Departments of Electrical and Computer Engineering and Computer Science, University of Virginia, Charlottesville, VA, United States

Preface

Medical imaging can be traced back at least to the year 1895, when Wilhelm Conrad Roentgen accidentally discovered X-rays. Over the years, a number of 2D, 2D+time, 3D, and 3D+time (or 4D) medical imaging modalities were discovered, developed, perfected, and added to the increasingly useful arsenal of diagnostic, clinical, and research tools. Invariably, medical imaging was used to produce printed or later displayed images for physicians to view and analyze visually. Over the years, the quality of medical imaging techniques, including their spatial and temporal resolution, coverage, and sensitivity, improved many-fold and helped revolutionize medicine.

With the widespread arrival of digital computers, medical images gradually became formed, processed, and displayed digitally, facilitating their computational analysis. Still, making them visually pleasing, denoising, enabling brightness and contrast adjustments, digital zooming, 3D flythroughs, playback, etc., were all first intended to improve their quality and facilitate visual and qualitative analysis. The computational character of medical images, however, also enabled their quantitative analysis, ranging from measurements of anatomical dimensions and quantification of anatomic changes and organ or lesion sizes to the description of tissue properties and as well as quantification of functional properties.

It has long been widely accepted that healthcare should be personalized, because what works for one person may not work for another, based on each person's genetic makeup as well as her or his phenotype. In this respect, quantitative analysis of medical images is in many cases a prerequisite to personalized medicine. Equally important, quantitative image analysis improves diagnostic accuracy, facilitates image-guided surgery and intervention, leads to early disease detection, and even enables patient-specific prediction of treatment outcomes.

Over the past at least 5 decades, hand in hand with the fast improvements of medical image acquisition techniques, tools, and devices, quantitative image analysis experienced similarly dramatic progress. The medical image acquisition and analysis community, which initially consisted of a small group of researchers with mainly signal processing and/or nuclear physics backgrounds, was joined by enthusiastic and enlightened physicians, with the earliest results emerging from in-depth workshops and conferences. Among such conferences, Information Processing in Medical Imaging (IPMI, since 1969), SPIE Medical Imaging (SPIE MedIm, since 1971), Computing in Cardiology (CinC, since 1974), the Medical Image Computing and Computer-Aided Intervention (MICCAI, since 1998), and the IEEE International Symposium on Medical Imaging (ISBI, since 2002) were arguably among the most influential ones. Increasingly, computational image analysis techniques and applications are also presented at medical and especially radiological scientific sessions, for example that of the Radiologic Society of North America (RSNA, since 1915). Technically, methodologically, and computationally influential and to this day most impactful are two highly regarded journals – *IEEE Transactions on Medical Imaging* (*IEEE TMI*, since 1982, impact factor = 11.0) and *Medical Image Analysis* (*MedIA*, since 1996, impact factor = 13.8). Needless to say, a number of other high-quality medical imaging journals exist today, contributing to the development of the community.

The field of medical imaging and medical image analysis has matured considerably. Despite continuous gains in the performance of application-oriented medical image analysis methods and tools, the foundational progress started to slow down about 2 decades ago, only to skyrocket again with the deep learning revolution that has affected each and every aspect of medical imaging starting around 2015. With the advances of computational power, the availability of digital data, and the improved access to data annotations and electronic medical records, powered by progress in machine learning, deep learning initiated a new era of dramatic progress in medical image analysis.

The deep learning-invoked changes to medical image acquisition (e.g., deeply learned tomographic reconstruction) as well as medical image analysis (e.g., segmentation convolutional neural networks) approaches are broad and widespread. Focusing on image analysis, deep neural nets quickly demonstrated performance better than that achieved previously for image segmentation, whole-image classification, and denoising, to name a few application areas. They have, however, equally quickly, revealed a number of associated problems – for example, methods that performed remarkably well when trained and applied to image data from one scanner failed equally remarkably when used on seemingly similar data from another scanner or on images acquired with even slightly different acquisition parameters. Equally problematic was and in many ways still is the overall black-box character of deep learning; if a solution fails, how can we understand why and how can we fix this problem? Yet another burning issue is the typical need for large-scale and highly diverse annotated data that are used for training. This is further associated with the understandable difficulty to include rare cases in sufficient numbers in the training sets, dealing with unbalanced data, and reaching sufficient levels of understanding to the internal decision-making mechanisms of the underlying deep networks.

Tremendous progress has been seen in the past 10 or so years with deep networks becoming architecturally more complex, adding long layer-to-layer connections, exploring and incorporating aspects of decision explainability, self-assessing the level of success, etc. One crucial direction of ongoing research is how to minimize the required training set sizes, how to minimize the effort of expert annotators, how to select and annotate images that are indeed different from those already available in annotated training data, how to train the networks incrementally, and how to improve training efficiency via assisted and smart annotation approaches. While progress has been made, these questions remain subjected to ongoing research. Similarly, following the initial impression of the overarching superiority of deep learning compared to traditional approaches, researchers are seeing more and more often that hybrid combinations of properly selected traditional and deep learning approaches can further enhance the performance. This fact is one of the major motivations that is reflected in the selection of topics of this book, enabling the reader to be well trained in both of these main medical image analysis directions and thus fostering development of novel hybrid approaches benefiting from the power of both traditional and machine learning analysis architectures.

An aspect, notable in its own right, is the democratization of medical image analysis research experienced following the arrival of deep learning methods. In the past, tight collaboration of medical experts with medical image computing experts was absolutely inevitable. With the availability of annotated medical image datasets, facilitated by widespread acknowledgment of the utility of medical image analysis challenges, image data and the so important "truth" (independent standards used for training and performance evaluations) are now available for all, not just a few. There are hundreds of newly emerging teams that focus on the development of new machine learning approaches for a variety of challenge-specific tasks. It certainly allows these new teams to contribute novel ideas and compete on a world stage that was not possible previously. As useful and door-opening these new opportunities are, it is also becoming very clear that not having the coveted tight collaboration with medical experts remains a limitation to achieving the full potential of machine learning in medical image analysis, and we want to caution that a lack of medical expertise and/or sufficient understanding of the underlying medical tasks may reduce the level and speed of achievements in this interdisciplinary field.

What we expect – and hope this book will help accomplish – is that novel combinations of traditional and machine learning-based approaches will continue improving quantitative, personalized medical image analysis. We also expect that these methods will not maintain their black-box character and that they will increasingly and uniformly include aspects of explainability, allow man–machine interactions, exhibit integrated quality control, and thus in many ways help maintain the needed level of "common sense" and perhaps even intuition that is so important in today's medicine.

This book is closing the gap that existed in the medical image analysis literature for some time and responds to the new developments brought on by the deep learning tidal wave. Yet, in agreement with and in support of the paradigm that hybrid traditional and deep learning approaches and their interlinkages bring yet another jump in performance, the book addresses this aspect by offering a comprehensive coverage of the topics and either describes or suggests their novel combinations.

The book is logically organized in eight main parts. The first four chapters present a foundation on which the rest of the book builds. Chapter 1 introduces the most relevant and frequently used medical imaging modalities. Chapter 2 provides a succinct reference and introductory explanation of essential mathematical preliminaries. Chapter 3 offers an overview of regression and classification methods that are critical for understanding the machine learning approaches to medical image analysis. Methods and strategies for data-based estimation and inference are given in Chapter 4.

Medical image data can be represented in many ways and image processing methods must consider underlying data representations. This is the topic of Part II. Chapter 5 introduces image representations that are suitable for signal processing-based image analysis approaches. Methods for image filtering, enhancement, and restoration as well as several noise models are presented in Chapter 6. Considering that medical images are acquired at different scales and resolutions and that different resolutions/scales may be relevant for different tasks, Chapter 7 is devoted to multiresolution and multiscale analysis strategies.

Part III presents approaches to medical image segmentation. Statistical shape models revolutionized medical image analysis when they were introduced and remain of great relevance, as discussed in Chapter 8. Approaches to medical image segmentation that use deformable models are presented in Chapter 9. Ways in which image segmentation tasks can be represented by graphs and solved via graph-cut optimization are presented in Chapter 10.

Part IV of the book is devoted to medical image registration. Chapter 11 introduces classic approaches to registering shapes defined by point sets and surfaces. Graph-based approaches to image registration are the topic of Chapter 12. Chapters 13 and 14 introduce parametric and non-parametric approaches to volumetric image registration together with the enabling mathematical concepts. Image mosaicking and associated image-based geometric transformations are covered in Chapter 15.

Machine-learning approaches to medical image analysis are a topic of Part V. Chapter 16 presents an overview of deep learning fundamentals. Chapter 17 provides theoretical foundations of deep learning approaches for vision and representation learning. The following Chapter 18 builds on the presented concepts and discusses deep learning specifically tailored for medical image segmentation. Novel machine learning methods for image registration are a focus of Chapter 19.

Several advanced topics are presented in Part VI. An overview of techniques for analysis of motion and object/organ deformation is given in Chapter 20. Chapter 21 describes strategies for imaging-based genetics.

The need for large-scale image databases is widely recognized and is the focus of Part VII. Chapter 22 discusses methods for the analysis of large-size microscopic histology images. Chapter 23 presents methods for image retrieval from large-scale databases.

Each and every automated medical image analysis approach must be thoroughly validated prior to its clinical or clinical research deployment. In the concluding Part VIII, Chapter 24 discusses methods and indices used for quantitative assessment of segmentation and registration performance.

This is indeed a long book with a number of diverse topics. As comprehensive as it is, this or any other book can never offer a complete coverage of the field. First, the number of available pages is always limited and included topics as well as the depth of coverage need to be determined, frequently in a process of elimination. Second, the medical image analysis field is very much alive and well – which means that new tasks, approaches, and solutions are appearing every day. Third, there are still way too many unsolved problems to tackle and it will undoubtedly take substantial, enduring interdisciplinary research effort that will continue being published in journals, presented at conferences and – yes – summarized in new books. It is our desire to help prepare a new generation of medical image analysis researchers, developers, and users that motivated us to bring this new book to you.

Alejandro F. Frangi
Jerry L. Prince
Milan Sonka
September 2023

Nomenclature

Nomenclature

$\alpha, \beta, \gamma, \omega_1, \omega_2$	Constant scalar values
a, b, c	Constant scalar values
x, y, z, u, v, w	Scalar values, typically used for continuous coordinates
i, j, m, n, k_1, k_2	Scalar values, typically used for discrete coordinates, summation indices, etc...
$\boldsymbol{x}, \boldsymbol{y}, \boldsymbol{z}$	Column deterministic vectors, typically used as variables
$\boldsymbol{\alpha}, \boldsymbol{\beta}$ ...	Constant column vectors
A, B, C ...	Set names
$\mathbf{A}, \mathbf{B}, \mathbf{C}, \mathbf{\Sigma}$	Matrices
$[\boldsymbol{x}]_k$	k-th component of vector $\boldsymbol{x}$ if expressed with vector notation; x_k is a natural way to denote this component
$\boldsymbol{x}_k$	k-th vector of a vector set
$[\mathbf{A}]_{i,j}$	Element on i-th row and j-th column of matrix $\mathbf{A}$
$\mathbf{x}, \mathbf{y}, \mathbf{z}$	Scalar random variables
$\mathbf{x}, \mathbf{y}, \mathbf{z}$	Vector random variables
$\mathbf{X}$	Matrix of vector random variables; each column is a random vector
f, g, h	Scalar functions
$\boldsymbol{f}, \boldsymbol{g}, \boldsymbol{h}$	Vector functions
$I(\boldsymbol{x}), J(\boldsymbol{x})$	(Scalar) image value at point $\boldsymbol{x}$
$\mathbb{E}$	Expectation operator
$\boldsymbol{t}(\boldsymbol{x})$	Spatial (vector) transformation at point $\boldsymbol{x}$
$\mathcal{F}(\boldsymbol{u})$	Fourier transform at location $\boldsymbol{u}$
$\mathcal{J}(\boldsymbol{x})$	Jacobian matrix at point $\boldsymbol{x}$

Frequent operations

x^*	Conjugate
$\boldsymbol{x}^T$	Transpose
$\boldsymbol{x}^H$	Conjugate transpose
$\boldsymbol{x} \cdot \boldsymbol{y}$	Dot product if used in isolation
$\boldsymbol{x}^T \boldsymbol{y}$	Dot product if expressed within an algebraic derivation (of H if complex)
$f * g$	Convolution
$f \circledast g$	Circular convolution
$< f, g >$	Inner product
$\boldsymbol{x} \odot \boldsymbol{y}$	Hadamard (pointwise) product between two vectors (or matrices)
$\boldsymbol{x} \otimes \boldsymbol{y}$	Kronecker product
$\|\boldsymbol{x}\|^2$	$L2$ norm
$\|\boldsymbol{x}\|$	Modulus of vector $\boldsymbol{x}$
$\|\mathbf{A}\|_F^2$	Frobenius norm of matrix $\mathbf{A}$
$\|\boldsymbol{x}\|_{\ell_p}$	Lp norm

$\arg\min_{\boldsymbol{x}}$	Value of $\boldsymbol{x}$ that minimizes ...
$\arg\max_{\boldsymbol{x}}$	Value of $\boldsymbol{x}$ that maximizes ...
$A \cup B$	Union of sets A and B
$\bigcup_{i=1}^{N} A_i$	Union of sets A_i through A_N
$A \cap B$	Intersection between sets A and B
$\bigcap_{i=1}^{N} A_i$	Intersection of sets A_i through A_N
$A \setminus B$	Difference between sets A and B
$\overline{A}$	Complement of set A
$\|A\|$	Cardinality of set A
$\frac{df(x)}{dx}$	First-order derivative of function f with respect to x
$f'(x)$ **or** f'	Shorthand for first-order derivative of function f with respect to x
$\frac{d^2 f(x)}{dx^2}$	Second-order derivative of function f with respect to x
$f''(x)$ **or** f''	Shorthand for second-order derivative of function f with respect to x
$\frac{d^n f(x)}{dx^n}$	nth-order derivative of function f with respect to x
$f^{('n)}(x)$ **or** $f^{('n)}$	Shorthand for nth-order derivative of function f with respect to x
$\frac{\partial \phi(x,y)}{dy}$	(First-order) partial derivative of function ϕ with respect to y
$\phi_y(x, y)$ **or** ϕ_y	Shorthand of partial derivative of function ϕ with respect to variable y
$\nabla f(\boldsymbol{x})$	Gradient of function f with respect to vector variable $\boldsymbol{x}$; column vector
$[\mathbf{H}(\boldsymbol{y}_0)]_{ij} = \frac{\partial^2 f(\boldsymbol{y})}{\partial y_i \partial y_j}\Big\|_{\boldsymbol{y}=\boldsymbol{y}_0}$	Hessian matrix of function $f(\boldsymbol{y})$ evaluated at point $\boldsymbol{y}_0$
$\Delta f(\boldsymbol{x})$	Laplacian of function $f(\boldsymbol{x})$
$\mathbf{div} \boldsymbol{f} = \nabla \cdot \boldsymbol{f}(\boldsymbol{x})$	Divergence of function $\boldsymbol{f}(\boldsymbol{x})$

Symbols (or notable functions)

$j = \sqrt{-1}$	Imaginary unit
$\delta(x), \delta(\boldsymbol{x})$	Dirac's delta $\boldsymbol{x} \in \mathbb{R}^d$
$\delta[m], \delta[m, n]$	Kronecker's delta ($m, n \in \mathbb{Z}$)
$H(x)$	Heaviside step function
$\Pi(x)$	Rectangular function

Acknowledgments

This is a comprehensive book and, as such, is the result of a collaboration among many medical imaging scholars, containing pedagogical insights and research contributions from many laboratories located in world-leading universities around the globe. Indeed, this book is the outcome of a true team effort and represents a consensus on the topics bringing essential skills from our field and having a lasting impact.

The book content and its many chapters resulted from in-depth discussions of author teams, colleagues, and students, and were influenced by attendees of international conferences—of which MICCAI (Medical Image Computing and Computer-Aided Intervention) not only played a primary role but is also the driving force of the Elsevier and MICCAI Society book series. Thank you all for directly or indirectly contributing to the book-writing effort and thus making this book possible.

We would especially like to acknowledge the role of the book's section editors—Carlos Alberola-López, Bulat Ibragimov, Ivana Išgum, Andreas Maier, Jon Sporring, and Gonzalo Vegas Sánchez-Ferrero—who carried the substantial responsibility of working closely with the chapter authors to deliver well-focused chapters which form the cohesive sections of this book. Similarly, each and every one of the chapter authors is to be thanked for working with the entire book-writing team to meet deadlines, collaborate with the section editors and the book editors, and thereby make the entire book a comprehensive and unified text rather than a set of individual topical chapters. We are grateful for the support that all the home institutions provided to our authors, helping the book to be born out of the many active research labs and diverse research projects that are continuously taking place within the scholarly environments of our community. Our cadre of authors, section editors and editors all bring substantial pedagogical experience in undergraduate and graduate courses that is directly reflected in these chapters.

The support we, as the book editors, received from the University of Sheffield (AFF), the University of Leeds (AFF), Johns Hopkins University (JLP), the University of Iowa (MS), and the editorial team at Elsevier made the entire project possible, and we are very grateful for that. From Elsevier, we want to mention, especially, Tim Pitts (Senior Acquisitions Editor), Maria Elaine D. Desamero (Editorial Project Manager), Prem K. Kaliamoorthi (Production Project Manager), Dinesh Natarajan (Copyrights Specialist) and Goutham Sampathkumar (Contracts Coordinator). Their trust, guidance and nudging (when necessary) have been essential for successfully completing this complex endeavor.

We thank the many undergraduate and graduate students who will hopefully read these pages and the faculty who will guide them through the content and assess the material. We would love to hear your feedback and receive constructive critiques on areas where the content could be expanded or may benefit from future revisions. We hope that the next editions of the book will expand and further consolidate into a broader pedagogical consensus on the contents and skills needed in our community.

We also hope that future editions will enable other key colleagues in our field to contribute their insights.

Lastly and perhaps most importantly, we acknowledge the entire medical image analysis community—past, present, and future members. To past and present members, we thank you for the support and encouragement to take on this book project, which has been both challenging and rewarding for all involved. To future members, we thank you for selecting this book as a vehicle to help you learn about medical image analysis and to generate new ideas, conduct new research, complete new relevant projects, and develop healthcare-oriented applications, all of which are critical to the continued progress of the medical imaging discipline. It will be our greatest reward to witness the advancement of our beloved field of medical image analysis as a result of our writing and editorial efforts.

Alejandro F. Frangi
Jerry L. Prince
Milan Sonka
September 2023

PART I

Introductory topics

CHAPTER

Medical imaging modalities

1

Mathias Unberath[a] and Andreas Maier[b]

[a]*Department of Computer Science, Johns Hopkins University, Baltimore, MD, United States*

[b]*Pattern Recognition Lab, Friedrich-Alexander-University Erlangen-Nuremberg, Erlangen, Germany*

Learning points

- Common medical imaging modalities
- Physics of imaging and imaging contrast mechanisms
- Exemplar clinical workflows and associated imaging modalities
- Modalities of importance in diagnosis and/or treatment

1.1 Introduction

In medical imaging, many different physical mechanisms are used to generate views of the body. Each of these so-called *modalities* is created for a specific diagnostic or therapeutic purpose. As imaging is driven by its utility, the physical effect that is best suited to achieve each diagnostic or therapeutic task is selected. As a result, diagnostic and interventional imaging is continuously evolving to optimally cover all aspects relevant to clinical diagnosis and treatment workflows.

From a high-level perspective, imaging can be summarized as follows: A pulse of energy is introduced in the system to be imaged and the response of the system is measured in a spatially and temporally resolved manner. This view of image formation reveals several key properties of medical images that determine their spatial, temporal, and *functional* resolution: First, the type of energy injected into the system (e.g., electromagnetic radiation) will determine the physical effects governing energy–tissue interaction, and thus the *contrast mechanism*. Consequently, the choice of energy source will alter the functional window into the human body and thus depend on the clinical task. Second, the temporal behavior of energy injection and interaction is highly diverse, ranging from quasi-instantaneous as for high-brilliance X-ray sources to several minutes as in radioactive decay. These properties will determine the temporal resolution of an imaging modality. Finally, the geometric arrangement of energy source and detector array will determine the imaging geometry (e.g., pinhole camera model) setting theoretical upper bounds on the ideally achievable spatial resolution.

Medical Image Analysis. https://doi.org/10.1016/B978-0-12-813657-7.00013-3

Contingent on the above properties, medical images of a particular modality will exhibit specific characteristics implying two conclusions: First, modality-specific image processing approaches become necessary to fully leverage a modality's potential; and second, multiple modalities are usually necessary in the clinical diagnosis and treatment workflow to optimally gain access to the required information.

In the remainder of this chapter, we will first discuss metrics to quantify and compare image quality to determine their clinical utility. Next, we briefly introduce common medical imaging modalities and reason about their core properties. Finally, we discuss how this panel of modalities is combined to derive clinically useful diagnostic and interventional workflows that optimally benefit the patient.

1.2 Image quality

This section briefly introduces concepts for objective image quality assessment, which are important to quantify the performance of any imaging system and investigate how image quality changes as components of the system are modified. For a more detailed discussion of related topics we refer the reader to [1] and [2].

1.2.1 Resolution and noise

Several aspects during image formation affect the spatial resolution of an imaging system. The fundamental lower bound on the spatial resolution that can be achieved with a particular configuration is determined by the Nyquist–Shannon sampling theorem, stating that only signals with f_s can be expressed without loss of actual information, where $2 \cdot f_s$ is the sampling rate, i.e., the inverse pixel size in case of images. In real systems, the resolution is typically limited due to imperfections and signal corruption during the image formation process, such as scattering or noise, rather than the pixel size. Fig. 1.1 shows this effect using a bar pattern. As a consequence, system response functions are used to describe the resolving power of an imaging system. The most common variants used are the point spread function (PSF) and the modulation transfer function (MTF). As depicted in Fig. 1.2, the PSF describes the response of an imaging modality to an infinitely small point source. Due to corruption processes during image formation, the point source will be blurred, the degree of which can be quantified representing the quality of an imaging system. While the PSF evaluates the system's response to an input that is singular in the spatial domain, the MTF assesses the response to a signal in the frequency domain, i.e., sinusoidal patterns. It is computed as the ratio of amplitudes at each sinusoidal frequency, and hence, in an idealistic scenario normalized to 100%. Consequently, the ideal MTF is unity for a perfect signal transfer and will decay to zero at the Nyquist–Shannon limit. In real systems, the resolution gradually decays towards the frequency limit, as shown in Fig. 1.2. Experimentally, both the PSF and MTF require prospective imaging of a dedicated test object, a so-called phantom, to quantify image quality. However, this approach may not always be feasible, suggesting the need for methods

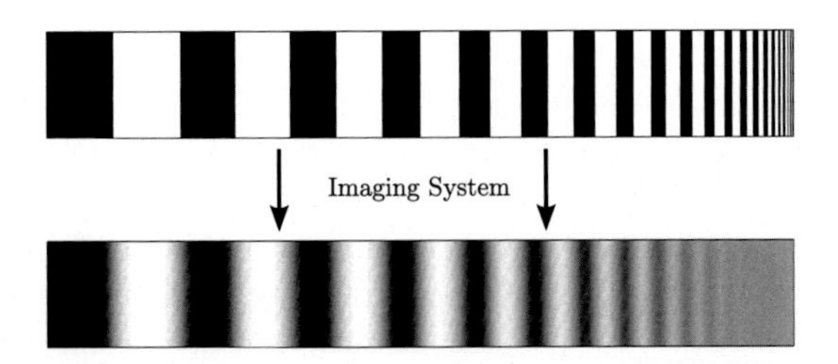

FIGURE 1.1

A bar pattern allows to visualize the loss of resolution that is introduced by an imaging system. The system blur reduces the contrast, which results in a loss of high spatial frequencies towards the right side of the bar pattern. Image reprinted from [3] with permission.

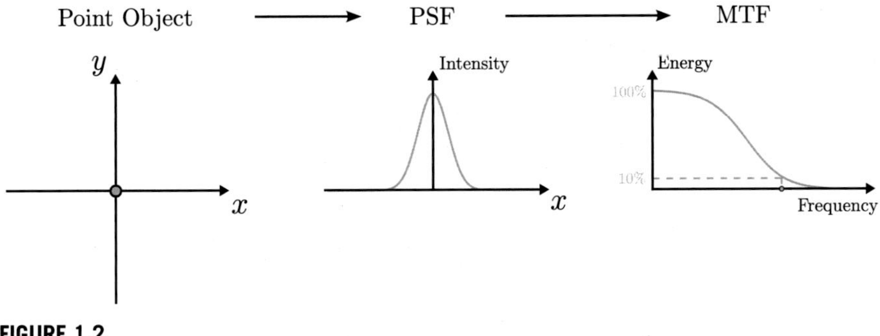

FIGURE 1.2

An ideal point object is used to measure the point spread function (PSF) of an imaging system. Its normalized Fourier transform is called the modulation transfer function (MTF). It measures the magnitude of each individual frequency. Image reprinted from [3] with permission.

that solely rely on the acquired image. One such metric is the signal-to-noise ratio (SNR), which is defined as ratio of the average signal value to its standard deviation computed within a homogeneous region of the image. Yet, the SNR does not fully characterize noise since it does not account for its texture, i.e., the spatial frequency distribution of noise. The noise power spectrum (NPS) describes both the magnitude and spatial frequency characteristics of image noise, and thus should be preferred when comparing and evaluating new imaging technologies. Unfortunately, its computation is not necessarily straightforward, since noiseless images usually cannot be obtained.

It is worth mentioning that (1) the list of aforementioned metrics to quantify image quality and resolving power is not exhaustive and (2) all these metrics are based on linear system theory, i.e., they assume linearity and shift-invariance. In fact, for most imaging systems, these quantities are not constant across the image; a popular example is the location-dependence of the PSF in wide field-of-view microscopy due to lens aberrations far from the optical axis.

1.2.2 Comparing image appearance

The quantitative metrics on image quality described above, such as the MTF or NPS, enable the comparison of image quality; however, they rely on metrics that are derived from the image to be evaluated only. If a template image with ideal or desirable characteristics is available, then relational metrics can be used that compare image appearance, e.g., before and after the application of a novel image processing algorithm. One can distinguish two classes of such measures, namely mathematically defined metrics and methods that consider characteristics of the human visual system to incorporate perceptual quality measures (see Section 1.2.3). Mathematically defined metrics are attractive since they exhibit low computational complexity and are independent of viewing conditions.

The most famous representative of mathematically defined measures is the root mean square error (RMSE), which, as the name suggests, computes the square root of the squared difference in image intensities averaged over the whole image. Since the RMSE follows a quadratic scoring rule, large errors are heavily penalized, suggesting that applying RMSE is useful if large errors are particularly undesirable. However, if this is not the case, the mean absolute error (MAE), which describes the average deviation, is more appropriate, since all individual differences receive equal weight. It is worth mentioning that both the RMSE and the MAE exhibit two fundamental limitations: First, they rely on intensity correspondence, i.e., pixel intensities across images must have the same magnitude in the ideal case, suggesting that cross-modality comparisons are difficult. Second, the magnitude of the metric itself is dependent on the value range and scaling of image intensities, suggesting that RMSE and MAD values are not easily interpretable. As an alternative, metrics that consider intensity correlations rather than deviations can be used. The normalized cross-correlation (NCC) measures the degree of linear correlation between the intensity values of two images with a dynamic range that is bound by $[-1, 1]$, and therefore is invariant to intensity scaling. Yet, even if two signals are linearly related there may be further distortions to luminance and contrast that can be assessed using the structural similarity index measure (SSIM), a perception-based technique. Image regions that are spatially close are assumed to have strong interdependencies that, if lost, will be perceived as a degradation of image quality. The SSIM is composed of three distinct terms that measure luminance, contrast, and structure. The universal quality index (UQI), SSIM's well-known predecessor, is a special case of SSIM that gives rise to unstable results if the intensity means or standard deviations are close to zero.

1.2.3 Task-based assessment

Evaluating and comparing image quality using the above metrics yields quantitative results that are meaningful overall; however, there are cases where straightforward computation-based analysis does not accurately reflect the clinical utility. A prominent example that is fiercely discussed is regularization as it is frequently used in image reconstruction problems in presence of insufficient measurements. In such cases, it is possible to obtain strikingly high SNR values with close to perfect SSIM, correlation coefficient, and RMSE to an image computed from sufficient samples. Unfortunately, these metrics fail to reflect that in most diagnostic settings overall image quality is secondary to detectability of pathology. Visual cues hinting at pathology, however, are usually small relative to the image content and as a consequence do not affect the aforementioned image quality metrics, setting the stage for task-based assessment of image quality, an active area of research on its own.

The traditional approach to task-based assessment considers reader studies. To this end, medical experts are recruited and presented with large numbers of images that have been processed with the method to be evaluated. In forced choice experiments, these experts must then rate images that may contain a pathologic cue. Every reader performing this task will exhibit a distinct true positive rate (TPR) and false positive rate (FPR), which largely depend on a reader's estimate of disease prevalence from their training and their aggressiveness in decision making. The TPR/FPR pair of a reader corresponds to a single operating point in receiver operating characteristic (ROC) space, suggesting that many readers must be recruited to populate the ROC diagram for medical imaging system evaluation and finally to draw representative conclusions on the system's performance. It is evident that this endeavor is time consuming and costly, and consequently, model observers (MOs) appear as promising surrogate for human readers in medical image quality assessment [4]. MOs are designed to detect a signal against a noisy background, rendering the problem as a two-exclusive-hypotheses test (signal present/absent). Many formulations for MOs exist, ranging from tasks where the signal is known exactly to tasks where the signal is only known statistically. Since the decision of MOs for either hypothesis depends on an adjustable threshold, evaluating the TPR and FPR for an MO at different values of this threshold populates the ROC space, and thus mimics human readers with different decision making characteristics.

1.3 Modalities and contrast mechanisms

This section provides an overview on typical imaging modalities and their respective contrast mechanisms. Understanding the contrast mechanism of any imaging modality offers insights into what anatomical structures can be imaged with high quality, which determines the clinical use cases for the modality. For each category, we present one table that captures the cornerstones of each modality. For a more extensive description, we refer the reader to [3].

1.3.1 X-ray transmission imaging

After the discovery of X-rays, transmission imaging was immediately found to be of diagnostic value. Today's most relevant X-ray-based modalities are reported in Table 1.1. All of them use high-energy X-rays that are able to penetrate soft and hard tissues in the body. Emitted from an X-ray source, the photons pass the body and are collected on the other side by a detector. Using the energy loss from source

Table 1.1 X-ray transmission imaging.

Name: Digital radiography **Energy range:** 50–150 kVp **Weight:** 200 kg **Max resolution:** 144 µm (@Detector) **Established in:** 1895		
Name: Mammography **Energy range:** 20–50 kVp **Weight:** 350 kg (stand + generator) **Max resolution:** 85 µm (@Detector) **Established in:** 1927		
Name: C-arm angiography **Energy range:** 70–120 kVp **Weight:** 665 kg (floor stand) **Max resolution:** 150 µm (@Detector) **Established in:** 1954		
Name: Computed tomography **Energy range:** 70–150 kVp **Weight:** 2.5 tons (gantry + couch) **Max resolution:** 450 µm (@Detector) **Established in:** 1971		
Name: X-ray phase contrast and dark-field **Energy range:** 40–100 kVp **Weight:** N.A. – research installation only **Max resolution:** 150 µm (@Detector) **Established in:** 2006		

Images from [3] and by courtesy of Siemens Healthcare AG.

to detector, contrast is generated. The more energy is lost, the higher the contrast. A typical application of digital radiography are chest X-ray images, as shown in Table 1.1. Here, we can clearly see the contrast difference between the lungs and the other soft tissues in the chest. Today, digital radiography is a standard modality to diagnose a wide range of applications from lung diseases to orthopedic applications.

Mammography is a modality that is specialized to the female breast. It is used in particular for breast cancer screening and diagnosis as it is cost-effective and widely available. The most common technique is projection imaging, which allows the detection of masses and dense tissues which have been shown to be risk factors towards development of breast cancer. With respect to the other modalities, also mammographic imaging allows the reconstruction of volumetric data from a series of projections. As the rotational range is limited, only incomplete data are sampled that are used to reconstruct slice images, which is commonly referred to as tomosynthesis. Given the same X-ray dose, tomosynthesis was shown to be superior for diagnosis of precursors of cancer. Yet, breast projection imaging is still predominantly used as the image data can be read much faster by experts than the volumetric data. Note that this is a significant cost factor in this high-throughput screening task.

C-arm systems also employ X-rays, but they are built with the aim of providing real-time guidance during minimally invasive treatments. In such a treatment a small wire that is referred to as catheter is inserted into the body through a blood vessel that is connected to the area of interest. Using real-time X-ray guidance, the catheter is navigated towards the actual target where the actual treatment is performed. Applications of C-arm angiography systems predominantly aim at treatment of the heart and the brain. To improve the visibility of vessels, iodine-based contrast agent is used such that the path towards the structure of interest can be visualized. C-arm systems are typically also equipped with a 3D option that allows to compute slice images from a rotational scan. In contrast to tomosynthesis, almost complete data can be acquired, which allows the reconstruction of high-resolution high-contrast images.

Computed tomography (CT) was a game changer when it was introduced as a medical product in 1974 as it allowed to virtually compute slice images of patients. Doing so, skulls could be virtually opened and its inner workings could be explored without actually having to perform surgery on a patient. As such the technology was quickly adopted in brain analysis to detect malformations and tumors already at early stages. Since then, CT has been continuously developed further and it is a routine diagnostic tool in today's clinical practice. With fast gantry rotation speeds of up to 4 Hz, the beating heart can be imaged with frame rates close to 75 ms. Full body scans can be performed in the range of about 20 seconds. As such CT is routinely used for emergency diagnostics.

The last X-ray-based modality that we present here is X-ray phase contrast imaging. It is still at an early state of development and is not routinely used in clinical practice. However, being able to not just measure X-ray absorption, but also X-ray phase and microangle scattering in the so-called darkfield signal adds further diagnostic value to any X-ray-based imaging modality. As demonstrated in Table 1.1, complicated measurement setups have to be designed in order to use this technology with clinically available X-ray sources and detectors. With these additional efforts, one is able to obtain this complementary information for each projection image. As the darkfield signal is dependent on the orientation of microscopic fibers, 3D darkfield imaging is able to reconstruct their orientation for each element of a volume.

1.3.2 Molecular imaging

Molecular imaging makes use of radioactive tracers that are used to explore the patient's metabolism. The tracers typically employ molecules that are known to be relevant for specific functional processes in the body. A well-known tracer that is quite commonly employed is fluorodeoxyglucose (FDG), which is a sugar that is

Table 1.2 Molecular imaging.

Name: Positron emission tomography (PET) **Physical effect:** β^+ decay **Weight:** 9 tons (PET/MR) **Max resolution:** 4 mm **Established in:** 1975	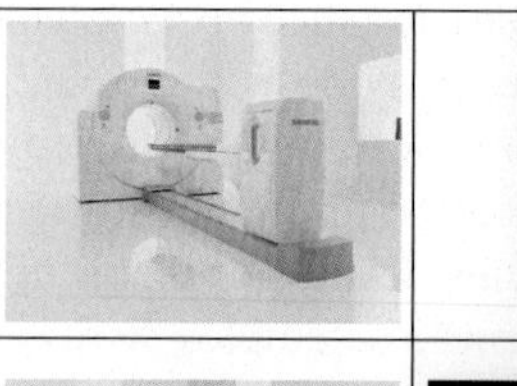	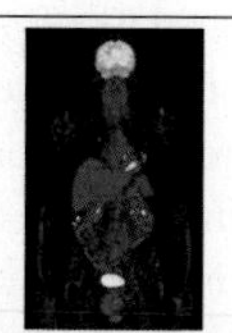
Name: Single-photon emission computed tomography (SPECT) **Physical effect:** γ decay **Weight:** 3 tons (SPECT/CT) **Max resolution:** 10 mm, 5.8 mm (iterative) **Established in:** 1968	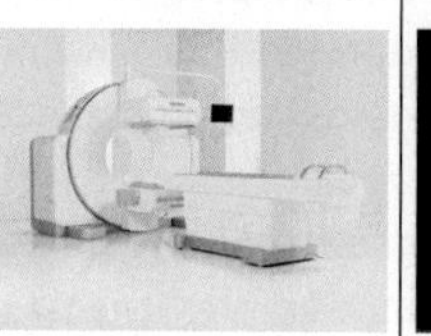	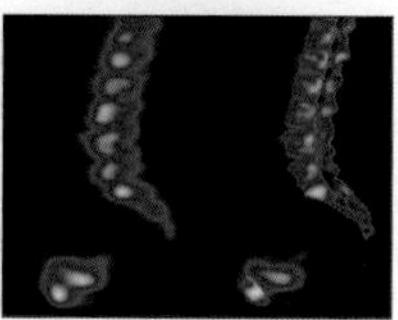

Images from [3].

marked with an ^{18}F fluorine atom. As ^{18}F is a radioactive isotope, it undergoes radioactive decay inside the patient's body. The goal of molecular imaging, or – as it is also called – emission tomography, is to reconstruct the distribution of the tracer inside the patient's body as it is indicative where the respective functional response is found. For FDG as in our example, the tracer is indicative of sugar consumption, which is often related to inflammation and immune response. Hence, an FDG image can be interpreted as a map of bodily activity.

In emission tomography, we distinguish two fundamental concepts. Positron emission tomography (PET) uses tracers that predominantly produce β^+ particles during decay. Once such a positron hits an electron, both annihilate and produce two γ quanta that are emitted exactly in opposite directions. A PET detector uses coincidence electronics to identify events that emerge from the same decay. Doing so, a set of rays can be determined and the final image is generated using an appropriate reconstruction algorithm.

In contrast to PET, single-photon emission computed tomography (SPECT) is based on γ decay. A single photon is directly produced by the radioactive isotope. In order to detect the direction of origin at the detector, large metal collimators are used to focus the acceptance angles for each pixel towards a certain orientation. Doing so, the shape of the collimator lamellae determines the projection geometry and resolution. As demonstrated in Table 1.2, the use of iterative, regularized reconstruction is very effective for emission tomography, which can incorporate prior knowledge on the scan to increase the maximal resolution of the system by almost a factor of 2.

For molecular imaging, typically very low amounts of radioactivity are introduced into the patient's body. As a result, activity maps are extremely sensitive to very low concentrations of tracers. However, there is also a price to pay for this high sensitivity: Typically emission tomography images are of much lower resolution than, e.g., X-ray-based images. As also noted in Table 1.2, most emission tomography systems that are sold today actually combine functional imaging with an additional structural modality such as CT or magnetic resonance imaging (MRI).

1.3.3 Optical imaging

All optical modalities that we discuss in this section operate either within the visible spectrum or at an infrared range that is close to the visible spectrum. Table 1.3 presents an overview on these modalities.

One of the earliest approaches to explore the inner workings of human beings is endoscopy. An optical system and a light source are used to transfer an image from within the body to the operator of the endoscope. Today, the optical systems are typically digital and can be operated in real-time. Applications are broad and range from diagnosis to interventions. The endoscope optic is either rigid as shown in Table 1.3 or flexible. Rigid endoscopes are often used in interventions in which the optic is inserted into the patient through a minimal incision that allows just enough space for the optics plus additional tools used during the intervention. Due to these restrictions, this type of intervention is often also referred to as *keyhole surgery*. Flexible endoscopes are able to follow human anatomy and allow the insertion using natural orifices and are often used in gastro- or colonoscopy.

Microscopes allow the optical magnification of small objects using visible light. They use a system of lenses in order to create high magnification factors so microscopic structures are visible at the eyepiece of the microscope. As such they are often used for biological analysis of cells, e.g., in Petri dishes. Clinically the analysis of live cells is not a standard procedure. However, microscopes play an important role in the analysis of tissues that are resected during surgery. After resection, the tissue is stained using particular dyes that indicate diagnostically relevant cellular structures in a certain color. Next, the specimen is fixed in a substrate that allows to cut the material into slices of few millimeters of thickness. These slices are then collected and transferred onto slides that are investigated using the microscope. In Table 1.3, we see a tissue sample stained with hematoxylin and eosin. This supports analysis of the image, e.g., as cell nuclei are shown in blue. This histological analysis gives important cues on the state of the tissue. For example, it allows to assess the aggressiveness of a cancer by counting the number of mitotic cells in the slide image as the number of dividing cells at a fixed time is related to the actual tumor growth.

Fundus imaging is a technique that explores the background of the eye – the retina. In principle, a normal camera plus an appropriate lens and light source is sufficient to acquire such images. Retinal imaging allows direct visualization of vessels inside the human body without having to perform any incision. Obvious applications are analysis of eye diseases such as glaucoma. Furthermore, fundus imaging also allows assessment of the vessel density and quality. In particular, in patients with a long history of high blood pressure, it can for example be seen that they typically exhibit higher curvature and tortuosity of their retinal vessels.

Optical coherence tomography (OCT) is a method to perform 3D imaging of the retina. In contrast to fundus imaging, not only a superficial image is scanned, but a full depth profile is acquired on each position of the retina. As such it allows for additional diagnostic analysis and in particular the thickness of different retinal layers was found to be an important factor that is often related to characteristic diseases. If multiple scanning is performed at each location, the analysis of the noise structure allows the

Table 1.3 Optical imaging.

Name: Endoscopy **Physical effect:** Visible light **Weight:** 4.9 kg **Wavelengths:** 400–700 nm **Established in:** 1853		
Name: Microscopy **Physical effect:** Visible light **Weight:** 6 kg **Wavelengths:** 400–700 nm **Established in:** 1609		
Name: Fundus imaging **Physical effect:** Visible light **Weight:** 24.5 kg **Wavelengths:** 400–700 nm **Established in:** 1851		
Name: Optical coherence tomography (OCT) **Physical effect:** Infrared light **Weight:** 65 kg **Wavelengths:** 800–1400 nm **Established in:** 1991		
Name: Depth imaging **Physical effect:** Infrared light **Weight:** 180 g **Wavelengths:** 800–1600 nm **Established in:** 2009		

Images from [3] and by courtesy of Talking Eyes GmbH.

visualization of vessels inside the retina without the use of contrast agent. This OCT angiography (OCTA) is an emerging technology that is speculated to have a high diagnostic value for several eye diseases. Table 1.3 shows a typical OCT scanner and one slice through the retina in which the different retinal layers are visible.

1.3.4 Large wavelengths and transversal waves

In this section, we present modalities that employ long wavelengths and transversal waves. One of the most important medical modalities is MRI. In MRI, a strong external magnetic field is used to align dipoles within the human body. One molecule that is of diagnostic importance is H_2O, i.e., water, which can be used for this so-called magnetic resonance effect. After alignment in direction of the strong external magnetic field, the dipoles can be excited using radio frequencies between 12 and

Table 1.4 Large wavelengths and transversal waves.

Name: Magnetic resonance imaging **Physical effect:** Radio waves and magnetic resonance effect **Weight:** 7.3 tons **Wavelengths:** 1–25 m **Established in:** 1974		
Name: Ultrasound **Physical effect:** Sound waves **Weight:** 10 kg **Wavelengths:** 35–400 µm **Established in:** 1942		
Name: Photoacoustic imaging **Physical effect:** Photoacoustic effect **Weight:** 10 kg **Wavelengths:** 300 µm **Established in:** 1993		

Images from [3,5]. Muyinatu Bell supplied the photograph of the photoacoustic imaging system used in her lab at Johns Hopkins University (https://pulselab.jhu.edu/).

300 MHz, depending on the strength of the external magnetic field. After excitation, these dipoles return to their equilibrium state and while doing so, they emit a matching radio frequency pulse corresponding to the one at which they have been excited. This resonance effect allows to determine the density of the affected dipoles in the volume under consideration. Using additional encoding steps, one is able to even resolve this density spatially which forms images and the one depicted in Table 1.4. The magnetic resonance effect is dependent on a multitude of tissue characteristics. As a result many different contrasts can be obtained to encode various effects from temperature to blood oxygenation and even Brownian motion and blood flow.

Ultrasound imaging uses a transceiver to emit and measure pressure waves into bodily tissues. At tissue boundaries these pressure waves are partially reflected. Measuring these reflections allows to reconstruct a depth profile along the direction of pressure wave propagation. If several of those directions are scanned in a linear manner, a slice image can be formed. This type of slice imaging is probably one of the most well-known imaging modalities and is used in many pregnancy screening tests. While 2D slice images are difficult to decipher for non-expert users, also 3D scan probes have emerged, as shown in Table 1.4. These 3D scans are popular with soon to be parents to get a first glimpse at their soon to be newborn child. Today, ultrasound is extremely accessible, as there are also hand-held devices available that have a total weight of less than 10 kg. Such devices can be used at sites of emergency and even inside ambulance cars.

Photoacoustic imaging relies on the same transceiver technology as ultrasound imaging for measuring pressure waves; however, the signal is not generated by emitting ultrasonic waves but by stimulating tissue using a high-energy density pulsed light source, i.e., a laser. Non-ionizing laser pulses are delivered to the region of interest where some of the energy will be absorbed and converted to heat, leading to thermoelastic expansion and emission of ultrasound. The magnitude of the photoacoustic signal is dependent on the optical absorption in the target tissue. As a consequence, photoacoustic imaging is primarily used for the visualization of blood vessels due to the absorption in hemoglobin that is several magnitudes higher than in surrounding tissue. However, exogenous contrast agents with well-known absorption spectra can be engineered to target specific anatomy. Photoacoustic imaging is usually integrated with ultrasound imaging platforms since both modalities share the detection mechanism, giving rise to hybrid imaging systems.

1.3.5 A historical perspective on medical imaging

The modalities that we described in this section have been developed within a time frame of millennia. Most relevant developments, however, happened in the time frame of the past 200 years. We can see an accelerating trend in terms of development over the last 50 years with relevant modalities still emerging after the year 2000.

Probably the oldest imaging modality presented here is endoscopy. Endoscopes that appear to be used in this context were already found in the ruins of Pompeii. Still, we can only date the first use of the term *endoscope* to 1953. In this year, Desormeaux developed a device for inspection of the urinary tract.

Also magnifying lenses have been used already since the 14th century. The term microscope and the first device were introduced by Galilei in 1609. Yet, it still was a rather simple design and it was improved numerous times to form the microscopes that we know today.

Shortly after the invention of photography in 1839, Helmholtz was the first to introduce this technology for the inspection of the eye in 1851, coining the term *ophthalmoscope*. Already in 1861 Maxwell introduced a color version of this technology.

The development of X-ray transmission-based modalities began with Röntgen's discovery on November 8, 1895. Within the short time frame of only roughly 6 weeks, Röntgen submitted a first paper on his observations on December 28, 1895 and shared his discovery with the rest of the world. As the new method was not protected by patents, quickly after the publication companies decided to create new products. In particular three companies, located in Hamburg, Erlangen, and Chicago, were amongst the first early adopters. Today these companies, after several merges and acquisitions, are still existing as parts of Philips Healthcare, Siemens Healthineers, and General Electric Healthcare.

Already right after the discovery of X-rays, X-ray movies or fluoroscopy was invented. Unfortunately, the idea took much longer to develop, as the rapid succession of X-ray images implied a high dose burden for the patient. As such it took until the discovery of the image intensifier in the 1940s/1950s to develop a useful medical-grade system. The first C-arm system that actually bore this name was created by Hugo Rost and colleagues in 1954 and forms the basis of today's C-arm angiography systems.

Mammography was already first applied in 1927 by Otto Kleinschmidt in Leipzig, Germany; however, it took the work of many others until the late 1960s to develop the approach into a valid breast screening method.

Ultrasound was already discovered by Spallanzani in 1794 as a possible means for navigation in bats. Its first medical use, however, was conducted in 1942 by Dussik, who attempted to detect brain tumors through the skull. In 1948, Ludwig developed A-mode scans to detect gall stones. Already in 1949, the first B-mode images were created by Howry and Holmes.

Already in 1962 the concept of emission tomography using radioactive isotopes was invented by Kuhl. The first human SPECT scans using an Anger Camera were conducted in 1968 using a rotating chair in order to observe the required information for 3D reconstruction by Muehllehner. Only in 1976, systems with a rotating SPECT gantry were introduced by Jaszczak and Keyes independently. For PET, the first 3D scans were conducted by Chesler in 1971. The first scans using the well-known FDG tracer were created by Ter-Pogossian and Phelps in 1975, which we deem as the founding year of today's clinical PET systems.

CT was first demonstrated by Hounsfield in 1969 in animal experiments using isotope sources. Only in 1971, the first medical exam could be performed in humans as the technology was deemed safe enough for human application.

The concepts of spin echo and free induction decay used for imaging in MRI were already discovered in 1950 by Hahn. In 1971, Lauterbur discovered a concept that would allow the encoding of spatial information into imaging gradients while Damadian was able to use nuclear magnetic resonance measurements to distinguish cancer from healthy tissues. In 1973 Lauterbur published the first MRI image, followed by the first image of a living mouse in 1974. Mansfield developed the concept of echoplanar imaging that helped to dramatically decrease the scan time in the late 1970s. Damadian and Minkoff were the first to acquire an in vivo human body image in 1977.

Already in the 1980s measurements of biological specimens were conducted using partial coherence or white-light interferometry. The term OCT and its widespread application in ophthalmology was mainly driven by research of Huang and Fujimoto in 1991. This formed the base for today's OCT developments.

The photophonic effect was already discovered by Bell and Tainter in 1880. Only in 1938 Veingerov developed a method to use the same effect for analysis of gas. Years later, in 1981, Bowen proposed to use the effect for imaging of tissue, showing thermoacoustic A-scans. The first in vivo imaging of a human finger was conducted, however, only in 1993 by Chen and colleagues. This work and others led to further popularization of the method for medical exploration since the 2000s.

Gating-based X-ray phase contrast imaging was first discovered using synchrotron radiation in 2002 and 2003 by David and Momose independently. However, only in 2006 Pfeiffer and colleagues identified an approach to apply the technique on regular X-ray sources. Widespread application of the technology is not yet in clinical use.

Time-of-flight imaging was already available for civil use in the early 2000s. The first prototype for time-of-flight endoscopy was documented in 2009 by Penne and colleagues using a low-resolution PMD chip with only 48 × 64 pixels.

1.3.6 Simulating image formation

For many tasks in medical image analysis it is desirable to have access to large amounts of data with precisely known properties. Unfortunately, such data are usually not readily available and the acquisition and annotation may be prohibitively expensive with respect to time, cost, or dose. A promising alternative is in silico simulation of image formation that has proved useful in a broad range of applications that range from the training of machine learning algorithms to the implementation of virtual clinical trials. Realistic simulation of image formation, however, is not trivial since it requires the modeling of complex physical interactions at the atomic level. If a 3D model of the anatomy to be imaged, i.e., the spatial distribution of materials and their properties, is known precisely, then Monte Carlo simulation can be used to simulate image formation with a very high degree of realism. In Monte Carlo simulation, small quanta of energy are propagated through the 3D model of anatomy in multiple stages. At each stage, a probability distribution function is evaluated that describes the physical interaction of energy with the tissue at this specific location to alter properties of the energy quanta (such as direction of travel due to scattering). The simulation stops when the quanta are registered in the simulated detector or otherwise absorbed. Since many quanta must be followed over many stages, Monte Carlo simulation is computationally very expensive and slow. We refer the reader to [6], [7], and [8] for more information on the use of Monte Carlo methods for the simulation of transmission, optical, and ultrasound imaging, respectively.

It is worth mentioning that for several imaging modalities, most physical effects that govern image formation can be expressed analytically. In this case, realistic simulation can be accelerated substantially without considerable loss of realism but will be highly specific to the imaging modality to be considered [9].

1.4 Clinical scenarios

From the above reasoning it should become clear that distinct medical imaging modalities will retrieve mutually exclusive information on the patient's anatomy. As a consequence, many complex clinical workflows rely on multiple imaging modalities at different stages throughout the course from diagnosis over treatment to follow-up. In the remainder of this chapter, we will describe three representative clinical scenarios that highlight the multimodal character of image-based decision making.

1.4.1 Stroke

Stroke is a severe emergency situation. In literature, two types of stroke are found that share similar acute symptoms. In case of the so-called hemorrhaging stroke, a vessel inside the brain ruptures and acute bleeding leads to loss of body control and acute spasm. Similar symptoms also emerge in ischemic stroke. Here, a small blood vessel is occluded by a clot that previously formed inside the blood of the patient. In this emergency situation, it is critical to treat the patient immediately. In case of a hemorrhage, the patient needs to undergo surgery and the bleeding needs to be stopped using a small metal clip. For the case of a blocked vessel, the blood clot needs to be removed. Small clots can be dissolved using anticoagulant agents. In case of larger clots, minimally invasive surgery is performed using a C-arm system to access the clot and to remove it using a screwdriver-shaped catheter, a so-called clot retriever.

Obviously, these are only two of the most common treatment options and we cannot describe all options in this clinical workflow at this point. However, we can already see that decisions need to be made quickly, as every minute that the patient is not treated properly about 2 million brain cells die. Furthermore, we need to determine quickly whether the patient is suffering from an ischemic or a hemorrhaging stroke. In case of a hemorrhage, the use of blood-thinning agents would be fatal. Hence, a typical stroke protocol first includes a CT scan of the brain to detect whether bleeding occurs. If no bleeding occurs, the patient is administered anticoagulant agent and in severe cases directly transferred to the angiography suite, where minimally invasive surgery can be performed. Presently the workflow is not standardized in many countries and alternatives exist, e.g., employing MRI instead of CT. Yet, only the combination of multiple, if not many modalities is key.

1.4.2 Oncology

Prostate cancer is one of the most common cancers in men. When a digital rectal exam or a PSA urine test reveals abnormal results, medical imaging can be used to further test whether prostate cancer is present. To this end, simultaneous PET/MRI imaging is a promising tool. Volumetric MRI images will reveal anatomic changes to the prostate such as abnormal growth, while simultaneous functional PET imaging uses prostate-specific membrane antigens, useful for evaluating the primary tumor and metastatic disease. If image-based tests also suggest the presence of prostate cancer, the diagnosis is confirmed with a biopsy. A needle is inserted into the prostate under transrectal ultrasound image guidance to retrieve a tissue sample. Finally, a pathologist will use an optical microscope to check the biopsy for presence of cancer cells, and if positive, assign a grade based on the Gleason score.

1.4.3 Osteonecrosis

Osteonecrosis, also referred to as avascular necrosis, describes the death of bone tissue due to a lack of blood supply. If not treated timely, it will eventually lead to bone collapse. If osteonecrosis occurs in joints, such as the femoral head, the potential collapse of the articular surface is particularly harmful since it will require total joint replacement. Since osteonecrosis exhibits high incidence in young adults, joint sparing surgery via core decompression is to be preferred. Osteonecrosis is first diagnosed using high-resolution MRI or PET scan (bone scan). Once the osteonecrotic region is identified, surgery is required to remove the lesion. To this end, a debridement tool is introduced into the affected bone under C-arm X-ray digital radiography to visualize the tool's position. After removal of the lesion, a bone graft is inserted to structurally support the bone.

1.5 Exercises

In this exercise, you will be working with X-rays and CT as a modality using the CONRAD (CONe-beam in RADiology) software. CONRAD is a medical image reconstruction toolkit that provides a powerful and flexible platform for research and development in X-ray and CT imaging. It has been specifically designed to support research in areas such as inverse problems, image processing, and computer-aided diagnosis. The software offers a wide range of features and functions, including advanced image reconstruction algorithms, support for different types of X-ray detectors and sources, and simulation capabilities for X-ray projections and CT slices. In addition, CONRAD provides a comprehensive set of tutorials and examples to help users get started quickly with using the software and exploring its many features and functions. By working with CONRAD in this exercise, you will gain valuable experience with a cutting-edge medical imaging toolkit and become familiar with its many capabilities and uses.

1. Download CONRAD from https://github.com/akmaier/CONRAD. Make sure to download the latest stable version of CONRAD for optimal performance.
2. Install CONRAD following the instructions on the CONRAD Tutorial pages available at https://www5.cs.fau.de/conrad/tutorials/user-guide/index.html.
3. Define a trajectory which allows reconstruction from X-ray views following the CONRAD Tutorials. This involves specifying the position of the X-ray source and detector for each projection. Follow the CONRAD tutorial on "Trajectory Configuration" that provides detailed instructions on how to define a trajectory.
4. Create simulated X-ray projections from the newly defined trajectory and a numerical phantom object. The tutorial on "Projection Generation" provides detailed instructions on how to create simulated X-ray projections.
5. Reconstruct CT slices using the simulated X-ray projections following the "Reconstruction" tutorial.
6. Explore further options of CONRAD like anthropomorphic phantoms, noise simulation, and photon counting detectors and reconstruction, which require use of the API. A tutorial on how to use the CONRAD API is available at https://www5.cs.fau.de/conrad/tutorials/api-tutorials/index.html.

References

[1] Chandler D.M., Seven challenges in image quality assessment: past, present, and future research, ISRN Signal Processing (2013) 1–53.

[2] Cunningham I.A., Applied linear-systems theory, in: Handbook of Medical Imaging, vol. 1, 2000, pp. 79–159.

[3] Maier A., Steidl S., Christlein V., Hornegger J., Medical Imaging Systems: An Introductory Guide, vol. 1, Springer, Cham, 2018.

[4] He X., Park S., Model observers in medical imaging research, Theranostics 3 (10) (2013) 774.

[5] Matsumoto Y., Asao Y., Yoshikawa A., Sekiguchi H., Takada M., Furu M., Saito S., Kataoka M., Abe H., Yagi T., Togashi K. , Toi M., Label-free photoacoustic imaging of human palmar vessels: a structural morphological analysis, Scientific Reports 8 (1) (2018) 786.

[6] Giersch J., Durst J., Monte Carlo simulations in X-ray imaging, Nuclear Instruments & Methods in Physics Research. Section A, Accelerators, Spectrometers, Detectors and Associated Equipment 591 (1) (2008) 300–305.

[7] Zhu C., Liu Q., Review of Monte Carlo modeling of light transport in tissues, Journal of Biomedical Optics 18 (5) (2013) 050902.

[8] Mattausch O., Makhinya M., Goksel O., Realistic ultrasound simulation of complex surface models using interactive Monte-Carlo path tracing, Computer Graphics Forum 37 (1) (2018) 202–213, Wiley Online Library.

[9] Unberath M., Zaech J.-N., Gao C., Bier B., Goldmann F., Lee S.C., Fotouhi J., Taylor R., Armand M., Navab N., Enabling machine learning in X-ray-based procedures via realistic simulation of image formation, International Journal of Computer Assisted Radiology and Surgery 14 (9) (2019) 1517–1528.

References

CHAPTER

2 Mathematical preliminaries

Carlos Alberola-López[a] **and Alejandro F. Frangi**[b,c,d,e]

[a]*Laboratorio de Procesado de Imagen (LPI), Universidad de Valladolid, Valladolid, Spain*

[b]*Department of Computer Science, School of Engineering, Faculty of Science and Engineering, The University of Manchester, Manchester, United Kingdom*

[c]*Division of Informatics, Imaging, and Data Sciences, School of Health Sciences, Faculty of Biology, Medicine, and Health, The University of Manchester, Manchester, United Kingdom*

[d]*Department of Electrical Engineering (ESAT), KU Leuven, Leuven, Belgium*

[e]*Department of Cardiovascular Sciences, KU Leuven, Leuven, Belgium*

Learning points

- Images: definitions and notation
- Relevant vector and matrix theory
- Linear processing in the image domain
- Images in the Fourier domain
- Introduction to calculus and variations
- Notions on shape analysis

2.1 Introduction

Imaging and image processing are intimately related to mathematics. Hence, in this chapter we will collect some well-known results that will be used throughout the book or that may be used when the reader begins doing processing themselves. We do not intend to be exhaustive in collecting results and enumerating them as a checklist. Instead, we have focused on selecting some topics and to include full derivations of some results so that the reader can gain insight into the results and develop the ability to derive new consequences.

The chapter begins with an introductory section, in which we provide some basic definitions and we introduce some quality and similarity metrics. Then we summarize some vector and matrix theory results. Linear processing and transformed (Fourier) domains are then analyzed and some transformations are analytically derived in several dimensions. Next, background concepts on calculus as well as some introductory material on calculus of variations are discussed. Finally, we conclude the chapter with an introductory section on shape analysis; we mainly concentrate on planar shapes so that the algebra does not obscure concepts.

Medical Image Analysis. https://doi.org/10.1016/B978-0-12-813657-7.00014-5

2.2 Imaging: definitions, quality and similarity measures

In this book image intensity will be represented by letters I and/or J. Images are typically considered as a function $I : \mathbb{Z}^N \to \mathbb{R}$, where $N = 2$ for 2D images and $N = 3$ for the 3D case. Higher dimensionality may exist if we include, for instance, a time dimension. In addition, images could be multichannel, either color images with three channels with/out a transparency channel or images with an arbitrary number of channels as is the case in multispectral images. Digital images are not only evaluated on a discrete grid but also have a finite number of pixels. However, this second feature will not be usually taken in consideration unless it is necessary for some applications.

Occasionally it may be useful to consider that images are functions of a continuous spatial variable $\boldsymbol{x}$, and hence they will be considered as $I : \mathbb{R}^N \to \mathbb{R}$. This may be convenient for some operations (for instance, the rotation property of the Fourier transform is easily derived in a continuous domain) or may be necessary for some others, such as finding the intensity of an image on non-grid points for a registration operation.

We may need to measure the quality of an image after our processing to analyze how good the algorithm was. To this end, several measures are at our disposal. We will follow [1] as a recent survey on the topic (see references therein for the original authors).

A popular measure is the mean square error (MSE), which is defined as follows. Consider $I[m,n]$ (the ideal image) and $J[m,n]$ (the processed image). Assume they are defined on a grid with dimensions M and N. The MSE is defined as

$$\text{MSE} = \frac{1}{MV} \sum_m \sum_n (I[m,n] - J[m,n])^2 .$$

If a square root is calculated we obtain the root MSE (RMSE). The peak signal-to-noise ratio is defined as

$$\text{MSE} = 10 \log_{10} \frac{L^2}{\text{MSE}},$$

where L is the dynamic range of the image (assuming the image consists of non-negative values). For 8-bit grayscale images, $L = 255$.

Many other measures have been defined in the past [2] in which further information is added. Normalized cross-correlation (NCC) is one of them; it is defined more easily by rearranging images $I[m,n]$ and $J[m,n]$ as vectors (say, a stack of columns from each image). Denoting this reordering as $I[m]$, $J[m]$, $1 \le m \le M \times N$, NCC is defined as

$$\text{NCC} = \frac{\sigma_{IJ}^2}{\sigma_I^2 \sigma_J^2}$$

with

$$\overline{I} = \frac{1}{MN} \sum_m I[m],$$

$$\sigma_I^2 = \frac{1}{MN-1}\sum_m \left(I[m]-\overline{I}\right)^2,$$

$$\sigma_{IJ}^2 = \frac{1}{MN-1}\sum_m (I[m]-\overline{I})(J[m]-\overline{J}),$$

and definitions for J are equivalent. The universal quality index (UQI) also takes in consideration intensity differences as well as contrast and correlation:

$$\mathrm{UQI} = \frac{4\sigma_{IJ}\overline{I}\cdot\overline{J}}{(\sigma_I^2\sigma_J^2)\left(\overline{I^2}+\overline{J^2}\right)},$$

with

$$\overline{I^2} = \frac{1}{MN}\sum_m I^2[m],$$

and equivalently for $\overline{J^2}$. We should highlight that $-1 \le Q \le 1$, and higher values indicate more similarity between I and J.

Two years after the UQI was defined, the same group defined the structural similarity index (SSIM):

$$\mathrm{SSIM} = \frac{(2\overline{I}\cdot\overline{J}+c_1)(2\sigma_{IJ}+c_2)}{((\overline{I})^2(\overline{J})^2+c_1)(\sigma_I^2+\sigma_J^2+c_2)},$$

where $c_1 = (K_1L)^2$ and $c_2 = (K_1L)^2$, with L being the dynamic range and $K_1, K_2 \ll 1$ ($K_1 = 0.01$ and $K_2 = 0.03$). UQI is the particular case of SSIM when $c_1 = c_2 = 0$, which is known to produce unstable results in some circumstances. The statistics defined above could also be calculated locally and the value returned as overall SSIM for the image would be the mean value of the local SSIMs. About this measure, $0 \le \mathrm{SSIM} \le 1$, where higher values are better. In this case, the correlation is denoted as *structure*.

2.3 Vector and matrix theory results

2.3.1 General concepts

This section is a summary of well-known results about vector and matrix theory. No specific references will be provided since the material hereinafter is quite basic; generally speaking most of the material has been extracted from [3].

Let $\boldsymbol{u}$ represent an N-component vector of either real or complex elements. The vector will be considered based on default column vectors, so

$$\boldsymbol{u} = \begin{bmatrix} u_1 & u_2 & \ldots & u_N \end{bmatrix}^T,$$

where T is the transposition operator that converts a row into a column and vice versa. Accordingly, the kth component of the vector will be denoted as $[\boldsymbol{u}]_k = u_k$. As for a matrix, let $\boldsymbol{A}$ denote a matrix of, say, N rows and M columns (i.e., of dimensions $N \times M$). The element of the matrix in row i and column j will be denoted by $[\mathbf{A}]_{i,i} = a_{ij}$ and, consequently, $[\mathbf{A}]^T_{i,j} = a_{ji}$. Occasionally, we may be interested in not only transposing but also conjugating the entries of a vector or a matrix. In this case, superscript H will be the operator, i.e., $[\mathbf{A}]^H_{i,j} = a^*_{ji}$, with the asterisk meaning "complex conjugate." Those matrices that are invariant with a transposition are called symmetric, while those that are invariant to operator H are called *Hermitian*.

Both addition and multiplication are defined for vectors and matrices. For the former, both elements must have the same dimensions and summation is performed elementwise. As for multiplication, focusing on a matrix, the operation is defined, for matrix $\boldsymbol{A}$ with N rows and P columns and matrix $\boldsymbol{B}$ with P rows and M columns, as

$$[\mathbf{AB}]_{i,j} = \sum_{p=1}^{P} [\mathbf{A}]_{i,p} [\mathbf{B}]_{p,j} .$$

Consequently, the number of columns of matrix $\boldsymbol{A}$ must coincide with the number of rows of matrix $\boldsymbol{B}$. The resulting matrix has N rows and M columns. Note that matrix multiplication does not commute, i.e., generally speaking $\boldsymbol{AB} \neq \boldsymbol{BA}$; actually, note that the existence of matrix $\boldsymbol{AB}$ does not guarantee that the multiplication $\boldsymbol{BA}$ is even defined. Note, however, that the following equality always holds:

$$(\boldsymbol{AB})^H = \boldsymbol{B}^H \boldsymbol{A}^H.$$

Matrix multiplication should not be confused with the Hadamard or the Kronecker product. The former is a componentwise operation, defined as follows:

$$[\mathbf{A} \odot \mathbf{B}]_{i,j} = [\mathbf{A}]_{i,p} [\mathbf{B}]_{p,j} .$$

As for the Kronecker product, assume that $\boldsymbol{A}$ has N rows and P columns and $\boldsymbol{B}$ has P rows and M columns; if this is the case, this product is defined as

$$\boldsymbol{A} \otimes \boldsymbol{B} = \begin{bmatrix} a_{11}\boldsymbol{B} & a_{12}\boldsymbol{B} & \cdots & a_{1P}\boldsymbol{B} \\ a_{21}\boldsymbol{B} & a_{22}\boldsymbol{B} & \cdots & a_{2P}\boldsymbol{B} \\ \vdots & \vdots & \cdots & \vdots \\ a_{N1}\boldsymbol{B} & a_{N2}\boldsymbol{B} & \cdots & a_{NP}\boldsymbol{B} \end{bmatrix},$$

where we have used the notation of a partition matrix, i.e., the insertion of a matrix into another matrix. Accordingly, the result of this product is another matrix with $N \cdot P$ rows and $P \cdot M$ columns.

As for vectors, two notable multiplication operations should be mentioned, namely, the inner and the outer product. The former is defined as

$$\boldsymbol{u}^H \boldsymbol{v} = \sum_{i=1}^{N} \left| u_i^* v_i \right|^2,$$

and both vectors should have the same number of components, giving rise to a scalar value. As for the latter, the vector dimensions may differ (say N for $\boldsymbol{u}$ and M for $\boldsymbol{v}$) and it is defined as

$$\boldsymbol{u}\boldsymbol{v}^H = \left[\begin{array}{cccc} v_1^* \boldsymbol{u} & v_2^* \boldsymbol{u} & \cdots & v_M^* \boldsymbol{u} \end{array} \right]. \tag{2.1}$$

As can be observed, the result is a matrix of dimensions $N \times M$. If the inner product is applied to just one vector we have

$$\boldsymbol{u}^H \boldsymbol{u} = \sum_{i=1}^{N} |u_i|^2 = \|\boldsymbol{u}\|_{\ell_2}^2,$$

which equals the (Euclidian) norm of the vector. By extension, this result can be generalized to

$$\|\boldsymbol{u}\|_{\ell_p} = \left(\sum_{i=1}^{N} |u_i|^p \right)^{\frac{1}{p}}.$$

Note that two vectors $\boldsymbol{u}$ and $\boldsymbol{v}$ are considered orthogonal if $\boldsymbol{u}^H \boldsymbol{v} = 0$. Vectors $\boldsymbol{u}_i$, $1 \leq i \leq Q \leq N$, with N being the dimension of the vectors, are considered orthonormal if the following condition holds:

$$\boldsymbol{u}_i^H \boldsymbol{u}_j = \delta[i - j],$$

where $\delta[n]$ is the Kronecker delta function. If this vector dataset is arranged as columns of matrix $\boldsymbol{A}$, then we have

$$\boldsymbol{A}^H \boldsymbol{A} = \boldsymbol{I}_N \tag{2.2}$$

and matrix $\boldsymbol{A}$ is referred to as orthonormal. Matrix $\boldsymbol{I}_N$ is the identity matrix, i.e., a matrix of the same size as $\boldsymbol{A}$, where all entries are null except those on the main diagonal, whose values are unity. The subscript indicates the dimension of the matrix, which may be omitted if context provides that information.

A popular matrix norm is the so-called *Frobenius norm*, which is defined, by extension, as follows:

$$\|\boldsymbol{A}\|_F = \left(\sum_{i=1}^{N} \sum_{j=1}^{M} \left| a_{ij} \right|^2 \right)^{\frac{1}{2}} = \left(\mathrm{Tr} \left\{ \boldsymbol{A}\boldsymbol{A}^H \right\} \right)^{\frac{1}{2}}.$$

The determinant of a square matrix is a scalar number (not necessarily real) that is obtained after

$$\det(\boldsymbol{A}) = |A| = \sum_{j=1}^{N} a_{ij} C_{ij},$$

with i being an arbitrary value, $1 \leq i \leq N$ (i.e., the index of any row in the matrix), and $C_{ij} = (-1)^{i+j} M_{ij}$, where M_{ij} is the determinant of another matrix, say $\boldsymbol{A}_{ij}$, which is obtained by deleting the ith row and the jth column. This matrix is called the minor of a_{ij}. The determinant has several properties, some of which are:

$$\det(\boldsymbol{AB}) = \det(\boldsymbol{A})\det(\boldsymbol{B}), \tag{2.3}$$

$$\det(\boldsymbol{A}^T) = \det(\boldsymbol{A}), \tag{2.4}$$

$$\det(\boldsymbol{A}^H) = (\det(\boldsymbol{A}))^*, \tag{2.5}$$

$$\det(k\boldsymbol{A}) = k^N \det(\boldsymbol{A}), \quad k \in \mathbb{C}. \tag{2.6}$$

The trace of a square matrix is the summation of the elements in the main diagonal, i.e.,

$$\mathrm{tr}(\boldsymbol{A}) = \sum_{i=1}^{N} [\mathbf{A}]_{i,i} .$$

The trace is invariant with matrix multiplication ordering (if well defined), i.e., $\mathrm{tr}(\boldsymbol{AB}) = \mathrm{tr}(\boldsymbol{BA})$.

The concept of a minor can be extended to any matrix (not necessarily square) as a square submatrix extracted by selecting some rows and columns. The order of the minor is its number of rows. With this idea in mind, the rank of a matrix is defined as the order of the greatest minor with non-null determinant. This has the consequence that the rank equals the number of rows (or columns) that are linearly independent. *Linear independence* means that, say, a row cannot be written as a linear combination of other rows, i.e., the row does not lie in the subspace spanned by the other rows. Obviously, $\mathrm{rank}(\boldsymbol{A}) \leq \min(N, M)$. Note that, by definition, a square matrix with rank different from the number of rows has null determinant.

The inverse of square matrix $\boldsymbol{A}$ is denoted as $\boldsymbol{A}^{-1}$ and satisfies $\boldsymbol{AA}^{-1} = \boldsymbol{A}^{-1}\boldsymbol{A} = \boldsymbol{I}$.

Several properties are satisfied by the inverse. Specifically:

$$\left(\boldsymbol{A}^T\right)^{-1} = \left(\boldsymbol{A}^{-1}\right)^T, \tag{2.7}$$

$$\left(\boldsymbol{A}^H\right)^{-1} = \left(\boldsymbol{A}^{-1}\right)^H, \tag{2.8}$$

$$(\boldsymbol{AB})^{-1} = (\boldsymbol{BA})^{-1}, \tag{2.9}$$

$$\det\left(\boldsymbol{A}^{-1}\right) = \frac{1}{\det(\boldsymbol{A})}. \tag{2.10}$$

Note that, following (2.2), the inverse of a square orthonormal matrix $\boldsymbol{A}$ equals $\boldsymbol{A}^H$.

A well-known and useful result is the *matrix inversion lemma*, which states the following:

$$(\boldsymbol{A}+\boldsymbol{BCD})^{-1} = \boldsymbol{A}^{-1} - \boldsymbol{A}^{-1}\boldsymbol{B}\left(\boldsymbol{DA}^{-1}\boldsymbol{B}+\boldsymbol{C}^{-1}\right)\boldsymbol{DA}^{-1}.$$

Matrix inversion arises naturally as the solution of a determined linear system, i.e., a linear system

$$\boldsymbol{Ax} = \boldsymbol{b} \tag{2.11}$$

in which $\boldsymbol{A}$ is a full-rank square matrix. In this case, since matrix $\boldsymbol{A}^{-1}$ exists, the solution is readily obtained as $\boldsymbol{x} = \boldsymbol{A}^{-1}\boldsymbol{b}$. However, matrix A may have dimensions $N \times M$ with $N \neq M$, i.e., the number of equations does not necessarily coincide with the number of unknowns. Let us assume that the matrix is full rank; in this case the inverse is generalized to the *pseudoinverse* (or *Moore–Penrose pseudoinverse*) as follows:

1. If the linear system is overdetermined, i.e., the number of equations N is larger than the number of unknowns M, then we have too many restrictions (N) for the number of degrees of freedom M available to satisfy (2.11) and hence an exact solution does not exist. So, the problem is approximated as

$$\arg\min_{\boldsymbol{x}} \|\boldsymbol{Ax}-\boldsymbol{b}\|^2, \tag{2.12}$$

whose solution turns out to be $\hat{\boldsymbol{x}} = \boldsymbol{A}^{\dagger}\boldsymbol{b}$, with

$$\boldsymbol{A}^{\dagger} = \left(\boldsymbol{A}^H\boldsymbol{A}\right)^{-1}\boldsymbol{A}^H.$$

In this case, it is easy to check that

$$\left(\boldsymbol{A}\hat{\boldsymbol{x}}-\boldsymbol{b}\right)^H\boldsymbol{A}\hat{\boldsymbol{x}} = 0. \tag{2.13}$$

2. Complementarily, if the system is undetermined, i.e., the number of unknowns M exceeds the number of restrictions N, then many solutions to (2.11) can be found. Then the solution that satisfies

$$\arg\min_{\boldsymbol{x}} \boldsymbol{x}^H\boldsymbol{x} \tag{2.14}$$

subject to making equation (2.11) hold turns out to be $\hat{\boldsymbol{x}} = \boldsymbol{A}^{\ddagger}\boldsymbol{b}$, with

$$\boldsymbol{A}^{\ddagger} = \boldsymbol{A}^H\left(\boldsymbol{AA}^H\right)^{-1}.$$

2.3.2 Eigenanalysis

Let $\boldsymbol{A}$ be a square matrix with N rows. Vector $\boldsymbol{v}$ is an *eigenvector* if the following equality holds:

$$\boldsymbol{A}\boldsymbol{v} = \lambda \boldsymbol{v},$$

where $\lambda \in \mathbb{C}$ is referred to as its *eigenvalue*. Note that a vector with null components is always an eigenvector. This equation can be rewritten as

$$\boldsymbol{A}\boldsymbol{v} - \lambda \boldsymbol{I}\boldsymbol{v} = \boldsymbol{0} \Longrightarrow (\boldsymbol{A} - \lambda \boldsymbol{I})\,\boldsymbol{v} = \boldsymbol{0},$$

i.e., as a linear system for which a solution (in $\boldsymbol{v}$) different from a null vector exists provided that the system is undetermined, i.e., making the equality

$$|\boldsymbol{A} - \lambda \boldsymbol{I}| = \boldsymbol{0}$$

hold, which gives rise to an N-grade polynomial

$$q(\lambda) = \sum_{i=0}^{N} \beta_i \lambda^i = 0, \tag{2.15}$$

and hence to N solutions for λ, which may all be different or some of them may be multiple. This polynomial is referred to as the characteristic polynomial of matrix $\boldsymbol{A}$.

Let α_i be the multiplicity of λ_i in the characteristic polynomial. The following results are of interest:

1. All eigenvalues are different ($\alpha_i = 1,\ 1 \le i \le N$): Let the columns of matrix $\boldsymbol{U}$ be the eigenvectors of matrix $\boldsymbol{A}$, with eigenvector $\boldsymbol{u}_i$ (the one associated to eigenvalue λ_i) positioned at column i, $1 \le i \le N$. Then matrix $\boldsymbol{U}$ turns out to be invertible and the following relation holds:

$$\boldsymbol{\Lambda} = \boldsymbol{U}^{-1}\boldsymbol{A}\boldsymbol{U}, \tag{2.16}$$

with $\boldsymbol{\Lambda} = \mathrm{diag}\begin{bmatrix} \lambda_1 & \lambda_2 & \cdots & \lambda_N \end{bmatrix}$, a matrix with all the entries zero except the ones in the main diagonal, the values of which are the eigenvalues. Note that the ordering of the eigenvalues in the matrix should coincide with ordering of the eigenvectors in the columns of matrix $\boldsymbol{U}$. As consequence, we can also write

$$\boldsymbol{A} = \boldsymbol{U}\boldsymbol{\Lambda}\boldsymbol{U}^{-1}. \tag{2.17}$$

2. Some eigenvalues have non-unitary multiplicity, i.e., there exists at least one index j for which the multiplicity s_j of λ_j is greater than 1. In this case, equation (2.16) holds, provided that matrix $\boldsymbol{A} - \lambda_j \boldsymbol{I}$ has rank equal to $N - s_j$.
3. Matrix $\boldsymbol{A}$ is Hermitian. Then all the eigenvectors will be different and real. This leads to an orthogonal set of eigenvectors, so an orthonormal matrix $\boldsymbol{U}$ can be

built. This can be easily seen by writing

$$\begin{aligned}
\boldsymbol{A}\boldsymbol{u}_i &= \lambda_i \boldsymbol{u}_i \rightarrow \boldsymbol{u}_j^H \boldsymbol{A}\boldsymbol{u}_i = \lambda_i \boldsymbol{u}_j^H \boldsymbol{u}_i, \\
\boldsymbol{A}\boldsymbol{u}_j &= \lambda_j \boldsymbol{u}_j \rightarrow \boldsymbol{u}_i^H \boldsymbol{A}\boldsymbol{u}_j = \lambda_j \boldsymbol{u}_i^H \boldsymbol{u}_j \rightarrow \boldsymbol{u}_j^H \boldsymbol{A}^H \boldsymbol{u}_i = \lambda_j^* \boldsymbol{u}_j^H \boldsymbol{u}_i.
\end{aligned}$$

Since $\boldsymbol{A}$ is Hermitian and all the eigenvalues are real, the last equation can be written

$$\boldsymbol{u}_j^H \boldsymbol{A}\boldsymbol{u}_i = \lambda_j \boldsymbol{u}_j^H \boldsymbol{u}_i,$$

and subtracting from the first equation, we can write

$$(\lambda_i - \lambda_j)\boldsymbol{u}_j^H \boldsymbol{u}_i = \boldsymbol{0},$$

which leads to $\boldsymbol{u}_j^H \boldsymbol{u}_i = \boldsymbol{0}$ since eigenvalues are different. Hence, eigenvectors are orthogonal, and they can be made orthonormal by normalization. If this is the case, since a matrix $\boldsymbol{U}$ composed of normalized eigenvectors is orthonormal as well, equations (2.16) and (2.17) can now be written

$$\boldsymbol{\Lambda} = \boldsymbol{U}^H \boldsymbol{A}\boldsymbol{U}, \tag{2.18}$$

$$\boldsymbol{A} = \boldsymbol{U}\boldsymbol{\Lambda}\boldsymbol{U}^H. \tag{2.19}$$

Matrix $\boldsymbol{U}$ is also called unitary. In addition, due to the orthonormality of $\boldsymbol{U}$, making use of equations (2.8) and (2.9) we can also write

$$\begin{aligned}
\boldsymbol{A}^{-1} &= \left(\boldsymbol{U}\boldsymbol{\Lambda}\boldsymbol{U}^H\right)^{-1} = \left(\boldsymbol{U}^H\right)^{-1} \boldsymbol{\Lambda}^{-1}\boldsymbol{U}^{-1} \\
&= \boldsymbol{U}\boldsymbol{\Lambda}^{-1}\boldsymbol{U}^H = \sum_{i=1}^{N} \frac{1}{\lambda_i} \boldsymbol{u}_i \boldsymbol{u}_i^H,
\end{aligned}$$

where $\boldsymbol{\Lambda}^p$ is a shortcut to denote

$$\boldsymbol{\Lambda} = \operatorname{diag}\left[\begin{array}{cccc} \frac{1}{\lambda_1^p} & \frac{1}{\lambda_2^p} & \cdots & \frac{1}{\lambda_N^p} \end{array}\right].$$

Considering the determinant of matrix $\boldsymbol{A}$, applying equations (2.5) and (2.10), we have

$$\begin{aligned}
\det(\boldsymbol{A}) &= \alpha, \\
\det\left(\boldsymbol{A}^H\right) &= \alpha^* = \det\left(\boldsymbol{A}^{-1}\right) = \frac{1}{\alpha} \Longrightarrow \\
\alpha^* &= \frac{1}{\alpha} \Longrightarrow \rho e^{-j\theta} = \frac{1}{\rho} e^{-j\theta} \Longrightarrow \rho = 1 \Longrightarrow |\alpha|^2 = 1,
\end{aligned}$$

i.e., the determinant of a unitary matrix has unitary modulus. This has the consequence, applying property (2.7),

$$\begin{aligned}\det(\boldsymbol{A}) &= \det(\boldsymbol{U})\det(\boldsymbol{\Lambda})\det\left(\boldsymbol{U}^H\right)\\ &= |\alpha|^2\left(\prod_{i=1}^{N}\lambda_i\right) = \left(\prod_{i=1}^{N}\lambda_i\right).\end{aligned}$$

A particular case of a Hermitian matrix is that of the covariance matrix of a vector of random variables. With the ideas we have described so far it is fairly simple to describe the grounds of the principal component analysis (PCA). Consider we have a zero-mean vector random variable x with covariance matrix $\boldsymbol{C}$ and let us accept that this matrix is full rank. Since this matrix is Hermitian (by definition) it can be expressed as

$$\boldsymbol{C} = \boldsymbol{U}\boldsymbol{\Lambda}\boldsymbol{U}^H, \tag{2.20}$$

where $\boldsymbol{U}$ is unitary. PCA is based on defining new variables y_i, $1 \leq i \leq Q$, which are the projections of the original variable onto the eigenvectors of the covariance matrix, i.e.,

$$\mathrm{y}_i = \left(\mathrm{x}^H\boldsymbol{u}_i\right)\boldsymbol{u}_i. \tag{2.21}$$

These transformations have the following consequences:

a. Variables y_i and y_j are uncorrelated. This is readily obtained by using the ideas we have just described:

$$\begin{aligned}\mathbb{E}\{\mathrm{y}_i\mathrm{y}_j^H\} &= \mathbb{E}\{\mathrm{y}_j^H\mathrm{y}_i\}\\ &= \mathbb{E}\left\{\boldsymbol{u}_j^H\left(\boldsymbol{u}_j^H\mathrm{x}\right)\left(\mathrm{x}^H\boldsymbol{u}_i\right)\boldsymbol{u}_i\right\}\\ &= \boldsymbol{u}_j^H\left(\boldsymbol{u}_j^H\mathbb{E}\left\{\mathrm{x}\mathrm{x}^H\right\}\boldsymbol{u}_i\right)\boldsymbol{u}_i = \boldsymbol{u}_j^H\left(\boldsymbol{u}_j^H\boldsymbol{C}\boldsymbol{u}_i\right)\boldsymbol{u}_i = 0,\end{aligned} \tag{2.22}$$

due to the orthogonality of $\boldsymbol{u}_i$ and $\boldsymbol{u}_j$.

b. The variance of the transformed variable turns out to coincide with λ_i (which are all positive due to the full rank of a Hermitian matrix). This is a direct consequence of equation (2.22) for $j = i$. Note that the term within braces for $j = i$ is a quadratic form that extracts the ith eigenvalue of matrix $\boldsymbol{\Lambda}$ in equation (2.16). Hence

$$\begin{aligned}\sigma_{\mathrm{y}_i}^2 &= \mathbb{E}\{\mathrm{y}_i^H\mathrm{y}_i\}\\ &= \boldsymbol{u}_i^H\left(\boldsymbol{u}_i^H\boldsymbol{C}\boldsymbol{u}_i\right)\boldsymbol{u}_i\\ &= \lambda_i\boldsymbol{u}_i^H\boldsymbol{u}_i = \lambda_i.\end{aligned}$$

c. Assume eigenvalues are ordered from high to low. The first $Q \leq N$ components y_i, $1 \leq i \leq Q$, will retain the maximum variance of the variable x. This variance will equal $\sum_{i=1}^{Q} \lambda_i$.

2.3.3 Singular value decomposition

Let $\boldsymbol{A}$ be a complex matrix of dimensions $N \times M$. This matrix can be alternatively expressed as

$$\boldsymbol{A} = \boldsymbol{V}\boldsymbol{\Sigma}\boldsymbol{U}^H,$$

where matrices $\boldsymbol{V}$ and $\boldsymbol{U}$ are unitary, i.e., $\boldsymbol{V}^H\boldsymbol{V} = \boldsymbol{I}_N$ and $\boldsymbol{V}^H\boldsymbol{V} = \boldsymbol{I}_M$, and matrix $\boldsymbol{\Sigma}$ is a diagonal (square) matrix of dimension r with singular values $\sigma_i > 0$. It can be interpreted as a generalization to non-square matrices of the eigenanalysis described so far, although, for a square matrix, the singular value decomposition (SVD) does not generally coincide with the eigendecomposition. It does when the matrix is Hermitian.

The SVD has nice geometrical interpretations for both the linear transformation $\boldsymbol{A}\boldsymbol{x}$ and the solution to equation (2.12). Specifically (in what follows $\boldsymbol{v}_i$ and $\boldsymbol{u}_i$ will denote the ith column of matrices $\boldsymbol{V}$ and $\boldsymbol{U}$, respectively):

1. For the linear transformation, we can write

$$\boldsymbol{A}\boldsymbol{x} = \boldsymbol{V}\boldsymbol{\Sigma}\boldsymbol{U}^H = \sum_{i=1}^{r} \left(\boldsymbol{u}_i^H \boldsymbol{x}\right)(\sigma_i \boldsymbol{v}_i),$$

i.e., the linear transform consists in projecting vector $\boldsymbol{x}$ onto the orthonormal vector set $\boldsymbol{u}_i$ defined on the departing linear space, i.e., the space in which $\boldsymbol{x}$ is defined. Then, each projection modulates the amplitude of each vector $\boldsymbol{v}_i$, defined on the output space, which has been previously modulated by the singular value σ_i. The transformed vector is obtained by summing the contribution of each modulated $\boldsymbol{v}_i$ vector.

2. Conversely, assume we want to solve equation (2.12). The solution will be

$$\begin{aligned}
\boldsymbol{A}^\dagger \boldsymbol{b} &= \left(\boldsymbol{A}^H\boldsymbol{A}\right)^{-1}\boldsymbol{A}^H \\
&= \left(\left(\boldsymbol{V}\boldsymbol{\Sigma}\boldsymbol{U}^H\right)^H \boldsymbol{V}\boldsymbol{\Sigma}\boldsymbol{U}^H\right)^{-1}\left(\boldsymbol{V}\boldsymbol{\Sigma}\boldsymbol{U}^H\right)^H \boldsymbol{V} \\
&= \left(\boldsymbol{U}\boldsymbol{\Sigma}\boldsymbol{V}^H\boldsymbol{V}\boldsymbol{\Sigma}\boldsymbol{U}^H\right)^{-1}\boldsymbol{U}\boldsymbol{\Sigma}\boldsymbol{V}^H\boldsymbol{b} \\
&= \left(\boldsymbol{U}\boldsymbol{\Sigma}^2\boldsymbol{U}^H\right)^{-1}\boldsymbol{U}\boldsymbol{\Sigma}\boldsymbol{V}^H\boldsymbol{b}.
\end{aligned}$$

The inverse to be calculated now resembles that in equation (2.19) but matrix $\boldsymbol{U}$ will not be, generally speaking, square, so the foregoing derivation is no longer valid for this case. However, it is simple to see that $\boldsymbol{u}_i$ is an eigenvector of the matrix to be inverted,

$$\boldsymbol{B}\boldsymbol{u}_i = \boldsymbol{U}\boldsymbol{\Sigma}^2\boldsymbol{U}^H\boldsymbol{u}_i = \sigma_i^2\boldsymbol{u}_i,$$

so $\boldsymbol{B}^{-1}\boldsymbol{u}_i$, provided that $\boldsymbol{B}^{-1}$ exists, will be equal to $\frac{1}{\sigma_i^2}\boldsymbol{u}_i$. Hence

$$\begin{aligned}
\boldsymbol{A}^{\dagger}\boldsymbol{b} &= \left(\boldsymbol{U}\boldsymbol{\Sigma}^2\boldsymbol{U}^H\right)^{-1}\boldsymbol{U}\boldsymbol{\Sigma}\boldsymbol{V}^H\boldsymbol{b} \\
&= \begin{bmatrix} \frac{1}{\sigma_1^2}\boldsymbol{u}_1 & \cdots & \frac{1}{\sigma_r^2}\boldsymbol{u}_r \end{bmatrix} \begin{bmatrix} \sigma_1\boldsymbol{v}_1^H\boldsymbol{b} \\ \vdots \\ \sigma_r\boldsymbol{v}_r^H\boldsymbol{b} \end{bmatrix} \\
&= \sum_{i=1}^{r} \frac{\boldsymbol{v}_i^H\boldsymbol{b}}{\sigma_i}\boldsymbol{u}_i.
\end{aligned}$$

In this case, note that the solution is obtained with a similar interpretation as in the forward case: the output vector $\boldsymbol{b}$ is projected onto each vector $\boldsymbol{v}_i$, this projection is normalized by σ_i, and the solution is obtained by summing the contribution of each modulated $\boldsymbol{u}_i$ vector.

3. Similar considerations can be made with respect to the problem defined by equation (2.14). Specifically:

$$\begin{aligned}
\boldsymbol{A}^{\ddagger}\boldsymbol{b} &= \boldsymbol{A}^H\left(\boldsymbol{A}\boldsymbol{A}^H\right)^{-1} \\
&= \left(\boldsymbol{V}\boldsymbol{\Sigma}\boldsymbol{U}^H\right)^H\left(\boldsymbol{V}\boldsymbol{\Sigma}\boldsymbol{U}^H\left(\boldsymbol{V}\boldsymbol{\Sigma}\boldsymbol{U}^H\right)^H\right)^{-1}\boldsymbol{b} \\
&= \boldsymbol{U}\boldsymbol{\Sigma}\boldsymbol{V}^H\left(\boldsymbol{V}\boldsymbol{\Sigma}\boldsymbol{U}^H\boldsymbol{U}\boldsymbol{\Sigma}\boldsymbol{V}^H\right)^{-1}\boldsymbol{b} \\
&= \boldsymbol{U}\boldsymbol{\Sigma}\boldsymbol{V}^H\left(\boldsymbol{V}\boldsymbol{\Sigma}^2\boldsymbol{V}^H\right)^{-1}\boldsymbol{b}.
\end{aligned}$$

Following the same reasoning as before, in this case, with respect to vectors $\boldsymbol{v}_i$, i.e.,

$$\boldsymbol{D}\boldsymbol{v}_i = \boldsymbol{V}\boldsymbol{\Sigma}^2\boldsymbol{V}^H\boldsymbol{v}_i = \sigma_i^2\boldsymbol{u}_i,$$

we have $\boldsymbol{v}_i^H\boldsymbol{D}^{-1} = \frac{1}{\sigma_i^2}\boldsymbol{v}_i^H$, so we can write

$$\boldsymbol{A}^{\ddagger}\boldsymbol{b} = \boldsymbol{U}\boldsymbol{\Sigma}\boldsymbol{V}^H\left(\boldsymbol{V}\boldsymbol{\Sigma}^2\boldsymbol{V}^H\right)^{-1}$$

$$= \begin{bmatrix} \sigma_1 \boldsymbol{u}_1 & \cdots & \sigma_r \boldsymbol{u}_r \end{bmatrix} \begin{bmatrix} \frac{\boldsymbol{v}_1^H \boldsymbol{b}}{\sigma_1^2} \\ \vdots \\ \frac{\boldsymbol{v}_r^H \boldsymbol{b}}{\sigma_r^2} \end{bmatrix}$$

$$= \sum_{i=1}^{r} \frac{\boldsymbol{v}_i^H \boldsymbol{b}}{\sigma_i} \boldsymbol{u}_i,$$

which coincides with the case of $\boldsymbol{A}^\dagger$.

2.3.4 Matrix exponential

2.3.4.1 Generalities

Given a square matrix $\boldsymbol{A}$, its matrix exponential is defined as

$$e^{\boldsymbol{A}} = \sum_{i=0}^{\infty} \frac{\boldsymbol{A}^i}{i!},$$

where $\boldsymbol{A}^0 = \boldsymbol{I}$. This is a particular case of a general matrix function, which is defined as

$$f(\boldsymbol{A}) = \sum_{i=0}^{\infty} \beta_i \boldsymbol{A}^i.$$

If the matrix is diagonalizable, i.e., if equation (2.16) holds, the exponential is readily calculated as

$$e^{\boldsymbol{A}} = \boldsymbol{U} e^{\boldsymbol{\Lambda}} \boldsymbol{A}^{-1} = \boldsymbol{U} \begin{bmatrix} e^{\lambda_1} & 0 & \dots & 0 \\ 0 & e^{\lambda_2} & \dots & 0 \\ \vdots & \vdots & \ddots & \vdots \\ 0 & 0 & \dots & e^{\lambda_N} \end{bmatrix} \boldsymbol{U}^{-1}.$$

Otherwise, the foregoing infinite power series expansion can be rewritten as a finite series by means of the Cayley–Hamilton theorem [4], which states that every matrix satisfies its characteristic polynomial. Recalling equation (2.15), we can write

$$\beta_0 \boldsymbol{I} + \beta_1 \boldsymbol{A} + \ldots + \beta_N \boldsymbol{A}^N = \boldsymbol{0}.$$

It turns out that any matrix function can be written as a finite power series as follows:

$$f(\boldsymbol{A}) = \sum_{k=0}^{N-1} \alpha_k \boldsymbol{A}^k. \tag{2.23}$$

The reason is the following. Assume $f_r(\lambda)$ is an rth-order power series, $r > N$. Dividing $f_r(\lambda)$ by $q(\lambda)$ we will obtain an $(r - N)$-grade polynomial $c_r(\lambda)$ and a remainder $o_r(\lambda)$, the grade of which will be $N - 1$. Hence, we can write

$$f_r(\lambda) = q(\lambda)c_r(\lambda) + o_r(\lambda).$$

Let $f(\lambda)$ denote the limit of function $f_r(\lambda)$ when $r \to \infty$. Similar notation is used for both $c_r(\lambda)$ and $o_r(\lambda)$. Then

$$f(\lambda) = q(\lambda)c(\lambda) + o(\lambda),$$

and evaluated on matrix $\boldsymbol{A}$ this gives rise to

$$f(\boldsymbol{A}) = q(\boldsymbol{A})\,c(\boldsymbol{A}) + r(\boldsymbol{A}) = r(\boldsymbol{A}),$$

since $q(\boldsymbol{A}) = \boldsymbol{0}$; hence equation (2.23) holds and the practical calculation of a matrix function is equivalent to the calculation of the N coefficients of the power series in equation (2.23). This can be easily done as follows:

1. All eigenvalues are different: If this is the case, we have N values λ_i, $1 \leq i \leq N$, for which $q(\lambda_i) = 0$. Then we can write

$$f(\lambda) = \gamma c(\lambda) \prod_{i=1}^{N} (\lambda - \lambda_i) + o(\lambda),$$

so we have a system of N equations with N unknowns as follows:

$$f(\lambda_i) = \sum_{k=0}^{N-1} \alpha_k \lambda_i^k, \quad 1 \leq i \leq N,$$

which can be easily solved.

2. Some eigenvalues have non-unitary multiplicity: For illustration purposes, we will assume that eigenvalue λ_j has multiplicity s_j. Then

$$f(\lambda) = \gamma c(\lambda)(\lambda - \lambda_j)^{s_j} \prod_{\substack{i=1 \\ i \neq j}}^{N-s_j} (\lambda - \lambda_i) + o(\lambda).$$

In this case, however, we will have only $N - s_j + 1$ equations for N, so we need $s_j - 1$ additional equations. These can be easily obtained by enlarging the system with the first $s_j - 1$ derivatives of the function above, i.e.,

$$f(\lambda_i) = \sum_{k=0}^{N-1} \alpha_k \lambda_i^k, \quad i \in (1, \ldots, N - s_j, j),$$

$$\frac{d^p f(\lambda)}{\lambda^p} = \sum_{k=0}^{N-1} \alpha_k \left. \frac{d^p \lambda^k}{d\lambda^p} \right|_{\lambda=\lambda_j}, \qquad 1 \leq p \leq s_j - 1.$$

2.3.4.2 An example which gives rise to a matrix exponential

Consider a problem of a dynamic system expressed as follows:

$$\frac{d\boldsymbol{x}(t)}{dt} = \boldsymbol{A}(t)\boldsymbol{x}(t) + \boldsymbol{B}(t)\boldsymbol{e}(t),$$

where $\boldsymbol{x}(t)$ is the system *state* at time t, $\boldsymbol{e}(t)$ is the known input to the system at that time instant, and both $\boldsymbol{A}(t)$ and $\boldsymbol{B}(t)$ are known system matrices.

The solution to this equation is known to be

$$\boldsymbol{x}(t) = \boldsymbol{\Phi}(t, t_0)\boldsymbol{x}_0 + \int_{t_0}^{t} \boldsymbol{\Phi}(t, \tau)\boldsymbol{B}(\tau)\boldsymbol{e}(\tau)d\tau, \quad t \geq t_0,$$

where the initial condition $\boldsymbol{x}(t_0) = \boldsymbol{x}_0$ has been applied. It turns out that matrix $\boldsymbol{\Phi}(t, \tau)$ is defined as a matrix exponential, according to

$$\boldsymbol{\Phi}(t, \tau) = e^{\int_{\tau}^{t} \boldsymbol{A}(\rho)d\rho}.$$

2.4 Linear processing and transformed domains

2.4.1 Linear processing. Convolution

The basis for linear processing is the convolution operation; this operation is used throughout the book and in this introductory chapter we will summarize some of its properties and recall what this operation becomes in the Fourier domain. To begin with, let $f[n]$, $g[n]$ and $h[n]$ be scalar functions of an integer variable, i.e., $f, g, h : \mathbb{Z} \to \mathbb{R}$. Then the (discrete) convolution operation is defined as

$$g[n] = \sum_{k=-\infty}^{\infty} f[k]h[n-k]. \tag{2.24}$$

This operation turns out to provide the output of a linear and invariant system, the impulse response of which is $h[n]$ and the input is $f[n]$ or, alternatively, the impulse response is $f[n]$ and the input is $h[n]$. Let us recall what these concepts mean. Assume a system provides the output $g_1[n]$ when the input is $f_1[n]$, and the output $g_2[n]$ when the input is $f_2[n]$:

- The system is linear if a linear combination of the inputs $af_1[n]+bf_2[n]$ provides the output $ag_1[n]+bg_2[n]$, $a, b \in \mathbb{R}$, i.e., it provides the same linear combination of the respective output.
- The system is said to be shift-invariant if the input $f_1[n-n_0]$ gives rise to the output $g_1[n-n_0]$, i.e., a shifted input gives rise to a shifted output, with the same shift as the input.
- Impulse response: Let $h[n]$ denote the system output for the input $\delta[n]$, with $\delta[n]$ being the Kronecker delta function, i.e., a function whose values are null $\forall n \neq 0$ and equal to 1 at $n=0$. Let $h_k[n]$ denote the system output for the shifted input $\delta[n-k]$. Note that $h_k[n]=h[n-k]$ if the system is linear (with $h_0[n]=h[n]$).

With these definitions, we can write

$$f[n]=\sum_{k=-\infty}^{\infty} f[k]\delta[n-k],$$

and provided that the system is both linear and shift-invariant, we can write

$$\begin{aligned} g[n] &= \sum_{k=-\infty}^{\infty} f[k]h_k[n] \\ &= \sum_{k=-\infty}^{\infty} f[k]h[n-k] = f[n]*h[n]. \end{aligned}$$

Similar conclusions can be derived for continuous variables; i.e., let $f, g, h : \mathbb{R} \to \mathbb{R}$. Then if $h(x)$ denotes the impulse response[1] of a linear and shift-invariant system, then the output $g(x)$ of such a system when the input is $f(x)$ can be expressed as

$$g(x)=\int_{-\infty}^{\infty} f(x-z)h(z)dz=\int_{-\infty}^{\infty} h(x-z)f(z)dz=f(x)*h(x). \tag{2.25}$$

Occasionally we will denote this operation by $g=f*h$ for notational simplicity.

Convolution, either discrete or continuous, satisfies several properties. We will concentrate on three of them:

- Convolution satisfies the commutative, associative, and distributive properties. With respect to the latter, it turns out that $(f+g)*h=f*h+g*h$.

[1] For a continuous variable, the impulse function $\delta(x)$ is the Dirac delta, the values of which are null $\forall t \neq 0$ and it also satisfies

$$\int_{-\infty}^{\infty} \delta(x)dx=1.$$

- Derivative of the convolution: It is easy to check that

$$\frac{dg(x)}{dx} = \frac{df(x)}{dx} * h(x) = f(x) * \frac{dh(x)}{dx},$$
$$\frac{d^n g(x)}{dx^n} = \frac{d^n f(x)}{dx^n} * h(x) = f(x) * \frac{d^n h(x)}{dx^n}.$$

The result is a direct consequence of the Leibniz rule (see Section 2.5.4).
- An extrapolation of the central limit theorem concerning convolutions: It is well known that the probability density function (PDF) of the summation of a large number of identically distributed random variables tends to be Gaussian. In addition, the PDF of the summation of independent random variables is the convolution of their PDFs. Hence, if a convolution kernel is applied in multiple convolutions, the effective convolution kernel tends to be Gaussian-shaped with as variance the variance of the convolution kernel times the number of convolutions carried out.

The convolution operator carries over to multiple dimensions, for functions defined on a discrete domain $f : \mathbb{Z}^N \rightarrow \mathbb{R}$ or a continuous domain $f : \mathbb{R}^N \rightarrow \mathbb{R}$. Specifically, for a discrete grid with $N = 2$, we could write

$$g[m,n] = \sum_{p,q} f[p,q]h[m-p,n-q] = f[m,n] * g[m,n].$$

2.4.2 Transformed domains

2.4.2.1 1D Fourier transform

Linear systems have a natural dual domain defined by means of the Fourier transform. By *dual* we mean a domain that has the same information as the original domain (so you can go forward and backward between both) but in which this information is shown differently and operations have alternative formulations. To get started, let $\mathcal{F}(u)$ denote the Fourier transform of function $f(x)$. Then the pair of direct and inverse transformations is

$$\mathcal{F}(u) = \int_{-\infty}^{\infty} f(x)e^{-j2\pi ux}dx, \tag{2.26}$$

$$f(x) = \int_{-\infty}^{\infty} F(u)e^{j2\pi ux}du. \tag{2.27}$$

These operations will be respectively denoted as $\mathbb{F}\{f(x)\}$ and $\mathbb{F}^{-1}\{\mathcal{F}(u)\}$. Sometimes this transformation is expressed as a function of another variable, say $\xi = 2\pi u$. If this is the case, a normalizing factor $\frac{1}{2\pi}$ multiplies the integral in the second equation above.

Fourier transforms satisfy a number of properties that are summarized in Chapter 5. Here we will compute some Fourier transform pairs to illustrate some take-home messages related to this operation. Consider a function $\delta(t)$. It is easy to see that

$$\int_{-\infty}^{\infty} \delta(x)e^{-j2\pi ux}dx = \int_{-\infty}^{\infty} e^{-j2\pi u0}\delta(x)dx = e^{-j2\pi u0}\int_{-\infty}^{\infty} \delta(x)dx = 1,$$

i.e., the Fourier transform of an infinitely narrow function has a constant value $\forall u \in \mathbb{R}$. Alternatively, the inverse Fourier transform of $\delta(u)$ is

$$\int_{-\infty}^{\infty} \delta(u)e^{j2\pi ux}du = e^{j2\pi 0x}\int_{-\infty}^{\infty} \delta(u)du = 1,$$

i.e., the same effect occurs but in the opposite direction. Consider now both $\delta(x - x_0)$ and the same function in the transformed domain $\delta(u - u_0)$. In this case, the respective direct and inverse Fourier transforms are

$$\int_{-\infty}^{\infty} \delta(x - x_0)e^{-j2\pi ux}dx = e^{-j2\pi ux_0}\int_{-\infty}^{\infty} \delta(x - x_0)dx = e^{-j2\pi ux_0},$$
$$\int_{-\infty}^{\infty} \delta(u - u_0)e^{j2\pi ux}du = e^{j2\pi u_0 x}\int_{-\infty}^{\infty} \delta(u - u_0)du = e^{j2\pi u_0 x}. \tag{2.28}$$

These results generalize to the shifting property, which can be expressed as

$$\mathbb{F}\{f(x - x_0)\} = e^{-j2\pi ux_0}\mathbb{F}\{f(x)\} = e^{-j2\pi ux_0}\mathcal{F}(u),$$
$$\mathbb{F}^{-1}\{\mathcal{F}(u - u_0)\} = e^{j2\pi u_0 x}\mathbb{F}^{-1}\{\mathcal{F}(u)\} = e^{j2\pi u_0 x}f(x). \tag{2.29}$$

The second property is of special interest in communication theory as we will mention shortly after.

This Fourier transform is linear, i.e., the Fourier transform of the linear combination of two functions is the same linear combination of the Fourier transforms of each function. With this in mind, we now calculate $\mathbb{F}\{\cos(2\pi u_0 x)\}$. To this end, we write

$$\begin{aligned}\mathbb{F}\{\cos(2\pi u_0 t)\} &= \mathbb{F}\left\{\frac{e^{j2\pi u_0 u} + e^{-j2\pi u_0 u}}{2}\right\}\\ &= \frac{1}{2}\{\delta(u - u_0) + \delta(u + u_0)\},\end{aligned}$$

in which we have used equation (2.28). As can be observed, a sinusoid of frequency – either spatial or temporal, depending on the physical meaning of variable x – consists of two Dirac delta functions in the Fourier domain. Actually, a complex exponential of frequency u_0 is a delta at that frequency in the Fourier domain. This lets us give a simple interpretation to equation (2.26); the operation carried out is the inner product of function $f(x)$ with the complex exponential $e^{j2\pi u_0 u}$; hence, equation (2.26) measures the contribution of frequency u to create function $f(x)$. Functions (signals) varying rapidly in the original domain will have higher values of $|\mathcal{F}(u)|$ at higher values of u than other functions with variations less pronounced.

We now get back to equation (2.29). Let us assume that we multiply $f(x)$ by the signal $\cos(2\pi u_0 x)$. Bearing in mind the linearity property of operator $\mathbb{F}\{\}$, we can write

$$\mathbb{F}\{f(x)\cos(2\pi u_0 x)\} = \frac{1}{2}\{\mathcal{F}(u-u_0)+\mathcal{F}(u+u_0)\},$$

which constitutes the basis for modulating a signal $f(x)$ to some central frequency u_0. This result generalizes to the modulation property, which states that the product of two functions in the original domain translates itself into a convolution in the transformed domain. Complementarily, the convolution property states that a convolution in the original domain results in a multiplication in the transformed domain. Let us check this is the case:

$$\begin{aligned}
\mathbb{F}\{f(x)*h(x)\} &= \mathbb{F}\left\{\int_{-\infty}^{\infty} f(x-z)h(z)dz\right\} \\
&= \int_{-\infty}^{\infty}\left[\int_{-\infty}^{\infty} f(x-z)h(z)dz\right]e^{-j2\pi ux}dx \\
&= \int_{-\infty}^{\infty}\left[\underbrace{\int_{-\infty}^{\infty} f(x-z)e^{-j2\pi ux}dx}_{\mathcal{F}(u)e^{-j2\pi uz}}\right]h(z)dz \\
&= \mathcal{F}(u)\int_{-\infty}^{\infty} h(z)e^{-j2\pi uz}dz = \mathcal{F}(u)\mathcal{H}(u),
\end{aligned}$$

where we have assumed we can change the order of integration. This is the case when the Dirichlet conditions of existence of the Fourier transform hold.

We now illustrate the property we mentioned above about a narrow function and how it translates in the Fourier domain. This can be easily generalized by means of transforming a Gaussian-shaped function. Specifically, consider the transformation of the PDF of a zero-mean Gaussian variable with variance σ^2, i.e.,

$$f(x) = \frac{1}{\sigma\sqrt{2\pi}}e^{-\frac{x^2}{2\sigma^2}},$$

which is defined $\forall x \in \mathbb{R}$. Remember that the higher σ^2 is, the wider is this function (and the smaller its maximum since the area under this function should equal 1). Hence

$$\begin{aligned}
\mathbb{F}\{f(x)\} &= \int_{-\infty}^{\infty}\frac{1}{\sigma\sqrt{2\pi}}e^{-\frac{x^2}{2\sigma^2}}e^{-j2\pi ux}dx \\
&= \int_{-\infty}^{\infty}\frac{1}{\sigma\sqrt{2\pi}}e^{-Q(x)}dx.
\end{aligned}$$

This integral can be readily solved by trying to find the area under a PDF; to this end, we work on the exponent as follows:

$$\begin{aligned}
Q(x) &= \frac{1}{2\sigma^2}\left(x^2 + j4\pi u\sigma^2 x\right) \\
&= \frac{1}{2\sigma^2}\left(x^2 + j4\pi u\sigma^2 x + (j2\pi u\sigma^2)^2 - (j2\pi u\sigma^2)^2\right) \\
&= \frac{1}{2\sigma^2}\left((x - j2\pi\sigma^2 u)^2 - (j2\pi u\sigma^2)^2\right).
\end{aligned}$$

Consequently[2]

$$\begin{aligned}
\mathbb{F}\{f(x)\} &= \int_{-\infty}^{\infty} \frac{1}{\sigma\sqrt{2\pi}} e^{-\frac{1}{2\sigma^2}\left((x-j2\pi\sigma^2 u)^2 - (j2\pi u\sigma^2)^2\right)} dx \\
&= e^{\frac{1}{2\sigma^2}(j2\pi u\sigma^2)^2} \underbrace{\int_{-\infty}^{\infty} \frac{1}{\sigma\sqrt{2\pi}} e^{-\frac{1}{2\sigma^2}(x-j2\pi\sigma^2 u)^2} dx}_{=1} \\
&= e^{-2\pi^2\sigma^2 u^2}.
\end{aligned} \tag{2.30}$$

As can be observed, the behavior of σ in the original and transformed domains is opposed. The greater σ, the wider $f(x)$ but the narrower $\mathcal{F}(u)$, and vice versa. It is noteworthy that a Gaussian-shaped function also has a Gaussian-shaped Fourier transform.

In the case that the function is defined on a discrete-valued variable, i.e., function $f : \mathcal{Z} \to \mathcal{R}$, the Fourier transform is adapted accordingly. All the properties mentioned above carry over to this case, although care must be taken about the periodic nature of this transformation. Let $\Omega = 2\pi u$ and let n be the independent variable of this function. Then

$$\mathcal{F}(\Omega) = \mathbb{F}\{f(n)\} = \sum_{n=-\infty}^{\infty} f[n] e^{-j\Omega n}, \tag{2.31}$$

$$f[n] = \frac{1}{2\pi}\int_{<2\pi>} \mathcal{F}(\Omega) e^{j\Omega n} d\Omega, \tag{2.32}$$

where $< 2\pi >$ means *integration within any interval of length* 2π. In this case, as stated above, $\mathcal{F}(\Omega + 2\pi k) = \mathcal{F}(\Omega)$, $k \in \mathcal{Z}$, due to the fact that

[2] This integral is unity irrespective of the location parameter of the Gaussian curve, even for the case of a complex parameter. This can be rigorously demonstrated by using the Cauchy-Goursat theorem. Similar results can be obtained by noting that $\mathbb{F}\{f(x)\} = E\{e^{-j2\pi ux}\}$ is the characteristic function $\Phi(\omega)$ of a Gaussian curve evaluated at $\omega = -2\pi u$.

$$
\begin{aligned}
\mathcal{F}(\Omega+2\pi k) &= \sum_{n=-\infty}^{\infty} f[n]e^{-j(\Omega+k2\pi)} \\
&= e^{-jk2\pi}\sum_{n=-\infty}^{\infty} f[n]e^{-j\Omega n} = \sum_{n=-\infty}^{\infty} f[n]e^{-j\Omega n} = \mathcal{F}(\Omega).
\end{aligned}
$$

This is the reason why the integration above can be carried out in $< 2\pi >$.

2.4.2.2 2D Fourier transform

The pair of direct and inverse Fourier transforms expressed in equations (2.26)–(2.27) carries over to the case $f : \mathcal{R}^2 \rightarrow \mathcal{R}$ as follows:

$$
\mathcal{F}(u,v) = \int_{-\infty}^{\infty}\int_{-\infty}^{\infty} f(x,y)e^{-j2\pi(ux+vy)}dxdy, \tag{2.33}
$$

$$
f(x,y) = \int_{-\infty}^{\infty}\int_{-\infty}^{\infty} F(u,v)e^{j2\pi(ux+vy)}dudv. \tag{2.34}
$$

As can be observed, the transformation kernel in this case is separable, so the transformation can be expressed as a cascade of 1D transformations,

$$
\begin{aligned}
\mathcal{F}(u,v) &= \int_{-\infty}^{\infty}\int_{-\infty}^{\infty} f(x,y)e^{-j2\pi(ux+vy)}dxdy \\
&= \int_{-\infty}^{\infty}\underbrace{\left[\int_{-\infty}^{\infty} f(x,y)e^{-j2\pi ux}dx\right]e^{-j2\pi vy}}_{H(u,y)}dy \\
&= \int_{-\infty}^{\infty} H(u,y)e^{-j2\pi vy}dy.
\end{aligned}
$$

In the case that the function $f(x,y)$ is separable, i.e., $f(x,y) = f_1(x)f_2(y), \forall x,y$, equation (2.33) is just the tensor product of two 1D Fourier transforms in, say, horizontal and vertical directions.

Let us get some insight into this, since this is of interest for image processing. Let x be a variable defined in the horizontal axis and let y be a variable defined in the vertical axis. We will find the Fourier transform of a pulse of height 1 and width Δ in the horizontal direction, and infinitely long in the vertical direction, i.e., $f(x,y) = \Pi\left(\frac{x}{\Delta}\right)$, with $\Pi(x)$ being a function whose value is 1 whenever $|x| < \frac{1}{2}$ and 0 otherwise. Accordingly,

$$
\begin{aligned}
\mathcal{F}(u,v) &= \int_{-\infty}^{\infty}\int_{-\infty}^{\infty} f(x,y)e^{-j2\pi(ux+vy)}dxdy \\
&= \int_{-\frac{\Delta}{2}}^{\frac{\Delta}{2}} e^{-j2\pi ux}dx \underbrace{\int_{-\infty}^{\infty} e^{-j2\pi vy}dy}_{=\delta(v)}
\end{aligned}
$$

$$
\begin{aligned}
&= \frac{1}{-j2\pi u} e^{-j2\pi ux}\Big|_{-\frac{\Delta}{2}}^{\frac{\Delta}{2}} \delta(v) = \frac{e^{\pi u\Delta} - e^{-\pi u\Delta}}{j2\pi u} \\
&= \frac{\sin(\pi u\Delta)}{\pi u}\delta(v) = \Delta \mathrm{sinc}(\Delta u)\delta(v),
\end{aligned}
$$

where function $\mathrm{sinc}(x) = \frac{\sin(\pi x)}{\pi x}$. If function $f(x, y)$ is interpreted as an image, you would see a vertical strip. However, its Fourier transform consists of a function that is non-null only in the horizontal direction and at $v = 0$ (because of the presence of $\delta(v)$). The weight of δ is a sinc function of the horizontal frequency variable, the main lobe of which becomes narrower as Δ increases, i.e., as the pulse in the original domain becomes wider, and vice versa.

It is also interesting to see the effect of a rotation of the function in the original domain on the Fourier transform. Assume that function $\mathcal{F}(u.v) = \mathbb{F}\{f(x, y)\}$. It will be more convenient to rewrite variables in their polar coordinates, i.e., $u = \xi \cos(\phi)$ and $v = \xi \sin(\phi)$, i.e.,

$$
\mathcal{F}_p(\xi, \phi) = \mathcal{F}(\xi \cos(\phi), \xi \sin(\phi)) = \int_{-\infty}^{\infty}\int_{-\infty}^{\infty} f(x, y)e^{-j2\pi(\xi \cos(\phi)x+\xi \sin(\phi)y)}dxdy.
$$

If we change variables within the integral $x = \rho \cos(\theta)$ and $y = \rho \sin(\theta)$, bearing in mind that the Jacobian of this transformation lets us write $dxdy = \rho d\rho d\theta$, we rewrite

$$
\begin{aligned}
\mathcal{F}_p(\xi, \phi) &= \mathcal{F}(\xi \cos(\phi), \xi \sin(\phi)) \\
&= \int_0^{\infty}\int_0^{2\pi} f(\rho \cos(\theta), \rho \sin(\theta))e^{-j2\pi(\xi \cos(\phi)\rho \cos(\theta)+\xi \sin(\phi)\rho \sin(\theta))}\rho d\rho d\theta \\
&= \int_0^{\infty}\int_0^{2\pi} f_P(\rho, \theta)e^{-j2\pi\xi\rho \cos(\phi-\theta)}\rho d\rho d\theta.
\end{aligned}
$$

If we aim to calculate the Fourier transform of the rotated function, say, $f_P(\rho, \theta + \alpha)$, then

$$
\begin{aligned}
\mathcal{F}_{\mathrm{rot}}(\xi, \phi) &= \int_0^{\infty}\int_0^{2\pi} f_P(\rho, \theta + \alpha)e^{-j2\pi\xi\rho \cos(\phi-\theta)}\rho d\rho d\theta \\
&= \int_0^{\infty}\int_0^{2\pi} f_P(\rho, \beta)e^{-j2\pi\xi\rho \cos(\phi-(\beta-\alpha))}\rho d\rho d\theta \\
&= \int_0^{\infty}\int_0^{2\pi} f_P(\rho, \beta)e^{-j2\pi\xi\rho \cos(\phi+\alpha-\beta)}\rho d\rho d\theta \\
&= \mathcal{F}_p(\xi, \phi + \alpha),
\end{aligned}
$$

where we have changed the variable $\beta = \theta + \alpha$.

Finally, for discrete variables, i.e., $f : \mathbb{Z}^2 \to \mathbb{R}$, straightforward extension of the definition in equation (2.31) applies, i.e.,

$$\mathcal{F}(\Omega_1, \Omega_2) = \sum_{m=-\infty}^{\infty} \sum_{n=-\infty}^{\infty} f[m,n] e^{-j\Omega_1 m + \Omega_2 n}. \tag{2.35}$$

Properties of the 1D case carry over to the 2D case.

2.4.2.3 *N*-dimensional Fourier transform

The definitions in the two previous sections generalize to the N-dimensional case as follows:

$$\mathcal{F}(\boldsymbol{u}) = \int f(\boldsymbol{x}) e^{-j2\pi \boldsymbol{u}^T \boldsymbol{x}} d\boldsymbol{x}, \tag{2.36}$$

$$f(\boldsymbol{u}) = \int \mathcal{F}(\boldsymbol{x}) e^{-j2\pi \boldsymbol{u}^T \boldsymbol{x}} d\boldsymbol{u}. \tag{2.37}$$

Properties also hold for this case. An example, we are going to calculate the Fourier transform of a multivariate Gaussian-shaped function, defined as

$$g(\boldsymbol{x}) = e^{-\frac{1}{2}\boldsymbol{x}^T \boldsymbol{\Delta} \boldsymbol{x}}, \tag{2.38}$$

where we have intentionally omitted normalizing constants. We will find the transform in two steps:

- Assume $\mathcal{F}(\boldsymbol{u}) = \{f(\boldsymbol{x})\}$. Let $g(\boldsymbol{x}) = f(\mathbf{A}\boldsymbol{x})$. We will find the Fourier transform of $g(\boldsymbol{x})$:

$$\begin{aligned}
\mathcal{G}(\boldsymbol{u}) &= \int f(\mathbf{A}\boldsymbol{x}) e^{-j2\pi \boldsymbol{u}^T \boldsymbol{x}} d\boldsymbol{x} \\
&= \int f(\boldsymbol{y}) e^{-j2\pi \boldsymbol{u}^T \mathbf{A}^{-1} \boldsymbol{y}} \frac{1}{|\det(\mathbf{A})|} d\boldsymbol{y} \\
&= \frac{1}{|\det(\mathbf{A})|} \int f(\boldsymbol{y}) e^{-j2\pi \left(\left(\left(\mathbf{A}^{-1}\right)^T \boldsymbol{u}\right)^T \boldsymbol{y}\right.} d\boldsymbol{y} \\
&= \frac{1}{|\det(\mathbf{A})|} \mathcal{F}\left(\left(\mathbf{A}^{-1}\right)^T \boldsymbol{u}\right),
\end{aligned} \tag{2.39}$$

  where we have changed variables $\boldsymbol{y} = \mathbf{A}\boldsymbol{x}$ so $|\det(\mathbf{A})| d\boldsymbol{x} = d\boldsymbol{y}$. We assume that $\mathbf{A}$ is a full-rank square matrix and, consequently, invertible.
- Recall equation (2.38). Matrix Δ is the inverse of a covariance matrix which is symmetric and positive definite for real variables. Hence, using equation (2.18) and defining $\mathbf{V} = \mathbf{U}^T$ and $\boldsymbol{y} = \mathbf{V}\boldsymbol{x}$ we can write

$$\begin{aligned} g(\boldsymbol{x}) &= e^{-\frac{1}{2}\boldsymbol{x}^T \boldsymbol{\Delta} \boldsymbol{x}} \\ &= e^{-\frac{1}{2}\boldsymbol{x}^T \mathbf{V}^T \boldsymbol{\Lambda} \mathbf{V} \boldsymbol{x}} \\ &= f(\mathbf{V}\boldsymbol{x}). \end{aligned}$$

Defining

$$f(\boldsymbol{y}) = e^{-\frac{1}{2}\boldsymbol{y}^T \boldsymbol{\Lambda} \boldsymbol{y}} = e^{-\frac{1}{2}\sum_{i=1}^{N} \lambda_i y_i^2} = \prod_{i=1}^{N} e^{-\frac{\lambda_i y_i^2}{2}},$$

its N-dimensional Fourier transform can be readily obtained by (1) taking in consideration that the function is separable in the variables, and so is the transformation kernel, and hence the N-dimensional transform will be the product of 1D Fourier transforms, and (2) making use of the result in equation (2.30) with $\lambda_i^2 = 1/\sigma^2$, and therefore

$$\mathcal{F}(\boldsymbol{u}) = \prod_{i=1}^{N} e^{-\frac{2\pi^2 u_i^2}{\lambda_i}} = e^{-2\pi^2 \boldsymbol{u}^T \boldsymbol{\Lambda}^{-1} \boldsymbol{u}}.$$

Finally, using equation (2.39) we end up with

$$\begin{aligned} \mathcal{G}(\boldsymbol{u}) = \mathcal{F}(\boldsymbol{u}) &= \frac{1}{|\det(\mathbf{V})|} \mathcal{F}\left(\left(\mathbf{V}^{-1}\right)^T \boldsymbol{u}\right) \\ &= e^{-2\pi^2 ((\mathbf{V}^{-1})^T \boldsymbol{u})^T \boldsymbol{\Lambda}^{-1} \mathbf{V}^{-1})^T \boldsymbol{u}} \\ &= e^{-2\pi^2 \boldsymbol{u}^T \mathbf{V}^T \boldsymbol{\Lambda}^{-1} \mathbf{V} \boldsymbol{u}} \\ &= e^{-2\pi^2 \boldsymbol{u}^T \boldsymbol{\Delta}^{-1} \boldsymbol{u}}, \end{aligned}$$

i.e., the Gaussian shape is maintained in the transformed domain albeit with inverse covariance behavior.

2.5 Calculus

2.5.1 Derivatives, gradients, and Laplacians

Let $f(\boldsymbol{x})$ be a scalar function of a vector variable, i.e., $f : \mathbb{R}^d \to \mathbb{R}$. The gradient of this function is a d-component vector, the components of which are the partial derivatives of the function with respect to each of the components of the input variable, i.e.,

$$\nabla f(\boldsymbol{x}) = \begin{bmatrix} \frac{\partial f(\boldsymbol{x})}{\partial x_1} & \frac{\partial f(\boldsymbol{x})}{\partial x_2} & \cdots & \frac{\partial f(\boldsymbol{x})}{\partial x_d} \end{bmatrix}^T,$$

where the *nabla* operator ∇ is a vector-differential operator defined as

$$\nabla = \sum_{i=1}^{d} \frac{\partial}{\partial x_i} \boldsymbol{n}_i$$

and $\boldsymbol{n}_i$ is the unit vector in the direction of each of the variables. The directional (or Gateaux) derivative of function $f(\boldsymbol{x})$ in the direction given by vector $\boldsymbol{n}$ is defined as follows:

$$\lim_{h \to 0} \frac{f(\boldsymbol{x} + h\boldsymbol{n}) - f(\boldsymbol{x})}{h},$$

which turns out to be equal to the dot product

$$< \nabla f(\boldsymbol{x}), \boldsymbol{n} >= [\nabla f(\boldsymbol{x})]^T \boldsymbol{n}.$$

The Gateaux derivative is usually represented as $\delta_{\boldsymbol{n}} f(\boldsymbol{x})$. The gradient operator is frequently used in optimization processes since a sufficient condition for a point, say $\boldsymbol{x}^*$, to be an extremum of a function (i.e., either a local maximum or local minimum) is

$$\nabla \; f(\boldsymbol{x})|_{\boldsymbol{x}=\boldsymbol{x}^*} = \boldsymbol{0}.$$

The Laplacian of a multivariate function is a scalar function obtained as the summation of the second-order partial derivatives of the function with respect to each component, i.e.,

$$\Delta f(\boldsymbol{x}) = \nabla \cdot (\nabla f(\boldsymbol{x})) = \sum_{i=1}^{d} \frac{\partial^2 f(\boldsymbol{x})}{\partial x_i^2}.$$

The divergence is defined on a vectorial function. Specifically, consider now that $\boldsymbol{f}(\boldsymbol{x})$ with $\boldsymbol{f} : \mathbb{R}^d \to \mathbb{R}^d$. Let f_i denote the ith component of $\boldsymbol{f}(\boldsymbol{x})$. Then, the divergence of $\boldsymbol{f}$ is defined as

$$\operatorname{div} \boldsymbol{f} = \nabla \cdot \boldsymbol{f}(\boldsymbol{x}) = \sum_{i=d}^{d} \frac{\partial f_i(\boldsymbol{x})}{\partial x_i}.$$

2.5.2 Calculus of variations

We may not only be interested in finding points that are local extrema of a function, but we may also be interested in finding a function itself that minimizes some cost function. This situation gives rise to the so-called calculus of variations which is, loosely speaking, an extension to infinite dimensions of the problem of finding local extrema of a function. We will first consider a functional $\Phi(y)$, i.e., a scalar function with argument a function y. Function $y : \mathcal{D} \subset \mathbb{R}^d \to \mathbb{R}$ is considered derivable to some order within some domain. In this scenario, a function y^* is considered an extremum of the functional if the Gateaux derivative

$$\delta_v \Phi(y)|_{y=y^*} = 0$$

$\forall v \in \mathcal{D}$.

We get started with a simple case. Assume we have a 1D problem in which we look for a curve $y(x)$ defined in the interval $[a, b]$ that should be the extremum of the following functional:

$$\Phi(y) = \int_a^b u(x, y(x), y'(x))dx,$$

where $y'(x)$ is a shortcut for $y'(x) = \frac{dy(x)}{dx}$. In this case, it is not difficult to apply the condition on the Gateaux derivative indicated above to obtain the *Euler–Lagrange* sufficient condition

$$\frac{d}{dx}\frac{\partial u(x, y, y')}{\partial y'} = \frac{\partial u(x, y, y')}{\partial y}$$

for $y(x)$ to be an extremum of the functional. Note that in the expressions above partial derivatives with respect to y and y' do not take into account dependence on x, i.e., y and y' are considered independent variables. The full derivative with respect to x takes this dependence into account, i.e., function $u()$ will, generally speaking, depend on x both by itself and through $y(x)$ and $y'(x)$.

Example. Find the 2D curve $y(x)$ that goes through points (x_1, y_1) and (x_2, y_2) and it is an extremum of the functional

$$\Phi(y) = \int_{x_1}^{x_2} \left((y')^2 - 3xy\right) dx.$$

The solution is found by applying the *Euler–Lagrange* sufficient condition just mentioned, taking into account that $u(x, y, y') = (y')^2 - 3xy$. Hence

$$\frac{\partial u}{\partial y} = -3x,$$
$$\frac{\partial u}{\partial y'} = 2y',$$
$$\frac{d}{dx}\left(\frac{\partial u}{\partial y'}\right) = 2y'',$$

so

$$2y'' = -3x,$$
$$y'' = -\frac{3}{2}x,$$

$$y' = -\frac{3}{4}x^2 + C,$$
$$y = -\frac{3}{12}x^3 + Cx + D.$$

The constants are found by applying the condition that the curve goes through the points indicated above, i.e.,

$$y_1 = -\frac{3}{12}x_1^3 + Cx_1 + D,$$
$$y_2 = -\frac{3}{12}x_2^3 + Cx_2 + D,$$
$$y_1 - y_2 = -\frac{3}{12}\left(x_1^3 - x_2^3\right) + C(x_1 - x_2) \rightarrow C = \frac{y_1 - y_2 + \frac{3}{12}\left(x_1^3 - x_2^3\right)}{x_1 - x_2},$$
$$D = y_1 + \frac{3}{12}x_1^3 - Cx_1.$$

Let us consider now more involved cases. Suppose we look for contour parameterized by s, and the contour is defined by its coordinates $(x(s), y(s))$. In this case, we will create a functional by considering derivatives up to order two, i.e.,

$$\Phi(x, y) = \int_0^1 u(s, x(s), y(s), x'(s), y'(s), x''(s), y''(s))ds.$$

In this case, the Euler–Lagrange equations take the form

$$\frac{\partial u}{\partial x} + \sum_{i=1}^{2}(-1)^i \frac{\partial^i}{\partial s^i}\frac{\partial u}{\partial x^{('q)}} = \frac{\partial u}{\partial x} - \frac{\partial}{\partial s}\frac{\partial u}{\partial x'} + \frac{\partial^2}{\partial s^2}\frac{\partial u}{\partial x''}, \tag{2.40}$$

$$\frac{\partial u}{\partial y} + \sum_{i=1}^{2}(-1)^i \frac{\partial^i}{\partial s^i}\frac{\partial u}{\partial y^{('q)}} = \frac{\partial u}{\partial y} - \frac{\partial}{\partial s}\frac{\partial u}{\partial y'} + \frac{\partial^2}{\partial s^2}\frac{\partial u}{\partial y''}. \tag{2.41}$$

Index i in the summation ranges up to the order of the derivatives that are used in the functional; in this case this order has been limited to two. We have used $x^{('q)}$ as a shortcut for the derivative of order q of function $x(s)$ (and similarly for $y(s)$).

Finally we consider the case of several variables. Assume we have variables $(x, y, z) \in \mathcal{D}$ and we assume that a function $\phi : \mathbb{R}^3 \rightarrow \mathbb{R}$ is also defined. The first-order partial derivatives of this function will be referred to as ϕ'_x, ϕ'_y, and ϕ'_z. This being the case, consider that we need to find an extremum of the following problem:

$$\Phi(\phi) = \int_{\mathcal{D}} u\left(x, y, z, \phi, \phi_x, \phi_y, \phi_z\right) dxdydz.$$

In this case, the Euler–Lagrange sufficient condition turns out to be

$$\frac{\partial u}{\partial \phi} - \frac{\partial}{\partial x}\left(\frac{\partial u}{\phi'_x}\right) - \frac{\partial}{\partial y}\left(\frac{\partial u}{\phi'_y}\right) - \frac{\partial}{\partial z}\left(\frac{\partial u}{\phi'_z}\right) = 0. \tag{2.42}$$

The solutions to equations (2.40), (2.41), and (2.42) are typically found by discretizing the image grid and applying numerical procedures to approximate the root of these equations. Specifically, consider equation (2.42) and let us accept a time dependence of ϕ. We write

$$\frac{\partial \phi}{\partial t} \approx \frac{\phi(t + \Delta t) - \phi(t)}{\Delta t} = \Psi(\phi(t)), \tag{2.43}$$

where $\Psi(\cdot)$ is the left-hand side term in equation (2.42). This gives rise to the iteration

$$\phi(t + \Delta t) = \phi(t) + \Delta t \Psi(\phi(t))$$

until a stable result is found.

2.5.3 Some specific cases

2.5.3.1 Laplace equation

A well-known equation is the so-called Laplace equation. Consider $\boldsymbol{x} \in \Omega \subset \mathbb{R}^d$ and $u : \mathbb{R}^d \rightarrow \mathbb{R}$. The Laplace equation is stated as follows:

$$\Delta u(\boldsymbol{x}) = 0, \tag{2.44}$$

i.e., as a function the Laplacian of which is null. This function is a particular case of the heat equation that we will describe in the next section (as well as the particular case of other equations). Solutions to this equation are called *harmonic functions*. The problem is completely defined when some conditions are added. Usual conditions are the *Dirichlet* conditions, which set the value of function $\Delta u(\boldsymbol{x})$ for $\boldsymbol{x} \in \delta\Omega$, where $\delta\Omega$ is the boundary of domain Ω.

Due to its simplicity, we can easily find a numerical scheme to approximate the solution to this equation. Consider that we have a discrete Cartesian grid, with points separated by the quantity h_i in each of the d dimensions of our problem. Consider also a canonical base $\boldsymbol{v}_i$, the value of which is h_i in the ith component $(1 \leq i \leq d)$ and 0 otherwise. Then, we approximate the partial derivative at point $\boldsymbol{x}^k$ as follows:

$$\left.\frac{\partial u(\boldsymbol{x})}{\partial x_i}\right|_{\boldsymbol{x}^k} \approx \frac{u(\boldsymbol{x}^k + \boldsymbol{v}_i) - u(\boldsymbol{x}^k)}{h_i}.$$

Likewise, second-order derivatives will be taken as *centered derivatives*, i.e.,

$$\begin{aligned}\left.\frac{\partial^2 u(\boldsymbol{x})}{\partial x_i^2}\right|_{\boldsymbol{x}^k} &\approx \frac{\left.\frac{\partial u(\boldsymbol{x})}{\partial x_i}\right|_{\boldsymbol{x}^k} - \left.\frac{\partial u(\boldsymbol{x})}{\partial x_i}\right|_{\boldsymbol{x}^k - \boldsymbol{v}_i}}{h_i} \\ &= \frac{u(\boldsymbol{x}^k + \boldsymbol{v}_i) - 2u(\boldsymbol{x}^k) + u(\boldsymbol{x}^k - \boldsymbol{v}_i)}{h_i^2}.\end{aligned}$$

With these definitions, we can approximate the Laplace equation at point $\boldsymbol{x}^k$,

$$\sum_{i=1}^{d} \frac{u(\boldsymbol{x}^k + \boldsymbol{v}_i) - 2u(\boldsymbol{x}^k) + u(\boldsymbol{x}^k - \boldsymbol{v}_i)}{h_i^2} = 0, \tag{2.45}$$

so we can solve for $u(\boldsymbol{x}^k)$ and find a solution through the iterative process

$$u^{(m+1)}(\boldsymbol{x}^k) = \frac{\sum_{i=i}^{d} \frac{u^{(m)}(\boldsymbol{x}^k+\boldsymbol{v}_i)+u^{(m)}(\boldsymbol{x}^k-\boldsymbol{v}_i)}{h_i^2}}{\sum_{i=1}^{d} \frac{2}{h_i^2}}.$$

In other to apply this procedure we need an initial solution, the Dirichlet conditions, and a visiting ordering. Then we iterate until convergence.

This is an explicit solution. Implicit solutions can also be devised. Recall from equation (2.45) that we have a linear relation among the values of $u(\boldsymbol{x}^k)$ and the neighbors involved in the derivatives. Actually, if equation (2.45) is extended to all the points within domain Ω, we get a linear system that we write as

$$\mathbf{A}\boldsymbol{u} = \boldsymbol{f},$$

where $\mathbf{A}$ is a sparse matrix with coefficients as indicated in equation (2.45), $\boldsymbol{u}$ is a vector of the unknown values of function $u()$ within the discrete set of points in domain Ω, and $\boldsymbol{f}$ is a vector containing zeroes and the values of $u()$ in $\delta\Omega$. The system can be solved with standard numerical methods.

2.5.3.2 Heat (or diffusion) equation

The heat equation is based on two principles. The first is the Fick law, which states that the gradient of the temperature $u(\boldsymbol{x})$ gives rise to a flow

$$\boldsymbol{\Phi} = -\mathbf{D}\nabla u,$$

where $\mathbf{D}$ is the diffusion tensor that models the possibility that diffusion does not take place in the same direction as the temperature gradient, i.e., the diffusion process is anisotropic. The isotropic case takes places when these two directions coincide, i.e., the tensor becomes a scalar value – which may still be a function of variable $\boldsymbol{x}$. In the case that this scalar turns out to be a constant, i.e., there is no spatial dependence, the diffusion process is referred to as homogeneous.

The second principle is the continuity equation, which states

$$\frac{\partial u}{\partial t} = -\mathrm{div}\,(\boldsymbol{\Phi}) = \mathrm{div}\,(\mathbf{D}\nabla u)\,.$$

Note that the Laplace equation is the particular case of this result for the stationary solution after an isotropic and homogeneous diffusion process.

2.5.4 Leibniz rule for interchanging integrals and derivatives

Assume function $\varphi(u)$ is a scalar function of a scalar variable, i.e., $\varphi : \mathbb{R} \to \mathbb{R}$, defined by means of the following integral expression:

$$\varphi(u) = \int_{\mu_i(u)}^{\mu_s(u)} G(\tau, u) d\tau.$$

Assume all the functions involved are derivable with respect to variable u. Then it turns out that the derivative of the equality above can be written as follows:

$$\begin{aligned}\frac{d\varphi(u)}{du} &= \frac{d}{du}\int_{\mu_i(u)}^{\mu_s(u)} G(\tau, u) d\tau \\ &= G(\mu_s(u), u)\frac{d\mu_s(u)}{du} - G(\mu_i(u), u)\frac{d\mu_i(u)}{du} + \int_{\mu_i(u)}^{\mu_s(u)} \frac{\partial G(\tau, u)}{\partial u} d\tau.\end{aligned} \tag{2.46}$$

2.6 Notions on shapes

In this section we will provide an introduction to shape analysis for the simple 2D case; this topic will be used in subsequent chapters for more general cases. Most of the material is inspired by [5], although notation has been adapted for this book and some mathematical derivations have been extended for better comprehension. Consider that a shape is defined by means of a set of 2D points, as many as k, and each of which will be represented by a complex scalar. Hence, if we have two shapes, they will be denoted by vectors $\boldsymbol{x}, \boldsymbol{y} \in \mathbb{C}^k$. Without loss of generality we will consider that these two shapes are centered; this means that their center of gravity is zero, i.e., $\mathbf{1}_k^T \boldsymbol{x} = \mathbf{1}_k^T \boldsymbol{y} = 0$, with $\mathbf{1}_k$ being a column vector with all its entries equal to 1.

2.6.1 Procrustes matching between two planar shapes

We will describe the planar Procrustes analysis. This analysis consists in finding the transformation that $\boldsymbol{y}$ requires so that the (Euclidean) distance between this vector and $\boldsymbol{x}$ is minimized. We will understand that the transformation consists in a rotation, a scaling, and a displacement, i.e.,

$$T(\boldsymbol{y}) = \beta e^{j\phi} \boldsymbol{y} + z\mathbf{1}_k,$$

where β is responsible for the scaling, the complex exponential carries out the rotation, and z is a complex number responsible of the displacement. The vectors entering the optimization will be normalized so that their norm is unity, i.e., $\boldsymbol{x}_N = \boldsymbol{x}/\sqrt{\boldsymbol{x}^H\boldsymbol{x}}$ and $\boldsymbol{y}_N = \boldsymbol{y}/\sqrt{\boldsymbol{y}^H\boldsymbol{y}}$. The problem is posed as follows:

$$\arg\min_{\beta,\phi,z} \mathcal{L}(\beta, \phi, z) = \arg\min_{\beta,\phi,z} \left\| \boldsymbol{x}_N - \beta e^{j\phi}\boldsymbol{y}_N + z\mathbf{1}_k \right\|^2.$$

This expression can be rewritten

$$\arg\min_{\beta,\phi,z} \mathcal{L}(\beta,\phi,z) = \arg\min_{\beta,\phi,z} \|\boldsymbol{x}_N - \boldsymbol{A}\boldsymbol{\theta}\|^2,$$

with

$$\boldsymbol{A} = \begin{bmatrix} \boldsymbol{y}_N & \boldsymbol{1}_k \end{bmatrix},$$
$$\boldsymbol{\theta} = \begin{bmatrix} \beta e^{j\phi} \\ z \end{bmatrix}.$$

Since the target function is a sum of squares, we have a least square problem which can be solved following equation (2.12) as follows:

$$\begin{aligned}
\hat{\boldsymbol{\theta}} &= \left(\boldsymbol{A}^H\boldsymbol{A}\right)^{-1}\boldsymbol{A}^H\boldsymbol{x}_N \\
&= \left(\begin{bmatrix} \boldsymbol{y}_N^H \\ \boldsymbol{1}_k \end{bmatrix}\begin{bmatrix} \boldsymbol{y}_N & \boldsymbol{1}_k \end{bmatrix}\right)^{-1}\begin{bmatrix} \boldsymbol{y}_N^H \\ \boldsymbol{1}_k \end{bmatrix}\boldsymbol{x}_N \\
&= \begin{pmatrix} \boldsymbol{y}_N^H\boldsymbol{y}_N & 0 \\ 0 & k \end{pmatrix}^{-1}\begin{bmatrix} \boldsymbol{y}_N^H\boldsymbol{x}_N \\ 0 \end{bmatrix} \\
&= \begin{bmatrix} \frac{\boldsymbol{y}_N^H\boldsymbol{x}_N}{\boldsymbol{y}_N^H\boldsymbol{y}_N} \\ 0 \end{bmatrix} = \begin{bmatrix} \beta e^{j\phi} \\ z \end{bmatrix},
\end{aligned}$$

so

$$T(\boldsymbol{y}) = \frac{\boldsymbol{y}_N^H\boldsymbol{x}_N}{\boldsymbol{y}_N^H\boldsymbol{y}_N}\boldsymbol{y}. \tag{2.47}$$

Hence, no displacement is needed, but just a translation and a scaling. With respect to function $\mathcal{L}(\cdot)$, recalling equation (2.13), we can write

$$\begin{aligned}
\mathcal{L}(\beta,\phi,z) &= \left(\boldsymbol{x}_N - \boldsymbol{A}\hat{\boldsymbol{\theta}}\right)^H\boldsymbol{x}_N \\
&= \boldsymbol{x}_N^H\boldsymbol{x}_N - \begin{bmatrix} \frac{\boldsymbol{x}_N^H\boldsymbol{y}_N}{\boldsymbol{y}_N^H\boldsymbol{y}_N} & 0 \end{bmatrix}\begin{bmatrix} \boldsymbol{y}_N^H \\ \boldsymbol{1}_k \end{bmatrix}\boldsymbol{x}_N \\
&= \boldsymbol{x}_N^H\boldsymbol{x}_N - \frac{\boldsymbol{x}_N^H\boldsymbol{y}_N\boldsymbol{y}_N^H\boldsymbol{x}_N}{\boldsymbol{y}_N^H\boldsymbol{y}_N} \\
&= 1 - \frac{\boldsymbol{x}^H\boldsymbol{y}\boldsymbol{y}^H\boldsymbol{x}}{\boldsymbol{y}^H\boldsymbol{y}\boldsymbol{x}^H\boldsymbol{x}}.
\end{aligned} \tag{2.48}$$

The last line has been written bearing in mind that $\boldsymbol{x}_N^H\boldsymbol{x}_N = 1$ and removing the normalization. The square root of the expression above is defined as the Procrustes distance between two shapes, i.e.,

$$d_P(\boldsymbol{x}, \boldsymbol{y}) = \sqrt{1 - \frac{\boldsymbol{x}^H \boldsymbol{y} \boldsymbol{y}^H \boldsymbol{x}}{\boldsymbol{y}^H \boldsymbol{y} \boldsymbol{x}^H \boldsymbol{x}}}.$$

2.6.2 Mean shape

Consider now that we have n shapes $\boldsymbol{x}_i$, $i = \{1, \dots, n\}$, which we assume are a random sample from a mean shape $\boldsymbol{\eta}$, which is then deterministically rotated, scaled, and shifted. The mean shape will be estimated as the shape that minimizes the sum of the Procrustes distances, i.e.,

$$\hat{\boldsymbol{\eta}} = \arg\min_{\boldsymbol{\eta}} \sum_{i=1}^{n} d_P^2(\boldsymbol{x}_i, \boldsymbol{\eta}).$$

Following equation (2.48), we can write

$$\begin{aligned}
\sum_{i=1}^{n} d_P^2(\boldsymbol{x}_i, \boldsymbol{\eta}) &= \sum_{i=1}^{n} 1 - \frac{\boldsymbol{\eta}^H \boldsymbol{x}_i \boldsymbol{x}_i^H \boldsymbol{\eta}}{\boldsymbol{x}_i^H \boldsymbol{x}_i \boldsymbol{\eta}^H \boldsymbol{\eta}} \\
&= n - \boldsymbol{\eta}^H \underbrace{\left[\sum_{i=1}^{n} \frac{\boldsymbol{x}_i \boldsymbol{x}_i^H}{\boldsymbol{x}_i^H \boldsymbol{x}_i} \right]}_{\boldsymbol{B}} \frac{\boldsymbol{\eta}}{\boldsymbol{\eta}^H \boldsymbol{\eta}}.
\end{aligned}$$

So minimizing the objective function is equivalent to

$$\hat{\boldsymbol{\eta}} = \arg\max_{\boldsymbol{\eta}} \boldsymbol{\eta}^H \boldsymbol{B} \boldsymbol{\eta} \qquad \text{s.t.} \quad \boldsymbol{\eta}^H \boldsymbol{\eta} = 1.$$

It is easy to see that the mean shape that maximizes the function above is the eigenvector associated to the largest eigenvalue of matrix $\boldsymbol{B}$. Note that $\boldsymbol{B}$ is Hermitian, so taking in consideration equation (2.19), we can write

$$\begin{aligned}
\boldsymbol{\eta}^H \boldsymbol{B} \boldsymbol{\eta} &= \boldsymbol{\eta}^H \left(\boldsymbol{U} \boldsymbol{\Lambda} \boldsymbol{U}^H \right) \boldsymbol{\eta} \\
&= \left(\boldsymbol{U}^H \boldsymbol{\eta} \right)^H \boldsymbol{\eta}^H \boldsymbol{\Lambda} \boldsymbol{U}^H \boldsymbol{\eta} = \boldsymbol{y}^H \boldsymbol{\Lambda} \boldsymbol{y} \\
&= \sum_{i=1}^{n} \lambda_i y_i^2 \leq \lambda_{\max} \sum_{i=1}^{n} y_i^2 = \lambda_{\max},
\end{aligned}$$

with $\lambda_{\max}$ being the maximum eigenvalue; equality is achieved when $\boldsymbol{\eta}$ is the eigenvector associated to that eigenvalue, say, $\hat{\boldsymbol{\eta}} = \boldsymbol{u}_{\max}$.

The coordinates of the Procrustes fit to the mean shape will be, following equation (2.47),

$$T(\boldsymbol{x}_i) = \frac{\boldsymbol{x}_i^H \hat{\boldsymbol{\eta}}}{\boldsymbol{x}_i^H \boldsymbol{x}_i} \boldsymbol{x}_i. \tag{2.49}$$

It is easy to check that the mean shape is, up to a scale factor, the average of the points in the fit. Specifically,

$$\begin{aligned} \frac{1}{n}\sum_{i=1}^{n} T(\boldsymbol{x}_i) &= \frac{1}{n}\sum_{i=1}^{n} \frac{\boldsymbol{x}_i^H \hat{\boldsymbol{\eta}}}{\boldsymbol{x}_i^H \boldsymbol{x}_i} \boldsymbol{x}_i = \sum_{i=1}^{n} \boldsymbol{x}_i \frac{\boldsymbol{x}_i^H \hat{\boldsymbol{\eta}}}{\boldsymbol{x}_i^H \boldsymbol{x}_i} \\ &= \frac{1}{n}\left[\sum_{i=1}^{n} \frac{\boldsymbol{x}_i \boldsymbol{x}_i^H}{\boldsymbol{x}_i^H \boldsymbol{x}_i}\right] \hat{\boldsymbol{\eta}} = \frac{1}{n}\boldsymbol{B}\hat{\boldsymbol{\eta}} = \frac{\lambda_{\max}}{n}\hat{\boldsymbol{\eta}}. \end{aligned} \tag{2.50}$$

Residuals are defined as the difference

$$T(\boldsymbol{x}_i) - \frac{1}{n}\sum_{i=1}^{n} T(\boldsymbol{x}_i) \tag{2.51}$$

and they could serve as the basis to calculate a sample covariance matrix y to determine variability out of the eigenanalysis of this matrix.

2.6.3 Procrustes analysis in higher dimensions

For the case that shapes have a higher dimension, say points in at least 3D space, the complex algebra we have used is no longer valid and we need to resort to other forms of rotation such as rotation matrices or quaternions. A rotation matrix is an orthogonal matrix, i.e., $\boldsymbol{\Lambda}^T\boldsymbol{\Lambda} = \boldsymbol{\Lambda}\boldsymbol{\Lambda}^T = \boldsymbol{I}$. With this in mind, let matrix $\boldsymbol{X}_i$ denote the ith shape in a population with k columns that represent the k points of the shape and m dimensions (so the matrix has m rows). The problem to solve in this case is

$$\min \sum_{i=1}^{n} \sum_{j=i+1}^{n} \left\| \left(\beta_i \boldsymbol{\Lambda}_i \boldsymbol{X}_i + \boldsymbol{d}_i \boldsymbol{1}_k^T\right) - \left(\beta_j \boldsymbol{\Lambda}_j \boldsymbol{X}_j + \boldsymbol{d}_j \boldsymbol{1}_k^T\right) \right\|^2, \tag{2.52}$$

where the free parameters are the scaling factor β_k, rotation matrices $\boldsymbol{\Lambda}_k$, and displacement vectors $\boldsymbol{d}_k$, $k = \{1, \dots, n\}$. In later chapters an algorithm to solve problems of this kind is analyzed.

2.7 Exercises

1. The discrete Fourier transform (DFT) of a sequence $f[n]$, $0 \le n \le N-1$ is a sampled version of the Fourier transform defined in Eq. (2.31). The DFT is defined

as

$$\mathcal{F}[k] = \sum_{n=0}^{N-1} f[n] e^{-j\frac{2\pi}{N}kn}, 0 \leq k \leq N-1$$

a. Defining

$$\mathcal{F} = \begin{bmatrix} F[0] & F[1] & \cdots & F[N-1] \end{bmatrix}^T$$
$$f = \begin{bmatrix} f[0] & f[1] & \cdots & f[N-1] \end{bmatrix}^T$$

find matrix $\boldsymbol{A}$ that guarantees that $\mathcal{F} = \boldsymbol{A}\boldsymbol{f}$.

b. Show that matrix $\boldsymbol{A}$ is orthogonal. Is it orthonormal?

c. Extending the DFT definition to the 2D case as follows

$$\mathcal{F}[k_1, k_2] = \sum_{n_1=0}^{N_1-1} \sum_{n_1=0}^{N_2-1} f[n_1, n_2] e^{-j2\pi\left(\frac{k_1 n_1}{N_1} + \frac{k_2 n_2}{N_2}\right)}, \quad 0 \leq k \leq N-1$$

and defining matrices $\boldsymbol{\mathcal{F}}$ and $\boldsymbol{f}$ such that $[\boldsymbol{\mathcal{F}}]_{k_1,k_2} = \mathcal{F}[k_1, k_2]$ and $[\boldsymbol{f}]_{n_1,n_2} = f[n_1, n_2]$, show that $\boldsymbol{\mathcal{F}} = \boldsymbol{A}\boldsymbol{f}\boldsymbol{B}$, with $\boldsymbol{A}$ and $\boldsymbol{B}$ properly defined.

d. Define

$$\mathcal{F}_k = \begin{bmatrix} F[0,k] & F[1,k] & \cdots & F[N_1-1,k] \end{bmatrix}^T$$
$$f_n = \begin{bmatrix} f[0,n] & f[1,n] & \cdots & f[N_1-1,n] \end{bmatrix}^T$$
$$\mathcal{F} = \begin{bmatrix} \mathcal{F}_0^T & \mathcal{F}_1^T & \cdots & \mathcal{F}_{N_2-1}^T \end{bmatrix}^T$$
$$f = \begin{bmatrix} f_0^T & f_1^T & \cdots & f_{N_2-1}^T \end{bmatrix}^T$$

Show that $\boldsymbol{\mathcal{F}} = \boldsymbol{C}\boldsymbol{f}$, with matrix $\boldsymbol{C}$ defined by means of a Kronecker product in which matrices $\boldsymbol{A}$ and $\boldsymbol{B}$ are involved.

e. Show that matrix $\boldsymbol{C}$ is orthogonal. Is it orthonormal?

f. Find the inverse transforms by using the orthogonality property of the matrices involved in the three transformations analyzed in this exercise.

2. Recall $\mathcal{F}(u)$ and $f(x)$ from equations (2.26) and (2.27) respectively, which are related by the operator $\mathbb{F}$. Then:

a. Show that $\mathbb{F}\{f(-x)\} = \mathbb{F}(-u)$

b. Show that $\mathbb{F}\{\mathcal{F}(u)\} = f(-x)$.

c. Let $\mathbb{F}^n$ denote the application of operator $\mathbb{F}$ n times, i.e., $\mathbb{F}^2$ means $\mathbb{F}\{\mathbb{F}\{f(x)\}\}$. With this convention, find a general expression for $\mathbb{F}^{2n}$.

3. Assume $\boldsymbol{A}$ is a real symmetric matrix.

a. Find an expression for $\boldsymbol{A}^n$, $n \in \mathbb{Z}^+$, with the elements of the eigenanalysis of the matrix.

b. Considering λ_{max} is the eigenvalue with the maximum modulus, justify an approximate expression of A^n for $n >> 1$, in which the only eigenvalue involved is λ_{max}.

4. The Bloch equations define the behavior of the magnetization vector under different circumstances within a magnetic resonance imaging (MRI) environment. A simplified version of them reduces to one equation which describes the free precession movement of the so-called longitudinal component, say $M_z(t)$, i.e., the behavior of this component when there is no external excitation or after this excitation has been removed. The equation is

$$\frac{dM_z(t)}{dt} = \frac{M_0 - M_z(t)}{T_1} \tag{2.53}$$

where M_0 is the equilibrium magnetization value, i.e., the magnetization value when the specimen to be imaged (a patient, a phantom, etc.) has only been inserted in the magnet. T_1 is a characteristic value of the tissue that plays a fundamental role in MRI. We intend to find the behavior of $M_z(t)$, $t \geq 0$, assuming that $M_z(0) = M_z^0$, which does not necessarily coincide with M_0. To this end:

a. This differential may be solved by first finding a general solution to the homogeneous equation. This equation is the one that results by nulling the term that does not contain either $M_z(t)$ or any of its derivatives. Show that this solution is $M_z(t) = Ae^{-t/T_1}$.

b. Next step is to find a particular solution to the complete equation. To this end we will use the so-called method of variation of constants. Specifically, we will write the previous solution as $M_z(t) = A(t)e^{-t/T_1}$, i.e., the constant becomes a function of time, and then insert this solution in Eq. (2.53) to find the value of $A(t)$ that satisfies the complete equation. Use this method to show that the final solution is

$$M_z(t) = M_z^0 e^{-\frac{t}{T_1}} + M_o\left(1 - e^{-\frac{t}{T_1}}\right)$$

5. Recall the Laplace equation (Eq. (2.44)). In this exercise we are going to explore its implicit solution for a specific geometry. Consider we have a grid of size $h_i = 0.5$, $i = \{1, 2\}$. We have two contours, namely, Γ_1 and Γ_2. In our discretized version Γ_1 consists of the points on the square with side length 1, centered on the origin and with sides parallel to the coordinates axes. Likewise, Γ_2 is a similar square albeit with side 3. Assume $u(\Gamma_1) = 1$ and $u(\Gamma_2) = 2$.

a. Let's arrange vector $\boldsymbol{u}$ of the implicit solution by piling up values $s(\boldsymbol{x})$ from columns in the grid, starting from the left end of the geometry. Taking in consideration, for simplicity, all the grid positions within the outer square, show that the solution can be found by solving the system

$$\begin{pmatrix} f_{-3} \\ f_{-2} \\ f_{-1} \\ f_0 \\ f_1 \\ f_2 \\ f_3 \end{pmatrix} = \begin{pmatrix} I & 0 & 0 & 0 & 0 & 0 & 0 \\ A & B & A & 0 & 0 & 0 & 0 \\ 0 & D & C & D & 0 & 0 & 0 \\ 0 & 0 & D & C & D & 0 & 0 \\ 0 & 0 & 0 & D & C & D & 0 \\ 0 & 0 & 0 & 0 & A & B & A \\ 0 & 0 & 0 & 0 & 0 & 0 & I \end{pmatrix} \begin{pmatrix} u_{-3} \\ u_{-2} \\ u_{-1} \\ u_0 \\ u_1 \\ u_2 \\ u_3 \end{pmatrix}$$

with $\boldsymbol{u}_j$ the vector of values $s(\boldsymbol{x})$ in the j-th column, $-3 \leq j \leq 3$, arranged from top to button. With this structure in mind, and making use of Eq. (2.45), find out the components of each of the matrices and vectors defined above.

b. Strictly speaking, the values $s(\boldsymbol{x})$, with $\boldsymbol{x}$ interior to the inner square, should not be part of the calculations since the boundary conditions apply, needless to say, to points belonging to $\Gamma_1 \cup \Gamma_2$. Redefine the matrix operations introduced above to remove the interior points (there is only one) from the calculations. Should results be affected?

c. Check with a computer the result of the calculations indicated in this exercise.

References

[1] Samajdar T., Quraishi M.I., Analysis and evaluation of image quality metrics, in: Information Systems Design and Intelligent Applications, Springer, 2015, pp. 369–378.

[2] Eskicioglu A.M., Fisher P.S., Image quality measures and their performance, IEEE Transactions on Communications 43 (12) (1995) 2959–2965.

[3] Van Trees H.L., Optimum Array Processing: Part IV of Detection, Estimation, and Modulation Theory, John Wiley & Sons, 2004.

[4] Szabo F., The Linear Algebra Survival Guide: Illustrated with Mathematica, Academic Press, 2015.

[5] Dryden I.L., Mardia K.V., Statistical Shape Analysis, Wiley Series in Probability and Statistics, John Wiley & Sons, Ltd, New York, NY, 1998.

CHAPTER

Regression and classification

3

Thomas Moreau and Demian Wassermann
Inria, Saclay Île-de-France, Palaiseau, France

Learning points

- Regression: definition and introduction
- Single and multidimensional regression
- Classification as a regression problem
- Prediction and its performance metrics
- Non-linear regression through linearizing strategies

3.1 Introduction

Most problems in medical imaging can be expressed as regression problems. A regression problem, where we try to find the relationship between a desired or measured noisy observation (y) and noiseless data or input (x), is a Swiss army knife formalism in uncountable applications. In this chapter, we will focus on the linear regression problem, that is, when the relationship between x and y is expected to be well captured through a linear relationship.

Linear models remain some of the most interesting tools to describe realities. Despite their limited expressive power, they have a large number of advantages: they are easy to solve; the quality of the solution is simple enough to assess; and they have the potential to use the fitted model parameters to explain the described phenomena. Several well-known problems in medical imaging, physics, and computer science find their solutions through reductions to linear models.

A typical example of the power of linear models for medical imaging is found in functional magnetic resonance imaging (MRI) in neuroscience. For such data, the scarcity of annotated data and the dimension of the problem make the linear models particularly efficient to capture interesting statistical effects such as part of the brain involved in certain cognitive tasks, also called functional networks. Using the NeuroQuery database [1] – a dataset for meta-analysis described in Section 3.2.1 – we illustrate this for an encoding task with a linear model in Fig. 3.1. The coefficients of the linear model can be interpreted as brain maps and the peak activities reveal areas that are in line with the functional networks described in the literature.

Medical Image Analysis. https://doi.org/10.1016/B978-0-12-813657-7.00015-7

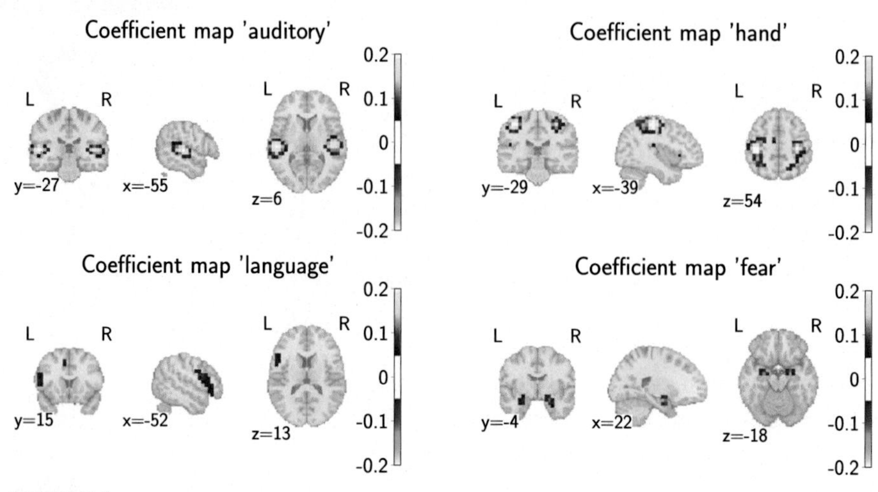

FIGURE 3.1

Encoding model based on the NeuroQuery meta-analysis database [1]. A linear model learns to predict activity in the brain from a term-frequency inverse document frequency (TF-IDF) representation of neuroscience publications. Then, for each word, we display the associated coefficient, which is in line with the functional network.

In this chapter we will introduce linear model regression as a minimization problem as well as a statistical inference one; we will show how to quantify the quality of an adjusted model to fit the original dataset, as well as to generalize it; and finally, we will show how to cast non-linear problems, such as classification, as instances of linear regression.

The following chapter-specific notation is used throughout this chapter:

Nomenclature

x, y, z	scalar values
X, Y, Z	column vectors and outputs respectively as matrices
$[x]_k$	If X is a vector, the kth scalar component
$\mathbf{x}, \mathbf{X}$	Random variables either scalar or vectorial
$x^{(i)}, X^{(i)}$	If $\mathbf{x}, \mathbf{X}$ are random variables, ith observation of the variable.
f, g, h	Functions
λ, α, β	function meta-parameters in scalar form, e.g., $f(X, \lambda)$ or $f_\lambda(X)$
a, b, c	scalar constants
A, B, C	constant vector, as column vectors where A_k is the kth scalar component
$\mathcal{D}, \mathcal{F}, \mathcal{G}$	Finite sets
$\langle X, Y \rangle_L$	Inner product between X and Y on the space L
$\|X\|_L^q$	q-norm for X on the space L
$\mathbf{Pr}_\beta(X\|Y)$	Probability of X conditional to Y with meta-parameter β
$\mathbb{E}_\beta[X\|Y]$	Expected value of X conditional to Y with meta-parameter β
$\mathbf{Var}_\beta[X\|Y]$	Variance value of X conditional to Y with meta-parameter β
$\mathcal{N}(\mu, \sigma^2)$	Gaussian distribution with mean μ and standard deviation σ

3.1.1 Regression as a minimization problem

The regression of Y from X is the task of estimating the relationship between the value of an output Y – also called observation – and the value of an input X – also called features or covariates. The most classical example of such regression is univariate linear regression. As all values are scalars here, we will consider an input $x \in \mathbb{R}$ and an output $y \in \mathbb{R}$. The relationship between the input and the output is given by the following model:

$$y = f(x, \boldsymbol{\beta}) + \epsilon = [\boldsymbol{\beta}]_0 + [\boldsymbol{\beta}]_1 x + \epsilon, \tag{3.1}$$

where ϵ is an additive (unknown) noise term that we will discuss later. Specifically, the parameter $[\boldsymbol{\beta}]_0$ encodes the intercept, i.e., the mean value of the outcome y when $x = 0$ and $[\boldsymbol{\beta}]_1$ is the slope of the relationship, meaning that an increase of 1 in x will result in an average increase of $[\boldsymbol{\beta}]_1$ in the outcome y. Through this chapter we present several synthetic and neuroscience-based examples for which we provide the corresponding code.

In the noiseless case, when $\epsilon = 0$, computing the values of $[\boldsymbol{\beta}]_0$ and $[\boldsymbol{\beta}]_1$ is very simple. Indeed, the generating model can be recovered using two points $(x^{(1)}, y^{(1)})$ and $(x^{(2)}, y^{(2)})$ generated by this model. The model (3.1) has two unknown variables, which we can resolve using two equations. We can compute the values of $[\boldsymbol{\beta}]_0$ and $[\boldsymbol{\beta}]_1$ with the formula

$$\begin{cases} y^{(1)} &= [\boldsymbol{\beta}]_0 + [\boldsymbol{\beta}]_1 x^{(1)}, \\ y^{(2)} &= [\boldsymbol{\beta}]_0 + [\boldsymbol{\beta}]_1 x^{(2)} \end{cases} \quad \Leftrightarrow \quad \begin{cases} [\boldsymbol{\beta}]_1 &= \dfrac{y^{(1)} - y^{(2)}}{x^{(1)} - x^{(2)}}, \\ [\boldsymbol{\beta}]_0 &= \dfrac{y^{(2)}x^{(1)} - y^{(1)}x^{(2)}}{x^{(1)} - x^{(2)}}. \end{cases} \tag{3.2}$$

When the noise ϵ is non-zero, it is no longer possible to use only to examples to reliably solve the regression. Indeed, with $(x^{(1)}, y^{(1)})$ and $(x^{(2)}, y^{(2)})$ corrupted by noise $\epsilon^{(1)}$ and $\epsilon^{(2)}$, respectively, the values of the coefficients $\widehat{[\boldsymbol{\beta}]}_0$ and $\widehat{[\boldsymbol{\beta}]}_1$ estimated with (3.2) would include a noise term, i.e.,

$$\widehat{[\boldsymbol{\beta}]}_0 = [\boldsymbol{\beta}]_0 + \frac{\epsilon^{(2)}x^{(1)} - \epsilon^{(1)}x^{(2)}}{x^{(1)} - x^{(2)}} \quad \text{and} \quad \widehat{[\boldsymbol{\beta}]}_1 = [\boldsymbol{\beta}]_1 + \frac{\epsilon^{(1)} - \epsilon^{(2)}}{x^{(1)} - x^{(2)}}. \tag{3.3}$$

The estimated coefficients are no longer exact and the noise can lead to large errors, as illustrated in Fig. 3.2. To obtain better estimate $[\boldsymbol{\beta}]_0$ and $[\boldsymbol{\beta}]_1$, it is necessary to aggregate the information from more than two points. If we have a set of N pairs $\{(x^{(1)}, y^{(1)}), \dots (x^{(N)}, y^{(N)})\}$ with noise, the linear system

$$\begin{cases} y^{(1)} = [\boldsymbol{\beta}]_0 + [\boldsymbol{\beta}]_1 x^{(1)} \\ \vdots \\ y^{(N)} = [\boldsymbol{\beta}]_0 + [\boldsymbol{\beta}]_1 x^{(N)} \end{cases} \tag{3.4}$$

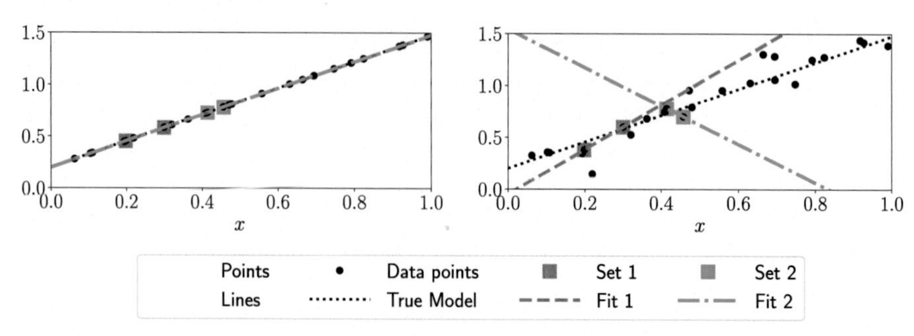

FIGURE 3.2

In the noiseless case (left), any two points give the same information and the correct coefficients $[\boldsymbol{\beta}]_0$, $[\boldsymbol{\beta}]_1$ can be recovered easily. In the noisy case (right), two random points can give very different coefficient estimates. It is necessary to use more than two points for the regression.

is overdetermined and in general does not admit an exact solution. A natural way to solve the regression is to select the model $f(\cdot, \boldsymbol{\beta})$ that minimizes the error on all the examples according to some criterion. For a given $\boldsymbol{\beta} = ([\boldsymbol{\beta}]_0, [\boldsymbol{\beta}]_1)$, the model predicts that the input value $x^{(i)}$ will result in the outcome $\widehat{y}^{(i)} = f(x^{(i)}, \boldsymbol{\beta})$. A cost is a function $\ell : \mathbb{R} \times \mathbb{R} \mapsto \mathbb{R}_+$ that will measure how far the predicted outcome $\widehat{y}^{(i)}$ is to the observed value $y^{(i)}$. Then, choosing the best parameters $\boldsymbol{\beta}$ can be reduced to solving the following minimization problem:

$$\boldsymbol{\beta}^* = \arg\min_{\boldsymbol{\beta}} \frac{1}{N} \sum_{i=1}^{N} \ell(y^{(i)}, \underbrace{f(x^{(i)}, \boldsymbol{\beta})}_{\widehat{y}^{(i)}}). \tag{3.5}$$

This is also known as empirical risk minimization (see Section 3.2.2).

Least squares The most common choice for the cost function ℓ is the squared distance, i.e., $\ell(\widehat{y}^{(i)}, y^{(i)}) = (y^{(i)} - \widehat{y}^{(i)})^2$. With this choice of distance, the regression can be cast as a minimization problem to find the parameters $\boldsymbol{\beta} = ([\boldsymbol{\beta}]_0, [\boldsymbol{\beta}]_1)$ which best fit the data

$$\boldsymbol{\beta}^* = \arg\min_{\boldsymbol{\beta}} \frac{1}{N} \sum_{i=1}^{N} (y^{(i)} - f(x^{(i)}, \boldsymbol{\beta}))^2. \tag{3.6}$$

This formulation of linear regression is known as the least squares regression (LSQ) or ordinary least squares (OLS) problem, as it minimizes the sum of the squared errors. This is the most common formulation but not the only one, as the goodness-of-fit function ℓ can be chosen to reflect different properties of the noise (see Section 3.2.3).

3.1.2 Regression from the statistical angle

In Section 3.1.1, we have formalized linear regression as a minimization problem for a given choice of loss measuring how well each example $(x^{(i)}, y^{(i)})$ is estimated by our model. The critical parameter in this framework is the choice of the loss ℓ, which drives the behavior of our model. The choice of this loss function is linked to the type of additive noise considered in the model (3.1).

Here, we consider a noise ϵ following a known probability law p and a linear generative model $f(\cdot, \boldsymbol{\beta})$ such that we have a random variable x and another random variable y distributed as

$$\mathrm{y} = f(x, \boldsymbol{\beta}) + \epsilon = [\boldsymbol{\beta}]_0 + [\boldsymbol{\beta}]_1 \, \mathrm{x} + \epsilon. \tag{3.7}$$

For a dataset $(x^{(i)}, y^{(i)})_{i=1}^{N}$ sampled independent and identically distributed (i.i.d.) from the previous model, the probability of observing the output $y^{(i)}$ given the input $x^{(i)}$ and the parameter of the model $\boldsymbol{\beta}$ can be written as

$$\Pr(\{\mathrm{y}^{(i)} = y^{(i)}\}_{i=1}^{n} | \{\mathrm{x}^{(i)} = x^{(i)}\}_{i=1}^{n}, \boldsymbol{\beta}) = \prod_{i=1}^{n} \Pr(\mathrm{y} = y^{(i)} | \mathrm{x} = x^{(i)}, \boldsymbol{\beta}). \tag{3.8}$$

In this case, we are conditioning that the features corresponding to observations, $x^{(i)}$, so that the randomness on the right-hand side of (3.7) comes only from the measurement errors ϵ. Then, under the hypothesis that the errors are i.i.d. and p-distributed, (3.7) becomes

$$\Pr(\{\mathrm{y}^{(i)} = y^{(i)}\}_{i=1}^{n} | \{\mathrm{x}^{(i)} = x^{(i)}\}_{i=1}^{n}, \boldsymbol{\beta}) = \prod_{i=1}^{n} p(y^{(i)} - f(x^{(i)}, \boldsymbol{\beta})) = \prod_{i=1}^{n} p(\epsilon^{(i)}). \tag{3.9}$$

Then $\hat{\beta}$ can be estimated through the principle of maximum likelihood, that is, selecting the model that gives the largest probability, or equivalently the smallest negative log-likelihood, i.e.,

$$\widehat{\beta} = \arg\min_{\boldsymbol{\beta}} \mathcal{L}(\boldsymbol{\beta}) = -\sum_{i=1}^{n} \log p(y^{(i)} - f(x^{(i)}, \boldsymbol{\beta})). \tag{3.10}$$

This optimization problem is equivalent to the one defined in equation (3.5). Indeed, choosing the loss function $\ell : \mathbb{R} \times \mathbb{R} \mapsto \mathbb{R}_+$ as $\ell(y, \widehat{y}) = -\log p(y - \widehat{y})$ yields the same solution. The choice of the error measurement function ℓ is thus intrinsically linked to the distribution of the noise p.

Gaussian white noise in the case where we assume that the additive noise is distributed according to $\mathcal{N}(0, \sigma^2)$, the probability density for ϵ is $p(\epsilon) = \frac{1}{\sqrt{2\pi\sigma^2}} e^{-\frac{\epsilon^2}{2\sigma^2}}$

and the negative log-likelihood becomes

$$\begin{aligned}\mathcal{L}(\boldsymbol{\beta}) &= -\sum_{i=1}^{n} \log(p(y^{(i)} - f(x^{(i)}, \boldsymbol{\beta}))) \\ &= \frac{1}{2\sigma^2}\sum_{i=1}^{n}(y^{(i)} - f(x^{(i)}, \boldsymbol{\beta}))^2 + \frac{1}{2}\log(2\pi\sigma^2) \\ &= \frac{1}{2\sigma^2}\sum_{i=1}^{N}(\epsilon^{(i)})^2 + \frac{1}{2}\log(2\pi\sigma^2).\end{aligned} \tag{3.11}$$

With this probabilistic perspective, we find an equivalent formulation to the LSQ formulated in equation (3.6).

3.2 Multidimensional linear regression

In the previous section, we have seen the basics of a linear model to predict an output y from a univariate input x. In this section, we explain how to extend this model to multidimensional inputs and how to evaluate its performance. Then, we explain how these models can be used to predict out-of-sample behavior and how we can assess the quality of such predictions.

In practice, it is often the case that we are interested in predicting the output y not only from one factor $x \in \mathbb{R}$ but from a variety of factors $\boldsymbol{x} \in \mathbb{R}^p$. Here, p denotes the number of features – or covariates – that are used to predict y and $[\boldsymbol{x}]_k$ denotes the kth feature in $\boldsymbol{x}$. In this case, the linear model presented in equation (3.1) can be extended with

$$y = f(\boldsymbol{x}, \boldsymbol{\beta}) + \epsilon = [\boldsymbol{\beta}]_0 + \sum_{k=1}^{p} [\boldsymbol{\beta}]_k [\boldsymbol{x}]_k + \epsilon, \tag{3.12}$$

where ϵ is an additive noise in $\mathbb{R}$. This model can also be denoted with vectorial notations $f(\boldsymbol{x}, \boldsymbol{\beta}) = [\boldsymbol{\beta}]_0 + \boldsymbol{\beta}^\top \boldsymbol{x}$, where $\boldsymbol{\beta}$ and $\boldsymbol{x}$ are both column vectors in $\mathbb{R}^p$. With this model, each feature k has an influence $[\boldsymbol{\beta}]_k$ – which can be positive or negative – on the output y. This is for instance the case when one wants to predict which voxels have been associated with a specific term through the neuroscience literature as we show in the examples in Fig. 3.1 and those described in Section 3.2.1.

The estimation of the model parameter $\boldsymbol{\beta}$ from a dataset of N pairs $\{(\boldsymbol{x}^{(i)}, y^{(i)})\}_{i=1}^N$ is performed with an optimization problem similar to the one in the univariate case (3.5),

$$\widehat{\beta} = \underset{\beta \in \mathbb{R}^p}{\arg\min} \frac{1}{N}\sum_{i=1}^{N} \ell(y^{(i)}, f(\boldsymbol{x}^{(i)}, \boldsymbol{\beta})). \tag{3.13}$$

The cost function ℓ measures the error between the observed value $y^{(i)}$ and the predicted output $f(\boldsymbol{x}^{(i)}, \boldsymbol{\beta})$ for a given sample i and a parameter $\boldsymbol{\beta} \in \mathbb{R}^p$. The optimization problem to fit the model's parameters (3.13) can be defined within the framework of risk minimization which we present in Section 3.2.2.

An important difference with the univariate model is that some of the covariates $[\boldsymbol{x}]_k$ might be correlated. In this case the number of effective parameters of the regression problem might be smaller than p. Specifically, the given parameter can then be expressed as a function of other parameters, $[\boldsymbol{x}]_j = [\boldsymbol{x}]_j\,([\boldsymbol{x}]_1, \ldots, [\boldsymbol{x}]_{j-1}, [\boldsymbol{x}]_{j+1}, \ldots, [\boldsymbol{x}]_p)$, and the linear model can become unstable. In Section 3.2.5.1 we present a specific strategy to deal with this issue called ridge regularization.

3.2.1 Direction of prediction

An interesting aspect of linear models is that the direction of prediction $\boldsymbol{\beta}$ can give interesting insights into the considered application. Indeed, the value of each $[\boldsymbol{\beta}]_k$ quantifies how much a feature $[\boldsymbol{x}]_k$ impacts the considered target on average, keeping the other features constant. A positive value means that when this feature grows, the target also grows while a negative value implies the opposite. Coupled with the distribution of the feature, this coefficient aims to quantify the *marginal* link $P(y|\,[\boldsymbol{x}]_k)$.

However, some pitfalls might make the interpretation of such direction not straightforward. First, when the features are not all homogeneous and have different scaling, the relative strength of the coefficients cannot be compared directly. To make all coefficients comparable, it is essential to standardize the features in the training data $\boldsymbol{x}$ in order to have a zero mean and a unit variance. Second, the linear model's coefficients do not account directly for marginal effects but for *conditional* links, which are marginal given that all the other features are the same. When the features are independent, this conditional link is the same as the marginal. However, when two features are correlated in the dataset used to learn the coefficients of the linear model, the quantified effect becomes the impact of a feature given that the other is constant. This effect can lead to a situation where, while a feature k has a positive marginal impact on the target, the associated coefficient is negative because a correlated feature l already accounts for this positive effect and given this extra feature l, the impact of $[\boldsymbol{x}]_k$ is negative.

To illustrate the direction of prediction for medical imaging, we consider an example in functional neuroscience. The NeuroQuery dataset is a collection of functional neuroscience studies aimed at constructing models for meta-analysis [1]. It contains the full texts of 13,459 neuroscience publications associated with peak activations in the brain: localizations of intensive brain activity that are reported in the study, in a common space. This dataset can be used to investigate the relationship between parts of the brain an some given tasks. In order to transform these data into vectorized space, each publication is encoded using the term-frequency inverse document frequency (TF-IDF) representation. This representation associates each word from the vocabulary with a feature $[\boldsymbol{x}]_k$ which is computed as the occurrence frequency of this word in the considered document multiplied by the inverse of the frequency of occurrence of this word in the publication corpus. Using these data, two tasks can be considered. The first is the *encoding* task, which regresses the brain activation from the TF-IDF representation of the document. The obtained coefficients quantify which part of the brain each word activates, as illustrated in Fig. 3.1. As both input and output are multidimensional, the reverse task can also be considered, where given the activation maps, one tries to predict how much a word will be used in the publication. This task is called the *decoding* task and will be illustrated in the following sections, e.g., in Fig. 3.9.

3.2.2 Risk minimization

The risk minimization framework aims at providing a global framework linking the observations, hypotheses on the distributions of $\mathbf{y}$ and $\mathbf{X}$, the loss function, and the choice of regression model $f(\cdot, \boldsymbol{\beta})$ with parameter $\boldsymbol{\beta}$ formally through the risk function $\mathcal{R}$. This framework allows for the characterization of different types of errors in the regression framework as well as the relationship between the number of observations and the quality of the regression problem resolution.

The risk function A first source of error for the estimator comes from the choice of estimation procedure. The aim of statistical learning is to be able to infer a model $f(\cdot, \boldsymbol{\beta})$ from a set of observations $\{(\boldsymbol{x}^{(i)}, y^{(i)})\}_{i=1}^{N}$ in order to predict the value of the output y from the observation of a novel input $\boldsymbol{x}$. To measure how well a model solves this task, we can define the risk function for choosing the parameter $\boldsymbol{\beta}$. If the dataset $\mathbf{X}$, y is distributed according to the joint probability law $P_{\mathbf{X},\mathrm{y}}$ and y follows the model in (3.12), the risk associated to a parameter $\boldsymbol{\beta}$ is defined as

$$\mathcal{R}(\beta) = \mathbb{E}_{\mathbf{X},\mathrm{y}}[\ell(\mathrm{y}, f(\mathbf{X}, \boldsymbol{\beta}))], \quad \mathbf{X}, \mathrm{y} \sim P_{\mathbf{X},\mathrm{y}}, \tag{3.14}$$

where loss ℓ can depend on assumptions on the noise that corrupts the observed output $\mathrm{y} = y$. In the case where the noise ϵ is additive with zero mean and variance σ^2, the most classical risk is the variance of the prediction error, i.e., choosing ℓ as the squared ℓ_2 distance $\ell_2^2(y_1, y_2) = \sum_i \left(y_1^{(i)} - y_2^{(i)}\right)^2$. Then, equation (3.14) reads

$$\mathcal{R}(\boldsymbol{\beta}) = \mathbb{E}[(\mathrm{y} - f(\mathbf{X}, \boldsymbol{\beta}))^2], \tag{3.15}$$

where the expectation, unless specified, is taken over all random variables. In this case, the risk can be decomposed with two concurrent terms, the bias and the variance. The risk function defined in equation (3.15) can be separated into three terms:

$$\begin{aligned}\mathcal{R}(\boldsymbol{\beta}) &= \mathbb{E}[(f(\mathbf{X},\boldsymbol{\beta}) - f(\mathbf{X},\boldsymbol{\beta}^*))^2] + \underbrace{\mathbb{E}[(y - f(\mathbf{X},\boldsymbol{\beta}^*))^2]}_{\text{is Var}(\epsilon)=\sigma^2}, \\ &= \underbrace{\left(\mathbb{E}[f(\mathbf{X},\boldsymbol{\beta})] - \mathbb{E}[f(\mathbf{X},\boldsymbol{\beta}^*)]\right)^2}_{Bias} + \underbrace{\mathbb{E}\left[\left(f(\mathbf{X},\boldsymbol{\beta}) - \mathbb{E}[f(\mathbf{X},\boldsymbol{\beta})]\right)^2\right]}_{Variance} + \sigma^2,\end{aligned} \tag{3.16}$$

where $\boldsymbol{\beta}^*$ is the true parameter for the regression problem. Both lines can be obtained by introducing an extra term in the square and expanding $(a+b)^2 = a^2 + b^2 + 2ab$. The cross terms $2ab$ disappear due to one of the term being 0 in each case. This relation is called the bias–variance decomposition. The first term – tagged bias – measures how far the expected prediction from the model with parameter $\boldsymbol{\beta}$ is from the true expected value with no noise. When the bias is 0, this means that in expectation, the model $f(\cdot, \boldsymbol{\beta})$ will output the correct prediction. The second term is named the variance. It measures the dispersion of the prediction made by the model. A model which would output a constant value would have a variance of 0, while a model whose value changes quickly as features change across datapoints would have a much larger variance.

As illustrated in Fig. 3.3(a), these two terms are not minimal at the same point. Thus, to minimize the risk, which is the sum of these two terms, it is necessary to find an equilibrium between these terms. This is the so-called **bias–variance trade-off**. This trade-off depends on the method chosen to estimate $\widehat{\boldsymbol{\beta}}$. In the case of OLS, the estimator is unbiased, i.e., its bias term is 0. Among all unbiased estimators, under some mild hypothesis, OLS is also the estimator with the smallest variance. This is explained in detail in the next section. Nonetheless, a different trade-off is needed in cases where the $\boldsymbol{\beta}$ parameters are correlated or the sample magnitude is too small with respect to the amount of noise. To handle these cases, regularized least squares techniques have been proposed. In these cases, the estimator becomes biased as the regularization changes parameter estimates but it also reduces the variance. Choosing the right regularization thus amounts to finding the right bias–variance trade-off to minimize the risk of the predictor. See Section 3.2.5 for more details on this topic.

3.2.2.1 Gauss–Markov theorem

The OLS problem is the canonical form of the linear regression. Indeed, the ℓ_2 distance is a very natural way to measure the error and it also appears naturally in the case where the noise is considered as Gaussian. Another explanation for its popularity is given by the Gauss–Markov theorem. We call an estimator unbiased when for any given dataset $\mathcal{D}$, the expected value of the estimator $\widehat{\boldsymbol{\beta}}$ is equal to the true parameter $\boldsymbol{\beta}^*$, i.e.,

$$\mathbb{E}_{\mathcal{D}}[\widehat{\boldsymbol{\beta}}] = \boldsymbol{\beta}^*. \tag{3.17}$$

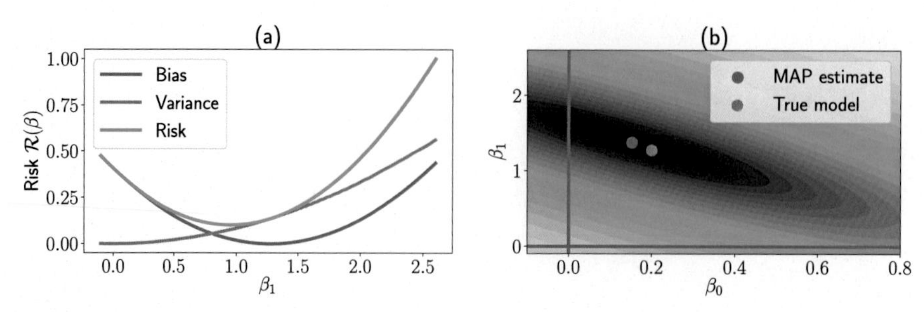

FIGURE 3.3

(a) Illustration of the bias–variance trade-off for a univariate linear model for $\mathrm{x} \sim \mathcal{U}([0, 1])$. Here, the value of $[\boldsymbol{\beta}]_0$ is fixed to the value of the true model. When the value of $[\boldsymbol{\beta}]_1$ increases, the variance in the prediction also increases as the distribution of predicted values is uniform in $[[\boldsymbol{\beta}]_0, [\boldsymbol{\beta}]_0 + [\boldsymbol{\beta}]_1]$. For the bias, the error decreases up to the optimal $\boldsymbol{\beta}^*$ and then increases again quadratically. The minimum of the risk is obtained in between these two extrema. (b) Illustration of the generalization error. The level set is the one of the empirical risk $\widehat{R}_N$ in a 1D regression. The point with the minimum risk (orange) is not located at the same position as the point which minimizes $\widehat{R}_N$ (blue).

Looking back at the bias–variance decomposition in equation (3.16), this implies that the bias term is null. Among the family of unbiased estimators, we say that one is the best unbiased estimator when it has the smallest variance term. Under mild assumptions, this theorem states that the OLS estimator $\widehat{\boldsymbol{\beta}}$ is the best linear unbiased estimator (BLUE). The assumptions on the p-distributed noise random variables $\epsilon^{(i)}$ are:

i The noise component of each observation is additive, $\epsilon^{(i)} = \mathrm{y}^{(i)} - f(\mathbf{X}, \boldsymbol{\beta}^*)^{(i)}$, and each element is independent: $\Pr(\epsilon^{(i)}, \epsilon^{(j)}) = p(\epsilon^{(i)})p(\epsilon^{(j)})$, for $i \neq j$.
ii The noise is centered: $\mathbb{E}[\epsilon^{(i)}] = 0$.
iii The noise is homoscedastic: The distribution of $\epsilon^{(i)}$ is independent of $\mathbf{X}$ and the specific observation i. In other words, $\mathrm{Var}[\epsilon^{(i)}] = \sigma^2$.

Empirical risk minimization Another source of error is the fact that we have a finite number of observations. In order to output a model that gives the best prediction on new observation according to the chosen loss ℓ, one would want to choose the one minimizing the risk function. However, it is not possible to compute the risk function as it would require an infinite number of samples. To estimate it, one can rely on the empirical risk $\widehat{\mathcal{R}}_N$, defined for a certain number N of observations $\mathcal{D}_N = \{\boldsymbol{x}^{(i)}, y^{(i)}\}_{i=1}^N$ of the random variables $\mathbf{x}$, y. The empirical risk corresponds to the empirical mean of the value of the risk for the samples in $\mathcal{D}_N$, i.e.,

$$\widehat{\mathcal{R}}_N(\boldsymbol{\beta}) = \frac{1}{N}\sum_{i=1}^{N} \ell(y^{(i)}, f(\boldsymbol{x}^{(i)}, \boldsymbol{\beta})). \tag{3.18}$$

One can verify that when N goes to infinity, the limit of the empirical risk $\widehat{\mathcal{R}}_N(\boldsymbol{\beta})$ is the true risk $\mathcal{R}(\boldsymbol{\beta})$; in fact, our regression problem can be framed as a risk minimization problem as

$$\boldsymbol{\beta}^* = \arg\min_{\boldsymbol{\beta}} \mathcal{R}(\boldsymbol{\beta}), \tag{3.19}$$

where $\boldsymbol{\beta}^*$ is the true parameter. Nonetheless, we do not have access to the true risk function nor to the true parameter. Hence, we generally look for an algorithm $\mathcal{A}: D_N \to \widehat{\boldsymbol{\beta}}$ implementing the regression such that the empirical risk converges, in expectation, to the true risk. Formally,

$$\mathbb{E}_{\mathcal{D}_N}[\mathcal{R}_N(\mathcal{A}(D_N))] \underset{N\to\infty}{\to} \mathcal{R}(\boldsymbol{\beta}^*). \tag{3.20}$$

The rate of convergence of different algorithms to the true risk is usually quantified through the sample complexity of the algorithm defined as

$$\sup_P \left\{\mathbb{E}_{\mathcal{D}_N}[\mathcal{R}_N(\mathcal{A}(D_N))] - \mathcal{R}(\boldsymbol{\beta}^*)\right\}, \tag{3.21}$$

where P is the probability distribution from which D_N has been drawn. We will not cover here the study of sample complexity for different algorithms and refer the interest reader to general textbooks on machine learning such as Hastie et al. [2].

Estimation error and link to generalization A final aspect to discuss in this section is the estimation error, which for a specific set of observations $\mathcal{D} = (\boldsymbol{X}, \boldsymbol{y})$ can be defined as

$$\text{Err}_{\mathcal{D}} = \mathbb{E}[\ell(\text{y}, f(\mathbf{x}, \widehat{\boldsymbol{\beta}}))|\mathcal{D}]. \tag{3.22}$$

If $\mathcal{D}$ was the dataset used to obtain $\widehat{\beta}$ we call this error the *in-sample error*. Nonetheless, the general goal of a regression problem is often to also have good *generalization* performance. This means that given a new sample $\mathcal{T} = (\boldsymbol{X}^{\mathcal{T}}, \boldsymbol{y}^{\mathcal{T}})$ drawn from the same probability distribution as $\mathcal{D}$ which has not been used to compute $\widehat{\boldsymbol{\beta}}$, we want the generalization error

$$\text{Err}_{\mathcal{T}} = \mathbb{E}[\ell(\mathbf{y}^{\mathcal{T}}, f(\mathbf{X}^{\mathcal{T}}, \widehat{\boldsymbol{\beta}}))|\mathcal{T}] \tag{3.23}$$

to be small. As the observations in $\mathcal{T}$ are not included in $\mathcal{D}$ we can use $\text{Err}_{\mathcal{T}}$ to quantify the out-of-sample performance

$$\mathbb{E}_{\mathcal{T}}\,\text{Err}_{\mathcal{T}} = \mathbb{E}_{\mathcal{T}}\mathbb{E}[\ell(\mathbf{y}^{\mathcal{T}}, f(\mathbf{X}^{\mathcal{T}}, \widehat{\boldsymbol{\beta}}))|\mathcal{T}]. \tag{3.24}$$

In the next sections we will provide specific measurements and techniques to quantify the in- and out-of-sample performance and the generalization, namely prediction, quality.

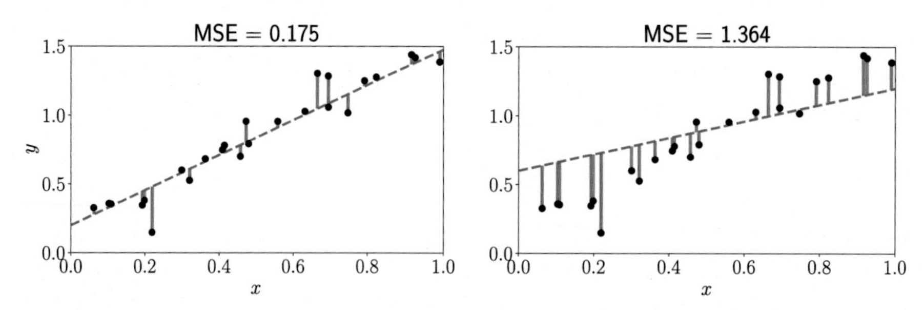

FIGURE 3.4

The core idea to measure the goodness-of-fit of a model is to quantify the errors between the prediction of the model and the training points. The quantification depends on the choice of metric to compare the prediction and the data point. In this figure we present two different datasets. The one on the left shows a better fit than the one of the right and this can be seen through the MSE as well as through the distance between the observations and the regressed line, shown dashed, which is shown through the orange bars.

3.2.3 Measures of fitting and prediction quality

As discussed in the previous section, the risk function plays a critical role, both for learning and to evaluate the model. While the most used risk function is the mean square error (MSE), defined in equation (3.15), choosing the correct error function ℓ is critical for many applications. In this section, we discuss the different properties of three of the most typically used loss functions for regression problems. These can be used to evaluate the quality of the fitting as well as that of the prediction of a model as shown in Fig. 3.4.

Mean square error The most classical way to measure a distance is to use the ℓ_2-norm. The MSE metric simply computes the average ℓ_2 distance between the predictions $\widehat{y}^{(i)}$ and the true targets $y^{(i)}$. Formally,

$$\mathrm{MSE}(y, \widehat{y}) = \frac{1}{N} \sum_{i=1}^{N} (y^{(i)} - \widehat{y}^{(i)})^2, \quad \text{where } \widehat{y}^{(i)} = f(\boldsymbol{x}^{(i)}, \widehat{\boldsymbol{\beta}}). \tag{3.25}$$

This metric is equivalent to the empirical risk for the least squares method. As described in Section 3.1.1, this risk is linked to the assumption that the noise ϵ is white and Gaussian. Thus, it can fail to distinguish between good and bad cases when these assumptions on the noise are broken. This is illustrated with the Anscombe Quartet [3] in Fig. 3.5, where four very different datasets achieve the same MSE for linear regression. Note that the reported MSE is the lowest one that can be achieved for a linear model, as it is achieved with the OLS estimator (see Section 3.2.2.1). Here, only the top left set of points seems to follow the linear model (3.12) with a

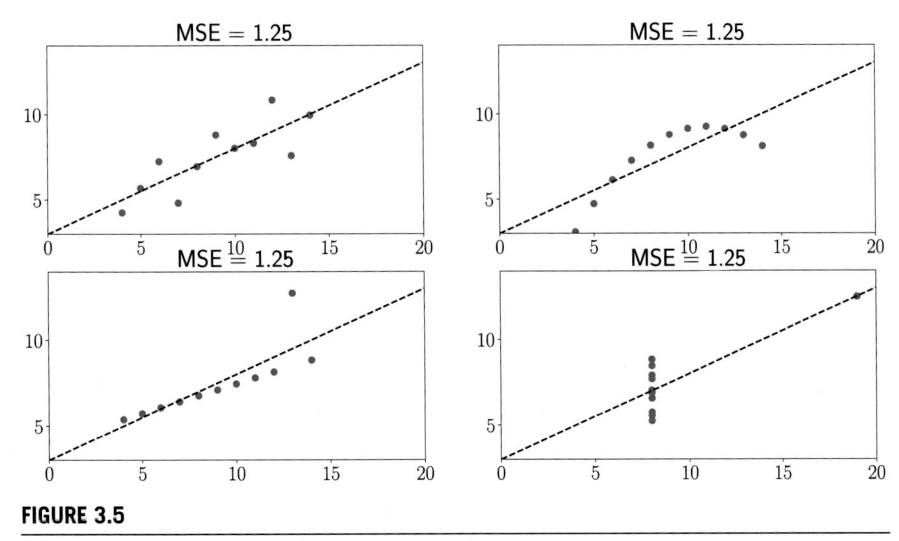

FIGURE 3.5

The Anscombe Quartet [3] is a set of four different datasets which are used to calculate the MSE for an ordinary linear regression. Specifically, that of the top left is the result one would expect, while all others display different behaviors where the fitted linear model does not accurately represent the dataset.

Gaussian white noise. The three other examples show different configurations, which challenges the chosen model. For the top right example, the linear assumption seems to be wrong as the points lie on a more complex curve which looks quadratic. The noise in the bottom left example is not Gaussian and the prediction is driven away from the true linear model by a single outlier. Finally, the bottom right example shows the importance of a good sampling in the input distribution. Indeed, the prediction is mainly driven by the farthest point, which could lead to low generalization quality if it is noisy.

Explained variance The MSE can be rescaled to take into account the variance of the original observation y. This is what is called the explained variance:

$$\mathrm{Var}_{\mathrm{explained}}(y, \widehat{y}) = \frac{\sum_{i=1}^{N}(y^{(i)} - \widehat{y}^{(i)})^2}{\sum_{i=1}^{N}(y^{(i)} - \bar{y})^2} = \frac{\mathrm{MSE}(y, \widehat{y})}{\mathrm{Var}(y)},$$

where $\bar{y}$ is the mean value of $y^{(i)}$. The interpretation of this value is the fraction of the variance of the original signal that can be explained by the chosen model. This rescaling is useful to compare the score of predictors on problems that have different scaling. This value is also linked to the coefficient of determination as $R^2 = 1 - \mathrm{Var}_{\mathrm{explained}}$.

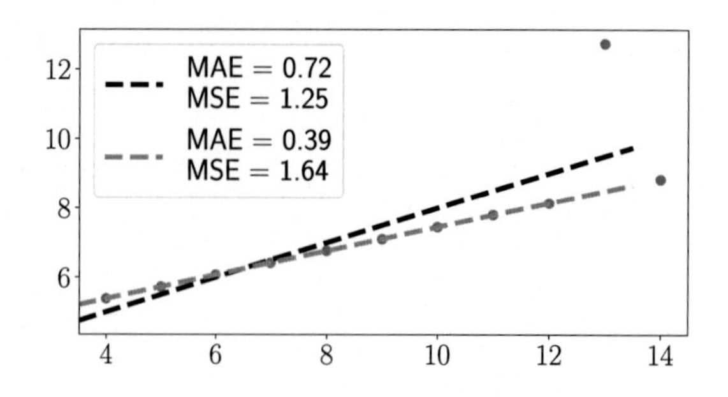

FIGURE 3.6

Illustration of the robustness of the MAE to outliers. Using the third dataset of the Anscombe Quartet, the least squares solution has a larger MAE than the model trained without the outlier, while this is not the case for the MSE.

Mean absolute error Another popular metric for regression is to use the ℓ_1-norm to quantify the difference between the predictions $\widehat{y}^{(i)}$ and the target $y^{(i)}$. Specifically,

$$\mathrm{MAE}(y, \widehat{y}) = \frac{1}{N}\sum_{i=1}^{N} |y^{(i)} - \widehat{y}^{(i)}|. \tag{3.26}$$

This choice of loss function ℓ can be linked to choosing noise, which implies the existence of outliers. Alternatively, the MAE metric is useful where the noise distribution is heavy-tailed, i.e., the probability of a sample as we reach the extremes of the random variable's support does not converge to 0. An example of such case is the Laplace distribution, where the probability of having a large deviation is larger than in the Gaussian setting, an example of what is usually called super-Gaussian noise. In this case, the negative log-likelihood from equation (3.10) becomes exactly equation (3.26). This metric is more balanced than the MSE. Indeed, the MSE tends to lower the importance of small deviations while it increases that of large deviations. For instance, returning to the lower left example from Fig. 3.5, the OLS estimator is biased toward the outlier point. The line passing by all points except this outlier thus has a larger MSE. In contrast, the MAE is twice as small as the MAE of the OLS estimator, showing that the importance of the outlier point is reduced in this case, which can be seen in Fig. 3.6.

3.2.4 Out-of-sample performance: cross-validation methods

A main issue in machine learning which can be very well analyzed in the linear model setting is that of generalizability. Specifically, once we have determined the parameter $\boldsymbol{\beta}$ obtaining the estimation $\hat{\boldsymbol{\beta}}$ in our regression problem (3.1), will $f(\boldsymbol{x}, \hat{\boldsymbol{\beta}})$ behave as expected for any given x? Said in statistical terms, "will my $\hat{\boldsymbol{\beta}}$ estimated from a sample $(\mathbf{x}^{(i)}, y^{(i)})_{i=1}^{N}$ generalize to the population?"

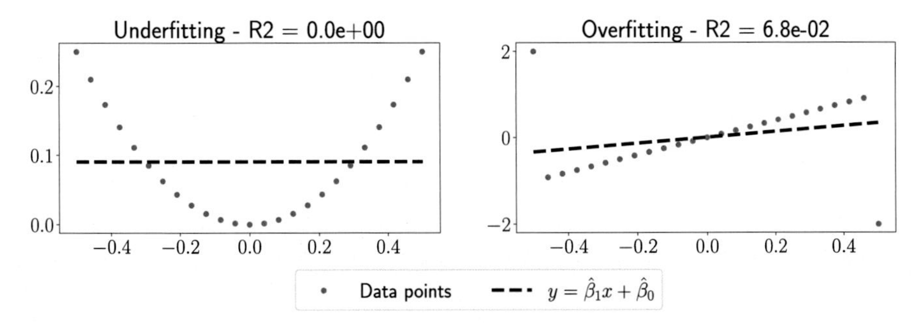

FIGURE 3.7

Illustration of the *underfitting* (left) and *overfitting* (right) phenomena. For the underfitting case, the data points have a rich structure that cannot be captured with a simple linear regression. A more complex model – such as polynomial regression – would be required to better account for this structure. In the overfitting case, the presence of two outliers makes the linear model deviate from the expected model. This behavior could be fixed using resampling techniques and regularization, which will detect models that fit the training data too well.

There are two main phenomena which can affect the generalization of our regression model. The first is *overfitting*, which means that $f(x, \hat{\boldsymbol{\beta}})$ will follow extremely well our training dataset but, nonetheless, it does not generalize. Specifically, the measurements quantifying the quality of the fit denote a very good performance while the same measurements on an out-of-sample set, such as the test set, are prone to show a very bad performance. In terms of the error measurements introduced in Section 3.2.2, the in-sample error $\mathrm{Err}_{\mathcal{D}}$ is small and the out-of-sample error $\mathrm{Err}_{\mathcal{T}}$ is large. An example of this is shown in Fig. 3.7 for a linear model with outliers and further in this chapter in a non-linear case in Section 3.3.2.2. The second phenomenon is *underfitting*, where our regression model does not have the capability of synthesizing the dataset nor of generalizing the behavior to out-of-sample observations, i.e., when both errors are large. One way to detect underfitting in a linear model is when the coefficient of determination R^2 of the fit is negative. This means that $\mathrm{MSE}(y, \hat{y}) > \mathrm{VAR}(y)$, or that the fitting error is larger than the variance of the data. An example of this case, trying to fit a parabola with a linear model, is shown in Fig. 3.7. Overfitting is harder to detect as it requires either strong hypotheses on the population or a data-rich scenario in which we can hold back observations, i.e., split our sample, to simulate confidently the behavior of our model in the whole population.

As discussed, a commonly used method to quantify the generalization capabilities of our model is to split our sample $(x^{(i)}, y^{(i)})_{i=1}^{N}$ into two segments. The *training* segment $(x_{\text{train}}^{(i)}, y_{\text{train}}^{(i)})_{i=1}^{N_{\text{train}}} = (X_{\text{train}}, Y_{\text{train}})$ is used to obtain $\hat{\beta}$, and the *test* segment $(x_{\text{test}}^{(i)}, y_{\text{test}}^{(i)})_{i=1}^{N_{\text{test}}} = (X_{\text{test}}, Y_{\text{test}})$ is used to quantify how well each $f(x_{\text{test}}^{(i)}, \hat{\boldsymbol{\beta}})$ resembles

$y_{\text{test}}^{(i)}$. Whether it is necessary to split our sample in this manner and how to split it is a matter of long discussion which we will take on later in this section.

The split-sample approach can work nicely, provided that the split is representative of the data and its noise and that the amount of data in the training sample is enough to fit the parameters of the regression model. Informally this is the case when both splits are i.i.d. and they have a similar coverage of the sampled domain. A strategy which is more robust to a wider spread of distributional assumptions of our sample, yet more computationally expensive, is through resampling methods. The main goal of these is to generate multiple pairs of training/test splits from the main sample. In the following, we denote $((\boldsymbol{x}_{\text{train}_k}, \boldsymbol{y}_{\text{train}_k}), (\boldsymbol{x}_{\text{test}_k}, \boldsymbol{y}_{\text{test}_k}))$ the kth split of the original sample $(\boldsymbol{X}, \boldsymbol{y})$ separated into training/test pairs. Then we can obtain K estimates for $\hat{\boldsymbol{\beta}}, \hat{\boldsymbol{\beta}}_1, \dots, \hat{\boldsymbol{\beta}}_K$, and characterize the distribution of the parameters $\{\hat{\boldsymbol{\beta}}_k\}_{k=1\dots K}$, as well as that of the predictions' error, for each of the K subsamples, $\{\ell(\boldsymbol{y}_{\text{test}_k}, f(\boldsymbol{X}_{\text{test}_k}; \hat{\boldsymbol{\beta}}_k))\}_{k=1\dots K}$, where the *score* function can be the MSE, R^2, etc. The generalization quality of our regression model can then be examined in light of the mean, standard deviation, or more general distributional analyses of the parameters and performance scores across all samples. We illustrate this in Fig. 3.8, where we have estimated the uncertainty that a specific voxel is related with a word mentioned in a neuroscience publication. It can be seen that there is less uncertainty for words directly associated with lower cognitive functions such as "auditory" than for those associated with higher cognitive functions such as "math." How to generate these samples and to guarantee that the obtained distribution of scores is a good representative of any out-of-sample score is an open area of research (see, e.g., [4–7]).

The accuracy of a resampling method to estimate the performance of the regression and the precision of the system's parameters $\widehat{\boldsymbol{\beta}}$ depends on the distributional characteristics of our annotated sample. For instance, it depends on whether the target observations are i.i.d. or whether the noise fits the hypotheses of the Gauss–Markov theorem described in Section 3.2.2.1. If the sample can be divided into K different independent sets while retaining the aforementioned conditions of i.i.d. observations and sample space coverage, then we can use *K-fold cross-validation* [8]. This means that we split the sample into K partitions and then use $K-1$ for training and 1 for testing. The exhaustive estimation of the parameter and score distributions then requires only K iterations. If this partition cannot be achieved but each sample point $(\boldsymbol{x}^{(i)}, y^{(i)})$ is exchangeable, meaning that the joint probability of all samples, $Pr[(\mathbf{x}^{(k_1)}, \mathrm{y}^{(k_1)}), \dots, (\mathbf{x}^{(k_N)}, \mathrm{y}^{(k_N)})]$, remains the same for any permutation $K = k_1, \dots, k_N$ of sample indices $i = 1, \dots, N$, then we can use a permutation strategy [9,10]. Estimation of the accuracy under permutations is implemented by performing a training/test procedure on each permutation K, which means that an exhaustive permutation procedure requires $N!$ training/test procedures, although under certain conditions the distributions of $\hat{\boldsymbol{\beta}}$ and $\ell(y, \hat{y})$ can be accurately estimated with much fewer operations. If none of these hypotheses are met, then we need to resort to the bootstrapping family of techniques [11], which generally require a much larger number of training/test procedures to accurately estimate the distributions of $\widehat{\boldsymbol{\beta}}$ and $\ell(y, \hat{y})$.

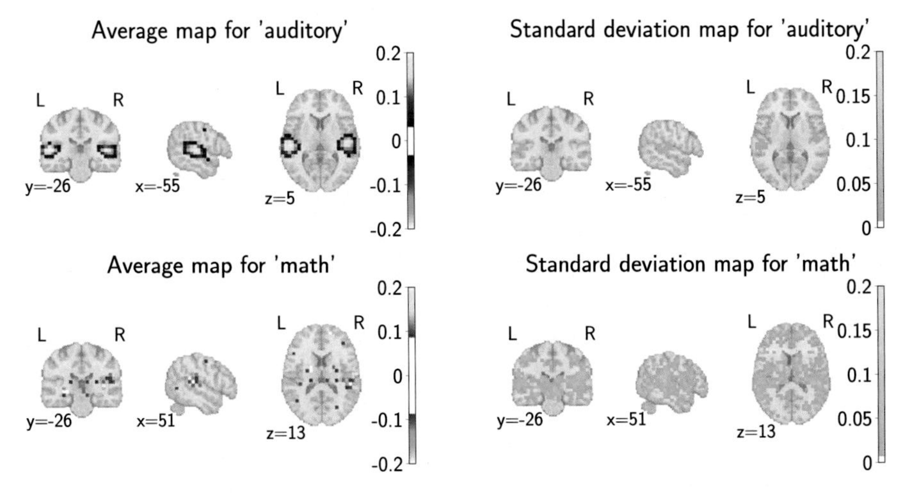

FIGURE 3.8

Using cross-validation to quantify uncertainty in inference tasks for decoding. Here, we learn the parameter of a linear model multiple times, average the learned β, and quantify the uncertainty. Using this technique, we can see that the functional network associated to the word "auditory" is much more stable than the one associated with "math". This is consistent with a quantification of the agreement in the community on the locations associated with such cognitive functions.

Finally, which method provides the most accurate estimation for $\widehat{\boldsymbol{\beta}}$ and $\ell(y, \hat{y})$ depends not only on the training sample characteristics, but also on the chosen regression function. Extensive analysis of such methods has been conducted in the context of a large number of applications and biomedical use cases (e.g., [7,10,12]).

3.2.5 Shrinkage methods

When the data are scarce or noisy, it is common to add prior information into the fitting algorithm to compensate for this. In terms of the bias–variance trade-off that we described in equation (3.16), this amounts to including a well-justified bias to reduce the variance in the data. Mathematically, we rewrite our minimization problem in equation (3.13) as

$$\widehat{\beta} = \underset{\beta \in \mathbb{R}^p}{\arg\min} \frac{1}{N} \sum_{i=1}^{N} \ell(y^{(i)}, f(\boldsymbol{x}^{(i)}, \boldsymbol{\beta})) + \lambda \operatorname{Reg}(\boldsymbol{\beta}), \tag{3.27}$$

where λ sets the trade-off between prioritizing a minimal loss function, i.e., reducing the bias, or the regularizer, i.e., reducing the variance. In general, these methods tend to reduce, or shrink, the value of $\widehat{\beta}$ under a certain norm, leading to the general name of "shrinkage methods." Depending on the hypothesis that we want to implement, the regularization function Reg can take different shapes.

3.2.5.1 Dealing with multicollinearity: ridge regression

If our problem is that our system results in highly correlated components in the optimal parameter $\widehat{\boldsymbol{\beta}}$, the solution becomes unstable and exhibits high variance due to the fact that different components of $\widehat{\boldsymbol{\beta}}$ can evolve together, providing a similar minimal value of the loss function. This can be alleviated by including a bias in which we push the magnitude of $\widehat{\boldsymbol{\beta}}$ to be small. Specifically, this defines our regularizer as $\text{Reg}(\boldsymbol{\beta}) = \sum_k [\boldsymbol{\beta}]_k^2$, leading to

$$\widehat{\boldsymbol{\beta}} = \underset{\boldsymbol{\beta} \in \mathbb{R}^p}{\arg\min} \frac{1}{N} \sum_{i=1}^{N} \ell(y^{(i)}, f(\boldsymbol{x}^{(i)}, \boldsymbol{\beta})) + \lambda \sum_{k=1}^{p} [\boldsymbol{\beta}]_k^2. \tag{3.28}$$

When our loss function ℓ is the MSE, this is easy to implement as our minimization function then becomes

$$\begin{aligned}
\widehat{\boldsymbol{\beta}} &= \underset{\boldsymbol{\beta} \in \mathbb{R}^p}{\arg\min} \frac{1}{N} \sum_{i=1}^{N} (y^{(i)} - \boldsymbol{x}^{(i)}\boldsymbol{\beta})^2 + \lambda \sum_{k=1}^{p} [\boldsymbol{\beta}]_k^2 \\
&= \underset{\boldsymbol{\beta} \in \mathbb{R}^p}{\arg\min} \frac{1}{N} \|\boldsymbol{y} - \boldsymbol{X}\boldsymbol{\beta}\|_2^2 + \lambda \|\boldsymbol{\beta}\|_2^2, \quad \boldsymbol{X} \in \mathbb{R}^{N \times p}, \ \boldsymbol{y} \in \mathbb{R}^N \\
&= \frac{1}{N} (\boldsymbol{X}^T \boldsymbol{X} + \lambda \, \text{Id})^{-1} \boldsymbol{X}^T \boldsymbol{y}.
\end{aligned} \tag{3.29}$$

In Fig. 3.9 we can see how the variance of the solution is increased and the parameters $[\boldsymbol{\beta}]_j$ become more disperse as λ decreases. This method is known as Tikhonov regularization, or ridge regression.

3.2.5.2 Dealing with sparsity: the LASSO

A different common issue in regression is whether we have enough independent data points $(y^{(i)}, \boldsymbol{x}^{(i)})_{i=1\ldots N}$ to fit all p parameters in our model. From a different perspective, we could also argue that not all parameters are important to achieve a minimal loss with good generalization. In other words, that $\widehat{\boldsymbol{\beta}}$ is *sparse*. To enforce this, Tibshirani et al. [13] proposed the *LASSO* (Least Absolute Shrinkage and Selection Operator). Instead of dealing with a selection problem, in which we should choose which components of $\widehat{\boldsymbol{\beta}}$ we keep, LASSO techniques relax this problem into one where we enforce the magnitude of a certain number of components to be close to 0. This is achieved through the following version of equation (3.27):

$$\widehat{\boldsymbol{\beta}} = \underset{\boldsymbol{\beta} \in \mathbb{R}^p}{\arg\min} \frac{1}{N} \sum_{i=1}^{N} \ell(y^{(i)}, f(\boldsymbol{x}^{(i)}, \boldsymbol{\beta})) + \lambda \sum_{k=1}^{p} |[\boldsymbol{\beta}]_k|, \tag{3.30}$$

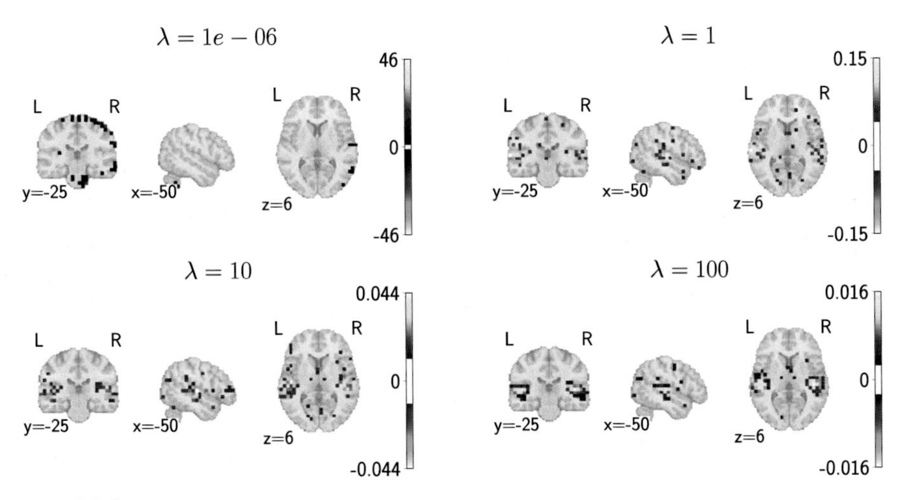

FIGURE 3.9

Decoding example with ridge regression for different regularization levels λ for the word "auditory." When the regularization is too low, the model overfits the data and learns large coefficients that are unrelated with the cognitive process. With larger regularization, the large coefficients capture the auditory cortices.

which when we deal with the MSE loss becomes

$$\begin{aligned}\widehat{\boldsymbol{\beta}} &= \underset{\boldsymbol{\beta}\in\mathbb{R}^p}{\arg\min}\, \frac{1}{N}\sum_{i=1}^{N}(y^{(i)} - \boldsymbol{x}^{(i)}\boldsymbol{\beta})^2 + \lambda\sum_{j=1}^{p}|[\boldsymbol{\beta}]_j| \\ &= \underset{\boldsymbol{\beta}\in\mathbb{R}^p}{\arg\min}\, \frac{1}{N}\|\boldsymbol{y} - \boldsymbol{X}\boldsymbol{\beta}\|_2^2 + \lambda\|\boldsymbol{\beta}\|_1, \quad \boldsymbol{X}\in\mathbb{R}^{N\times p},\ \boldsymbol{y}\in\mathbb{R}^N.\end{aligned} \tag{3.31}$$

Differently from the ridge-regularized regression problem presented in Section 3.2.5.1, this problem does not have a closed-form solution which can be easily implemented. Hence, an extensive number of minimization techniques have been proposed to solve the LASSO problem in different settings (see, e.g., [14,15]). In Fig. 3.10 we show the effect of this regularization in the decoding task. In this figure it can be observed how, differently from the ridge, the LASSO technique pushed the solutions to be more sparse, thus reducing the number of voxels recognized as being associated with the term "auditory" without the need for post-analysis thresholding.

3.2.5.3 Other general regularizers

A more general expression for our regularized regression problem can be written as follows:

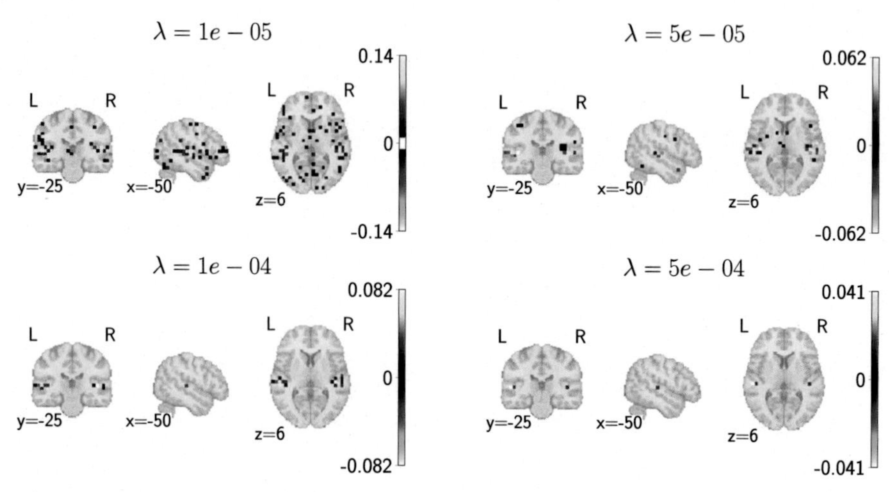

FIGURE 3.10

Decoding example with ℓ_1-regularization for different regularization levels λ for the word "auditory." The ℓ_1-regularization promotes sparse solutions for any level of regularization. For low regularization levels, an overfitting effect similar to the one in the ridge example can be observed. For higher regularization levels, only a few voxels are activated in the brain, leading to a much more localized decoding.

$$\widehat{\boldsymbol{\beta}} = \arg\min_{\beta} \sum_{i=1}^{N} \ell(y^{(i)}, f(\boldsymbol{x}^{(i)}, \boldsymbol{\beta})) + \sum_{k=1}^{K} [\boldsymbol{\lambda}]_k \, \| [\boldsymbol{M}]_k \, \boldsymbol{\beta} \|_{p_k, q_k},$$
$$\text{where } \| [\boldsymbol{M}]_k \, \boldsymbol{\beta} \|_{p,q} \triangleq \left(\sum_j \left(\sum_i |[[\boldsymbol{M}]_k \, \boldsymbol{\beta}]_{ij}|^p \right)^{\frac{q}{p}} \right)^{\frac{1}{q}}. \tag{3.32}$$

This general form of matrix norm-based regularizers has been used to develop a range of regression approaches such as elastic net [16],

$$\widehat{\boldsymbol{\beta}} = \arg\min_{\beta} \sum_{i=1}^{N} \ell(y^{(i)}, f(\boldsymbol{x}^{(i)}, \boldsymbol{\beta})) + \underbrace{\lambda_1 \|\boldsymbol{\beta}\|_{2,2}^2}_{\text{Ridge}} + \underbrace{\lambda_2 \|\boldsymbol{\beta}\|_{1,1}}_{\text{LASSO}}, \tag{3.33}$$

which balances between reducing the variance of the parameters β in a Ridge setting and the sparsity imposition of LASSO. Other cases such as compressive sensing (see, e.g., [17]) and graph-net [18] can also be expressed in terms of equation (3.32). We show some examples in Fig. 3.11, where it can be seen that the Total Variation-regularizer (TV) privileges larger constant areas with the same values.

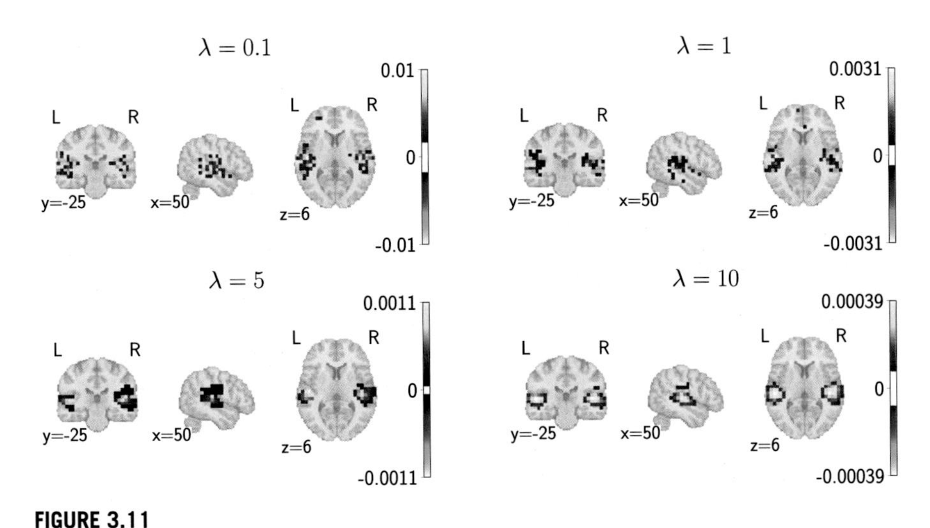

FIGURE 3.11

Decoding example with TV-regularization for different regularization level λ for the word "auditory." The TV-regularization promotes localized solution and even for low regularization levels, it selects regions that are coherent with the literature.

3.3 Treating non-linear problems with linear models

The linear regression model can be extended in a simple manner to non-linear settings. Here we show as examples two specific cases. First, we discuss the case where the noise does not meet the Gauss–Markov requirements, which means that if we keep the right-hand side of equation (3.1) constant, the relationship between the linear model on the right-hand side and the measured signal stops being a simple equality. In a non-negligible set of cases, this can be achieved by transforming the observations with a *link function* (see, e.g., [19]). Specifically, we address the problem of classification through logistic regression. Second, we address the same problem but by transforming our features X, specifically in two cases, one where the features are categorical variables and another where y is a non-linear function which can be expressed by harnessing the power of harmonic analysis and choosing discretized functional bases as features.

3.3.1 Generalized linear models: transforming y

In Section 3.1.2 we described linear regression, in its more usual setting, as corresponding to the estimation of the parameters of an output Y which is a Gaussian-distributed random variable. Although this scenario is one of the most common ones, it is by no means universal. An example of this case can be found in the binary classification task, where the output Y is one of two labels. Specifically, and without loss of generality, we have $y^{(i)} \in \{0, 1\}$, which does not follow a Gaussian probability law, invalidating the algorithms and assumptions we have introduced in previous sections of this chapter. Writing this in a formal manner, we now have the problem of finding θ such that

$$\mathrm{y}^{(i)} = f(\boldsymbol{x}^{(i)}, \theta; \epsilon) \sim \mathcal{D}(\theta), \quad \epsilon \sim P_\epsilon, \tag{3.34}$$

where the randomness on the right-hand side comes from the, usually implicit, noise represented by the random variable ϵ. However, there is a solution to this conundrum if we can "translate" the problem of performing the regression on Y to perform it on a transformed $g(Y)$, which is a Gaussian-distributed random variable. Then, we only need to be able to translate the regressed parameters β back to the original setting. More formally, if we can produce an invertible *link* function g such that

$$\mathrm{y}^{(i)} \sim \mathcal{D}(\theta^{(i)}), \tag{3.35}$$

$$g(\mathrm{y}^{(i)}) \sim \mathcal{N}(X^{(i)}\beta, \sigma^2), \tag{3.36}$$

$$g^{-1}(\boldsymbol{x}^{(i)}\boldsymbol{\beta}) = \theta^{(i)}, \tag{3.37}$$

then we can apply g to $Y^{(i)}$, use the methods introduced in Section 3.1.2 to solve our problem of finding β such that

$$\begin{cases} g(\mathrm{y}^{(1)}) = \boldsymbol{\beta}\boldsymbol{x}^{(1)} + \epsilon^{(1)}, \\ \vdots \\ g(\mathrm{y}^{(N)}) = \boldsymbol{\beta}\boldsymbol{x}^{(N)} + \epsilon^{(N)}, \end{cases} \quad \text{where} \quad \epsilon^{(i)} \sim \mathcal{N}(0, \sigma^2), \tag{3.38}$$

and obtain our desired result by applying $\theta^{(i)} = g^{-1}(\boldsymbol{x}^{(i)}\boldsymbol{\beta})$.

Although this approach is used in many different cases, the most commonly used version of generalized linear models is a particular brand of regression for classification problems usually dubbed *logistic regression*, which was developed and coined by Pierre-François Verhulst in the 19th century.

3.3.1.1 Classification as a regression problem: logistic regression

In the case of binary classification, each input $\boldsymbol{x}^{(i)}$ is associated with one of two classes, denoted 0 and 1. In this case, the output $\mathbf{y}^{(i)}$ does not live in $\mathbb{R}$ but in $\{0, 1\}$. Using a linear regression would lead to counting the same error for predicting $\widehat{\mathbf{y}} = 0$ and $\widehat{\mathbf{y}} = 2$ when the true value is $\mathbf{y} = 1$. If the value $\widehat{\mathbf{y}}$ is projected in $\{0, 1\}$ by taking the indicator function $\mathbf{1}\{\widehat{\mathbf{y}} > 0.5\}$, it would be much better to predict $\widehat{\mathbf{y}} = 2$ than the other. In order to avoid this problem, one can use *logistic regression*, which can be seen as a slight modification of linear regression.

Logistic regression is a linear model that regresses a continuous random variable y in the $]0, 1[$ open interval from a continuous variable x. The core idea is to assume that y is Bernoulli-distributed and use a linear model to predict a value $l = g(\mathrm{y})$ in $\mathbb{R}$

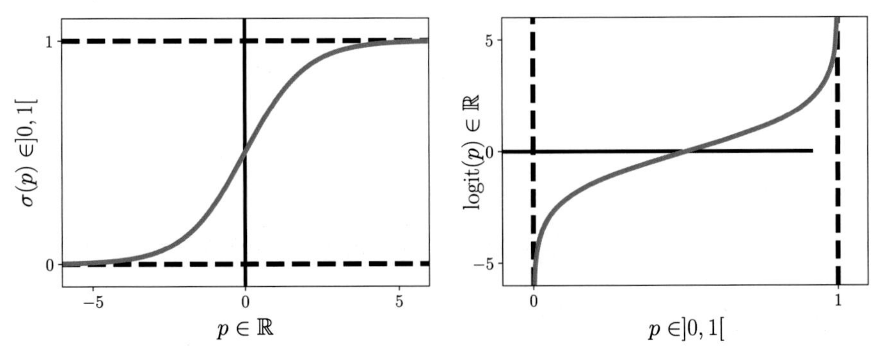

FIGURE 3.12

The sigmoid function (left) σ transforms the space of real numbers $\mathbb{R}$ in the interval $]0,1[$. The logit function (right) is the inverse transformation.

from x,

$$l = g(\mathrm{y}) = [\boldsymbol{\beta}]_0 + [\boldsymbol{\beta}]_1 x \quad \text{with} \quad \mathrm{y} \sim \text{Bernoulli}(\theta^{(i)}), \tag{3.39}$$

$$g : y \mapsto \log\left(\frac{y}{1-y}\right). \tag{3.40}$$

This value is often called the *logit* of y. Then, it is transformed from $\mathbb{R}$ to $]0,1[$ using the sigmoid as inverse link function $g^{-1} : x \mapsto \frac{1}{1+e^{-x}}$ (see Fig. 3.12). The full model thus reads

$$\begin{cases} \mathrm{y}^{(1)} \sim \text{Bernoulli}(\theta(\boldsymbol{x}^{(1)})), \\ \vdots \\ \mathrm{y}^{(N)} \sim \text{Bernoulli}(\theta(\boldsymbol{x}^{(N)})), \end{cases} \Leftrightarrow \begin{cases} g(\mathrm{y}^{(1)}) = \boldsymbol{\beta}\boldsymbol{x}^{(1)} + \epsilon^{(1)}, \\ \vdots \\ g(\mathrm{y}^{(N)}) = \boldsymbol{\beta}\boldsymbol{x}^{(N)} + \epsilon^{(N)}, \end{cases}$$
$$\text{where} \quad \epsilon^{(i)} \sim \mathcal{N}(0,\sigma^2), \quad \theta(\boldsymbol{x}^{(i)}) = g^{-1}(\boldsymbol{\beta}\boldsymbol{x}^{(i)}). \tag{3.41}$$

With this transform, the linear model aims to output the largest logit possible for a target $Y = 1$ and the lowest for $\mathbf{y} = 0$, and we can solve our problem for a given set of observations $\mathrm{y}^{(i)}$ as

$$\widehat{\boldsymbol{\beta}} = \underset{\boldsymbol{\beta}}{\arg\min} \left\| \begin{pmatrix} g\left(\mathrm{y}^{(1)}\right) \\ \vdots \\ g\left(\mathrm{y}^{(N)}\right) \end{pmatrix} - \begin{pmatrix} [\boldsymbol{x}]_1^{(1)} & \cdots & [\boldsymbol{x}]_1^{(N)} \\ \vdots & \ddots & \vdots \\ [\boldsymbol{x}]_p^{(1)} & \cdots & [\boldsymbol{x}]_p^{(N)} \end{pmatrix} \boldsymbol{\beta} \right\|_2^2, \quad \widehat{\boldsymbol{\theta}}^{(i)} = g^{-1}(\widehat{\boldsymbol{\beta}}\boldsymbol{x}^{(i)}). \tag{3.42}$$

This derivation is equivalent to a maximum likelihood estimation, with the same hypothesis that each $\mathrm{y}^{(i)}$ is Bernoulli-distributed.

Derivation as a maximum likelihood estimation Logistic regression can also be derived as a probabilistic model. The log-odds of a model is the logarithm of the ratio between the probability $P(\mathrm{y}=1|\mathrm{x}) = p$ of having $\mathrm{y}=1$ and the probability $P(\mathrm{y}=0|\mathrm{x}) = 1 - P(\mathrm{y}=1|\mathrm{x}) = 1 - p$ of having $\mathrm{y}=0$. Logistic regression is based on the assumption that there exists a linear relationship between the input x and the log-odds of the event $\mathrm{y}=1$, i.e.,

$$\log \frac{p}{1-p} = g(p) = [\boldsymbol{\beta}]_0 + [\boldsymbol{\beta}]_1\, x. \tag{3.43}$$

This corresponds to the *logit* described in equation (3.40), which is another name for the log-odds. Interestingly, the log-odd function $p \mapsto \frac{p}{1-p}$ is the inverse of the sigmoid σ. Thus, to recover the value of $P(\mathrm{y}=1|\mathrm{x}=x)$, one can use the sigmoid with the relationship (3.43),

$$P(\mathrm{y}=1|\mathrm{x}=x) = g^{-1}([\boldsymbol{\beta}]_0 + [\boldsymbol{\beta}]_1\, x), \tag{3.44}$$

$$P(\mathrm{y}=0|\mathrm{x}=x) = 1 - P(\mathrm{y}=1|\mathrm{x}=x) = 1 - g^{-1}([\boldsymbol{\beta}]_0 + [\boldsymbol{\beta}]_1\, x). \tag{3.45}$$

Given a set of N i.i.d. samples $\{(\boldsymbol{x}^{(1)}, \mathrm{y}^{(1)}) \ldots (x^{(N)}, \mathrm{y}^{(N)})\}$, the likelihood of a model $\boldsymbol{\beta}$ can be written as

$$L(\beta|x) = P(\mathrm{y}|x, \boldsymbol{\beta}) = \prod_{i=1}^{N} P(\mathbf{y} = y^{(i)}|\mathrm{x} = x^{(i)}; \boldsymbol{\beta}) \tag{3.46}$$

$$= \prod_{i=1}^{N} P(\mathrm{y}=1|\mathrm{x}=x^{(i)}; \boldsymbol{\beta})^{y^{(i)}}\, P(\mathrm{y}=0|\mathrm{x}=x^{(i)}; \boldsymbol{\beta})^{1-y^{(i)}}. \tag{3.47}$$

Substituting the relationship (3.44) in this equation and taking the negative log of this expression, we obtain the classical negative log-likelihood function

$$\mathcal{L}(\boldsymbol{\beta}|x) = -\log(L(\boldsymbol{\beta}|x))$$

$$= \sum_{i=1}^{N} y^{(i)} \log(1 + e^{-[\boldsymbol{\beta}]_0 - [\boldsymbol{\beta}]_1 x^{(i)}}) + (1 - y^{(i)}) \log(1 + e^{[\boldsymbol{\beta}]_0 + [\boldsymbol{\beta}]_1 x^{(i)}}). \tag{3.48}$$

This function can be optimized using gradient descent to find the value of $\boldsymbol{\beta}$.

Note that for classification tasks, the performance evaluation is slightly different compare to the regression problem. This matter is discussed at length in Chapter 24, and in particular in Section 24.3.

3.3.2 Feature spaces: transforming X

By generalizing the feature matrix, linear regression can be used to express a more general set of problems than a linear relationship between real numbers and a matrix in $\mathbb{R}^N$. Examples of these are functional basis-driven, such as the Fourier transform, or categorical variable representation. In this section, we focus on two cases.

3.3.2.1 Categorical variables

Another conundrum which we face when working on regression models are categorical variables. These variables, which live in a discrete, usually finite support, space, are non-trivial to represent in a linear regression setting. Nonetheless, these appear in many medical applications when representing gender, age groups, and disease prevalence, amongst other cases. The general way to deal with categorical variables in a linear model is through encoding them as numerical ones. Depending on the type of categorical variable and the model we are designing, several options arise. The simplest one is usually the ordinal case, where categories can be sorted in an unequivocal importance order, such as a star-based ranking. This can be encoded as an integer variable with consecutive numbers for the different consecutive categories (for instance, 1 for one-star hotels, 2 for two-star hotels, etc.). When the categories do not represent a particular ordered set but different exclusive options, we can use one-hot encoding. In this case we add one column to X per value of the categorical variable; for instance, a binary gender variable with M/F options will be encoded as two binary columns, one for those tagged as M and one for those tagged as F. This solution is very simple to capture; however, there is an issue: it adds to the system one more parameter per category. A simple regression model for a gender-exclusive effect will then look like

$$y = [\boldsymbol{\beta}]_M \, [\boldsymbol{x}]_M + [\boldsymbol{\beta}]_F \, [\boldsymbol{x}]_F + [\boldsymbol{\beta}]_0 \, . \tag{3.49}$$

There is a problem with this encoding: the number of parameters increases with the number of categories, which increases the rank of the matrix X, and hence the number of data points required to fit the regression problem. A simple option to reduce the matrix rank is to define one category as the baseline; for instance, we could define F as baseline in the model in (3.49):

$$y = [\boldsymbol{\beta}]_M \, [\boldsymbol{x}]_M + [\boldsymbol{\beta}]_F \, . \tag{3.50}$$

In this case, which has one parameter less than the one in (3.49), the intercept $[\boldsymbol{\beta}]_F$ encodes also the female effect and $[\boldsymbol{\beta}]_M$ encodes the difference in the contribution to the value of Y that comes from being tagged as a male.

3.3.2.2 Linearizing non-linear regression: functional bases

Several families of non-linear functions are expressible as linear models: we define a family, or basis, of functions and then we express our regressed function $y = f(x; [\boldsymbol{\beta}]_0, \ldots, [\boldsymbol{\beta}]_J)$ as a linear combination of these basis functions. For instance, let us assume that our sample $(x^{(i)}, y^{(i)})_{i=1}^N$ is equally sampled over a bounded domain, e.g., $x^{(i)} \in [0, 2\pi]$ has been sampled at frequency $2\pi i/N$. Then, instead of stating the usual regression problem described in equation (3.4), we can state a regression problem such that $y^{(i)} = \sum_{j=0}^{J-1} [\boldsymbol{\beta}]_j \cos(j \cdot x^{(i)})$. Our linear regression problem will be able to fit any non-linear function which can be approximated by this family of cosine-based functions. Generally, this is achieved when $f(x)$ can be expressed as

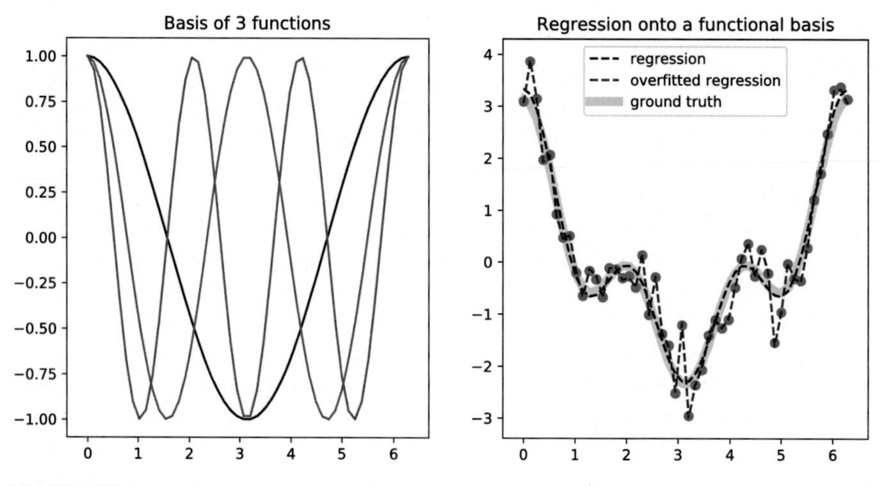

FIGURE 3.13

Functional basis regression. A cosine-based family of functions, as seen on the left panel, can be used to represent non-linear functions through linear regression, as shown in the right panel.

a linear combination of different functions, rendering our regression problem in the following general setting:

$$y = \sum_{j=0}^{J-1} [\boldsymbol{\beta}]_j \, f_j(x) + \epsilon, \tag{3.51}$$

which, when y and x are discretely sampled as in our example, can be stated as

$$\begin{aligned} y^{(1)} &= \sum_{j=0}^{J-1} [\boldsymbol{\beta}]_j \, f_j(x^{(1)}) + \epsilon^{(1)}, \\ &\vdots \\ y^{(N)} &= \sum_{j=0}^{J-1} [\boldsymbol{\beta}]_j \, f_j(x^{(N)}) + \epsilon^{(N)}, \end{aligned} \tag{3.52}$$

and this can be solved as a least squares problem as described in Section 3.1.1. We show the cosine-based example in Fig. 3.13. It is worth noting that non-linear representations are more prone to overfitting than regular linear models. In the example case of Fig. 3.13, it is simple to see that when we use too many basis functions, i.e., J is large, we are bound to fit high-frequency noise. This is shown in Fig. 3.13, where, on the right panel, it is observable that the red curve represents overfitted regression.

3.3.3 Going further

Regression and classification problems are a "Swiss army knife" paradigm to frame several of the issues in medical imaging. Specifically, most supervised problems, such as denoising, registration (Chapter 19), or object detection (Chapter 22), can be framed as regression problems.

Furthermore, here we have but touched the basic scenarios and tools regarding the power of regression. Currently a large number of deep learning models are used to perform non-linear regression (Sections 19.4, 22.3).

3.4 Exercises

1. In Fig. 3.7 we exemplify underfitting by using a linear regression to fit $y = x^2$ in the interval $x \in [-0.5, 0.5]$. Re-evaluate the experiment in the interval $x \in [0.2, 0.5]$. Is the linear model still underfitting? If this is the case, how does the underfitting compare with that presented in the original case?
2. In Fig. 3.8, we explore the results of a decoding experiment for the word "auditory". Take random subsamples of the NeuroQuery database at the following ratios: 10%, 25%, 50%, 75%, and 100% of the total number of documents. Run the regression analysis plot average map across cross-validation experiments for each sample size. How does it change? Compute the 99-percentile of the standard deviation (std) across cross-validation experiments and plot it against sample size. How does it change with sample size?
3. In exercise 2, we explored the capacity of a linear regressor to predict foci reported by studies that mention the 'auditory' word. We achieve this by examining the algorithm's behavior when used with different sample sizes. Reproduce exercise 2 using a Ridge regularization, as seen in Section 3.2.5.1. How do the average and standard deviations change? Does the 99th percentile of the standard deviation across cross-validation experiments decrease faster than without regularization? Why?

References

[1] Dockès J., Poldrack R., Primet R., Gözükan H., Yarkoni T., Suchanek F., Thirion B., Varoquaux G., NeuroQuery, comprehensive meta-analysis of human brain mapping, eLife 9 (Mar. 2020) e53385.

[2] Hastie T., Tibshirani R., Friedman J., Elements of Statistical Learning, Data Mining, Inference, and Prediction, Springer, Feb. 2011.

[3] Anscombe F.J., Graphs in statistical analysis, American Statistician (ISSN 0003-1305) 27 (1) (Feb. 1973) 17, https://doi.org/10.2307/2682899.

[4] Efron B., Tibshirani R., Improvements on cross-validation: the .632+ bootstrap method, Journal of the American Statistical Association 92 (438) (1997) 548–560.

[5] Bates S., Hastie T., Tibshirani R., Cross-validation: what does it estimate and how well does it do it?, arXiv:2104.00673 [math, stat], Apr. 2021.

[6] Wager S., Cross-validation, risk estimation, and model selection: comment on a paper by Rosset and Tibshirani, Journal of the American Statistical Association 115 (529) (Jan. 2020) 157–160, https://doi.org/10.1080/01621459.2020.1727235, ISSN: 0162-1459, 1537-274X.

[7] Varoquaux G., Cross-validation failure: small sample sizes lead to large error bars, NeuroImage (June 2017), https://doi.org/10.1016/j.neuroimage.2017.06.061.

[8] Geisser S., The predictive sample reuse method with applications, Journal of the American Statistical Association 70 (350) (1975) 320–328, https://doi.org/10.1080/01621459.1975.10479865.

[9] Freedman D., Lane D., A nonstochastic interpretation of reported significance levels, Journal of Business & Economic Statistics 1 (4) (Jan. 1983) 292–298.

[10] Winkler A.M., Ridgway G.R., Webster M.A., Smith S.M., Nichols T.E., Permutation inference for the general linear model, NeuroImage (ISSN 1053-8119) 92 (May 2014) 381–397, https://doi.org/10.1016/j.neuroimage.2014.01.060.

[11] Efron B., Estimating the error rate of a prediction rule: improvement on cross-validation, Journal of the American Statistical Association 78 (382) (June 1983) 316–331, https://doi.org/10.1080/01621459.1983.10477973, ISSN: 0162-1459, 1537-274X.

[12] Molinaro A.M., Simon R., Pfeiffer R.M., Prediction error estimation: a comparison of resampling methods, Bioinformatics 21 (15) (Aug. 2005) 3301–3307, https://doi.org/10.1093/bioinformatics/bti499, ISSN: 1367-4803, 1460-2059.

[13] Tibshirani R., Regression shrinkage and selection via the lasso, Journal of the Royal Statistical Society, Series B, Methodological 58 (1) (Jan. 1996) 267–288, https://doi.org/10.2307/2346178.

[14] Meier L., Van De Geer S., Bühlmann P., The group lasso for logistic regression, Journal of the Royal Statistical Society, Series B, Statistical Methodology (ISSN 1369-7412) 70 (1) (Jan. 2008) 53–71, https://doi.org/10.1111/j.1467-9868.2007.00627.x.

[15] Meinshausen N., Relaxed lasso, Computational Statistics & Data Analysis (ISSN 0167-9473) 52 (1) (Sept. 2007) 374–393, https://doi.org/10.1016/j.csda.2006.12.019.

[16] Zou H., Hastie T., Regularization and variable selection via the elastic net, Journal of the Royal Statistical Society, Series B, Statistical Methodology 67 (2) (Apr. 2005) 301–320, https://doi.org/10.1111/j.1467-9868.2005.00503.x.

[17] Donoho D.L., Compressed sensing, IEEE Transactions on Information Theory 52 (4) (Mar. 2006) 1289–1306, https://doi.org/10.1109/TIT.2006.871582.

[18] Grosenick L., Klingenberg B., Katovich K., Knutson B., Taylor J.E., Interpretable whole-brain prediction analysis with GraphNet, NeuroImage 72 (May 2013) 304–321, https://doi.org/10.1016/j.neuroimage.2012.12.062.

[19] Searle S.R., McCulloch C.E., Generalized Linear, and Mixed Models, Wiley-Interscience, Jan. 2001, ISBN: 978-0-471-19364-7, 0-471-19364-X.

CHAPTER

Estimation and inference

4

Gonzalo Vegas Sánchez-Ferrero[a] and Carlos Alberola-López[b]

[a]*Department of Radiology, Harvard Medical School, Brigham and Women's Hospital, Boston, MA, United States*

[b]*Laboratorio de Procesado de Imagen (LPI), Universidad de Valladolid, Valladolid, Spain*

Learning points

- What is estimation?
- Sampling distributions
- Data-based estimation methods
- Bayesian estimation methods
- Monte Carlo estimation methods

4.1 Introduction: what is estimation?

The field of medical imaging is full of random sources of variation: radiation of X-rays in computed tomography, random spatial distribution of scatters in ultrasound imaging, and thermal noise in the k-space of magnetic resonance acquisitions are some examples of the intrinsic randomness of medical imaging.

Perhaps the most evident example is the presence of noise that corrupts the acquisition, reduces the contrast, or may even introduce biases in the measures obtained from the image. A proper characterization of the random phenomena leads to a better understanding of the information provided by imaging.

Probability theory is the mathematical approach commonly applied to deal with random events. It provides a solid framework to analyze the properties of random events. In this framework, we can model the stochastic events through random variables with known probabilistic distributions. In this way, if the random variable is completely characterized – its probability distribution is known – several valuable results can be derived, i.e., "the distribution of intensities in a magnetic resonance image follows a Rician distribution," or "the granular pattern observed in blood chambers of a cardiac ultrasound images follows a Rayleigh distribution."

It is clear, however, that the study of random events and the fitting of probabilistic models require the analysis of samples. In this case, the goal is to obtain information about the probability law of a real phenomenon from the analysis of samples. This is precisely the aim of statistical inference: the analysis of data to obtain the underlying probability distribution that governs the random event. Note that this goal is

Medical Image Analysis. https://doi.org/10.1016/B978-0-12-813657-7.00016-9

beyond the aims of the probability theory, which assumes the probability distribution to be known from the beginning. So, when we talk about *estimation* we will refer to the parameters of the underlying probabilistic distribution that model a physical phenomenon.

In the context of medical imaging, estimation theory provides a powerful tool to obtain clinically relevant information. To give some examples, the characterization of the distributions of samples may lead to the detection of pathological processes such as vulnerable plaque in intravascular ultrasound imaging, or the description of anatomical structures such as the geometry of heart chambers in cardiac ultrasound. In magnetic resonance imaging (MRI), the characterization of noise leads to a whole family of noise reduction filters that not only reduces the effect of noise but also removes biases in the measures.

Obviously, any estimation obtained from a sample will be a function of that random sample, which in turn is a new random variable. The idea of inferring the parameters from the estimator (a function of the samples) requires a link between the sample distribution and the distribution of the physical phenomenon studied.

In this chapter, we aim to provide the essential tools to establish the links between the samples we can acquire from a physical phenomenon and the characteristic parameters that might describe the nature of that phenomenon. For this purpose, we give an overview of the main techniques to analyze sampling distributions through the sampling cumulative distribution function (CDF) and the sampling density function. These tools will provide well-grounded evidence supporting the statistical characterization of the phenomena studied. Then, we give an overview of estimation methods to calculate the parameters of the underlying distributions of our observations. Along with the explanations, we will provide detailed examples to illustrate the concepts.

4.2 Sampling distributions

This section aims to show why we can obtain relevant statistical information from a random sample when the knowledge of the probabilistic nature of the sample is imprecise or unknown. To do so, we will study how the sampling probability distribution function behaves as the number of samples increases, and we will provide the fundamentals that, in fact, the sampling distribution converges to the true distribution describing the phenomenon under study.

We also discuss a metric to compare the empirical distribution to some candidate distribution when they are both continuous, the Kolmogorov–Smirnov statistic. The probabilistic distribution of this statistic is known when both distributions are equal and it has the advantage of being independent of the distribution. Therefore, it provides a suitable way to check how likely is the distance observed between the sampling and the candidate distributions. This will lead to the definition of the Kolmogorov–Smirnov test.

This section also studies the behavior of the probability density function (PDF). This function offers a more detailed description of the behavior of the random variable, since the mode is easily observed, the skewness becomes clearer, and the general shape is more distinctive. The common way to estimate the PDF is through histograms. We provide a detailed description of the PDF behavior as the number of samples increases and also discuss how to set a suitable step size in the histogram. We describe a simple rule to define an appropriate step size that minimizes the integral mean square error (IMSE) of the sampling PDF, the Scott's rule.

Finally, we introduce the Pearson goodness-of-fit test. This test compares the differences of observed frequencies to the expected frequencies of a candidate distribution. The probability distribution of the statistic employed in the comparison of frequencies when both distributions are the same is known and, as in the case of the Kolmogorov–Smirnov statistic, we can check how likely is the statistic observed and conclude if both distributions are comparable or not.

4.2.1 Cumulative distribution function

Let us consider a phenomenon described by a certain random variable X with probability distribution function F. In this case, if we consider a simple random sample where all the samples are independent, all the samples follow the same distribution F. If we approximate the distribution function with the sampling distribution of n samples $\mathbf{X} = (X_1, X_2, \ldots, X_n)$ as the events detected for a certain interval $(-\infty, x]$, we get

$$F_n(x) = \frac{\#\text{ samples within }(-\infty, x]}{n} = \frac{1}{n}\sum_{i=1}^{n} \mathbf{1}_{(-\infty,x]}(X_i), \tag{4.1}$$

where $\mathbf{1}_A(X)$ is the indicator function that indicates the membership of an element X in a set $A \in \mathbb{R}$. Note that for each x, $F_n(x)$ is a random variable that results from the sum of n independent Bernoulli random variables $\mathbf{1}_{(-\infty,x]}(X_i)$, i.e., the probability of X_i being within $(-\infty, x]$ is $P\{\mathbf{1}_{(-\infty,x]}(X_i) = 1\} = P\{X_i \leq x\} = F(x)$. Therefore, $nF_n(x) = \sum_{i=1}^{n} \mathbf{1}_{(-\infty,x]}(X_i)$ follows a binomial distribution $\mathcal{B}(n, F(x))$ with probability distribution [1]:

$$P\left\{F_n(x) = \frac{k}{n}\right\} = \binom{n}{k} F(x)^k (1 - F(x))^{n-k}, \tag{4.2}$$

with mean $E\{F_n(x)\} = F(x)$ and variance $\mathrm{Var}\{F_n(x)\} = F(x)(1 - F(x))/n$.

This result shows that the sampling distribution is centered in its theoretical value $F(x)$ for each x. Furthermore, since $nF_n(x)$ is a sum of independent and identically distributed (i.i.d.) random variables, the central limit theorem applies. So, the asymptotic distribution follows a normal distribution $\mathcal{N}\left(F(x), \sqrt{F(x)(1 - F(x))/n}\right)$ for each x as $n \to \infty$ (note that the variance decreases as n increases). Similarly, the strong law of large numbers [2] states that the sample average *converges almost surely* to its theoretical value: $F_n(x) \to F(x)$ (almost sure convergence means that the probability of events in which $F_n(x)$ do not converge to $F(x)$ is 0).

This result is of special interest since it shows that we can approximate the true distribution of the underlying phenomenon with arbitrary precision (considering an arbitrary number of samples). Note, however, that we stressed the fact that this convergence is always for *every fixed* x, meaning that there may be different convergence rates for each x (known as pointwise convergence). Glivenko and Cantelli proved that, in fact, $F_n \to F$ uniformly:

$$\sup_{x\in\mathbb{R}} |F_n(x) - F(x)| \to 0 \qquad \text{almost surely.} \tag{4.3}$$

The uniform convergence established with this result strongly supports the analysis of unknown distributions through the observation of the empirical CDF (ECDF). In the following section we also provide an appropriate metric to measure the discrepancies between the empirical distribution F_n and a candidate distribution F.

4.2.2 The Kolmogorov–Smirnov test

The Glivenko–Cantelli theorem proves that the ECDF is a powerful tool to describe the underlying distribution of a physical phenomenon just by the analysis of a sample. Let us suppose we are interested in evaluating a family of probabilistic distributions to model the stochastic behavior of a certain tissue in a clinical image. In that case, the Glivenko–Cantelli theorem states that the ECDF calculated in equation (4.1) converges to the true distribution function. Therefore, the supremum norm of the difference between a candidate distribution function,

$$D_n = \sup_x |F_n(x) - F(x)|, \tag{4.4}$$

offers a good metric to evaluate the suitability of the candidate distribution. Furthermore, a surprising result proved by Kolmogorov showed that if the cumulative function F is continuous, the distribution of D_n follows a probability distribution which is independent of F: the Kolmogorov distribution.

The independence of the Kolmogorov distribution and F enables the construction of a goodness-of-fit test by establishing critical values from which the value D_n can be considered as too far from what is expected if the data follows the distribution F.

In Fig. 4.1, we show an example of the ECDF calculated for an increasing number of samples following a Rayleigh distribution with scale parameter $s = 1$ jointly with the theoretical distribution. We also indicate the locations at which the Kolmogorov metric D_n is reached for both examples. Note that as the number of samples increases, the ECDF fits better the theoretical CDF, resulting in a dramatic decrease of D_n (from 0.34 to 0.02 in this example).

This example evidences the suitability of the supremum norm as a metric and offers a statistical test to evaluate the fitting of the candidate distribution. The Kolmogorov–Smirnov test has, however, a subtle yet essential detail. The independence between F and D_n is lost when the parameters of the theoretical distribution

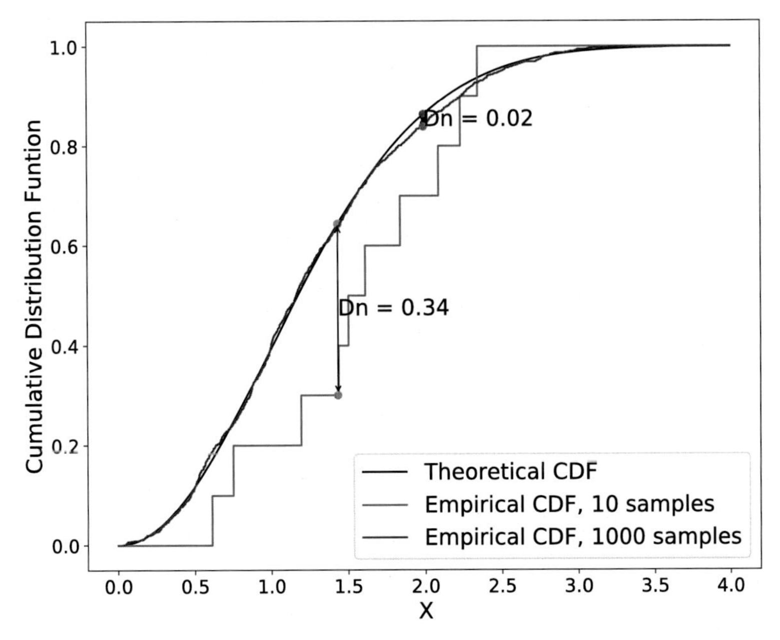

FIGURE 4.1 Empirical cumulative distribution function of Rayleigh-distributed random samples.

Empirical cumulative distribution function of Rayleigh-distributed random samples with scale parameter $s = 1$.

are estimated from the samples. So, although the metric D_n can still be used to check the fitting with an estimated distribution, the statistical test cannot be performed in its original form due to the dependence between F and D_n.

4.2.3 Histogram as probability density function estimate

Sometimes we are interested in analyzing and estimating the density function of a random phenomenon described by a continuous random variable. This is, however, not a trivial task since the PDF is the derivative of the CDF, $f(x) = F'(x)$, and the calculation of a derivative of $F_n(x)$ is not continuous and will strongly depend on the samples.

A reasonable way to analyze the behavior of the PDF is to study the normalized histogram of the sample (i.e., the area under the histogram is equal to 1). However, is this a good way of estimating the PDF of a random variable?

To answer this question we can perform a similar analysis as the one presented in Section 4.2.1 to evaluate the suitability of the histogram approach to the PDF estimation. We can represent the histogram as a sum of indicator functions as we did for the CDF as follows:

$$f_n(x) = \frac{F(x_{k+1}) - F(x_k)}{x_{k+1} - x_k} = \frac{1}{n(x_{k+1} - x_k)} \sum_{k=1}^{n} \mathbf{1}_{(x_k, x_{k+1}]}(X_i) = \frac{n_k}{n(x_{k+1} - x_k)}, \tag{4.5}$$

where $x \in (x_k, x_{k+1}]$ and $n_k = \sum_{k=1}^{n} \mathbf{1}_{(x_k, x_{k+1}]}(X_i)$ is the number of samples in $(x_k, x_{k+1}]$.

As previously mentioned, the random variable n_k is the sum of n Bernoulli random variables and, therefore, it follows a binomial distribution $\mathcal{B}(n, p_k)$, where $p_k = F(x_{k+1}) - F(x_k)$. So, similarly to what we obtained for the CDF, the mean and variance of $f_n(x)$ for $x \in (x_k, x_{k+1}]$ are

$$E\{f_n(x)\} = \frac{p_k}{x_{k+1} - x_k}, \tag{4.6}$$

$$\text{Var}\{f_n(x)\} = \frac{p_k(1 - p_k)}{n(x_{k+1} - x_k)^2}. \tag{4.7}$$

To make the analysis simpler, let us consider a homogeneous spacing in the histogram $h = x_{k+1} - x_k$ for all k. So, when $h \to 0$,

$$\lim_{h \to 0} E\{f_n(x)\} = \lim_{h \to 0} \frac{F(x_{k+1}) - F(x_{k+1} - h)}{h} = F'(x), \tag{4.8}$$

since $x \in (x_{k+1} - h, x_{k+1}]$.

Therefore, as the step size of the histogram becomes smaller, the estimate of the PDF becomes the true PDF on average. However, it is worth to note that the variance also depends on the step size, and as we reduce the step size, fewer values fall within each interval, causing an increase in the variance per interval. Consequently, there is a trade-off in the selection of the number of bins and the step size. Unfortunately, there is not a straightforward solution to select the best step size since it depends on the shape of the PDF to be estimated.

A good choice is the rule proposed in 1979 by David Scott [3], who suggests to use the step size h as the one that minimizes the integral mean square error (IMSE) of $f_n(x)$, when $f(x)$ is a Gaussian distribution, as follows:

$$\text{MSE}(f_n(x)) = E\left\{(f(x) - f_n(x))^2\right\} \tag{4.9}$$

$$= E\left\{(f(x) - E\{f_n(x)\} + E\{f_n(x)\} - f_n(x))^2\right\} \tag{4.10}$$

$$= E\left\{(f(x) - E\{f_n(x)\})^2\right\} + E\left\{(f_n(x) - E\{f_n(x)\})^2\right\} \tag{4.11}$$

$$+ 2(f(x) - E\{f_n(x)\}) \underbrace{E\{f_n(x) - E\{f_n(x)\}\}}_{=0} \tag{4.12}$$

$$= (f(x) - E\{f_n(x)\})^2 + E\left\{(f_n(x) - E\{f_n(x)\})^2\right\}. \tag{4.13}$$

So, considering the results of equation (4.7), we get

$$\mathrm{MSE}(f_n(x)) = \left(f(x) - \frac{p_k}{h}\right)^2 + \frac{p_k(1-p_k)}{nh^2}. \tag{4.14}$$

Now, since $f(x)$ has two continuous and bounded derivatives, the Taylor expansion for $p_k(x)$ at $x \in (x_k, x_{k+1}]$ is

$$\begin{aligned} p_k(x) &= \int_{x_k}^{x_k+h} f(t)dt = \int_{x_k}^{x_k+h} \left(f(x) + f'(x)(t-x) + O\left(h^2\right)\right) dt \\ &= f(x)h - f'(x)(x-x_k)h + \frac{1}{2}f'(x)h^2 + O(h^3). \end{aligned} \tag{4.15}$$

Then, plugging equation (4.15) in equation (4.14) and integrating in the real line, we get the IMSE:

$$\mathrm{IMSE}(f_n(x)) = \frac{1}{nh} + \frac{h^2}{4}\int_{-\infty}^{\infty} f'(x)^2 dx + \int_{-\infty}^{\infty} f'(x)^2(x-x_k)^2 \tag{4.16}$$

$$-h\int_{-\infty}^{\infty} f'(x)^2(x-x_k) + O\left(\frac{1}{n} + h^3\right). \tag{4.17}$$

Now, noting that

$$\int_{-\infty}^{\infty} f'(x)^2(x-x_k)^2 = \sum_{k=-\infty}^{\infty}\int_{x_k}^{x_k+h} f'(x)^2(x-x_k)^2 \tag{4.18}$$

$$= \sum_{k=-\infty}^{\infty}\int_0^h f'(y+x_k)^2 y^2 dy = \sum_{k=-\infty}^{\infty}\int_0^h (f'(x_k)^2 + O(h))y^2 dy \tag{4.19}$$

$$= \frac{h^3}{3}\sum_{k=-\infty}^{\infty} f'(x_k)^2 + O(h^4) = \frac{h^2}{3}\int_{-\infty}^{\infty} f'(x)^2 + O(h^3), \tag{4.20}$$

where the last term was obtained from the left Riemann sum approach,

$$\int_{-\infty}^{\infty} f(x)dx = h\sum_{k=-\infty}^{\infty} f(x_k) + O(h^2), \tag{4.21}$$

we get

$$\int_{-\infty}^{\infty} f'(x)^2(x-x_k)^2 dx = \frac{h^2}{3}\int_{-\infty}^{\infty} f'(x)^2 dx + O(h^3), \tag{4.22}$$

$$h\int_{-\infty}^{\infty} f'(x)^2(x-x_k) dx = -\frac{h^2}{2}\int_{-\infty}^{\infty} f'(x)^2 dx + O(h^3). \tag{4.23}$$

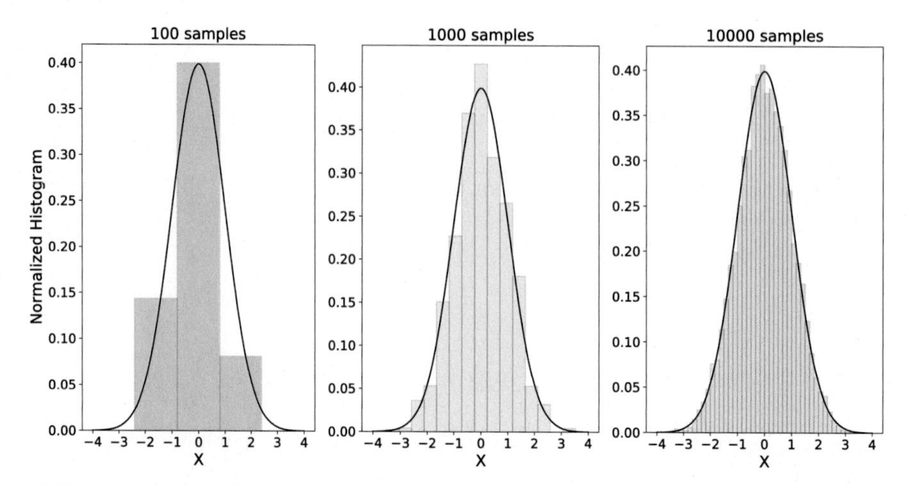

FIGURE 4.2 Normalized histograms of a standard Gaussian random variable with an increasing number of samples.

Normalized histograms of a standard Gaussian random variable with an increasing number of samples.

Therefore, plugging equations (4.22) and (4.23) in equation (4.17) we get

$$\mathrm{IMSE}(f_n(x)) = \frac{1}{nh} + \frac{h^2}{12}\int_{-\infty}^{\infty} f'(x)^2 dx + O\left(\frac{1}{n} + h^3\right). \tag{4.24}$$

So, minimization of the first two terms, $\hat{h} = \arg\min_h IMSE(f_n(x))$, results in

$$\hat{h} = \left(\frac{6}{n\int_{-\infty}^{\infty} f'(x)^2 dx}\right)^{\frac{1}{3}}. \tag{4.25}$$

Finally, considering that, for a Gaussian distribution, $\int_{-\infty}^{\infty} f'(x)^2 dx = 1/\left(4\sqrt{\pi}\sigma^3\right)$, Scott's rule becomes

$$\hat{h} = 24^{1/3}\pi^{1/6}\sigma n^{-1/3} \approx 0.5\sigma n^{-1/3}. \tag{4.26}$$

The effect of Scott's rule can be observed in Fig. 4.2, where the fitting of the normalized histograms to a standard Gaussian random variable is depicted. Note that the number of bins and their width adapt to the samples.

4.2.4 The chi-squared test

Similarly to the Kolmogorov–Smirnov test, the analysis of a distribution through its sample density function (histogram) offers a way to evaluate the goodness-of-fit to a certain distribution. Pearson in 1900 proposed a statistical test to evaluate whether the difference between observed frequencies and expected frequencies comes from chance. What Pearson proposed is to consider the weighted squared differences between observed and expected observations within each bin of the histogram:

$$D = \sum_{i=1}^{k} \frac{1}{np_i}(N_i - np_i)^2, \tag{4.27}$$

where n is the number of samples, k is the number of bins considered, and p_i is the theoretical probability of events within the ith interval of the histogram. This metric has a remarkable property: it asymptotically converges to a chi-squared distribution with $k-1$ degrees of freedom, χ^2_{k-1}. Since the asymptotic convergence depends on the number of samples n, a lower limit in the number of samples is usually considered to assume a good approximation. This limit depends on the theoretical probability p_i for each bin. The number of observations commonly accepted is $np_i \geq 5$ for each bin. Obviously, this constraint may affect the way the bins are selected.

In contrast to the Kolmogorov–Smirnov test, the chi-squared test allows us to estimation of parameters of the theoretical distribution. The only consequence of the estimation of parameters is a reduction in the number of degrees of freedom of the χ^2 distribution. So, if we assume the data come from an m-parameter distribution and we estimate the m parameters from the samples we have, the chi-squared goodness-of-fit test can be realized by considering a χ^2_{k-m-1} distribution.

4.3 Estimation. Data-based methods

4.3.1 Definition of estimator and criteria for design and performance measurement

In this section we will provide an overview on how to determine the value of the parameters of the underlying distribution that our observations have been drawn from. To this end, the symbol $\mathbf{x}$ will be used as an N-component data vector and the symbol $\boldsymbol{\theta}$ will denote the p-component parameter vector to be identified. In the case that this vector turns out to be a scalar, we will use the symbol θ. The vector $\mathbf{x}$ will also be referred to as *sample*; we assume that this vector is a sample of the distribution $p(\mathbf{x}; \boldsymbol{\theta})$, and the dependence of the sample on the parameter $\boldsymbol{\theta}$ through the function $p()$ is precisely the source of information that makes our estimation plausible. The presence of a semicolon within the braces of function $p()$ is noteworthy. This symbol means that the function $p()$ is parameterized by $\boldsymbol{\theta}$, so the only random variable involved in that function is $\mathbf{x}$; consequently, the parameter $\boldsymbol{\theta}$ is assumed to be an *unknown deterministic constant*. In Section 4.5 we will depart from a different assumption.

The estimator is a function of the data vector, say $\hat{\boldsymbol{\theta}}(\mathbf{x})$, about which we have the expectation that it will take on a value close to the actual value $\boldsymbol{\theta}$. However, the definition of *closeness* is a key ingredient in the way the estimator will be derived; we should bear in mind that the estimator $\hat{\boldsymbol{\theta}}(\mathbf{x})$, since it is a function of the random variable $\mathbf{x}$, is a random variable itself, so closeness should be appropriately defined for this particular situation. The dependence of $\mathbf{x}$ in the estimator $\hat{\boldsymbol{\theta}}$ will occasionally be made explicit, but most often will only be implied.

Specifically, a sensible criterion for the case of a scalar parameter θ is to calculate an estimator such that its associated mean square error (MSE) is as small as possible, i.e., defining the estimation error as $\epsilon = \hat{\theta} - \theta$, it would make sense to calculate

$$\hat{\theta} = \arg\min_{\hat{\theta}} E\{\epsilon^2\} = \arg\min_{\hat{\theta}} E\{(\hat{\theta} - \theta)^2\}. \tag{4.28}$$

However, things are slightly more complicated than that, because of the behavior of the two components involved in the equation above. Let us add and subtract the quantity $E\{\hat{\theta}\} = \eta_{\hat{\theta}}$ in the expression above; doing some algebra we obtain

$$E\{(\hat{\theta} - \theta)^2\} = E\{(\hat{\theta} - \eta_{\hat{\theta}} + \eta_{\hat{\theta}} - \theta)^2\} \tag{4.29}$$

$$= E\{(\hat{\theta} - \eta_{\hat{\theta}})^2\} + E\{(\eta_{\hat{\theta}} - \theta)^2\} \tag{4.30}$$

$$= \sigma_{\hat{\theta}}^2 + b^2(\theta), \tag{4.31}$$

i.e., the MSE turns out to consist of two components, namely, the variance of the estimator and the square of the estimator bias. The latter is the square value of the difference between the mean value of the estimator and the actual parameter value. What is the point in this? The point is twofold:

1. The bias is usually a function of θ. This being the case, estimators built on the basis of equation (4.28) usually turn out to be a function of the unknown parameter, which makes the estimators non-writable as $\hat{\theta}(\mathbf{x})$ but as $\hat{\theta}(\mathbf{x};\theta)$, i.e., as a function of the quantity that we mean to estimate. That makes estimators non-realizable and hence this criterion is typically not practical.
2. This being the case, a customary criterion is to select unbiased estimators (i.e., those for which $b(\theta) = 0, \forall\theta$) and within that group search for the estimator with minimum variance. At first glance this may seem equivalent to minimizing equation (4.28), but note there is a difference: we select a subset of estimators, so we do not permit an interchange of bias and variance to minimize that equation.

So, to summarize, we will accept as our initial criterion to design the estimators so that they satisfy

$$\hat{\theta} = \arg\min_{\hat{\theta}} \sigma_{\hat{\theta}}^2 \quad \text{s.t.} \quad E\{\hat{\theta}\} = 0, \tag{4.32}$$

where "s.t." stands for *subject to*.

In the case that our parameter turns out to be a vector, i.e., $\boldsymbol{\theta}$, this criterion is applied to each of its components.

4.3.2 A benchmark for unbiased estimators: the Cramer–Rao lower bound

Ideally, the estimator that satisfied equation (4.32) for all values of θ would be referred to as the minimum variance unbiased estimator (MVUE), for obvious reasons. However, there is no guarantee that such an estimator exists. What we do have at our disposal is a benchmark for unbiased estimators that can tell us how close (or far) we are to optimality. This benchmark is known as the Cramer–Rao lower bound (CRLB).

Before we get into the details, recall that we indicated in the previous section that our source of information (that makes our estimation plausible) is $p(\mathbf{x}; \boldsymbol{\theta})$, i.e., the dependence of our sample on $\boldsymbol{\theta}$ through $p()$. So, apparently, the more sensitive this function is to $\boldsymbol{\theta}$, the better the estimation should be. This turns out to be the case.

The CRLB theory departs from requiring that the following regularity condition holds:

$$E\left\{\frac{\partial \log p(\mathbf{x}; \boldsymbol{\theta})}{\partial \boldsymbol{\theta}}\right\} = \mathbf{0}, \tag{4.33}$$

where the expectation is taken with respect to the distribution $p(\mathbf{x}; \boldsymbol{\theta})$ using the actual value of the parameter.

This condition is not very stringent and it typically holds for most of the distributions used in practice as long as the parameters involved are not those defining intervals in which the variable $\mathbf{x}$ can take on values.

We now define some quantities that are involved in CRLB theory. Specifically,

$$[\mathbf{I}(\boldsymbol{\theta})]_{(i,j)} = E\left\{-\frac{\partial^2 \log p(\mathbf{x}; \boldsymbol{\theta})}{\partial \theta_i \partial \theta_j}\right\}, \tag{4.34}$$

where $\mathbf{I}(\boldsymbol{\theta})$ is the $p \times p$ *Fisher information* matrix, the (i, j) element of which is defined as just indicated. In this situation, it turns out that the variance of any unbiased estimator of the ith component of $\boldsymbol{\theta}$ satisfies

$$\sigma^2_{\hat{\theta}_i} \geq \left[\mathbf{I}^{-1}(\boldsymbol{\theta})\right]_{(i,j)}. \tag{4.35}$$

This is, consequently, the CRLB. What does this expression show? We can point out two ideas:

1. First, consider the case that $\boldsymbol{\theta}$ is a scalar, so the Fisher information matrix becomes a scalar as well. In this case, the inverse of the matrix is just the inverse of the expectation of the second derivative. Clearly, the more sensitive function $p(\mathbf{x}; \theta)$ (actually, the expectation of its logarithm) is with respect to θ, the lower is its inverse, so the lower is the achievable variance of an unbiased estimator. This seems pretty much intuitive.
2. Generally speaking, the more parameters need to be estimated (i.e., the larger p), the worse the estimation variance; this is due to the fact that we have the same data (i.e., the Nth component vector $\mathbf{x}$) to estimate not just one, but p parameters,

so our information must be shared, and generally speaking, this will lead to a performance decrease. Mathematically, this is due to the following inequality:

$$\left[\mathbf{I}^{-1}(\boldsymbol{\theta})\right]_{(i,i)} \geq \frac{1}{[\mathbf{I}(\boldsymbol{\theta})]_{(i,i)}}, \tag{4.36}$$

where the term on the right-hand side would be the variance of the ith component of vector $\hat{\boldsymbol{\theta}}$ provided the other components were known (and hence, do not need to be estimated).
Note that this will not be the case when the Fisher information matrix is diagonal. In Section 4.4 we show a situation where this occurs.

One additional consequence should be borne in mind with respect to this result. Let us refer to an estimator that achieves the CRLB as the efficient estimator; if this estimator exists, it will obviously be the MVUE. Let us accept that an efficient estimator exists; then, the following result holds:

$$\frac{\partial \log p(\mathbf{x};\boldsymbol{\theta})}{\partial \boldsymbol{\theta}} = \mathbf{I}(\boldsymbol{\theta})(\mathbf{g}(\mathbf{x}) - \boldsymbol{\theta}). \tag{4.37}$$

This factorization exists if and only if the efficient estimator exists; note that if the factorization holds, then $\hat{\boldsymbol{\theta}} = \mathbf{g}(\mathbf{x})$ would be the efficient estimator. Note that the efficient estimator is unbiased (as a direct consequence of equation (4.37) and equation (4.33)) and its covariance matrix is

$$\mathbf{C}_{\hat{\boldsymbol{\theta}}} = \mathbf{I}^{-1}(\boldsymbol{\theta}). \tag{4.38}$$

4.3.3 Maximum likelihood estimator

The principle of the maximum likelihood estimator (MLE) is based on the concept of *likelihood function*. What is that function? This function, nominally, is the same as $p(\mathbf{x};\boldsymbol{\theta})$ but with a subtle difference. When one speaks about a PDF, the variable in the function is assumed to be $\mathbf{x}$, while the parameter vector $\boldsymbol{\theta}$ remains fixed. For the likelihood function, roles are interchanged, i.e., the variable is $\boldsymbol{\theta}$ and the sample, i.e., the observed quantity $\mathbf{x}$, remains fixed. In the latter case, we measure a quantity proportional to the probability that the random variable takes on value $\mathbf{x}$ as a function of $\boldsymbol{\theta}$.

This being the case, it makes sense to build an estimator on the basis of choosing the value of $\boldsymbol{\theta}$ that makes the observation $\mathbf{x}$ the most probable. This is the MLE.

So, formally, the MLE of parameter $\boldsymbol{\theta}$ is defined as

$$\hat{\boldsymbol{\theta}} = \arg\max_{\boldsymbol{\theta}} p(\mathbf{x};\boldsymbol{\theta}) = \arg\max_{\boldsymbol{\theta}} \log p(\mathbf{x};\boldsymbol{\theta}), \tag{4.39}$$

where the second equality is due to the monotonicity of the log function.

Two interesting properties make the MLE of special interest. These properties are:

1. If the efficient estimator exists, the MLE will give rise to it. This is a direct consequence of equation (4.37) and the definition of the MLE, as expressed in equation (4.39). This second equation is maximized by nulling the derivative of any of the two functions involved (either the likelihood function or its logarithm). Nulling the second function means nulling equation (4.37), and the estimator that makes the latter null is precisely $\hat{\boldsymbol{\theta}} = \mathbf{g}(\mathbf{x})$, i.e., the efficient estimator.
2. The MLE has nice asymptotic properties, i.e., properties that are verified for a large data record ($N \to \infty$). Specifically, irrespective of the original distribution of the data sample $\mathbf{x}$ (as long as a few mild conditions are satisfied, such as the existence of derivatives and the regularity condition defined in equation (4.33)), it turns out that

$$\hat{\boldsymbol{\theta}} \overset{a}{\sim} N\left(\boldsymbol{\theta}, \mathbf{I}^{-1}(\boldsymbol{\theta})\right), \tag{4.40}$$

where the symbol $\overset{a}{\sim}$ means "is distributed asymptotically as." So, as we can see, the MLE is asymptotically unbiased and asymptotically efficient. This property is often used by software packages to simplify the calculation of confidence intervals (CIs) for point estimators, by providing the approximate value

$$\hat{\boldsymbol{\theta}}_i \pm \lambda\sqrt{\mathbf{I}^{-1}(\hat{\boldsymbol{\theta}}_i)}, \tag{4.41}$$

where λ is the $1 - 0.05/2$ quantile of the standard normal distribution (i.e., $\lambda \approx 1.96$) for an approximate CI at level $\alpha = 0.05$.

These two properties provide undoubtable evidence that the MLE is a preferred choice whenever it is tractable.

4.3.4 The expectation-maximization method

In some circumstances we might be interested in estimating the parameters of a probability distribution that depends on unobserved latent variables. This is the case, for example, in the statistical characterization of the echomorphology of vulnerable plaque in intravascular ultrasound imaging, where the patterns observed result from the contribution of different echogenic components of the plaque that follow different probabilistic distributions. In this situation, we know the family of distributions to be considered – we will consider Gamma distributions [4] – but we do not know which one corresponds to each pixel. This is precisely the unobserved latent variable and the pixel intensity is the observed variable.

The expectation-maximization method [5] allows us to find the maximum likelihood estimates of the parameters of the distributions iteratively. We will explain the details of the expectation-maximization method within the context of the characterization of vulnerable plaque in ultrasound imaging with a Gamma mixture model.

Let us assume we have $\mathbf{X} = \{x_i\}$, $1 \leq i \leq N$, a set of observations (pixel intensities) of a given region of the ultrasound image. To keep things simpler, we will consider these samples as i.i.d. random variables. This assumption can be taken if a downsampling is performed to reduce the potential correlation between neighboring pixels. The mixture model considers that these variables result from the contributions of J distributions:

$$p(x|\Theta) = \sum_{j=1}^{J} \pi_j f_\Gamma(x|\Theta_j), \tag{4.42}$$

where Θ is a vector of the parameters of the Gamma Mixture Model $\{\pi_1, \cdots, \pi_J, \Theta_1, \cdots, \Theta_J\}$ and Θ_j are the parameters of the PDF (in our case the shape and scale parameters of a Gamma distribution are represented as α_j and β_j, respectively). The Gamma PDF is defined as

$$f_\Gamma(x|\alpha, \beta) = \frac{x^{\alpha-1}}{\beta^\alpha \Gamma(\alpha)} e^{-\frac{x}{\beta}}, \quad x \geq 0 \text{ and } \alpha, \beta > 0, \tag{4.43}$$

where $\Gamma(x)$ is the Euler Gamma function. Note that the condition $\sum_{j=1}^{J} \pi_j = 1$ must hold to guarantee that $p(x_i|\Theta)$ is a well-defined probability distribution.

The joint distribution of i.i.d. samples is given by

$$p(\mathbf{X}|\Theta) = \prod_{i=1}^{N} p(x_i|\Theta). \tag{4.44}$$

The expectation-maximization method maximizes the log-likelihood function considering the unobserved variables as discrete random variables, $\mathbf{Z} = \{Z_i\}$. These random variables take values in $\{1, \ldots, J\}$, so that $Z_i = j$ means that the sample x_i belongs to the distribution class j.

We will denote the estimate of the mixture parameters in the nth iteration as $\Theta^{(n)}$. The expectation step is performed by calculating the expected value of the log-likelihood $\mathcal{L}(\Theta|\mathbf{X}, \mathbf{Z})$ considering the previous estimation of parameters $\Theta^{(n)}$:

$$\mathcal{Q}(\Theta|\Theta^{(n)}, \mathbf{X}) = E_{\mathbf{Z}|\Theta^{(n)}, \mathbf{X}}\{\mathcal{L}(\Theta|\mathbf{X}, \mathbf{Z})\}. \tag{4.45}$$

In the maximization step, the new estimate $\Theta^{(n)}$ is obtained by maximizing the expectation of the likelihood function $\mathcal{Q}(\Theta|\Theta^{(n)}, \mathbf{X})$. These steps are iterated until a stop criterion is reached. A common criterion can be $\|\Theta^{(n+1)} - \Theta^{(n)}\| < \text{Tol}$ for some pre-established tolerance (Tol).

The expectation of the likelihood function with respect to the latent variables when samples $\mathbf{X}$ and the previous estimate $\Theta^{(n)}$ are known is

$$
\begin{aligned}
&\mathcal{Q}(\Theta|\Theta^{(n)}, \mathbf{X}) = E_{Z|\Theta^{(n)},\mathbf{X}}\{\mathcal{L}(\Theta|\mathbf{X}, \mathbf{Z})\} = \\
&\sum_{i=1}^{N} E_{Z_i|\Theta^{(n)},x_i}\{\log p(x_i|\Theta_{z_i}, Z_i = z_i) + \log p(Z_i = z_i|\Theta)\} = \\
&\sum_{i=1}^{N}\sum_{j=1}^{J} p(Z_i = j|x_i, \Theta^{(n)}) \left(\log p(x_i|\Theta_j) + \log \pi_j\right),
\end{aligned}
\tag{4.46}
$$

where $p(Z_i = j|\Theta)$ is the probability of x_i to belong to class j, denoted as π_j. On the other hand, $p(Z_i = j|x_i, \Theta^{(n)})$ can be derived by the Bayes theorem as follows:

$$
p(Z_i = j|x_i, \Theta^{(n)}) = \frac{p(x_i|\Theta_j^{(n)})p(Z_i = j|\Theta^{(n)})}{f_\Gamma(x_i|\Theta^{(n)})}, \tag{4.47}
$$

where, as in equation (4.42),

$$
\sum_{j=1}^{J} p(x_i|\Theta_j^{(n)})p(Z_i = j|\Theta^{(n)}) = \sum_{j=1}^{J} \pi_j f_\Gamma(x_i|\Theta_j^{(n)}). \tag{4.48}
$$

Since equation (4.46) is composed of two independent terms, the maximization step can be done to each term separately. For the term depending on π_j, optimization via *Lagrange multipliers* can be performed in a straightforward way, where the constraint is the well-defined probability condition $\sum_{j=1}^{J} \pi_j = 1$. The Lagrange method of multipliers guarantees a necessary condition for optimality in this problem. The *Lagrange function* with the Lagrange multiplier, λ, is the following:

$$
\Lambda(\pi, \lambda) = \sum_{i=1}^{N}\sum_{j=1}^{J} \gamma_{i,j} \log \pi_j + \lambda \left(\sum_{j=1}^{J} \pi_j - 1\right), \tag{4.49}
$$

where $\gamma_{i,j} = p(Z_i = j|x_i, \Theta^{(n)})$ to make notation simpler.

Now, we can calculate the derivative with respect to each π_j and equaling to 0, the following expression is derived:

$$
\sum_{i=1}^{N} \gamma_{i,j} = -\lambda \pi_j. \tag{4.50}
$$

By summing both terms of the equation over j, we obtain $\lambda = -N$, since $\sum_{j=1}^{J} \gamma_{i,j} = 1$ for each $i = 1, \ldots, N$. Finally, the values of $\hat{\pi}_j$ that maximize the Lagrange function (and the likelihood term) are

$$
\hat{\pi}_j = \frac{1}{N}\sum_{i=1}^{N} \gamma_{i,j} = \frac{1}{N}\sum_{i=1}^{N} p(Z_i = j|\Theta). \tag{4.51}
$$

Now, we can maximize the term of equation (4.46) which depends on $\Theta_j = (\alpha_j, \beta_j)$:

$$\frac{\partial}{\partial \beta_j}\left\{\sum_{i=1}^{N}\sum_{j=1}^{J}\gamma_{i,j}\log p(x_i|\Theta_j)\right\} = 0, \tag{4.52}$$

where the log-likelihood of $p(x_i|\Theta_j)$ is

$$\log p(x_i|\Theta_j) = (\alpha_j - 1)\log x_i - \frac{x_i}{\beta_j} - \alpha_j \log(\beta_j) - \log(\Gamma(\alpha_j)). \tag{4.53}$$

This results in the following expression:

$$\sum_{i=1}^{N}\gamma_{i,j}\left(\frac{x_i}{\beta_j^2} + \frac{\gamma_{i,j}}{\alpha_j}\right) = 0, \tag{4.54}$$

which gives the following result:

$$\beta_j = \frac{1}{\alpha_j}\frac{\sum_{i=1}^{N}\gamma_{i,j}x_i}{\sum_{i=1}^{N}\gamma_{i,j}}. \tag{4.55}$$

Now, plugging equation (4.55) into equation (4.46) and deriving with respect to α_j,

$$\frac{\partial}{\partial \alpha_j}\left\{\sum_{i=1}^{N}\sum_{j=1}^{J}\gamma_{i,j}\log p\left(x_i|\alpha_j, \frac{1}{\alpha_j}\frac{\sum_{i=1}^{N}\gamma_{i,j}x_i}{\sum_{i=1}^{N}\gamma_{i,j}}\right)\right\} = 0. \tag{4.56}$$

The result is

$$\sum_{i=1}^{N}\gamma_{i,j}\log(x_i) - \sum_{i=1}^{N}\gamma_{i,j}\log\left(\frac{\sum_{k=1}^{N}\gamma_{k,j}x_k}{\sum_{k=1}^{N}\gamma_{k,j}}\right) + \sum_{i=1}^{N}\gamma_{i,j}\log(\alpha_j) - \sum_{i=1}^{N}\gamma_{i,j}\psi(\alpha_j) = 0, \tag{4.57}$$

where $\psi(x)$ is the *Digamma* function defined as $\psi(x) = \frac{\Gamma'(x)}{\Gamma(x)}$.

Finally, reordering terms, the following equality is derived:

$$\log(\alpha_j) - \psi(\alpha_j) = \log\left(\frac{\sum_i^N \gamma_{i,j}x_i}{\sum_i^N \gamma_{i,j}}\right) - \frac{\sum_i^N \gamma_{i,j}\log x_i}{\sum_i^N \gamma_{i,j}}. \tag{4.58}$$

This expression has no closed solution. However, we can ensure a unique solution since the function $f(x) = \log(x) - \psi(x)$ is well behaved (continuous, positive, and monotonically decreasing in $(0, \infty)$) and, by virtue of Jensen's inequality, the right part of the equation (4.58) is always positive. This solution can be easily obtained by any root finding algorithm.

From the estimated value $\hat{\alpha}_j$ that maximizes the log-likelihood, the estimate of $\hat{\beta}$ is directly obtained from equation (4.55).

4.4 A working example

Assume that we have an image with a suspicion of a malignant tumor in it. We are given a number of samples of that image within the area that is suspected to be affected by the tumor; assume that samples have been obtained from pixels that are sufficiently far from each other, so that they can be considered as approximately i.i.d. normal variables. Due to previous measures with this imaging modality, the tumor is considered malignant if the mean density is η_0. We are asked to provide a justified answer whether, in our case, the evidence tells us that the tumor is malignant.

We will solve this simple exercise as follows. First, we will estimate the two unknown parameters of the original distribution $\boldsymbol{\theta} = [\ \eta \quad \sigma^2\ ]^T$; to this end, we will calculate the MLE of these two parameters. Then, we will find a CI of the mean; if this interval contains the value η_0 we will conclude that the tumor is malignant, according to the evidence. As for the MLE calculation, since we now know that the MLE will give rise to the efficient estimator, provided it exists, we will first check whether the factorization in equation (4.37) is possible in our case. So we first write

$$p(\mathbf{x}, \boldsymbol{\theta}) = \prod_{n=0}^{N-1} p(x_n, \boldsymbol{\theta}) = \prod_{n=0}^{N-1} \left(\frac{1}{\sigma^2 2\pi}\right)^{\frac{N}{2}} \exp\left(-\sum_{n=0}^{N-1} \frac{(x_n - \eta)^2}{2\sigma^2}\right), \tag{4.59}$$

$$\log p(\mathbf{x}, \boldsymbol{\theta}) = -\frac{N}{2} \log \sigma^2 - \frac{N}{2} \log 2\pi - \frac{1}{2\sigma^2} \sum_{n=0}^{N-1} (x_n - \eta)^2. \tag{4.60}$$

In order to avoid confusion with the exponent that accompanies the symbol σ we will write $\tau = \sigma^2$, i.e.,

$$\log p(\mathbf{x}, \boldsymbol{\theta}) = -\frac{N}{2} \log \tau - \frac{N}{2} \log 2\pi - \frac{1}{2\tau} \sum_{n=0}^{N-1} (x_n - \eta)^2. \tag{4.61}$$

Now we can calculate the first-order derivatives, i.e.,

$$\frac{\partial \log p(\mathbf{x}, \boldsymbol{\theta})}{\partial \eta} = \frac{1}{\tau} \sum_{n=0}^{N-1} (x_n - \eta), \tag{4.62}$$

$$\frac{\partial \log p(\mathbf{x}, \boldsymbol{\theta})}{\partial \tau} = -\frac{N}{2\tau} + \frac{1}{2\tau^2} \sum_{n=0}^{N-1} (x_n - \eta)^2, \tag{4.63}$$

and with these expressions, the regularity conditions (equation (4.33)) are clearly satisfied since $E\left\{\sum_{n=0}^{N-1}(x_n-\eta)\right\}=0$ and $E\left\{\sum_{n=0}^{N-1}(x_n-\eta)\right\}=N\tau=N\sigma^2$. The second-order derivatives can now be written:

$$\frac{\partial^2 \log p(\mathbf{x},\boldsymbol{\theta})}{\partial \eta^2}=\frac{N}{\tau}, \tag{4.64}$$

$$\frac{\partial^2 \log p(\mathbf{x},\boldsymbol{\theta})}{\partial \eta \partial \tau}=-\frac{1}{\tau^2}\sum_{n=0}^{N-1}(x_n-\eta)^2, \tag{4.65}$$

$$\frac{\partial^2 \log p(\mathbf{x},\boldsymbol{\theta})}{\partial \eta \partial \tau}=-\frac{1}{\tau^2}\sum_{n=0}^{N-1}(x_n-\eta)^2, \tag{4.66}$$

$$\frac{\partial^2 \log p(\mathbf{x},\boldsymbol{\theta})}{\partial \tau^2}=\frac{N}{2\tau^2}-\frac{1}{\tau^3}\sum_{n=0}^{N-1}(x_n-\eta)^2. \tag{4.67}$$

Calculating expectations we now write the Fisher information matrix (keeping in mind that $\tau=\sigma^2$ and using equation (4.34))

$$\mathbf{I}(\boldsymbol{\theta})=\begin{bmatrix} \frac{N}{\sigma^2} & 0 \\ 0 & \frac{2\sigma^4}{N} \end{bmatrix}. \tag{4.68}$$

However, in this case there is no way to apply equation (4.37) since in equation (4.63) we have a polynomial in τ. We can then conclude that the efficient estimator does not exist so we have to explicitly calculate the MLE. To this end, nulling equation (4.62), solving for η, and inserting this result in equation (4.63), we obtain

$$\hat{\eta}=\frac{1}{N}\sum_{n=0}^{N}x_n, \tag{4.69}$$

$$\hat{\tau}=\hat{\sigma^2}=\frac{1}{N}\sum_{n=0}^{N}\left(x_n-\hat{\eta}\right)^2. \tag{4.70}$$

In order to give the CI, we may directly apply the asymptotic result in equation (4.41). However, for this specific case we can calculate an exact CI by using well-established theory. Specifically, recall that the MLE is asymptotically unbiased, but this does not necessarily mean that it is actually unbiased for a particular sample size. This being the case, instead of using equation (4.70) we are going to use the customary unbiased version, namely,

$$\hat{s}^2=\frac{1}{N-1}\sum_{n=0}^{N}\left(x_n-\hat{\eta}\right)^2. \tag{4.71}$$

Now, we will make use of the fact that

$$\frac{\hat{\eta} - \eta}{\frac{\hat{s}}{\sqrt{N}}} \sim t_{N-1}, \tag{4.72}$$

i.e., such a statistic (with η being the actual mean of the distribution) is distributed as a Student t with $N-1$ degrees of freedom; consequently, the CI is defined as

$$\hat{\eta} \pm t_{N-1,1-\frac{\alpha}{2}} \frac{\hat{s}}{\sqrt{N}}, \tag{4.73}$$

where $t_{N-1,1-\frac{\alpha}{2}}$ is the $1-\frac{\alpha}{2}$ quantile of this distribution, for a CI at level α. So, as a conclusion, the tumor will be considered malignant if η_0 falls within the CI defined in equation (4.72) and benign otherwise.

4.5 Estimation. Bayesian methods

4.5.1 Definition of Bayesian estimator and design criteria

In our previous discussions, as explicitly stated in Section 4.3.1, we considered that the unknown parameter $\boldsymbol{\theta}$ was a deterministic constant. The Bayesian philosophy is drastically different from this since the departing assumption is that the unknown $\boldsymbol{\theta}$ is sampled from a random variable.

This makes things quite different; in our previous estimators we only relied on one source of information, namely, $p(\mathbf{x}; \boldsymbol{\theta})$, which models how the data $\mathbf{x}$ is distributed as a function of $\boldsymbol{\theta}$. The Bayesian philosophy, on the other side, integrates two sources of knowledge; the first source is the same as in the previous case, but now we write it as $p(\mathbf{x}|\boldsymbol{\theta})$, meaning how the data behaves *conditioned* on the particular realization of the variable $\boldsymbol{\theta}$. The second source of information is how the variable behaves irrespective of the observations, i.e., the Bayesian assumption assumes knowledge of a function $p(\boldsymbol{\theta})$. This latter piece of information is referred to as the *a priori* density function or *the prior*, for short, since this information is known before the realization of the experiment that gives rise to the observation $\mathbf{x}$. This function, consequently, allows us to include our prior knowledge about the behavior of the parameter. Note that this function may also have some parameters; these parameters are called *hyperparameters* of the prior, and should be assumed known.

The cornerstone of the Bayesian philosophy is the Bayes theorem,

$$p(\boldsymbol{\theta}|\mathbf{x}) = \frac{p(\mathbf{x}, \boldsymbol{\theta})}{p(\mathbf{x})} = \frac{p(\mathbf{x}|\boldsymbol{\theta})p(\boldsymbol{\theta})}{p(\mathbf{x})} = \frac{p(\mathbf{x}|\boldsymbol{\theta})p(\boldsymbol{\theta})}{\int p(\mathbf{x}, \boldsymbol{\theta})d\boldsymbol{\theta}} \tag{4.74}$$

$$= \frac{p(\mathbf{x}|\boldsymbol{\theta})p(\boldsymbol{\theta})}{\int p(\mathbf{x}|\boldsymbol{\theta})p(\boldsymbol{\theta})d\boldsymbol{\theta}}, \tag{4.75}$$

and allows us to update our knowledge about $\boldsymbol{\theta}$ once $\mathbf{x}$ has been observed, i.e., the Bayes theorem gives a way to construct the *posterior* density function $p(\boldsymbol{\theta}|\mathbf{x})$ out of our two pieces of information. The main Bayesian estimators, as we describe in the next section, are derived from the posterior.

4.5.2 Design criteria for Bayesian estimators

In Section 4.3.1 we pointed out that a sensible criterion to build a scalar estimator could be minimizing the MSE, i.e., minimizing the mean square value of the error $\epsilon = \hat{\theta} - \theta$, although, as described, the estimators derived from this criterion could be non-realizable because of a bias issue. In the Bayesian case, as we now show, these types of estimators are perfectly realizable. Note, however, that in this case, the expectation (to calculate the mean square value) is twofold, since we now have two sources of randomness.

Keeping this in mind, recall the definition of MSE:

$$\mathrm{MSE}(\hat{\theta}) = E_{\mathbf{x},\theta}\{\epsilon^2\} = E_{\mathbf{x},\theta}\{(\hat{\theta}(\mathbf{x}) - \theta)^2\} = \int (\hat{\theta}(\mathbf{x}) - \theta)^2 p(\mathbf{x}, \theta) d\mathbf{x} d\theta \tag{4.76}$$

$$= \int \left[\int (\hat{\theta}(\mathbf{x}) - \theta)^2 p(\theta|\mathbf{x}) d\theta \right] p(\mathbf{x}) d\mathbf{x}. \tag{4.77}$$

Clearly, minimizing $\mathrm{MSE}(\hat{\theta})$ can be achieved by minimizing the integral within brackets in equation (4.77) since the functions involved are non-negative. This being the case, if we derive that integral with respect to $\hat{\theta}$ and solve for it, we will come up with

$$\hat{\theta}(\mathbf{x}) = \int \theta p(\theta|\mathbf{x}) d\theta = E\{\theta|\mathbf{x}\}. \tag{4.78}$$

So, as can be appraised, the estimator that minimizes the MSE turns out to be the mean of the posterior distribution. This is called the minimum MSE estimator (MMSEE).

This result is just a particular case of a more general formulation. The Bayesian philosophy is grounded on the definition of both an error measure, which is referred to as a cost, and a Bayesian risk. The latter is the mean of the cost with respect to the random variables involved. In the case that the cost is defined as $C(\epsilon) = \epsilon^2$ we obtain the MMSEE. But there are other criteria.

One popular criterion is the hit-or-miss cost, namely, $C(\epsilon) = 1$ if $|\epsilon| > \delta$ and 0 otherwise. With this cost function, it is simple to derive the resulting Bayesian estimator when $\delta \to 0$, which turns out to be

$$\hat{\theta}(\mathbf{x}) = \arg\max_{\theta} p(\theta|\mathbf{x}), \tag{4.79}$$

i.e., the mode of the posterior distribution. This estimator is known as the *maximum a posteriori* (MAP) estimator.

Two additional comments apply when the parameter $\boldsymbol{\theta}$ is a vector:

1. For the MMSEE, the reason in equations (4.76) and (4.76) can be applied as such when other components are involved. This is due to the fact that in the Bayesian case we can integrate out the variables that are not explicitly used in the subintegral function. The result is that the MMSEE for a vector parameter is the vector extension of the scalar estimator of equation (4.78), i.e.,

$$\hat{\boldsymbol{\theta}}(\mathbf{x}) = E\{\boldsymbol{\theta}|\mathbf{x}\} = \begin{bmatrix} E\{\theta_1|\mathbf{x}\} \\ \vdots \\ E\{\theta_p|\mathbf{x}\} \end{bmatrix}. \tag{4.80}$$

2. For the scalar MAP estimator, one of the main advantages is that the function $p(x)$ is not needed, which makes things far simpler. The reason why this is true can be observed in equation (4.74): certainly, maximizing $p(\boldsymbol{\theta}|\mathbf{x})$ is equivalent to maximizing $p(\mathbf{x}|\boldsymbol{\theta})p(\boldsymbol{\theta})$, since the denominator of the second equality is non-negative.
   However, when the parameter is a vector we do need to calculate $p(\mathbf{x})$ to finally derive $p(\theta_i|\mathbf{x})$. Consequently, the MAP estimator for a vector parameter is usually redefined as

$$\hat{\boldsymbol{\theta}}(\mathbf{x}) = \arg\max_{\boldsymbol{\theta}} p(\boldsymbol{\theta}|\mathbf{x}) = \arg\max_{\boldsymbol{\theta}} p(\mathbf{x}|\boldsymbol{\theta})p(\boldsymbol{\theta}). \tag{4.81}$$

   As a rule, this estimator does not coincide with the one defined by extending equation (4.79) to every component.

4.5.3 Performance measurement

As is the case of all estimators, better performance means a higher concentration of error values around zero; the point is that performance now has to consider the fact that the parameter is a random variable as well.

The case of the MMSEE makes characterization of the error variable especially simple. Let us get started with a scalar parameter. We have

$$E\{\epsilon\} = E_{\mathbf{x},\theta}\{\theta - \hat{\theta}\} = E\{\theta\} - E_{\mathbf{x}}\{E\{\theta|\mathbf{x}\}\} \tag{4.82}$$

$$= E\{\theta\} - \int \theta p(\theta|\mathbf{x})p(\mathbf{x})d\theta d\mathbf{x} \tag{4.83}$$

$$= E\{\theta\} - \int \theta p(\theta)d\theta = 0. \tag{4.84}$$

With respect to the variance of the error, since it is zero mean, we can write

$$E\{\epsilon^2\} = E_{\mathbf{x},\theta}\{(\theta - \hat{\theta})^2\}, \tag{4.85}$$

which coincides with the definition of MSE as written in equation (4.76).

If we now move on to the vector case, we may calculate the error covariance matrix

$$\mathbf{C}_\epsilon = E_{\mathbf{x},\theta}\{\epsilon\epsilon^T\} = E_{\mathbf{x},\theta}\left\{(\boldsymbol{\theta} - E\{\boldsymbol{\theta}|\mathbf{x}\})(\boldsymbol{\theta} - E\{\boldsymbol{\theta}|\mathbf{x}\})^T\right\} \tag{4.86}$$

$$= E_{\mathbf{x}}\left\{E_{\boldsymbol{\theta}|\mathbf{x}}\left\{(\boldsymbol{\theta} - E\{\boldsymbol{\theta}|\mathbf{x}\})(\boldsymbol{\theta} - E\{\boldsymbol{\theta}|\mathbf{x}\})^T\right\}\right\} \tag{4.87}$$

$$= E_{\mathbf{x}}\left\{\mathbf{C}_{\boldsymbol{\theta}|\mathbf{x}}\right\}. \tag{4.88}$$

This is an interesting result, since, as we will promptly see, for the Gaussian case, the covariance of the posterior distribution, $\mathbf{C}_{\boldsymbol{\theta}|\mathbf{x}}$, is not a function of $\mathbf{x}$ and hence both covariance matrices coincide.

4.5.4 The Gaussian case

Assume that variables $\mathbf{x}$ and $\boldsymbol{\theta}$ are jointly Gaussian, with means $E\{\mathbf{x}\}$ and $E\{\boldsymbol{\theta}\}$ and covariance matrices $\mathbf{C}_\mathbf{x}$ and $\mathbf{C}_{\boldsymbol{\theta}}$, respectively. Their cross-covariance matrices are $\mathbf{C}_{\mathbf{x}\boldsymbol{\theta}}$ and $\mathbf{C}_{\boldsymbol{\theta}\mathbf{x}}$ (recall that in the covariance matrix, the first subscript refers to the variable that enters the expectation operator as a column vector while the second refers to one that enters as a row vector).

If these variables are jointly Gaussian, then their joint PDF can be written

$$p(\mathbf{x}, \boldsymbol{\theta}; E\{\mathbf{x}\}, E\{\boldsymbol{\theta}\}, \mathbf{C}) = \frac{1}{(2\pi)^{\frac{N+p}{2}} |\mathbf{C}|^{\frac{1}{2}}} \times \exp\left(-\frac{1}{2}\begin{bmatrix} \mathbf{x} - E\{\mathbf{x}\} \\ \boldsymbol{\theta} - E\{\boldsymbol{\theta}\} \end{bmatrix}^T \mathbf{C}^{-1} \begin{bmatrix} \mathbf{x} - E\{\mathbf{x}\} \\ \boldsymbol{\theta} - E\{\boldsymbol{\theta}\} \end{bmatrix}\right), \tag{4.89}$$

where matrix $\mathbf{C}$ is a block matrix defined as

$$\mathbf{C} = \begin{bmatrix} \mathbf{C}_\mathbf{x} & \mathbf{C}_{\mathbf{x}\boldsymbol{\theta}} \\ \mathbf{C}_{\boldsymbol{\theta}\mathbf{x}} & \mathbf{C}_{\boldsymbol{\theta}} \end{bmatrix}. \tag{4.90}$$

Then, it turns out that:

1. Both $\mathbf{x}$ and $\boldsymbol{\theta}$ are also Gaussian, i.e., their respective PDFs are $p(\mathbf{x}; E\{\mathbf{x}\}, \mathbf{C}_\mathbf{x})$ and $p(\boldsymbol{\theta}; E\{\boldsymbol{\theta}\}, \mathbf{C}_{\boldsymbol{\theta}})$.
2. The posterior is also Gaussian, i.e., $p\left(\boldsymbol{\theta}; E\{\boldsymbol{\theta}|\mathbf{x}\}, \mathbf{C}_{\boldsymbol{\theta}|\mathbf{x}}\right)$, where the posterior parameters are

$$E\{\boldsymbol{\theta}|\mathbf{x}\} = E\{\boldsymbol{\theta}\} + \mathbf{C}_{\boldsymbol{\theta}\mathbf{x}}\mathbf{C}_\mathbf{x}^{-1}(\mathbf{x} - E\{\mathbf{x}\}), \tag{4.91}$$

$$\mathbf{C}_{\boldsymbol{\theta}|\mathbf{x}} = \mathbf{C}_{\boldsymbol{\theta}} - \mathbf{C}_{\boldsymbol{\theta}\mathbf{x}}\mathbf{C}_\mathbf{x}^{-1}\mathbf{C}_{\mathbf{x}\boldsymbol{\theta}}. \tag{4.92}$$

As can be observed from equation (4.92), the posterior covariance matrix is not a function of $\mathbf{x}$, so its expectation coincides with itself, and hence $\mathbf{C}_\epsilon = \mathbf{C}_{\boldsymbol{\theta}|\mathbf{x}}$ for the Gaussian case. The error ϵ is then fully characterized, since it will also be Gaussian because it is a linear function of jointly Gaussian variables.

4.5.5 Conjugate distribution and conjugate priors

As we have described in equation (4.79), Bayes theory relies on the Bayes theorem, and, specifically, is a matter of how to move from the prior to the posterior, which is clearly expressed in equation (4.75).

Occasionally, these two distributions turn out to belong to the same family. If this is the case, they are called *conjugate distributions* and the prior distribution, referred to as *conjugate prior* with respect to the likelihood function entered in equation (4.75), gives rise to a posterior of the same family as the prior.

The reader is invited to explore Section 4.7 for several examples of these distributions.

4.5.6 A working example

Numerous examples can be shown about the use of the Bayesian paradigm for random field segmentation [6]; this research field was very active in the beginning of the 2000s. However, we will resort to a simple example to illustrate the concepts that we have introduced in this section.

Consider we pursue tumors in a certain image modality; consider, as well, that we have a notion of how extensive these tumors are at our working resolution; so we have a discrete probability function, say $p(a, b)$, that tells us how probable it is to find a tumor of a pixels wide and b pixels high. This function runs in a sensible range for both dimensions. This is our prior information.

Now, let us accept that pixels within the tumor can be modeled as independent Gaussian random variables with mean η_t and σ_t. On the other hand, pixels in the background will be modeled as independent Gaussian random variables with mean η_b and σ_b.

We will inspect our image in sliding windows of $M \times M$ pixels; M should be chosen large enough to give room to the highest values of the tumor in our prior. Our purpose is to obtain the MAP estimate of a and b, i.e., the width and height of the tumor, respectively.

To this end, we will calculate the posterior distribution $p(a, b|\mathbf{x})$; this will be accomplished by using the Bayes theorem (recall equation (4.74)), which we now write as

$$p(a, b|\mathbf{x}) = \frac{p(\mathbf{x}|a, b)p(a, b)}{p(\mathbf{x})} = \frac{p(\mathbf{x}|a, b)p(a, b)}{\sum_i \sum_j p(\mathbf{x}|a_i, b_j)p(a_i, b_j)}. \tag{4.93}$$

Note that in order to calculate the MAP estimator we do not need to calculate the denominator of the expression above since it is common for every choice of the pair $(a_l b_m)$. So, the problem reduces to calculating the function $p(\mathbf{x}|a_l, b_m)$ for every choice of the variables and choose that combination that maximizes the numerator in equation (4.93). With the modeling assumptions indicated in the problem statement, this can be easily accomplished by

$$p(\mathbf{x}|a_l, b_m) = \prod_{n=1}^{a_l b_m} p(x_n; \eta_t, \sigma_t) \prod_{m=1}^{M^2 - a_l b_m} p(x_m; \eta_b, \sigma_b), \tag{4.94}$$

where x_n are the values within the tentative tumor mask, with as many pixels as $a_l b_m$, while x_m are the values outside the mask but within the sliding inspection window, consisting of $M^2 - a_l b_m$ pixels. Note the parameters $(\eta_t, \sigma_t, \eta_b, \sigma_b)$ can be easily estimated from the data using an MLE approach, similar to that described in Section 4.4.

So, MAP estimation will be achieved by

$$(a^*, b^*) = \arg\max_{a_l, b_m} p(a_l, b_m|\mathbf{x}) = \arg\max_{a_l, b_m} p(\mathbf{x}|a_l, b_m) p(a_l, b_m). \tag{4.95}$$

Needless to say, we have assumed that the tumor indeed exists in the data. We could test this hypothesis at this point, by means, for instance, of the maximum likelihood ratio for that choice of (a^*, b^*), i.e., the tumor will be considered present at the location of the sliding window if

$$\frac{p(\mathbf{x}|a^*, b^*)}{p(\mathbf{x}|a = 0, b = 0)} > 1. \tag{4.96}$$

We may want to raise our threshold to avoid false positives just because of chance due to a large number of comparisons (since decisions are pixelwise); but this is beyond our scope. The interested reader may consult other sources [7] for deeper insight on testing correction (False Discovery Rate and multiple comparisons procedures).

4.6 Monte Carlo methods

4.6.1 A non-stochastic use of Monte Carlo

Let us assume that we intend to solve the problem

$$\int_a^b \Psi(x) dx, \tag{4.97}$$

where function $\Psi(x)$ is sufficiently complex so as to make the analytical solution intractable. A very simple solution to this problem is to use a Monte Carlo method. Specifically, we rewrite the foregoing equation as

$$(b - a) \int_a^b \Psi(x) \frac{1}{b - a} dx, \tag{4.98}$$

where we can interpret the subintegral function as a product of function $\Psi(x)$ with the PDF of a uniform random variable within the interval (a, b). With this interpretation, we may write

$$(b-a)\int_a^b \Psi(x)\frac{1}{b-a}dx = (b-a)E\{\Psi(X)\}, \quad (4.99)$$

where X denotes the uniform random variable. Therefore, we may obtain the approximate result

$$\int_a^b \Psi(x)dx = (b-a)E\{\Psi(X)\} \approx \frac{b-a}{N}\sum_{n=0}^{N-1}\Psi(x_n), \quad (4.100)$$

as the Monte Carlo method dictates, with x_n independent samples of a uniform random variable in the interval (a, b). For *sufficiently* large N, results should be stable and accurate.

4.7 Exercises

1. Assume your observations are $x_0, \ldots, x_{N-1}$, where each x_n is a sample of mean η (common for all albeit unknown) and known standard deviation σ_n (which may vary through the dataset). The observations are assumed uncorrelated. An estimator for the mean is written as

$$\hat{\eta} = \sum_{n=0}^{N-1} a_n x[n].$$

  a. Find the coefficients $a_n, 0 \le n \le N-1$, so that $E\{\hat{\eta}\} = \eta$ and $\sigma_{\hat{\eta}}^2$ is minimum. Do your results make sense?
  b. For the case that the observations are Gaussian, find out whether the estimator just calculated coincides with the MLE.
  c. Find the CRLB for $\hat{\eta}$.

2. Assume you have a dataset $x_0, \ldots, x_{N-1}$, where the observations are i.i.d. We know that $P(x_n = k) = \binom{N}{k}p^k(1-p)^{N-k}$, $0 \le k \le N$, $E\{x_n\} = Np$, and $\sigma_{x_n}^2 = Np(1-p)$, but the actual value of the parameter p is unknown.
  a. Determine the CRLB associated to $\hat{p}$.
  b. Determine whether an efficient estimator exists and, if so, find it.
  c. Find the MLE of p as well as its asymptotic distribution.

3. Assume your observations $x_0, \ldots, x_{N-1}$ are i.i.d. exponential random variables with parameter $\lambda > 0$. This unknown parameter is modeled as a random variable, the density function of which is

$$p(\lambda; \alpha, \beta) = \frac{\beta^{-\alpha}}{\Gamma(\alpha)}\lambda^{\alpha-1}e^{-\frac{\lambda}{\beta}}u(\lambda),$$

where $u(\lambda)$ is the unit step function, $\alpha > 0$, $\beta > 0$, and $\Gamma()$ guarantees that $p(\lambda; \alpha, \beta)$ is correctly defined. We wish to find the following:

a. $\hat{\lambda}_{MLE}$.
b. $\hat{\lambda}_{MAP}$.
c. Discuss the values of α and β for which $\hat{\lambda}_{MAP}$ should tend to $\hat{\lambda}_{MLE}$.

Hint. The last bullet can be answered without knowing the results of the other two questions. Just analyze the behavior of function $p(\lambda; \alpha, \beta)$ with the variation of α and β.

4. The dataset $x_0, \ldots, x_{N-1}$ consists of N i.i.d. samples from a Poisson random variable with parameter $\lambda > 0$, i.e.,

$$P(x_n = k)\frac{\lambda^k}{k!}e^{-\lambda}, \ k \geq 0.$$

The unknown parameter λ is modeled as a random variable with density function

$$p(\lambda; \alpha, \beta) = \frac{\beta^{-\alpha}}{\Gamma(\alpha)}\lambda^{\alpha-1}e^{-\frac{\lambda}{\beta}}u(\lambda),$$

where $u(\lambda)$ is the unit step function, $\alpha, \beta > 0$ (known), and $\Gamma(\alpha)$ is defined as

$$\Gamma(p) = k^p \int_0^\infty \tau^{p-1} exp(-k\tau)d\tau, \ \mathrm{Re}(p), \mathrm{Re}(k) > 0.$$

According to this, calculate:
a. $\hat{\lambda}_{MAP}$.
b. Assuming that is $p(\mathbf{x})$ known (so you do not need to calculate it), obtain $\hat{\lambda}_{MMSE}$.

5. Your dataset consists of the observations $x_0, \ldots, x_{N-1}$, which are N i.i.d. samples of the distribution

$$p(x_n; x_{min}, \alpha) = \frac{\alpha x_{min}^\alpha}{(x_n)^{\alpha+1}} u(x[n] - x_{min}) = \mathcal{P}(x_n; x_{min}, \alpha),$$

where $\alpha > 0$, x_{min} is an arbitrary real number, and $u(x)$ is the unit step function. We wish to find the following:
a. MLE of parameters α and x_{min}.
b. Assuming now that x_{min} is known, find the CRLB of $\hat{\alpha}$.
c. Consider that the observations $y[0], \ldots, y[N-1]$ are i.i.d. samples from a uniform distribution in the interval $(0, \theta)$. Assume $\theta \sim \mathcal{P}(\theta; \theta_{min}, \alpha)$. Provided that both $\alpha > 0$ and $\theta_{min} > 0$ are known, find $\hat{\theta}_{MAP}$.

6. Observations of the dataset $x_0, \ldots, x_{N-1}$ lead us to assume that they follow the model

$$x_n = A + Bn + w_n,$$

where w_n is a sequence of i.i.d. Gaussian zero-mean random variables with variance σ_w^2. A and B are also modeled as Gaussian random variables which are uncorrelated to each other and with w_n, $0 \leq n \leq N-1$. Let

$$\mathbf{\Theta} = \begin{bmatrix} A \\ B \end{bmatrix}.$$

We wish to find:

a. $\hat{\mathbf{\Theta}}_{MMSE}$.

b. If $\boldsymbol{\epsilon} = \mathbf{\Theta} - \mathbf{\Theta}_{MMSE}$, find $\mathbf{C}_{\boldsymbol{\epsilon}}$.

c. $\hat{\mathbf{\Theta}}_{MAP}$.

References

[1] Papoulis A., Unnikrishna Pillai S., Probability, Random Variables, and Stochastic Processes, Fourth ed., McGraw Hill, Boston, 2002, http://www.worldcat.org/search?qt=worldcat_org_all&q=0071226613.

[2] Billingsley P., Probability and Measure, Wiley Series in Probability and Statistics, Wiley, ISBN 9780471007104, 1995, https://books.google.com/books?id=z39jQgAACAAJ.

[3] Scott D.W., On optimal and data-based histograms, Biometrika (ISSN 0006-3444) 66 (3) (1979) 605–610, https://doi.org/10.1093/biomet/66.3.605, https://academic.oup.com/biomet/article-lookup/doi/10.1093/biomet/66.3.605.

[4] Saba L., Sanches J.M., Pedro L.M., Suri J.S., Multi-Modality Atherosclerosis Imaging and Diagnosis, 1st ed., Springer Publishing Company, Incorporated, ISBN 9781493942893, 2016.

[5] Dempster A.P., Laird N.M., Rubin D.B., Maximum likelihood from incomplete data via the EM algorithm, Journal of the Royal Statistical Society: Series B 39 (1) (1977) 1–38.

[6] Cheng H., Bouman C.A., Multiscale Bayesian segmentation using a trainable context model, IEEE Transactions on Image Processing 10 (4) (2001) 511–525.

[7] Rosner B.A., Fundamentals of Biostatistics, Brooks/Cole, Cengage Learning, Boston, 2011.

PART

Image representation and processing

CHAPTER

Image representation and 2D signal processing

5

Santiago Aja-Fernández[a], Gabriel Ramos-Llordén[b,c], and Paul A. Yushkevich[d]

[a]*Image Processing Lab. (LPI), Universidad de Valladolid, Valladolid, Spain*

[b]*Athinoula A. Martinos Center for Biomedical Imaging, Department of Radiology, Massachusetts General Hospital, Charlestown, MA, United States*

[c]*Harvard Medical School, Boston, MA, United States*

[d]*Penn Image Computing and Science Laboratory, University of Pennsylvania, Philadelphia, PA, United States*

Learning points

- Image representation: continuous, digital, and multichannel images
- Images as two-dimensional signals
- Linear space- and circular invariant systems
- Convolution
- Frequency representations: Fourier, Discrete Fourier, and Discrete Cosine Transforms
- Sampling and interpolation
- Image quantization

5.1 Image representation

In medical imaging, when we talk about *images* we usually refer to digital images, and specifically to raster or bitmap images. A digital image is a 2D numerical representation of a picture of reality. See the example in Fig. 5.1: an ideal continuous image is captured by a device and it is transformed into a digital image through the following procedure:

1. The spatial extent that can be captured by the imaging device is limited.
2. The original image is sampled in order to obtain a limited and discrete number of values. This is related with the resolution of the image. The basic unit of image, which usually corresponds to one sampled value, is known as *pixel*.
3. The range of values that a pixel can take is reduced to a discrete set. This procedure is known as quantification.

Medical Image Analysis. https://doi.org/10.1016/B978-0-12-813657-7.00018-2

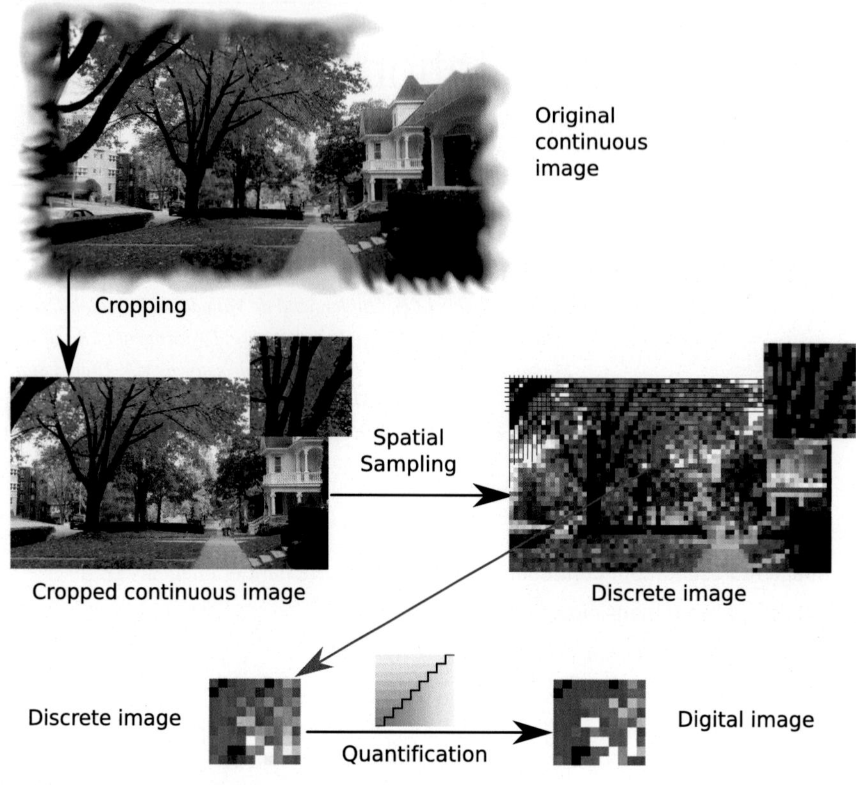

FIGURE 5.1 From a continuous to a digital image.

The original continuous image when acquired by a device is (1) limited in space, (2) spatially sampled, and (3) quantized.

Although this procedure is described in Fig. 5.1 for real-world images, it can be easily translated to the different modalities of medical imaging.

Information in bitmap images is stored inside every pixel (see Fig. 5.2). The value of that pixel can correspond to an intensity value, usually in the range [0, 1], or it can be an index to a color map, where the intensity values are stored. While both methods are equivalent for grayscale images, color images usually store indexed information (see Fig. 5.2, bottom). If an image is coded using three color components (red, green, and blue, for instance), a pixel stores the indices to the three color components. There may be images encoded with more than three components. Indeed, multicomponent data appear in (diffusion/functional) magnetic resonance imaging (MRI), as well as in microscopy. Nevertheless, in this chapter we focus on single-component images, which is still the most traditional way to acquire images in the medical field.

In most medical imaging modalities, the information stored in a single pixel is related to the acquisition units of every specific modality (see Fig. 5.3 for illustration). The acquired data are usually given to the user as a digital image. However, the val-

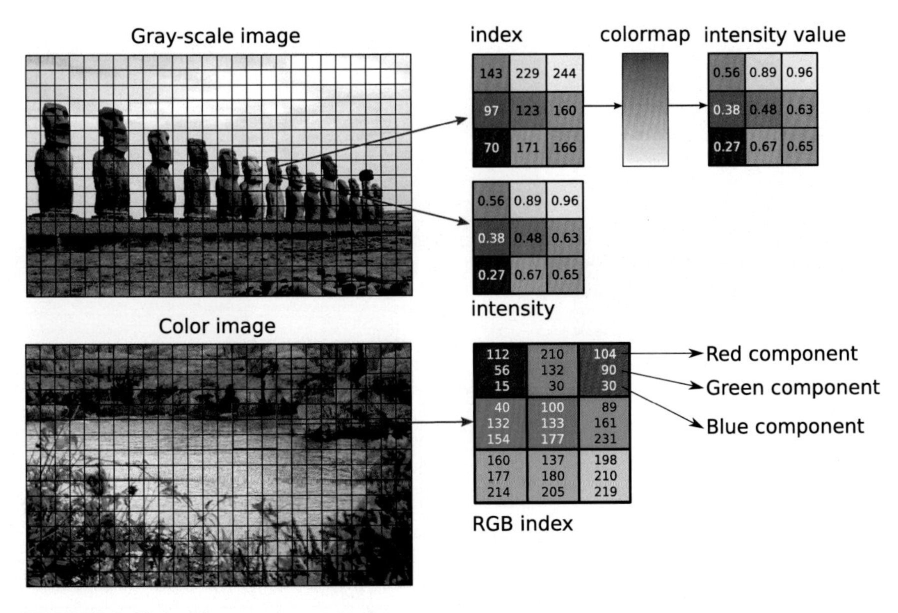

FIGURE 5.2 Digital image representation.

Top: A grayscale image can store the intensity information or an index that points to the color map with the intensity information. Bottom: Example of the representation of a color image using an RGB color component.

ues from each pixel can vary: computed tomography (CT) images provide units in the Hounsfield scale, where air areas have a value of -1000, water has a value of 0, and bone values range from 200 to 3000. The scale is a linear transformation of the original attenuation coefficient measured in the scanner. It is quantitative and therefore it is totally related to the structure of the materials and tissues. In the example of the ultrasound image, in this case an echocardiography image, the data provided by the vendor have been normalized in the range $[0, 255]$, so that every pixel is represented by an integer number (one byte). Last, the values provided by the MRI slice are related to the scanner and acquisition features, so the values are not related to tissue properties. In this example, the information is given by the visual contrast between regions.

In this chapter we will consider the images as 2D signals, so we can use the whole signal processing theory for them. The extension to 3D or multidimensional imaging is straightforward. The original continuous image would be denoted as $f(x, y)$, where $x, y \in \mathbb{R}$ are the spatial coordinates. The sampled (discrete) version of the image is defined as $f[m, n]$, where $n, m \in \mathbb{Z}$. Finally, the digital image is the quantized version of the sampled signal, and it can be represented as $\tilde{f}[m, n]$. However, for the sake of simplicity, digital images are usually simply pictured as $f[m, n]$, or even as $f(x, y)$ (in this case $x, y \in \mathbb{Z}$). In that case, the difference between continuous and digital images must be derived from the context.

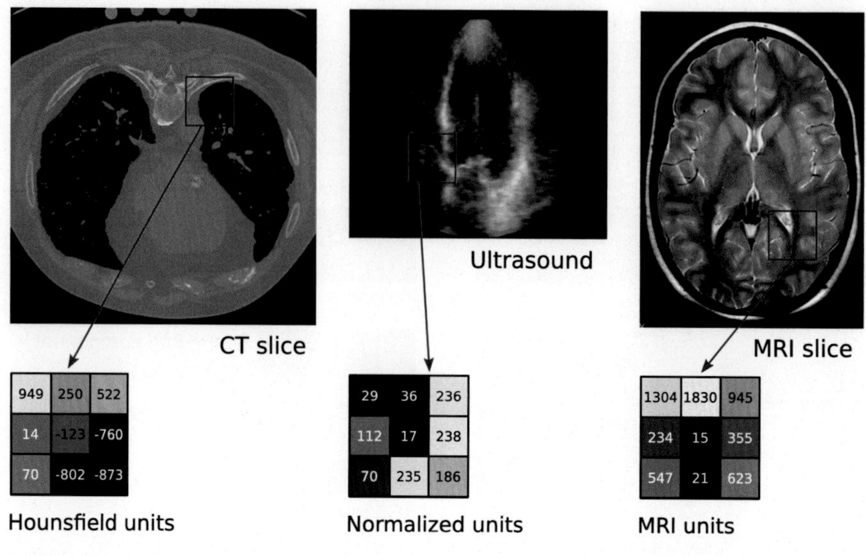

FIGURE 5.3 Medical imaging modalities.

Example of the intensity values provided by three different modalities of medical images: computed tomography, ultrasound, and magnetic resonance imaging.

5.2 Images as 2D signals

From a signal processing perspective, an image can be seen as a 2D signal that depends on two spatial components, width and height. The *ideal* original image (which actually does not exist) would be a continuous 2D signal defined as $f(x, y)$, where $x, y \in \mathbb{R}$ and

$$f : \mathbb{R}^2 \rightarrow \mathbb{R}.$$

In reality, the images we usually deal with are modified versions of the ideal continuous signal: the image is sampled, quantified, and limited in space (the image is limited by a border). The discrete version of the image is $f[m, n]$, where $n, m \in \mathbb{Z}$ and

$$f : \mathbb{Z}^2 \rightarrow \mathbb{R}.$$

Finally, the digital image $\widehat{f}[m, n]$ is the quantized version of the discrete signal.

Digital and discrete images rely on a discrete grid as a coordinate system (see Fig. 5.4). Fig. 5.4(a) shows a general coordinate system for a 2D signal $f[m, n]$. This system can be directly translated to an image coordinate system (Fig. 5.4(b)), where the origin is set to the center of the image. However, in order to avoid negative indices, it is useful to set the origin in one of the corners of the image (see Fig. 5.4(c)).

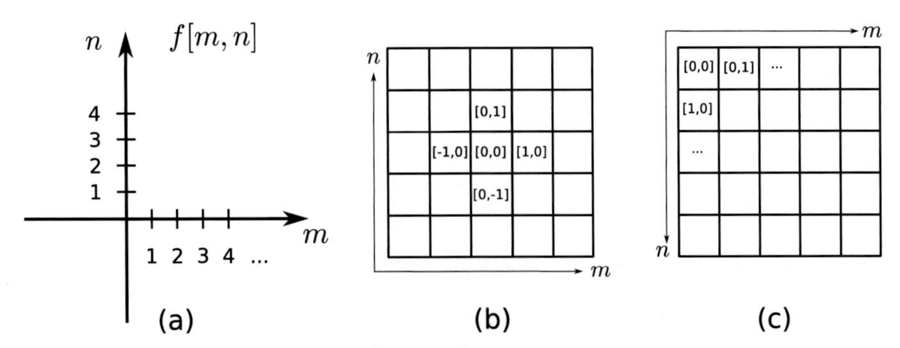

FIGURE 5.4 Representation of the coordinate system for a 2D discrete signal.

(a) Coordinate system for 2D signals $f[m,n]$, with $n, m \in \mathbb{Z}$. (b) Coordinate systems for digital images considering the center of the image as the center of the space. (c) Alternative coordinate system using one corner of the image as the origin of the space.

5.2.1 Linear space-invariant systems

The concept of signal is usually linked to the concept of a *system*. Although the word *system* can be used in many different contexts to denote many different processes, in signal/image processing theory a system is a process by which signals (e.g., images) are transformed into other signals (e.g., images). When an image arrives, the system can modify the input image into an output image or, alternatively, the system can react to the input, producing a specific behavior. Under the concept of *system*, we will include the various image processing filters defined throughout the book. Mathematically, a system is then a mapping $T[\cdot]$ that operates on an image $f(x, y)$ and produces an output image $g(x, y)$, which is given by $g(x, y) = T[f(x, y)]$. Among all the possible definitions for $T[\cdot]$, a special class of systems are those that fulfill two features, linearity and spatial invariance. Those concepts are described below.

Linearity: The relationship between the input and output images is a linear map. That is, if we define $f_1(x, y)$ and $f_2(x, y)$ so that their outputs are $g_1(x, y) = T[f_1(x, y)]$ and $y_2(t) = T[f_2(x, y)]$, a linear combination of those inputs, $f(x, y) = \alpha_1 f_1(x, y) + \alpha_2 f_2(x, y)$, will produce the same linear combination of outputs: $g(x, y) = T[f(x, y)] = \alpha_1 g_1(x, y) + \alpha_2 g_1(x, y)$.

Spatial (shift-)invariance: If the input is shifted by a given amount $[x_0, y_0]$, the output is shifted by the same amount: $g(x - x_0, y - y_0) = T[f(x - x_0, y - y_0)]$. This is equivalent to saying that the location of the origin of the coordinate system is irrelevant.

Those systems that fulfill both properties are known as linear space-invariant (LSI) systems, and they are characterized by their unitary impulse response, $h(x, y) = T[\delta(x, y)]$ for continuous systems and $h[m, n] = T\,[\delta[m, n]]$ for discrete systems, where $\delta(x, y)$ is the Dirac delta function in 2D and $\delta[m, n]$ is its discrete counterpart.

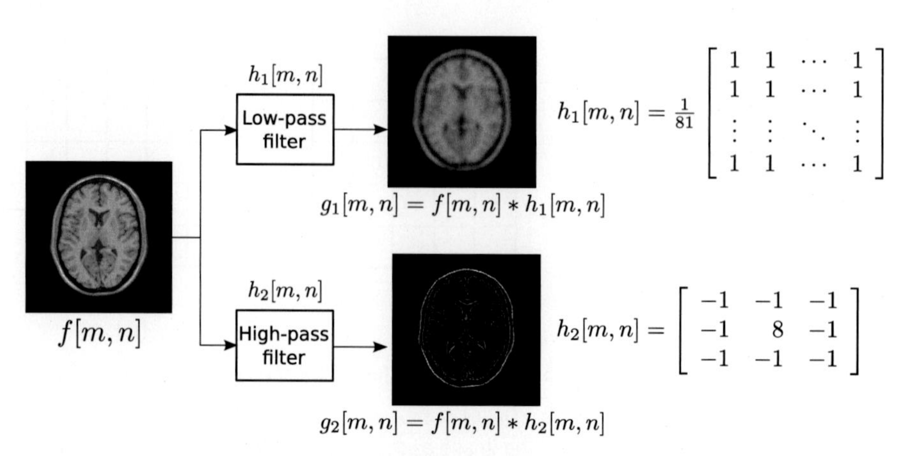

FIGURE 5.5 Example of image processing using LSI systems.

Two different impulse responses have been considered. (1) $h_1[m,n]$ corresponds to a 9×9 low-pass filter. (2) $h_2[m,n]$ is a high-pass filter.

This impulse response completely characterizes the system so that the output for any input image can be calculated. Sometimes systems are defined using continuous notation. That is why in what follows we will keep both notations in parallel. However, practical implementations nowadays are mostly done using discrete (digital) systems. The output of a continuous system to an input image $f(x, y)$ is given by the convolution with the impulse response of the system:

$$g(x, y) = f(x, y) * h(x, y) \tag{5.1}$$

$$= \int_{x'} \int_{y'} f(x', y')h(x - x', y - y')dx'dy'. \tag{5.2}$$

Similarly, for a discrete signal, the output is defined as

$$g[m, n] = f[m, n] * h[m, n] \tag{5.3}$$

$$= \sum_{m'} \sum_{n'} f[m', n']h[m - m', n - n']. \tag{5.4}$$

In Fig. 5.5 an example of LSI filtering is depicted: a discrete image $f[m, n]$ is processed using two different LSI systems via convolution.

Properties of 2D convolution

Convolution gives a product that satisfies the following properties:

Commutativity: $f(x, y) * h(x, y) = h(x, y) * f(x, y)$.
Associativity: $f(x, y) * (h(x, y) * g(x, y)) = (f(x, y) * h(x, y)) * g(x, y)$.
Distributivity: $f(x, y) * (h(x, y) + g(x, y)) = f(x, y) * h(x, y) + f(x, y) * g(x, y)$.

Identity element: $f(x, y) * \delta(x, y) = f(x, y)$.
Differentiation: $(f(x, y) * h(x, y))' = f'(x, y) * h(x, y) = f(x, y) * h'(x, y)$.
Stability: A system defined by an impulse response $h(x, y)$ or $h[m, n]$ is stable if

$$\int_x \int_y |h(x, y)| dx dy < \infty,$$

$$\sum_m \sum_n |h[m, n]| < \infty.$$

5.2.2 Linear Circular Invariance systems

In practical situations, images are not infinite, but limited in space. Digital images are known to have a limited number of rows and columns. The limitation of an $M \times N$ discrete signal is equivalent to multiplying a limitless image $f[m, m]$ by a rectangular function $r[m, n]$:

$$f_d[m, m] = f[m, m] \cdot r[m, n] = \begin{cases} f[m, n] & 0 \leq m \leq M-1,\ 0 \leq n \leq N-1, \\ 0 & \text{otherwise}, \end{cases}$$

where

$$r[m, n] = \begin{cases} 1 & 0 \leq M-1,\ 0 \leq n \leq N-1, \\ 0 & \text{otherwise}. \end{cases}$$

If LSI systems are used to process a signal $f_d[m, m]$, we must be aware that outside the borders of the image, the signal value is 0. As an effect of this area of zero value, the different procedures may fail near the borders. In addition, in image processing, one of the requirements of the processing systems is that the output has the exact same size as the input. In an LSI system, if the input $f[m, n]$ is an $M_1 \times N_1$ signal and the impulse response is $M_2 \times N_2$, the output $g[m, n] = f[m, n] * h[m, n]$ will be an $(M_1 + M_2 - 1) \times (N_1 + N_2 - 1)$ signal, which is greater than the original size. If only *effective* values of the convolution are taken into account (those in which no pixels from the zero-value area are involved), the size would be smaller: $(M_1 - M_2 + 1) \times (N_1 - N_2 + 1)$.

In order to avoid these issues, a periodic extension of the image can also be assumed. A periodic expansion of the $M \times N$ discrete signal $f_d[m, m]$ is defined as

$$f_p[m + k_1 M, n + k_2 N] = f_p[m, n] = f[m, n], \quad (k_1, K_2, M, N) \in \mathcal{Z}.$$

This extension of the image assumes that all the operations are done in a space of periodic signals. The systems involved are no longer LSI but LCI: linear and circular shift-invariant. LCI systems are also completely characterized by the unitary impulse response, $h[m, n]$, and the output can be calculated using the circular discrete convo-

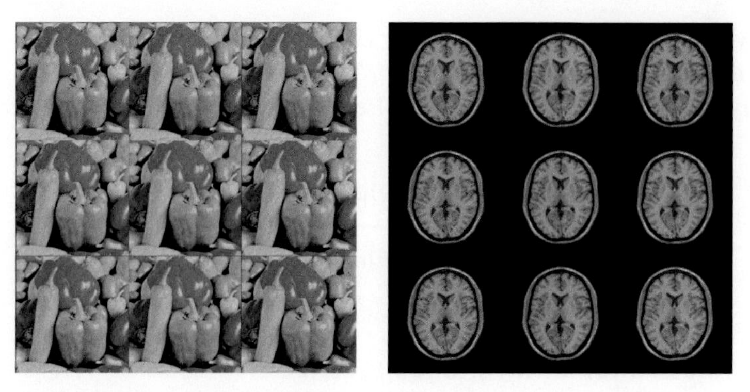

FIGURE 5.6 Periodic expansion of digital image $f[m,n]$.

This expansion is used in LCI systems and implicitly assumed when DFT is used.

lution

$$g[m,n] = f_p[m,n] \circledast h[m,n] \tag{5.5}$$

$$= \sum_{m'=0}^{M-1} \sum_{n'=0}^{N-1} f_p[m',n'] h[(m-m')\text{modM}, (n-n')\text{modM}]. \tag{5.6}$$

It can also be seen as the periodic expansion of the discrete convolution. If

$$g_0[m,n] = f[m,n] * h[m,n],$$

then

$$g[m,n] = \sum_{k=-\infty}^{\infty} \sum_{l=-\infty}^{\infty} g_0[m+kM, n+lN].$$

As we will see in the next section, LCI systems and the circular convolution naturally arise when dealing with the discrete Fourier transform (DFT).

In Fig. 5.6 we show the periodic expansion of two different images. Note that in the first one, the periodic expansion creates discontinuities in the border of the image. This effect can produce undesired behaviors when using certain filtering procedures. However, note that in some medical images, like the MRI slice in Fig. 5.6, due to the black background present this is no longer an issue.

5.3 Frequency representation of 2D signals

One powerful technique in image processing is to use alternative representations of the signal that allow a better representation of certain features of the image. One of the most used transformations is the Fourier transform, which is a representation of the distributions of frequencies within the image.

5.3.1 Fourier transform of continuous signals

The 2D Fourier transform of a continuous signal $f(x, y)$ is defined as

$$F(u, v) = \int_{-\infty}^{\infty}\int_{-\infty}^{\infty} f(x, y)e^{-j2\pi(ux+vy)}dxdy \quad \text{(analysis equation)}, \tag{5.7}$$

$$f(x, y) = \int_{-\infty}^{\infty}\int_{-\infty}^{\infty} F(u, v)e^{j2\pi(ux+vy)}dudv \quad \text{(synthesis equation)}, \tag{5.8}$$

where u and v are the spatial frequency components. The Fourier transform can also be computed over angular frequencies $\omega_i = 2\pi\xi_i$, and therefore

$$f(x, y) = \left(\frac{1}{2\pi}\right)^2 \int_{-\infty}^{\infty}\int_{-\infty}^{\infty} F(\omega_1, \omega_2)e^{j(\omega_1 x+\omega_2 y)}d\omega_1 d\omega_2. \tag{5.9}$$

For some applications it is useful to define the Fourier transform using polar coordinates:

$$F_p(\xi, \phi) = \int_{-\infty}^{\infty}\int_{-\infty}^{\infty} f(x, y)e^{-j2\pi(\xi\cos\phi x+\xi\sin\phi y)}dxdy \tag{5.10}$$

$$= \int_{0}^{\infty}\int_{0}^{2\pi} f_p(\rho, \theta)e^{-j2\pi(\xi\cos\phi\rho\cos\theta+\xi\sin\phi\rho\sin\theta)}\rho d\rho d\theta. \tag{5.11}$$

Properties of the Fourier transform

Some of the main properties of the 2D Fourier transform with special relevance for image processing are the following:

1. Uniqueness: If $f(x, y)$ is square-integrable, the Fourier transform is unique and reversible.
2. Linearity:

$$\alpha f(x, y) + \beta g(x, y) \overset{\mathfrak{F}}{\longleftrightarrow} \alpha F(u, v) + \beta F(u, v).$$

3. Shift:

$$f(x - x_0, y - y_0) \overset{\mathfrak{F}}{\longleftrightarrow} e^{j2\pi(x_0 u+y_0 v)}F(u, v).$$

4. Separability: The 2D Fourier transform is equivalent to two consecutive 1D Fourier transforms in normal directions:

$$F(u, v) = \int_{-\infty}^{\infty}\left[\int_{-\infty}^{\infty} f(x, y)e^{-j2\pi vy}dy\right]e^{-j2\pi ux}dx.$$

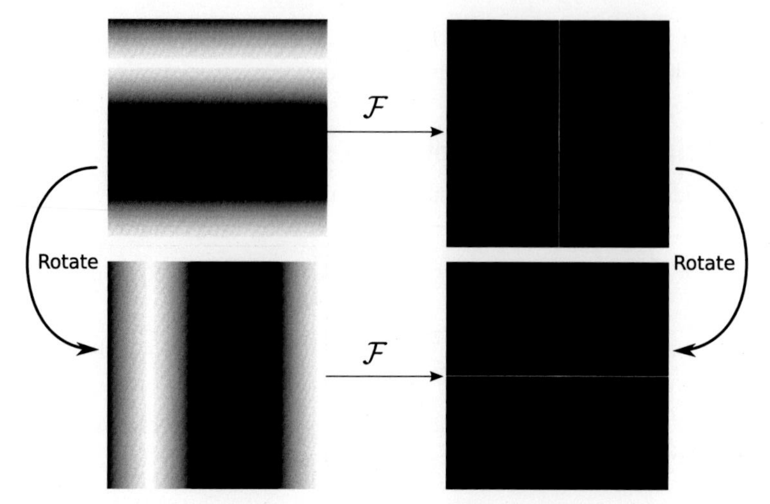

FIGURE 5.7 The rotation of an image implies the rotation of its Fourier transform.

The rotation of an image implies the rotation of its Fourier transform.

5. Convolution: The Fourier transform of a convolution is a product:

$$g(x, y) = f(x, y) * h(x, y) \stackrel{\mathfrak{F}}{\longleftrightarrow} G(u, v) = F(u, v)H(u, v).$$

6. Rotation: The rotation of $f(x, y)$ implies a rotation of its Fourier transform (see Fig. 5.7):

$$\mathfrak{F}\left[f_P(\rho, \theta + \alpha)\right] = F_P(\xi, \phi + \alpha).$$

5.3.2 Discrete-space Fourier transform

For a discrete signal $f[m, n]$ we can define the 2D Fourier transform as

$$F(\Omega_1, \Omega_2) = \sum_m \sum_n f[m, n] e^{-j(\Omega_1 m + \Omega_2 n)} \quad \text{(analysis equation)},$$

$$x[m, n] = \left(\frac{1}{2\pi}\right)^2 \int_{-\pi}^{\pi} \int_{-\pi}^{\pi} F(\Omega_1, \Omega_2) e^{j(\Omega_1 m + \Omega_2 n)} d\Omega_1 d\Omega_2 \quad \text{(synthesis equation)}.$$

We call this transform the discrete-space Fourier transform (DS-FT). Note that the DS-FT is a periodic continuous signal, with period 2π in both frequency directions.

The computational alternative to the DS-FT is the 2D DFT, where the signal and its transform are both discrete.

5.3.3 2D discrete Fourier transform

Let us assume that $f[m,n]$ is an $M \times N$ finite discrete image. The 2D DFT of that signal is computed as

$$F[k_1,k_2] = \sum_{m=0}^{N-1}\sum_{n=0}^{N-1} f[m,n]e^{-j\frac{2\pi}{N}(k_1 m + k_2 n)} \quad \text{(analysis equation)}, \tag{5.12}$$

$$f[m,n] = \frac{1}{N^2}\sum_{k_1=0}^{N-1}\sum_{k_2=0}^{N-1} F[k_1,k_2]e^{j\frac{2\pi}{N}(k_1 m + k_2 n)} \quad \text{(synthesis equation)}. \tag{5.13}$$

Properties of the 2D DFT

1. Periodicity: The DFT assumes a periodic extension of image $f[m,n]$ so that $f[m+M,n+N] = f[m,n]$ and a periodic expansion of the DFT: $F[u+M, v+N] = F[u,v]$.
2. The values of the DFT are samples of the DS-FT:

$$\left.\begin{array}{l} f[m,n] \overset{\mathcal{FS-FT}}{\longleftrightarrow} F_d(\Omega_1,\Omega_2) \\ f[m,n] \overset{\mathcal{DFT}}{\longleftrightarrow} F[k_1,k_2] \end{array}\right\} F[k_1,k_2] = F_d\left(\frac{2\pi}{M}k_1, \frac{2\pi}{N}k_2\right).$$

3. The DFT can be implemented using fast computational solutions, like the fast Fourier transform, with a considerable reduction in the number of operations.
4. Conjugate symmetry: $F[k_1,k_2] = F^*[M-k_1, N-k_2]$.
5. (Circular) convolution: The DFT of the circular convolution is a product:

$$f[m,n] \circledast h[m,n] \overset{\mathcal{DFT}}{\longleftrightarrow} F[k_1,k_2] \cdot F[k_1,k_2].$$

6. Product:

$$f[m,n] \cdot h[m,n] \overset{\mathcal{DFT}}{\longleftrightarrow} F[k_1,k_2] \circledast F[k_1,k_2].$$

Note that the LCI systems previously defined implicitly arise when working with DFT. When dealing with limited discrete signals and DFT, we are assuming the periodic expansion of the images that makes the traditional convolution unfit for the problem.

5.3.4 Discrete cosine transform

One of the problems of the DFT is that the periodicity that it implies creates non-natural transitions in images. In order to obtain a better behavior in the borders, the discrete cosine transform (DCT) can alternatively be used. The DCT of an $N \times N$

$f[m, n]$

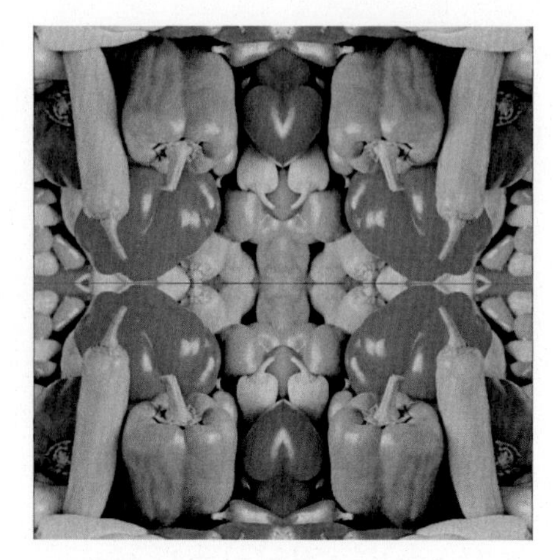

$f_e[m, n]$

FIGURE 5.8 Symmetric expansion implicitly assumed by the DCT.

The discontinuities in the border of the image are reduced with respect to the periodic expansion in Fig. 5.6.

image $f[m, n]$ is equivalent to the DFT of an extension of the image into $2N \times 2N$ (see Fig. 5.8), $f_e[m, n]$:

$$\mathrm{DCT}\{f[m, n]\} = \mathrm{DFT}\{f_e[m, n]\} = F[k_1, k_2].$$

The DCT is a separable linear transformation. For a 2D signal it is defined as

$$F[k_1, k_2] = \alpha(k_1)\alpha(k_2) \sum_{m=0}^{N-1} \sum_{n=0}^{N-1} f[m, n] \cos\left(\frac{(2m+1)k_1\pi}{2M}\right) \cos\left(\frac{(2n+1)k_2\pi}{2N}\right)$$

$$\text{(analysis equation)}, \tag{5.14}$$

$$f[m, m] = \sum_{k_1=0}^{N-1} \sum_{k_2=0}^{N-1} \alpha(k_1)\alpha(k_2) F[k_1, k_2] \cos\left(\frac{(2m+1)k_1\pi}{2M}\right) \cos\left(\frac{(2n+1)k_2\pi}{2N}\right)$$

$$\text{(synthesis equation)}, \tag{5.15}$$

where

$$\alpha(k_1) = \begin{cases} \frac{1}{\sqrt{M}} & \text{if } k_1 = 0, \\ \sqrt{\frac{2}{M}} & \text{if } k_1 = 1, \cdots, M-1, \end{cases}$$

$$\alpha(k_2) = \begin{cases} \frac{1}{\sqrt{N}} & \text{if } k_2 = 0, \\ \sqrt{\frac{2}{N}} & \text{if } k_2 = 1, \cdots, M-1. \end{cases}$$

5.4 Image sampling

5.4.1 Introduction

So far, medical images have been modeled as 2D or 3D continuous functions, and algorithms for medical image analysis are often conceived for such a continuous representation. Nevertheless, in practice, modern medical image processing and storing require images to be available in digital form, that is, as arrays of finite lengths of binary words [1]. Transforming a continuous image into a digitized one is comprised of (at least) two steps: image sampling and image quantization. Both processes are described in this chapter, with a special emphasis on the former step. For simplicity, we will focus on the 2D case, the extension to 3D being straightforward.

Image sampling can be seen as a discretization of a continuous image in the spatial domain. That is, given a continuous image $f(x, y)$ with $(x, y) \in \mathbb{R}^2$, the output of image sampling is a 2D array of points $f[m, n]$ with $(m, n) \in \mathbb{Z}^2$, where each point represents a particular sample $f(x, y)$, i.e.,

$$f[m,n] = f(m \cdot \Delta_x, n \cdot \Delta_y), \tag{5.16}$$

where Δ_x and Δ_y are the so-called sampling rates in the x- and y-directions. Points $(m, n) \in \mathbb{Z}^2$ are often referred to as pixels in the 2D case and as voxels in the 3D case.

The fundamental question in image sampling is the following: Given samples $f[m, n]$ of $f(x, y)$, *is it possible to exactly reconstruct* $f(x, y)$ *by solely using information from* $f(m, n)$? And if so, *which conditions should the image* $f(x, y)$ *fulfill? How should we then choose the values of* Δ_x *and* Δ_y? The answers to these questions are provided by a branch of mathematical analysis called sampling theory.

5.4.2 Basics on 2D sampling theory

Let us consider an ideal image sampling function, which is a 2D infinite array of Dirac delta functions located on a rectangular grid with spacing Δ_x and Δ_y (see Fig. 5.9), i.e.,

$$\delta_p(x, y) = \sum_{m=-\infty}^{\infty} \sum_{n=-\infty}^{\infty} \delta(x - m\Delta_x, y - n\Delta_y). \tag{5.17}$$

Its Fourier transform is another train of Dirac functions with spacing $1/\Delta_x$ and $1/\Delta_y$,

$$\hat{\delta}_p(u, v) = \frac{1}{\Delta_x \Delta_y} \sum_{m=-\infty}^{\infty} \sum_{n=-\infty}^{\infty} \delta(u - m/\Delta_x, v - n/\Delta_y). \tag{5.18}$$

The sampled image is thus defined as the product of the original function and the sampling function,

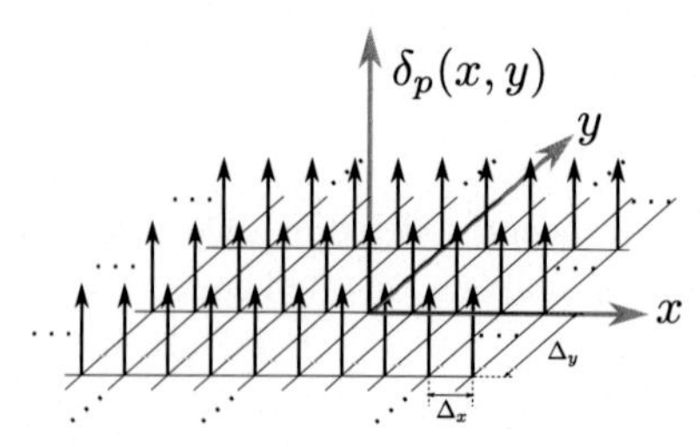

FIGURE 5.9 Sampling function.

2D infinite array of Dirac delta functions.

$$f_P(x,y) = f(x,y) \cdot \delta_P(x,y) \tag{5.19}$$

$$= \sum_m \sum_n f(m\Delta_x, n\Delta_y)\delta(x - m\Delta_x, y - n\Delta_y)$$

$$= \sum_m \sum_m f[m,n]\delta(x - m\Delta_x, y - n\Delta_y), \tag{5.20}$$

and its Fourier transform is given by

$$F_P(u,v) = F(u,v) * \hat{\delta}_P(u,v) \tag{5.21}$$

$$= \frac{1}{\Delta_x \Delta_y} \sum_m \sum_n F(u,v) * \delta(u - m/\Delta_x, v - n/\Delta_y)$$

$$= \frac{1}{\Delta_x \Delta_y} \sum_m \sum_n F(u - m/\Delta_x, v - n/\Delta_y). \tag{5.22}$$

Important to note, $F_P(u,v)$ contains infinite overlapped replicas of the original spectrum of $f(x,y)$, that is, $F(u,v)$, so in principle, it is impossible to recover $F(u,v)$ and hence $f(x,y)$ from $F_P(u,v)$. However, let us assume that image $f(x,y)$ is band-limited, that is, $F(u,v)$ has finite support. As an illustration we assume that the support of $F(u,v)$ is rectangular, i.e., it is zero for $|u| > B_x$, $|v| > B_y$. We will further assume that $2B_x \leq 1/\Delta_x$, $2B_y \leq 1/\Delta_y$. If that is so, it is straightforward to prove that the replicas of $F(\xi_x, \xi_y)$ in equation (5.22) do not overlap, and hence,

$$F(u,v) = \Delta_x \Delta_y F_P(u,v) \Pi\left(\frac{u}{2B_x}\right) \Pi\left(\frac{v}{2B_y}\right), \tag{5.23}$$

where $\Pi\left(\frac{u}{2B_x}\right)$ and $\Pi\left(\frac{v}{2B_y}\right)$ are rectangular functions with width $2B_x$ and $2B_y$, respectively. The inverse Fourier transform equation (5.23) yields

$$f(x,y) = r_x r_y f_s(x,y) * \mathrm{sinc}(2B_x x)\,\mathrm{sinc}(2B_y y), \tag{5.24}$$

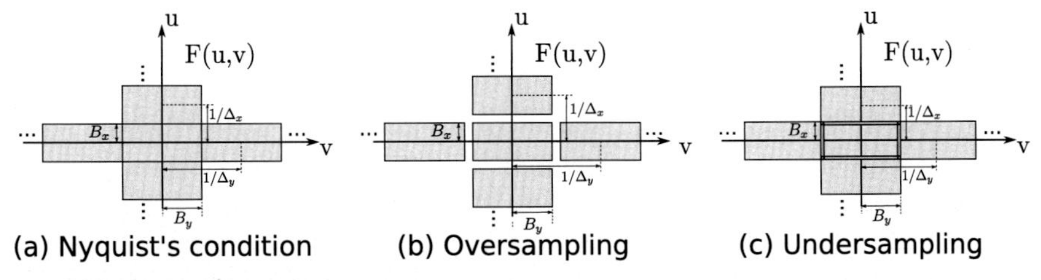

FIGURE 5.10 Sampling and aliasing.

Three scenarios depending on the choice of Δ_x and Δ_y. The Nyquist regime (a) and oversampling (b) allow exact reconstruction, whereas this is no longer possible when undersampling occurs (c).

where $r_x \leq 1$ and $r_y \leq 1$ are the sampling rate parameters, defined as

$$r_x = \frac{2B_x}{1/\Delta_x}, \quad r_y = \frac{2B_y}{1/\Delta_y}. \tag{5.25}$$

Substituting equation (5.20) into equation (5.24), we arrive at the following expression:

$$f(x, y) = r_x r_y \sum_{m=-\infty}^{\infty} \sum_{n=-\infty}^{\infty} f[m, n] \operatorname{sinc}(2B_x x - r_x m) \operatorname{sinc}(2B_y y - r_y n). \tag{5.26}$$

If $1/\Delta_x$ and $1/\Delta_y$ are chosen to be equal to $2B_x$ and $2B_y$, then $r_x = 1$ and $r_y = 1$, and equation (5.26) becomes the celebrated cardinal series, also called the Whittaker–Shannon–Kotelnikov sampling theorem formula, which can be found in any basic book of image processing. Frequencies $2B_x$ and $2B_y$ are called the Nyquist frequencies or Nyquist rates. If $1/\Delta_x$ and $1/\Delta_y$ are higher than the Nyquist frequencies, i.e., sampling rate parameters are strictly smaller than 1, we are dealing with *oversampling*. In both situations, exact reconstruction is possible with equation (5.26). If $1/\Delta_x$ and $1/\Delta_y$ are smaller than the Nyquist frequencies, the replicas in equation (5.22) do overlap. As a result, the derivation that we followed to arrive at the expression in equation (5.26) is not valid. This case is referred to as *undersampling*. Reconstruction is no longer possible in this situation. These three scenarios are illustrated in Fig. 5.10.

In summary, a band-limited image $f(x, y)$ with rectangular bandwidth $2B_x$ and $2B_y$ can be reconstructed without error provided $2B_x\Delta_x \leq 1$ and $2B_y\Delta_y \leq 1$. This condition is called the Nyquist condition.

5.4.2.1 Inexact reconstruction

Though exact reconstruction is possible if the Nyquist conditions hold, it should be noted that, in practice, real-world images are barely band-limited. Moreover, instead of an infinite set of samples, the number of image samples is always finite. These two cases violate the assumptions we have made above. Therefore, equation (5.26) is not directly applicable. We are then in a scenario where inexact reconstruction is the best we can achieve. These two types of errors are described below.

1. **Aliasing:** When images are not band-limited, the Nyquist conditions never hold. In this undersampling scenario, spurious spatial frequency components will be introduced into the reconstruction. This effect is called aliasing. The effects of aliasing in an actual image are shown in Fig. 5.11. In this example, the starting point is a sampled image of a brick wall (Fig. 5.11(a)). The original continuous image is assumed to be sampled fulfilling the Nyquist condition. To mimic the effect of sampling with a different sampling step, we decimate the discrete image by a factor of 5 and 10, respectively. Images are then zoomed in at the same size of the original image. Observe that when the sampling frequency is reduced by a factor of 5, the particular pattern of the brick wall is still noticeable, whereas it is impossible to distinguish when the sampling frequency is reduced 10-fold. In the latter case, the Nyquist conditions do not hold, and therefore aliased frequencies are present in the image. These types of artificial low-frequency spatial components are known in the image jargon as moiré patterns.
   Aliasing errors can be substantially reduced by low-pass filtering the image before sampling. This attenuates the spectral foldover that appears when replicas of the sampled image overlap. In Fig. 5.12, the same image was sampled before it was first low-pass filtered and thus sampled with a sampling frequency reduced by a factor of 10.
   Obviously, attenuating high spatial frequencies to avoid undersampling introduces a loss of information in the sampled image. As a result, there is always a trade-off between sampled image resolution and aliasing error. In a practical design, consideration must be given to the choice of the low-pass filter as well as the degree of aliasing error that is acceptable. It may be the case that a small aliasing error is preferable instead of a substantial loss of resolution. For a more elaborated discussion, the reader is referred to [2].
2. **Truncation error:** Even if Nyquist conditions hold, exact reconstruction is not possible, since the number of samples is finite and thus equation (5.26) is not realizable. A typical line of action is then to calculate equation (5.26) with a finite number of samples N, thereby truncating the summand. This has the effect of loss of resolution, manifested in terms of blurring.

In the previous example, it has been assumed that the finite support of $F(u, v)$ is rectangular. Clearly, the most efficient sampling scheme is achieved when $2B_x\Delta_x = 1$ and $2B_y\Delta_y = 1$, since replicas do not overlap and all the space (u, v) is filled in. This gives the biggest Δ_x and Δ_y that we may choose to reconstruct the sampled image. A smaller Δ_x and Δ_y, i.e., turning in more relaxed sampling conditions, will necessarily end up in aliasing. This is so because the replicas of $F(u, v)$ are maximally packed.

It should be noted, though, that continuous images that are band-limited can have frequency support with any arbitrary shape. Let us assume, for example, that the support of $F(u, v)$ is circular, that is, $F(u, v)$ is zero for $u^2 + v^2 > r^2$, with r be-

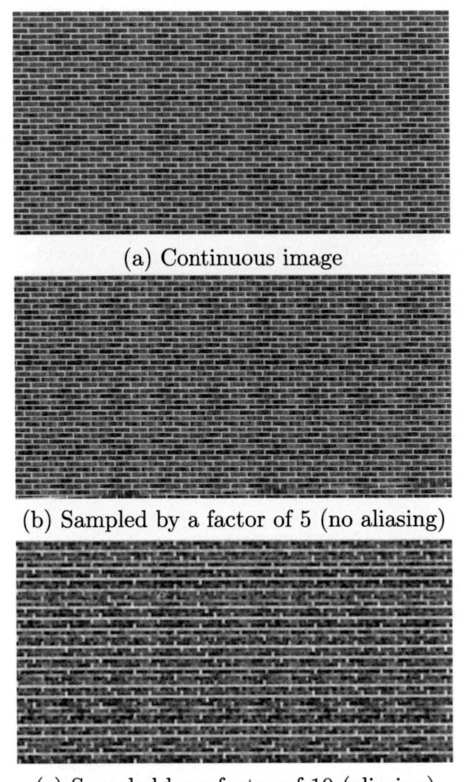

FIGURE 5.11 Example of the effect of sampling.

A continuous image of a brickwall (a) is sampled at a factor of 5x, respecting Nyquist conditions. When sampling frequency is reduced a factor of ten, aliasing occurs, and is manifested in the image domain as artificial texture patterns (moiré patterns).

ing the radius. Clearly, if $2r\Delta_x < 1$ and $2r\Delta_y < 1$, the replicas of $F(u, v)$ do not overlap (Fig. 5.13), and reconstruction is theoretically possible with the same formula as equation (5.26). Nevertheless, the space (u, v) is not completely filled in, and therefore the reader may wonder whether it would be possible to relax the sampling condition by trying to fill in the spectral gaps that are shown in Fig. 5.14.

Evidently, it is impossible to fill in the space (u, v) with a circular spectral support if a rectangular sampling scheme is implemented since replicas are repeated in a rectangular fashion. The branch of geometry that studies how to arrange circles in such a way no overlapping occurs is named circle packing. Since complete packing is impossible, the interesting question is to find the arrangement that obtains the maximum proportion of space covered. Translating in terms of image sampling, the optimal arrangement determines the optimal sampling scheme. It is well known that in a 2D Euclidean space, the optimal arrangement for a circle is a hexagonal tiling.

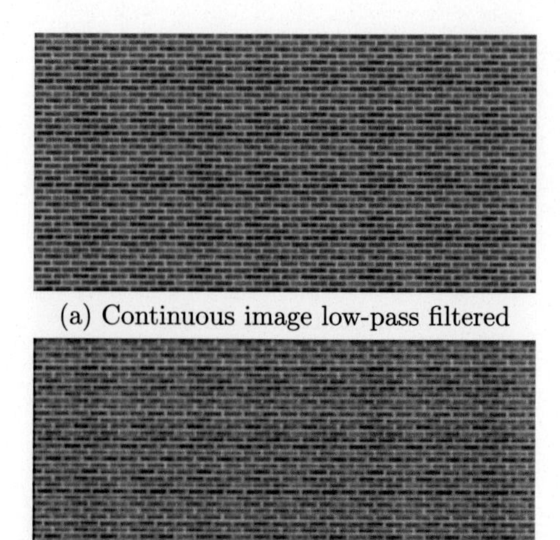

(a) Continuous image low-pass filtered

(b) Sampled by a factor of 10 (no aliasing)

FIGURE 5.12 Example of the effects of sampling and filtering.

When a continuous image is low-pass filtered (a), high-frequency components are left out, and the effect of aliasing is severely reduced when reducing the sampling frequency by a factor of 10 (b).

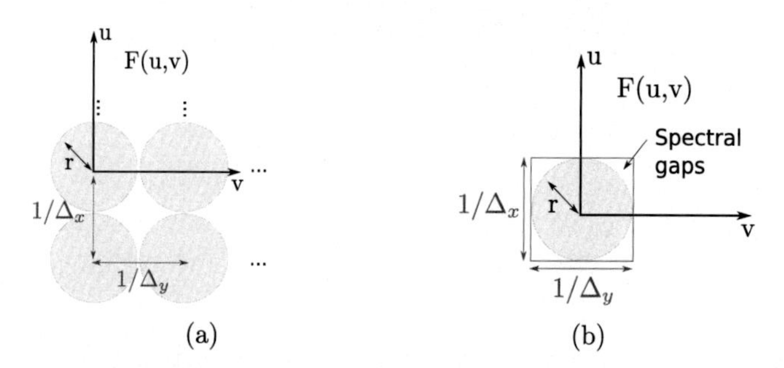

FIGURE 5.13 Effects of sampling on the spectral domain.

(a) Rectangular sampling of an image with circular spectral support. (b) Spectral gaps that appear with rectangular sampling and circular spectral support.

The sampling scheme that produces this type of tiling for a continuous image with circular spectral support is named hexagonal sampling.

In what follows, we will generalize the sampling theory presented above for any non-rectangular sampling scheme with arbitrary spectral support. Then, we will illustrate the general case with both rectangular and circular support, and the concept of Nyquist density will be introduced. Let us start by defining a generalized version

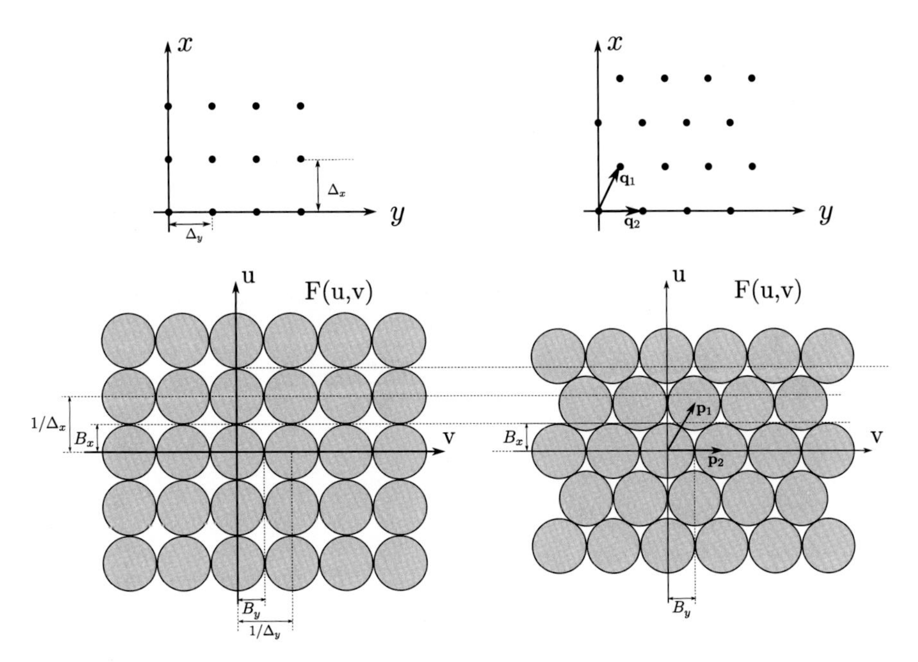

FIGURE 5.14 Rectangular and hexagonal sampling.

Comparison between rectangular sampling (left) and hexagonal sampling (right). Note that circles are maximally packaged with hexagonal sampling.

of the 2D array of Dirac delta functions of equation (5.17):

$$\delta_p(\boldsymbol{x}) = \sum_{m=-\infty}^{\infty} \sum_{n=-\infty}^{\infty} \delta(\boldsymbol{x} - \boldsymbol{Q}\begin{pmatrix} m \\ n \end{pmatrix}), \tag{5.27}$$

where $\boldsymbol{x} = (x, y)^T$, $\boldsymbol{Q} = (\boldsymbol{q}_1, \boldsymbol{q}_2)$ is the so-called sampling matrix, and the 2D column vectors $\boldsymbol{q}_1$ and $\boldsymbol{q}_2$, are the sampling directions. Observe that, indeed, this is a generalization of equation (5.17) since the latter is a special case of equation (5.27) with $\boldsymbol{q}_1 = (\Delta_x, 0)^T$ and $\boldsymbol{q}_2 = (0, \Delta_y)^T$.

The Fourier transform of $\delta_p(\boldsymbol{x})$ is [3]

$$\hat{\delta}_p(\boldsymbol{u}) = |\boldsymbol{P}| \sum_{m=-\infty}^{\infty} \sum_{n=-\infty}^{\infty} \delta(\boldsymbol{u} - \boldsymbol{P}\begin{pmatrix} m \\ n \end{pmatrix}), \tag{5.28}$$

where $\boldsymbol{u} = (u, v)^T$, $\boldsymbol{P} = \boldsymbol{Q}^{-T}$, and $|\boldsymbol{P}|$ is the determinant of $\boldsymbol{P}$. Let us assume as well that the continuous image $f(\boldsymbol{x})$ is band-limited in the sense of $F(\boldsymbol{u}) = 0$ if $\boldsymbol{u} \notin C$, where $F(\boldsymbol{u})$ is its Fourier transform and C is its spectral support. Multiplying

$f(\boldsymbol{x})$ by $\delta_p(\boldsymbol{x})$ and then applying the Fourier transform, it is possible to prove that

$$F_p(\boldsymbol{u}) = |\boldsymbol{P}| \sum_{m=-\infty}^{\infty} \sum_{n=-\infty}^{\infty} F(\boldsymbol{u} - \boldsymbol{P}\begin{pmatrix} m \\ n \end{pmatrix}). \tag{5.29}$$

Therefore, $F(\boldsymbol{u})$ is replicated in the $\boldsymbol{u} = (u, v)^T$ space on tiles with a periodicity matrix $\boldsymbol{P} = (\boldsymbol{p}_1, \boldsymbol{p}_2)$, where vectors $\boldsymbol{p}_1$ and $\boldsymbol{p}_2$ determine the directions of replication of the spectral support $\mathcal{C}$.

Similar to equation (5.23), we can write

$$F(\boldsymbol{u}) = \frac{F_p(\boldsymbol{u})\Pi_{\mathcal{C}}(\boldsymbol{u})}{|\boldsymbol{P}|}, \tag{5.30}$$

with $\Pi_{\mathcal{C}}(\boldsymbol{u})$ defined as

$$\Pi_{\mathcal{C}}(\boldsymbol{u}) = \begin{cases} 1 & \text{if } \boldsymbol{u} \in \mathcal{C}, \\ 0 & \text{otherwise.} \end{cases} \tag{5.31}$$

The inverse Fourier transform equation (5.30) gives

$$f(\boldsymbol{x}) = \sum_{m=-\infty}^{\infty} \sum_{n=-\infty}^{\infty} f[m,n] s_{\mathcal{C}}(\boldsymbol{x} - \boldsymbol{Q}\begin{pmatrix} m \\ n \end{pmatrix}), \tag{5.32}$$

where $f[m,n] = f(\boldsymbol{Q}\begin{pmatrix} m \\ n \end{pmatrix})$ and $s_{\mathcal{C}}(\boldsymbol{x})$ is given by

$$s_{\mathcal{C}}(\boldsymbol{x}) = |\boldsymbol{Q}| \int_{\boldsymbol{u} \in \mathcal{C}} e^{-2\pi i \langle \boldsymbol{u}, \boldsymbol{x} \rangle} d\boldsymbol{u}. \tag{5.33}$$

5.4.3 Nyquist sampling density

A sampling scheme is efficient when the number of samples per unit area is kept low. The number of samples per unit area, or sampling density (SD), is defined by means of the inverse of the area of the parallelogram that is defined by vectors $\boldsymbol{q}_1$ and $\boldsymbol{q}_2$. Such an area is given by $|\boldsymbol{Q}|$; thus, the SD is defined as [3]

$$\text{SD} = \frac{1}{|\boldsymbol{Q}|} = |\boldsymbol{P}| \text{ samples/unit area.} \tag{5.34}$$

Given a particular image $f(\boldsymbol{x})$ with spectral support $\mathcal{C}$, the minimum value of SD that can be achieved with a given sampling scheme without aliasing is known as the *Nyquist sampling density*. There may exist different sampling schemes that achieve *Nyquist sampling density*. If the spectral support $\mathcal{C}$ is rectangular, we have already stated that rectangular sampling achieves the *Nyquist sampling density* if $2B_x\Delta_x = 1$ and $2B_y\Delta_y = 1$. This choice is not unique and other different tilings can be shown to reach the minimum possible density.

In case of circular $\mathcal{C}$, the gain of using hexagonal sampling over rectangular sampling can be substantial. Indeed, if $\mathcal{C}$ is circular with radius r and rectangular sampling is used with $\Delta_x = \frac{1}{2r}$ and $\Delta_y = \frac{1}{2r}$, then

$$\boldsymbol{Q} = \begin{pmatrix} \Delta_x & 0 \\ 0 & \Delta_y \end{pmatrix} = \begin{pmatrix} \frac{1}{2r} & 0 \\ 0 & \frac{1}{2r} \end{pmatrix} \tag{5.35}$$

and

$$\boldsymbol{P} = \begin{pmatrix} \frac{1}{\Delta_x} & 0 \\ 0 & \frac{1}{\Delta_y} \end{pmatrix} = \begin{pmatrix} 2r & 0 \\ 0 & 2r \end{pmatrix} \tag{5.36}$$

given a value of SD of

$$\mathrm{SD}_{\mathrm{rect}} = 4r^2. \tag{5.37}$$

If hexagonal sampling is applied, which is defined through the sampling matrix

$$\boldsymbol{Q} = \begin{pmatrix} T & -T \\ \frac{T}{\sqrt{3}} & \frac{T}{\sqrt{3}} \end{pmatrix} \tag{5.38}$$

yielding

$$\boldsymbol{P} = \begin{pmatrix} \frac{1}{2T} & -\frac{1}{2T} \\ \frac{\sqrt{3}}{2T} & \frac{\sqrt{3}}{2T} \end{pmatrix} \tag{5.39}$$

with $T = \frac{1}{2r}$, then the SD can be shown to be

$$\mathrm{SD}_{\mathrm{hex}} = 2\sqrt{3}r^2. \tag{5.40}$$

The reduction of SD with respect to rectangular sampling is

$$r = \frac{\mathrm{SD}_{\mathrm{hex}}}{\mathrm{SD}_{\mathrm{rect}}} = \frac{2\sqrt{3}r^2}{4r^2} = \frac{\sqrt{3}}{2} = 0.866. \tag{5.41}$$

That is, hexagonal sampling reduces the SD by 13.4% over rectangular sampling. For a given value r there is no other sampling scheme that gives a SD such that

$$\mathrm{SD} \leq \mathrm{SD}_{\mathrm{hex}}. \tag{5.42}$$

Thus $\mathrm{SD}_{\mathrm{hex}}$ is the Nyquist SD.

5.5 Image interpolation

At the beginning of Section 5.4, it was mentioned that sampling theory provides the conditions and rules to choose the sampling settings in order to exactly reconstruct any point of an image $f(x, y)$ from a discrete set of points $f[m, n]$. Such a question is intimately linked to interpolation theory. Interpolation is the task of constructing new points of a given mathematical function from knowledge of a discrete set of points. Image interpolation is thus a subfield of mathematical analysis which deals with the generation of new pixel intensities from a given set of values.

In most of thc cases, any interpolation method can be modeled as [4]

$$\hat{f}(x, y) = \sum_{m=-\infty}^{\infty} \sum_{n=-\infty}^{\infty} f[m, n] h\left(\frac{x - m\Delta_x}{\Delta_x}, \frac{y - n\Delta_y}{\Delta_y}\right), \tag{5.43}$$

where $h(x, y)$ is the interpolation kernel. Observe that $\hat{f}(x, y) \neq f(x, y)$, except for very singular cases, where $h(x, y)$ is then called an ideal interpolation kernel. For example, kernel $h(x, y)$ defined as (see equation (5.26))

$$h(x, y) = r_x r_y \operatorname{sinc}(r_x x) \operatorname{sinc}(r_y y) \tag{5.44}$$

is an ideal interpolation kernel for images with spectral support confined to a rectangle with width B_x and height and B_y. As we have seen, ideal interpolators are impractical since the conditions imposed on the image may not hold, and perhaps more importantly, ideal interpolators do not have limited spatial support, thus, infinite samples are required.

There are multiple options to define $h(x, y)$; however, interpolation methods are required to fulfill a certain list of conditions, which ultimately determine the shape of $h(x, y)$. Here, we just mention two of them [5]:

1. *An image should not be modified if it is interpolated in the same grid.*
   Mathematically, this means that $\hat{f}(m\Delta_x, n\Delta_y) = f[m, n]$ for every m and n. The reader can easily check that such a condition is guaranteed if $h(0, 0) = 1$ and $h(m, n) = 0$ for $|m| = 1, 2, \ldots$ and $|n| = 1, 2, \ldots$. Observe that equation (5.44) meets this condition, as the sinc function is null when evaluated at integer points, except at zero, where its value is 1.
2. *Direct current (DC) preservation: DC-constant interpolator.*
   It is desirable that the DC component of the image is not amplified during the interpolation process. Mathematically, this condition is guaranteed if

$$\sum_{m=-\infty}^{\infty} \sum_{n=-\infty}^{\infty} h(d_x + m, d_y + n) = 1 \tag{5.45}$$

for any displacement $0 \leq d_x, d_y \leq 1$. This condition is also known as the partition of unity condition. It can be demonstrated that $H(0, 0) = 1$ and $H(u, v) = 0$ for $|u| = 1, 2, \ldots$ and $|v| = 1, 2, \ldots$ are necessary conditions for the partition of unity condition to hold.

5.5.1 Typical interpolator kernels

In this section, we briefly describe the most common interpolation methods in medical imaging. Most of the kernels $h(x, y)$ are assumed to be separable, that is, $h(x, y) = h_x(x)h_y(y)$. Hence, we focus on the 1D case, i.e., the shape of $h_x(x)$ or $h_y(y)$. For ease of notation, we call this 1D interpolation kernel $h(x)$, no matter if it refers to the x- or y-coordinate.

5.5.1.1 Windowed sinc

Implementation of the ideal sinc interpolator requires an infinite number of samples. Hence, a straightforward approach is truncating the summand. Truncation of the summand is equivalent to multiplication of the sinc function with a rectangular function. In the frequency domain, this is equivalent to convolution with a sinc function. As a result, truncation of the ideal interpolator produces ringing effects in the frequency domain. The rectangular function is a just a particular example of a window or tapered function, which is a mathematical function that is zero outside of a given interval. More elaborated window functions can be considered to implement the sinc interpolator. Such window functions produce significantly better results. An example of a window function is the Blackman–Harris window function. For $N = 6$ samples, the Blackman–Harris window function $w(x)$ is given by

$$w(x) = \begin{cases} 0.42323 + 0.49755\cos\left(2\pi\frac{x}{N}\right) + 0.07922\cos\left(2\pi\frac{2x}{N}\right) & 0 \leq |x| \leq N/2, \\ 0 & \text{otherwise.} \end{cases} \tag{5.46}$$

A windowed sinc interpolator with the previous window can be shown to be a DC-constant interpolator. The profile of the filter is shown in Fig. 5.15.

Other typical windows functions are Hamming, Hanning, and Lanczos.

5.5.1.2 Nearest neighbor interpolation

Nearest neighbor interpolation is probably the simplest interpolation one can consider. The value $\hat{f}(x, y)$ at point (x, y) is given by the sample $f[m, n]$ whose associated discrete point (m, n) is the closest to (x, y). The way closeness is determined is given by the distance that is chosen. In the conventional nearest neighbor interpolation, the l_1-norm or Manhattan distance is used. With this distance, the closest point (m, n) is simply the discrete point whose components m and n are the nearest integers to x and y, respectively. Nearest neighbor interpolation, which is a DC-constant interpolator and fulfills property 1 of Section 5.5, corresponds to the kernel $h(x)$ (Fig. 5.16) defined as

$$h(x) = \begin{cases} 1 & 0 \leq |x| < 0.5, \\ 0 & \text{otherwise.} \end{cases} \tag{5.47}$$

Strong aliasing and blurring effects are prominent with nearest neighbor interpolation.

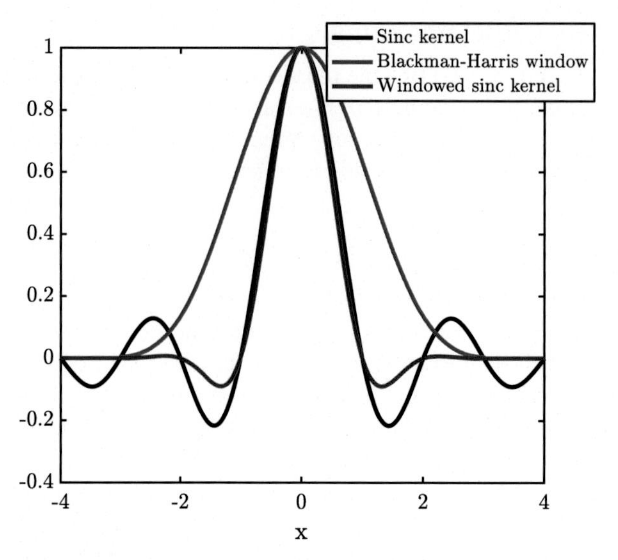

FIGURE 5.15 Interpolation kernel.

Windowed sinc kernel with Blackman–Harris window ($N = 6$).

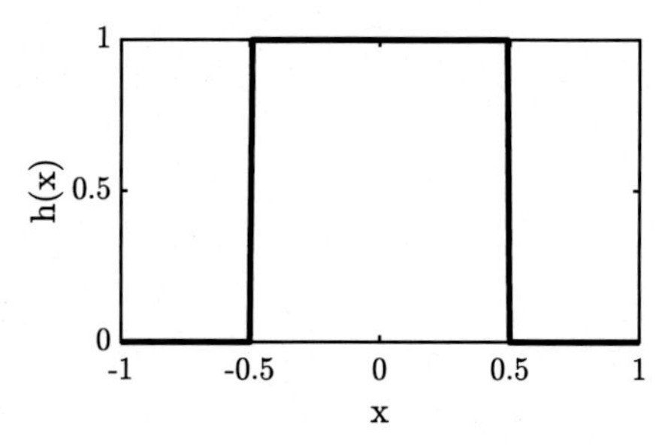

FIGURE 5.16 Interpolation kernel.

Kernel of nearest neighbor interpolation.

5.5.1.3 Linear interpolation

In a linear interpolation scheme, the values of direct neighbors (x and y) are weighted by their distance (absolute value) to the opposite point of interpolation [5]. The associated kernel $h(x)$ is a triangular function (Fig. 5.17):

$$h(x) = \begin{cases} 1 - |x| & 0 \leq |x| < 1, \\ 0 & \text{otherwise.} \end{cases} \tag{5.48}$$

Linear interpolation is a DC-constant interpolation method and obviously meets $h(m) = 0$ for $|m| = 1, 2, \ldots$. Linear interpolation corresponds to a low-pass filter in the frequency domain. Consequently, it attenuates the high-frequency components,

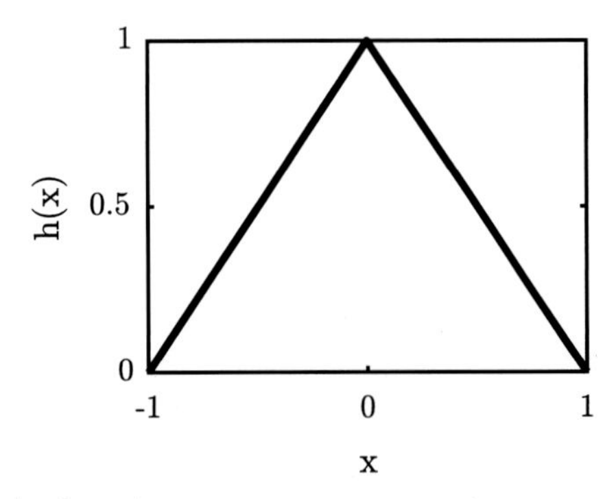

FIGURE 5.17 Interpolation kernel.

Kernel of linear interpolation.

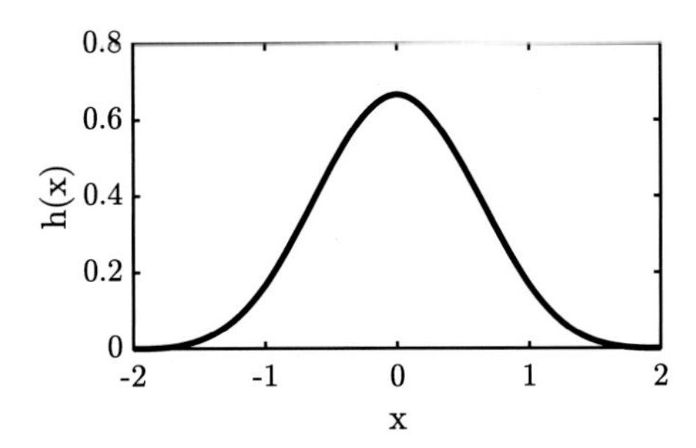

FIGURE 5.18 Interpolation kernel.

Kernel of cubic interpolation.

manifesting as blurring effects in the image domain. In the 2D case, bilinear interpolation is the common nomenclature.

5.5.1.4 Cubic interpolation

Cubic interpolation is defined through the following piecewise cubic polynomial (Fig. 5.18):

$$h(x) = \begin{cases} 1/2|x|^3 - |x|^2 + 2/3 & 0 \leq |x| < 1, \\ -1/6|x|^3 + |x|^2 - 2|x| + 4/3 & 1 \leq |x| < 2, \\ 0 & \text{otherwise}. \end{cases} \tag{5.49}$$

Note that though cubic interpolation is a DC-constant kernel, it does not meet the first property of interpolation kernels. Indeed, $h(0) \neq 1$ and $h(m) \neq 0$, $|m| = 1, 2, \ldots$. Consequently, cubic interpolation causes severe blurring effects. Nevertheless, it presents a favorable stopband response and allows the attenuation of unwanted high-frequency noise in the image [5].

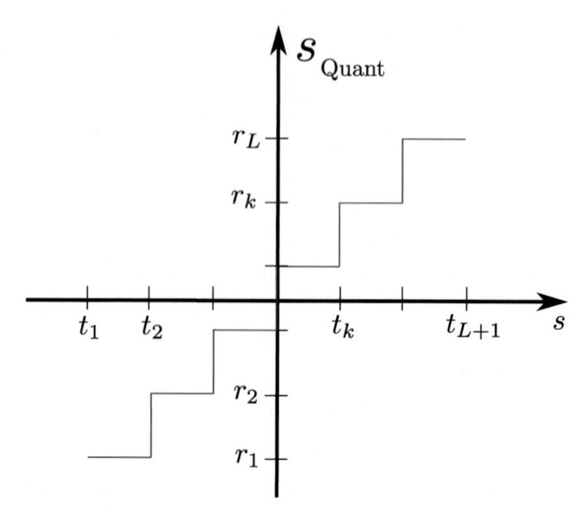

FIGURE 5.19 An image quantizer.

An image quantizer.

5.6 Image quantization

Pixel intensities are represented by bits. Since the number of bits is limited, the continuous range of intensity should be quantized, that is, it should be transformed into a discrete set. This process is called image quantization. Thus, a quantizer transforms the continuous variable that corresponds to the intensity, let us say s, into a discrete variable. Such a discrete variable can only take a finite set of numbers $\{r_1, r_2, \ldots, r_L\}$. A quantizer is determined by the number of levels, that is, the discrete set of numbers that the variable can take, as well as by the decision levels. Given L decision levels $\{t_k, k = 1, \ldots, L+1\}$, the quantizer rule is defined as follows: if s lies in the interval $[t_k, t_{k+1})$, the quantized value is r_k. An example of a quantizer is shown in Fig. 5.19.

If the intervals $[t_k, t_{k+1})$ have the same length, the quantizer is said to be uniform. Clearly, an image quantizer with non-uniform intervals may have a better performance. Indeed, if intensity values u are likely to lie in a given range, it makes sense to design a quantizer with much more decision levels/intervals in that region. As the error is proportional to the length of the intervals, the error is reduced in the selected region. If the probability density function of the random variable u is known, the optimal quantizer in terms of mean square error can be derived analytically. This image quantizer is called Lloyd–Max quantizer. An example of a quantized image is shown in Fig. 5.20. For a more comprehensive treatment of image quantizers, we refer the reader to the excellent reference [1].

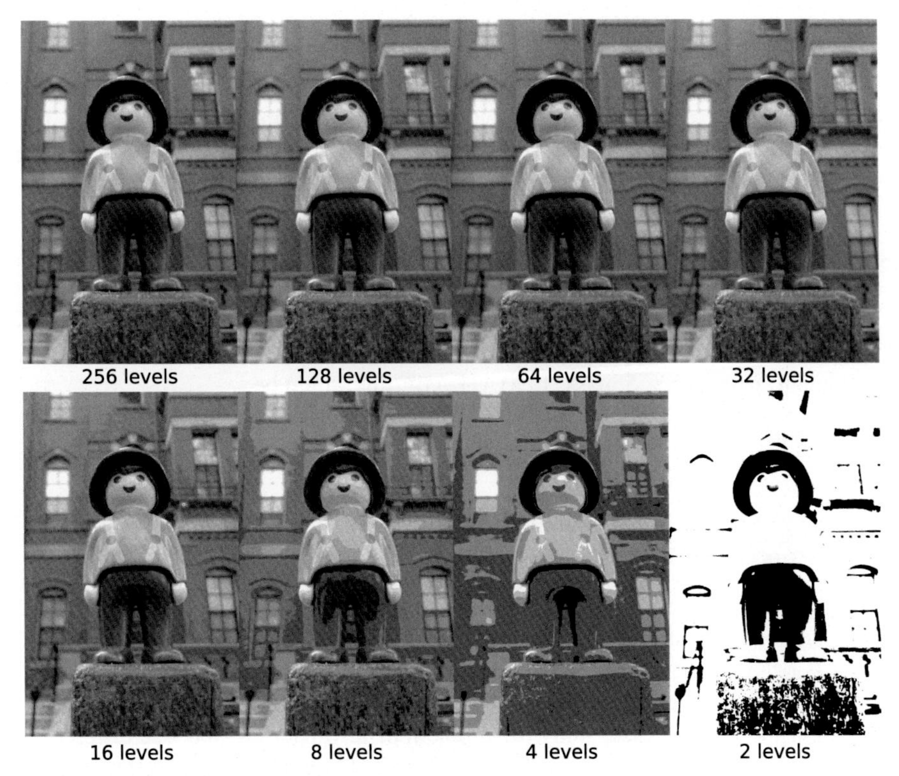

FIGURE 5.20 Example of quantization of a digital image.

The same image is shown with different quantization levels, from 256 (8 bits) to 2 (1 bit).

5.7 Further reading

2D signal processing is a very well-known field inside image processing. The main handbooks on image processing have chapters devoted to these two procedures. We recommend some of them for further reading:

- Basics on signal processing (1D): [6,7].
- 2D signal processing: [1,2,8,9].
- Quantization: [1].
- Fourier analysis: [6,7,10,11].

5.8 Exercises

1. Define a function (using any programing language) that deals with problem of edges caused by the 2D convolution. You are asked to give different options to deal with the discontinuities arising at the edges:

- Use a mirror version of the image.
- Replicate the final line/column.

2. Define a function that calculates the local mean of an image using a mask. The mean must only be done in those regions in which the mask is one. You must take into account (and reduce) any effect that could happen in the edges of the image. Different types of neighborhoods must be considered (square, Gaussian...). For instance:

```
M=local_mean2v(Image,[window size],'square');
```

3. From an image $I(x, y)$ calculate its local mean by convolution (in the spatial domain) with a 15×5 mask. In MATLAB® this would be

```
h = ones([15,5])/75;
If = filter2(h,I);
```

4. Perform filtering in the Fourier domain (by multiplication) and compare the results. Specifically:

   - Study theoretically the problem for the case of a continuous $I(x, y)$ signal. Define the continuous filter in the time and frequency domain.
   - For the discrete case, calculate its theoretical continuous Fourier transform. Plot the sampled theoretical transform versus the FFT.
   - For the discrete case, study the problem of zero-padding, the size differences between the signal in space and in Fourier, the edge effect and the position of the center in the Fourier transform. Also study the edge effect of having 2 limited signals.
   - Describe the complete process to be carried out so that the filtering in space and in frequency are tomally equivalent.

5. Design a function that calculates the gradient of a 1D discrete signal continuously, using the derivative of a linear interpolator. Define an expansion of the function to 2D. Implement the function and compare it to the discrete gradient.

References

[1] Jain A.K., Fundamentals of Digital Image Processing, Prentice Hall, Englewood Cliffs, NJ, 1989.

[2] Pratt W.K., Digital Image Processing, 3rd ed., Wiley, New York, NY, USA, 1991.

[3] Marks R.J., Handbook of Fourier Analysis & Its Applications, Oxford University Press, 2009.

[4] Keys R., Cubic convolution interpolation for digital image processing, IEEE Transactions on Acoustics, Speech, and Signal Processing 29 (6) (1981) 1153–1160.

[5] Lehmann T.M., Gonner C., Spitzer K., Survey: Interpolation methods in medical image processing, IEEE Transactions on Medical Imaging 18 (11) (1999) 1049–1075.

[6] Oppenheim A.V., Willsky A.S., Nawab S.H., Signals & Systems, Prentice-Hall, Inc, Upper Saddle River, NJ, USA, 1996.
[7] Haykin S., Van Veen B., Signals and Systems, 2nd ed., Wiley, 2002.
[8] Gonzalez R.C., Woods R.E., Digital Image Processing, 4th ed., Pearson, 2018.
[9] Lim J.S., Two-Dimensional Signal and Image Processing, Prentice Hall, Englewood Cliffs, NJ, 1990.
[10] Beerends R.J., ter Morsche H.G., van den Berg J.C., van de Vrie E.M., Fourier and Laplace Transforms, 7th ed., Cambridge University Press, 2003.
[11] Tolstov G.P., Fourier Series, Dover Publications Inc., Mineola, NY, 1977.

CHAPTER

Image filtering: enhancement and restoration

6

Santiago Aja-Fernández[a], Ariel H. Curiale[b,c], and Jerry L. Prince[d]

[a]*Image Processing Lab. (LPI), Universidad de Valladolid, Valladolid, Spain*
[b]*Applied Chest Imaging Laboratory, Brigham and Women's Hospital - Harvard Medical School, Boston, MA, United States*
[c]*Medical Physics Department, CONICET - Bariloche Atomic Center, Río Negro, Argentina*
[d]*Electrical and Computer Engineering, Johns Hopkins University, Baltimore, MD, United States*

Learning points

- Image enhancement and restoration
- Point-to-point transformations: e.g., contrast enhancement and histogram equalization
- Spatial operations: e.g., image smoothing filters, border detection
- Operations in transformed domains: linear and homomorphic filtering
- Noise models and image restoration: inverse and Wiener filters

6.1 Medical imaging filtering

The first thing to consider when dealing with filtering in medical imaging, regardless of the modality, is that the data under consideration can incorporate very sensitive information. The knowledge contained within the intensity pattern that conforms the image has not been acquired with esthetic purposes but with a clinical or research aim. Therefore, special care must be taken not to eliminate or modify that information: no filtering procedure should be done with simple artistic or esthetic purposes. Although this premise is clearly shared by most medical imaging researchers, it is sometimes left aside when validating new filtering schemes using *visual comparison*.

From a practical viewpoint, filtering in medical imaging must be *conservative* under the following terms [1]:

1. No significant information present in the image must be erased or modified. For instance, aggressive filtering can eliminate small calcifications in a mammogram,

Medical Image Analysis. https://doi.org/10.1016/B978-0-12-813657-7.00019-4

which could be a risk for diagnosis. In the same way, some filtering methodologies can alter the edges on the image, causing a distortion of objects' sizes, which in the end may cause incorrect volume or distance measures.

2. All information relevant to the physicians must be kept. In many occasions noisy patterns have information useful for the expert. Before *cleaning* a specific area of the image, we must ensure the visual role of noise in diagnosis. For instance, in ultrasound imaging, very aggressive filtering that removes the speckle of the image can also remove valuable information about the mobility of certain structures.
3. No information must be added. Filtering artifacts can appear as a side effect of certain techniques. Sometimes these artifacts can be interpreted as anatomical features, and a false diagnosis can be derived.

Thus, the rule of thumb would be "*if you cannot keep all the important information, do not filter*." Most of the approaches in literature are usually validated via spectacular visual results. However, the success of a filtering procedure is not to produce good-looking pictures, but to ensure that no relevant information is removed.

With this strong requirement in mind, the next step is to consider the final purpose of the filtering of the data. Every proposal in literature will present some advantages and disadvantages, and there is no method that is suitable for all purposes. Thus, image processing must be done only after considering the final use of the filtered image. Let us consider some possible scenarios:

1. *Visual quality*, where the purpose of the filtering is to improve its visual quality. The processing must not only produce *visually pleasing* pictures, but also ease the visual understanding of the data by an expert.
2. *Further processing,* where the purpose is to improve the response of algorithms that will be used to extract information from the filtered images. In this case, the quality of a filtered result is not related to the appearance of the filtered images but instead to its ability to improve the performance of algorithms that use them. Some significant applications are segmentation, measurement of geometrical dimensions, and numerical processing.

There is no all-purpose filter that will perform well in all situations. Also, very simple filtering techniques are often appropriate for the application.

In this chapter, the basic procedures to improve the quality of an image are reviewed. We will make a distinction between image enhancement and restoration:

1. *Image enhancement:* This term includes all those techniques that seek to (1) improve the visual appearance of an image or (2) transform the image to a different shape that is more suitable for analysis. Image enhancement techniques do not assume an underlying model for the image acquisition, and therefore their task is not to improve the fidelity to the original image. Some examples of these techniques are contrast manipulation, histogram processing, and border detection,

2. *Image restoration:* Image restoration is a process that seeks to estimate the original image that has been degraded in the acquisition step. A degradation model is needed and it usually involves blurring of the image and noise. Many different degradation models can be used, and they differ among different modalities.

In this chapter, the first sections are devoted to image enhancement: point-to-point operations, spatial operations, and operations in the transform domain. The second part is focused on image restoration. Finally, some examples of medical image filtering techniques are given.

6.2 Point-to-point operations

Point-to-point operations are commonly used to enhance low-contrast images as well as high-luminance images, but they are not only restricted to these tasks. In these operations, each point (pixel or voxel) is modified according to a particular transformation, T, that only depends on the image intensity at each point (see Fig. 6.1). For example, in a 2D grayscale image, $I(\boldsymbol{x})$, the processed image, $J(\boldsymbol{x})$, is defined by the transformation, T, as follows:

$$J(\boldsymbol{x}) = T(I(\boldsymbol{x})), \tag{6.1}$$

where $\boldsymbol{x}$ corresponds to the spatial pixel location, $\boldsymbol{x} = (x, y)$, the image intensity belongs to $[0, \cdots, L-1]$, and the enhanced image intensity is in the range of $[0, \cdots, L'-1]$, where L and L' are the maximum intensity levels. It is straightforward to make an extension of equation (6.1) to color images or higher dimensions.

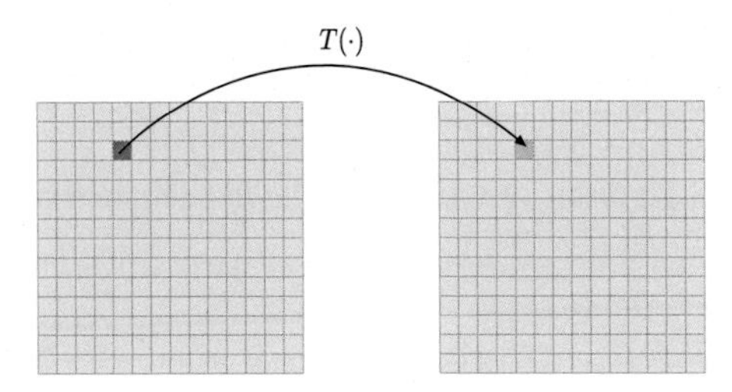

FIGURE 6.1

Illustration of a point-to-point transformation.

6.2.1 Basic operations

Some of the simpler point-to-point operations that can be performed over images are the following ones:

Basic operation: Pointwise addition, subtraction, multiplication, and division:

$$T(I_1(\boldsymbol{x}), I_2(\boldsymbol{x})) = I_1(\boldsymbol{x})\langle\cdot\rangle I_2(\boldsymbol{x}), \qquad \langle\cdot\rangle \in \{+, -, \times, /\}. \tag{6.2}$$

Negative image: The negative of an image is calculated as

$$T(I(\boldsymbol{x})) = (L-1) - I(\boldsymbol{x}), \qquad I \in [0, \cdots, L-1]. \tag{6.3}$$

Operations on color images: A color image is usually defined by three intensity levels. Each of these levels represents an intensity color defined in a specific color mode. The most used color models are RGB (red, green, and blue), YCM (yellow, cyan, and magenta), HSL (hue, saturation, and lightness), and HSV (hue, saturation, and value). Color point-to-point transformations can combine the color intensities, or they can be applied for each color. For example, the following point-to-point operation transforms an RGB image into a grayscale:

$$L = 0.299\,\mathrm{R} + 0.587\,\mathrm{G} + 0.114\,\mathrm{B}, \tag{6.4}$$

where R, G, and B represent the input image intensity for each of the RGB channels and L corresponds to the output image intensity or luminance. An example of this transformation can be seen in Fig. 6.2, where also the negative of the grayscale image according to equation (6.3) is shown.

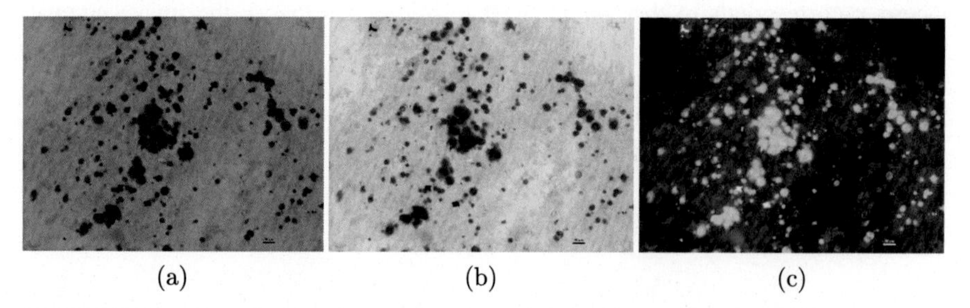

(a) (b) (c)

FIGURE 6.2 Examples of grayscale and negative point-to-point transformations.

(a) Original mesenchymal stem cell. (b) Grayscale image of (a). (c) Negative of the grayscale image in (b).

-Original image courtesy of Dr. Diego Bustos, Cell Signal Integration Lab – IHEM CONICET UNCUYO, Mendoza, Argentina.

6.2.2 Contrast enhancement

In digital images, the dynamic range refers to the ratio between the highest and lowest intensity values, and it is commonly associated with the image contrast. When images with high dynamic ranges are displayed on a monitor, the highest values dominate on the screen. Hence, details around lower-intensity values disappear. This is because

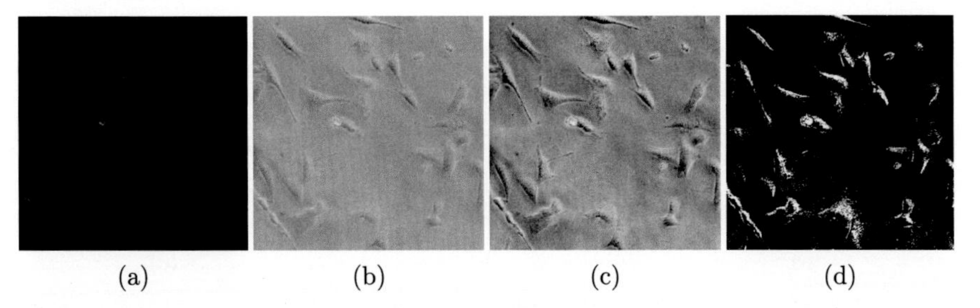

FIGURE 6.3 Point-to-point transformations of a grayscale image.

(a) Dark confocal microscopy of stem cells with high dynamic range. (b) Result of using a log transformation. (c) Result of contrast stretching applied to the image in (b). (d) Result of a threshold applied to the image in (b).

-Original image courtesy of Dr. Diego Bustos, Cell Signal Integration Lab – IHEM CONICET UNCUYO, Mendoza, Argentina.

image intensity values are typically linearly scaled to a fixed number of bits – e.g., 8 bits – when they are acquired or displayed. To avoid this issue and allow us to better distinguish details in bright or dark regions, a logarithm transformation can be applied to compress the dynamic range as follows:

$$T(I(\boldsymbol{x})) = c \cdot \log(1 + I(\boldsymbol{x})), \tag{6.5}$$

where c is a constant. As illustrated in Fig. 6.3, cells cannot be easily distinguished in the dark image (Fig. 6.3(a)) because the image has a very high dynamic range. Note that the intensity values around the air bubble are extremely high compared with those belonging to the stem cells, which makes it quite difficult to identify the cells. However, when a dynamic range compression is carried out, the stem cells show up (Fig. 6.3(b)). Even after this operation, however, it can be seen that the cells still have a low contrast, i.e., their image intensity values are too similar.

Low-contrast images might be acquired for many reasons including poor illumination and incorrect dynamic range in the sensors. Image normalization or contrast stretching aims to improve the contrast in an image by expanding a narrow range of input intensity values into a wide (stretched) range of output intensity values (usually the full range of gray values that can be displayed). The contrast stretching function shown in Fig. 6.4(a) is defined as follows:

$$T(I(\boldsymbol{x})) = \frac{1}{1 + \left(\frac{m}{I(\boldsymbol{x})}\right)^k}, \tag{6.6}$$

where k controls the slope of the function and m corresponds to the intensity value where the stretching will be performed. The result of applying this transformation is a high-contrast image as shown in Fig. 6.3(c). In the limit case, $k \to \inf$, equation (6.6) becomes just a threshold function (Fig. 6.4(b)) and the output results in a binary image (see Fig. 6.3(d)).

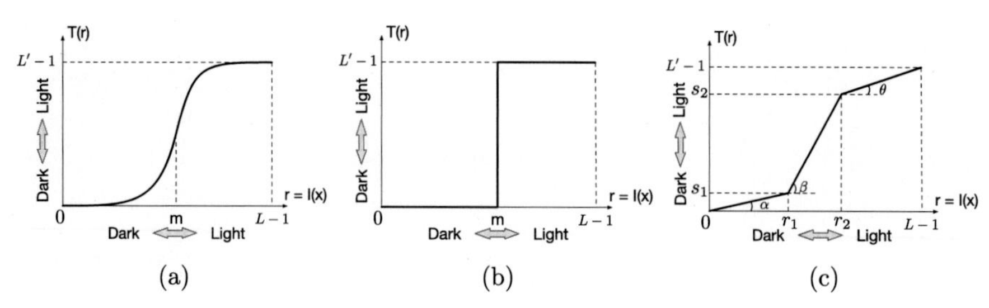

FIGURE 6.4 Common histogram transformation functions.

(a) Contrast stretching transformation. (b) Threshold transformations. (c) Piecewise transformation.

Different contrast stretching functions can be used for contrast enhancement. However, one of the simplest is the piecewise linear function depicted in Fig. 6.4(c). Piecewise transformations use different linear functions to modify the output intensity levels. In fact, a piecewise transformation can be seen as a general contrast stretching transformation. The shape of the transformation is controlled by the number of linear transformations used and its connection points. Fig. 6.4(c) shows a piecewise transformation where three linear transformations are combined at points (r_1, s_1) and (r_2, s_2). If $r_1 = s_1$ and $r_2 = s_2$, then the piecewise transformation reduces to a linear transformation and no change occurs in the image contrast. Also, if $L = L'$ there is no change at all in the image intensity. If $r_1 = r_2$, $s_1 = 0$, and $s_2 = L' - 1$, then it reduces to the threshold function. The following equation is used to perform the piecewise transformation depicted in Fig. 6.4(c):

$$T(r) = \begin{cases} \tan(\alpha)\, r & : 0 \le r < r_1, \\ \tan(\beta)(r - r_1) + s_1 & : r_1 \le r < r_2, \\ \tan(\theta)(r - r_2) + s_2 & : r_2 \le r < L - 1, \end{cases} \tag{6.7}$$

where $\tan(\alpha) = s_1/r_1$, $\tan(\beta) = (s_2 - s_1)/(r_2 - r_1)$, and $\tan(\theta) = (L' - 1 - s_2)/(L - 1 - r_2)$. An example of this transformation with $r_1 = 5$, $r_2 = 100$, $s_1 = 0$, $s_2 = 200$ can be seen in Fig. 6.5(c).

Another useful family of grayscale transformations is commonly named power-law transformations. Power-low transformations have the basic form

$$T(I(\boldsymbol{x})) = c\,(\epsilon + I(\boldsymbol{x}))^{\gamma}, \tag{6.8}$$

where c, γ, and ϵ are constants. The constant ϵ provides an offset which can be used for display calibration, but they are usually set to zero. Therefore, equation (6.8) is usually simplified to $T(I(\boldsymbol{x})) = c\, I(\boldsymbol{x})^{\gamma}$.

Many devices used for capturing and displaying medical images have non-linearities in their response to the input stimulus. The luminance non-linearities introduced by these devices can often be described using a power-law relationship, and the process used to correct this phenomenon is called gamma correction. Images

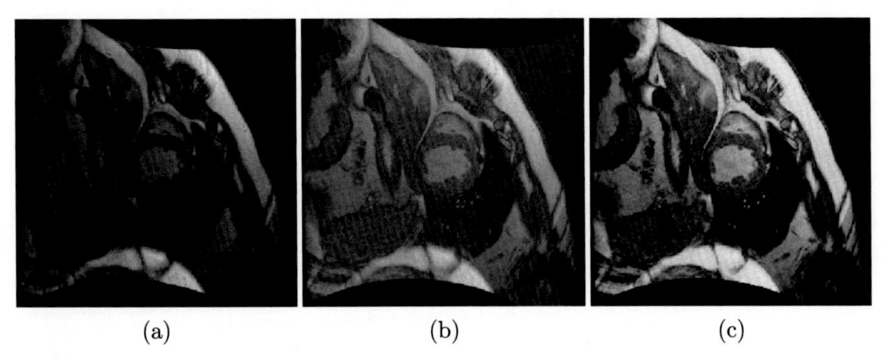

FIGURE 6.5 Power-law and piecewise transformation.

(a) Original cardiac magnetic resonance imaging. (b) The result of using a power-law transformation. (c) The result of using a piecewise transformation.

-Image courtesy of the Cardiac Atlas Project.

that are not properly corrected can look too dark. In addition to gamma correction, power-law transformations are useful for general-purpose contrast manipulation [2]. Fig. 6.5(b) shows an example of a power-law transformation on a cardiac magnetic resonance image with $c = 1$ and $\gamma = 0.6$.

6.2.3 Histogram processing

In what follows we will describe some useful point-to-point operations based on modifying the image histogram. A histogram is an accurate and simple way to estimate the *probability density function* (PDF) for a random variable. It was first introduced by Karl Pearson in [3]. The histogram of an image, I, with L gray levels (i.e., $I(\boldsymbol{x}) \in [0, \cdots, L-1]$) is a discrete function $h(r_k) = n_k$ that describes the relative frequency, n_k, for a particular gray level intensity r_k. In particular, if the histogram is normalized by dividing each of its values by the total number of pixels/voxels in the image, it gives an estimate that a pixel has a particular gray level and $\sum_{r_k=0}^{L-1} h(r_k) = 1$.

The histogram of an image is a powerful tool for describing basic gray level characteristics such as dark, light, low contrast, and high contrast. Fig. 6.6 shows the normalized histogram for these four characteristic images. The vertical axis of each histogram represents the relative number of pixels (it is normalized) with a particular gray level intensity. Dark and bright images tend to have more pixels concentrated in the low- and high-gray level regions, as the histograms show in Fig. 6.6(a and b). Thus, the histogram gives us information about the spatial distribution of the gray levels in an image. It can also provide information about the shape of this distribution. For example, the histogram of low-contrast images shows that most of the pixels are in a narrow range of gray levels (see Fig. 6.6(c)). On the contrary, the gray levels in a high-contrast image tend to be spread into a wide range of gray intensities (Fig. 6.6(d)).

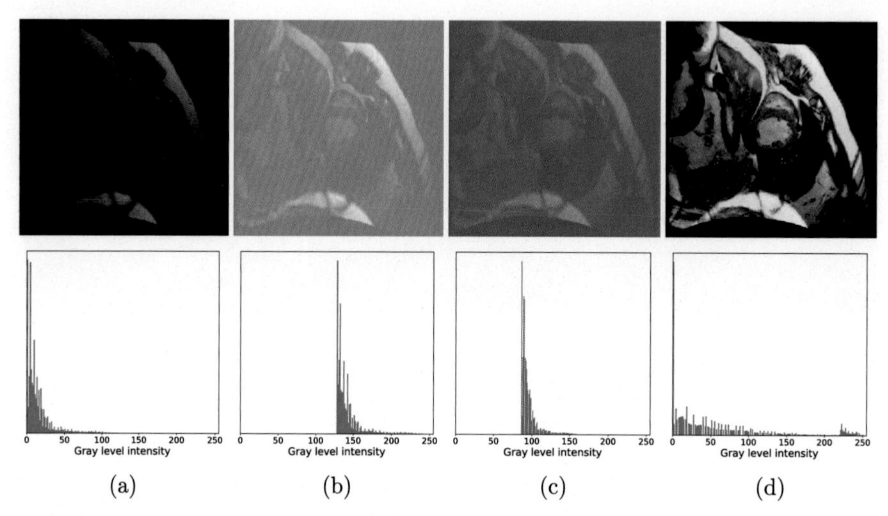

FIGURE 6.6 Basic types of images and their histograms.

(a) Dark cardiac magnetic resonance image. (b) Bright cardiac magnetic resonance image. (c) Low-contrast cardiac magnetic resonance image. (d) High-contrast cardiac magnetic resonance image.

-Original image courtesy of the Cardiac Atlas Project.

Histogram equalization

Histogram equalization aims to find a transformation, T, that properly distributes the input image intensity, where the most frequent intensity levels will be assigned to new intensity levels with higher dynamic range. In fact, the main goal of histogram equalization is to transform the input image intensity in such a way that the intensity levels transformed will be uniformly distributed. Hence, the image contrast is enhanced.

As introduced before, the image intensity level can be seen as a random variable in the range of $[0, L-1]$. Let p_x be the PDF of a random variable x that represents the input intensity level of an image. Now, let p_y be the PDF of a random variable $y \in [0, L-1]$ that represents the output intensity level for a transformation $y = T(x)$ defined as the cumulative distribution function (CDF):

$$y = T(x) = (L-1)\int_0^x p_x(w)dw. \tag{6.9}$$

It is easy to prove that $p_y(y) \sim U[0, L-1]$. A basic result from probability theory is that if $p_x(x)$ and $T(x)$ are known and $T^{-1}(y)$ is single-valued and monotonically increasing in the interval $0 \le y \le L-1$, then the PDF $p_y(y)$ is defined as

$$
\begin{aligned}
p_Y(y) &= p_X(x)\left|\frac{dx}{dy}\right| \\
&= p_X(x)\left|\frac{1}{\frac{dy}{dx}}\right| \\
&= p_X(x)\left|\frac{1}{\frac{dT(x)}{dx}}\right| = \frac{1}{L-1}, \quad y \in [0, L-1], \\
p_Y(y) &\sim U[0, L-1],
\end{aligned}
$$

where

$$\frac{dT(x)}{dx} = (L-1)\, p_X(x). \tag{6.10}$$

It is important to note that unlike its continuous counterpart, the discrete CDF

$$y = T(x) = (L-1)\sum_{j=0}^{x} p_X(j), \quad j \in [0, L-1], \tag{6.11}$$

where

$$p_X(x) = \frac{n_x}{n}, \quad x \in [0, L-1], \tag{6.12}$$

cannot produce a discrete equivalent of a uniform PDF, which would be a uniform histogram. However, as shown in Fig. 6.7, the use of equation (6.11) does have the general tendency of spreading the histogram of the input image, which enhances the image contrast, as expected. Note that the CDF of the equalized image (i.e., histogram equalization) plotted in red over the histogram in Fig. 6.7 is similar to a uniform distribution.

Histogram specification

Histogram equalization automatically determines a transformation that distributes the input image intensity to make a new image with a uniform histogram. However, there are some applications in which attempting to perform such transformation is not the best approach for image enhancement. In fact, we sometimes find it useful to *specify* the shape of the histogram that we want to get. The method used for this task is known as *histogram specification* or *matching*.

Using the same notation as before, let p_X be the PDF of a continuous random variable $x \in [0, L-1]$ that represents the image intensity level. Now, let p_Z be the specified PDF of another continuous random variable, $z \in [0, L-1]$, that represents the desired image intensity level. Suppose next that we define the transformation $y = T(x)$ as the CDF in a similar way as for histogram equalization:

$$y = T(x) = \int_0^x p_X(w)dw. \tag{6.13}$$

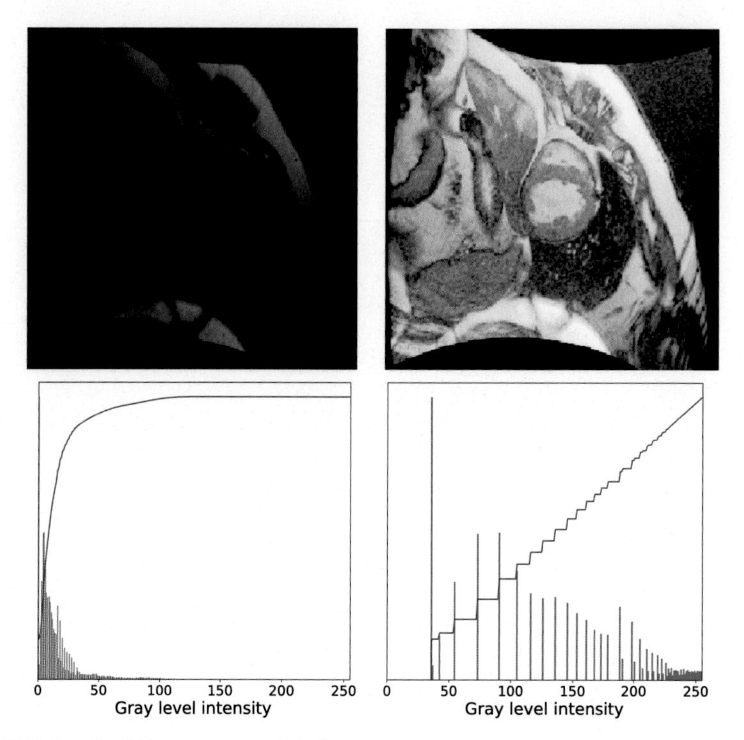

FIGURE 6.7 Result of histogram equalization.

Dark cardiac magnetic resonance image (left) and the result of a histogram equalization (right), where the cumulative distribution function is plotted in red over the histogram (bottom).

-Original image courtesy of the Cardiac Atlas Project.

It is important to note that now $y \in [0, 1]$ instead of $[0, L-1]$. Then, we define the transformation $G(z)$ as the CDF of p_z as

$$G(z) = \int_0^z p_z(w)dw \tag{6.14}$$

and $y = G(z) = T(x)$. Therefore, z must satisfy the following condition:

$$z = G^{-1}(y) = G^{-1}(T(x)). \tag{6.15}$$

Assuming that G^{-1} exists and that it is single-valued and monotonically increasing, it is possible to transform the image gray levels from the original image to get an image according to the specified PDF p_z by using G^{-1} and T. Note that the problem of finding G^{-1} is considerably simplified for the discrete case. However, in this case, only an approximation of the desired histogram is achieved. Since the gray levels in the image are integers, a simple approach can be used to estimate $\hat{z} = G^{-1}(y_k)$. Indeed, a good approximation is obtained by finding the smallest integer $\hat{z} \in [0, L-1]$ such that

$$(G(\hat{z}) - y_k) \geq 0, \quad k = 0, \ldots, L-1. \tag{6.16}$$

Fig. 6.8 shows an example of the histogram specification approach (Fig. 6.8(d)) for a desired p_z (plotted in red in Fig. 6.8(e)). In particular, this example shows that histogram equalization is not the best approach for improving the contrast (Fig. 6.8(b)). Furthermore, it shows that histogram specification avoids the saturation issue introduced when histogram equalization is performed.

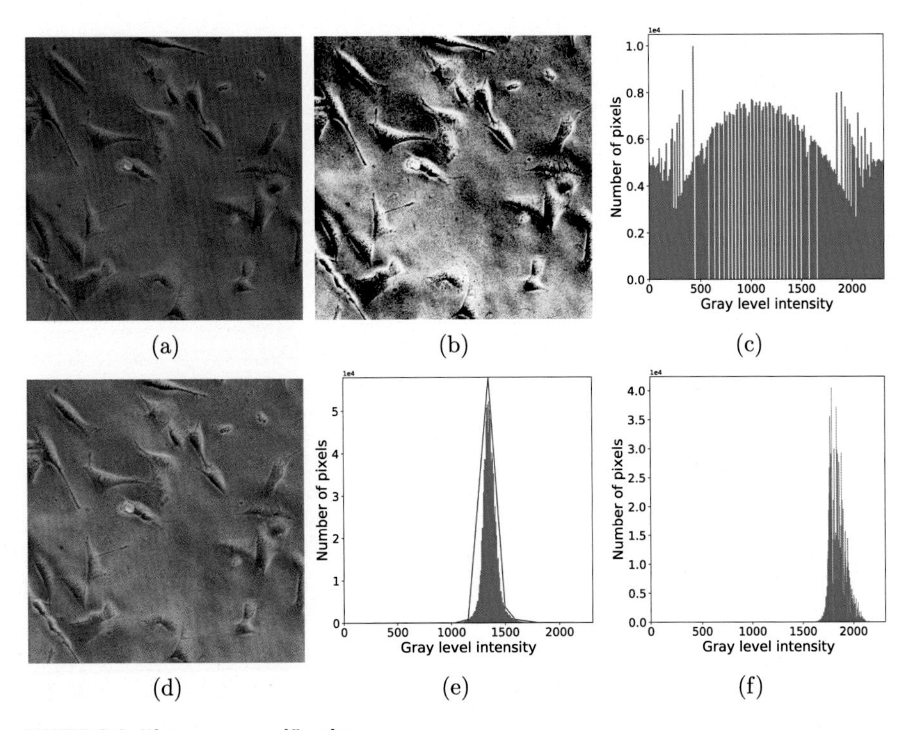

FIGURE 6.8 Histogram specification.

(a) Original confocal microscopy of stem cells. (b) Result of a histogram equalization. (c) Histogram of (b). (d) Result of histogram specification according to the desired distribution plotted in red on the image histogram (e). (f) Histogram of the specified histogram image.

-Original image courtesy of Dr. Diego Bustos, Cell Signal Integration Lab – IHEM CONICET UNCUYO, Mendoza, Argentina.

6.3 Spatial operations

The terms *spatial operations*, *spatial transformations*, and *spatial filtering* are described by operations that take multiple pixels in an image to yield a single value or multiple values. Approaches in this category commonly use information of a neighborhood around a pixel to be transformed, as depicted in yellow in Fig. 6.9. This neighborhood is commonly called a *mask*, *filter*, or *kernel*, and their values are referred to as *coefficients*. Following the same notation that we have introduced in point-to-point operations, each point (pixel or voxel) in the input image is transformed according to a spatial transformation, T, that takes into account the information of the point and their neighborhood according to the values defined in the filter or kernel as follows:

$$J(\boldsymbol{x}) = T(I(\boldsymbol{x}), W(\boldsymbol{x})), \tag{6.17}$$

where $I(\boldsymbol{x})$ is the intensity level of the point $\boldsymbol{x}$ and $W(\boldsymbol{x})$ corresponds to the spatial kernel to be used to process the neighborhood of $\boldsymbol{x}$.

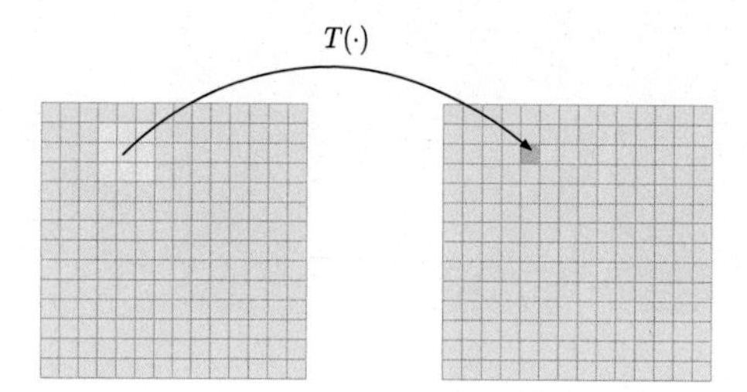

FIGURE 6.9

Schematic spatial transformation.

6.3.1 Linear filtering

Most of the spatial operations discussed in this section can be formulated as linear operations between the input image $I(\boldsymbol{x})$ and the kernel $W(\boldsymbol{x})$ as follows:

$$J(x, y) = \sum_{i=-n}^{n} \sum_{j=-m}^{m} W(i, j) \cdot I(x+i, y+j), \tag{6.18}$$

where $W(x, y)$ is a 2D kernel with a size of $(2n+1, 2m+1)$. An example of a generic 3×3 kernel is shown in Fig. 6.10.

From a signal processing point of view, this processing is equivalent to a convolution[1] between the signal – image $I(\boldsymbol{x})$ – and the impulse response of the

[1] Theoretically, it will only be a *convolution* if the kernel $W(\boldsymbol{x})$ is symmetrical. Otherwise, it will be a *correlation*.

$w(-1,-1)$	$w(-1,0)$	$w(-1,1)$
$w(0,-1)$	$w(0,0)$	$w(0,1)$
$w(1,-1)$	$w(1,0)$	$w(1,1)$

FIGURE 6.10 Generic 3 × 3 filter mask.

Generic 3 × 3 filter mask.

system – kernel $W(\boldsymbol{x})$ –

$$J(\boldsymbol{x}) = I(\boldsymbol{x}) * W(\boldsymbol{x}), \tag{6.19}$$

where "$*$" stands for the convolution. In this way, a filter mask is sometimes called a *convolution mask* or *convolution kernel*. This will be further discussed in Section 6.4.

Regarding implementation details, it is important to consider what happens when the center of the filter approaches the borders of the image. When the convolution mask reaches the image borders – where points in $I(x+i, y+j)$ might be found outside the image domain – one possible action is to restrict the center of the mask to be a distance no less than $(n-1)/2$, where n is the mask size in the dimension of the border that is reached. The resulting image will be smaller than the original, but the processed pixels will have been computed with the full mask. Other possible solutions are to complete the pixels outside the image with 0 or to repeat the border values beyond the image domain. This operation is commonly called padding. Usually, the best option depends on both the problem that is being addressed and the particular filter that is being used. For example, for most linear spatial operations, zero-padding is commonly used. However, for the median filter, which is a non-linear filter, the use of zero-padding can introduce artifacts along the image borders.

In what follows, we introduce some of the most common spatial operations used for smoothing, border detection, and sharpening. Furthermore, we will describe some non-linear spatial transformation for similar purposes.

Smoothing filters

Smoothing filters are designed to remove small details from images. These filters are commonly used in image processing for object detection or noise reduction. The mean filter, which simply assigns the average value of its neighborhood to each point in an image, is the most intuitive and simplest such filter. Fig. 6.11 shows an example

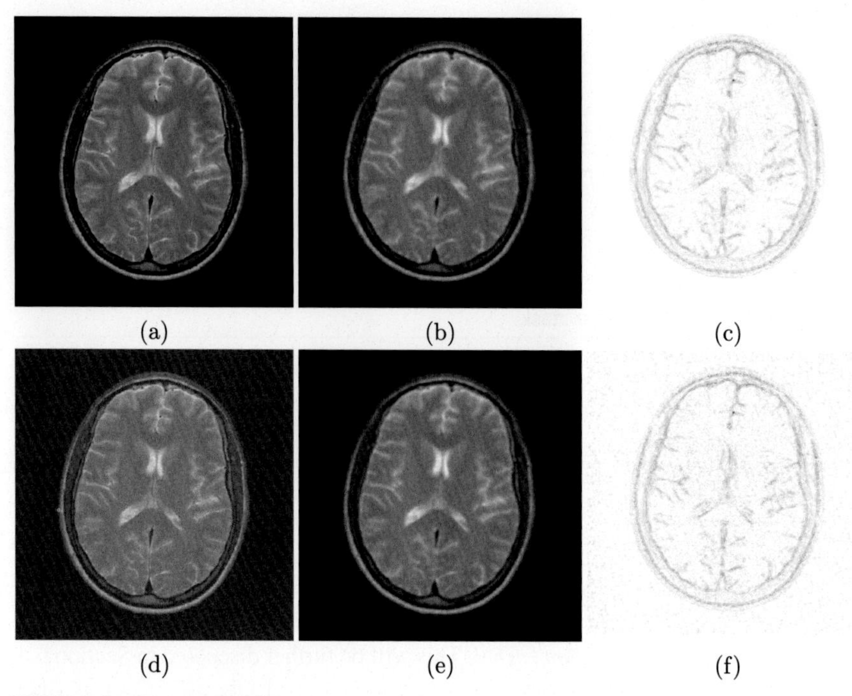

FIGURE 6.11 Average filtering.

(a) Original brain magnetic resonance image. (b) Result of smoothing the image in (a) with a squared mask of 5×5. (c) Differences between (a) and (b). (d) Original image corrupted with white noise. (e) Result of smoothing the noisy image in (d) with a squared mask of 5×5. (f) Differences between the original image (a) and the smoothed noise image (e).

-Original image courtesy of Hospital Clínico Universitario de Valladolid, Spain.

of white noise[2] reduction by using a mean filter with the following mask:

$$W(\boldsymbol{x}) = \frac{1}{25}\begin{bmatrix} 1 & 1 & 1 & 1 & 1 \\ 1 & 1 & 1 & 1 & 1 \\ 1 & 1 & 1 & 1 & 1 \\ 1 & 1 & 1 & 1 & 1 \\ 1 & 1 & 1 & 1 & 1 \end{bmatrix},$$

which can be rewritten as follows:

$$J(x, y) = \frac{1}{25} \sum_{i=-2}^{2} \sum_{j=-2}^{2} I(x+i, y+j). \tag{6.20}$$

[2] Details on noise models can be found in Section 6.5.

It is important to note that all coefficients in the mean filter contribute in the same way to the result. As shown in Fig. 6.11(e), this filter is effective in removing white noise; however, it also tends to remove small objects or details (Fig. 6.11(f)), which may not be desirable. To address this, a weighted mean filter might be used. In this variation, not all the pixels in the neighborhood contribute to the result in the same way. For example, the mask

$$W(\boldsymbol{x}) = \frac{1}{16}\begin{bmatrix} 1 & 1 & 1 \\ 1 & 8 & 1 \\ 1 & 1 & 1 \end{bmatrix}$$

yields a result in which the central pixel is weighted eight times higher than the others. A special case of a weighted mean operator is Gaussian smoothing, where the kernel is defined by the Gaussian function

$$W(\boldsymbol{x}) = \frac{\exp(-\frac{1}{2}(\boldsymbol{x}-\boldsymbol{\mu})^T \Sigma^{-1}(\boldsymbol{x}-\boldsymbol{\mu}))}{\sqrt{(2\pi)^2|\Sigma|}}, \tag{6.21}$$

where $\boldsymbol{\mu}$ and Σ correspond to the multidimensional mean and covariance matrix, respectively. In fact, if the dimensions are uncorrelated (i.e., Σ is a diagonal matrix), the n-dimensional Gaussian filter can be generated with only a 1D Gaussian filter. Note that the kernel should be carefully designed to include at least 2σ (or 3σ) of the distribution values. In this sense, its size must increase with increasing σ to maintain the Gaussian nature of the filter. Fig. 6.12 shows an example of a 2D Gaussian convolution kernel with $\mu_x = \mu_y = 0$ and $\sigma_x = \sigma_y = 1$, where 3σ of the distribution values are included. Additionally, results of applying a 2D Gaussian smoothing filter for different isotropic $\sigma \in [1, 2, 4]$ are depicted in Fig. 6.13.

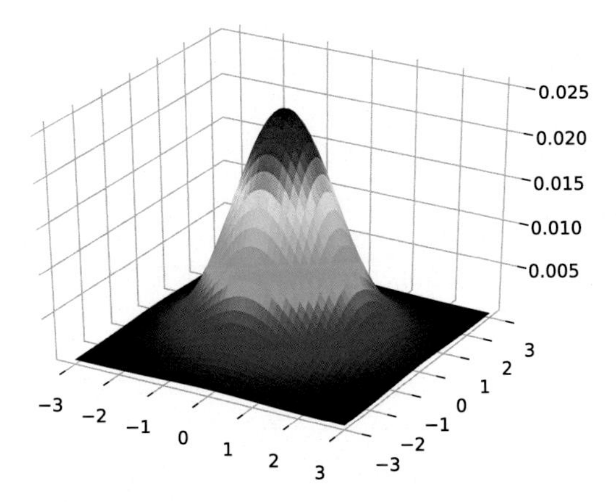

FIGURE 6.12 2D Gaussian kernel.

Example of a 2D Gaussian kernel with $\mu_x = \mu_y = 0$ and $\sigma_x = \sigma_y = 1$.

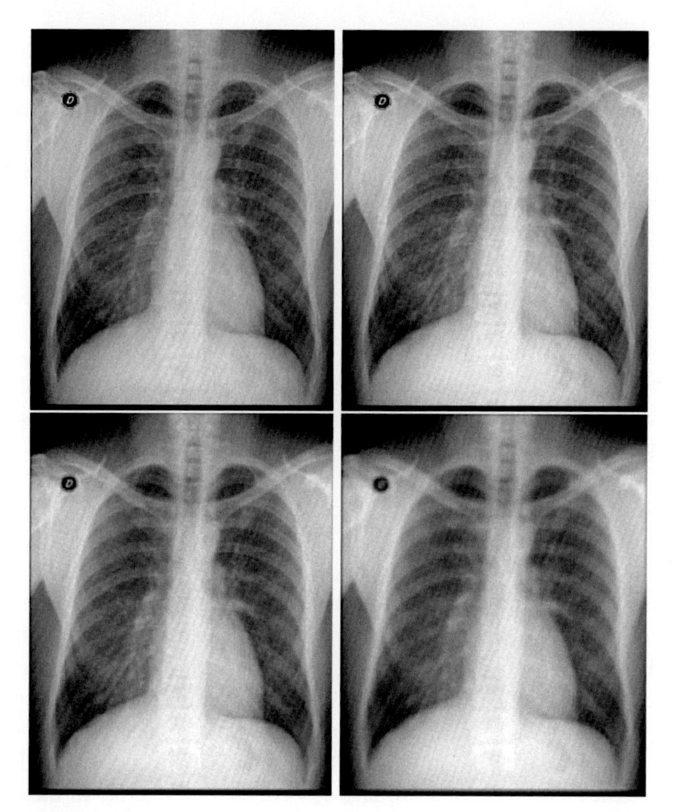

FIGURE 6.13 Gaussian filtering.

Results of applying a Gaussian smoothing filter to a chest X-ray image (upper left) for different isotropic $\sigma \in [1, 2, 4]$.

-Original image courtesy of Hospital Universitario UNCUYO, Mendoza, Argentina.

Highlighting borders and small details

The purpose of the spatial filters described in this section is to highlight the boundaries of objects and small details in the images. These filters can be thought of as the opposite of spatial smoothing filters and are based on first- and second-order derivatives, especially the gradient and Laplacian

$$\nabla I(\boldsymbol{x}) = \left(\frac{\partial I(\boldsymbol{x})}{\partial x_1}, \cdots, \frac{\partial I(\boldsymbol{x})}{\partial x_n} \right), \tag{6.22}$$

$$\Delta I(\boldsymbol{x}) = \nabla^2 I(\boldsymbol{x}) = \nabla \cdot \nabla I(\boldsymbol{x}) = \sum_i^n \frac{\partial^2 I(\boldsymbol{x})}{\partial^2 x_i}, \tag{6.23}$$

where $\boldsymbol{x} \in \mathbb{R}^N$.

The spatial derivatives of an image provide an easy way to identify where the image intensities show high variations. Possible definitions of the first-order spatial derivatives of a 2D image are the finite differences

$$\frac{\partial I(x,y)}{\partial x} \approx I(x+1,y) - I(x,y), \tag{6.24}$$

$$\frac{\partial I(x,y)}{\partial y} \approx I(x,y+1) - I(x,y). \tag{6.25}$$

There are different convolution masks that implement equations (6.24) and (6.25), but the most well known are the Prewitt, Roberts, and Sobel masks (see Fig. 6.14).

$$\underset{(a)}{\begin{bmatrix} -1 & 1 \end{bmatrix}} \quad \underset{(b)}{\begin{bmatrix} 1 \\ -1 \end{bmatrix}} \quad \underset{(c)}{\begin{bmatrix} 1 & 0 \\ 0 & -1 \end{bmatrix}} \quad \underset{(d)}{\begin{bmatrix} 0 & 1 \\ -1 & 0 \end{bmatrix}}$$

$$\underset{(e)}{\begin{bmatrix} -1 & -1 & -1 \\ 0 & 0 & 0 \\ 1 & 1 & 1 \end{bmatrix}} \quad \underset{(f)}{\begin{bmatrix} -1 & 0 & 1 \\ -1 & 0 & 1 \\ -1 & 0 & 1 \end{bmatrix}} \quad \underset{(g)}{\begin{bmatrix} -1 & -2 & -1 \\ 0 & 0 & 0 \\ 1 & 2 & 1 \end{bmatrix}} \quad \underset{(h)}{\begin{bmatrix} -1 & 0 & 1 \\ -2 & 0 & 2 \\ -1 & 0 & 1 \end{bmatrix}}$$

FIGURE 6.14 Examples of kernel masks used for border detection.

Different masks used for computing a 2D gradient vector. (a, b) Partial derivatives. (c, d) Roberts operators. (e, f) Prewitt kernel filters. (g, h) Sobel convolution masks.

Similarly, we can define the second-order derivative as

$$I_{xx} = \frac{\partial^2 I(x,y)}{\partial x^2} \approx I(x+1,y) - 2I(x,y) + I(x-1,y), \tag{6.26}$$

$$I_{yy} = \frac{\partial^2 I(x,y)}{\partial y^2} \approx I(x,y+1) - 2I(x,y) + I(x,y-1). \tag{6.27}$$

As in the computation of the gradient, there are several 2D masks that can be used to approximate the Laplacian (equation (6.23)) of an image, as shown in Fig. 6.15.

$$\underset{(a)}{\begin{bmatrix} 0 & 1 & 0 \\ 1 & -4 & 1 \\ 0 & 1 & 0 \end{bmatrix}} \quad \underset{(b)}{\begin{bmatrix} 0 & -1 & 0 \\ -1 & 4 & -1 \\ 0 & -1 & 0 \end{bmatrix}} \quad \underset{(c)}{\begin{bmatrix} 1 & 1 & 1 \\ 1 & -8 & 1 \\ 1 & 1 & 1 \end{bmatrix}} \quad \underset{(d)}{\begin{bmatrix} -1 & -1 & -1 \\ -1 & 8 & -1 \\ -1 & -1 & -1 \end{bmatrix}}$$

FIGURE 6.15 Laplacian masks.

Laplacian masks without (a, b) and with (c, d) diagonal elements.

The main goal of boundary detection is to highlight transitions between object boundaries in the image. A simple way to highlight these transitions is by using the magnitude of a first-order derivative, as shown in Fig. 6.16(b). In the same way, to

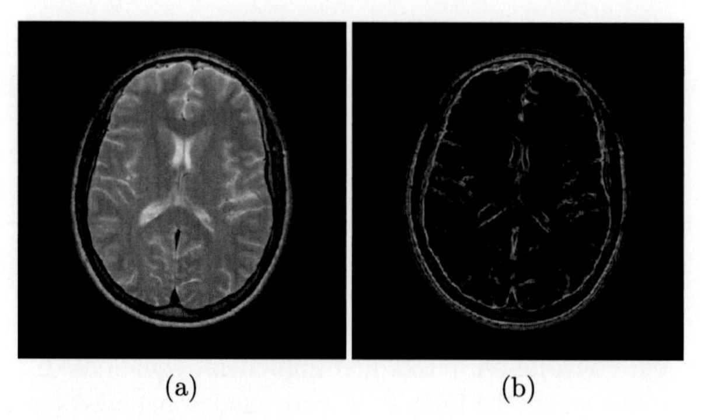

FIGURE 6.16 Example of border detection.

(a) Original brain magnetic resonance image. (b) Sobel gradient of (a).

-Original image courtesy of Hospital Clínico Universitario de Valladolid, Spain.

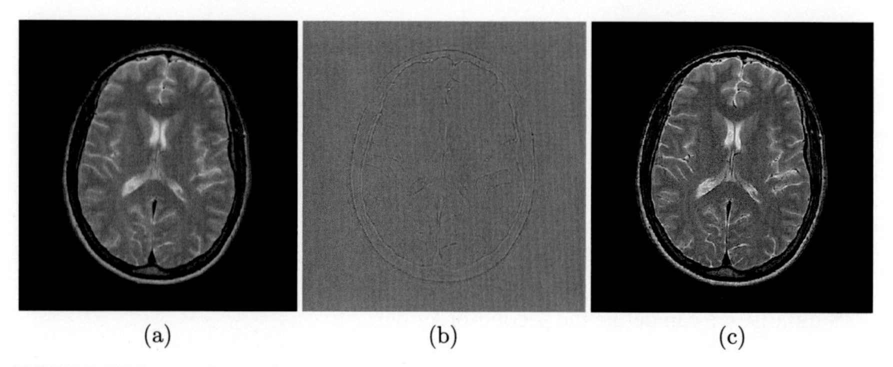

FIGURE 6.17 Image sharpening.

(a) Original brain magnetic resonance image. (b) Laplacian of (a). (c) Sharpening of (a).

-Original image courtesy of Hospital Clínico Universitario de Valladolid, Spain.

highlight small details in the image (Fig. 6.17(c)) it is common to add a second-order derivative such as the Laplacian to the original image as

$$J(\boldsymbol{x}) = I(\boldsymbol{x}) - c\,\nabla^2 I(\boldsymbol{x}) \tag{6.28}$$

when the center coefficient of the Laplacian mask is negative or as

$$J(\boldsymbol{x}) = I(\boldsymbol{x}) + c\,\nabla^2 I(\boldsymbol{x}) \tag{6.29}$$

when it is positive. For both cases, the sharpening effect is controlled by the constant c. Another option to highlight small details in images is known as high-boost filtering, which is defined using a blurred version of the original image, $\overline{I}$, as follows:

$$J(\boldsymbol{x}) = a\,I(\boldsymbol{x}) - \overline{I}(\boldsymbol{x}), \quad a \geq 1. \tag{6.30}$$

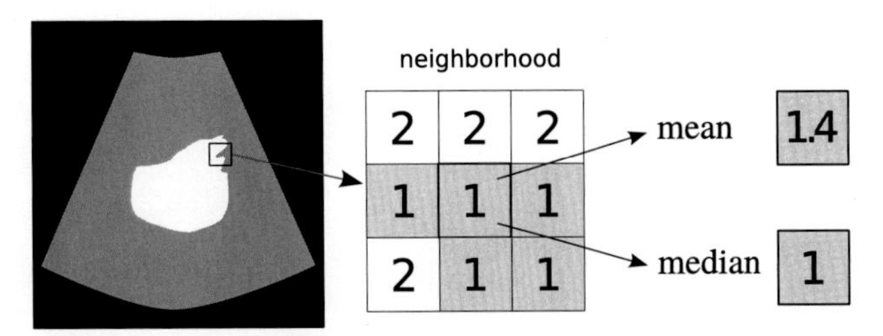

FIGURE 6.18 Schematic median filtering example.

Example of median filtering in a 3 × 3 neighborhood compared to a mean filter of the same area.

6.3.2 Non-linear filters

Non-linear spatial filters such as max, min, median, variance, or thin-structure filters also operate on neighborhoods, and the mechanism of sliding a mask over the image is the same. However, the transformation T that yields a pixel value from itself and its neighborhood is non-linear, unlike convolution.

Median filter

The transformation T corresponding to the median filter simply replaces the pixel value with the median value of the pixel and its neighborhood, as in

$$J(\boldsymbol{x}) = \underset{\boldsymbol{k}\in W(\boldsymbol{x})}{\text{median}} \{I(\boldsymbol{k})\}, \tag{6.31}$$

where $W(\boldsymbol{x})$ is a neighborhood centered in $\boldsymbol{x}$. Assuming an $n \times n$ window, the median filter works as follows:

1. The n^2 values of the pixels in the neighborhood are extracted.
2. These values are ordered.
3. The output of the filter corresponds to the value placed in the $\frac{n^2+1}{2}$ position.

The median filter is a powerful tool for impulsive noise reduction and removal of isolated outliers.

An illustration of the action of the median filter in comparison to the mean filter is shown in Fig. 6.18. It can be observed in Fig. 6.18 that in contrast to the mean filter (and any other linear filter), the output values of the median filter are always values that are already present in the image. This property has two advantages: first, no new values are introduced into the image; second, there is no smoothing or blurring across edges. The visual impact of this property can be observed in Fig. 6.19, which compares a Gaussian filter (which is a linear filter) with the median filter. A better preservation of the edges can be seen in the output of the median filter.

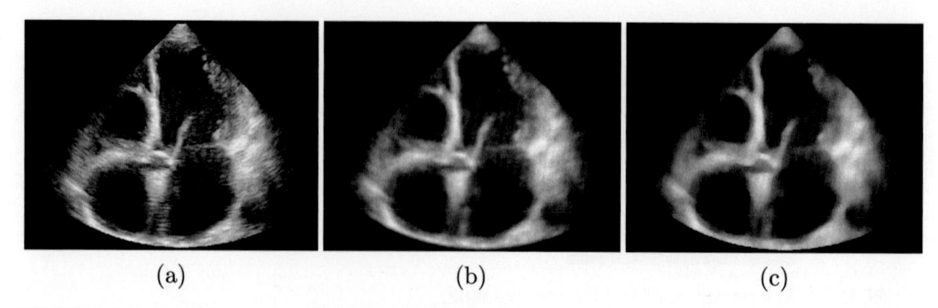

FIGURE 6.19 Gaussian vs. median filtering example.

(a) Four-chamber cardiac ultrasound image. (b) Gaussian smoothing filter. (c) Median filter.

-Original image courtesy of Dr. T. Perez Sanz, Río Hortega, Valladolid, Spain.

Pseudomedian filter

One of the main drawbacks of the median filter is its computational cost, with the number of operations growing exponentially with the size of the window. A computational alternative would be the so-called *pseudomedian* filter [4]. If $\{M_N\}$ is a sequence of elements $m_1, m_2, \cdots, m_N$, the pseudomedian of the sequence is defined as

$$\text{pmed}\{M_N\} = \frac{\text{maximin}\{M_N\} + \text{minimax}\{M_N\}}{2}, \tag{6.32}$$

where

$$\text{maximin}\{M_N\} = \max\{\min(m_1, \cdots, m_L), \min(m_2, \cdots, m_{L+1}), \\ \cdots, \min(m_{N-L+1}, \cdots, m_N)\}$$

and

$$\text{minimax}\{M_N\} = \min\{\max(m_1, \cdots, m_L), \max(m_2, \cdots, m_{L+1}), \\ \cdots, \max(m_{N-L+1}, \cdots, m_N)\},$$

with $L = (N+1)/2$.

Only when the data present a symmetric distribution, the pseudomedian coincides with the median.

6.4 Operations in the transform domain

Many of the operations previously introduced can be carried out in the transform domain as opposed to the presented spatial domain. Different transforms can be considered, but in this chapter we will briefly consider two of them: Fourier and homomorphic transforms. Other transforms that are of paramount importance in image processing are the wavelet and cosine transforms.

6.4.1 Linear filters in the frequency domain

As seen in previous chapters, there is an equivalence between convolution in the spatial domain and multiplication of the Fourier transforms of the signals. The linear processing of an image using a kernel as described in equation (6.18) can be written as a convolution,

$$J(x,y) = I(x,y) * h(x,y),$$

where $h(x, y)$ is the impulse response of the system, i.e., the convolution kernel. If continuous signals are assumed, the Fourier transform of the convolution is carried out by the product of every term:

$$J(u,v) = I(u,v) \cdot H(u,v). \tag{6.33}$$

However, in practical situations the images are band limited and discrete signals, and the discrete Fourier transform (DFT) is used. In this case, the equivalent of this operation in the frequency domain implicitly assumes a periodic expansion of the signal, a linear and circular shift-invariant (as seen in the previous chapter) system, and a circular convolution instead:

$$J(x,y) = I(x,y) \circledast h(x,y).$$

The DFT of this convolution is

$$J[k_1,k_2] = I[k_1,k_2] \cdot H[k_1,k_2], \tag{6.34}$$

where $W[k_1, k_2]$ is the DFT of $h(x, y)$. As a consequence, processing can alternatively be carried out into the frequency domain. Note that in the Fourier domain, the spatial operation becomes a point-to-point operation.

The process of filtering in the frequency domain involves the following steps:

1. Calculate the DFT of the original image $I(x,y) \xrightarrow{\mathcal{F}} I[k_1,k_2]$.
2. Define a filter $H[k_1, k_2]$ in the frequency domain, i.e., the transfer function. Note that this filter must be of exactly the same size as image $I[k_1, k_2]$.
3. The output image in the Fourier domain is obtained by a pointwise multiplication,

$$J[k_1,k_2] = I[k_1,k_2] \cdot H[k_1,k_2].$$

4. Calculate the output image in the spatial domain $J[k_1,k_2] \xrightarrow{\mathcal{F}^{-1}} J(x,y)$.
5. The output image $J(x, y)$ could be a complex image. Most of the times it will be necessary to obtain the real part (of the absolute value).

The process of designing a filter in the frequency domain is a relatively easy and intuitive task. For instance, for a cutoff frequency Ω_c we can define:

- **Low-pass filter:**

$$H[k_1, k_2] = \begin{cases} 1, & \text{if } \sqrt{k_1^2 + k_2^2} \leq \Omega_c, \\ 0, & \text{otherwise.} \end{cases} \tag{6.35}$$

- **High-pass filter:**

$$H(u, v) = \begin{cases} 1, & \text{if } \sqrt{k_1^2 + k_2^2} \geq \Omega_c, \\ 0, & \text{otherwise.} \end{cases} \tag{6.36}$$

6.4.2 Homomorphic processing

A very useful enhancement scheme can result from considering the image as the product of two components:

$$f(\boldsymbol{x}) = i(\boldsymbol{x}) \cdot r(\boldsymbol{x}), \tag{6.37}$$

where $i(\boldsymbol{x})$ is the *illumination* component and $r(\boldsymbol{x})$ the *reflectance*. The illumination is related to the overall brightness of the image while the reflectance defines the objects in the image, i.e., their boundaries and textures. As a consequence, $i(\boldsymbol{x})$ is a low-pass signal and $r(\boldsymbol{x})$ is a high-pass signal. If these signals can be separated, then it becomes possible to correct for uneven illumination, which is important in some medical imaging modalities like radiography and magnetic resonance imaging (MRI) (although the phenomenon is better described as "intensity inhomogeneity" since light is not used in these modalities).

In homomorphic processing, the logarithm is first used to yield a sum of the two components as follows:

$$\log f(\boldsymbol{x}) = \log i(\boldsymbol{x}) + \log r(\boldsymbol{x}). \tag{6.38}$$

Then a simple linear filter (low-pass or high-pass) is used to separate a component, which can then be processed separately. As an example, in Fig. 6.20(a) a scheme for local contrast enhancement is presented. Homomorphic processing is also used to separate multiplicative noise from the original signal.

6.5 Model-based filtering: image restoration

Image restoration aims to improve the image when for some reason it has been degraded. This degradation can occur for multiple reasons, but noise and blurring are the most common causes. Noise is inherent to all medical image acquisition systems because of both the stochastic nature of the physical processes themselves and the addition of noise through hardware use to acquire the images. It can be thought of as random variations from the truth at each spatial position. Images can also be corrupted by artifacts, which can be considered as random, but are typically more

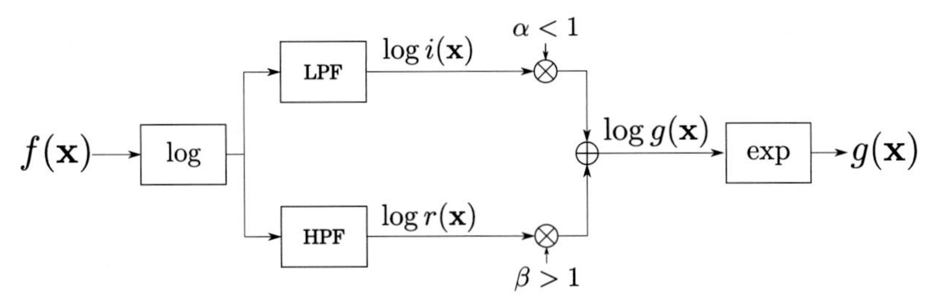

FIGURE 6.20 Schematic homomorphic processing.

Example of homomorphic processing: local contrast enhancement scheme.

difficult to characterize by simply probability models, and are therefore usually considered to be different from noise. To measure the power of the noise in images, an important measure from signal theory, the *signal-to-noise ratio* (SNR), is used. In particular, if this ratio is too small, it means that the noise level is high compared to the image intensity, and the meaningful information will be lost in the noise.

Blurring is a form of image degradation that is different from noise. It can be produced by patient motion (which is a random event) or by fixed or known characteristics of the imaging system itself. Like noise, blurring is typically unavoidable, but unlike noise it is often systematic rather than random. Typical sources of blurring include finite detector sizes in computed tomography and radiography, anti-aliasing filters in MRI, and transducer element size and physical field patterns in ultrasound. A simple general model (Fig. 6.21) that takes into account the noise and image degradation present in the observed image, I, can be given as

$$I(\boldsymbol{x}) = g(\boldsymbol{x}) + \eta(\boldsymbol{x}), \tag{6.39}$$

where η corresponds to the noise model. The image formation process, g, is modeled by a linear and space-invariant system where its impulse response, h, is convolved with the original image f as follows:

$$g(\boldsymbol{x}) = h(\boldsymbol{x}) * f(\boldsymbol{x}). \tag{6.40}$$

Additionally, the noise model, η, could be *signal-dependent* or *signal-independent* ($\eta_1 \neq 0$ vs. $\eta_1 = 0$),

$$\eta(\boldsymbol{x}) = g(\boldsymbol{x}) \times \eta_1(\boldsymbol{x}) + \eta_2(\boldsymbol{x}), \tag{6.41}$$

where η_1 and η_2 are two particular noise PDFs. The reader is referred to [5] for more details on a possible generalization of an image observation model.

Depending on the image acquisition process and transmission system used for imaging the body, equations (6.39)–(6.41) can be simplified by only considering η_1

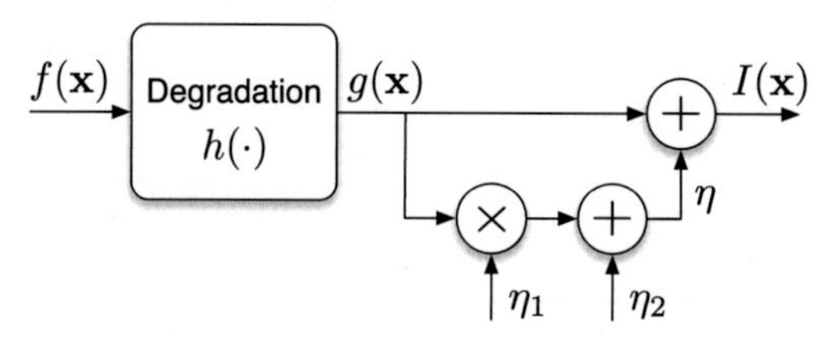

FIGURE 6.21 Model degradation.

Simple image degradation model.

or η_2 as follows:

$$I(x) = g(x) + \eta \quad \text{(signal-independent noise model)}, \tag{6.42}$$

$$I(x) = g(x) + g(x) \times \eta \quad \text{(signal-dependent noise model)}, \tag{6.43}$$

where I refers to the observed image, g corresponds to the detected image, and η is a particular noise PDF.

6.5.1 Noise models

The first thing to take into account when dealing with a restoration procedure is to properly model the features of noise in your image. There are many noise models in computer vision and medical images, depending on the formation of the image and the subsequent processing. Some of the most common models in image processing are *Poisson noise*, *Gaussian noise*, *salt & pepper*, *Rayleigh*, *Rician*, and *Gamma*. Poisson noise could be found, for instance, in those images related to photon-based acquisition, where the variation of photons collected by a digital sensor over a given time interval can be modeled as signal-dependent (equation (6.43)) with a Poisson PDF.

On the other hand, in computer vision, it is common to assume that the image is corrupted with uncorrelated white noise which is signal-independent (equation (6.42)) and is modeled following a zero-mean Gaussian PDF. This is the case with thermal and electronic noise. The impulse noise, also called *salt & pepper*, is often used for modeling malfunctioning of sensor cells, memory cell failure, or synchronization errors in image digitalization or transmission, for example. Impulsive noise has only two possible intensity levels, a and b. If $b > a$, gray level b will appear as a light value in the image. Conversely, gray level a will appears as a dark one. For an 8-bit/pixel image, the typical intensity values for a (pepper noise) are close to 0 and for b (salt noise) they are close to 255 [2].

Nevertheless, in order to select a proper noise model for a particular imaging modality, we recommend to consult specialized literature on that particular modality, because (1) models are constantly evolving and improving and (2) models are domain-specific. Ultrasound imaging, for instance, is usually modeled using Rayleigh or Gamma distributions, while MRI is usually considered to follow a Rician distribution [1].

Fig. 6.22 shows images and histograms resulting from different image degradation models.

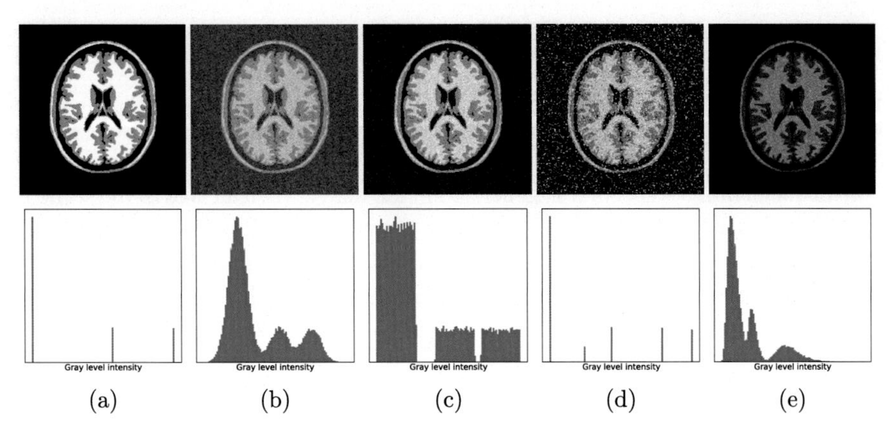

FIGURE 6.22 Examples of noise in images.

Images and histogram resulting from adding (b) signal-independent Gaussian, (c) uniform, (d) impulsive, and (e) signal-dependent Gamma noise to a synthetic brain image (a).

6.5.2 Point spread function

As previously stated, the degradation of an image can be modeled by the impulse response function of a linear system, $h(\boldsymbol{x})$, that in this particular problem is also known as *point spread function* (PSF). If we do not consider the noise component, we can write equation (6.39) as

$$g(x, y) = h(x, y) * f(x, y). \tag{6.44}$$

The equivalent in the frequency domain as introduced in Section 6.4 becomes a multiplication:

$$G(u, v) = H(u, v)\, F(u, v), \tag{6.45}$$

where H and F are the Fourier transform of the corresponding terms described before. These two equations provide the basis for image restoration for linear space-invariant systems.

The main hypothesis in restoration is that the same *kind* of blur equally affects all the points of the image, i.e., the blur is invariant to space and it can be modeled with an impulse response. The most common PSF used in restoration are listed below.[3]

Linear motion blur: The blur in the image is caused by a relative translation between the camera and the scene at constant velocity v_h during a time t_h. The PSF is defined by

$$h(x,y)=\begin{cases}\frac{1}{L_h}, & \text{if } \sqrt{x^2+y^2}\leq\frac{L_h}{2}; \frac{y}{x}=\phi,\\ 0, & \text{otherwise,}\end{cases} \tag{6.46}$$

where ϕ is the angle in which the translation takes place and $L_h=v_h\cdot t_h$ is the length of the movement.

Out-of-focus blur: This blur is generated in the image when an area of the scene is *out of focus* and therefore small details cannot be perceived. It is related with the so-called circle of confusion (COC), and the PSF that models this distortion is a uniform circle defined as

$$h(x,y)=\begin{cases}\frac{1}{\pi R^2}, & \text{if } \sqrt{x^2+y^2}\leq R,\\ 0, & \text{otherwise,}\end{cases} \tag{6.47}$$

where R is the radius of the COC and depends on acquisition parameters.

Gaussian blur: In some cases the blur may depend on a large number of parameters, making it difficult to properly model. That is the case, for instance, in atmospheric turbulence. In those cases, if the exposition time is long enough, the PSF may be accurately described by a Gaussian:

$$h(x,y)=C\exp\left(-\frac{x^2+y^2}{2\sigma_G^2}\right), \tag{6.48}$$

where σ_G controls the *amount* of blur.

Current restoration techniques avoid the prior usage of a PSF like the ones here proposed. On the contrary, the aim is to precisely estimate the *actual* PSF from the blurred image. However, these advanced methods are beyond the scope of this chapter.

6.5.3 Image restoration methods

The restoration process can be seen as the estimation of the original image $f(x,y)$ from the distorted one, $g(x,y)$. If we assume that the PDF is known *a priori*, the

[3] Note that these PSFs are common in natural image processing and computer vision. However, in medical imaging, distortion comes from the specificity of the different acquisition processes and, for different modalities, the PSF may differ from these.

estimation of the original signal is given by

$$\hat{f}(x, y) = g(x, y) * r(x, y), \tag{6.49}$$

where $h_r(x, y)$ is the restoration filter. In the frequency domain, we can write

$$\hat{F}(u, v) = G(u, v) \cdot R(u, v). \tag{6.50}$$

There are different ways to define the restoration filter $r(x, y)$. Let us review some of them.

Inverse filter: From a signal processing point of view, if $r(x, y)$ is the inverse of $h(x, y)$, then

$$[f(x, y) * h(x, y)] * r(x, y) = f(x, y),$$

and therefore

$$h(x, y) * r(x, y) = \delta(x, y). \tag{6.51}$$

In the frequency domain this is equivalent to

$$H(u, v) \cdot R(u, v) = 1. \tag{6.52}$$

The inverse filter can be defined as

$$R(u, v) = \frac{1}{H(u, v)}. \tag{6.53}$$

The restoration takes place in the frequency domain;

$$\hat{F}(u, v) = R(u, v) \cdot G(u, v)$$

If we consider that the image is not corrupted by noise, we simply obtain

$$\begin{aligned} \hat{F}(u, v) &= \frac{1}{H(u, v)} (F(u, v) \cdot H(u, v)) \\ &= F(u, v). \end{aligned}$$

However, if a component of noise is added to the image, the restoration process yields

$$\begin{aligned} \hat{F}(u, v) &= \frac{1}{H(u, v)} (F(u, v) \cdot H(u, v) + N(u, v)) \\ &= F(u, v) + \frac{N(u, v)}{H(u, v)}, \end{aligned}$$

where $N(u, v)$ is additive noise in the frequency domain. The main advantage of this filter is its simplicity. However, it also presents some important drawbacks, the main one being that the PSF could be non-invertible. For instance, $H(u, v)$ may have zero-values. As a consequence, some instability may be introduced into the solution. But, even if the PSD is invertible, the term $\frac{N(u,v)}{H(u,v)}$ may become dominant in the equation, amplifying the noise.

Pseudoinverse filter: In order to avoid the problems caused by the small values of $H(u, v)$, the pseudoinverse filter is introduced:

$$R(u,v) = \begin{cases} \dfrac{1}{H(u,v)}, & \text{if } |H(u,v)| > T_h, \\ 0, & \text{if } |H(u,v)| \le T_h, \end{cases} \tag{6.54}$$

where T_h is a given threshold.

Wiener filter: The Wiener filter assumes the PSF and the noise to be random variables. In order to estimate $\hat{f}(x, y)$, it uses a Bayesian approach that seeks to minimize the mean square error:

$$\text{error} = E\{(f^2(x,y) - \hat{f}^2(x,y))\}. \tag{6.55}$$

This minimization corresponds to a linear minimum mean square error (LMMSE) estimation. If we assume that signal and noise are not correlated, we can achieve a closed-form equation for the restoration process:

$$\hat{F}(u,v) = \left[\frac{H^*(u,v)S_f(u,v)}{S_f(u,v)|H(u,v)|^2 + S_\eta(u,v)}\right] G(u,v), \tag{6.56}$$

where $H^*(u, v)$ is the complex conjugate of $H(u, v)$, $S_f(u, v) = |F(u, v)|^2$ is the spectral density of the original image $f(x, y)$, and $S_\eta(u, v) = |N(u, v)|^2$ is the spectral density of noise. Note that equation (6.56) can be written as

$$\hat{F}(u,v) = \left[\frac{1}{H(u,v)} \frac{|H(u,v)|^2}{|H(u,v)|^2 + S_\eta(u,v)/S_f(u,v)}\right] G(u,v).$$

When $S_\eta(u, v)$ is small compared to $S_f(u, v)$, i.e., when the SNR is high, the term $S_\eta(u, v)/S_f(u, v) \to 0$ and the Wiener filter becomes the pseudoinverse filter.

The main drawback of this method is the necessity to estimate the spectral density of the original image, $S_f(u, v)$, and the spectral density of noise. The latter becomes easier if the noise is known to be Gaussian, since $S_\eta(u, v)$ becomes a constant. Many methods that allow an accurate estimation of this parameter have been reported in literature. On the other hand, in order to avoid the estimation of $S_f(u, v)$, some alternatives can be used:

1. Approximate the Wiener filter by the expression

$$\hat{F}(u,v) = \left[\frac{1}{H(u,v)} \frac{|H(u,v)|^2}{|H(u,v)|^2 + K}\right] G(u,v), \tag{6.57}$$

in which the term $S_\eta(u, v)/S_f(u, v)$ is replaced by a tunable parameter K. This term can be adjusted in order to control the filter behavior.

2. The spectral density of the signal can be roughly estimated from the degraded image, simply assuming that

$$S_f(u, v) \approx S_g(u, v) - \sigma_{eta}^2,$$

where σ_{eta}^2 is the spectral density of noise assuming uncorrelated Gaussian noise.

Restoration iterative schemes: Alternatively, there is a whole family of iterative methods that work in the spatial domain. These methods are good approaches when the PSF is only roughly estimated or when the degradation is spatially dependent. The basic iterative scheme is the following one:

$$\hat{f}_{i+1}(x, y) = \hat{f}_i(x, y) - \beta \left[g(x, y) - h(x, y) * \hat{f}_i(x, y) \right], \tag{6.58}$$

where $\hat{f}_i(x, y)$ is the image estimated after the ith iteration and β is a parameter that controls the estimation procedure. In more complex methods, the PSF $h(x, y)$ could also change in each iteration. When the number of iterations is large, the solution of equation (6.58) tends to the inverse filter.

For the sake of illustration, an example of restoration is shown in Fig. 6.23. A synthetic image (Fig. 6.23(a)) is blurred using the Gaussian PSF in Fig. 6.23(b). The original image is then restored using a pseudoinverse filter (Fig. 6.23(d)) and the Wiener approach (Fig. 6.23(e)). In this case, the inverse filter result is not shown since the instability in the solution does not produce a proper image. Note that even in the absence of noise, the Wiener filter produces a better result than the pseudoinverse, avoiding the spurious borders that appear in the latter. When Gaussian noise is also added (Fig. 6.23(f)), the pseudoinverse filter (Fig. 6.23(g)), since it behaves like a high-pass filter, amplifies the noise. The Wiener filter (Fig. 6.23(h)) has been implemented using the approach in equation (6.57), selecting a value of K that ensures a good balance between noise filtering and edge enhancement.

6.6 Further reading

Image filtering and restoration are very well-known fields within image processing. The main textbooks on image processing have chapters devoted to these two procedures. We recommend some of them to delve deeper into the topic:

- Some general image processing handbooks with great chapters about image filtering and restoration: [2,4–6].
- Some reading about noise in medical imaging: MRI [1], computed tomography [7], and speckle and ultrasound [8–10].
- Advanced filtering design [11].
- The statistics of the Wiener filter and LMMSE estimation can be found in [6,12]. For the application to image processing, you can read [6].

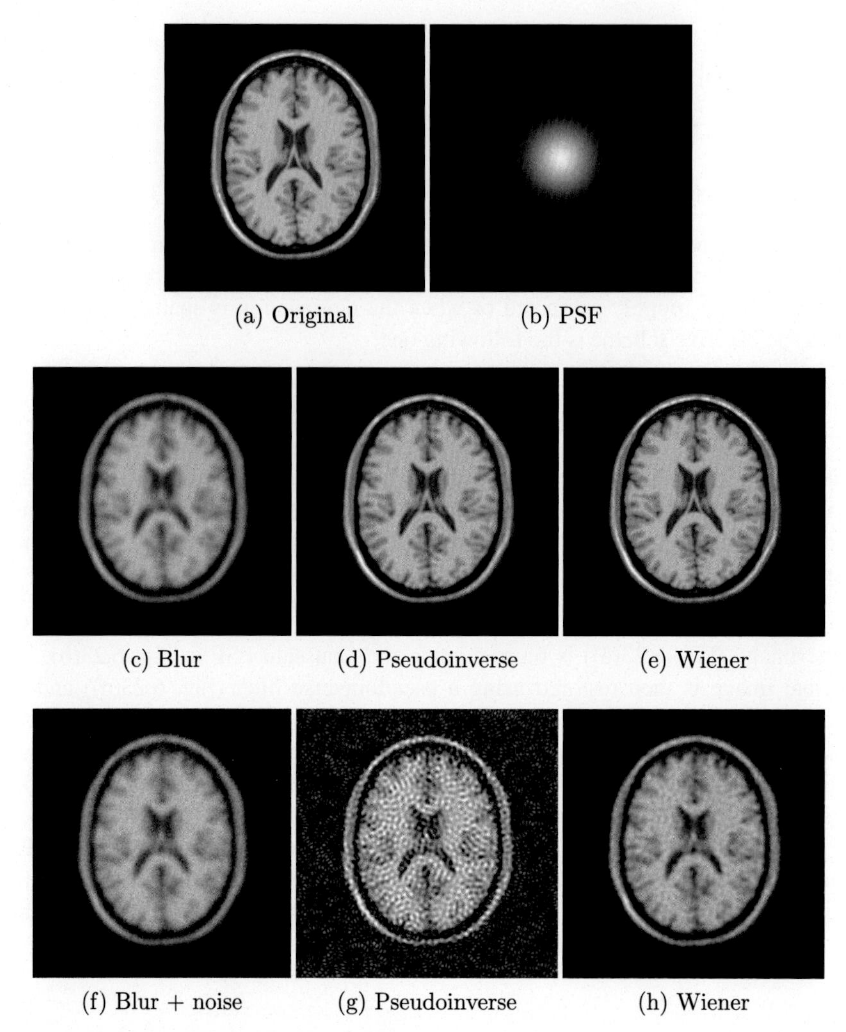

FIGURE 6.23 Restoration of distorted images.

(a) Original image. (b) Gaussian PSF in the frequency domain, $H(u, v)$. (c) Original image blurred with the PSF (no noise is added). (d) Image restored with a pseudoinverse filter (no noise). (e) Image restored with a Wiener filter (no noise). (f) Original image degraded with a PSF and Gaussian noise. (g) Image restored with a pseudoinverse filter (blur and noise). (h) Image restored with a Wiener filter (blur and noise).

- If you are interested in restoration techniques specially applied to medical imaging, you can check [13], where you can also find another interesting chapters on filtering.

6.7 Exercises

1. For a grayscale image, create a function (from scratch) that is able to vary brightness and contrast.
2. Create from scratch a function that automatically equalizes the histogram.
3. Define a function that adds different types of noise to an image.
4. Generate a system that performs Unsharp Masking of an image. Make it so that the output signal is an integer with the same dynamic range that the input.
5. Select an x-ray image. Use a thresholding method to separate the bones from the background and tissue. Repeat the process trying to equalize the non-uniform illumination by homomorphic processing.
6. Define a multilevel thresholding of x-ray and MRI data in order to separate background, soft tissue and bone. For this purpose study the histograms of the images and apply various thresholding techniques. Study the effect of applying some preprocessing (homomorphic processing, filtering, etc.).
7. Maximum likelihood estimation. Assume that an image has complex multiplicative noise:

$$I_c(\mathbf{x}) = I(\mathbf{x}) \times \left(N_1(0, \sigma^2) + j \cdot N_2(0, \sigma^2)\right)$$

   where $I(\mathbf{x})$ is the original image and $N_1(0, \sigma^2)$ is Gaussian noise with zero mean and variance σ^2. When considering the modulus of the signal:

$$M(\mathbf{x}) = |I_c(\mathbf{x})|$$

   the resulting signal $M(\mathbf{x})$ follows a Rayleigh distribution, the σ parameter of the distribution being the original Gaussian parameter multiplied by the value of the signal at that point.

   a. From a synthetic image, add multiplicative Rayleigh noise.
   b. Generate 50 realizations of the image (each with a different noise realization).
   c. Derive the Maximum likelihood estimator of the sigma parameter of a Rayleigh.
   d. Estimate the original signal (assume that the sigma value of the Gaussians is known).
8. Blur an image with a Gaussian lowpass filter of $\sigma = 1.5$ and size 11×11. Calculate the inverse and pseudoinverse filter in the frequency domain. Check the results of applying these two methods to correct the blurring.

References

[1] Aja-Fernández S., Vegas-Sánchez-Ferrero G., Statistical Analysis of Noise in MRI, Springer International Publishing AG, 2016.

[2] González R., Woods R., Digital Image Processing, 3rd ed., Addison & Wesley, 2011.

[3] Pearson K., Contributions to the mathematical theory of evolution. II. Skew variation in homogeneous material, Philosophical Transactions of the Royal Society of London 186 (Part I) (1895) 343–424.
[4] Pratt W.K., Digital Image Processing, 3rd ed., Wiley, New York, NY, USA, 1991.
[5] Jain A.K., Fundamentals of Digital Image Processing, Prentice Hall, Englewood Cliffs, NJ, 1989.
[6] Lim J.S., Two-Dimensional Signal and Image Processing, Prentice Hall, Englewood Cliffs, NJ, 1990.
[7] Vegas-Sánchez-Ferrero G., Ledesma-Carbayo M.J., Washko G.R., San José Estépar R., Statistical characterization of noise for spatial standardization of CT scans: enabling comparison with multiple kernels and doses, Medical Image Analysis 40 (2017) 44–59.
[8] Burckhardt C.B., Speckle in ultrasound B-mode scans, IEEE Transactions on Sonics and Ultrasonics 25 (1) (1978) 1–6.
[9] Thijssen J.M., Ultrasonic speckle formation, analysis and processing applied to tissue characterization, Pattern Recognition Letters 24 (4–5) (2003) 659–675.
[10] Shankar P.M., Reid J.M., Ortega H., Piccoli C.W., Goldberg B.B., Use of non-Rayleigh statistics for the identification of tumors in ultrasonic B-scans of the breast, IEEE Transactions on Medical Imaging 12 (4) (1993) 687–692.
[11] Granlund G.H., Knutsson H., Signal Processing for Computer Vision, Kluwer Academic Publishers, Dordrecht, Netherlands, 1994.
[12] Kay S.M., Fundamentals of Statistical Signal Processing. Volume I: Estimation Theory, Prentice-Hall, 1993.
[13] Epstein C.L., Introduction to the Mathematics of Medical Imaging, SIAM, 2007.

CHAPTER

Multiscale and multiresolution analysis

7

Jon Sporring

Department of Computer Science, University of Copenhagen, Copenhagen, Denmark

Learning points

- What is scale-space?
- Representation in image pyramids and Gaussian scale-space
- Gaussian scale-space, differential geometry of images
- Scale selection: blob and edge detection
- Scale-space histogram: three fundamental imaging scales

7.1 Introduction

For typical image processing tasks, the size of objects is unknown, and therefore, we must design algorithms that are versatile to work for all sizes. Consider the oversimplified case of trying to find a flower in an image of some unknown size, as shown in Fig. 7.1. In the example, the model is a small example of a flower, while the image contains an identical flower, but with a different size. Direct comparison of the model with image patches will not find the flower due to the size difference. However, if we resize either the model or the image, then there will be a scaling factor, which gives a perfect match. Imagining similar cases with other objects such as houses or dogs, one may realize that the resizing operation of either the model or the image is an operation which allows us to extend arbitrary models at a fixed size to a range of sizes. Further, the range of sizes, we can possibly consider, will be from the size of a single pixel to the size of the entire image. We will use the terms size and scale interchangeably, thus we say that the pixel size defines the inner scale and the image size defines the outer scales. While it is arbitrary whether we change the scale of the model or the image, we prefer the latter, and augmenting an image with the full range of scales is called a scale-space.

7.2 The image pyramid

Many scale-spaces have been designed for various purposes. One of the earliest suggested is the image pyramid [1], which is a sequence of images, starting with the

Medical Image Analysis. https://doi.org/10.1016/B978-0-12-813657-7.00020-0

FIGURE 7.1

Finding flowers. Left: a model of a flower. Right: An image containing a flower with a different size than the model.

original and followed by downsampled versions of the previous. For example, a pyramid of an original 2D image with 1024×1024 pixels could consist of the sequence of images of the same content but at resolutions of 1024×1024, 512×512, 256×256, $128 \times 128, \ldots$. This is illustrated in Fig. 7.2. The pyramid extends until the image is 1×1, but in practice, the pyramid is terminated earlier. Similarly for 3D images, each level consists of a downsampled 3D image, for example, $1024 \times 1024 \times 1024$, $512 \times 512 \times 512, \ldots$.

Each image in the pyramid is called a pyramid level, and upwards in the pyramid, we say that the resolution is coarser, while downwards we say that the resolution is finer. In the example, a downsampling factor of 2 was used, which is common, and in which case a pixel at some level in the pyramid approximately corresponds to an averaging of 2×2 pixels in the level below for 2D images and of $2 \times 2 \times 2$ pixels for 3D images. However, other downsampling factors larger than 1 could be used as well.

With a downsampling factor of 2, the added storage cost is not great, that is, for an original 2D image of size N^2, the number of pixels in the pyramid is

$$N^2 + \frac{1}{4}N^2 + \left(\frac{1}{4}\right)^2 N^2 + \left(\frac{1}{4}\right)^3 N^2 + \cdots = N^2\left(1 + \frac{1}{4} + \left(\frac{1}{4}\right)^2 + \left(\frac{1}{4}\right)^3 + \ldots\right), \tag{7.1}$$

where there is one term for each level in the pyramid. This sequence is bounded by the geometric series with infinite number of terms

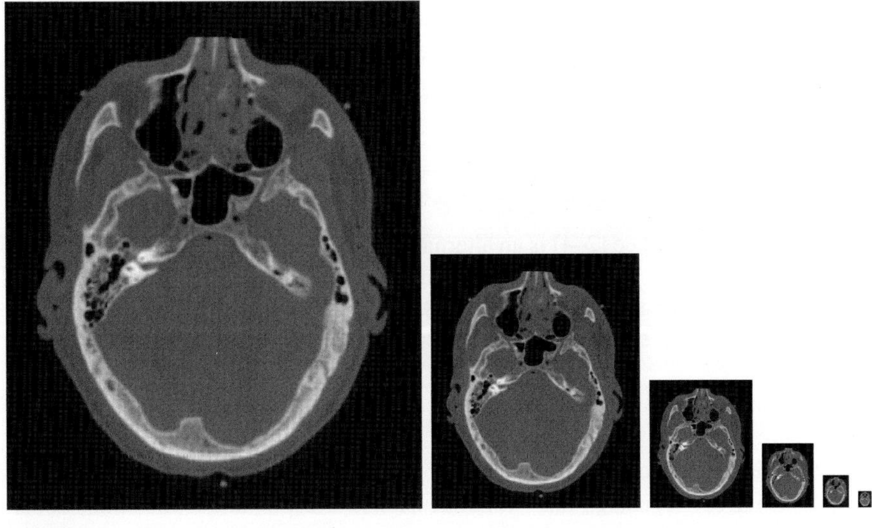

FIGURE 7.2

A pyramid is a sequence of downsampled images. Top left: The original. From left to right: Increasingly downsampled versions with factors $\left(\frac{1}{2}\right)^i$, $i = 1 \ldots 5$. The original image was obtained from the Visible Human Project at the U.S. National Library of Medicine.

$$s = 1 + \frac{1}{4} + \left(\frac{1}{4}\right)^2 + \left(\frac{1}{4}\right)^3 + \ldots. \tag{7.2}$$

For the geometric series, the value of s may be found by first multiplying the left- and right-hand sides by $\frac{1}{4}$,

$$\frac{1}{4}s = \frac{1}{4} + \left(\frac{1}{4}\right)^2 + \left(\frac{1}{4}\right)^3 + \left(\frac{1}{4}\right)^4 + \ldots, \tag{7.3}$$

subtracting (7.3) from (7.2), and solving for s,

$$s - \frac{1}{4}s = 1 \Rightarrow s = \frac{4}{3}. \tag{7.4}$$

For a finite number of levels, the pyramid including the original image will have no more than $\frac{4}{3}N^2$ pixels. For a 3D image of size N^3, a level contains $\frac{1}{8}$ of the number of pixels in the level below, and by similar considerations, the storage pyramid will have no more than $\frac{8}{7}N^3$ pixels.

The image pyramid is an effective representation of the image at a finite set of scales, but the relation between structures at various levels is discrete due to the resampling of the image. Also, the downsampling phase is typically chosen in relation to the image coordinates, which are almost always unrelated to the image content.

7.3 The Gaussian scale-space

The Gaussian scale-space [2–4] is an alternative structure that like the image pyramid gradually reduces the spatial resolution but in contrast, avoids downsampling, and does so in a smooth manner. The basis of Gaussian scale-space is convolution of the original image with a Gaussian kernel,

$$f_\sigma = f * g_\sigma, \tag{7.5}$$

where "$*$" is the convolution operator, g_σ is an isotropic Gaussian kernel with standard deviation $\sigma > 0$, and g_σ has with dimensions corresponding to the dimensionality of f,

$$g_\sigma(x) = \frac{1}{\sqrt{2\pi\sigma^2}} \exp\left(\frac{-x^2}{2\sigma^2}\right), \tag{7.6}$$

$$g_\sigma(x, y) = g_\sigma(x) g_\sigma(y), \tag{7.7}$$

$$g_\sigma(x, y, z) = g_\sigma(x) g_\sigma(y) g_\sigma(z). \tag{7.8}$$

In the limit, $\lim_{\sigma \to 0} f_\sigma = f$ and $\lim_{\sigma \to \infty} f_\sigma = \bar{f}$, where $\bar{f}$ is the average value of f. The parameter σ is considered the scale parameter, corresponding to the degree of downsampling: the larger the value of σ, the larger the corresponding downsampling. Levels in Gaussian scale-space are shown in Fig. 7.3.

The Gaussian kernel is defined in the continuous domain, while images are almost always samples on a regular grid. This is both an advantage and a challenge, since on the one hand, many of the properties of the Gaussian scale-space are nicely proven in the continuous domain, but it is not clear to what extent such properties extend to the discrete domain. For example, the 1D definition in (7.6) integrates to 1,

$$\int_{-\infty}^{\infty} g_\sigma(x)\, dx = 1, \tag{7.9}$$

while a sample version surely does not sum to 1,

$$\sum_i g_\sigma(x_i) \neq 1, \tag{7.10}$$

regardless of how closely we sample the kernel. The unit integral is essential for the kernel to be an averaging filter; hence, convolving with sampled Gaussian kernels, $\tilde{g}_\sigma$,

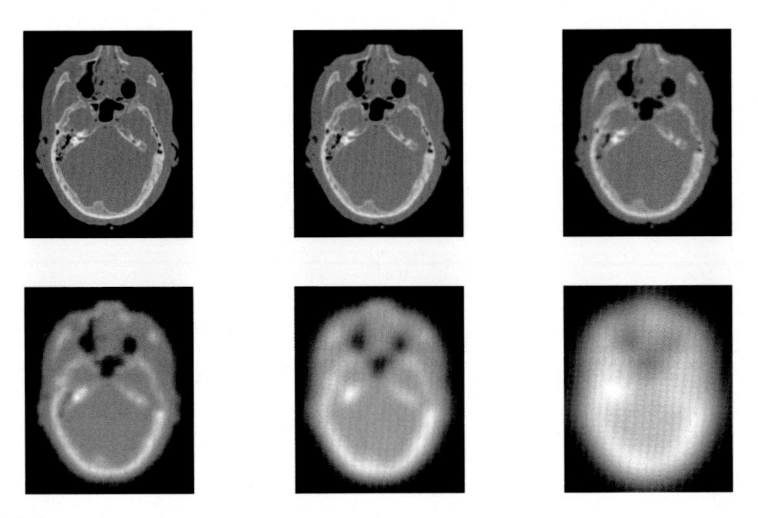

FIGURE 7.3

Levels from a Gaussian scale-space. Bottom: In reading order: the original image followed by downscaled versions with factors $2^i/\sqrt{12}$, $i = 1\dots5$. The original image was obtained from the Visible Human Project at the U.S. National Library of Medicine.

will not converge to the average value: $\lim_{\sigma\to\infty} \tilde{g}_\sigma * f \neq \bar{f}$. A simple normalization will correct this, $k_\sigma = \sum_i \tilde{g}_\sigma(x_i)$, in which case $\lim_{\sigma\to\infty} \tilde{g}_\sigma/k_\sigma * f = \bar{f}$. Nevertheless, it is my experience that it is advantageous to consider algorithms on discrete images as approximations of algorithms on continuous images, since the continuous domain offers more possibilities for simplifications and conclusions.

There exists no exact equivalence between the Gaussian scale-space and the image pyramid; one way to align them is by choosing values of σ which correspond to the standard deviation of a box function implicitly used in the downsampling operation. For example, the downsampling value of pixels in 1D is similar to the values of applying the box kernel,

$$\mathrm{box}_a(x) = \begin{cases} \frac{1}{2a} & \text{if } |x| < a, \\ 0 & \text{otherwise,} \end{cases} \tag{7.11}$$

with $a = 1$. The variance of $\mathrm{box}_a(x)$ is

$$\int_{-\infty}^{\infty} \mathrm{box}_a(x) x^2 \, dx = \int_{-a}^{a} \frac{1}{2a} x^2 \, dx = a^2/3. \tag{7.12}$$

Thus, the Gaussian scale-space may be aligned with the image pyramid by setting $\sigma = \frac{a}{\sqrt{3}}$ with $a = 1, 2, 4, \dots$. For $a = 1$ this is illustrated in Fig. 7.4. Further, in 1D,

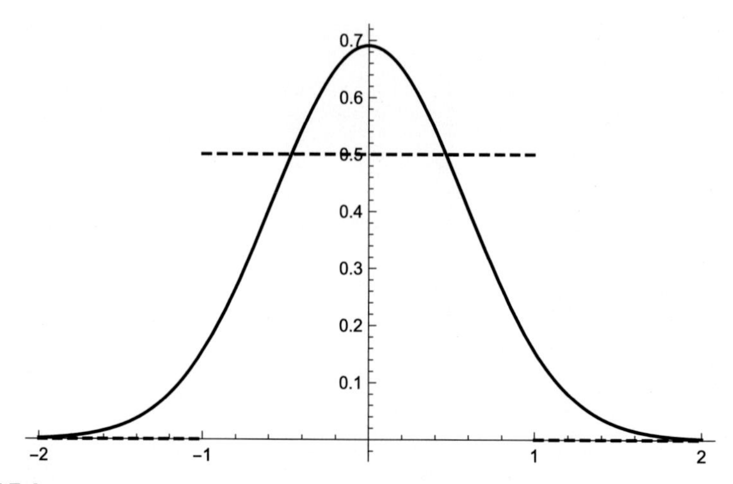

FIGURE 7.4

Aligning a Gaussian kernel (full line) to a box kernel (dotted line) by equal variance.

the mass of the Gaussian corresponding to the width of the box kernel is

$$\int_{-a}^{a} g_{a/\sqrt{3}}(x)\, dx = 0.917, \tag{7.13}$$

i.e., independent of a. In this perspective, the qualitative information content in scale-space is similar, as demonstrated in Fig. 7.5.

7.4 Properties of the Gaussian scale-space

The Gaussian scale-space is a useful construction particularly for studying differential geometric structures across scales. In the following, a number of properties and perspectives will be highlighted.

7.4.1 The frequency perspective

By the convolution theorem, we know that the Fourier transformation of a convolution of two functions is equal to the product of two Fourier transformations of these functions. Thus, we may gain insight into the Gaussian scale-space by studying (7.5) in the Fourier domain. The Fourier transform of the 1D Gaussian distribution is

$$\mathcal{F}\{g_\sigma(x)\} = \int_{-\infty}^{\infty} \frac{1}{\sqrt{2\pi\sigma^2}} \exp\left(-\frac{x^2}{2\sigma^2}\right) \exp(-i2\pi x u)\, dx \tag{7.14}$$

$$= \exp\left(-2\pi^2\sigma^2 u^2\right), \tag{7.15}$$

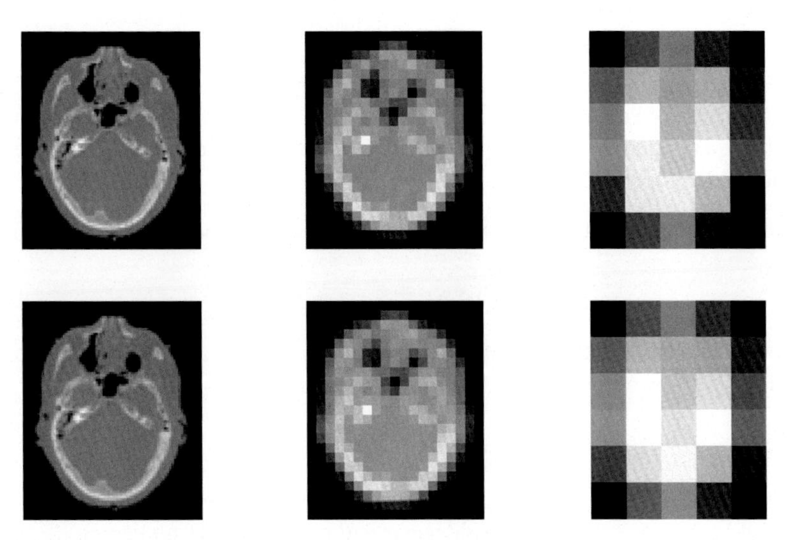

FIGURE 7.5

Comparing the pyramid values and the corresponding Gaussian levels downsampled for the purpose of illustration. Top and bottom rows show the corresponding pyramid and downsampled scale-space level. We used $\sigma = \frac{a}{\sqrt{3}}$ with $a = 2^{i-1}$, $i = 1, 3, 5$, corresponding to the downsampling factors $\left(\frac{1}{2}\right)^i$, $i = 1, 3, 5$. The original image was obtained from the Visible Human Project at the U.S. National Library of Medicine.

where u is the frequency parameter corresponding to x. By separability, we find that

$$\begin{aligned}\mathcal{F}\{g_\sigma(x, y)\} &= \mathcal{F}\{g_\sigma(x)g_\sigma(y)\} && (7.16)\\ &= \mathcal{F}\{g_\sigma(x)\}\mathcal{F}\{g_\sigma(y)\} && (7.17)\\ &= \exp\left(-\left(u^2+v^2\right)2\pi^2\sigma^2\right), && (7.18)\end{aligned}$$

where v is the frequency parameter corresponding to y, and that

$$\begin{aligned}\mathcal{F}\{g_\sigma(x, y, z)\} &= \mathcal{F}\{g_\sigma(x)g_\sigma(y)g_\sigma(z)\} && (7.19)\\ &= \exp\left(-\left(u^2+v^2+w^2\right)2\pi^2\sigma^2\right), && (7.20)\end{aligned}$$

where w is the frequency parameter corresponding to z.

Thus, the maximum frequency of the Gaussian filter is infinite, but it is essentially band-limited, since its amplitude converges exponentially fast to zero as the frequency increases. Further, when σ increases, the width of the Gaussian kernel increases in the spatial domain, but decreases in the frequency domain, and therefore, increasing the scale removes high frequencies or, equivalently, fine detail.

7.4.2 The semi-group property

Mathematically, the Gaussian scale-space may be described as a group, that is, if we consider the set of all possible Gaussian kernels $g_\sigma \in G$ and we consider convolution, $*$, as the binary operator, then the pair $(G, *)$ is a semi-group, or equivalently, we have the following properties.

Closure: Convolving two Gaussian kernels results in another Gaussian kernel,

$$g_\sigma * g_\tau = g_{\sqrt{\sigma^2+\tau^2}}. \tag{7.21}$$

Associative: The order in which three Gaussian kernels are convolved does not matter,

$$(g_\sigma * g_\tau) * g_\upsilon = g_\sigma * (g_\tau * g_\upsilon). \tag{7.22}$$

The associative rule is a consequence of the convolution operator and not special to Gaussian kernels, but the closure rule is particular to the Gaussian kernel.
Identity element: It is common to extend the family of Gaussian kernels with an identity element δ, also known as Dirac's delta function,

$$\delta = \lim_{\sigma \to 0} g_\sigma, \tag{7.23}$$

which has the property that

$$\delta * g_\tau = g_\tau * \delta = g_\tau. \tag{7.24}$$

However, it is not possible to define an inverse element, i.e., it is not possible to undo the result of a Gaussian convolution.

7.4.3 The analytical perspective

The Gaussian kernel is smooth, i.e., all derivatives $\frac{d^i}{dx^i} g_\sigma(x)$ exist. Assuming the Leibniz rule for interchanging integrals and derivatives and assuming that the convolution integral exists, then, e.g., all derivatives of 1D Gaussian smoothed functions exist:

$$\frac{d^i}{dx^i} f_\sigma(x) = \frac{d^i}{dx^i} (f * g_\sigma)(x) \tag{7.25}$$

$$= \frac{d^i}{dx^i} \int_{-\infty}^{\infty} f(\alpha) g_\sigma(x-\alpha)\, d\alpha \tag{7.26}$$

$$= \int_{-\infty}^{\infty} f(\alpha) \frac{d^i}{dx^i} g_\sigma(x-\alpha)\, d\alpha \tag{7.27}$$

$$= \left(f * \frac{d^i}{dx^i} g_\sigma \right)(x). \tag{7.28}$$

Equivalently, in higher dimensions,

$$\frac{\partial^{i+j}}{\partial x^i \partial y^j} f_\sigma(x, y) = \left(f * \frac{\partial^{i+j}}{\partial x^i \partial y^j} g_\sigma \right)(x, y) \tag{7.29}$$

and

$$\frac{\partial^{i+j+k}}{\partial x^i \partial y^j \partial z^k} f_\sigma(x, y, z) = \left(f * \frac{\partial^{i+j+k}}{\partial x^i \partial y^j \partial z^k} g_\sigma \right)(x, y, z). \tag{7.30}$$

The implication is that any differential operator can be evaluated on any image, as soon as it has been smoothed with a Gaussian kernel regardless of the size of the kernel.

Further, since the Taylor series of the Gaussian kernel converges, so will the Taylor series of any smoothed function; for example, in 2D,

$$f_\sigma(x+h, y+k) = f_\sigma(x+h, y+k) + h\frac{\partial}{\partial x} f_\sigma(x, y) + k\frac{\partial}{\partial y} f_\sigma(x, y) + \dots. \tag{7.31}$$

7.4.4 The heat diffusion perspective

The final perspective on the Gaussian scale-space, we will give, is diffusion of heat. Considering the image f as a heatmap, then f_σ is the solution to the heat diffusion equation,

$$\frac{\partial f}{\partial t} = \frac{1}{2} \nabla f, \tag{7.32}$$

where t is the diffusion time, ∇ is the Laplacian operator, i.e., $\nabla = \frac{\partial^2}{\partial x^2} + \frac{\partial^2}{\partial y^2} + \dots$, with terms matching the dimensionality of the image, and $t = \sigma^2$. This relation to the heat diffusion equation has inspired a number of non-linear scale-spaces, where smoothing is set to depend on the image structures being smoothed. Further discussion of this is, however, outside the scope of this chapter.

7.5 Scale selection

One of the key properties of Gaussian scale-space is its ability to compare structures across scales. A useful added construct is the γ-normalized derivative [5],

$$\sigma^{(i+j)\gamma/2} \frac{\partial^{i+j}}{\partial x^i \partial y^j} f_\sigma(x, y), \tag{7.33}$$

where γ is a parameter related to the structure. The intuition is that a structure like a solid disk responds differently to scaling than an edge structure. The γ-normalized derivatives allow for elegant algorithms to detect sizes of structures according to their type and how they respond to smoothing. We will look at two scale-space detectors, the blob detector and the edge detector.

7.5.1 Blob detection

Blobs are regions in images that have a structure similar to a solid disk. The blob detector relies on the γ-normalized Laplacian operator, where $\gamma = 2$:

$$L_\sigma(x, y) = \sigma^2 \left(\frac{\partial^2 f_\sigma(x, y)}{\partial x^2} + \frac{\partial^2 f_\sigma(x, y)}{\partial y^2} \right). \tag{7.34}$$

The blob detector is defined as

$$\nabla L_\sigma(x, y) = 0, \tag{7.35}$$

where ∇ is the Jacobian operator, i.e., $\nabla L_\sigma(x, y) = \left[\frac{\partial}{\partial x} L_\sigma(x, y), \frac{\partial}{\partial y} L_\sigma(x, y), \frac{\partial}{\partial \sigma} L_\sigma(x, y) \right]^T$. Note that we use the usual convention in image processing to write the Jacobian vector as a column vector, when obtained using the ∇ operator. As an example, Fig. 7.6 shows the 100 highest and 10 lowest points in the microscope image.

To further analyze the Blob detector, consider the image of a Gaussian function

$$f(x, y) = \frac{1}{2\pi\tau^2} \exp\left(-\frac{x^2 + y^2}{2\tau^2} \right). \tag{7.36}$$

Due to the closure property, we can analytically write up the Gaussian scale-space of this image as

$$f_\sigma(x, y) = \frac{1}{2\pi(\sigma^2 + \tau^2)} \exp\left(-\frac{x^2 + y^2}{2(\sigma^2 + \tau^2)} \right). \tag{7.37}$$

The γ-normalized Laplacian of f_σ is found to be

$$L_\sigma(x, y) = -\frac{\sigma^2(2\sigma^2 + 2\tau^2 - x^2 - y^2)}{(\sigma^2 + \tau^2)^2} f_\sigma(x, y). \tag{7.38}$$

Since L has a "Mexican hat" shape, we quickly realize that it has an extremum at $[x, y] = 0$, so we can restrict our attention to the development of L as a function of σ. In Fig. 7.7 the values of L_σ for $x = y = 0$ and $\tau = 1$ are plotted against σ. Solving for where $\nabla L_\sigma(0, 0) = 0$ we find three solutions: $\sigma = 0$, $\sigma = -\tau$, and $\sigma = \tau$. Of the three solutions only $\sigma = \tau$ is within our solution domain, and we conclude that the γ-normalized Laplacian has an extremum at the center of the original function f and at the scale that exactly matches the original size of f. Hence, we can detect not only the positions of blobs but also their sizes as extrema of L.

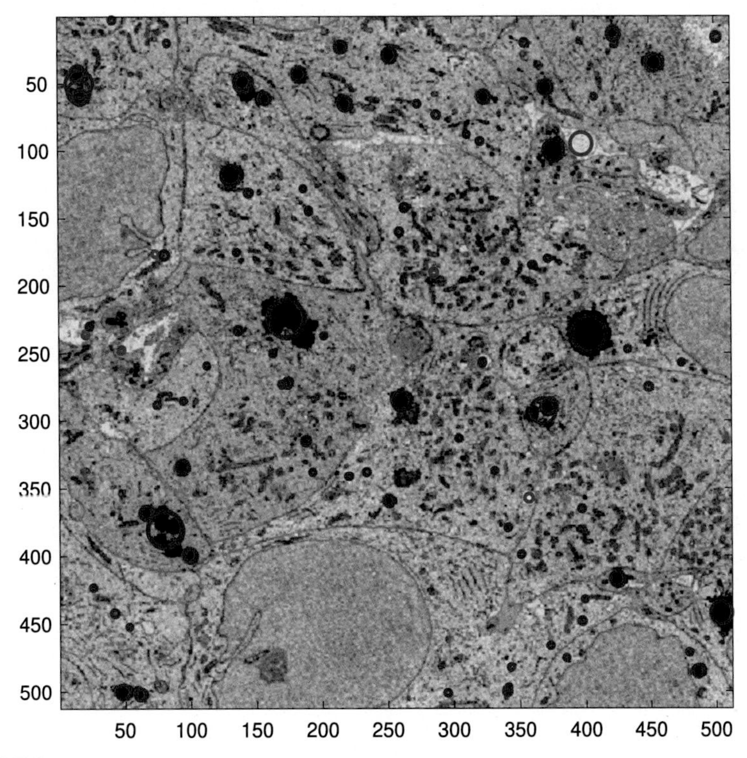

FIGURE 7.6

The first 100 maxima and 10 minima detected in a microscope image. Circles have a radius corresponding to the detection scale. Image courtesy of J.Midtgaard, K. Qvortrup, & M. Møller, Core Facility for Integrated Microscopy, Medical and Health Sciences Faculty, Copenhagen University.

7.5.2 Edge detection

Edges are regions, where image functions have a high slope, and such a structure is similar to a line. An indication of the location of an edge is the squared gradient magnitude with $\gamma = 1$,

$$E_\sigma(x, y) = \sigma^2 \left(\left(\frac{\partial f_\sigma(x, y)}{\partial x} \right)^2 + \left(\frac{\partial f_\sigma(x, y)}{\partial y} \right)^2 \right), \tag{7.39}$$

and the location of the edge is found by

$$\nabla E_\sigma(x, y) = 0. \tag{7.40}$$

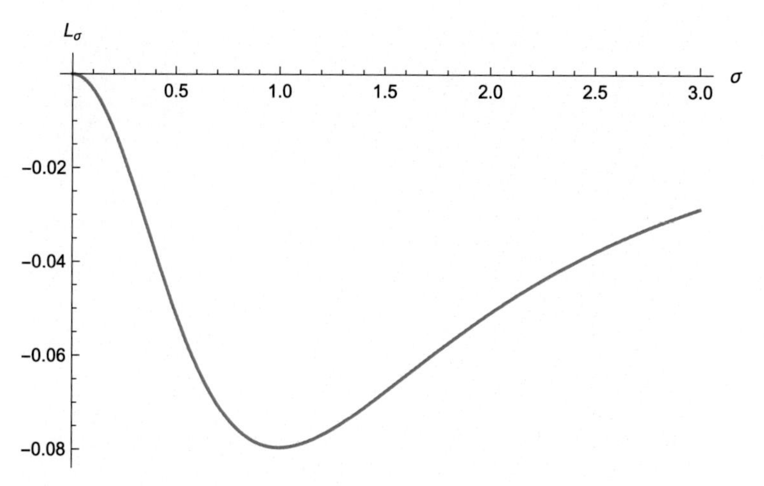

FIGURE 7.7

The γ-normalized Laplacian (7.38) with $\tau = 1$ and as a function of σ. Optima are found in $\sigma = -1, 0, 1$.

To further analyze the edge detector, consider a prototypical example of an edge: the integral of a 1D Gaussian function,

$$g_\tau(x) = \frac{1}{\sqrt{2\pi\tau^2}} \exp -\frac{x^2}{2\tau^2}, \tag{7.41}$$

$$f(x, y) = \int_{-\infty}^{x} g_\tau(\alpha)\, d\alpha. \tag{7.42}$$

We like to call f for a Gauss edge. Note that f is the edge, which has a sigmoidal shape in the x-direction and is constant in the y-direction. An example of this is shown in Fig. 7.8. The Gaussian scale-space of the image of this edge is

$$f_\sigma(x, y) = f * g_\sigma, \tag{7.43}$$

which in 2D has the following closed-form solution

$$f_\sigma(x, y) = \int_{-\infty}^{x} g_{\sqrt{\sigma^2+\tau^2}}(\alpha)\, d\alpha. \tag{7.44}$$

To prove this, we will need 4 properties: First f is constant in the y-direction. Second, consider $f(x, y) = \int_{-\infty}^{x} g(\alpha)\, d\alpha$ and the limit,

$$\frac{\partial f(x, y)}{\partial x} = \lim_{h \to 0} \frac{f(x+h, y) - f(x-h, y)}{2h} \tag{7.45}$$

$$= \lim_{h \to 0} \frac{1}{2h} \int_{x-h}^{x+h} g(\alpha)\, d\alpha \tag{7.46}$$

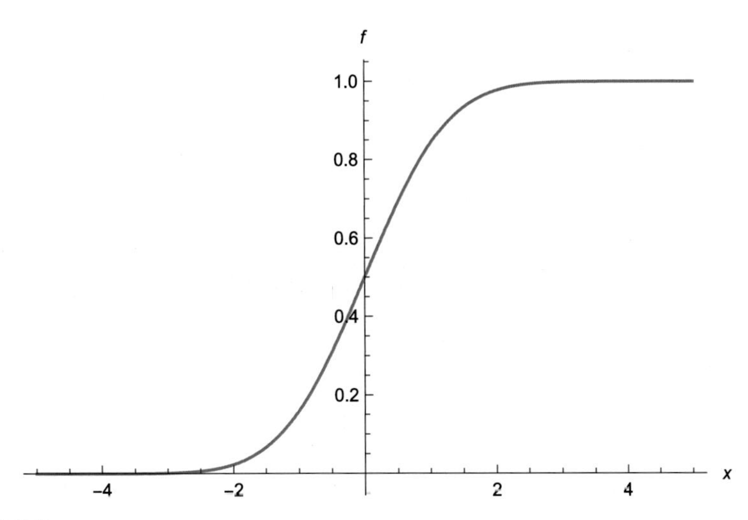

FIGURE 7.8

A prototypical example of an edge: the integral of a Gaussian of standard deviation $\tau = 1$.

$$= \lim_{h\to 0} \frac{g(x)}{2h} \int_{x-h}^{x+h} 1\, d\alpha \tag{7.47}$$

$$= \lim_{h\to 0} \frac{g(x)}{2h} \left[\alpha\right]_{x-h}^{x+h} \tag{7.48}$$

$$= g(x). \tag{7.49}$$

Third, convolution and differentiation commutes, i.e.,

$$\frac{\partial}{\partial x}(f * g)(x) = \left(\left(\frac{\partial}{\partial x} f\right) * g\right)(x) = \left(f * \left(\frac{\partial}{\partial x} g\right)\right)(x), \tag{7.50}$$

Thus,

$$\frac{\partial}{\partial x}(f * g_\sigma)(x, y) = \left(\left(\frac{\partial}{\partial x} f\right) * g_\sigma\right)(x, y) = g_\tau(x) * g_\sigma(x, y), \tag{7.51}$$

And due to the closure property, we conclude that

$$\frac{\partial}{\partial x}(f * g_\sigma)(x, y) = g_{\sqrt{\tau^2+\sigma^2}}(x). \tag{7.52}$$

and by (7.45)-(7.49) we have found the closed-form solution for the Gaussian scale-space of Gauss-edges.

The γ-normalized squared gradient magnitude of a Gauss edge is

$$E_\sigma(x, y) = \sigma^2 g^2_{\sqrt{\sigma^2+\tau^2}}(x, y). \tag{7.53}$$

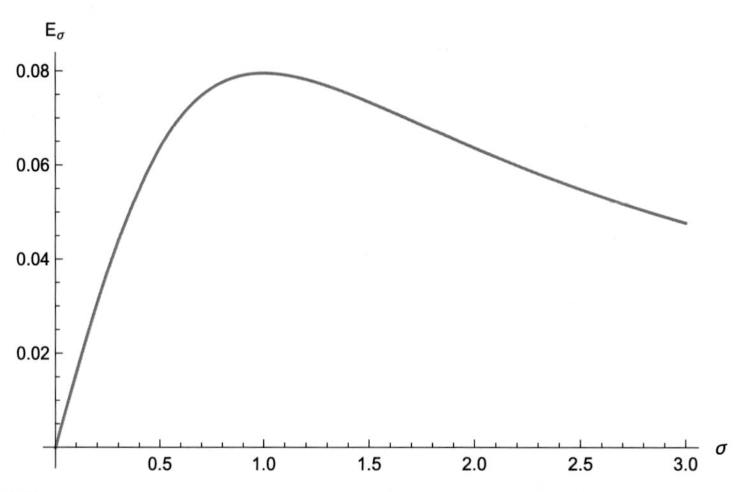

FIGURE 7.9

The γ-normalized squared gradient magnitude at $(x, y) = (0, 0)$ of a prototypical edge with $\tau = 1$.

In Fig. 7.9 the values of E plotted against σ are shown. Solving for where $\nabla E_\sigma(0, 0) = 0$ we realize that the derivative of E in the y-direction is zero, since the image is constant in that direction, and in the x-direction it is also zero, since E is proportional to a Gaussian kernel squared with its center in 0. In the σ-direction, we find that

$$\frac{\partial}{\partial \sigma} E_\sigma(0, 0) = \frac{-\tau^2 + \frac{1}{2}(\sigma^2 + \tau^2)}{\pi(\sigma^2 + \tau^2)^2}, \tag{7.54}$$

and solving for $\frac{\partial}{\partial \sigma} E_\sigma(0, 0) = 0$, we find two solutions: $\sigma = -\tau$ and $\sigma = \tau$. Of these solutions only $\sigma = \tau$ is within our solution domain, and we conclude that the γ-normalized squared gradient magnitude has an extremum at the center of the original function f and at the scale that exactly matches the original width of f. Hence, we can detect not only the position of edges, but also their sizes.

7.6 The scale-space histogram

In this section, we will introduce the Gaussian scale-space histogram, as previously defined [6].

A histogram of an image is a summary of its intensity distribution. Consider a discrete image $I : \Omega \to \Gamma$, where $\Gamma = \{v | v_0 < v < v_1\}$, and where Ω is a subset of $\mathbb{R}^2$ or $\mathbb{R}^3$. A normalized histogram is estimated as the count of intensity values in intensity bins. Parameterizing the bins by their left edges, $v_0 = b_0 < b_1 < \ldots < b_n = v_1$, the value of bin i is proportional to the count of pixels, where $b_i < I(\vec{x}) \leq b_{i+1}$.

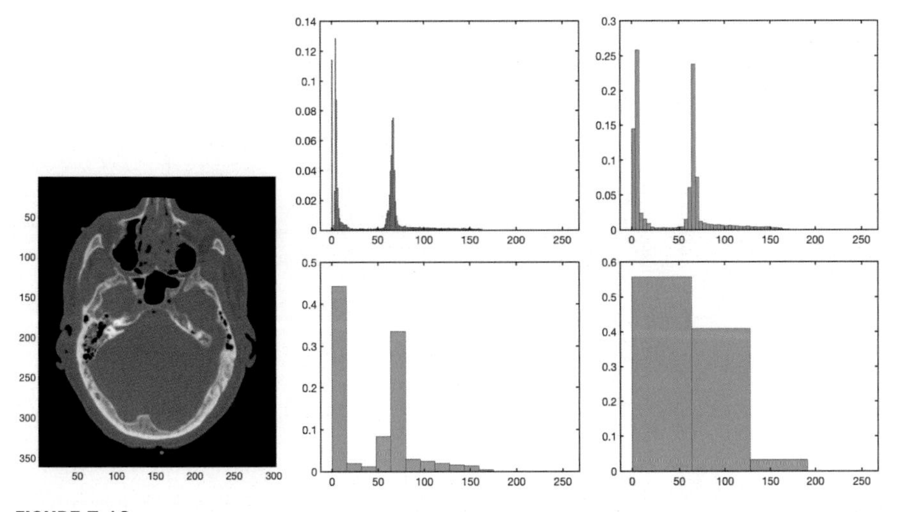

FIGURE 7.10

A histogram is estimated using different bin sizes. Normalized histograms of the top image are shown with 256, 64, 16, and 4 bins corresponding to Fig. 7.13. The original image was obtained from the Visible Human Project at the U.S. National Library of Medicine.

In Fig. 7.10 examples of normalized histograms for different bin sizes of constant width $w = b_{i+1} - b_i$ are shown. We see that the expression of the image content varies with bin width. Further, for constant bin width, we may vary the offset of the bins, i.e., translate all bin edges with a constant. Examples of some offsets are shown in Fig. 7.11.

In the example, the original image has 2^8 different possible intensity values, and thus, we may consider a range of bin widths from 1 to 2^8 or equivalently from 2^8 bins to 1. As for the Gaussian scale-space, we will call these two boundaries the inner and outer intensity scales. The value 2^8 is often chosen as the number of intensity values for an image, since this is the approximate number of intensity values the human eye can distinguish; however, it is most often not related to the scene or object being imaged. Hence, as in the Gaussian scale-space, we must consider all possible bin widths. One solution for a scale-space of histograms is to smooth the histogram at the inner scale to coarser levels. Writing h_0 as the histogram with bins at the inner scale, e.g., $b_{i+1} = b_i + 1$ when the inner scale is 1, we have

$$h_\beta(i) = h_0(i) * g_\beta. \tag{7.55}$$

Examples of Gaussian smoothed histograms are shown in Fig. 7.12. Note that normalized histograms transform as densities; hence, to compare a sampled, smoothed histogram directly with Fig. 7.10, we must either integrate or multiply with $w\sqrt{12}$.

The bin widths correspond to making a pyramid of images, where the number of gray values is reduced, as shown in Fig. 7.13. Such a figure illustrates the informa-

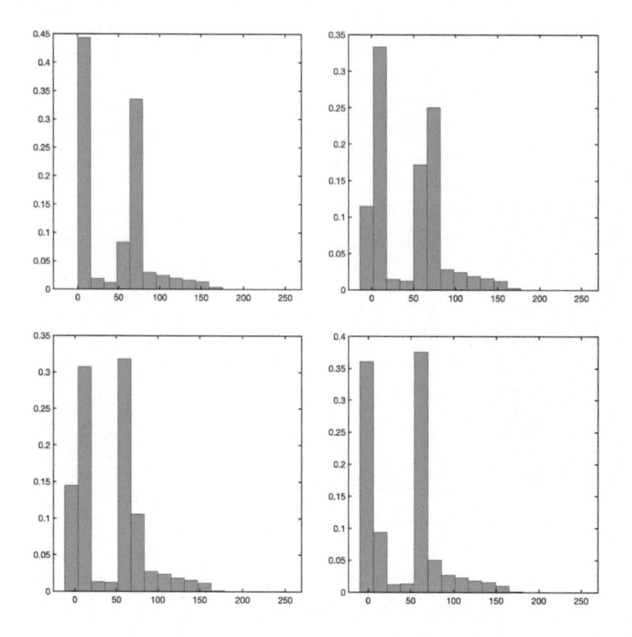

FIGURE 7.11

The histogram depends on the offset of the bin edges. A histogram of the image in Fig. 7.10 with a fixed bin width of 15.9 and varying offsets. Reading from top left, the offsets are −15.9, −13.9, −11.9, and −9.9.

tion content analyzed in the histograms with corresponding bin widths. Conceptually, we may consider the reduction in the number of intensities as an increase in the isophote's thickness. In the figure, the corresponding thicknesses are 1, 4, 16, and 64 in the original intensity values. In the intensity scale-space, the equivalent is a soft isophote defined as

$$p_{f,i,\beta}(\vec{x}) = \exp\left(\frac{-(f(\vec{x}) - i)^2}{2\beta^2}\right). \tag{7.56}$$

In the image from Fig. 7.10, the bone has an approximate value of 144. Some corresponding soft isophotes are shown in Fig. 7.14.

Local histograms may be evaluated by extracting a subimage and calculate the histogram of a subimage. For the general problem and as for space and intensity, we will investigate all possible subimages at all possible locations, and we will use a soft windowing function with width α. Thus, the local normalized histogram for a region centered at $\vec{x}$ and with a width scale α for an image with a spatial scale σ and an intensity scale β is

$$h_{\vec{x},\alpha,\beta,\sigma}(i) = \left(g_\alpha * p_{f_\sigma,i,\beta}\right)(\vec{x}). \tag{7.57}$$

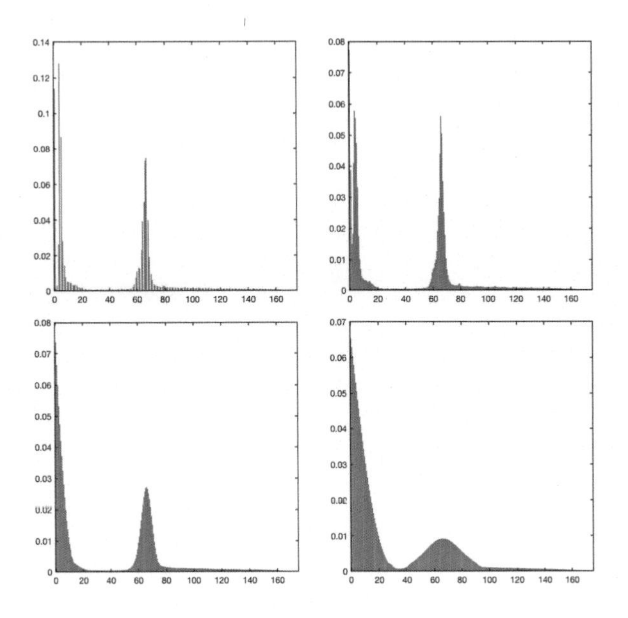

FIGURE 7.12

A Gaussian scale-space of a histogram. The widths have been set to align with the corresponding graphs in Fig. 7.10, i.e., scales $\beta = w/\sqrt{12}$.

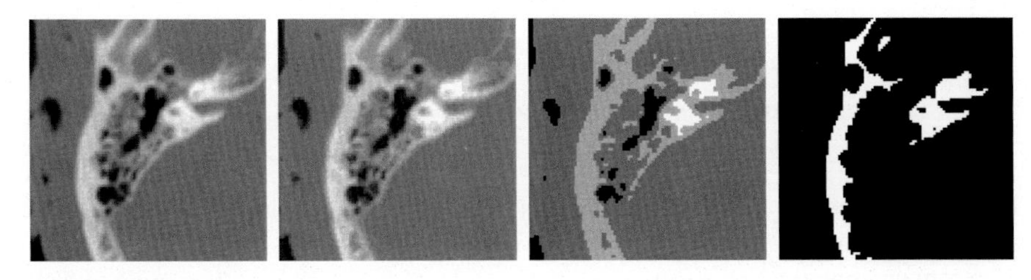

FIGURE 7.13

An image may be viewed using a different number of intensity values. Reading from top right, the images have 45, 12, 4, and 2 intensity levels. The original image was obtained from the Visible Human Project at the U.S. National Library of Medicine.

We call σ, β, and α for the intrinsic scales. No simple relations between σ, β, and α exist, but we can make simple interpretations of the relation between local histograms for varying values of these scale-space parameters. For $\sigma, \beta, \alpha \to 0$ a local histogram will be a delta function at the value of $I(\vec{x})$. Increasing σ makes the image smoother and reduces the number of intensity values in the image, and in the limit of $\sigma \to \infty$, all local histograms converge to the delta function at the mean image intensity. Increasing β makes the histogram smoother, and in the limit of $\beta \to \infty$, all local histograms converge to the constant histogram. Finally, increasing α makes the

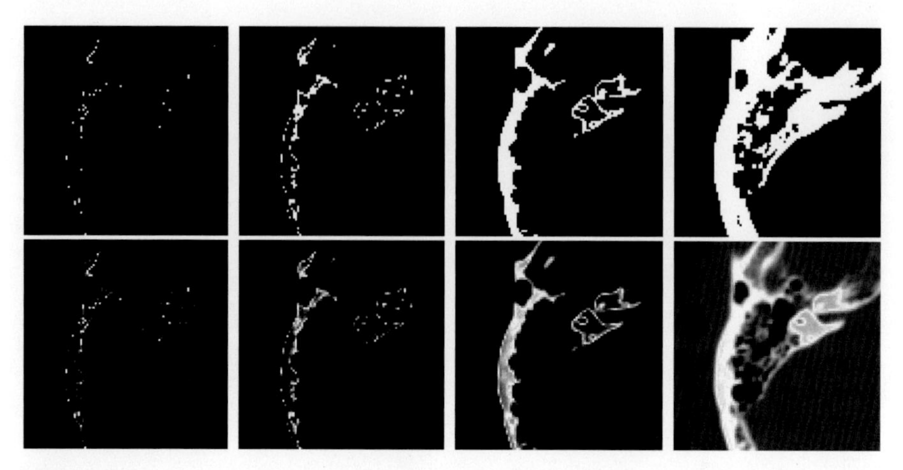

FIGURE 7.14

The bin width of a histogram corresponds to a thick isophote. Here isophote 144 is shown with thickness 2, 8, 32, and 128 (top row) and corresponding $\beta = 0.6, 0.9, 2.3, 37.0$ (bottom row) of a subimage from Fig. 7.10.

Table 7.1 A verbal description of the scales used in Fig. 7.16. Although applied differently, the σ- and α-parameters are both spatial scales, and thus have the same description.

Parameter	1.0	2.5	6.3	16.0
σ, α	$\sim$ inner scale	fine bone	skull	nose
β	$\sim$ inner scale	$\sim$ noise level	bone values	bone cavity and bone

local histograms less local, and in the limit of $\alpha \to \infty$, all histograms converge to the global histogram.

From an analytical perspective, all derivatives of $h_{\vec{x},\alpha,\beta,\sigma}(i)$ with respect to $\vec{x}$, α, β, σ, and i exist, and the Taylor series of h converges.

Examples of local histograms are shown in Fig. 7.15. The three intrinsic scales influence the resulting histogram in different manners, which is illustrated in Fig. 7.16. As should be expected, the figure shows that there is no simple relation between the values of σ, β, and α. The position (97, 188) is in the middle of two bright bone areas, and thus, a verbal description of the different variations in scales is given in Table 7.1.

7.7 Exercises

1. Write a function `gaussSmooth` in your favorite programming language, which takes an image $I : \mathbb{R}^2 \to \mathbb{R}$ and a standard deviation σ returns the image smoothed with a Gaussian with standard deviation σ.

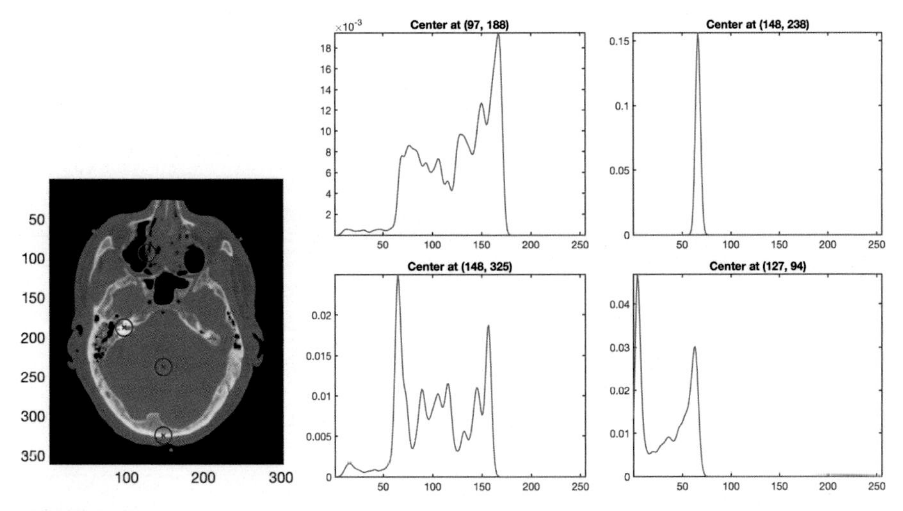

FIGURE 7.15

Using local histograms to summarize local image content. Top shows the image with circles indicating where the local histograms have been taken. The scale values are $\sigma = 1$, $\beta = 2.5$, and $\alpha = 6.3$.

2. Use `gaussSmooth` from Exercise 1., and make a function `scaleSpace` in your favorite programming language, which takes an image I and a sequence of standard deviations $\sigma List$ and returns a sequence or stack of smoothed images according to $\sigma List$.
3. Given a function in terms of its Fourier transform $f(x) = \int_{-\infty}^{\infty} F(u) \exp(2\pi iux)\, du$, where $i = \sqrt{-1}$, show that $\frac{d^n}{dx^n} f(x) = \int_{-\infty}^{\infty} (2\pi iu)^n F(u) \exp(2\pi iux)\, du$.
4. Extend the result in Exercise 3. to 2 dimensions and show that differentiation $\frac{\partial^{m+n}}{\partial x^m \partial y^n} f(x, y)$ is equivalent to filtering with a kernel, whose Fourier transform is $(2\pi iu)^m (2\pi iv)^n$, where (u, v) are the Fourier transformation variables corresponding to (x, y).
5. Make a function `scale` in your favorite programming language which takes an image $I : \mathbb{R}^2 \to \mathbb{R}$, a standard deviation σ and the pair (m, n) and returns $\frac{\partial^{m+n}}{\partial x^m \partial y^n} I(x, y)$ smoothed with a Gaussian kernel of standard deviation σ. The implementation must implement the smoothing in the Fourier domain and differentiation using multiplication with the monomials from Exercise 4..
6. Make an alternative algorithm for calculating derivatives of an image by first smoothing the image with `gaussSmooth` from Exercise 1. followed by finite differences to for estimating derivatives, such as the central difference $df(x, y) \simeq \frac{f(x+1,y)-f(x-1,y)}{2}$. Compare the result with `scale` with the same degree of smoothing and w.r.t. how the two algorithms reduce noise and the computation speed for a range of σ-values.

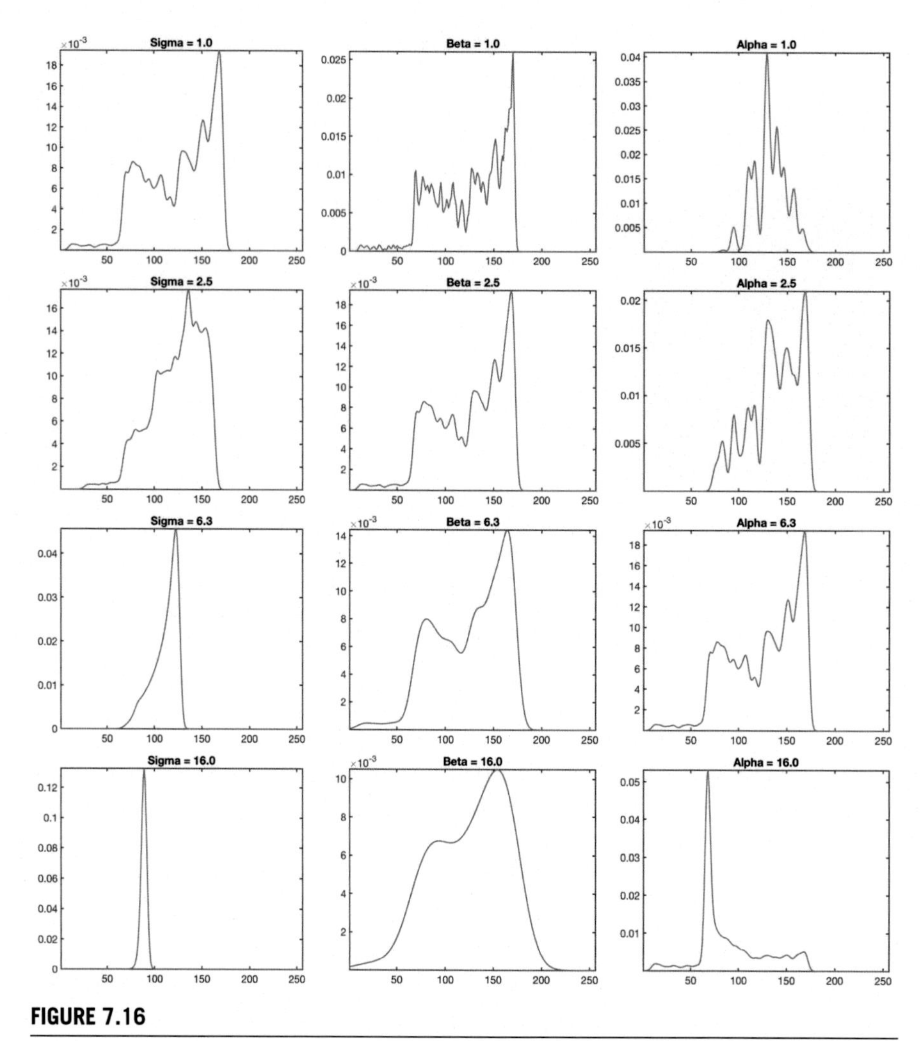

FIGURE 7.16

The local histogram at (97, 188). The scales were $\sigma = 1$, $\beta = 2.5$, and $\alpha = 6.3$ except in the first column we varied σ, in the second column we varied β, and in the third column we varied α. The top first, middle second, and bottom third histogram are intentionally set to use identical parameters.

7. Implement blob detection using an exponential list of scales $k^j \sigma_0$, $k > 1$, $j = 0, 1, \ldots$, and detect extrema by comparing neighbors in (x, y, σ) space. Test your algorithm first on a simple image with a range of blob-like objects, e.g., a synthetic image, and then on a real image. What do you observe about the extrema for which L in (7.38) is largest?
8. Make a function `histogram` which takes an image $I : \mathbb{R}^2 \to \mathbb{R}$, the set of scales (α, β, σ) and returns the scale-space histogram $h_{I,\alpha,\beta,\sigma}(x, y, i)$.

9. Choose or generate an image with a range of different textures situated next to each other. Calculate the scale-space histogram of it and evaluate summary measures of each local histogram such as the variance or the entropy. Choose a measure and the scale parameters (α, β, σ) which separates the textures, and calculate the gradient magnitude of this measure-image.

10. Consider the variance of a local histogram for fixed scale parameters (α, β, σ), and calculate the analytically the (spatial) gradient magnitude squared of the variance. What are the advantages and disadvantages of this analytical expression as compared to an algorithm, where the gradient magnitude squared is estimated using finite differences of the image of variances?

References

[1] Burt P.J., Adelson E.H., The Laplacian pyramid as a compact image code, IEEE Transactions on Communications COM-31 (4) (1983) 532–540.

[2] Iijima T., Basic theory of pattern observation, Tech. rep. (in Japanese), Technical Group on Automata and Automatic Control, IECE, Japan, 1959.

[3] Witkin A.P., Scale-space filtering, in: Proc. 8th Int. Joint Conf. Art. Intell., Karlsruhe, Germany, 1983, pp. 1019–1022.

[4] Koenderink J.J., The structure of images, Biological Cybernetics 50 (1984) 363–370.

[5] Lindeberg T., Feature detection with automatic scale selection, International Journal of Computer Vision 30 (2) (1998) 77–116.

[6] Koenderink J.J., Van Doorn A.J., The structure of locally orderless images, International Journal of Computer Vision 31 (2) (1999) 159–168.

PART

Medical image segmentation

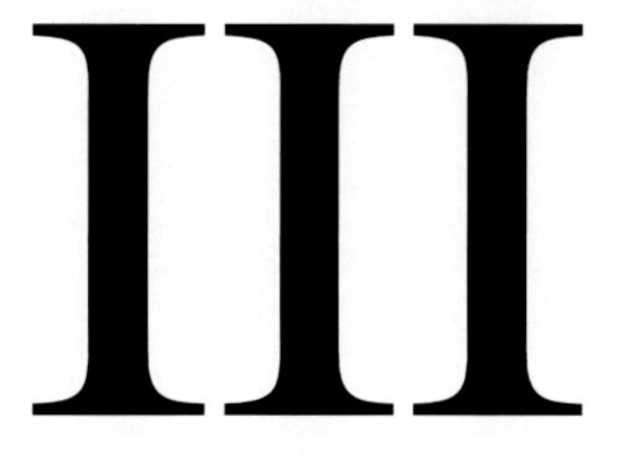

CHAPTER

Statistical shape models

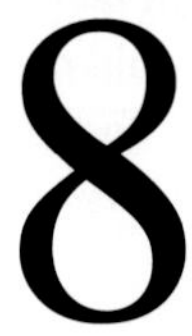

Tim Cootes

Division of Informatics, Imaging and Data Sciences, The University of Manchester, Manchester, United Kingdom

Learning points

- Explicit shape descriptions
- Statistical models of shape
- Statistical models of appearance
- Automatic methods for fitting models to images

8.1 Introduction

In many medical applications the shape of an object in an image is important. Sometimes one is interested in geometric measurements (distances, angles), but more often one wishes to understand the *variation* in shape, particularly when looking for abnormalities. For instance, human bones such as the femur vary from one person to another, but changes outside the normal range may be signs of disease.

This chapter focuses on representing shape using a set of labeled *landmark* points, and how one can use such a representation to build models of the variation in shape across a set. Such models can be used to compactly encode examples from a class of shapes and to help locate new examples in previously unseen images.

We assume that for each landmark on each example there is an equivalent landmark on every other example. This means that the methods described are only appropriate for classes of objects in which any individual example can be thought of as a deformed version of some reference template. For instance, the outlines of most normal left femur bones have roughly the same shape and one could generate any of them by modest deformations of any other. Thus the techniques can be applied to many bones and organs in the body. However, the methods will not work for things such as tree structures, where there is not necessarily equivalence between branches from one example to another.

Whether or not a meaningful correspondence exists between points on two shapes depends on the resolution of the measurements being made. For instance, if points are placed roughly every millimeter on the boundary of a bone, it is reasonable to find equivalent points on another bone. However, if one placed points at the center of each

Medical Image Analysis. https://doi.org/10.1016/B978-0-12-813657-7.00022-4

cell making up the bone, there could not be a sensible correspondence as other bones would not have equal numbers of cells.

In the following we will focus on 2D shape models initially, but there is a natural extension to 3D.

8.2 Representing structures with points

We represent the geometric information in a structure using a set of user-defined *landmark* points. These define the location of interesting parts of the structure (such as boundaries or the centers of substructures).

The position of each landmark should be well defined so that it can be located consistently on each example. Thus the same number of landmarks must be placed on each example of the structure, and each defines a *correspondence* between the images containing the structures (see Fig. 8.1).

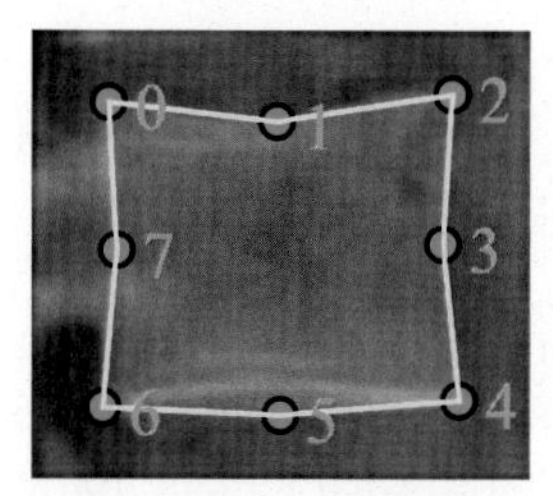

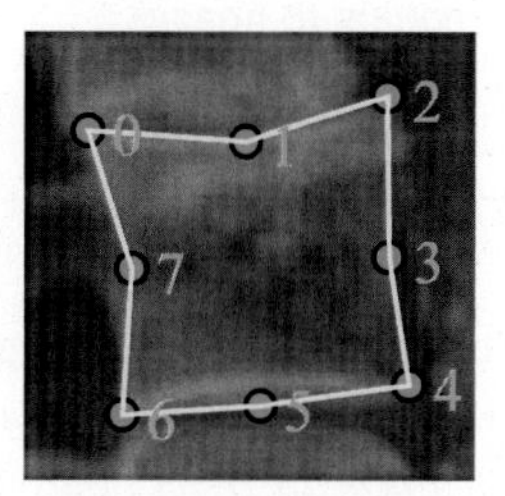

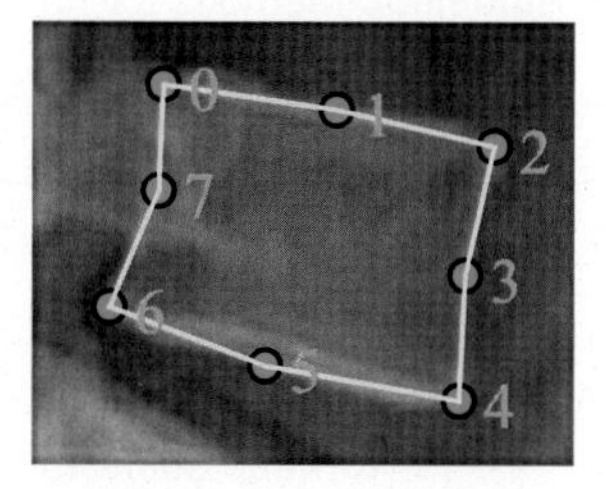

FIGURE 8.1 Point annotations.

Points used to represent the outline of a vertebral body. Each point is numbered and defines a particular position on the object. The same numbers are used on each example. Corners $(0, 2, 4, 6)$ are well defined in 2D (Type 1). Others are used to define the boundary in more detail (Type 2).

In 2D there are three types of such landmarks:

1. points which are well defined in 2D (such as corners, centers of small structures, etc.),
2. points which are well defined in 1D (such as points on boundaries),
3. points which are weakly defined (such as points at the intersections of occluding boundaries).

Most shapes consist of Types 1 and 2 only. A shape is defined by some well-defined corners, together with points along boundaries between them. The latter are only present to define the curve of the boundary, so their exact position along the curve can vary (see Fig. 8.1). Typically such boundary points are equally spaced between Type 1 landmarks.

An important advantage of the point-based representation is that it can be used for structures containing multiple parts – for instance, the model in Fig. 8.1 could be extended with more points to describe a neighboring vertebra – the model would then enable study of both the shapes of the vertebrae and their relative positions.

8.3 Comparing shapes

In 2D an individual shape is represented by the set of n points, $\{(x_i, y_i)\}$, which can be encoded in a single vector

$$\mathbf{x} = (x_1, y_1, \ldots, x_n, y_n)^T. \tag{8.1}$$

Two shapes are considered to be the same if one can translate, rotate, and scale one so that it exactly matches the other. For instance, a square remains a square if it is moved, scaled, or translated.

This observation leads to a mathematical definition of shape (due to Kendall) as *that geometric information which is invariant to some class of transformations.*

Usually we work with similarity transformations (translation, rotation, scale), but in some cases one may use translations only, rigid transformations (translation+rotations), dilations (translation+scale), or affine transformations.

Suppose $T(\mathbf{x}; \mathbf{t})$ applies a transformation to the points defined in $\mathbf{x}$ with parameters $\mathbf{t}$. For instance, if $T()$ is the 2D similarity transformation,

$$T(\mathbf{x}; \mathbf{t}) = s\mathbf{R}_\theta \mathbf{x} + \begin{pmatrix} t_x \\ t_y \end{pmatrix} = \begin{pmatrix} a & -b \\ b & a \end{pmatrix} \mathbf{x} + \begin{pmatrix} t_x \\ t_y \end{pmatrix}, \tag{8.2}$$

where s is a scaling factor and $\mathbf{R}_\theta$ is a matrix which applies a rotation by θ, then linear parameterization is defined by $\mathbf{t} = (a, b, t_x, t_y)^T$.

To compare two shapes we can compute the sum of square distances between their points.

If $\mathbf{x}_i$ is point i from shape $\mathbf{x}$, then the sum of square distances between two shapes, $\mathbf{x}$ and $\mathbf{z}$, is given by

$$d = \sum_{i=1}^{n} |\mathbf{x}_i - \mathbf{z}_i|^2 = |\mathbf{x} - \mathbf{z}|^2. \tag{8.3}$$

Two shapes $\mathbf{x}$ and $\mathbf{z}$ can be considered to be the same under a particular class of transformations if one can be transformed to exactly match the other, so that

$$\min_{\mathbf{t}} |\mathbf{z} - T(\mathbf{x}; \mathbf{t})|^2 = 0. \tag{8.4}$$

8.4 Aligning two shapes

To align shape $\mathbf{x}$ to $\mathbf{z}$ we find the transformation that maps $\mathbf{x}$ as closely as possible to $\mathbf{z}$. This involves finding the parameters $\mathbf{t}$ to minimize

$$Q(\mathbf{t}) = |\mathbf{z} - T(\mathbf{x}; \mathbf{t})|^2. \tag{8.5}$$

When $T()$ is a similarity transformation there is a simple linear solution to this.

Let the vector $\mathbf{x}_x = (x_1, ..., x_n)^T$ be the x-values from $\mathbf{x}$ and let $\mathbf{x}_y = (y_1, ..., y_n)^T$ be the y-values from $\mathbf{x}$. Similarly, let $\mathbf{z}_x$ and $\mathbf{z}_y$ be the x- and y-values from shape $\mathbf{z}$.

Let us also assume that $\mathbf{x}$ has been translated so that its center of gravity is at the origin. This can be achieved by applying a translation of $(-\mathbf{x}_x.\mathbf{1}/n, -\mathbf{x}_y.\mathbf{1})^T/n$.

For the case of 2D similarity transformations, the optimal choice of parameters to minimize (8.5) is given by

$$\begin{aligned} t_x &= \mathbf{z}_x.\mathbf{1}/n, \quad a = \mathbf{x}.\mathbf{z}/|\mathbf{x}|, \\ t_y &= \mathbf{z}_y.\mathbf{1}/n, \quad b = (\mathbf{x}_x.\mathbf{z}_y - \mathbf{x}_y.\mathbf{z}_x)/|\mathbf{x}|. \end{aligned} \tag{8.6}$$

When aligning in cases where some points may be corrupted with outliers, one must find the parameters that minimize a robustified version of the cost function

$$d = \sum_{i=1}^{n} \rho(|\mathbf{z}_i - T(\mathbf{x}_i)|), \tag{8.7}$$

where $\rho(.)$ is some suitable robust kernel.

8.5 Aligning a set of shapes

To analyze a set of N shapes, $\{\mathbf{x}_j\}$, we need to first align them into a common reference frame, a process known as generalized Procrustes analysis. The aim is to remove the effects of the general pose transformation (such as translation, rotation, and scaling) before applying statistical analysis. If this is not done, then the shape model will include translations, rotations, and scaling effects which may be rather arbitrary and thus limit the ability of the model to generalize to new examples.

When building a linear shape model, we seek the transformations that minimize

$$\sum_{j=1}^{N} |\mathbf{x}_j - T_j(\bar{\mathbf{x}})|^2, \quad \text{where} \quad \bar{\mathbf{x}} = \frac{1}{N}\sum_{j=1}^{N} T_j^{-1}(\mathbf{x}_j). \tag{8.8}$$

This can be interpreted as finding a set of transformations that best map the discovered mean to match each of the original target shapes (see Fig. 8.2).

The aligned shapes are given by applying the inverse of the transformation to each $T_j^{-1}(\mathbf{x}_j)$.

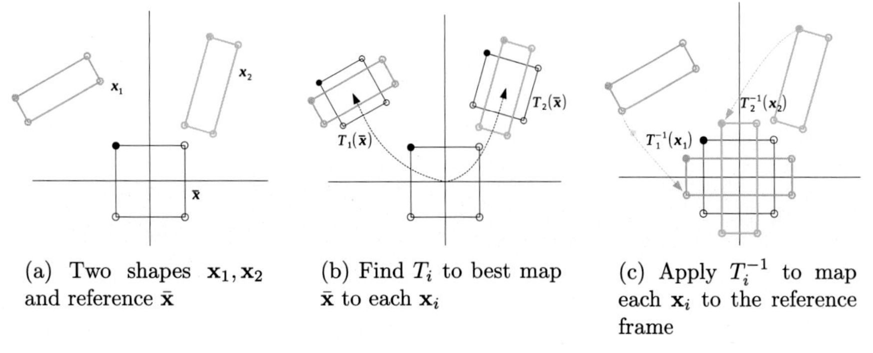

(a) Two shapes $\mathbf{x}_1, \mathbf{x}_2$ and reference $\bar{\mathbf{x}}$

(b) Find T_i to best map $\bar{\mathbf{x}}$ to each $\mathbf{x}_i$

(c) Apply T_i^{-1} to map each $\mathbf{x}_i$ to the reference frame

FIGURE 8.2 Aligning shapes.

Shapes are aligned into a common reference frame.

8.5.1 Algorithm for aligning sets of shapes

A range of algorithms have been proposed for aligning sets of shapes [1–4]. A simple iterative approach is as follows:

1. Choose one shape to define the orientation, say $\mathbf{x}_r = \mathbf{x}_0$.
2. Center $\mathbf{x}_r$ so its center of gravity is at the origin.
3. Scale $\mathbf{x}_r$ to have unit size ($|\mathbf{x}_r| = 1$).
4. Set initial estimate of mean $\bar{\mathbf{x}} = \mathbf{x}_r$.
5. Repeat until convergence:
   a. For each shape compute T_j to minimize $|T_j(\bar{\mathbf{x}}) - \mathbf{x}_j|^2$.
   b. Update estimate of mean: $\bar{\mathbf{x}} = \frac{1}{N}\sum_{j=1}^{N} T_j^{-1}(\mathbf{x}_j)$.
   c. Align $\bar{\mathbf{x}}$ to $\mathbf{x}_r$ to fix rotation, then scale so that $|\bar{\mathbf{x}}| = 1$.

   This usually converges in a few iterations.

8.5.2 Example of aligning shapes

Fig. 8.3 shows examples from a set of images of a face with 68 landmark points. Fig. 8.4 shows the scatter of all the points before and after aligning to a reference frame with different transformations. Note that when the faces are just translated to align them there is significantly more scatter about the mean positions.

8.6 Building shape models

The points of the aligned shapes often form approximately Gaussian clusters. Thus a simple model would be to treat a new shape as a displacement from the mean shape, where each point has an independent Gaussian distribution with covariance estimated from the training set. However, such a model would not take account of the correla-

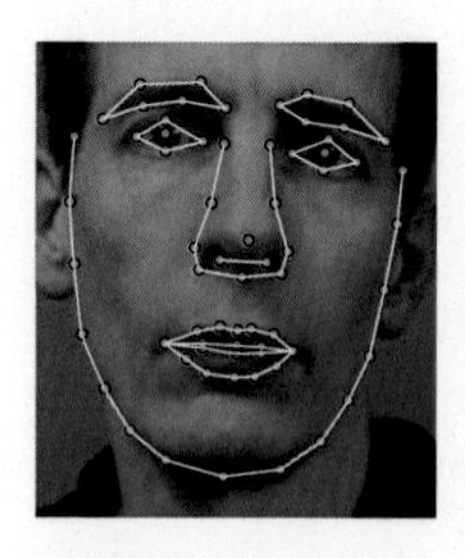
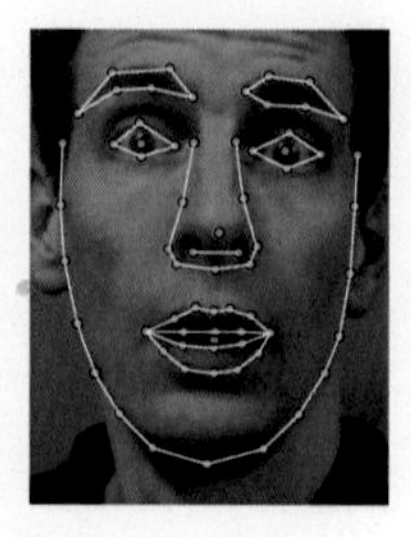
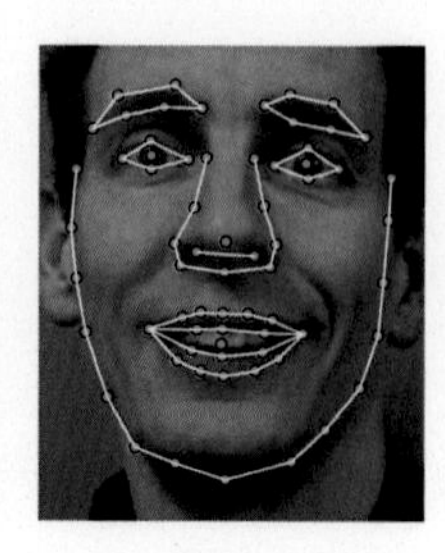
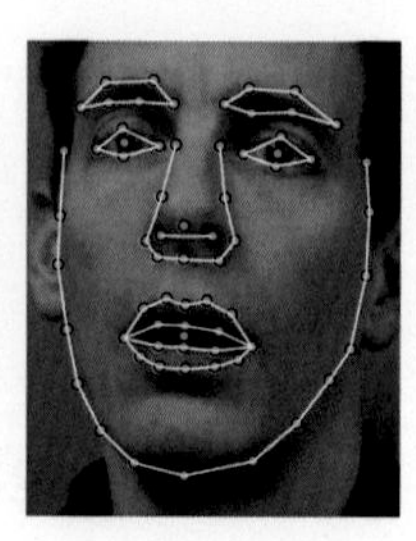

FIGURE 8.3 Face examples.

Examples of a face with 68 landmark points.

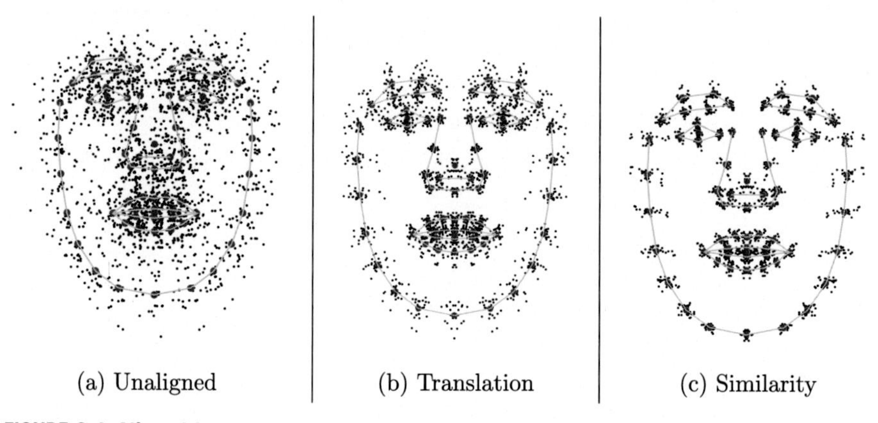

FIGURE 8.4 Aligned faces.

Results of aligning a set of faces with different transformations. Mean points are shown in red.

tions between the points. Nearby points on boundaries tend to move together, and shapes can exhibit symmetries which limit the independence of points. For instance, the approximate symmetry of the human face leads to strong correlations between points on the left and their equivalents on the right.

Principal component analysis (PCA) is a technique for taking account of such correlations. It enables the construction of a simple linear model with linearly independent parameters. Given a set of aligned shapes $\{\mathbf{x}_i\}$, we build a linear statistical shape model by applying PCA using the following steps:

1. Compute the mean of the set $\bar{\mathbf{x}} = \frac{1}{n}\sum_i \mathbf{x}_i$.
2. Compute the covariance about the mean, $\mathbf{S} = \frac{1}{n-1}\sum_i (\mathbf{x}_i - \bar{\mathbf{x}})(\mathbf{x}_i - \bar{\mathbf{x}})^T$.
3. Compute the eigenvectors $\{\mathbf{p}_j\}$ and associated eigenvalues λ_j of $\mathbf{S}$.
4. Choose the number of modes t to retain in the model.

The final model has the form

$$\mathbf{x} = \bar{\mathbf{x}} + \mathbf{P}\mathbf{b}, \tag{8.9}$$

where $\mathbf{P} = (\mathbf{p}_1|\ldots|\mathbf{p}_t)$ is a matrix whose columns are the t eigenvectors associated with the largest eigenvalues.

Note that when there are many points and few examples there are more efficient methods of computing the eigenvectors and eigenvalues than creating the whole covariance matrix.

8.6.1 Choosing the number of modes

The number of modes (eigenvectors) retained, t, defines the amount of flexibility in the model and will affect the performance of any system which uses the resulting model. If too few modes are retained, the model will not be able to represent the full variation seen in the training shapes. However, usually the modes associated with small eigenvalues are effectively just noise, so are not useful.

The eigenvalues, λ_j, describe the variance of the training data projected onto the associated eigenvector, $\mathbf{p}_j$. The total variance about the mean in the training set is given by the sum of all eigenvalues, $v_T = \sum_j \lambda_j$. A widely used approach is to choose t to represent a particular proportion of the total variance, p_t, such as 0.95 or 0.98;

$$t = \arg\min_{t'} \sum_{j=1}^{t'} \lambda_j > p_t v_T, \tag{8.10}$$

where the eigenvalues are sorted, $\lambda_j \geq \lambda_{j+1}$.

This ensures that most of the main modes of variation are encoded in the model, which is usually sufficient for most applications. The optimal number of modes retained will depend on the application, and is best found by experiment.

8.6.2 Examples of shape models

Fig. 8.5 shows the effect of changing the first three parameters of a 68-point shape model of the face. Modes 1 and 3 approximate rotations of the head. Procrustes analysis can correct for rotations in the image plane, but not those out of the plane – these lead to modes such as those shown. Mode 2 varies between narrow and wide faces. Later modes capture changes in expression. Note that the order of the modes is based on the total variance exhibited in the training set and may vary for other sets. Although the modes can often appear to capture particular types of variation or expression, they are not always "pure". For instance, face mode 3 includes a hint of a smile as well as a nodding effect, suggesting that people looking down in the training set were more likely to be smiling.

Fig. 8.6 shows the effect of changing the first three parameters of a shape model of the pelvis, where 74 points were used to represent the outline of the structure in a

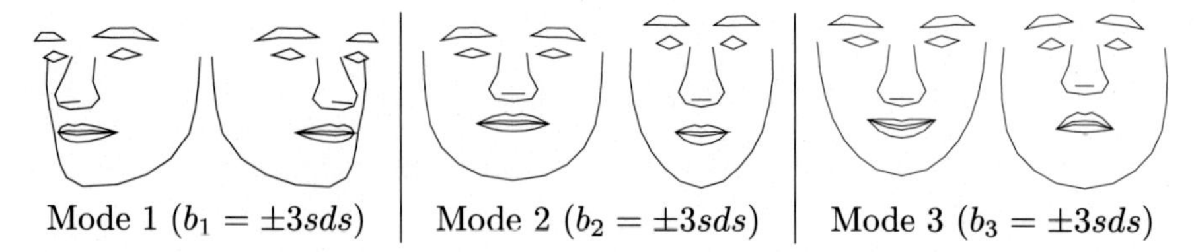

FIGURE 8.5 Face shape models.

Modes of a shape model of the face trained on different people and expressions.

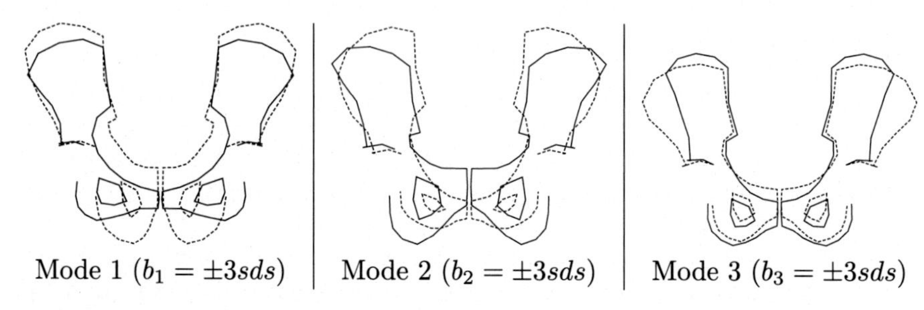

FIGURE 8.6 Pelvis shape models.

First three modes of a shape model of the pelvis.

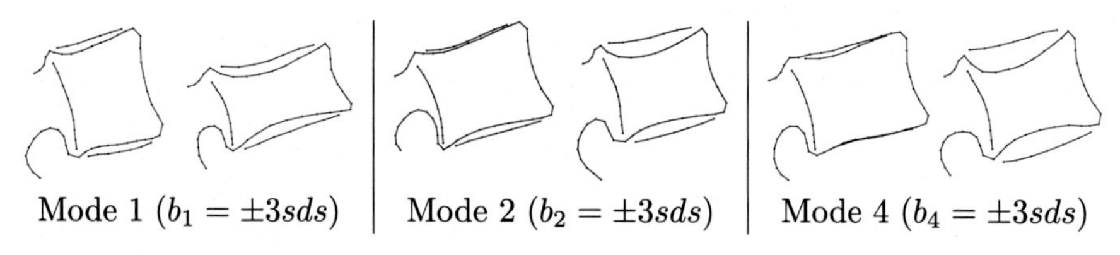

FIGURE 8.7 Vertebra shape models.

Three modes of a shape model of a vertebra.

radiograph.[1] The first two modes show large shape changes, which are related to the differences between men and women.

Fig. 8.7 shows three modes from a shape model of a vertebra as it appears in a lateral radiograph. Each vertebra was represented by 69 points. The first mode corresponds to the effect of a "crush" fracture.

8.6.3 Matching a model to known points

Given a set of points $\mathbf{z}$, one often wishes to find the best approximation with the shape model, enabling one to estimate the shape parameters representing $\mathbf{z}$. In the simplest

[1] Thanks to C. Lindner and R. Ebsim for the annotations.

case one finds the shape and pose parameters to minimize

$$Q_2(\mathbf{b},\mathbf{t}) = |T(\bar{\mathbf{x}} + \mathbf{P}\mathbf{b};\mathbf{t}) - \mathbf{z}|^2. \tag{8.11}$$

A fast alternating algorithm for this is as follows.

Initialize $\mathbf{b}$, for instance $\mathbf{b} = 0$

Repeat

- Find the optimal pose, $\mathbf{t}$, to minimize $Q_2(\mathbf{b},\mathbf{t})$;
- Project $\mathbf{z}$ into the model space: $\mathbf{z}' = T^{-1}(\mathbf{z};\mathbf{t})$;
- Estimate the shape using $\mathbf{b} = \mathbf{P}^T(\mathbf{z}' - \bar{\mathbf{x}})$;

Until converged.

This will usually converge in a small number of iterations. Minimizing equation (8.11) is equivalent to finding the most likely parameters if (a) the uncertainties on all the points in $\mathbf{z}$ are equal isotropic Gaussians and (b) the prior distribution of shape and pose parameters is flat. A more general case is to assume that each point in $\mathbf{z}$, $\mathbf{z}_i$, has a Gaussian uncertainty with an inverse covariance matrix, $\mathbf{W}_i$.

In this case the optimal parameters are found by minimizing

$$Q_w(\mathbf{b},\mathbf{t}) = 0.5\sum_{i=1}^{n}(\mathbf{x}_i(\mathbf{b},\mathbf{t}) - \mathbf{z}_i)^T\mathbf{W}_i(\mathbf{x}_i(\mathbf{b},\mathbf{t}) - \mathbf{z}_i) - \log p_b(\mathbf{b}), \tag{8.12}$$

where $\mathbf{x}_i(\mathbf{b},\mathbf{t})$ is the position of model point i after transformation and $p_b(\mathbf{b})$ is the prior distribution of the shape parameters.

In cases where candidates for point positions are not available (such as due to occlusion or failure of a detector) the corresponding inverse covariance matrix, $\mathbf{W}_i$, can be set to zero, indicating no knowledge about its position. The position will be estimated from those of the other points.

8.7 Statistical models of texture

We can use a similar methodology to model patterns of intensity (texture). However, to do this we must place pixels in correspondence so that we are comparing intensities at equivalent positions on different images. We wish to sample the intensity at a dense set of positions in a consistent way. For instance, the 117th sample point might be in the middle of the cheek, and should be so placed on every image. The landmarks define a sparse set of correspondences – we need to interpolate between them to generate a dense set of sampling positions.

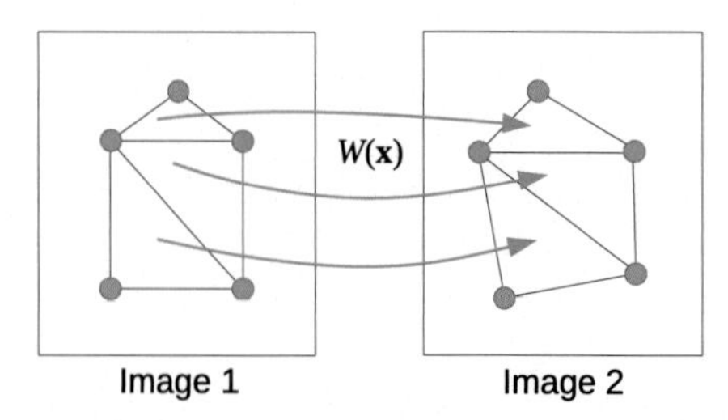

FIGURE 8.8 Deformation function.

The function $W(\mathbf{x})$ maps from image 1 to image 2.

This is equivalent to warping the image (deforming it as if printed on an elastic sheet) so that the landmarks are placed in consistent positions in a reference frame (typically chosen to be the mean point positions), and all other pixels end up at roughly corresponding positions. Once the images are warped into this common frame we can compare intensities in equivalent pixels directly, as each will correspond to the same position on the original objects. This enables the study of the pattern of intensities across a region. Warping each example into a reference frame removes variations due to shape, leaving variations due only to the intensity changes.

Suppose that we have two images, $I_1(\mathbf{x})$ and $I_2(\mathbf{x})$, with corresponding points $\{\mathbf{x}_{1,i}\}$ and $\{\mathbf{x}_{2,i}\}$. We can construct a deformation function, $W(\mathbf{x})$, which maps points from image 1 to image 2, so that

$$W(\mathbf{x}_{1i}) = \mathbf{x}_{2i} \quad \forall \quad i = 1 \ldots n. \tag{8.13}$$

W can either be a thin plate spline [5] or a piecewise affine function (which maps triangles in one image to equivalent triangles in the other using a linear transformation) (see Fig. 8.8).

Given such a function, to warp image 2 into the frame of image 1, we scan through all desired pixels (i, j) in the new image, I_{new}, and interpolate their values from image 2 using

$$I_{new}(i, j) = I_2(W(i, j)). \tag{8.14}$$

Given a set of annotated images, we can construct a shape model as described above. This gives us a mean shape $\bar{\mathbf{x}}$ which can define a reference frame in which we can study the texture.

Typically the mean shape is of unit size ($|\bar{\mathbf{x}}| = 1$) because of the normalization steps in the model construction. Since it is convenient to use integer indexed pixels, we must construct a set of reference points as a scaled version of the mean, $\bar{\mathbf{x}}_r = s\bar{\mathbf{x}}$, where the scale is chosen so that the resulting shape has a particular width (or area). This defines the resolution of the model (the number of pixels to be used) (see Fig. 8.9).

Let $\mathbf{g}_i$ be a vector containing the intensity values from the pixels sampled from image i (i.e., all the pixels in the reference frame that have been interpolated from

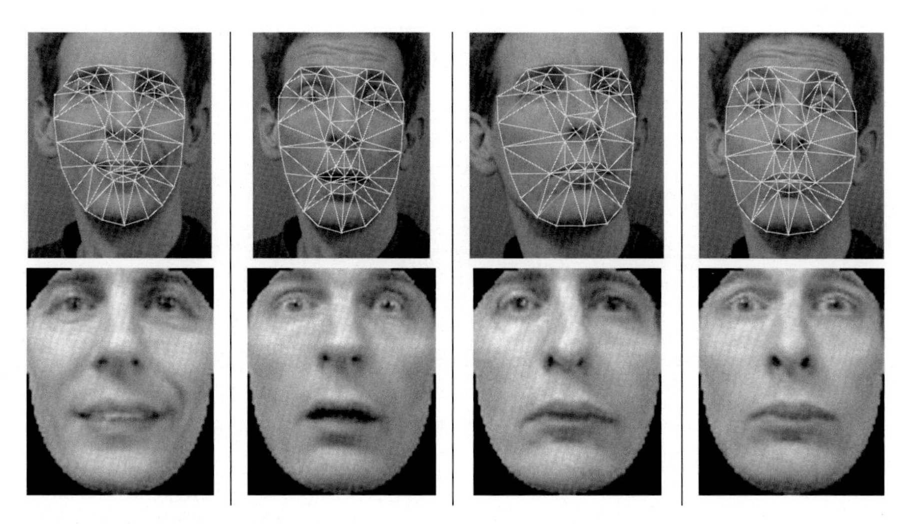

FIGURE 8.9 Warped examples.

Examples of faces warped to the reference frame (defined by the mean). Piecewise affine transformations based on the triangles in the mesh were used to represent the deformation.

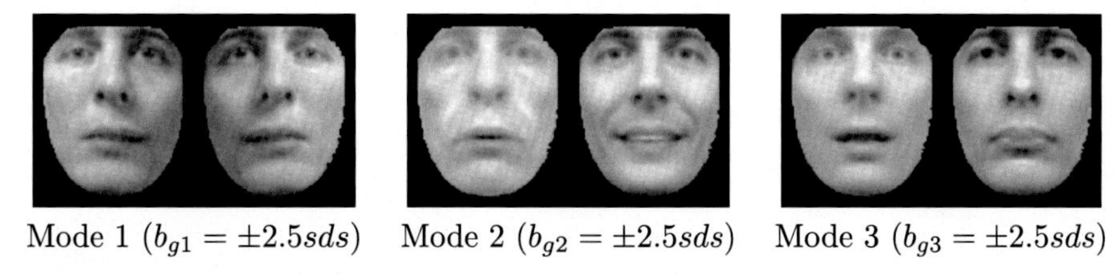

FIGURE 8.10 Facial texture modes.

First three modes of a texture model of a face.

image I_i). If necessary we may normalize this vector by subtracting the mean of all values (so that the mean is then zero) and then dividing by the length to produce a unit vector ($|\mathbf{g}_i| = 1$). This accounts for lighting variation. We can then apply PCA to the set of vectors, $\{\mathbf{g}_i\}$, to learn a linear model of the intensities with the form

$$\mathbf{g} = \bar{\mathbf{g}} + \mathbf{P}_g \mathbf{b}_g, \tag{8.15}$$

where $\bar{\mathbf{g}}$ is the mean, $\mathbf{P}_g$ is a matrix with the main eigenvectors, and $\mathbf{b}_g$ are the parameters which control the weights on each mode.

Fig. 8.10 shows the effect of varying each of the first three parameters of a model of the face by ± 2.5 standard deviations from the mean. Note that the shape is the same in each case, only the pattern of intensities changes.

8.8 Combined models of appearance (shape and texture)

The shape and texture of any example can be summarized by the parameter vectors $\mathbf{b}$ and $\mathbf{b}_g$. Since there may be correlations between the shape and texture variations, we apply a further PCA to the data as follows. For each example we generate the concatenated vector

$$\mathbf{b}_c = \begin{pmatrix} w_s\mathbf{b} \\ \mathbf{b}_g \end{pmatrix} = \begin{pmatrix} w_s\mathbf{P}^T(\mathbf{x}-\bar{\mathbf{x}}) \\ \mathbf{P}_g^T(\mathbf{g}-\bar{\mathbf{g}}) \end{pmatrix}, \tag{8.16}$$

where w_s is a scaling factor applied to the shape parameters allowing for the difference in units between the shape and intensity models. By applying PCA on these vectors we obtain a model combining the shape and intensity components

$$\mathbf{b}_c = \mathbf{P}_c\mathbf{c}, \tag{8.17}$$

where $\mathbf{P}_c$ are the eigenvectors and $\mathbf{c}$ is a vector of *appearance* parameters controlling both the shape and intensities of the model. By construction, $\mathbf{c}$ has zero mean across the training set.

The linear nature of the model enables the shape and intensities to be computed directly from $\mathbf{c}$:

$$\begin{aligned} \mathbf{x} &= \bar{\mathbf{x}} + \mathbf{Q}_s\mathbf{c}, \\ \mathbf{g} &= \bar{\mathbf{g}} + \mathbf{Q}_g\mathbf{c}, \end{aligned} \tag{8.18}$$

where

$$\begin{pmatrix} \mathbf{Q}_s \\ \mathbf{Q}_g \end{pmatrix} = \begin{pmatrix} w_s^{-1}\mathbf{P} & 0 \\ 0 & \mathbf{P}_g \end{pmatrix} \mathbf{P}_c \tag{8.19}$$

An example image can be synthesized for a given $\mathbf{c}$ by generating the shape-free intensity image from the vector $\mathbf{g}$ and warping it using the control points described by $\mathbf{x}$.

Fig. 8.11 shows the effect of varying each of the first three combined model parameters on the face synthesized by the system. Changing the parameters varies both the shape and the pattern of intensities. Fig. 8.12 shows the first three modes for a model of the pelvis as it appears in radiographs.

8.9 Image search

The techniques described above assume that the landmark point positions are known. When presented with a new image, one often wants to locate the landmarks with minimal human intervention. Many different techniques have been proposed for using shape models to help locate landmarks in images automatically. Most systems

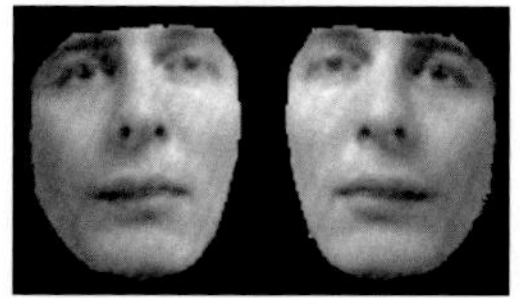
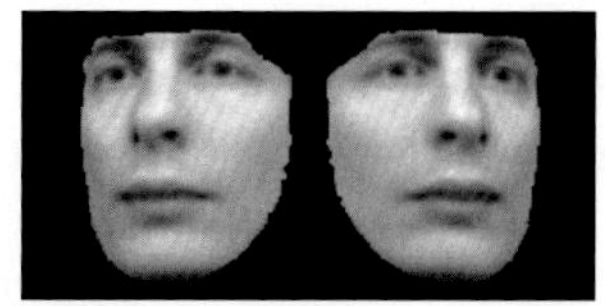
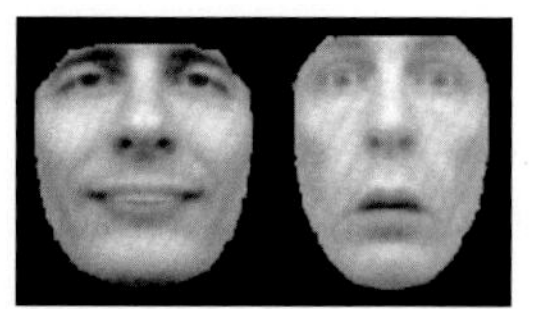
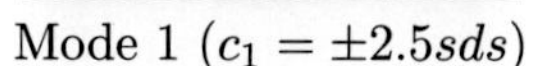

Mode 1 ($c_1 = \pm 2.5sds$) Mode 2 ($c_2 = \pm 2.5sds$) Mode 3 ($c_3 = \pm 2.5sds$)

FIGURE 8.11 Facial combined modes.

First three modes of a combined shape and texture model of a face.

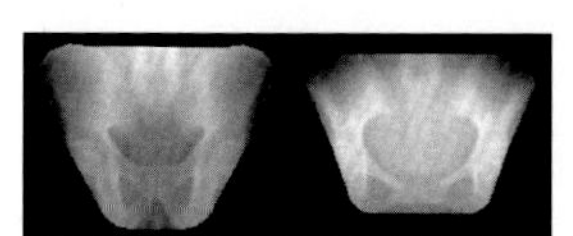
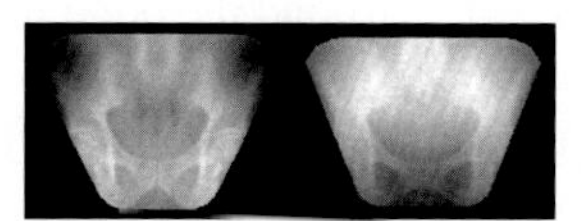
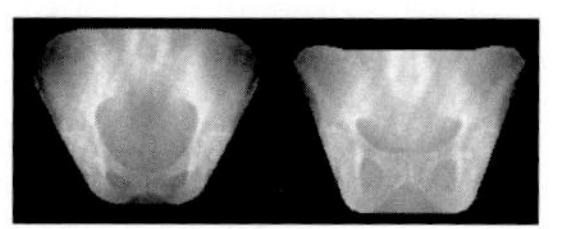
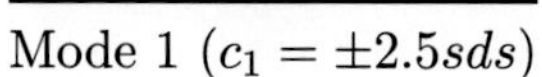

Mode 1 ($c_1 = \pm 2.5sds$) Mode 2 ($c_2 = \pm 2.5sds$) Mode 3 ($c_3 = \pm 2.5sds$)

FIGURE 8.12 Pelvis combined modes.

First three modes of a combined shape and texture model of a pelvis in radiographs.

first use a *global search* to locate the approximate pose (position, scale, and orientation) of the object by scanning the whole image. They then use a *local search* to find the model points accurately within this region. The local search techniques usually need a reasonably good estimate of the pose to converge correctly.

Here we focus on the local search step, though where necessary the methods described for exhaustive search in Section 8.10 can be applied to the whole image to estimate the positions of a few points, which is usually sufficient to initialize the local model.

There are three broad classes of approach for the local search:

1. **Exhaustive search**, in which each point is searched for independently in an image region.
2. **Alternating approaches**, in which candidates for each point are found with independent local searches, and then a shape model is used to constrain the set of points to a valid shape (such as the Active Shape Model [4]).
3. **Iterative approaches**, in which the model parameters are modified using estimates of the optimal update vector computed from image features sampled from around the current points.

In most cases it is quicker and more robust to use a multistage technique, in which initial models are trained and run on low-resolution versions of the image to get initial approximations to point positions. These are then improved by using models and information from higher-resolution images.

8.10 Exhaustive search

We first consider the case in which each individual point is located independently, as the approaches can also be used in the alternating algorithms. The goal is to search a region of interest (which may be the whole image) for the most likely position of a target landmark. We will assume that the approximate scale and orientation of the target is known.[2] An effective approach is to scan the region of interest with a model, which evaluates at every candidate position how likely that position is to be the true landmark location. The point with the highest response is then chosen as the best candidate. A simple but effective version of this is to use normalized correlation with a template based on the image patch around the point in one example (or the average appearance over a training set).

A generalization of this approach is to compute a feature vector describing the local image structure at every point (such as the response to a set of filters) and then to use a cost function on this vector to evaluate how likely each point is to be the target point. More specifically, for each position, $\mathbf{x}$, we can calculate a feature vector, $\mathbf{f}(\mathbf{x})$, either by sampling pixels in the area around $\mathbf{x}$ or by applying filters (such as Gabor filters, Haar wavelets, or suitably trained convolutional neural networks) to the image and sampling the response images. In the latter case each filter gives one element of the final feature vector.

One of the simplest features to use is the difference between intensities at the chosen positions within a patch [6]. More robust features can be computed from the means of intensities of rectangular patches, which can be computed efficiently from integral images (Haar features [7–9]).

The feature vector is usually normalized, for instance by subtracting the mean value from each element and scaling so that it is of unit length.

To evaluate a feature vector we must first train a model to distinguish good matches from poor matches.

Suppose we have a set of N training images, I_i, with a single point annotated in each (at $\mathbf{x}_i$). We can compute feature vectors sampled around the point in each training image, $\{\mathbf{f}_i\}$ $(i = 1 \ldots N)$. From this set we can build a model of what is expected at the point, such as estimating the distribution of the feature vectors, $p_f(\mathbf{f})$. For instance, if we assume $p_f(\mathbf{f})$ is Gaussian we would estimate the mean and covariance matrix from the set of example vectors.

To search for the best match in a new image, feature vectors are computed from samples taken around each candidate position in the region of interest and evaluated (by computing $p_f(\mathbf{f})$) to find that which is most similar to the training examples.

Alternatively, by sampling features both at the true point (positive examples) and away from the true point (negative examples), a classifier can be trained to discriminate the target point from its surroundings. Again the candidate position giving the strongest positive response would be chosen as the most likely target point.

[2] One can search at multiple scales and orientations to find the best pose if necessary.

8.10.1 Regression voting

Rather than attempt to classify each pixel we can use the features to vote for where the target point is likely to be. This has been found to be more robust than using classifiers at each pixel. The idea is that we train a regressor to predict the offset between the pixel position and the true landmark location, $(\delta_x, \delta_y) = \mathbf{r}(\mathbf{f})$. During search we scan the region of interest. If the feature vector sampled at point $\mathbf{x}$ is $\mathbf{f}$, we compute $(\delta_x, \delta_y) = \mathbf{r}(\mathbf{f})$ and then use this to vote into an accumulator array at $\mathbf{x} + (\delta_x, \delta_y)$.

The regressor is trained on a set of features taken at known (random) displacements about the true points as follows.

For each i

1. Generate a set of m random displacements $\mathbf{d}_{ij}$ with $|\mathbf{d}_{ij}| < r$
2. Generate small random scale, orientation displacement poses T_{ij}
3. Sample the feature vectors $\mathbf{f}_{ij}$ centered at $\mathbf{x}_i + \mathbf{d}_{ij}$ in image I_i with pose T_{ij}

The pairs $\{(T_{ij}^{-1}(-\mathbf{d}_{ij}), \mathbf{f}_{ij})\}$ are used to train a regressor, $\mathbf{d} = \mathbf{r}(\mathbf{f})$. The transformation of the displacement by T_{ij}^{-1} calculates the displacement in the transformed frame (the landmark position as seen from the sampling reference frame).

Random forests [8,10] are widely used for this regression task because of their efficiency and flexibility. A forest contains a set of decision trees. Each decision node in each tree applies a threshold to a particular feature measured from the image patch.

Training a tree. To train a tree in a random forest, one selects a feature and a threshold at each node so as to best split the data arriving at that node from its parent. When the tree is to be used to predict a location, one seeks to encourage the set of displacement vectors $\{\mathbf{d}_k\}$ which reach a node to be as compact (similar to one another) as possible. Let the samples arriving at the node be pairs $\{(\mathbf{d}_k, \mathbf{v}_k)\}$. Let $g(\mathbf{v})$ compute a scalar feature from the vector. Then during training one tries a random selection of possible features g and thresholds t, choosing the one that minimizes

$$G_T(t) = G(\{\mathbf{d}_k : g(\mathbf{v}_k) < t\}) + G(\{\mathbf{d}_k : g(\mathbf{v}_k) \geq t\}), \tag{8.20}$$

where $G()$ is an entropy measure $G(\{\mathbf{d}_k\}) = N \log|\Sigma|$, where Σ is the covariance matrix of the N samples.

During training one recursively selects features and thresholds to add new nodes to a tree until either a maximum depth is reached or the number of samples reaching a node falls below a threshold. The resulting leaf nodes then record (a) the mean of the displacement vectors reaching them, $\bar{\mathbf{d}}$, and (b) a weight $w = \min(0.5, |\Sigma|^{-0.5})$. The latter is used to reduce the importance of ambiguous cases (where the samples reaching the node are widely spread, with a large covariance, Σ).

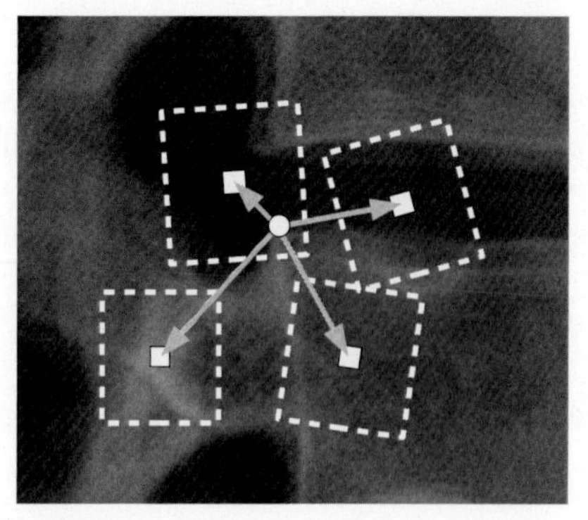

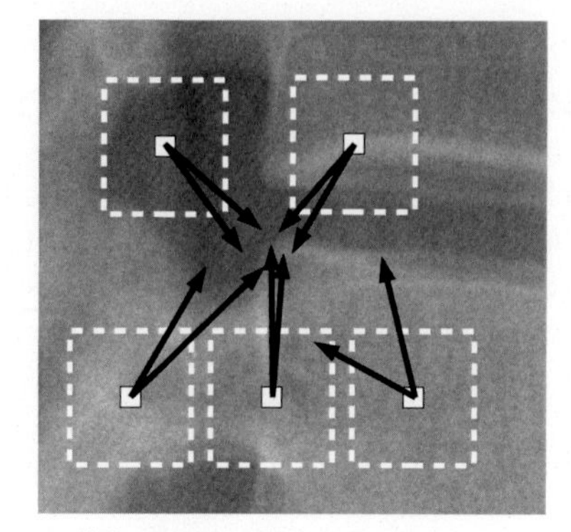

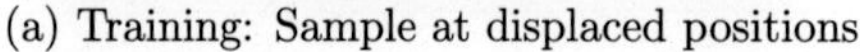

(a) Training: Sample at displaced positions (b) Search: One vote per tree

FIGURE 8.13 Regression voting.

During training, random displacements in position, scale, and orientation are used to improve generalization. During search, samples are taken at a fixed scale and orientation at regular grid positions.

A common way to combine the results of multiple trees is to compute the mean of the output of each. However, it is better to allow each tree to vote independently (with a weight given by the value in the leaf, w), combining the results in the accumulator array (see Fig. 8.13(b)).

8.11 Alternating approaches

Alternating approaches are variants of the Active Shape Model algorithm [4], which involves iterating through three steps (see also Fig. 8.14).

Initialize shape and pose parameters, $\mathbf{b}, \mathbf{t}$.

Repeat

1. Estimate point positions $\mathbf{x} = T(\bar{\mathbf{x}} + \mathbf{P}\mathbf{b}; \mathbf{t})$
2. Search the image around each point $\mathbf{x}_i$ for a better position $\mathbf{x}'_i$
3. Update the shape and pose parameters to fit to the new points $\mathbf{x}'$

Until converged.

The third step effectively applied a regularization, ensuring that points are always in a valid configuration and correcting cases in which the search (in step 2) has found a false match for one or more of the points.

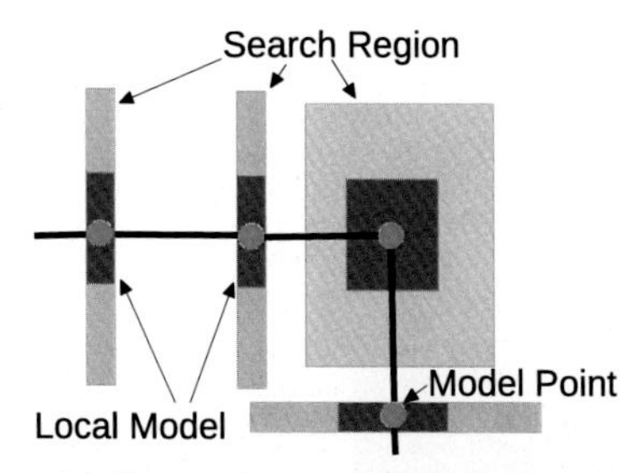

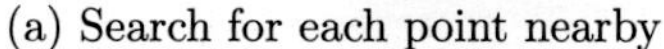
(a) Search for each point nearby

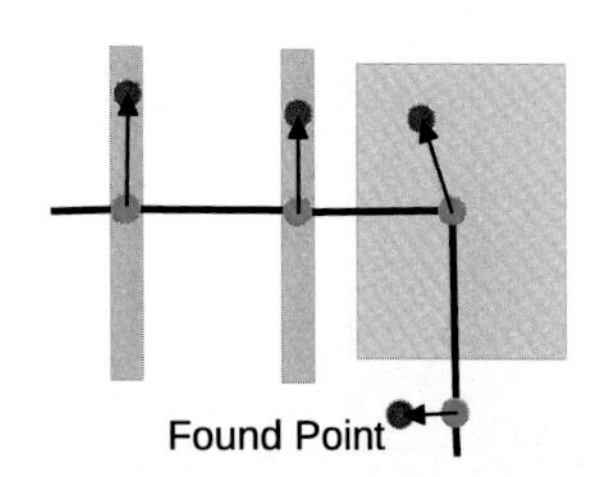

(b) Move towards best local matches

FIGURE 8.14 Active Shape Model steps.

The Active Shape Model alternates between finding good local matches and updating the shape model parameters.

When searching for a better position for each point (step 2) one can either:

- search along a line normal to the boundary at the point (which is efficient and suitable for cases where the target point is on a smooth boundary);
- search in a region around the point (which is slower than just searching along a line, but is more suitable when the target is a corner or the initial point is far from the true point).

8.11.1 Searching for each point

Alternating approaches require a method for searching a region to find the most likely position for a given point. Any of the techniques for searching for a single point described in Section 8.10 can be applied to the regions of interest around each point, though it is important to take into account the local scale and orientation using the current estimate of the shape.

Let $\mathbf{x}_j$ be the position of point j in image I and let $\mathbf{u}_j$ be a vector defining the scale and orientation of a sampling region at that point; $\mathbf{u}_j$ can either be derived from the pose, $\mathbf{t}$, using $\mathbf{u}_j = T((\delta,0);\mathbf{t}) - T((0,0);\mathbf{t})$ (where δ is the sample step size in the reference frame), it or can be a suitably scaled normal to a boundary at the point. Let $\mathbf{v}_j$ be $\mathbf{u}_j$ rotated by 90 degrees.

The intensities in a patch around the point can be obtained by sampling in a grid around the point using

$$P(a,b) = I_i(\mathbf{x}_j + a\mathbf{u}_j + b\mathbf{v}_j), \tag{8.21}$$

where a, b are integers in the range $-w_a \leq a \leq w_a$, $-w_b \leq b \leq w_b$.

The patch $P(a,b)$ is then a $(2w_a+1)\times(2w_b+1)$ image, which can be scanned with a model to find the best point. Suppose this is at point (a,b) within patch P.

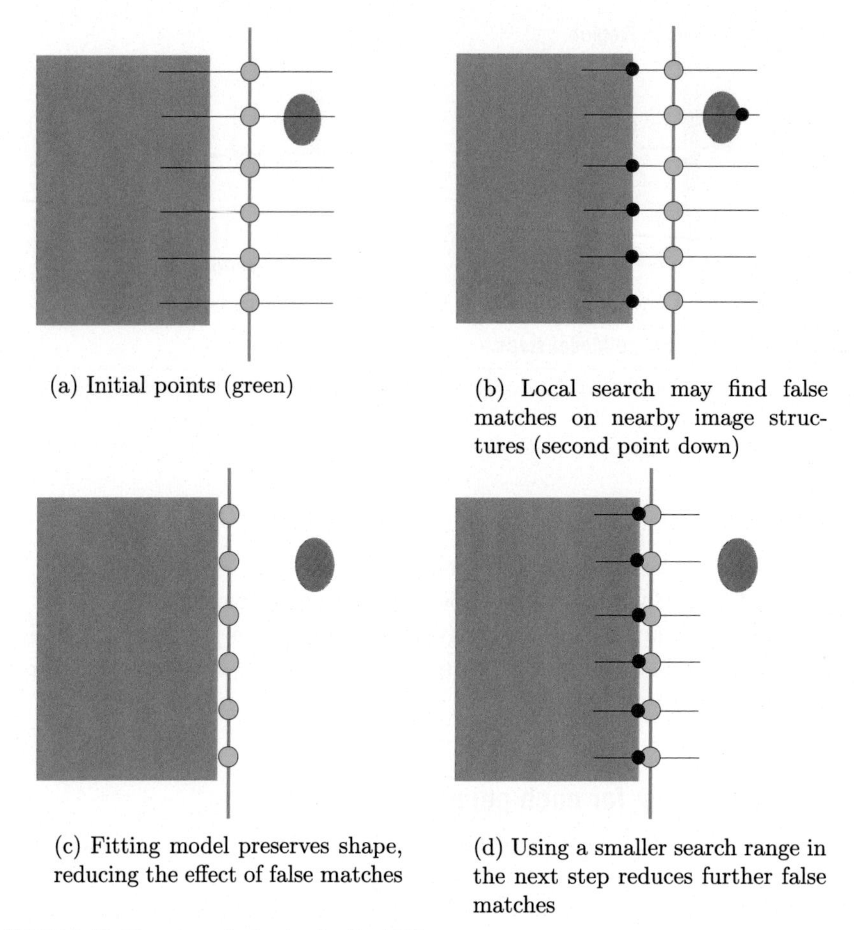

(a) Initial points (green)

(b) Local search may find false matches on nearby image structures (second point down)

(c) Fitting model preserves shape, reducing the effect of false matches

(d) Using a smaller search range in the next step reduces further false matches

FIGURE 8.15 Shape regularization in the ASM.

Fitting the shape model reduces the effect of false matches.

Then the resulting point corresponds to the position $\mathbf{x}_j + a\mathbf{u}_j + b\mathbf{v}_j$ in the image. This can be done independently for each model point to obtain new candidate positions.

8.11.2 Shape model as regularizer

Fitting the shape model to the found point positions enables the system to minimize the impact of false matches (see Fig. 8.15). As long as the majority of the matches are close to the right place the model points will move to the correct object. Reducing the search range after each iteration helps reduce further outlier matches. Alternatively robust methods can be used to fit the shape to the found points and explicitly identify the outliers.

8.12 Constrained local models

The Active Shape Model involves selecting a candidate for each point independently, which loses information if it chooses the wrong point (due to noise or inherent ambiguity in the image). A more effective alternative (the "Constrained Local Model" [CLM] approach) is to create "score images" for each point, indicating the quality of fit for that point for every pixel in a region around the original point position. The shape model is then used to constrain a search to find the overall optimum position of all the points together [9,11].

The individual shape model points can be computed using

$$\mathbf{x}_j = T(\bar{\mathbf{x}}_j + \mathbf{P}_j\mathbf{b}; \mathbf{t}), \tag{8.22}$$

where $\bar{\mathbf{x}}_j$ is the mean position of the point in the model frame and $\mathbf{P}_j$ defines the effect of each shape mode on the point.

One iteration of the CLM algorithm involves searching around each point $\mathbf{x}_j$ to compute a score image $S_j(\mathbf{x}_j)$ and then finding the pose and shape parameters $(\mathbf{t}, \mathbf{b})$ to maximize

$$Q_c(\mathbf{t}, \mathbf{b}) = \sum_{j=1}^{n} S_j(T(\bar{\mathbf{x}}_j + \mathbf{P}_j\mathbf{b}; \mathbf{t})) - R(\mathbf{b}), \tag{8.23}$$

where $R(\mathbf{b})$ is an optional regularization term used to encourage plausible shapes. The cost function $Q_c(\mathbf{t}, \mathbf{b})$ can then be optimized either using a general-purpose optimizer [11] or using a mean-shift approach [12].

Computing the score images, S_j, is best done in a standardized reference frame to make the algorithm invariant to the scale and orientation of the object in an image. A rectangular region around the current shape should be sampled to warp it into the standardized frame (see Fig. 8.16(a and b)).

The width (in pixels) of the mean shape in the standardized frame indicates the resolution at which the search will be done. Let s_r be a scaling factor chosen so that the scaled mean, $s_r\bar{\mathbf{x}}$, has a particular width (say 100 pixels). A point (x, y) in the image can be mapped into the shape model reference frame by applying the inverse of the pose transformation: $T^{-1}(x, y : \mathbf{t})$. Thus it can be mapped to the standardized frame as $(a, b) = s_r T^{-1}(x, y : \mathbf{t})$. For convenience, we can summarize this transformation as $s_r \circ \mathbf{t}^{-1}$. Similarly a point in the standardized frame, (a, b) maps to $T((a/s_r, b/s_r) : \mathbf{t})$ in the image. We can summarize this transformation as $\mathbf{t} \circ s_r^{-1}$. We can thus fill each integer pixel position (a, b) in the standardized frame by interpolating from the target image, $S(a, b) = I(T((a/s_r, b/s_r) : \mathbf{t}))$. One iteration of the search algorithm involves updating the initial estimates of pose, $\mathbf{t}$, and shape, $\mathbf{b}$, parameters using the following steps (see also Fig. 8.16).

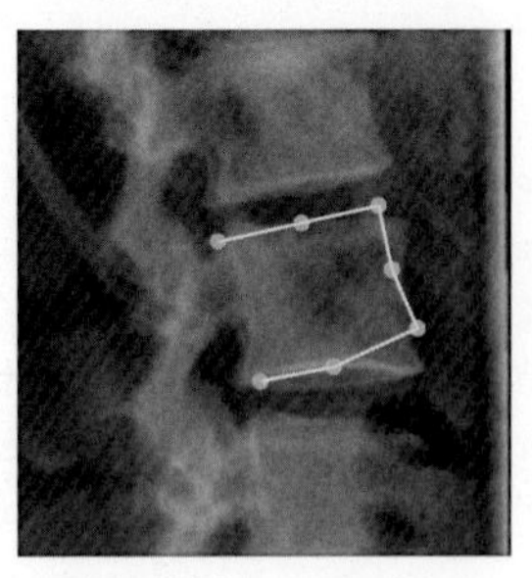

(a) Initial points on original image

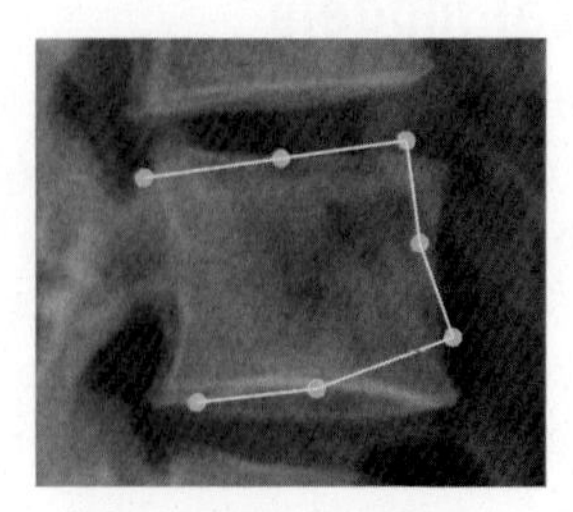

(b) Image sampled to standardized frame

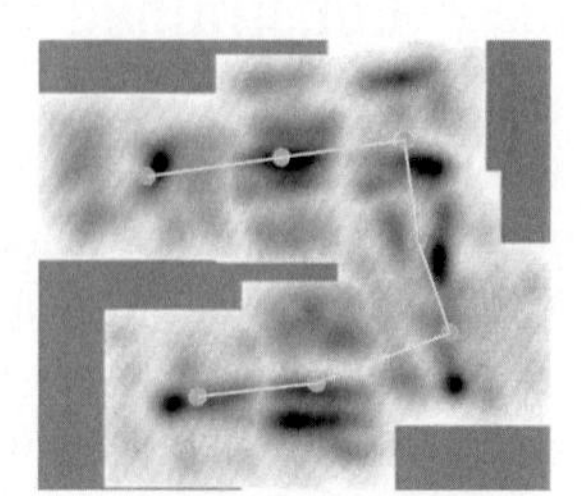

(c) Response images computed for each point

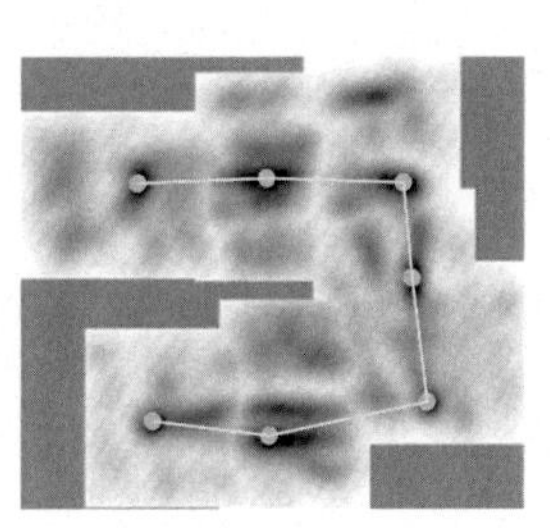

(d) Shape fitted to responses

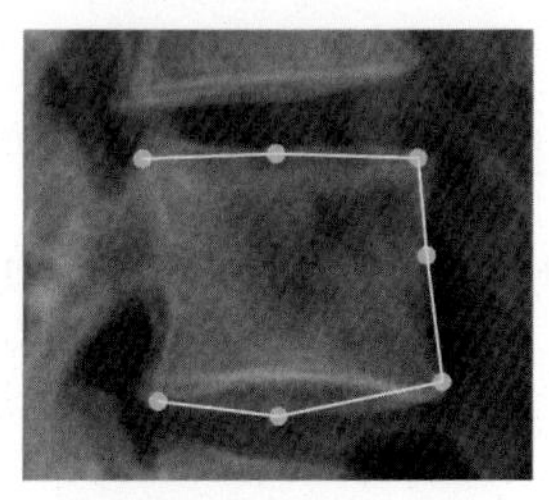

(e) Result in standardized frame

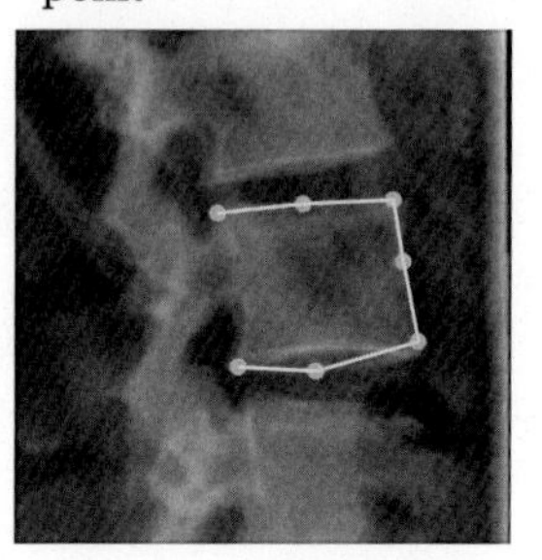

(f) Points mapped to original image

FIGURE 8.16 Constrained local model steps.

The CLM performs search in a standardized frame.

1. $S(a,b) = I(T((a/s_r, b/s_r) : \mathbf{t}))$ to sample the image into a standardized frame
2. Search about each point j in this sampled image to create score images $\{S_j\}$
3. Find the parameters $\mathbf{t}_r, \mathbf{b}$ to optimize $Q_c(\mathbf{t}_r, \mathbf{b})$ in (8.23)
4. Update the pose estimate to $\mathbf{t}_{new} = \mathbf{t} \circ s_r^{-1} \circ \mathbf{t}_r$

Fig. 8.16(c) shows examples of score images produced by voting with random trees based on Haar features. Corner points are well localized in 2D. Scores for points along boundaries tend to give good values along the boundary – here the shape model constraints help to locate the best position (Fig. 8.16(d)).

Usually a sequence of such models are created, the first having relatively low resolution in the standardized frame (with, for instance, 50 pixels across the reference shape), the last having higher resolution (perhaps 200 pixels across). One or two iterations of the search algorithm are applied at each stage.

8.12.1 Iteratively updating parameters

A wide range of approaches involve directly estimating an update to the current model parameters at each step, either using methods that approximate the optimal direction to minimize a cost function or using regression functions to predict the update directly.

If $\mathbf{p}$ is a vector combining all the parameters (for instance the shape and pose vectors), one iteration on an image I has the following steps.

1. Sample features from image based on current points, $\mathbf{v} = \mathbf{f}(I, \mathbf{x})$
2. Estimate updates to parameters, $\delta\mathbf{p} = \mathbf{h}(\mathbf{v})$
3. Modify current parameters, $\mathbf{p} \rightarrow \mathbf{p} + \delta\mathbf{p}$

Algorithms can thus be characterized by:

- how they extract features from the image (the function $\mathbf{f}(I, \mathbf{x})$);
- how they estimate the updates (the function $\mathbf{h}(\mathbf{v})$).

8.12.2 Extracting features

Extracting features based on the points effectively has two steps.

1. Sample from the target image into a reference frame
2. Apply filters to compute image features

The first step is important to ensure features are computed consistently, introducing independence to the overall scale and orientation of the object as it appears in an image.

It can be done either by warping a region with a triangulated mesh (as for texture models in Section 8.7), or rectangular regions can be sampled about each point using an estimate of the local scale and orientation (as described by equation (8.21)).

A feature vector can then be constructed by, for instance, applying filters to the sampled reference image and then sampling the output at predefined positions. The simplest approach is just to sample the intensity at every pixel in the reference frame to form a feature vector. However, better results are usually obtained by applying suitable filters first – for instance, computing gradients at each point.

8.12.3 Updating parameters

Given the features sampled from the image, $\mathbf{v}$, an update is calculated using some function $\delta\mathbf{p} = \mathbf{h}(\mathbf{v})$. The function can be either computed as a gradient descent step (where an explicit cost function is being optimized) or trained as a regressor.

Explicit cost functions. Suppose that for a particular structure we have a generative model which computes the feature vector given a set of parameters, $\mathbf{q}$, $\hat{\mathbf{v}} = \mathbf{v}(\mathbf{q})$. An example are statistical appearance models which apply PCA to feature vectors sampled from a training set, modeled as $\hat{\mathbf{v}} = \bar{\mathbf{v}} + \mathbf{P}_v\mathbf{q}$ [13]. The quality of fit of a model with parameters $\mathbf{p} = (\mathbf{b}^T, \mathbf{t}^T, \mathbf{q}^T)^T$ could be given by a sum of square error term, such as $Q(\mathbf{p}) = |\mathbf{d}(\mathbf{p})|^2$, where function $\mathbf{d}()$ computes the vector of differences between sampled and reconstructed features,

$$\mathbf{d}(\mathbf{p}) = \mathbf{v}(\mathbf{b}, \mathbf{t}) - V(\mathbf{q}). \tag{8.24}$$

The quadratic form of Q means that near the minimum, a step towards the minimum is given by

$$\delta\mathbf{p} = -(\mathbf{J}_D)^{-1}\mathbf{d}(\mathbf{p}), \tag{8.25}$$

where $\mathbf{J}_D$ is the Jacobian of $D(\mathbf{p})$ with respect to $\mathbf{p}$.

This general approach can be improved by careful choice of reference frame and the use of compositional updates, leading to the inverse compositional algorithm and its variants [14,15], which enable efficient and accurate model fitting.

Regression-based updates. Regression techniques aim to directly estimate good parameter updates directly from the current feature values. Suppose we have a training set of N images, I_i, and the best shape and pose parameters to match to the labeled points in the image are $\mathbf{p}_i$. We can construct a set of training pairs $\{(\mathbf{v}_{ij}, \delta\mathbf{p}_{ij})\}$ as follows.

For all images I_i with associated points $\mathbf{x}_i$

1. Generate a set of m random displacements $\delta\mathbf{p}_{ij}$
2. Sample the feature vectors $\mathbf{v}_{ij} = \mathbf{f}(I_i, \mathbf{p}_i + \delta\mathbf{p}_{ij})$

This set is then used to train a regression function, $\delta\mathbf{p} = \mathbf{h}(\mathbf{v})$.

This regression function estimates the parameter displacements required to get to the target position from any current position, and thus forms the basis of an iterative update scheme. The accuracy of the prediction depends on (i) how similar the current offset is to that used in training, (ii) how broad the training set was, and (iii) the form of the regression function.

In general, models trained on small displacements will give good predictions when applied to cases where the current position is near the true answer, but will break down when the current position is far from the right one. Models trained on large displacements are less accurate, but may give a useful step toward the right position.

Thus a useful strategy is to train a series of models, the first trained on large displacements, with later models trained on increasingly small ranges. For instance, the supervised descent method [16] uses a sequence of linear predictors, each leading to a more accurate estimate of the target position.

An example of a very fast technique is to use ensembles of trees to generate sparse binary vectors from input features, followed by sparse linear regression on these vectors to predict the parameter updates [6]. During training, a set of trees is trained to predict the offset of each point. However, rather than use the outputs stored in the leaves, each tree returns a binary vector which indicates which leaf was reached. The resulting binary vectors from all trees and all points are concatenated together and a linear regressor is trained to predict the actual parameter updates. The use of the trees builds non-linearity into the system, enabling it to deal with a wide range of variation in appearance.

8.13 3D models

The modeling and matching techniques described above are equally valid for 3D models. One of the main challenges is the initial annotation of the training images. In 2D it is practical to have human experts place all the points (often with semi-automatic support). However, to represent 3D surfaces requires hundreds, or even thousands of landmarks, which cannot be placed by hand.

Usually the 3D volumetric image is segmented (often with slice-by-slice manual editing) to create binary masks for the structures of interest. The challenge then is to place landmarks on the surfaces of each such structure so as to define useful correspondences. There are two broad approaches to this:

1. Convert each mask to a surface. Place points on each surface and then slide them around to optimize some measure of correspondence [17].
2. Use volumetric non-linear registration to deform each image (and mask) into a reference frame, place landmarks on the mean in this frame, and then propagate them back to the training images [18–20].

8.14 Recapitulation

The take-home messages from this chapter are:

- The shape of an object can be represented by a set of landmark points, which define correspondences across examples.
- After aligning shapes we can apply PCA in order to learn the mean shape and main modes of variation.

- An example shape can be approximated using this model, and the shape parameters give a useful compact representation of the deformation from the mean.
- Effective methods exist to use such models to locate landmarks on previously unseen images, enabling automatic analysis of large datasets.

8.15 Exercises

1. **Aligning shapes.** Given two shapes, represented by vectors $\mathbf{x}_1$ and $\mathbf{x}_2$, we can align them by finding the parameters of the transformation $T(\mathbf{x})$ which minimizes $|T(\mathbf{x}_1) - \mathbf{x}_2|^2$.
   a. Show that if T is a translation, then the optimal parameters are equal to the difference between the centers of gravity of the two shapes.
   b. Suppose T applies a rotation and scaling, encoded using two parameters ($a = s\cos\theta,\ b = s\sin\theta$) in the matrix $\begin{pmatrix} a & b \\ -b & a \end{pmatrix}$. Derive the equation giving the optimal values for a, b to align two shapes.
2. **PCA modeling.** Given a model of the form $\mathbf{x} = \bar{\mathbf{x}} + \mathbf{P}\mathbf{b}$, where $\mathbf{P}$ are the eigenvectors corresponding to the t largest eigenvalues of a covariance matrix, prove:
   a. The parameters which minimize the sum of square distances of a model instance to a new shape, $\mathbf{x}$, are given by $\mathbf{b} = \mathbf{P}^T(\mathbf{x} - \bar{\mathbf{x}})$.
   b. The sum of square distances between the best model fit and the target $\mathbf{x}$ is given by $\|\mathbf{x} - \bar{\mathbf{x}}\|^2 - \|\mathbf{b}\|^2$.
3. **Fitting to incomplete shapes.** Suppose we only know the positions of a subset of the points and wish to find the best shape model fit to this subset. Let $\mathbf{W}$ be a diagonal matrix with diagonal elements set to 1 for elements corresponding to the coordinates of the known points and 0 for the coordinates of the unknown points.
   a. Demonstrate that the shape parameters, $\mathbf{b}$, which best fit the model to a partially defined shape $\mathbf{x}$, in which unknown points are set to zero, are given by the solution to the linear equation $\mathbf{P}^T\mathbf{W}\mathbf{P}\mathbf{b} = \mathbf{P}^T\mathbf{W}(\mathbf{x} - \bar{\mathbf{x}})$.

References

[1] Kent J.T., The complex Bingham distribution and shape analysis, Journal of the Royal Statistical Society B 56 (1994) 285–299.

[2] Dryden I.L., Mardia K.V., Statistical Shape Analysis, Wiley, 1998.

[3] Goodall C., Procrustes methods in the statistical analysis of shape, Journal of the Royal Statistical Society B 53 (2) (1991) 285–339.

[4] Cootes T.F., Taylor C.J., Cooper D.H., Graham J., Active shape models - their training and application, Computer Vision and Image Understanding 61 (1) (1995) 38–59.

[5] Bookstein F.L., Principal warps: thin-plate splines and the decomposition of deformations, IEEE Transactions on Pattern Analysis and Machine Intelligence 11 (6) (1989) 567–585.

[6] Ren S., Cao X., Wei Y., Sun J., Face alignment via regressing local binary features, IEEE Transactions on Image Processing (ISSN 1057-7149) 25 (3) (2016) 1233–1245, https://doi.org/10.1109/TIP.2016.2518867.
[7] Viola P., Jones M., Rapid object detection using a boosted cascade of simple features, in: CVPR, vol. 1, 2001, pp. 511–518.
[8] Criminisi A., Shotton J. (Eds.), Decision Forests for Computer Vision and Medical Image Analysis, Springer, 2013.
[9] Lindner C., Bromiley P.A., Ionita M.C., Cootes T.F., Robust and accurate shape model matching using random forest regression-voting, IEEE Transactions on Pattern Analysis and Machine Intelligence (ISSN 0162-8828) 37 (9) (2015) 1862–1874, https://doi.org/10.1109/TPAMI.2014.2382106.
[10] Breiman L., Random forests, Machine Learning 45 (2001) 5–32.
[11] Cristinacce D., Cootes T.F., Automatic feature localisation with constrained local models, Pattern Recognition 41 (10) (2008) 3054–3067.
[12] Saragih J.M., Lucey S., Cohn J.F., Deformable model fitting by regularized mean-shifts, International Journal of Computer Vision 91 (2011) 200–215.
[13] Cootes T.F., Edwards G.J., Taylor C.J., Active appearance models, IEEE Transactions on Pattern Analysis and Machine Intelligence 23 (6) (2001) 681–685.
[14] Matthews I., Baker S., Active appearance models revisited, International Journal of Computer Vision 60 (2) (2004) 135–164.
[15] Tzimiropoulos G., Pantic M., Fast algorithms for fitting active appearance models to unconstrained images, International Journal of Computer Vision (ISSN 1573-1405) 122 (1) (2017) 17–33, https://doi.org/10.1007/s11263-016-0950-1.
[16] Xiong X., De la Torre F., Supervised descent method and its application to face alignment, in: IEEE CVPR, 2013.
[17] Davies R.H., Twining C.J., Cootes T.F., Waterton J.C., Taylor C.J., A minimum description length approach to statistical shape modelling, IEEE Transactions on Medical Imaging 21 (2002) 525–537.
[18] Duchesne S., Collins D.L., Analysis of 3D deformation fields for appearance-based segmentation, in: Proc. MICCAI, 2001, pp. 1189–1190.
[19] Frangi A.F., Rueckert D., Schnabel J.A., Niessen W.J., Automatic construction of multiple-object three-dimensional statistical shape models: application to cardiac modeling, IEEE Transactions on Medical Imaging 21 (2002) 1151–1166.
[20] Cootes T.F., Twining C.J., Petrovic V.S., Babalola K.O., Taylor C.J., Computing accurate correspondences across groups of images, IEEE Transactions on Pattern Analysis and Machine Intelligence 32 (11) (2010) 1994–2005.

CHAPTER

Segmentation by deformable models

Jerry L. Prince
Electrical and Computer Engineering, Johns Hopkins University, Baltimore, MD, United States

Learning points

- Implicit boundary representations
- Implicit boundary evolution. Markers and level sets
- External, internal, and pressure forces. Speed functions
- Numerical implementation of parametric and geometric shape models
- Extensions to topological changes, multiple scales and higher dimensions
- Extensions to multiple objects and incorporating prior shape information

9.1 Introduction

Segmenting objects in medical images by finding their boundaries is a classical approach which might be considered a "dual" to finding the objects themselves. In this chapter, we explore the use of moving object boundaries – so-called deformable models – to segment objects. Unlike many image segmentation approaches, which label many disparate pixels throughout an image as belonging to an object, deformable models generally focus on finding a single connected object within the image (though we consider the multiple-object case later in the chapter as well). In this sense, deformable models have more in common with regions growing from a single seed pixel than segmentation by graph cuts, k-means, or deep convolutional neural networks. Two broad classes of deformable models are described in this chapter: (1) parametric deformable models, in which connected points or markers that represent a boundary are moved around within the image until they conform (ideally) to the desired object boundary, and (2) geometric deformable models, in which a function defined on the entire image domain is modified until its zero level set conforms (ideally) to the desired object boundary. As we will show, these two approaches are equivalent in certain ways but quite different in other ways. This chapter will consider both types of models together so that it will be easier to compare their representations and computational strategies and thereby set the stage for a better understanding of the trade-offs one might consider in choosing one approach over the other.

Medical Image Analysis. https://doi.org/10.1016/B978-0-12-813657-7.00023-6

9.2 Boundary evolution

We consider only closed boundaries that contain a single object. Despite the fact that most medical image analysis takes place in 3D, for the sake of brevity and clarity, we will consider in detail only 2D images and objects. In places, we briefly consider extension to 3D merely to take note of its potential increase in complexity.

9.2.1 Marker evolution

The boundary of an object in the plane can be described by a function $\mathbf{X}(s)$, $s \in [0, 1]$, where $\mathbf{X}(0) = \mathbf{X}(1)$ and $\mathbf{X}(s)$ represents the 2D spatial point given by the pair $(x(s), y(s))$. Here, s does not represent arclength, but is merely an independent scalar that indexes a particular point on the curve. Any deformable model based on this explicit representation of a curve is called a parametric deformable model because the boundary position is described by its parameter s. A sampled version of this curve might be described by the set of points $\mathbf{X}(s_i)$, $i = 0, \dots, N-1$, where N is the number of points on the curve and s_i are consecutive samples within the interval $[0, 1]$ – e.g., $s_i = i/N$. Therefore, the curve can also be represented as a pair of vectors $\mathbf{x} = [x(s_0)\ x(s_1)\ \dots\ x(s_{N-1})]^T$ and $\mathbf{y} = [y(s_0)\ y(s_1)\ \dots\ y(s_{N-1})]^T$. Each point $\mathbf{X}(s_i)$ represents a so-called "marker" that will be caused to move around with the aim to reach the true or desired boundary position. This representation of a deformable model, where the neighboring markers are connected with line segments, is not the only finite representation of a continuous curve; for example, one can replace line segments with basis function expansions using Fourier or spline basis functions. But this sampling representation is the simplest to explain and implement and is by far the most commonly used in practice (see also [1]).

Since the goal of deformable models is to start in an arbitrary position within the image domain and deform to (ideally) the object's true boundary, we need another independent variable representing time or iteration. For this, we extend our previous notation to $\mathbf{X}(s, t)$, where $t \geq 0$ is a (pseudo) time variable. Time is discrete in any practical algorithm, so we give time at iteration k the notation $t^{(k)}$. Accordingly, the discrete deformable model is described by the time-dependent pair of vectors $(\mathbf{x}^{(k)}, \mathbf{y}^{(k)})$, as shown in Fig. 9.1. A parametric deformable model must describe how these vectors evolve to find an object boundary. To initiate motion, there must be something to move the markers around, and this is done by defining "forces" $\mathbf{F}$ that act on the points. In particular, the markers will move according to the following simple partial differential equation:

$$\gamma \frac{\partial \mathbf{X}}{\partial t} = \mathbf{F}, \tag{9.1}$$

where γ is a scaling parameter that determines how far the markers move in response to the forces. This simple equation is fundamental to all parametric deformable models.

There are a few practical issues to consider when using markers. First, one should ask how many markers to use in representing a boundary. Using too few markers

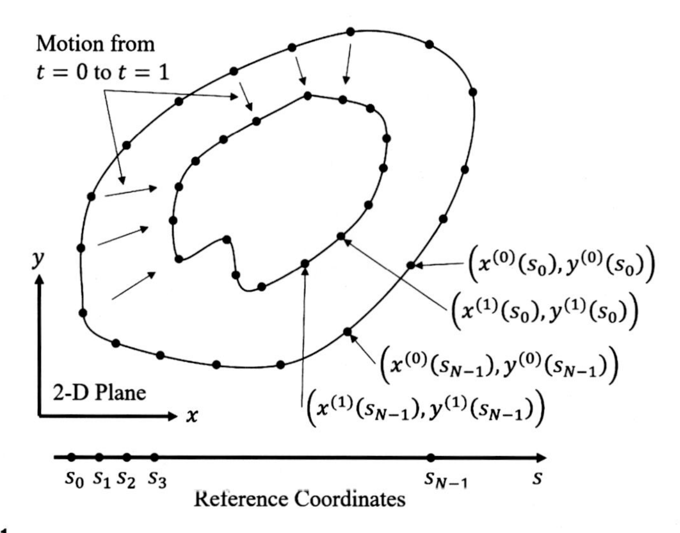

FIGURE 9.1

Marker representation of a deformable contour. The outer markers represent a starting shape at $t = 0$. After a motion step from $t = 0$ to $t = 1$, the markers move to form the inner shape. Each marker has a 2D coordinate that is determined by a mapping from a reference coordinate s for that particular marker. To maintain a closed curve, it is assumed that the reference coordinate s_{N-1} is circularly adjacent to the reference coordinate s_0.

forces one to omit boundary details should they be present and use of too many markers requires excessive computational burden. As well, if the digital representation of the object itself is coarse – i.e., the number of pixels in the image is small – then the curve can logically be represented by fewer markers. Second, markers can move farther apart or closer together during deformation, even simultaneously at different parts of the boundary. Because of this, resampling to better space out markers is typically required during deformation, which adds an additional computational burden. Third, markers that move independently or in an unconstrained manner can result in a boundary that crosses itself, i.e., a self-intersecting boundary, which is unrealistic and should be avoided. Thus, parametric deformable models often require extra computations to avoid self-intersecting boundaries.

9.2.2 Level set evolution

Consider a function $\phi(\mathbf{x})$ defined on the plane having both positive and negative values. The zero level set of this function defines a collection of (x, y) points which, without further qualification, do not have any particular relationship or designation as a curve. If, however, the function ϕ is taken to be the *signed distance function* of the contour $\mathbf{X}(s, t)$, where the function is negative on the inside of the contour and positive on the outside (a very common but arbitrary assumption), then its zero level set defines the shape represented by $\mathbf{X}(s, t)$. In fact, if ϕ also becomes a function of

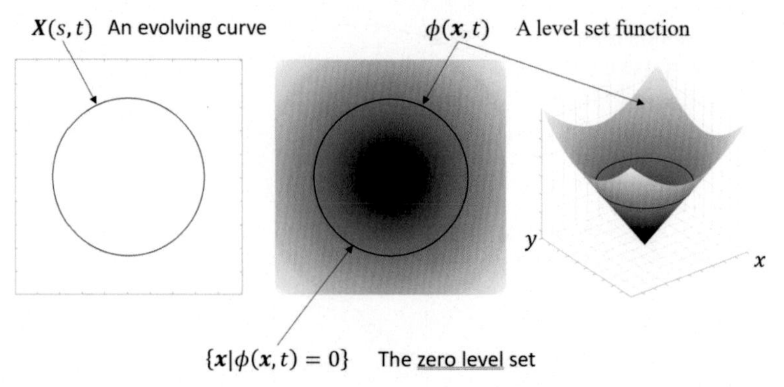

FIGURE 9.2

Relation between a contour and its level set representation. The curve $\mathbf{X}(x,t)$ defines an explicit set of points in the plane, whereas the level set function $\phi(\mathbf{x},t)$ provides an implicit representation of the same curve through its zero level set. By convention, a level set function is defined for every point $\mathbf{x}$ in the plane and is negative on the inside and positive on the outside of the closed curve.

time t – e.g., $\phi(\mathbf{x},t)$ – then the zero level set of ϕ is a curve that moves in time and can therefore be used as a deformable model for image segmentation. The relationship between the contour and the level set function is shown in Fig. 9.2 (see also [2]).

The fact that $\mathbf{X}(s,t)$ defines a collection of points explicitly while $\phi(\mathbf{x},t)$ only defines the curve implicitly via its zero level set forces us to view the movements of their defined shapes differently. In particular, imagine a force that moves the points in $\mathbf{X}(s,t)$ tangentially at every point on the curve. This does not change the shape of the object, but it does cause $\mathbf{X}(s,t)$ to change its parameterization of the curve. So, in the parametric representation, tangential forces cause a reparameterization of the curve while not changing its shape. Now consider the effect of tangential forces on the level set representation; there is no change in $\phi(\mathbf{x},t)$ in the presence of tangential forces. This happens because there is no parameter to identify the location within a curve in the level set representation. Because of this, forces that would act in a tangential direction on the curve do not affect the level set function that represents the curve. Of course, such forces do not affect the shape of the curve in either representation, so we can ignore the tangential component of forces that act to move a deformable contour from both a practical and theoretical standpoint.

Now consider a curve $\mathbf{X}(s,t)$ evolving only due to normal forces – i.e., forces acting in a direction $\mathbf{N}$ that is normal to the curve. This implies that the curve's partial derivative with respect to t can be written as

$$\frac{\partial \mathbf{X}}{\partial t} = V\mathbf{N}, \tag{9.2}$$

where V is called the *speed function*, which depends on both the spatial position and the curvature κ of the curve.

By the definition, the associated level set function must be zero at every point of the curve, which gives

$$\phi(\mathbf{X}(s,t),t) = 0. \tag{9.3}$$

By taking the total derivative of equation (9.3), it can be shown that (see Exercise 1.)

$$\frac{\partial \phi}{\partial t} = V|\nabla \phi|. \tag{9.4}$$

This is the fundamental non-linear partial differential equation, called the *eikonal equation*, that governs all geometric deformable models. It is deceptively simple, but actually somewhat tricky to solve, as we will see. Like the forces, which are key to designing effective parametric deformable models, the speed function V is the key to designing effective geometric deformable models.

9.3 Forces and speed functions

In parametric deformable models, we move markers by applying (pseudo) forces to the points, which mimics what a real object might experience except that these forces are completely artificial and derived from both intuition and mathematical principles. The forces $\mathbf{F}$ are vector quantities, in general, having both a direction and a magnitude as in real forces. The behavior and therefore the performance of deformable models is mostly dependent on the creativity used to design the forces that are used to move the model towards its desired position. Generally, the forces we use are divided into three categories: external forces, internal forces, and pressure forces:

- *External forces* are derived from the image or features computed from the image.
- *Internal forces* act on the points defining the object and are generally designed to maintain a cohesive structure independent of the underlying image.
- *Pressure forces* are those that act to grow or shrink the model and may or may not depend on properties of the underlying image.

The key to creating an effective deformable model is to define these forces in a suitable manner for the specific problem that is to be solved. We now expand on the most common mathematical definitions of these forces in the following sections.

9.3.1 Parametric model forces

Here, we focus on finding object boundaries where the objects themselves are generally defined by sharp changes in intensities.

External forces. We assume that images $f(x,y)$ are scaled so that $f(x,y) \in [0,1]$, which helps to ensure that parameters selected for one type of image – e.g., a cardiac image – are likely to also work in another type of image – e.g., a brain image. Object edges should have large gradient magnitudes. Therefore, we can define an edge map as $e(x, y) = |\nabla f(x, y)|$. Given that $e(x, y)$ can be expected to peak on object edges, the gradient of $e(x, y)$ will generally point towards the location of edges. Therefore, a suitable external force, specifically designed to find edges, can be defined as

$$\mathbf{v}_{\text{ext}}(x, y) = \nabla|\nabla f(x, y)|. \tag{9.5}$$

This is a very rudimentary external force that illustrates the basic idea but is not often used because it is not robust to image noise, has a magnitude that is highly dependent on the strength of the edge, and tends to disappear on so-called weak edges (parts of the object boundary where the gradient is not very large).

To address the above issues, it is common to define more complex forces. To do this, we first define the concept of *edge potential*, which is a function $p(x, y)$ that is large away from edges and approaches zero on edges. With a potential defined in this way, the goal is to move the markers so that they fall into potential "wells" that are approximately zero, which implies that the markers reside on edges. Clearly, $p(x, y) = -e(x, y)$ is such a potential given this definition, and therefore a possible force might be defined by

$$\mathbf{v}_{\text{ext}}(x, y) = -\nabla p(x, y). \tag{9.6}$$

A weakness of this force is that it depends directly on the size of the image edge, which can result in good convergence in some areas and poor convergence in others – e.g., at the weak edges.

To address the issues with variable edge strength, we can augment the edge potential defined above. First, to address noise, we smooth the image with a Gaussian kernel and then take the gradient, which yields a more robust edge map $e(x, y) = |\nabla(G(x, y) * f(x, y))|$. To create a potential, we could negate the robust edge map and then add a constant, but this would create potential wells that are not all near zero. Alternatively, we could invert $e(x, y)$ so that large values are close to zero; but this would be unstable when the gradient is zero (which happens in all regions that have constant intensities). But that is the right idea. The most common edge potential in parametric deformable models is given by

$$p(x, y) = \frac{1}{1 + |\nabla(G(x, y) * f(x, y))|}, \tag{9.7}$$

which is nearly zero on edges and equal to unity in constant regions. Another benefit of this potential is that it is inherently scaled to the range [0, 1], which is useful in calibrating constants that might otherwise need to be different for different types of images – e.g., cardiac versus brain or X-ray versus magnetic resonance images. With this potential, the image forces are found using equation (9.6).

Pressure forces. All of the forces defined so far are considered to be external forces. If these are the only forces that are used, then the markers would act independently and move along the negative gradient directions of the potential. This might be appropriate in some cases but, generally, it is important to find a way to keep the markers moving in a coordinated fashion. The coordinated movement of all points can be encouraged by using pressure forces. Pressure forces, in their simplest form, act to encourage the deformable model to either move inward or outward. Since pressure acts on a contour in the normal direction – i.e., it has no tangential component – it can be defined by a simple scalar multiple of **N**, i.e.,

$$\mathbf{v}_{\text{pres}} = w_{\text{pres}}\mathbf{N}. \tag{9.8}$$

The sign of w_{pres} coupled with the definition of **N** determines in which direction the active contour is programmed to move under pressure. Getting the sign wrong is a common mistake in specifying parameters for a deformable model. Often, pressure is used at the beginning of evolution to help the active contour rapidly reach the region around the target object, called the *capture range*, where the potential starts to drop from unity to zero. In this case we would write the scalar w_{pres} as $w_{\text{pres}}(t)$. It is also possible that the pressure depends on the marker – i.e., $w_{\text{pres}}(s,t)$ – though this is uncommon.

Internal forces. The final types of forces that must be considered are the internal forces. These forces consider only the shape of the contour and are oblivious to the image itself; therefore, they are thought to be a type of prior information on shape. They can be very complicated, in general, perhaps derived from a prior collection of typical shapes, but we will consider only the most basic types here, those based on curvature. From the study of curves in calculus, we know that $d\mathbf{X}/ds$ yields a tangent vector (which would be a unit vector if s were arclength). The second derivative $d^2\mathbf{X}/ds^2$ yields a normal vector whose length is proportional to curvature. We can define a curvature force to be one that tends to straighten out curves implying a reduction in curvature. In this way, straighter curves are favored, which is a type of prior shape knowledge. With a little reasoning about planar curves, it can be shown that the normal force

$$\mathbf{v}_{\text{int}} = \alpha\frac{\partial^2\mathbf{X}}{\partial s^2} \tag{9.9}$$

with α positive will tend to straighten the curve and reduce curvature.

Putting all three forces together yields a comprehensive total force

$$\mathbf{F} = \mathbf{v}_{\text{ext}} + \mathbf{v}_{\text{pres}} + \mathbf{v}_{\text{int}}, \tag{9.10}$$

which when used in equation (9.1) causes an active contour to move toward object edges while keeping a smooth shape and, in the absence of other forces, tends to move either outward or inward depending on the sign of the pressure forces.

Illustrative example. Fig. 9.3 shows a simple example from a mediolateral oblique mammogram (left). A common initial step in the automatic analysis of mammograms is to limit subsequent analyses to the breast tissue, and a deformable model can be used to provide a smooth boundary. Since the edge of the breast does not form a sharp discontinuity between bright and dark regions, a breast mask was first created using Otsu's binarization method. With this simple binarization approach, the label

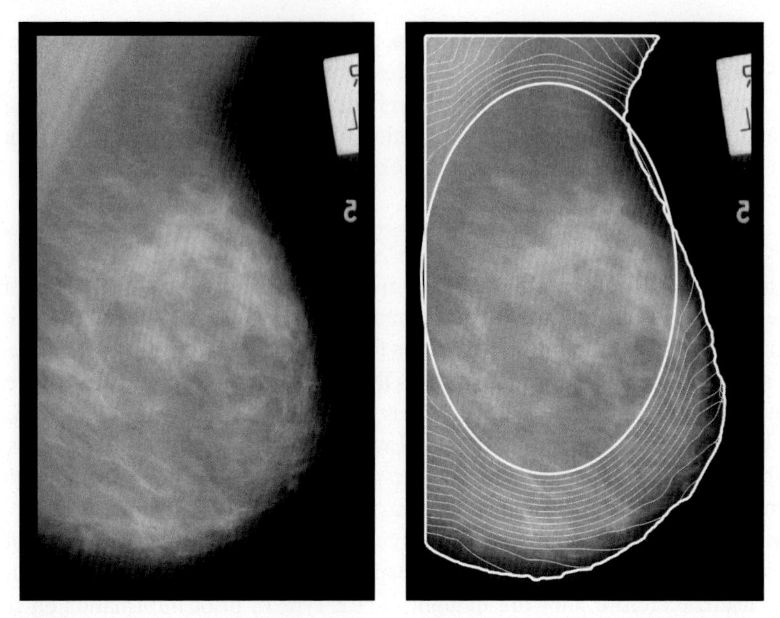

FIGURE 9.3

Mediolateral oblique mammogram (left). Evolution of a parametric deformable model starting from an ellipse (right).

in the upper right corner of this image was also included and the breast boundary was not smooth. To find a smoother representation of just the breast region itself, an ellipse made from a set of markers (a snake) was initialized in roughly the expected orientation and size of the breast tissue. External forces were created using gradient vector flow (GVF) (see [5]) and a small amount of internal forces from equation (9.9) was applied to smooth out the rough edges. Several steps showing the evolution of this deformable contour are shown in Fig. 9.3 (right). The initial and final contours are shown using thicker curves.

9.3.2 Geometric model speed functions

We already know that we must define speed functions instead of forces for geometric deformable models. Since speed functions determine the size and sign of vector forces acting in the normal direction of a curve, we could (naively) define V to be the projection of $\mathbf{F}$ in the normal direction – i.e., $V = \mathbf{F} \cdot \mathbf{N}$. However, there is a problem with this approach since curvature, which is key to defining internal forces, is defined with knowledge of the parametric form of the curve, which is a representation that is not present in geometric deformable models. It turns out that this form can be useful for external forces, as we will see a bit later, but for now let us start by building up, intuitively, what might be a good speed function from the elements that we have already defined.

Most importantly, we have defined a potential function p that is large away from edges and close to zero at edges. It seems, therefore, that a geometric model could use p as a speed function, since it will stop moving at the edges (where the speed is close to zero). So this could replace the external force in parametric models. A speed function for pressure is very natural since speeds are already intended to relate to forces in the normal direction. All we need is a constant speed term (which can depend on time) which is either positive or negative depending on whether the curve should move outward or inward. What would make a speed term that replaces internal forces? Suppose we could compute curvature directly from the level set function and did not require a parametric form of the curve. Then if curvature is large, we would want to encourage the curve to straighten out or, equivalently, to reduce curvature. Therefore, curvature κ also seems to be a good term for speed.

Putting the above notions together could lead to a wide variety of possibilities for a viable speed term, and it is somewhat of an "art" in specifying speed functions for any given application. Like the basic forces for parametric models described above, the following speed function has emerged as something of a prototype for all speed functions,

$$V_{\text{proto}}(x, y) = p(x, y)(\kappa(x, y) + w_{\text{pres}}). \tag{9.11}$$

Here, we can see that when the edge potential is zero, the curve will stop moving and when it is not zero, it will slow down when the curvature is smaller. The pressure will apply whenever $p \neq 0$, causing the curve to move in a specified inward or outward direction. It is critical, however, that the signs of all the terms in this speed function are chosen appropriately. Imagine that the curvature is large and that this causes the curve to move rapidly toward making the curve even curvier; this would be the opposite of the smoother curves that we are trying to achieve.

To determine the correct signs, let us start by assuming that $p = 1$ and $\kappa = 0$, which implies that $V = w_{\text{pres}}$ (dropping the explicit position notation and subscript for convenience). Now consider equation (9.4), the governing evolution equation for geometric deformable models, $\phi_t = V|\nabla\phi|$ (where $\phi_t = \partial\phi/\partial t$), and look at Fig. 9.4(a). It is clear that setting $w_{\text{pres}} > 0$ will increase ϕ, which has the consequence of shrinking the curve. Likewise, the curve will expand if $w_{\text{pres}} < 0$.

Now let us look at curvature. Customarily, curvature is defined without a sign – i.e., it is positive whether the curve bends inward or outward. But we need a signed curvature here to tell the curve in which direction to move in order to straighten itself. Since κ needs to be defined throughout the image domain, its definition is tied to the level set function, not the curve (which only exists on a subset of points), except that it must agree with the curvature of the curve on points that the curve passes through. It turns out that the (signed) curvature of any level set curve is given by the divergence of its normalized gradient (see Exercise 3.),

$$\kappa = \nabla \cdot \frac{\nabla\phi}{|\nabla\phi|}. \tag{9.12}$$

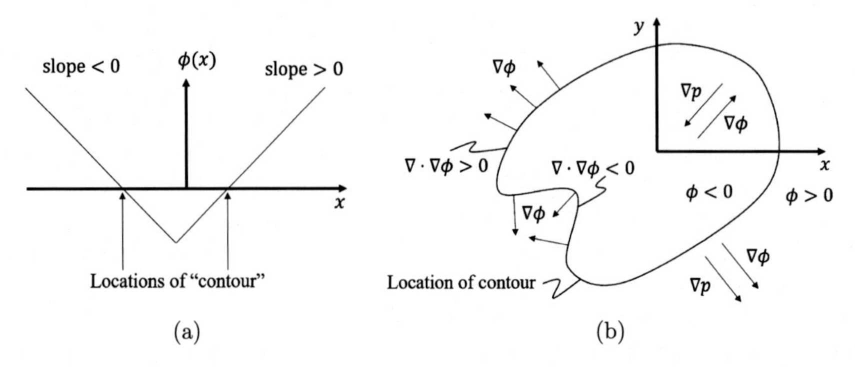

FIGURE 9.4

Illustrations of level set functions in (a) 1D and (b) 2D. In 1D, a well-defined level set function has a slope of ± 1 and the "contour" is defined by the points at which it crosses zero. In 2D, a well-defined level set function has unit gradient vectors and the contour is defined by the collection of points at which its value is zero.

This can be seen intuitively by looking at Fig. 9.4(b). On the lower left, the gradient vectors are converging – hence, the divergence is negative – as one moves along the curve, while on the upper right, the gradient vectors are diverging, and the divergence is positive. It should be clear that all of these signs are dependent on our choice of level set function and the sign conventions of differential calculus. Getting the signs wrong is a common source of error in implementations of deformable models. Note that because of the way it is defined, κ is a function of spatial position and could well be written as $\kappa(\mathbf{x})$. Thus, there is curvature defined for every level set of ϕ, not just the zero level set, where our curve resides.

We have not considered the sign of p to this point, and here we see a problem. By our previous definition, p is always positive. Consider a circle as a potential initial curve. In the absence of pressure, $k > 0$ and $p > 0$; thus $\phi_t > 0$ and our curve will shrink. If the desired boundary is outside the initial curve, we would then require $w_{\text{pres}} < 0$ to make the curve move outward to find the desired boundary. This is awkward since it requires prior knowledge that we may not have. Clearly, it would be better to distinguish outside from inside automatically while evolving the curve. This can be done by considering the interaction of the gradients of both ϕ and p. In particular, consider the inner product $\nabla p \cdot \nabla \phi$ and look at Fig. 9.4(b). For this exercise, imagine that the depicted curve is the desired curve. The inner product satisfies $\nabla p \cdot \nabla \phi > 0$ for a point on the outside and $\nabla p \cdot \nabla \phi < 0$ for a point on the inside. The sign of this inner product tells the contour exactly which way to move and can be directly added to the velocity as follows:

$$V_{\text{better}}(x, y) = p(x, y)(\kappa(x, y) + w_{\text{pres}}) + \nabla p(x, y) \cdot \frac{\nabla \phi(x, y)}{|\nabla \phi(x, y)|}. \tag{9.13}$$

A quick check and glance at Fig. 9.4(b) verifies that the sign of this extra term is correct. Clearly the term that we have added acts as an additional component of the external forces.

Plugging the speed function in equation (9.13) into the fundamental geometric deformable model evolution equation (9.4) yields

$$\frac{\partial \phi(x, y, t)}{\partial t} = p(x, y)(\kappa(x, y) + w_{\text{pres}})|\nabla \phi(x, y, t)| + \nabla p(x, y) \cdot \nabla \phi(x, y, t), \tag{9.14}$$

which is probably the most common evolution equation for geometric deformable models. It should be noted that positive coefficients multiplying the different terms can be used to trade off the relative influence of the internal, external, and pressure speeds as desired.

Illustrative example. Fig. 9.5 shows an example of deformable contour evolution using a geometric deformable model applied to a microscopic picture of cell nuclei. After staining, multiple cell nuclei stand out separately from the cytoplasm and membrane in this microscopy image (left). Because there is an unknown number of nuclei in such an image, use of a geometric deformable model is preferred over a parametric model because the geometric model can split into multiple pieces without having to detect the need for and then deliberately create a new object. Objects can be destroyed and created just by moving the level set function up and down, respectively, regionally as determined by the values and distribution of the external speed function. In this case, a single object was initialized as a rectangle near the border of the image. A speed function was specified to penalize high curvatures, the length of the curve, and the enclosed area, while seeking to minimize the edge potential computed according to equation (9.7) (see [4]). With these terms, the curve shrinks in the absence of image gradients, but stops at edges. Because the curve continues to shrink in zero-gradient areas, it leaves behind segmented cells until it no longer moves. The thicker lines in Fig. 9.5 (right) show both the starting and ending contours while the thinner lines show curves at a variety of intermediate stages; all were found using a zero level isocontour algorithm. It should be noted that the initial and intermediate contours are implicit functions of the evolving level set function and are not ordinarily computed.

9.3.3 Non-conservative external forces

Up to this point we have derived external forces (or speeds) entirely from edge potential functions and their gradients. Vector fields that can be written as the gradient of a scalar potential are called conservative, and they do not constitute the full breadth of possible vector fields. Conservative fields, in particular, are curl-free due to the vector identity $\nabla \times \nabla p = 0$. As part of the creativity in the development of external forces and for practical reasons, researchers have invented external forces that have a component of curl and are therefore not conservative.

In the case of parametric deformable models, non-conservative forces are easily incorporated into the general definition of forces in equation (9.10) by simply

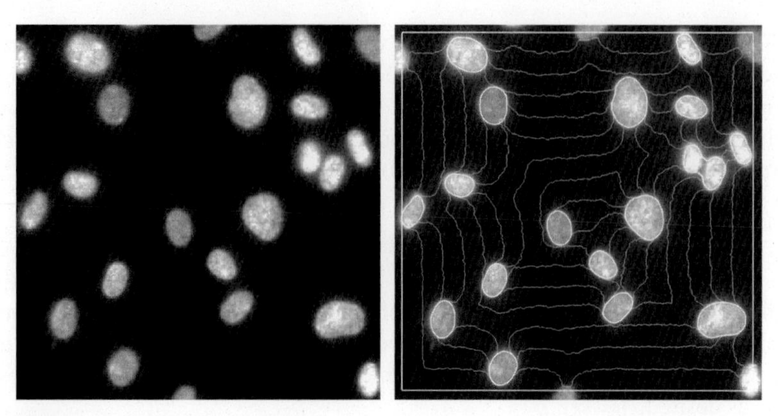

FIGURE 9.5

Microscopic picture of cell nuclei (left). Evolution of a geometric deformable model starting from a square (right).

replacing $\mathbf{v}_{\text{ext}}$ with whatever force is defined, curl-free or not. In the case of geometric deformable models, we should return to our original notion of creating speeds by simply projecting all the defined forces onto the normal vector of the curve. We had originally rejected this idea because of the requirement to define a normal vector using the parametric form of the curve. However, we have just developed the approach of projecting ∇p onto the gradient of ϕ, as written out in equation (9.13). In fact, the gradient of ϕ must point in the normal direction of its level sets throughout the plane; this is a fundamental concept of vector calculus. Because of our definition of the normal direction in parametric deformable models and the definition of our level set function, we have the following:

$$\mathbf{N} = -\frac{\nabla\phi}{|\nabla\phi|}. \tag{9.15}$$

With this, we have a clear and simple relationship between the normal direction on the curve and the level set function itself. In fact, this tells us that if we have an external force $\mathbf{v}_{\text{ext}}$ defined in any way, we can create a speed by projecting it in the normal direction and then add this to any other speed function already derived. Since our external forces point toward the desired boundary while ∇p points away from the desired boundary, we would need to use a minus sign and therefore write

$$V_{\text{general}}(x, y) = p(x, y)(\kappa(x, y) + w_{\text{pres}}) - \mathbf{v}_{\text{ext}}(x, y) \cdot \frac{\nabla\phi(x, y)}{|\nabla\phi(x, y)|}. \tag{9.16}$$

Here, $\mathbf{v}_{\text{ext}}$ could incorporate ∇p or replace it by something entirely different. Two very popular forces that are often used for this purpose are GVF (see [5]) and vector field convolution (VFC) (see [6]).

9.4 Numerical implementation

In this section, we develop some basic approaches for numerical implementation of both parametric and geometric deformable models. This development serves to illustrate some of the issues that must be considered when implementing deformable models but does not introduce many details that are often critical to successful implementation in real-world situations. To gain further insight, additional literature must be consulted (for example, see [3]).

9.4.1 Parametric model implementation

We consider implementation of the marker representation of parametric deformable models (also called active contours or snakes) in 2D. We start with the simple expression equation (9.1), repeated here:

$$\gamma \frac{\partial \mathbf{X}}{\partial t} = \mathbf{F}. \tag{9.17}$$

Let the active contour be represented by points $\mathbf{X}_i = (X_i, Y_i) = (X(ih), Y(ih))$, $i = 1, \ldots, N$. Then the temporal derivative is approximated by

$$\mathbf{X}_t \approx \frac{\mathbf{X}_i^n - \mathbf{X}_i^{n-1}}{\Delta t}, \tag{9.18}$$

where the superscripts indicate time or iteration counter. The external forces are simply evaluated at each point $\mathbf{X}_i^n$ as in $\mathbf{v}_{\text{ext}}(\mathbf{X}_i^n)$, which will require interpolation in general. The pressure forces act in a normal direction and therefore require computation of a unit normal $\mathbf{N}_i^n$, which can be computed by first defining the unit tangent vector at point i as

$$\mathbf{T}_i = \frac{\mathbf{X}_{i+1} - \mathbf{X}_i}{|\mathbf{X}_{i+1} - \mathbf{X}_i|}. \tag{9.19}$$

Then, the unit normal vector approximation is

$$\mathbf{N}_i = \frac{\mathbf{T}_i - \mathbf{T}_{i-1}}{|\mathbf{T}_i - \mathbf{T}_{i-1}|}, \tag{9.20}$$

which means that the pressure forces are $w_{\text{pres}}\mathbf{N}$.

The curvature forces are generally very grossly approximated using simple finite differences wherein the sampled points are assumed to be approximately equally spaced. Accordingly, we can approximate the derivatives along the curve using

$$\dot{\mathbf{X}}_i^n = \mathbf{X}_{i+1} - \mathbf{X}_i, \tag{9.21a}$$

$$\ddot{\mathbf{X}}_i^n = \mathbf{X}_i - \mathbf{X}_{i-1} \tag{9.21b}$$

$$= \mathbf{X}_{i+1} - 2\mathbf{X}_i + \mathbf{X}_{i-1}, \tag{9.21c}$$

which implies that the basic internal force can be written as $\mathbf{v}_{\text{int}}^n = \alpha \ddot{\mathbf{X}}_i^n$.

Now let $\tau = \Delta t/\gamma$, a single parameter that controls the step size in time. Let us consider the entire collection of points together as one vector $\mathbf{X}^n$, which is a bit of abuse of notation. We can write this vector as the collection of x-positions for $i = 1, \ldots, N$ followed by the collection of y-positions for $i = 1, \ldots, N$. Then we can write the evolution equation as

$$\frac{\mathbf{X}^n - \mathbf{X}^{n-1}}{\tau} = \alpha \mathbf{A} \mathbf{X}^n + \mathbf{F}(\mathbf{X}^{n-1}), \tag{9.22}$$

where the matrix $\mathbf{A}$ provides the finite differences that are necessary to compute the internal forces and $\mathbf{F}$ is a function that provides the x- and y-components of both the external and pressure forces acting on each of the points of the active contour. This equation can be rearranged to yield

$$\mathbf{X}^n = (\mathbf{I} - \tau \alpha \mathbf{A})^{-1} [\mathbf{X}^{n-1} + \tau \mathbf{F}(\mathbf{X}^{n-1})], \tag{9.23}$$

which is a matrix form for active contour evolution that includes smoothing, pressure, and external forces that push an active contour towards a boundary.

The intuition behind equation (9.23) is straightforward. Consider the position $\mathbf{X}^{n-1}$ of the active contour at time $n-1$. In what direction should the active contour move? Clearly, it should try to move in the direction pointed to by the external and pressure forces, $\mathbf{F}(\mathbf{X}^{n-1})$. How far should it move the points? Well, that is a numerical consideration that is governed by the step size in time, τ. The term in the brackets, therefore, makes sense. The application of $(\mathbf{I} - \tau \mathbf{A})^{-1}$ implements a smoothing operation since it is the inverse of a term containing a differential operator $\mathbf{A}$. The degree of smoothing depends on both the time step τ and the positive coefficient α.

9.4.2 Geometric model implementation

We consider the general geometric deformable model given by

$$\frac{\partial \phi}{\partial t} = (\rho \kappa + w_{\text{pres}}) |\nabla \phi| - \mathbf{v}_{\text{ext}} \cdot \nabla \phi. \tag{9.24}$$

Slightly different from equation (9.16), this model decouples the pressure speed term from the external force for convenience. There are four terms that must be computed to implement this equation numerically. We will now show how to implement these terms in numerically stable ways using finite differences.

Temporal derivative. We follow convention and use a forward difference to compute the temporal for the left-hand side of (9.24), yielding

$$\frac{\partial \phi}{\partial t} \approx \frac{\phi_{ij}^{n+1} - \phi_{ij}^n}{\Delta t}. \tag{9.25}$$

Here, i and j indicate x- and y-positions in the image. (Note that we are assuming standard laboratory coordinates; i and j are not indexing the row and column of a matrix, though this may be a good way to implement the equations.) This equation will immediately provide a way to update ϕ at time $n+1$ given functions at time n. Therefore, all remaining quantities on the right-hand side of equation (9.24) are to be evaluated at time n. If the notation is missing, it is assumed to be at time n.

Curvature term. In the term $p\kappa|\nabla\phi|$, we assume that both the image potential p and the curvature κ can be directly evaluated at each point (i, j) in the image using, for example, equations (9.7) and (9.12). In computing κ using equation (9.12), care should be taken to use central differences for the evaluation of both the gradient and the divergence so that there is no shift in its evaluation. Because κ can be both positive and negative, there is no natural way that information flows – that is, there is no way *a priori* to know whether the level set is moving inward or outward – at any given point. Therefore, central differences should be used to evaluate the gradient. These can be represented as follows:

$$D_{ij}^{0x} = \frac{\phi_{i+1,j}^n - \phi_{i-1,j}^n}{2\Delta x} \quad \text{and} \quad D_{ij}^{0y} = \frac{\phi_{i,j+1}^n - \phi_{i,j-1}^n}{2\Delta y}. \tag{9.26}$$

Putting this together, we find that the curvature term in equation (9.24) can be approximated by

$$p\kappa_{ij}^n|\nabla\phi| = p\kappa_{ij}^n\sqrt{\phi_x^2+\phi_y^2} \approx p\kappa_{ij}^n\left[(D_{ij}^{0x})^2+(D_{ij}^{0y})^2\right]^{1/2}. \tag{9.27}$$

Advection term. The external force $\mathbf{v}_{\text{ext}}$ is computed in advance from the image and is not dependent on ϕ. Therefore, the direction in which a curve should move in response to this force is known, and information should flow in that direction. This means that forward and backward differences should be judiciously used to guarantee the information flow direction; this is guaranteed by satisfying the so-called *upwind condition*, as described next.

We define the following forward and backward finite differences for ϕ:

$$D_{ij}^{-x} = \frac{\phi_{i,j}^n - \phi_{i-1,j}^n}{\Delta x}, \quad D_{ij}^{+x} = \frac{\phi_{i+1,j}^n - \phi_{i,j}^n}{\Delta x}, \tag{9.28}$$

$$D_{ij}^{-y} = \frac{\phi_{i,j}^n - \phi_{i,j-1}^n}{\Delta y}, \quad D_{ij}^{+y} = \frac{\phi_{i,j+1}^n - \phi_{i,j}^n}{\Delta y}. \tag{9.29}$$

Since $\mathbf{v}_{\text{ext}} \cdot \nabla\phi = u\phi_x + v\phi_y$, there should be just two terms in this approximation. The choice of which two to use depends on the signs of u and v, i.e., which direction the information is flowing at pixel (i, j). Rather than writing out an "if then else" type of statement, it is convenient to write this with "switches" as follows:

$$\begin{aligned}\mathbf{v}_{\text{ext}} \cdot \nabla\phi \approx{}& \max(u_{ij}^n, 0)D_{ij}^{-x} + \min(u_{ij}^n, 0)D_{ij}^{+x} \\ &+ \max(v_{ij}^n, 0)D_{ij}^{-y} + \min(v_{ij}^n, 0)D_{ij}^{+y}.\end{aligned} \tag{9.30}$$

The min and max functions act as switches that choose the correct terms to use in the computation. Recall that all these numerical differences depend on ϕ at time n even though the notation does not explicitly include these variables.

Pressure term. The sign of w_{pres} in the pressure term, $w_{\text{pres}}|\nabla\phi|$, could depend on the location in the image. Region forces, for example, are derived from a prior pixelwise segmentation or classification. If inside the object, the pressure should push the curve outward, but if outside the object, the pressure should push the curve inward. Like advection forces, these are known beforehand, but the computation must be sensitive to the possibility of either a positive or negative pressure. Accordingly, using the max and min switches, we approximate the pressure term using

$$w_{\text{pres}}|\nabla\phi| \approx \max(w_{\text{pres},ij}, 0)\nabla^- + \min(w_{\text{pres},ij}, 0)\nabla^+, \tag{9.31}$$

where

$$\nabla^+ = \left[\max(D_{ij}^{-x}, 0)^2 + \min(D_{ij}^{+x}, 0)^2 + \max(D_{ij}^{-y}, 0)^2 + \min(D_{ij}^{+y}, 0)^2\right]^{1/2}, \tag{9.32}$$

$$\nabla^- = \left[\max(D_{ij}^{+x}, 0)^2 + \min(D_{ij}^{-x}, 0)^2 + \max(D_{ij}^{+y}, 0)^2 + \min(D_{ij}^{-y}, 0)^2\right]^{1/2}. \tag{9.33}$$

Like the curvature term, the pressure term requires the magnitude of $\nabla\phi$, which is why the individual terms are squared and added and the square root is applied on each sum.

Full geometric deformable model numerical approximation. The individual terms can now be put together to reveal an iterative solution to the geometric deformable model evolution equation,

$$\begin{aligned}\phi_{ij}^{n+1} = \phi_{ij}^n + \Delta t \{&\max(w_{\text{pres},ij}, 0)\nabla^- + \min(w_{\text{pres},ij}, 0)\nabla^+ \\ &\max(u_{ij}^n, 0)D_{ij}^{-x} + \min(u_{ij}^n, 0)D_{ij}^{+x} + \\ &\max(v_{ij}^n, 0)D_{ij}^{-y} + \min(v_{ij}^n, 0)D_{ij}^{+y} + \\ &\rho\kappa_{ij}^n\left[(D_{ij}^{0x})^2 + (D_{ij}^{0y})^2\right]^{1/2}\}.\end{aligned} \tag{9.34}$$

Although this equation appears complicated, the reasons for each term should be clear from the previous development and, overall, the computation is not overly difficult to implement.

Unfortunately, despite its complexity, equation (9.34) represents only a basic implementation of a geometric deformable model; although direct implementation of this equation works, it is both inefficient and numerically unstable. Its lack of efficiency arises primarily from the need to update ϕ at every pixel of the image even though the desired contour resides only at the zero level set. This is typically addressed using a so-called *narrow band* method (see [7]), which keeps track of the approximate zero level set and computes updates only in a narrow band of pixels around it. This approach requires some "book-keeping" but ends up being much more efficient. There are even faster methods that are available, many based on the fact that the equation to implement is nothing more than a partial differential equation for which there are many numerically efficient approaches.

The numerical instability arises from the fact that all level sets, not just the zero level set, tend to move toward the desired boundary. Thus, ϕ tends to get steeper and steeper around the desired contour over time, which tends to make all of the finite difference approximations that we just presented inaccurate. Ideally, keeping the magnitude of the gradient of ϕ close to unity is best; this means that ϕ should be a signed distance function, ideally. One way to handle this problem is to recompute ϕ from the zero level set every 10–12 iterations or so. This is accomplished using a fast marching algorithm (see [3]), which is very fast, so it is not too much of a burden. There are other ways to accomplish this during the update itself.

9.5 Other considerations

The above presentation has outlined only the most basic considerations in deformable models. It is important to be aware of some of the issues that researchers have also addressed in the past.

Topology. It may not be immediately apparent from the preceding discussion that geometric deformable models have great flexibility in addressing the segmentation of multiple objects and objects with arbitrary topologies while this creates a problem in parametric deformable models. It should be pretty clear that parametric models require a fixed representation – a 2D curve is what we presented above – at the outset. The markers that are used in the numerical implementation are assumed to be connected to each other in a fixed way; thus, there is a fixed topology in the model and in what it will represent in every iteration until convergence. There exist parametric deformable models that will sense the need to adjust their topologies and create new objects, for example, but one can imagine the additional complexity in their implementation. On the other hand, geometric deformable models can create and remove objects simply by having parts of ϕ become negative or positive, respectively. The number of objects and their topologies are decided only at the end when the zero level set is computed. This additional flexibility is highly desired in some situations. In situations where the topology of the object(s) to be segmented is known and the use of a geometric deformable model is desired for computational (or other) reasons, there exist methods to preserve topology over the iterations – at the cost of additional computations.

Three dimensions. The boundary of a 3D object is represented by a surface rather than a curve. The concept of an "active contour" is no longer quite appropriate, but one can still say "deformable model" and be correct. Although the parametric representation using markers can still be used, the connections between markers are no longer simple line segments. Instead, the most common representation is a triangle mesh – i.e., the markers are connected to each other by triangular surface elements. The implementation of external forces remains straightforward, but pressure forces require a more involved computation of surface normals and internal forces must consider multiple curvatures as defined using numerical approximations of the principal curvatures on the surface. In contrast, both the mathematical representation and the implementation of geometric deformable models in 3D are more straightforward. In particular, the vector calculus notation that we used in deriving the evolution equations of geometric deformable models is largely unchanged. Accordingly, implementation of geometric deformable models in 3D is also much more straightforward.

Multiple objects. This chapter has focused on the problem of estimating the boundary of one object, but suppose there are multiple objects. We have already noted that geometric deformable models can adapt their topologies without any specific intervention. Because of this property, segmentation of multiple objects whose boundaries are similarly defined is quite straightforward using geometric deformable models. However, when the objects to be segmented are to be separately labeled – i.e., as object 1, object 2, etc. – then either multiple parametric models or multiple level set functions must be used. Interactivity of the objects must then be specified; issues such as connectivity, intersection, and object topology often must be tracked or specified. A great deal of effort has been put into the development of computationally efficient ways to handle multiple objects in the framework of deformable models, and there is no single widely accepted approach.

Prior shape information. If knowledge of the approximate shape of the object is available or if knowledge of the space of possible or likely shapes is available, then one must think about ways to specify prior shape information on the deformable model. There are many such methods that have been proposed in the literature, some of which are described in Chapter 8. Overall, one can think of prior shape knowledge as being implemented as another internal force term that is updated as the shape of the deformable model evolves. Clearly, the desired result will be a trade-off between the prior shape, which will tend to cause the model to evolve towards the mean object shape, and the image data, which will cause the model to evolve towards nearby strong edges in the image. Typically, there will be a regularization coefficient that weights this trade-off and computational strategies that try to identify a globally optimal result.

Multiple scales. Evolution towards features that are close to the current model is an ever-present feature of deformable models. If strong features coming from noise or another object are attractive to the model, then it will be difficult for the evolving model to find the desired object. This undesirable behavior is often said to happen because the model converged to a local minimum. The concept of a local minimum has a specific interpretation when the deformable model is derived as one that minimizes a cost function, which is an alternative formulation of deformable models that we did not emphasize in this chapter. But whether formulated as a force balance equation, as we have done here, or as a cost function minimization problem, it is highly desirable to avoid such local minima. One very common approach is to use multiple scales in which to evolve the model, typically starting with a coarse scale at the beginning and shifting to finer and finer scales over time. The meaning of scale can be varied and is often highly problem-dependent. Using a larger Gaussian blurring function in the definition of external forces at the beginning is one simple way to implement multiple scales. Using active contours with different numbers of markers starting with a smaller number at first and increasing the number over time is another way. Working on downsampled versions of the image – often using a Gaussian scale space – at the beginning is probably the most common way to implement multiple scales. In the case of geometric deformable models, this has the great benefit of reducing the number of pixels that require level set function updating at the beginning, thus substantially reducing computation time.

9.6 Recapitulation

The take-home messages from this chapter are:

- There are two fundamental ways to represent deformable models: parametric and geometric.
- Both methods yield iterative approaches to describe how the contour moves to find the desired boundary.
- There are three fundamental forces, which are known as speeds in geometric models: internal, external, and pressure.
- Both methods can be implemented using finite difference schemes that change the representation iteratively until convergence.

9.7 Exercises

1. Given

$$\frac{\partial \mathbf{X}}{\partial t} = V\mathbf{N},$$

take the total derivative of $\phi(\mathbf{X}(s,t),t) = 0$ to show that

$$\frac{\partial \phi}{\partial t} = V|\nabla \phi|.$$

2. Suppose you wanted to penalize higher-order properties of a curve's shape and augment the internal forces by considering $\beta \partial^4 \mathbf{X}/\partial s^4$ with $\beta > 0$. To make the curve smoother, should you add or subtract this term?

3. (a) Prove that the curvature of level sets of ϕ is given by

$$\kappa = \nabla \cdot \frac{\nabla \phi}{|\nabla \phi|}.$$

(b) Show that a circle has positive curvature.

4. Weak edges are a problem in deformable models, as the curve tends to move right through them instead of stopping. The second term on the right-hand side of this geometric deformable model was added to try to address this problem. Note that $\mathbf{X}$ is the position of a point on the active contour as measured from the origin, which is generally assumed to be contained within the contour. We have

$$\frac{\partial \phi}{\partial t} = \lambda\left(c\kappa|\nabla\phi| + \nabla c \cdot \nabla\phi\right) + \left(c + \frac{1}{2}\mathbf{X} \cdot \nabla c\right)|\nabla\phi|.$$

(a) Explain how this second term works. Provide a couple of illustrations to bolster your words.
(b) What happens if the origin is not included in the active contour?

References

[1] Kass M., Witkin A., Terzopoulos D., Snakes: Active contour models, International Journal of Computer Vision 1 (4) (1988) 321–331.

[2] Caselles V., Kimmel R., Sapiro G., Geodesic active contours, in: Proceedings of IEEE International Conference on Computer Vision, IEEE, June 1995, pp. 694–699.

[3] Sethian J.A., Level Set Methods and Fast Marching Methods: Evolving Interfaces in Computational Geometry, Fluid Mechanics, Computer Vision, and Materials Science (vol. 3), Cambridge University Press, 1999.

[4] Li C., Xu C., Gui C., Fox M.D., Level set evolution without re-initialization: a new variational formulation, in: 2005 IEEE Computer Society Conference on Computer Vision and Pattern Recognition (CVPR'05) (vol. 1), IEEE, June 2005, pp. 430–436.

[5] Xu C., Prince J.L., Snakes, shapes, and gradient vector flow, IEEE Transactions on Image Processing 7 (3) (1998) 359–369.

[6] Li B., Acton S.T., Active contour external force using vector field convolution for image segmentation, IEEE Transactions on Image Processing 16 (8) (2007) 2096–2106.

[7] Mille J., Narrow band region-based active contours and surfaces for 2D and 3D segmentation, Computer Vision and Image Understanding 113 (9) (2009) 946–965.

CHAPTER

Graph cut-based segmentation

10

Jens Petersen[a,b], **Ipek Oguz**[c], **and Marleen de Bruijne**[a,d]

[a]*Department of Computer Science, University of Copenhagen, Copenhagen, Denmark*
[b]*Department of Oncology, Rigshospitalet, Copenhagen, Denmark*
[c]*Department of Electrical Engineering and Computer Science, Vanderbilt University, Nashville, TN, United States*
[d]*Department of Radiology and Nuclear Medicine, Erasmus MC - University Medical Center Rotterdam, Rotterdam, the Netherlands*

Learning points

- Graph cut methods for energy minimization
- Segmentation tasks as energy minimization problems
- Graph cuts for segmentation of complex structures
- Limitations and advantages of using graph cuts

10.1 Introduction

The purpose of this chapter is to give an introduction to graph cut-based methods in medical image segmentation with a particular focus on vessel-like structures. Graph cut methods can be used to compute solutions – optimal solutions in some cases and good approximate solutions in other cases – to problems that can be modeled using Markov random fields (MRF). As such they have found relatively wide use in medical image analysis for tasks such as segmentation, registration, and reconstruction.

The chapter first introduces the needed graph theory and then describes how MRF models can be used for image segmentation, followed by a description of ways of computing solutions using graph cuts for different classes of problems. As examples, we show how graph cuts can be used to segment coronary arteries in computed tomography angiography (CTA) and the interior and exterior surface of airways in computed tomography (CT) images.

Medical Image Analysis. https://doi.org/10.1016/B978-0-12-813657-7.00024-8

10.2 Graph theory

10.2.1 What is a graph?

A graph G consists of vertices $\mathcal{V}$ and edges $(u, v) \in \mathcal{E}$ between pairs of vertices $u, v \in \mathcal{V}$. Edges can be directed or undirected. A graph with directed edges is called a directed graph, whereas a graph where the edges are bidirectional is called an undirected graph.

10.2.2 Flow networks, max flow, and min cut

A flow network is a directed graph in which each edge has a non-negative capacity $c(u, v)$ for $(u, v) \in \mathcal{E}$ and the node set $\mathcal{V}$ has two special nodes, the source s and the terminal (or sink) t.

A feasible flow $f(u, v)$ in a flow network has to satisfy the following requirements:

$$f(u, v) = -f(v, u), \tag{10.1}$$

$$f(u, v) \leq c(u, v), \text{ and} \tag{10.2}$$

$$\sum_{v \in \mathcal{V}} f(v, u) = 0 \quad \text{for all } v \in \{\mathcal{V} \setminus \{s, t\}\}. \tag{10.3}$$

The value of the flow is the amount of flow flowing from the source s to the sink t through the network:

$$\sum_{v \in \mathcal{V}} f(v, t) \quad \text{for all } v \in \mathcal{V}. \tag{10.4}$$

The maximum flow problem is the problem of finding the maximum value of feasible flows.

A cut in a flow network is a partition of the vertices of the graph into two disjoint sets. Of particular relevance to the content of this chapter is the s-t-cut, where the source and sink vertices are in different sets, called respectively the source set and the sink set. The cost of an s-t-cut is the total capacity of edges going from the source set to the sink set. The minimum cut problem is the problem of finding the minimum cost s-t-cut.

The min-cut max-flow theorem states that the cost of the minimum cut in a flow network is equal to the maximum flow in the same network.

10.3 Modeling image segmentation using Markov random fields

Suppose we are given an image, that is, several pixel values $\mathbf{x} = \{x_0, x_1, \cdots, x_{n-1}\}$, and we are interested in segmenting this image into foreground and background, that is, we are interested in variables (or labels) $\mathbf{y} = \{y_0, y_1, \cdots y_{n-1}\}$, where $y_i \in \{0, 1\}$ for $i \in \{0, 1, \cdots, n-1\}$.

An MRF can be used to model such problems. MRFs are a type of graphical model, in which the conditional probability of observing a given variable y_i is independent of all other variables given a neighborhood N_i of i, that is,

$$P(y_i|\mathbf{y} \setminus y_i) = P(y_i|N_i). \tag{10.5}$$

Such dependencies can be described using an undirected graph, with a vertex for each variable and an edge for each dependency. With images, neighborhoods consisting of the set of directly neighboring pixels are often used, for instance in 3D pixels lying immediately above, below, left, right, front, and back. In such cases, the model is saying that the label y_i of a particular pixel i is independent of all other pixel labels given the neighboring pixels N_i. This is of course not always correct, but it turns out to be a useful model in many cases.

Given an MRF it is possible to write the joint distribution over all variables $\mathbf{y}$ as a Gibbs distribution:

$$P(\mathbf{y}) = \exp(-E(\mathbf{y})/T)/Z, \tag{10.6}$$

where $E(\mathbf{y})$ is called the energy, T is called the temperature, and

$$Z = \sum_{i\in\{0,1,\cdots,n-1\}} \exp(-E(\mathbf{y})/T)$$

is the partition function or normalization constant. A complete subgraph, that is, a subgraph where each vertex is connected to every other vertex, is called a clique. The energy $E(\mathbf{y})$ can be written as a sum of clique potentials, $V_c(\mathbf{y})$, over all possible maximum cliques $c \in C$:

$$E(\mathbf{y}) = \sum_{c\in C} V_c(\mathbf{y}). \tag{10.7}$$

By defining the clique potentials, one specifies the degree to which the variables that are part of the cliques are dependent. With the example neighborhood system given above, where each pixel label is dependent on the labels of the pixels above, below, left, right, front, and back, the maximum clique size is two and the model can thus be specified through pairwise potentials:

$$E(\mathbf{y}) = \sum_{(i,j)\in C} V_{i,j}(y_i, y_j), \tag{10.8}$$

where $(i, j) \in C$ are each of the pairwise cliques in the model.

It is typically not feasible to compute the probability distribution $P(\mathbf{y})$ directly, due to the large number of possible combinations of labels. Moreover, in segmentation we want to determine what is the most likely labeling given the observed pixel values. In such cases, we do not need the full distribution $P(\mathbf{y})$, and what is sought is the maximum a posteriori (MAP) estimate of the labels given the data, $\hat{\mathbf{y}}$, where

$$\hat{\mathbf{y}} = \arg\max_{\mathbf{y}} P(\mathbf{y}|\mathbf{x}). \tag{10.9}$$

Using Bayes' rule this can be rewritten as

$$\hat{\mathbf{y}} = \arg\max_{\mathbf{y}} P(\mathbf{x}|\mathbf{y})P(\mathbf{y}) \tag{10.10}$$

$$= \arg\max_{\mathbf{y}} \prod_{i=0}^{n-1} P(x_i|y_i)P(\mathbf{y}) \tag{10.11}$$

$$= \arg\max_{\mathbf{y}} \sum_{i=0}^{n-1} \log\left(P(x_i|y_i)\right) + \log\left(P(\mathbf{y})\right), \tag{10.12}$$

where we have assumed the conditional probabilities are independent and therefore allow $P(\mathbf{x}|\mathbf{y})$ to be factorized over its individual variables and computed the logarithm.

Inserting the value of $P(\mathbf{y})$ from equation (10.6) with energy from equation (10.8) gives

$$\hat{\mathbf{y}} = \arg\max_{\mathbf{y}} \sum_{i=0}^{n-1} \log\left(P(x_i|y_i)\right) - \sum_{(i,j)\in C} V_{i,j}(y_i, y_j)/T + log(Z) \tag{10.13}$$

$$= \arg\min_{\mathbf{y}} \sum_{i=0}^{n-1} -\log\left(P(x_i|y_i)\right) + \sum_{(i,j)\in C} V_{i,j}(y_i, y_j)/T. \tag{10.14}$$

Note that the normalization constant Z is left out because it is independent of $\mathbf{y}$. The segmentation problem – obtaining the MAP – can now be solved by solving the energy minimization of equation (10.14). We can simplify the notation a bit to make clear that the MAP solution can be computed by minimizing the sum of a unary term and a pairwise term:

$$\hat{y} = \arg\min_{\mathbf{y}} \sum_{i=0}^{n-1} f_i(y_i) + \sum_{(i,j)\in C} g_{i,j}(y_i, y_j), \tag{10.15}$$

where $f_i(y_i) = -\log\left(P(x_i|y_i)\right)$ and $g_{i,j}(y_i, y_j) = V_{i,j}(y_i, y_j)/T$. The unary term depends on the data and can be thought of as the cost of observing the combination of pixel value x_i and label y_i. The pairwise term similarly describes the cost of observing labels y_i and y_j at the same time.

10.4 Energy function, image term, and regularization term

Suppose we assume a model of the form described in Section 10.3, in which the neighborhood of each pixel or voxel is equivalent to the pixels immediately connected to it, that is, to the left, right, up, and down in 2D and also front and back in 3D. The energy function can be written in the following form:

$$E(\mathbf{y}) = \sum_{v \in \mathcal{V}} f_v(y_v) + \sum_{(u,v) \in \mathcal{E}} g_{u,v}(y_u, y_v), \tag{10.16}$$

where $\mathbf{y} = \{y_v | v \in \mathcal{V}\}$ is a given configuration, $f_v(y_v)$ gives the unary cost in energy of the variable y_v, and $g_{u,v}(y_u, y_v)$ gives the pairwise cost in energy of the variables y_v and y_u.

10.4.1 Image term

As described, the segmentation class that a particular pixel or voxel belongs to is modeled with the variable y_v. For example, if we are interested in segmenting vessels in volumetric CT images, then $y_v = 1$ might signify that the voxel referenced by v belongs to a vessel and $y_v = 0$ would conversely signify that the voxel referenced by v belongs to the background or non-vessel class. In such cases, the function $f_v(y_v)$ models the cost of assigning a particular class to a particular voxel. Finding a suitable function $f_v(y_v)$ for a particular problem can be difficult. In the simplest cases one could derive $f_v(y_v)$ based on models of the intensity distributions of the classes, for instance using Gaussian mixture models. In the segmentation of tubular structures such as vessels, one can also derive suitable functions based on Hessian eigenvalue analysis with the widely used vesselness filter [1] or learned appearance models based on the output of machine learning classifiers.

10.4.2 Regularization term

The pairwise term in equation (10.16) models dependencies between labels of neighboring pixels. Typically, we expect labels to be somewhat spatially coherent, and we can encode this by assigning higher energies (higher costs) to configurations in which neighboring pixels have different labels. One such example is the Potts model:

$$g_{u,v}(y_u, y_v) = \begin{cases} \lambda & \text{if } y_u \neq y_v, \\ 0 & \text{else.} \end{cases} \tag{10.17}$$

Here, an extra energy of $\lambda > 0$ is imposed for each neighboring pair of pixels that have different labels. The parameter λ can be tuned to give the level of smoothness that fits the particular problem best. As λ increases, the optimal solutions will tend to have fewer and larger regions of similar values.

10.5 Graph optimization and necessary conditions

This section describes how to find the configuration of $\mathbf{y}$ that minimizes the energy in equation (10.16) using a minimum cut algorithm. To begin with, we will only consider problems where each y_v can take on one of two values, that is, $y_v \in \{0, 1\}$ for all $v \in \mathcal{V}$. With this, binary segmentation problems can be solved. Multiclass problems are discussed in Section 10.7.

10.5.1 Energy minimization and minimum cuts

We will follow the method described in [2] and construct a graph with the vertices $\mathcal{V}_{s,t}$ consisting of the vertices $\mathcal{V}$ and a source s and a sink t. The goal is to do this in such a manner that the s-t-cut in this graph gives us the optimal configuration in terms of the energy $E(\mathbf{y})$. Specifically, we will do this such that vertices in the source set S get the label 0 and the vertices in the sink set get the label 1.

Note that the cost of the minimum cut does not have to equal the minimum energy and it would not be possible for many energies, because we cannot have negative edge capacities. All that is needed is for the cost of a cut to be a constant from the energy of the configuration given by the cut. The minimum cut will then also minimize energy.

10.5.2 Necessary conditions

For equation (10.16) to be minimized by graph cuts it has to be graph-representable. This is the case for so-called submodular energies/functions [2]. A necessary and sufficient condition for an energy function with binary variables to be submodular is that the pairwise terms satisfy the following property:

$$g_{u,v}(0,0) + g_{u,v}(1,1) \leq g_{u,v}(0,1) + g_{u,v}(1,0). \tag{10.18}$$

A central part of the construction by Kolmogorov et al. [2] is the so-called additivity theorem, which says that the sum of two graph-representable functions is itself graph-representable. In particular, if we have two functions each represented by a graph with the same number of (corresponding) vertices, then simply adding the edge capacities of the two graphs results in a new graph representing the sum of the two functions. If edges are missing in one of the graphs, we may simply assume that they are there with zero capacity.

10.5.3 Minimum cut graph construction

The minimum cut graph $G_{s,t}$ consists of the set of vertices $\mathcal{V}_{s,t} = \{\mathcal{V}, s, t\}$ and directed edges $\mathcal{E}_{s,t}$. Each directed edge has a capacity, and $c(u, v) > 0$ denotes that there is an edge from vertex u to v with capacity $c(u, v)$. Because of the additivity theorem, it suffices to look at ways of representing $f_v(y_v)$ and $g_{u,v}(y_u, y_v)$ for individual and pairs of variables, respectively. For a quick overview of the needed edges and capacities, the reader is referred to Boxes 10.1 and 10.2. The following will explain why these edge capacities work in detail.

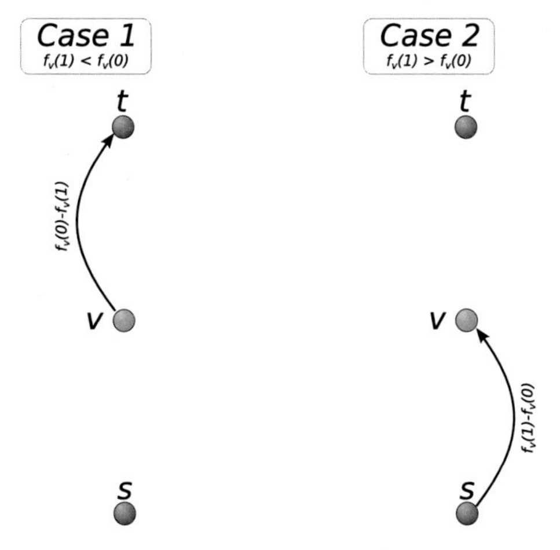

FIGURE 10.1 Unary term edges of graph $G_{s,t}$.

The edges implementing the unary terms of equation (10.16) for variable y_v and the two different cases of problems.

For each variable y_v there are two possible configurations $y_v = 0$ and $y_v = 1$ and two associated costs $f_v(0)$ and $f_v(1)$. If $f_v(0) > f_v(1)$, an edge is added, which represents the extra cost of choosing $y_v = 0$ over $y_v = 1$, that is, an edge with capacity $f_v(0) - f_v(1)$. This edge goes from the vertex v to the sink t. If, on the other hand, $f_v(1) > f_v(0)$, then an edge is added from the source s to v with capacity $f_v(1) - f_v(0)$.

BOX 10.1 (Capacities of edges implementing a unary term).
For each $v \in \mathcal{V}$,

$$c(v,t) = f_v(0) - f_v(1) \quad \text{if } f_v(0) > f_v(1) \text{ and}$$
$$c(s,v) = f_v(1) - f_v(0) \quad \text{if } f_v(0) < f_v(1).$$

Clearly, the added edges have a capacity larger than 0 and although the cost of a cut is not equal to the energy (equation (10.16)) of the particular variable assignment, it deviates only a constant from this. This constant is equal to $-\min(f_v(0), f_v(1))$. The two possible cuts for each of the two cases can be seen in Fig. 10.2.

For pairs of variables, y_u and y_v, there are four possible configurations with costs given by $g_{u,v}(0,0)$, $g_{u,v}(0,1)$, $g_{u,v}(1,1)$, and $g_{u,v}(0,0)$. We can write these costs as

$$g_{u,v}(0,0) = g_{u,v}(0,0), \tag{10.19}$$

$$g_{u,v}(1,0) = g_{u,v}(0,0) + \big(g_{u,v}(1,0) - g_{u,v}(0,0)\big), \tag{10.20}$$

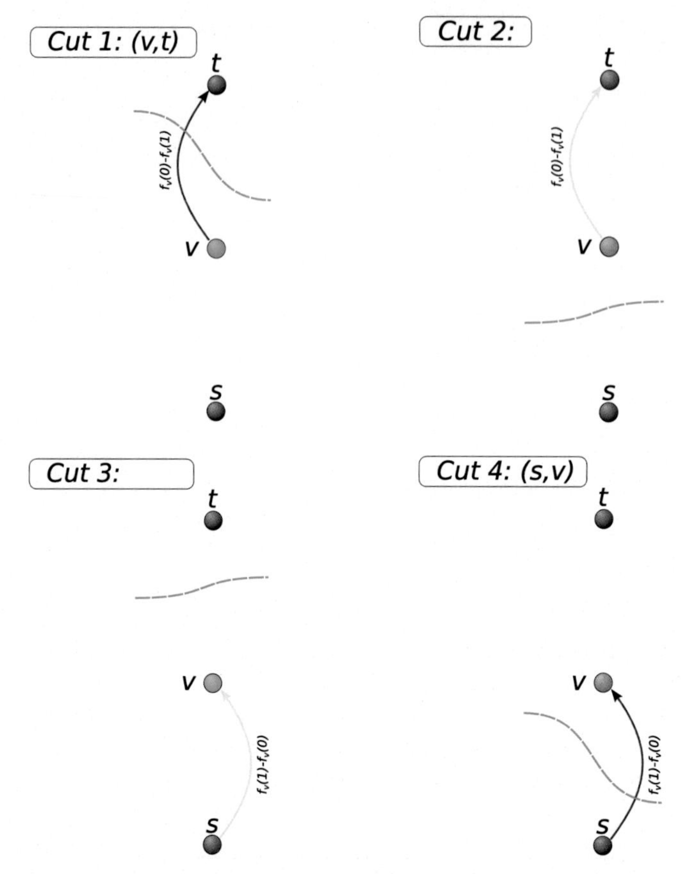

FIGURE 10.2 Cut cost of the unary edge configurations.

Cut edges are highlighted in red. The cost of the corresponding cuts (green punctuated line) are for Case 1, top left: $f_v(0) - f_v(1) = f_v(0) - C$ and top right: $0 = f_v(1) - C$, where $C = f_v(1)$, and for Case 2: bottom left $0 = f_v(0) - C$ and bottom right: $f_v(1) - C$, where $C = f_v(0)$. It is clear that the cost of the cut is only a constant from the corresponding unary term given in equation (10.16).

$$g_{u,v}(1,1) = g_{u,v}(0,0) + \big(g_{u,v}(1,0) - g_{u,v}(0,0)\big) + \big(g_{u,v}(1,1) - g_{u,v}(1,0)\big), \tag{10.21}$$

$$\begin{aligned} g_{u,v}(0,1) = g_{u,v}(0,0) &+ \big(g_{u,v}(1,1) - g_{u,v}(1,0)\big) \\ &+ \big(g_{u,v}(0,1) + g_{u,v}(1,0) - g_{u,v}(0,0) - g_{u,v}(1,1)\big). \end{aligned} \tag{10.22}$$

One $g_{u,v}(0,0)$ is added to every configuration, so we can disregard that as a constant cost. Whether $g_{u,v}(1,0) - g_{u,v}(0,0)$ or $g_{u,v}(1,1) - g_{u,v}(1,0)$ is added to the cost of a particular configuration depends only on one of the variables each, respectively, $y_u = 1$ and $y_v = 1$. So these can be added as edges to and from the source and sink. More specifically, if $g_{u,v}(1,0) > g_{u,v}(0,0)$, then an edge from the source to u is added with a capacity of $g_{u,v}(1,0) - g_{u,v}(0,0)$. If, on the other hand, $g_{u,v}(0,0) > g_{u,v}(1,0)$, then an edge is added from u to t with the cost of $g_{u,v}(0,0) - g_{u,v}(1,0)$. The same applies for $g_{u,v}(1,1) - g_{u,v}(1,0)$ and the vertex v. The remaining term $g_{u,v}(0,1) + g_{u,v}(1,0) - g_{u,v}(0,0) - g_{u,v}(1,1)$ depends on the configuration of both y_u and y_v, however, because the energy function is submodular. We know that $g_{u,v}(0,1) + g_{u,v}(1,0) - g_{u,v}(0,0) - g_{u,v}(1,1) \geq 0$, and the term can thus be implemented with a single directed edge from u to v. The complete set of edges implementing the pairwise term is shown in Box 10.2.

BOX 10.2 (Capacities of edges implementing the pairwise term).
For each $(u, v) \in \mathcal{E}$,

$$
\begin{aligned}
c(s,u) &= g_{u,v}(1,0) - g_{u,v}(0,0) \quad \text{if } g_{u,v}(1,0) > g_{u,v}(0,0),\\
c(u,t) &= g_{u,v}(0,0) - g_{u,v}(1,0) \quad \text{if } g_{u,v}(1,0) < g_{u,v}(0,0),\\
c(s,v) &= g_{u,v}(1,1) - g_{u,v}(1,0) \quad \text{if } g_{u,v}(1,1) > g_{u,v}(1,0),\\
c(v,t) &= g_{u,v}(1,0) - g_{u,v}(1,1) \quad \text{if } g_{u,v}(1,1) < g_{u,v}(1,0),\\
c(u,v) &= g_{u,v}(0,1) + g_{u,v}(1,0) - g_{u,v}(0,0) - g_{u,v}(1,1).
\end{aligned}
$$

It is helpful to look at a few examples. Assuming we are given a problem where $g_{u,v}(1,0) < g_{u,v}(0,0)$ and $g_{u,v}(1,1) > g_{u,v}(1,0)$, that is, Case 1 in Fig. 10.3, the four possible configurations and cuts can be seen in Fig. 10.4. On the other hand, if we are given a problem where $g_{u,v}(1,0) > g_{u,v}(0,0)$ and $g_{u,v}(1,1) > g_{u,v}(1,0)$, that is, Case 4 in Fig. 10.3, the four possible configurations and cuts can be seen in Fig. 10.5. Note how the costs of the four possible cuts are only a constant, which is different from case to case, from the corresponding pairwise energy of the configurations.

10.5.4 Limitations (and solutions)

There are several limitations due to both the model and its optimization using graph cuts.

Large memory requirements and difficulty to parallelize. Medical images can be both 3D and large. To represent every voxel as a vertex and itself and every neighboring voxel as edges thus leads to large graphs and therefore large memory requirements. Although there have been works looking at solving such large graph problems by splitting them into subproblems that can be solved separately and perhaps even in parallel, they either have poor worst-case running times or rely on heuristics.

Metrification artifacts. One way to limit graph size is to limit the neighborhood size, and this is perhaps part of the reason relatively small neighborhoods are often used. Assuming a voxel label is dependent only on the label of its four or six nearest neighbors is an approximation at best and this approximation tends to lead to results that

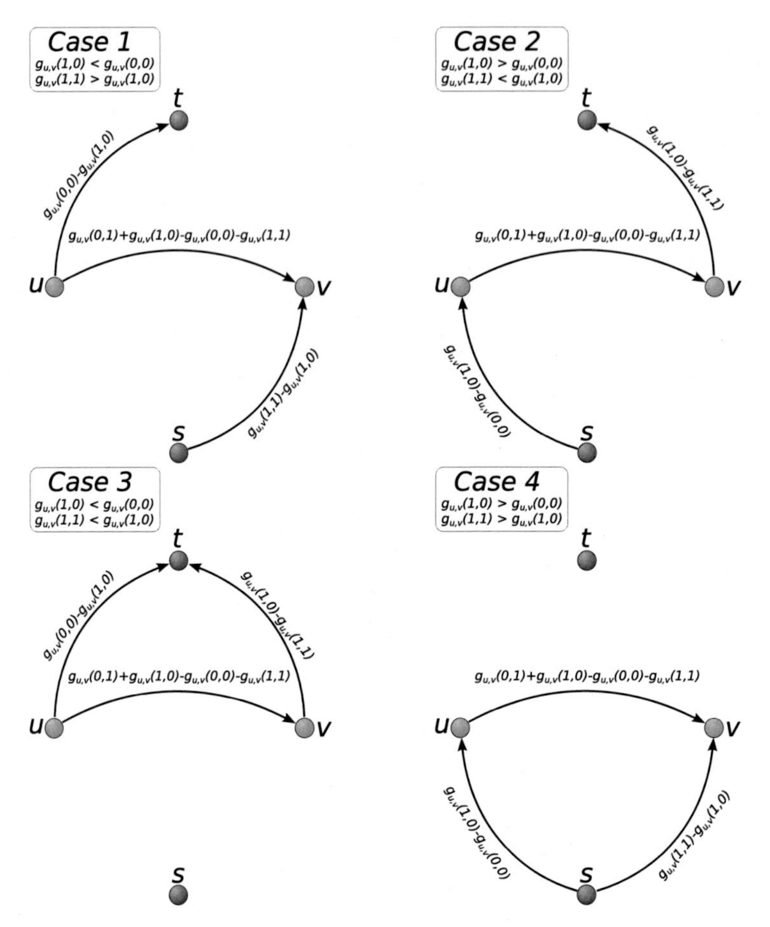

FIGURE 10.3 Pairwise term edges of graph $G_{s,t}$.

The edges implementing the pairwise terms of equation (10.16) for the variables y_u and y_v and the four different cases of problems.

are blocky in appearance. This less than desirable result is known as metrification artifacts. Longer-range dependencies are a solution, but the number of edges added to the graph grows quadratically in 2D and cubically in 3D with the radius of the neighborhood.

10.6 Interactive segmentation

MRF models and inference using graph cuts can be used to build fully automated approaches for medical image segmentation, and while automation is ideal when it

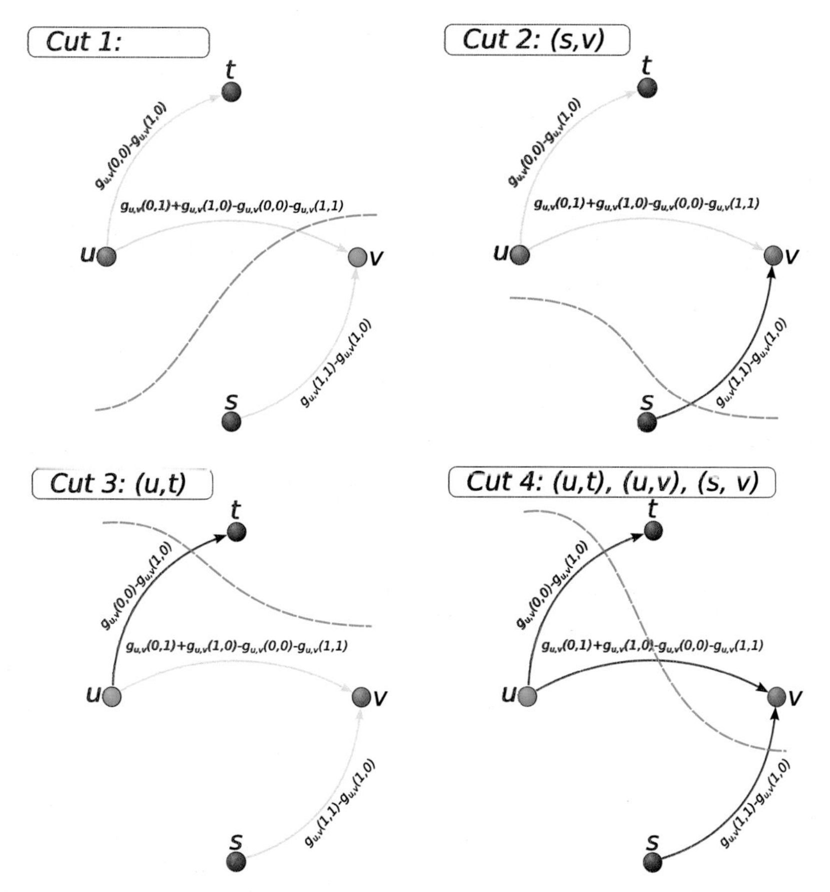

FIGURE 10.4 Cut cost of the four possible configurations in Case 1.

Cut edges are highlighted in red. The cost of the corresponding cuts (green punctuated line) are top left: $0 = g_{u,v}(1,0) - C$, top right: $g_{u,v}(1,1) - g_{u,v}(1,0) = g_{u,v}(1,1) - C$, bottom left: $g_{u,v}(0,0) - C$, and bottom right: $g_{u,v}(0,0) - g_{u,v}(1,0) + g_{u,v}(0,1) + g_{u,v}(1,0) - g_{u,v}(0,0) - g_{u,v}(1,1) + g_{u,v}(1,1) - g_{u,v}(1,0) = g_{u,v}(0,1) - C$, where $C = g_{u,v}(1,0)$. It is clear that the cost of the cut is only a constant from the corresponding pairwise term given in equation (10.16).

works, deciding which part of an image belongs to a given anatomical structure or pathology can not always be done just based on image information alone, nor can it always be accurately done using the relatively simple models described in this chapter. Correcting mistakes in segmentation outputs can be very time consuming, and in some cases as time consuming as complete manual segmentation. An advantage of graph cuts in this respect is that they allow for an elegant way of incorporating expert knowledge through user interaction.

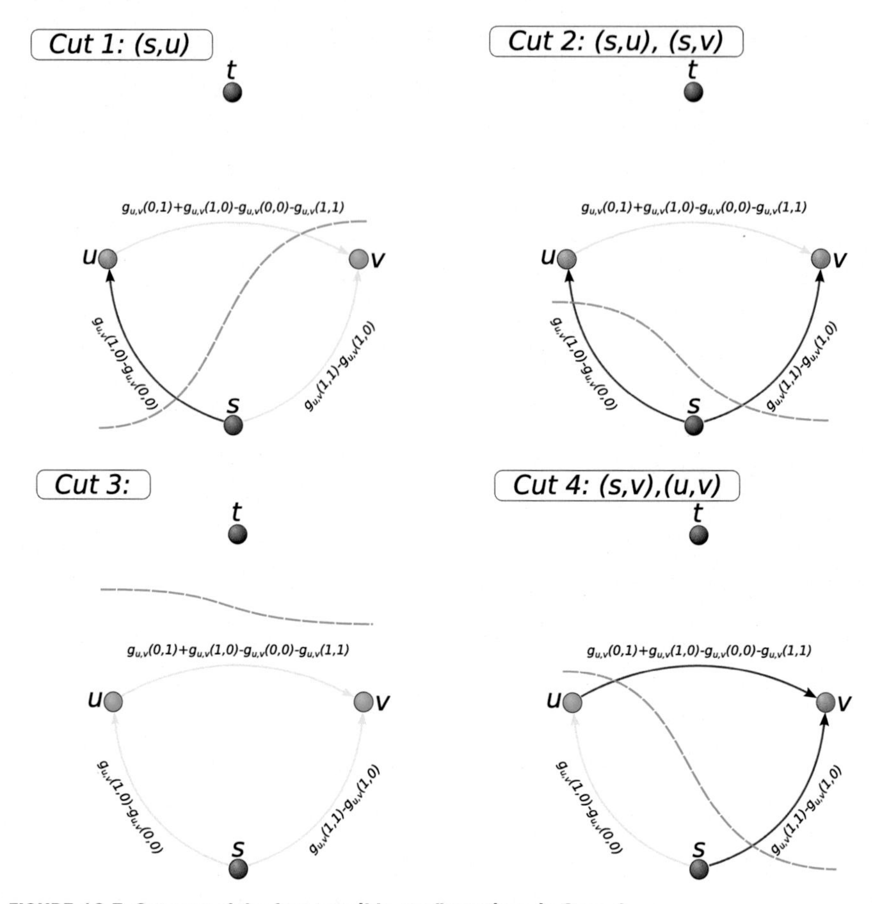

FIGURE 10.5 Cut cost of the four possible configurations in Case 4.

Cut edges are highlighted in red. The cost of the corresponding cuts (green punctuated line) are top left: $g_{u,v}(1,0) - C$, top right: $g_{u,v}(1,0) - g_{u,v}(0,0) + g_{u,v}(1,1) - g_{u,v}(1,0) = g_{u,v}(1,1) - C$, bottom left: $0 = g_{u,v}(0,0) - C$, and bottom right: $g_{u,v}(1,1) - g_{u,v}(1,0) + g_{u,v}(0,1) + g_{u,v}(1,0) - g_{u,v}(0,0) - g_{u,v}(1,1) = g_{u,v}(0,1) - C$, where $C = g_{u,v}(0,0)$. It is clear that the cost of the cut is only a constant from the corresponding pairwise term given in equation (10.16).

10.6.1 Hard constraints and user interaction

Assuming we know that a given variable y_k has a particular label, let us say 1, this may be enforced by setting the cost of any other possible option to infinity. Given a binary problem, the cost of $y_k = 0$ would thus be $f_k(0) = \infty$. Looking at Box 10.1 and Fig. 10.1, it can be seen that the consequence of this is that an edge with infinite capacity from vertex k to the sink vertex t is added. This will ensure that the vertex v_k will always be part of the sink set and thus have the label 1. In practice, dealing with infinite values in programming may be difficult and a similar effect can be achieved by just choosing a large enough value. This value needs to be larger than the cost of all the alternative configurations. Given the MRF property (equation (10.5)), only the cost of the configuration of y_v and its neighborhood N_k needs to be considered. That is, if $f_k(0) > f_k(1) + \max_{y_i, i \in N_k} \sum_{i \in N_k} g_{k,i}(1, y_i)$, then for any choice of neighborhood configuration, the cost will be minimized only if $y_v = 1$.

10.6.2 Example: coronary arteries in CT angiography

Coronary arteries are part of the cardiovascular system. Diseases affecting the coronary arteries, such as coronary artery disease, are a leading cause of death globally. Diagnosis can be done using coronary CTA, which requires an injection of a radiocontrast material to highlight the coronary arteries and subsequent imaging using a CT scanner. There is a large interest in automated and objective quantification of the dimensions of the coronary arteries. In the following example, we will assume that we are provided with a limited number of points as well as an indication for each of them whether they are inside (foreground) or outside (background) the artery lumen. The artery lumen is the volume of the artery that transports the contrast-enhanced blood. Such points can be interactively provided by manually selecting and classifying points in the image or using automated approaches. The task is then to segment the complete coronary artery lumen volume.

A straightforward approach could be to estimate the distribution of the voxel intensity values of the points within the coronary artery lumen and background to arrive at a unary term. This on its own, however, will not work well, as there is a large overlap between the intensities seen in the lumen and the background. To improve, we may try to regularize the result using the Potts model (equation (10.17)); however, in the Potts model, the energy is proportional to the number of neighboring variables that have different values. In a segmentation problem, where the variables correspond to pixels or voxels with a given value of 0 or 1 indicating whether they belong to background or vessel lumen, the energy of the Potts model is proportional to the segmentation surface area, and because coronary vessels are thin elongated structures with a relatively large surface area-to-volume ratio, minimizing Potts energy will more often than not tend to remove rather than bring forth vessels.

Coronary artery lumen surfaces are typically characterized as regions with a relatively high image gradient, and we may instead formulate the problem as one of looking for a closed surface in high-gradient regions which contains all of the given coronary artery centerline points and none of the given background points. This can be done by creating a pairwise term that has small values for label differences in high-gradient regions and large values for label differences in low-gradient regions [3]:

$$g_{u,v}(y_u, y_v) = \begin{cases} e^{-\frac{(I_u - I_v)^2}{\sigma^2}} & \text{if } y_u \neq y_v, \\ 0 & \text{else,} \end{cases} \tag{10.23}$$

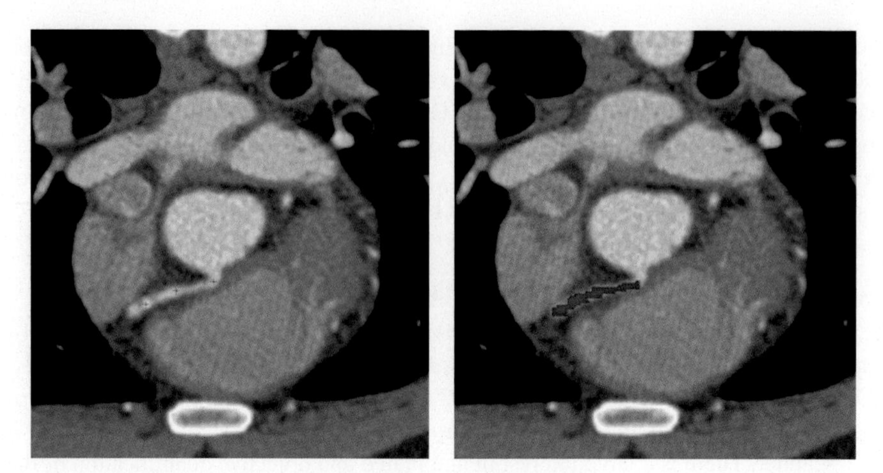

FIGURE 10.6 Segmentation of coronary artery using graph cuts.

An axial slice view of 3D scan overlaid (red) with the input centerline points (left) and the segmentation (right).

where I_u and I_v are the image intensity values at the vertex u and v, respectively. The σ-value controls the speed with which the pairwise energy cost goes to zero as a function of intensity differences. Such a pairwise term may of course be combined with a unary term, but if no unary term is used, except for the infinite values at the known vessel lumen or background points, then because the pairwise term is always non-zero, minimizing the energy will result in a segmentation with components that all contain at least one given input point. The exception is of course if no input values are given, as the entire input image will then be one component (background or foreground). In the example (scan data coming from [4]) in Fig. 10.6, the vessel centerline was given and the points on it were defined as foreground. The background was defined as all voxels more than 10 mm from any given foreground point. The σ-parameter was set to 7.5.

10.7 More than two labels

When the number of labels $|\mathcal{L}|$ is greater than two, minimizing $E(\mathbf{y})$ is in general NP-hard [2], but good approximate solutions exist when the pairwise energy term g is a semi-metric, which is the case if

$$g(\alpha, \beta) = g(\beta, \alpha) \text{ and} \tag{10.24}$$

$$g(\alpha, \beta) = 0 \Leftrightarrow \alpha = \beta, \tag{10.25}$$

for $\alpha, \beta \in \mathcal{L}$.

10.7.1 Move-making algorithm(s)

Some of the most useful methods are generally referred to as move-making algorithms. These minimize energy through "moves" that progressively lessen energy by solving suitably formulated binary labeling problems until convergence.

The α-β-expansion method. Given a current configuration of $\mathbf{y}$, a graph G, and a possible label set $\mathcal{L}$, α-β-expansion loops through all pairs of labels α and β in the set $\mathcal{L}$ and for each pair computes the minimum energy configuration that can be obtained by possibly interchanging α and β labels of the individual variables in the current configuration. The remaining variables that have labels of neither α or β are not changed by the move. The looping repeats until all unique unordered pairs of α and β have been cycled through once with no reduction in energy.

An α-β-expansion move can be computed as a graph cut on the G' graph consisting of the subset of vertices $\mathcal{V}' \subseteq \mathcal{V}$, where the corresponding variables in the current configuration have the label α or β, that is, $\mathcal{V}' = \left\{v | y_v \in \{\alpha, \beta\}, v \in \mathcal{V}\right\}$. The edges of this graph $\mathcal{E}' \subseteq \mathcal{E}$ correspond to the subset of the complete edge set that connects vertices in $\mathcal{V}'$, that is, $\mathcal{E}' = \{(u, v) | (u, v) \in \mathcal{E}, y_u, y_v \in \mathcal{V}'\}$. Dependencies on variables that are not in the graph G' because they have labels that are not α or β in the current configuration are handled by changing the unary term to include a summation of their cost:

$$f'_v(y'_v) = f_v(y'_v) + \sum_{(u,v)\in\mathcal{E}, y_u \in \left\{\mathcal{L}-\{\alpha,\beta\}\right\}} g_{u,v}(y_u, y'_v), \tag{10.26}$$

for $v \in \mathcal{V}'$ and $y'_v \in \{\alpha, \beta\}$. The pairwise term, on the other hand, is identical to the overall pairwise term:

$$g'_{u,v}(y'_u, y'_v) = g_{u,v}(y'_u, y'_v), \quad \text{for } (u, v) \in \mathcal{E}', \tag{10.27}$$

for $u, v \in \mathcal{V}'$ and $y'_u, y'_v \in \{\alpha, \beta\}$.

The α-expansion method. If the pairwise term is a metric, that is, if in addition to being a semi-metric the triangle inequality also holds,

$$g(\alpha, \beta) \leq g(\alpha, \gamma) + g(\gamma, \beta), \tag{10.28}$$

for any $\alpha, \beta, \gamma \in \mathcal{L}$, then the α-expansion method can be shown to be faster than α-β-expansion.

Given a current configuration $\mathbf{y}$, a graph G, and a possible label set $\mathcal{L}$, α-expansion loops through each label $\alpha \in \mathcal{L}$ and computes the minimum energy configuration when any variable is allowed to change its label to α. The looping repeats until all $\alpha \in \mathcal{L}$ have been cycled through once with no reduction in energy.

An α-expansion move can be computed as a graph cut in the graph G, where each variable y'_v for $v \in \mathcal{V}$ can take on one of two labels, α or y_v. The α-expansion move unary cost is given by

$$f'_v(y'_v) = \begin{cases} \infty & \text{if } y_v = \alpha,\ y'_v \neq \alpha, \\ f_v(y'_v) & \text{else.} \end{cases} \tag{10.29}$$

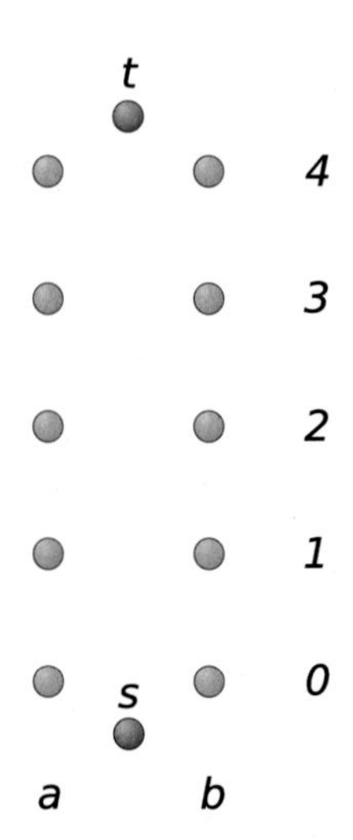

FIGURE 10.7 Vertices of graph $G_{s,t}$.

The vertices of graph $G_{s,t}$ are arranged into columns – one column for each variable and one vertex in each column for each possible label. Source s and sink t nodes are also added.

The α-expansion move pairwise term is identical to the overall pairwise term

$$g'_{u,v}(y'_u, y'_v) = g_{u,v}(y'_u, y'_v), \quad \text{for } (u, v) \in \mathcal{E}', \tag{10.30}$$

for $u, v \in \mathcal{V}'$ and $y'_u, y'_v \in \{\alpha, y_v\}$.

10.7.2 Ordered labels and convex priors

If we assume that the possible set of labels $L = \{0, 1, \cdots, k\}$, where $k \in \mathbb{N}$ for the variables y_v, where $v \in \mathcal{V}$ can be ordered and the pairwise cost is convex, increasing, and determined from differences in this order, that is, if $g_{u,v}$ can be written in the form $g_{u,v}(y_u, y_v) = \hat{g}_{u,v}(|y_u - y_v|)$, then the solution can be found in low-order polynomial time using graph cuts by an algorithm due to H. Ishikawa [5].

The minimum cut graph, $G_{s,t}$, in this case in addition to the source s and sink t, includes a column of vertices for each $v \in \mathcal{V}$ associated with each of the possible labels $y_v \in L$. In other words, $\mathcal{V}_{s,t} = \{s, t\} \cup \{v_l | v \in \mathcal{V}, l \in L\}$. Fig. 10.7 shows a simple example of the columns in $\mathcal{V}_{s,t}$, where $\mathcal{V} = \{a, b\}$, $\mathcal{E} = \{(a, b)\}$, and $k = 4$. The goal of the construction is to form $G_{s,t}$ in such a manner that the minimum cut in $G_{s,t}$ intersects each column at a position indicating the optimal label for each variable according to equation (10.16). For a quick overview of the needed edges and capacities, the reader is referred to Boxes 10.3, 10.4, and 10.5.

To enforce that the minimum cut intersects each column exactly once, infinite capacity edges are added from the source s to the bottommost vertices of each column and from each vertex in each column to the vertex immediately below it (Box 10.3).

BOX 10.3 (Capacity of edges enforcing a single cut per column).
For each $v \in \mathcal{V}$, we have

$$c(s, v_0) = \infty,$$
$$c(v_l, v_{l-1}) = \infty, \quad \text{for } 0 < l \le k.$$

To implement the unary cost of equation (10.16), edges are added from each vertex in each column to the vertex immediately above it, with a capacity given by the relevant unary term (Box 10.4).

BOX 10.4 (Capacity of edges implementing the unary term).
For each $v \in \mathcal{V}$, we have

$$c(v_k, t) = f_v(v_k),$$
$$c(v_l, v_{l+1}) = f_v(v_l), \quad \text{for } 0 \le l < k.$$

These edges are illustrated for the simple example graph mentioned above in Fig. 10.8. With these edges, the minimum cut has to intersect each column once because it is a path from the source to the sink. It also cannot intersect it more than once, as this would mean an infinite cost edge would be cut.

A minimum cut in columns constructed in the above manner can be interpreted as assigning a label to the variable the column represents. Correspondingly, a minimum cut in $G_{s,t}$ can be interpreted as a configuration of the MRF. In particular, the assigned label is given by the source set vertex with the largest index

$$y_v = \arg\max_{v_l \in S} l, \tag{10.31}$$

for all $v \in \mathcal{V}$. Note that for a particular configuration the cost of the minimum cut is equivalent to the sum of the unary terms in equation (10.16).

To implement the pairwise terms of equation (10.16), edges are added to $G_{s,t}$ between columns whose corresponding vertices in G are connected by an edge. To do this a new function is defined from $g_{u,v}$:

$$\tilde{g}_{u,v}(y) = \begin{cases} 0 & \text{if } y < 0, \\ g_{u,v}(1) & \text{if } y = 0, \\ g_{u,v}(y+1) - 2g_{u,v}(y) + g_{u,v}(y-1) & \text{if } y > 0. \end{cases}$$

Note we assume without loss of generality that $g_{u,v}(0) = 0$. If this is not the case, we can simply shift the function until it does become the case, as this will not change the configuration minimizing the energy.

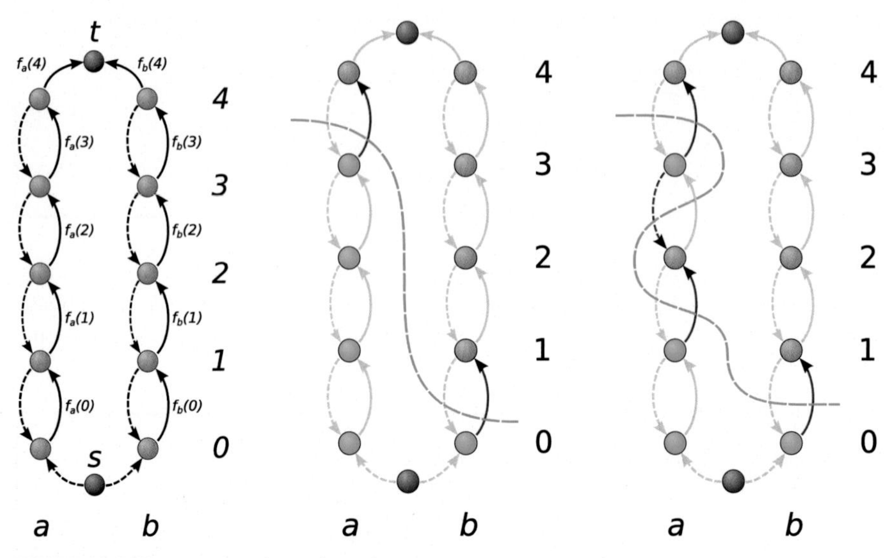

FIGURE 10.8 Unary term edges of graph $G_{s,t}$.

The edges implementing the unary terms of equation (10.16). Left: directed edges are either of infinite capacity (punctuated arrows) or of a specific finite capacity (fully drawn arrows). Middle: showing a feasible cut as a green punctuated curve, the source set vertices in green and the edges that are cut in red. The corresponding configuration is given by $y_a = 3$ and $y_b = 0$. Note the cost of the cut $f_a(3) + f_b(0)$ is equivalent to unary part of the energy of the configuration (equation (10.16)). Right: showing an infeasible cut in which column a is cut multiple times and as a consequence an infinite capacity edge is cut (blue).

Next, for each vertex v and connected vertex u in G, that is, $(u, v) \in \mathcal{E}$, directed edges are added to $\mathcal{E}_{s,t}$ going from every vertex in column u to every vertex in column v. The capacity of these edges is given by $\tilde{g}_{u,v}(l - m)$, where l and m are the indices of the particular vertices in column u and v, respectively (Box 10.5).

BOX 10.5 (Capacity of edges implementing the pairwise term).
For each $(u, v) \in \mathcal{E}$, we have

$$c(u_l, v_m) = \tilde{g}_{u,v}(l - m), \quad \text{for } l, m \in L.$$

Note that because $\tilde{g}_{u,v}(y) = 0$ for $y < 0$, approximately half of these edges are not actually added to $\mathcal{E}_{s,t}$. How this works can be illustrated with a simple example graph. In Fig. 10.9 all the above edges with non-zero capacity are drawn.

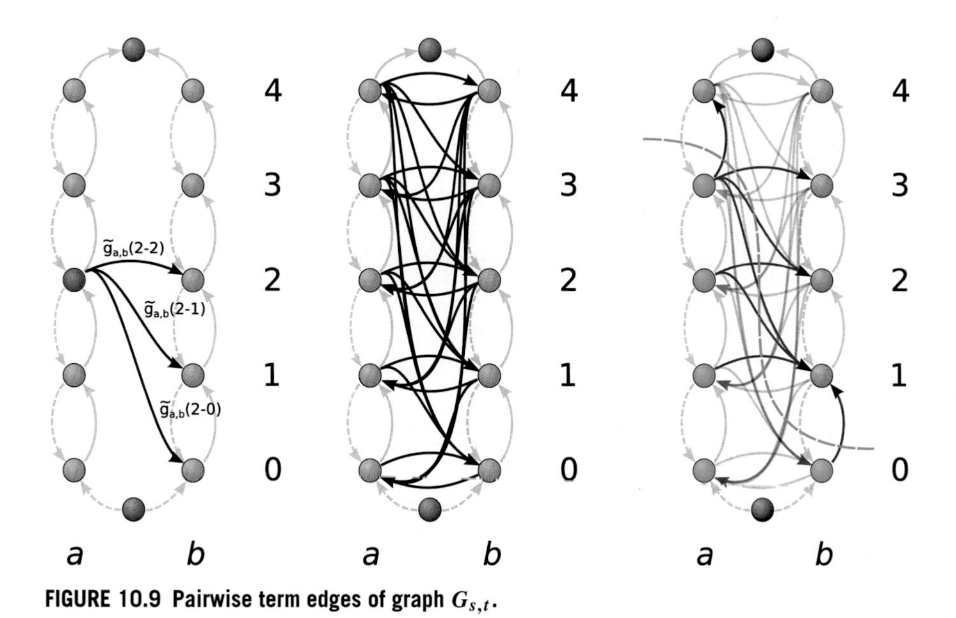

FIGURE 10.9 Pairwise term edges of graph $G_{s,t}$.

The edges implementing the pairwise terms of equation (10.16). Left: showing the directed edges added from vertex a_2 and their capacity as listed in Box 10.5. Edges where the capacity is zero are left out. Middle: the resulting graph, when such directed edges are added from all vertices. Right: showing an example cut ($y_a = 3$ and $y_b = 0$) as a green punctuated curve and the corresponding source set vertices in green. The unary and pairwise term edges that are cut are highlighted in red and blue, respectively. The cost of the cut $f_a(3) + f_b(0) + 3\tilde{g}_{a,b}(0) + 2\tilde{g}_{a,b}(1) + \tilde{g}_{a,b}(2)$ is equivalent to the energy of the configuration (equation (10.16)) $f_a(3) + f_b(0) + g_{a,b}(3)$.

Note that if the second-order derivative of $g_{u,v}$ is zero, then the number of intra-column edges added to $G_{s,t}$ is proportional to k, as $\tilde{g}_{u,v}(y) = 0$ for $y > 0$. If, on the other hand, the second-order derivative of $g_{u,v}$ is non-zero, then the number of intra-column edges added to $G_{s,t}$ is proportional to k^2. Performance and memory requirements may thus not be satisfying if k is large and the pairwise terms are non-linear.

10.7.3 Optimal surfaces

Graph cuts with ordered labels and convex priors are used to find surfaces and in particular layered surfaces in medical imaging and have been published under names such as graph search and optimal (net) surfaces [6]. In these models, one can think of the graph G as being a representation of the surface to be found. The surface topology is specified through the set of edges in $\mathcal{E}$. The ordered set of labels represent possible positions for each of the vertices in the sought surface. Fig. 10.10 shows how this is done. The unary terms $f_v(y_v)$ in equation (10.16) represent the cost of placing the

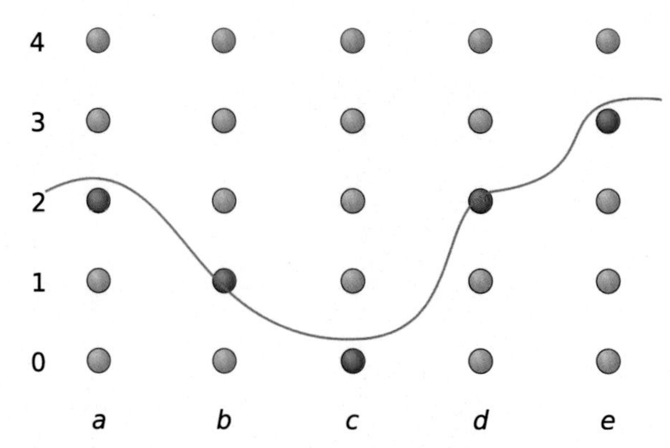

FIGURE 10.10 Optimal surfaces – surface positions as labels.

In optimal surfaces the possible labels for a given variable represent possible positions for a point on the surface. In this example the contour represents the continuous surface that we want to find. The graph columns discretize the space of possible positions for the surface and allows it to be found with graph cuts.

surface at a given position and the pairwise terms $g_{u,v}(y_u, y_v)$ are used to regularize the solutions, such that more smooth surfaces are found.

Forming the graph G and the columns in $G_{s,t}$ is not always straightforward. The topology of the surface must be known beforehand and while the surface positions can be 2D, 3D, ..., the labels are integer indices into these positions. Because $g_{u,v}(y_u, y_v)$ has to be convex and increasing in $|y_u - y_v|$, differences in labels must reflect some useful property of the underlying surface positions. If G is thought of as an initial representation of the surface to be found, then the labels at a given position on this surface can be a measure of how much this representation has to be deformed relative to neighboring labels in a certain direction to match the given input. In some cases the sought surfaces are terrain-like and approximately aligned with the axes of the image and the graph G and corresponding columns can thus be formed easily based on image dimensions alone. If topology and approximate surface positions are not known beforehand, we may have to rely on some other method to give us this information first. An example of this can be seen in Section 10.7.4.

Multiple surfaces. Multiple surfaces can be found simultaneously by constructing a subgraph for each of the surfaces (Fig. 10.11). Just as dependencies between neighboring points on the surfaces can be modeled by using pairwise terms, so can inter-surface dependencies. This can be used to, for instance, model that surfaces have a tendency to be at a certain distance from each other or that they form layers without intersections.

Hard constraints. Pairwise hard constraints can be implemented by adding infinite cost edges between columns. We can for instance enforce that $y_v - y_u \geq \delta_{u,v}$, where

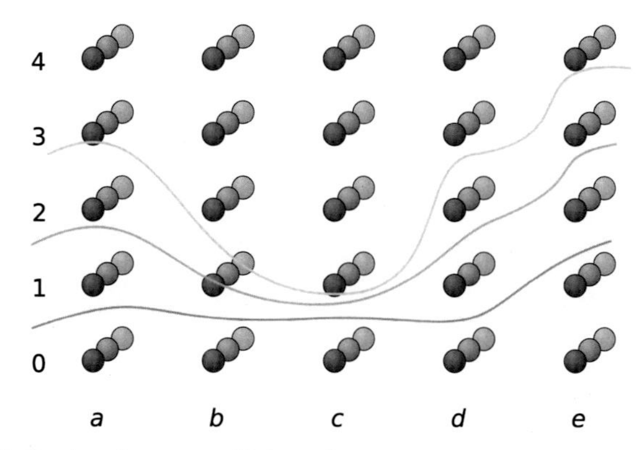

FIGURE 10.11 Optimal surfaces – multiple surfaces.

Multiple surfaces can be modeled together by stacking multiple single surface graphs.

$-k \geq \delta_{u,v} \leq k$. This is done by adding the following infinite capacity edges:

$$c(u_l, v_{l+\delta_{u,v}}) = \infty \text{ for } l \in \{\max(0, -\delta_{u,v}), \max(0, -\delta_{u,v}) + 1, \cdots, \min(k - \delta_{u,v}, k)\},$$
$$c(u_l, t) = \infty \text{ for } l \in \{k - \delta_{u,v} + 1, k - \delta_{u,v} + 2, \cdots, k\}.$$

When modeling surfaces, this can for instance be used to enforce hard smoothness constraints, such that the found surfaces never vary more than a certain amount $\eta \geq 0$ between two neighboring points $(u, v) \in \mathcal{E}$. This is equivalent to adding the infinite capacities corresponding to $\delta_{u,v} = \delta_{v,u} = \eta$ for all $(u, v) \in \mathcal{E}$.

In multisurface problems, a purpose of hard constraints is to enforce that surfaces are layered, such that one surface is always a certain number of indices, δ, where $0 \geq \delta < k$ to one side of the other. To do this, suppose $v_j, v_{j+1} \in \mathcal{V}$ represent corresponding points on surface j and $j+1$ and infinite capacity edges implementing $\delta_{v_j, v_{j+1}} = \delta$ are added. Examples of both of these constructions can be seen in Fig. 10.12.

Note that because hard constraints limit the range of possible solutions, they can quite drastically limit the number of edges in $\mathcal{E}_{s,t}$ and therefore improve performance. For instance, edges from u_i and v_j can be left out of $\mathcal{E}_{s,t}$ if $\delta_{u,v} < i - j$. This can be used to reduce the number of intra-column edges added to $G_{s,t}$ so that the number is proportional to k even in cases with non-linear $g_{u,v}$. One can get an impression of the reduction in edge count by comparing the number of edges (including non-highlighted edges) in Fig. 10.12 with the full edge set seen in Fig. 10.9.

10.7.4 Example: airways in CT

The human airways form a branching tree-like structure involved in respiration by conducting air between the mouth and nasal cavities and the alveoli within the lungs, where gas exchange happens. In CT images, the airway branches appear as a bright tubular wall containing a darker lumen. The wall appears bright because it consists

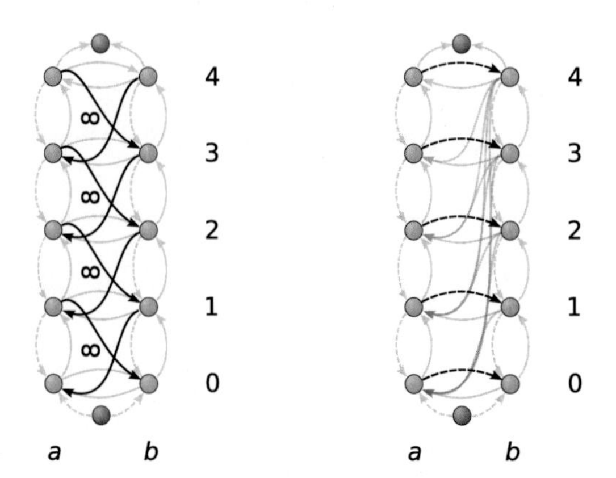

FIGURE 10.12 Optimal surfaces – hard constraints.

Hard constraints can be incorporated with infinite capacity edges (punctuated arrows). Left: Feasible solutions of variable y_a and y_b are within one ($\eta = 1$) label from each other, that is, $|y_a - y_b| \leq 1$. Right: Feasible solutions always have variable y_b larger than or equal to y_a ($\delta_{a,b} = 0$).

of tissues such as cartilage and smooth muscle that are similar in density to water, with a Hounsfield value of approximately 0. The lumen is dark because it conducts air and thus has a Hounsfield value of around -1000. In practice, however, many other things affect airway voxel values, such as disease, noise, and partial volume effects, and observed airway voxel values may therefore be quite far from these ideal values. Most of the airways lie within the lungs. The lung background parenchyma is a sponge-like structure filled with air and consists of alveoli, alveolar ducts, and respiratory bronchioles. The alveoli, alveolar ducts, and respiratory bronchioles are all well below the typical resolution of clinical CT and due to partial volume effects they can be observed as voxels with Hounsfield values that vary with the inspiration level of the scan. Values in the range of -900 to -800 are normal in scans taken at maximum inspiration, but disease can greatly affect these values. The lungs are also filled with blood vessels, which form another branching tubular structure. The blood vessels are filled with blood and thus are a little bit denser than water, and voxel values are typically around 50.

Diseases such as asthma, chronic obstructive pulmonary disease, and cystic fibrosis affect the airways and the airway wall and there is therefore a clinical interest in quantifying the dimensions of these structures. Because of the higher contrast between the airway lumen and the wall than between the wall and the lung background, the task of segmenting the lumen is easier than the task of segmenting the wall. It has often been approached as a two-step process in which the lumen is first segmented and then the resulting lumen model is used to help segment and/or measure the wall. The recent developments in deep learning models, which are capable of learning and exploiting interdependencies of multiple classes, are perhaps going to replace this two-step approach; however, it is nonetheless the strategy we will follow below.

In some cases, an acceptable segmentation of the airway lumen can be achieved with something as simple as region growing, in which the region of interest is iteratively grown from a single seed point by continuously adding neighboring voxels to the lumen if they are below a given voxel value threshold for the lumen–wall boundary (Fig. 10.14). However, this is sensitive to noise and partial volume effects and fails catastrophically whenever the region growing leaks into the background lung. Better performance may be achieved using machine learning techniques. Suppose we are given some images where regions of airway lumen and background have been indicated; we can then train a machine learning classifier to recognize airway lumen voxels and apply it to as of yet unseen images. Using features based on spatial image derivatives and eigenvalues of the Hessian matrix combined with a classifier such as K-nearest neighbors, we can get results such as shown in the middle row of Fig. 10.13. Region growing can still be used to remove disconnected false positives in this result, as can be seen in the bottom row of the same figure and in Fig. 10.14. These results may then be used as an initial guess at the interior surface which we can simultaneously refine and use to find the exterior surface.

Suppose we represent this initial interior surface or input surface as the graph G, for instance constructed by using a meshing algorithm like Marching Cubes, to give us vertices ($\mathcal{V}$) on the surface and edges ($\mathcal{E}$) between neighboring vertices. The problem of both refining the input surface and finding the exterior surface can be posed as one optimal surface problem with two surfaces. A column representing the possible positions for every vertex in $\mathcal{V}$ needs to be constructed. The most straightforward construction is to orient the column orthogonal to the input surface and sample it equidistantly. In this way, a column for a combination of a particular point $v \in \mathcal{V}$ on the input surface and sought surface is the set of possible positions for the sought surface at this particular point (Fig. 10.15). This set is ordered by the signed distance along the column from the input surface. Straight columns may intersect and therefore can cause output surfaces to be self-intersecting; to prevent this, one can instead form columns defined from gradient vector flow [7] or flow lines in a smoothed version of the lumen segmentation [8] (see Fig. 10.15). Choosing the sampling density of both the lumen surface and the columns is a trade-off between surface precision and computation cost.

Once the columns have been computed it is only a matter of defining $f_v(y_v)$ and $g_{u,v}(y_u, y_v)$, and the graph $G_{s,t}$ can be constructed as described in Section 10.7.2 and the minimum cut can be computed. Remember the function $f_v(y_v)$ gives the cost of assigning a particular label (in this case indexed column position) to y_v. This could for instance be based on the output of a machine learning classifier; however, in this example, we will simply use the negative image gradient along the outward column direction for the interior surface and the positive image gradient for the exterior surface. Why this works should be clear from the description of the airway appearance in CT above – the column starts inside the airway lumen where intensity is ideally around -1000, then transitions the interior surface boundary to the airway

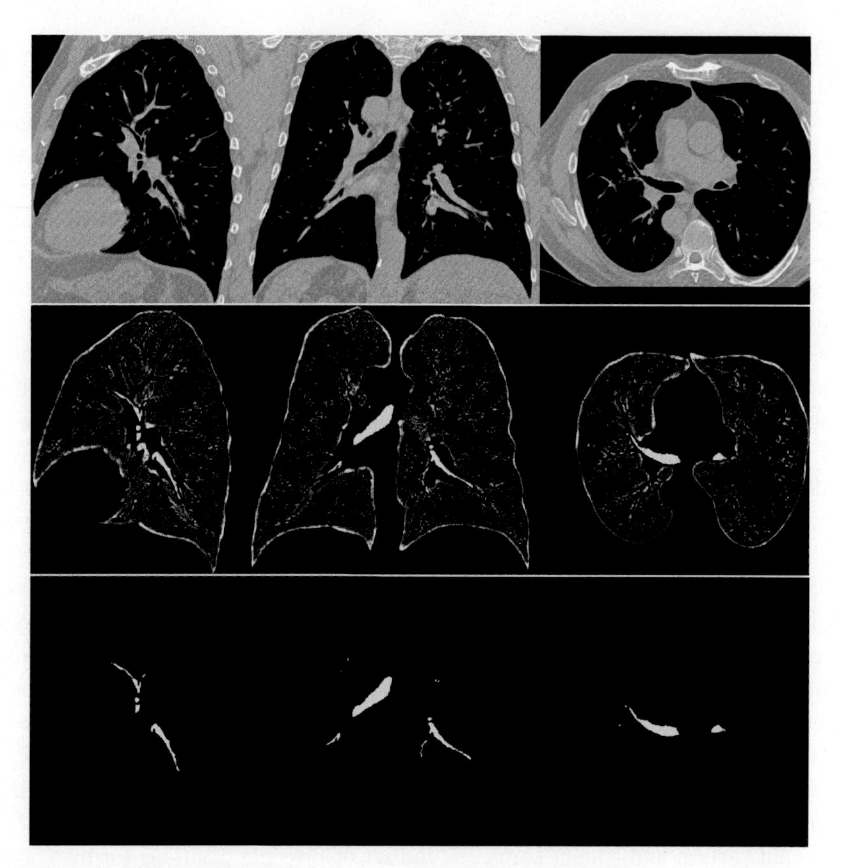

FIGURE 10.13 KNN classifier trained to recognize airway lumen in CT.

Top: CT of the lung. Middle: Processed using a K-nearest neighbor classifier and features based on spatial image derivatives and eigenvalues of the Hessian matrix. Colors from red to yellow to white indicate increasingly higher probabilities of lumen according to the classifier output. Note the false positive outputs at the lung boundary. Bottom: The result is post-processed with region growing using a single seed point in the trachea and a probability threshold of 0.5.

wall where intensity is around 0, and then transitions the exterior surface boundary where the intensity is around −900 to −800. We thus have a large positive gradient at the interior surface and a large negative gradient at the exterior surface, and since we are minimizing cost, these values are inverted. An example can be seen in Fig. 10.16.

We have a little less flexibility in choosing $g_{u,v}(y_u, y_v)$ as it has to be convex and increasing. In this example, we will use a linear cost of $p_m|y_v - y_u|$, where $m \in \{\text{interior, exterior}\}$ and $p_m \geq 0$. Fig. 10.17 shows visualizations of the results.

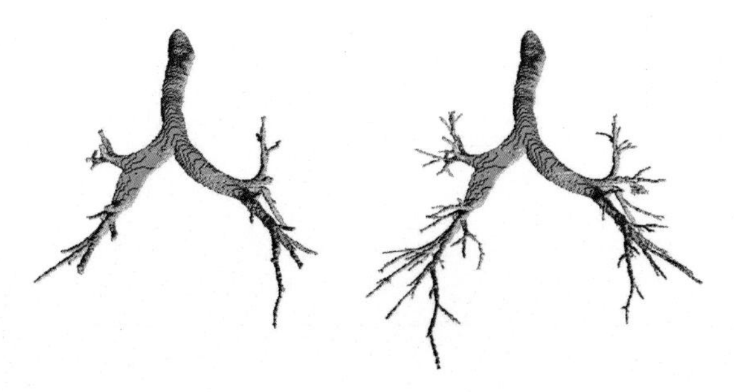

FIGURE 10.14 Region growing on image intensity or classifier output.

The result of using region growing on the CT attenuation values directly with a threshold of −900 (left) or on the output of a K-nearest neighbor classifier. Machine learning models can help to identify more of the smaller branches.

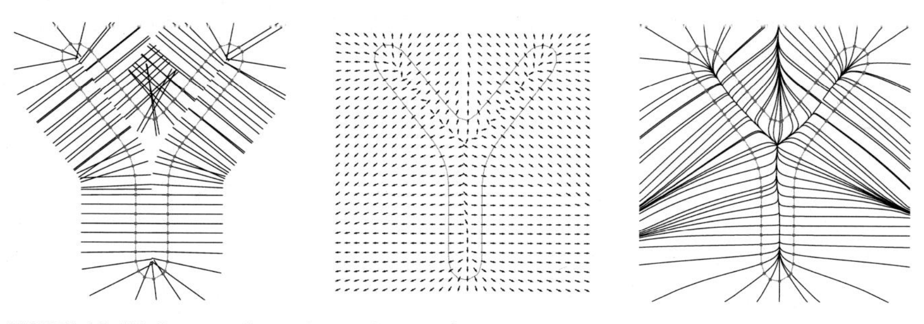

FIGURE 10.15 Constructing columns from an input surface.

Columns can be formed along the surface normal direction (left), but this may lead to poor results in high-curvature regions. Alternatively a vector field (middle) can be computed from the input surface and used to form columns that follow flow lines in this field (right). This improves results in high-curvature regions.

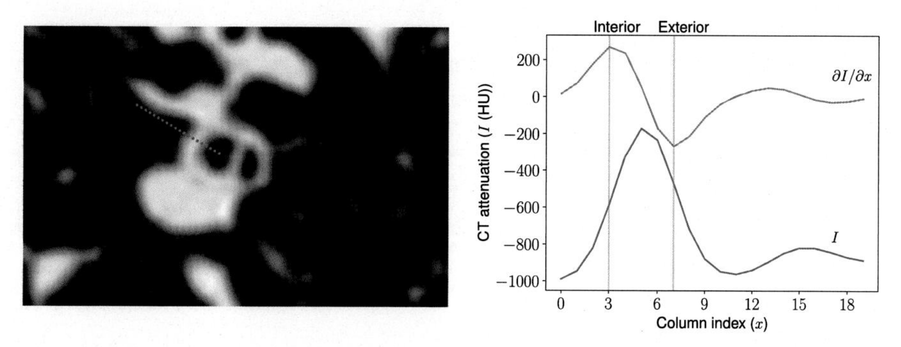

FIGURE 10.16 Column and intensity profile.

Left: An example column overlaid on a cross-section of an airway in CT. Right: CT attenuation along this column (blue). The magnitude of the first-order derivative (orange) can be used as a simple edge detector.

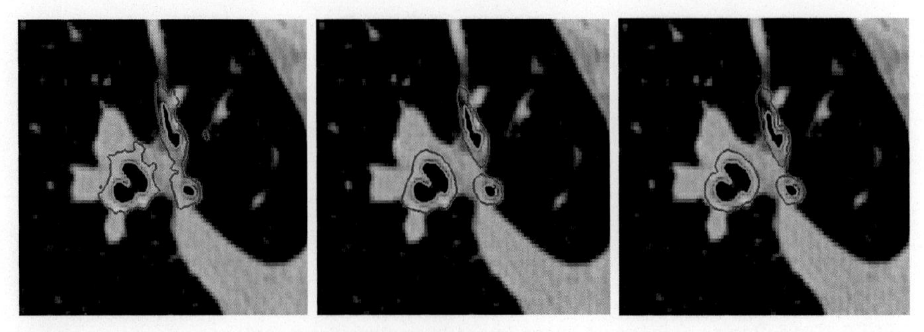

FIGURE 10.17 Regularizing effect of the pairwise term.

Increasing the cost of differences in labels between neighboring columns showing: too little regularization (left), a good choice of regularization (middle), and too much regularization (right). Note that with too little regularization the unary term dominates and particularly the exterior surfaces will become irregular due to nearby structures such as vessels that have similar attenuation values as airway walls. With too much regularization, the pairwise term dominates and the output surfaces will tend to have similar shapes as the input surface.

10.8 Recapitulation

This chapter has provided an overview of image segmentation using graph cuts, covering:

- how image segmentation can be modeled as an energy minimization problem with specific energy terms modeling image and regularization aspects;
- some basic concepts from graph theory, such as flows and minimum cuts;
- how and under what conditions an energy minimization problem can be solved efficiently using minimum cut algorithms;
- ways of handling more difficult cases involving more than two labels using optimal surface and move-making algorithms;
- two examples of medical image segmentation problems solved with graph cuts – segmentation of coronary arteries in CTA and airways in CT.

10.9 Exercises

1. What are the possible clique sizes with a 3D voxel neighborhood consisting of the 26 nearest neighbors?
2. Draw the graph $G_{s,t}$ for the problem given by $\mathcal{V} = \{a, b, c\}$, $\mathcal{E} = \{(a, b), (b, c)\}$, $f_a(1) = f_b(0) = f_c(1) = 1$, and $f_a(0) = f_b(1) = f_c(0) = -1$ and a Potts model pairwise term with $\lambda = 1$. Draw the minimum cut.
3. Repeat Exercise 2. with $\lambda = 2$.

4. Draw the four possible cuts in Cases 2 and 3 (Fig. 10.3) and compute the associated costs. Verify that these are indeed a constant from the energy of the corresponding configurations (equation (10.16)).
5. Draw $G_{s,t}$ for $\mathcal{V} = \{a, b\}$, $\mathcal{E} = \{(a, b)\}$, and $k = 4$, with a hard constraint of $|y_a - y_b| \leq 2$. Indicate the capacity of all non-zero capacity edges.

References

[1] Frangi A.F., Niessen W.J., Vincken K.L., Viergever M.A., Multiscale vessel enhancement filtering, in: International Conference on Medical Image Computing and Computer-Assisted Intervention, Springer, 1998, pp. 130–137.

[2] Kolmogorov V., Zabih R., What energy functions can be minimized via graph cuts?, IEEE Transactions on Pattern Analysis and Machine Intelligence 26 (2) (2004) 147–159.

[3] Boykov Y., Jolly M.-P., Interactive organ segmentation using graph cuts, in: International Conference on Medical Image Computing and Computer-Assisted Intervention, Springer, 2000, pp. 276–286.

[4] Baskaran L., Al'Aref S.J., Maliakal G., Lee B.C., Xu Z., Choi J.W., Lee S.-E., Sung J.M., Lin F.Y., Dunham S., Mosadegh B., Kim Y.-J., Gottlieb I., Lee B.K., Chun E.J., Cademartiri F., Maffei E., Marques H., Shin S., Choi J.H., Chinnaiyan K., Hadamitzky M., Conte E., Andreini D., Pontone G., Budoff M.J., Leipsic J.A., Raff G.L., Virmani R., Samady H., Stone P.H., Berman D.S., Narula J., Bax J.J., Chang H.-J., Min J.K., Shaw L.J., Data from: Automatic segmentation of multiple cardiovascular structures from cardiac computed tomography angiography images using deep learning, https://doi.org/10.5061/dryad.9s4mw6mc9, 2021.

[5] Ishikawa H., Exact optimization for Markov random fields with convex priors, IEEE Transactions on Pattern Analysis and Machine Intelligence 25 (10) (2003) 1333–1336.

[6] Wu X., Chen D.Z., Optimal net surface problems with applications, in: International Colloquium on Automata, Languages, and Programming, Springer, 2002, pp. 1029–1042.

[7] Xu C., Prince J.L., Snakes, shapes, and gradient vector flow, IEEE Transactions on Image Processing 7 (3) (1998) 359–369.

[8] Petersen J., Nielsen M., Lo P., Nordenmark L.H., Pedersen J.H., Wille M.M.W., Dirksen A., de Bruijne M., Optimal surface segmentation using flow lines to quantify airway abnormalities in chronic obstructive pulmonary disease, Medical Image Analysis 18 (3) (2014) 531–541.

PART

Medical image registration

IV

CHAPTER

Points and surface registration

11

Shan Cong and Li Shen

Department of Biostatistics, Epidemiology and Informatics, University of Pennsylvania Perelman School of Medicine, Philadelphia, PA, United States

Learning points

- Procrustes analysis for linear registration of multiple corresponding point sets
- Quaternions in rigid registration of two corresponding point sets
- Iterative Closest Point (ICP) for rigid points registration in general cases
- Thin plate spline for non-rigid alignment of two corresponding point sets
- Surface mesh representations, modeling and parameterization
- Surface registration SPHARM and related extensions

11.1 Introduction

Registration aims to discover the spatial correspondence between two geometric configurations by seeking an optimal global or local transformation that best aligns one configuration to the other in a homogeneous fashion. This chapter focuses on two types of geometric configurations: one is defined by a set of points; the other is described by a surface mesh representation. We discuss methods for points registration in Section 11.2 and methods for surface registration in Section 11.3.

The essential capability of points and surface registration is needed in various medical image analysis applications, biomedical research, and clinical practice. For example, in modern radiotherapy, registration techniques play a critical role in treatment planning, delivery, and monitoring. Registration is also a popular topic in medical image processing such as atlas-based segmentation and object matching across multiple medical imaging modalities. In addition, registration is commonly treated as a fundamental step for morphological analyses and longitudinal studies. Below we provide a brief review of example biomedical applications of points and surface registration.

First of all, in the preparation of radiotherapy, a motion evaluation and correction model can be built to register a preoperative medical image (MRI, PET, etc.) to the intraoperative 3D ultrasound image. The correspondence between the preop-

Medical Image Analysis. https://doi.org/10.1016/B978-0-12-813657-7.00026-1

erative image and intraoperative image can be established by aligning the points or normal vectors sampled from the surfaces of a tissue or organ. For example, in order to guide transrectal ultrasound imaging procedures which are widely adopted for prostate cancer biopsy and therapies, the motion evaluation and correction model is applied to predict and therefore compensate the prostate gland motion, primarily caused by the movement of the ultrasound probe. The estimated deformation then can locate the tumor, identified from magnetic resonance imaging (MRI) or positron emission tomography (PET) images, in the ultrasound imaging coordinates, which can be targeted for biopsy or localized treatment during procedures.

Second, 3D dental surface models acquired by optical digitizers can help describe dental regions in different angles, and thus are extensively applied in orthodontic planning to simulate and predict treatments. To achieve a certain angle of view, a rigid transformation is often applied to the 3D surface model, which is a process of surface alignment. Similar to the applications in radiotherapy, accurate target localization and verification is essential in orthognathic surgery due to the development of target position. A template or reference in image registration is a portion of the preoperative model which remains fixed during the registration process.

Third, automatic segmentation of complex organs, such as the heart and brain subcortical structures, is sometimes formulated as a fitting process from a prior model to the target images. The prior models contain the segmentation information and are used to guide the segmentation procedure, and the fitting procedure is designed to find spatial correspondence between the prior model and target image. For example, in atlas-based segmentation, a prior model (atlas) contains information of the shape and essential texture features of the images, and the fitting process from the prior to the target image is the key to propagate the segmentation such that the correspondence can be found by aligning 3D point clouds, surface normal vectors, or manually placed landmarks in a rigid or non-rigid manner.

Fourth, statistical shape analysis of 3D surface objects aims to quantify statistical differences and their significance in groups using statistical models. It is an analysis of the geometrical properties of some given set of surface shapes. It involves anatomical surface modeling and statistical comparisons on anatomy and highly relies on surface registration techniques. For example, in computational neuroanatomy, surface feature measures reconstructed from MRI scans have been used to study group differences in brain structure and also to predict clinical outcomes.

Fifth, quantitative prediction of physical defects and disease risks can provide valuable information for the preparation of a specific treatment or even assist in preventive surgery. For example, bone quality examination in osteoporosis patients and fracture risk assessment can be performed by aligning landmark points placed on 2D or 3D bone images to a reference. Another application is the detection of prenatal facial dysmorphism using 3D ultrasound. Statistical surface shape modeling and registration are used in quantitative extraction and assessment of fetal face features; thus, genetic effects on facial shape variability and pathological bone shapes can be understood.

Another application is dose assessment. Dose assessment and adaptive radiotherapy using automatic contours propagation can be achieved by surface registration. To monitor up-to-date estimates of anatomical changes and deliver a proper dose, a series of surface-aligned longitudinal image data can provide comparisons of accumulated doses taken over the course of therapy. This information can help detect structural and functional changes that might suggest changes in the treatment plan. A practical example is the determination of the delivered dose to the spinal cord in head-and-neck cancer radiotherapy.

At last, longitudinal analysis is designed to measure the trajectory of a disordered object such as a breast or lung tumor throughout fractionated radiotherapy. Points and surface registration play an important role in longitudinal shape analysis, which aligns tissue or organ shape of all visits together.

11.2 Points registration

This section presents four fundamental approaches for points registration. First, we discuss Procrustes analysis for aligning multiple corresponding point sets together, where shape information can be extracted after removing the effects of translation, scaling, and rotation. Second, we discuss a quaternion-based method for registering two corresponding point sets via a rigid body transformation. Third, we discuss the iterative closest point (ICP) algorithm, which extends the quaternion method to handle points registration in general cases. Finally, we discuss the thin plate spline (TPS) method for non-rigid alignment of two corresponding point sets coupled with non-linear warping of the underlying image domain.

11.2.1 Procrustes analysis for aligning corresponding point sets

Procrustes analysis [1] is a points registration method which aims to align a set of N geometric configurations (say, X_i for $i \in [1, N]$), where each configuration is defined by a set of corresponding landmark points. Its goal is to extract the shape information via removing the translational, rotational, and scaling effects. To remove the translational effect, each landmark set is centered by moving its center of gravity to the origin. After that, to remove rotational and scaling effects, Procrustes analysis seeks for a transformation T such that $T(X_i)$ best aligns the point set to landmarks on the mean configuration $\overline{X} = \frac{1}{N}\Sigma_{i=1}^{N} X_i$. The way of achieving best alignment is to minimize the summed square distance over the corresponding landmark points. This process is conducted in an iterative manner, where each landmark set is repeatedly aligned with the mean configuration until the mean converges.

Procrustes analysis can easily be understood graphically. Fig. 11.1(a) shows two sets of corresponding landmark points marked in red and blue, respectively, which need to be registered together. First, we compute the centroid of each set and move

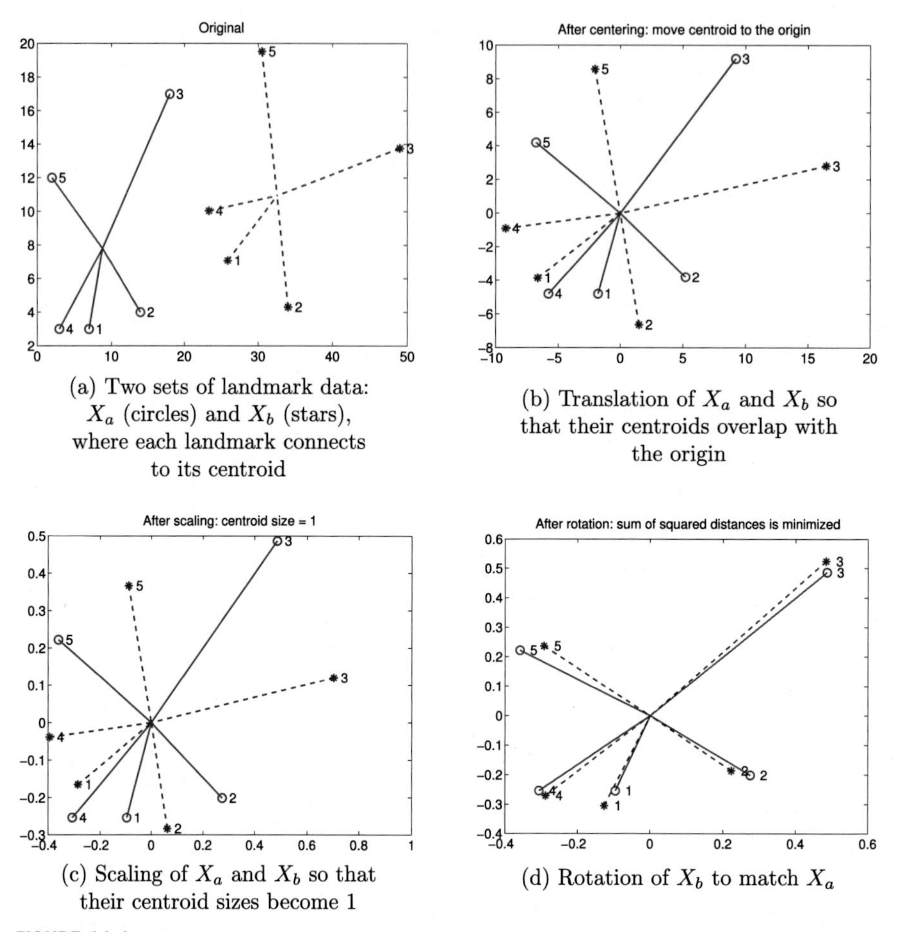

(a) Two sets of landmark data: X_a (circles) and X_b (stars), where each landmark connects to its centroid

(b) Translation of X_a and X_b so that their centroids overlap with the origin

(c) Scaling of X_a and X_b so that their centroid sizes become 1

(d) Rotation of X_b to match X_a

FIGURE 11.1

Procrustes analysis. (a) Two example landmark sets. (b) Centering of the landmark sets. (c) Scaling of the landmark sets. (d) Rotation of one to match the other (i.e., minimizing the summed square distances of corresponding landmarks).

the centroids together (i.e., to the origin, see Fig. 11.1(b)). Second, we rescale them so that the sum of square distances shown is fixed at unity (Fig. 11.1(c)). The original root sum of squares, the scaling factor here, is usually called centroid size in the shape analysis literature [2]. It is the same as the square root of the net moment of the original set of landmarks around their centroid. Finally, we superimpose one of the scaled forms over the other at their centroids and spin it with a unique rotation that minimizes the sum of squares of the distances between corresponding landmarks (Fig. 11.1(d)). The squared Procrustes distance between the forms is the sum of squares of those residual distances at its minimum.

11.2.2 Quaternion algorithm for registering two corresponding point sets

The quaternion-based algorithm described in [3] is another popular method for registering two corresponding point sets. It utilizes the property that every unit quaternion represents a rotation of objects in 3D. The quaternion-based algorithm plays an essential role in applying quaternions to perform 3D rotations [4]. For example, it aligns two corresponding point sets in a least squares sense and computes a rigid body transformation to align one set to the other. Let $P = \{\vec{p}_1, \cdots, \vec{p}_n\}$ be a measured data point set to be aligned with a model point set $X = \{\vec{x}_1, \cdots, \vec{x}_n\}$, where both sets have the same number of elements (denoted as $N_x = N_p$) and where each point $\vec{p}_i$ corresponds to the point $\vec{x}_i$ with the same index. P and X are defined as two landmark configurations, where $\vec{p}_i$ corresponds to $\vec{x}_i$ for $i \in \{1, \cdots, n\}$. To align P to X, we minimize the following objective function by using a rotation matrix $\mathbf{R}$ and a translation vector $\mathbf{T}$:

$$f(\mathbf{R}, \mathbf{T}) = \frac{1}{n} \sum_{i=1}^{n} \|\vec{x}_i - \mathbf{R}\,\vec{p}_i - \mathbf{T}\|^2. \tag{11.1}$$

A quaternion-based algorithm has been presented by Besl et al. [3] to achieve the above least squares rotation and translation for the problem of corresponding point set registration. The algorithm is described below.

A^T is used to denote the transpose of a matrix A. The *unit quaternion* is a column vector containing four components,

$$\vec{q}_R = (q_0, q_1, q_2, q_3)^T, \tag{11.2}$$

where $q_0 \geq 0$ and $q_0^2 + q_1^2 + q_2^2 + q_3^2 = 1$. The unit quaternion is one way to represent rotation [5]. The 3×3 rotation matrix generated by a unit rotation quaternion is given by

$$\mathbf{R}(\vec{q}_R) = \begin{bmatrix} q_0^2 + q_1^2 - q_2^2 - q_3^2 & 2(q_1q_2 - q_0q_3) & 2(q_1q_3 + q_0q_2) \\ 2(q_1q_2 + q_0q_3) & q_0^2 + q_2^2 - q_1^2 - q_3^2 & 2(q_2q_3 - q_0q_1) \\ 2(q_1q_3 - q_0q_2) & 2(q_2q_3 + q_0q_1) & q_0^2 + q_3^2 - q_1^2 - q_2^2 \end{bmatrix}. \tag{11.3}$$

Let $\vec{q}_T = (q_4, q_5, q_6)^T$ be a translation vector. Thus, the complete transformation vector $\vec{q}$ could be denoted

$$\vec{q} = \begin{pmatrix} \vec{q}_R \\ \vec{q}_T \end{pmatrix} = (q_0, q_1, q_2, q_3, q_4, q_5, q_6)^T. \tag{11.4}$$

Now the objective function to be minimized becomes

$$f(\vec{q}) = \frac{1}{n} \sum_{i=1}^{n} \|\vec{x}_i - \mathbf{R}(\vec{q}_R)\,\vec{p}_i - \vec{q}_T\|^2. \tag{11.5}$$

We use $\vec{\mu}_p$ and $\vec{\mu}_x$ to denote the centers of mass of the point sets P and X, respectively:

$$\vec{\mu}_p = \frac{1}{n}\sum_{i=1}^{n} \vec{p}_i \quad \text{and} \quad \vec{\mu}_x = \frac{1}{n}\sum_{i=1}^{n} \vec{x}_i. \tag{11.6}$$

Thus, the cross-covariance matrix Σ_{px} of the sets P and X is given by

$$\Sigma_{px} = \frac{1}{n}\sum_{i=1}^{n}((\vec{p}_i - \vec{\mu}_p)(\vec{x}_i - \vec{\mu}_x)^T) = \frac{1}{n}\sum_{i=1}^{n}(\vec{p}_i \vec{x}_i^T) - \vec{\mu}_p \vec{\mu}_x^T. \tag{11.7}$$

The cyclic components of the anti-symmetric matrix

$$A_{ij} = (\Sigma_{px} - \Sigma_{px}^T)_{ij} \tag{11.8}$$

are used to form a column vector,

$$\Delta = (A_{23}, A_{31}, A_{12})^T. \tag{11.9}$$

Now a symmetric 4×4 matrix can be formed as

$$Q(\Sigma_{px}) = \begin{bmatrix} \mathrm{trace}(\Sigma_{px}) & \Delta^T \\ \Delta & \Sigma_{px} + \Sigma_{px}^T - \mathrm{trace}(\Sigma_{px})\,\mathbf{I}_3 \end{bmatrix}, \tag{11.10}$$

where $\mathbf{I}_3$ is the 3×3 identity matrix and $\mathrm{trace}(\Sigma_{px})$ is the sum of the diagonal elements of the matrix Σ_{px}. Then, the *optimal rotation* is determined by the unit eigenvector

$$\vec{q}_R = (q_0, q_1, q_2, q_3)^T \tag{11.11}$$

corresponding to the maximum eigenvalue of the matrix $Q(\Sigma_{px})$. The *optimal translation* vector is given by

$$\vec{q}_T = \vec{\mu}_x - \mathbf{R}(\vec{q}_R)\,\vec{\mu}_p. \tag{11.12}$$

This quaternion-based algorithm is simple and effective. Fig. 11.2 shows a sample result of applying the algorithm in aligning two point sets, and the result is desired.

11.2.3 Iterative closest point algorithm for general points registration

In Section 11.2.2, we assume $N_x = N_p$. But in general cases, the numbers of landmark points N_x and N_p are not necessarily the same. Given no correspondence on the two point sets, it is impossible to determine the optimal relative transformation such as translation, rotation, and scaling as we discussed in Section 11.2.2. To overcome this limitation, the quaternion-based registration can be extended to the ICP algorithm [3] for aligning two point sets in general cases.

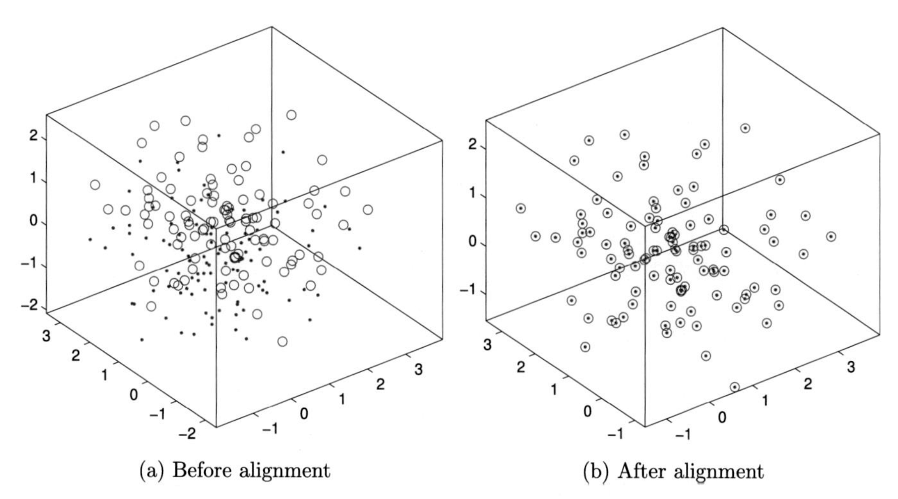

FIGURE 11.2

Aligning corresponding point sets. (a) Blue circles are formed by rotating and translating red dots by a certain amount. (b) The quaternion-based algorithm can perfectly align red dots to blue circles.

Let $P = \{\vec{p}_1, \cdots, \vec{p}_m\}$ be a measured data point set to be aligned with a model point set $X = \{\vec{x}_1, \cdots, \vec{x}_n\}$. A notable difference between the previously discussed quaternion method and ICP is that in CIP there is no known point correspondence between $\vec{p}_i$ and $\vec{x}_j$. Thus, the first step of ICP is to find the correspondence.

The ICP algorithm finds a correspondence between $\vec{p}_i$ and $\vec{x}_j$ by matching the closest points corresponding with each other. For each point $\vec{p}_i$ in $P = \{\vec{p}_1, \cdots, \vec{p}_m\}$, ICP searches for the closest point $\vec{x}_j$ in the model shape X; usually the distance is measured by Euclidean distance, which is defined (in the 3D case) as

$$d\left(p_i, x_j\right) = \left\|p_i - x_j\right\| = \sqrt{\left(p_{xi} - x_{xj}\right)^2 + \left(p_{yi} - x_{yj}\right)^2 + \left(p_{zi} - x_{zj}\right)^2}, \quad (11.13)$$

and the goal is to find

$$d\left(p_i, X\right) = \min_{j=1,\ldots N_x} d\left(p_i, x_j\right) = \min_{j=1,\ldots N_x} \left\|p_i - x_j\right\|.$$

The closest point $x_k \in X$ satisfies $d\left(p_i, x_k\right) = d\left(p_i, X\right)$, where k is the index of the closest point p_i such that

$$k = \underset{j=1,\ldots N_X}{\arg\min}\, d\left(p_i, x_j\right).$$

The closest point in the model set X that yields the minimum Euclidean distance is denoted as s_i, so we have $s_i = x_k$ such that $d(p_i, s_i) = d(p_i, X)$; thus, the correspondence between p_i and s_i is identified. Let S denote the resulting set of closest points.

We define C as the closest point operator such that $S = C(P, X)$ and $S \subset X$. We can align $p_i \in P$ to $s_i \in S$ by applying the quaternion registration method discussed in Section 11.2.2. This process is iterative and converges if the starting position is mostly approached. The algorithm can be summarized in the following steps:

1. Initialize the registration error to ∞.
2. Calculate correspondence $S = C(P, X)$.
3. Calculate the rigid transformation matrix.
4. Apply quaternion registration.
5. Update the registration error.
6. If error > threshold, go back to Step 2.
7. Converge.

11.2.4 Thin plate spline for non-rigid alignment of two corresponding point sets

The quaternion algorithm discussed in Section 11.2.2 aims to identify a rigid transformation to align two corresponding point sets. Of note, rigid transformations are global transformations. When a local patch needs to be aligned, the global transformation usually cannot achieve the best registration results. To overcome this limitation, TPS [1] can be used to yield a non-rigid registration for matching two corresponding point sets and subsequently warping their underlying image domains accordingly.

In a 2D setting, the TPS function $f(P)$ has heights (values) h_i at points $P_i = (x_i, y_i)$, i.e., $f(P_i) = h_i$, for $i = 1, \ldots, k$. This indicates that f interpolates the heights h_i at landmarks P_i. In addition, the function f has the minimum bending energy of all functions that interpolate heights h_i in that way:

$$\min\left(\iint_{\mathbf{R}^2}\left(\left(\frac{\partial^2 f}{\partial x^2}\right)^2 + 2\left(\frac{\partial^2 f}{\partial x \partial y}\right)^2 + \left(\frac{\partial^2 f}{\partial y^2}\right)^2\right)\right), \tag{11.14}$$

where the integral is taken over the entire 2D plane. Detailed definition of the 2D TPS is available in [1].

Fig. 11.3 shows example TPS applications to the registration of 2D landmark data. The red diamonds are the control points in the 2D domain. The goal is to match blue stars to the control points and warp the underlying mesh (e.g., image content) accordingly. We compute two TPS functions: one f_x, where the height corresponds to the x-coordinate, and the other f_y, where the height corresponds to the y-coordinate. Thus, f_x and f_y provide the interpolated x- and y-coordinates, respectively. The resulting map $(f_x(P), f_y(P))$ is a deformed version of the original one that matches the corresponding landmarks perfectly with the minimum bending energy.

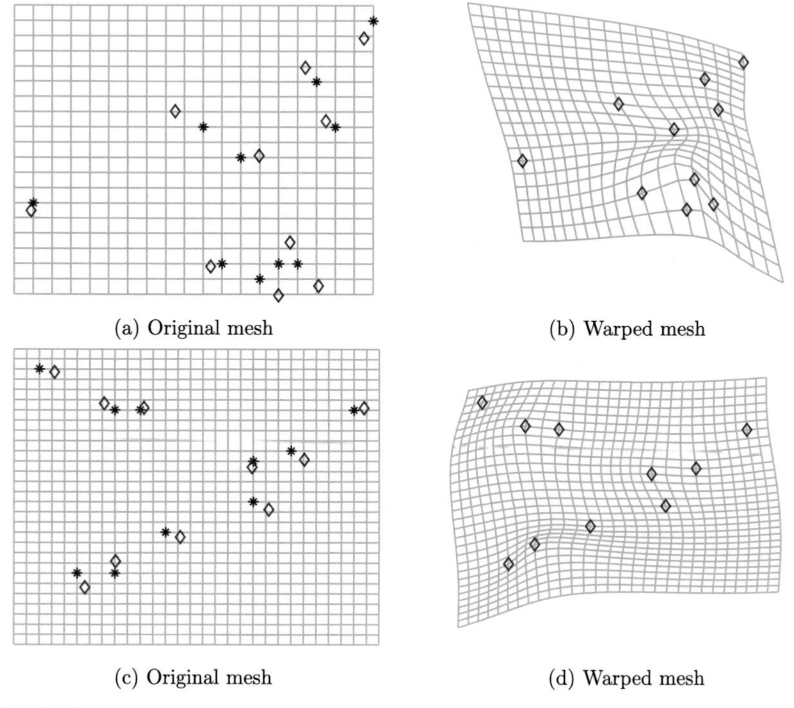

FIGURE 11.3

Example thin plate spline (TPS) applications to the registration of 2D landmark data. The goal is to match blue stars to the control points and warp the underlying mesh (e.g., image content) accordingly.

11.3 Surface registration

While image registration is deemed as one of the most important research topics in image processing and analysis, surface registration plays a likewise significant role in medical imaging, computer graphics, and shape analysis. Surface registration aims to find correspondence between two surfaces, which could be guided by geometric information such as curvatures and thickness or surface signals such as landmarks or regional parcels. This technique meets vast clinical needs, as partly described in Section 11.1, it enables groupwise comparisons by aligning 3D dental surfaces [6]; moreover, it helps to identify deformations by aligning 3D cortical surfaces [7]; it also helps in adjusting head-and-neck positions in radiotherapy [8]; in practise, the patient-to-image surface registration technique is a crucial factor that affects the accuracy of image-guided ear–nose–throat and neurosurgery systems [9], etc. In this section, we first describe how to represent surfaces using polygonal meshes, then dis-

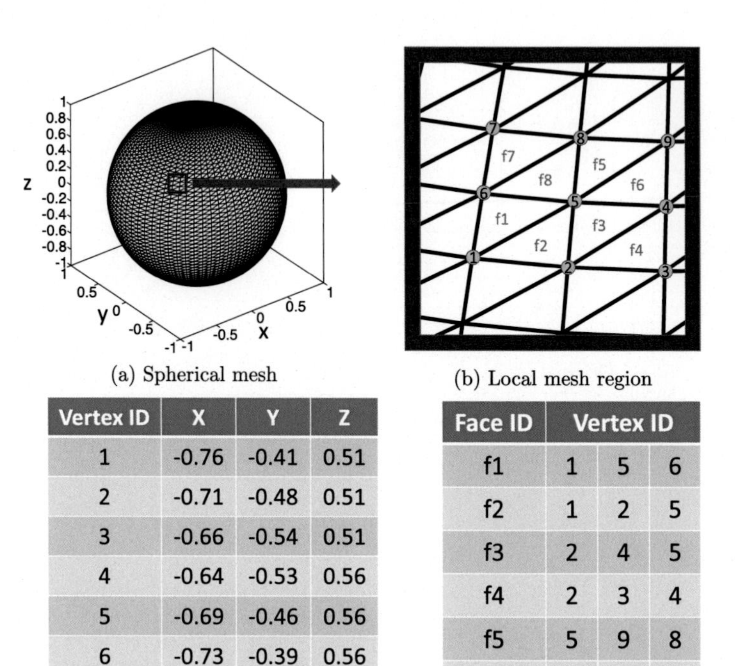

(a) Spherical mesh

(b) Local mesh region

Vertex ID	X	Y	Z
1	-0.76	-0.41	0.51
2	-0.71	-0.48	0.51
3	-0.66	-0.54	0.51
4	-0.64	-0.53	0.56
5	-0.69	-0.46	0.56
6	-0.73	-0.39	0.56
7	-0.71	-0.38	0.6
8	-0.67	-0.45	0.6
9	-0.62	-0.51	0.6

(c) Vertex list

Face ID	Vertex ID		
f1	1	5	6
f2	1	2	5
f3	2	4	5
f4	2	3	4
f5	5	9	8
f6	5	4	9
f7	6	7	8
f8	6	5	8

(d) Face list

FIGURE 11.4

Example surface mesh structure. (a) Mesh representation of the surface of a unit sphere. (b) A zoom-in view of a local mesh region. (c) A list of vertices from the local mesh region. (d) A list of faces from the local mesh region.

cuss methods for surface parameterization, and finally present several algorithms for surface registration.

11.3.1 Surface mesh representation

A surface is typically described by a polygonal mesh including a set of vertices, edges, and faces. As shown in Fig. 11.4, an example surface mesh consists of vertices, edges (i.e., links between vertices), and faces (i.e., closed areas bounded by edges). Surface models are usually represented by triangular meshes and sometimes by quadrilateral or other polygonal meshes.

For example, Fig. 11.4(a) shows a triangular mesh describing the surface of a unit sphere. For a local mesh region shown in Fig. 11.4(b), we take a close look at nine example vertices labeled with vertex IDs from 1 to 9 as well as eight related faces

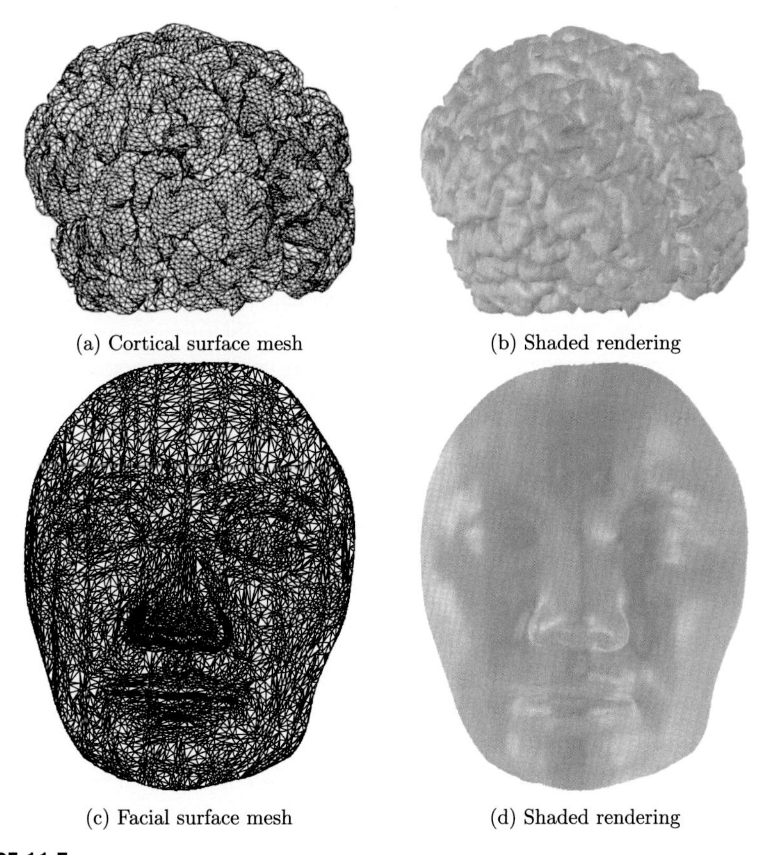

FIGURE 11.5

Surface mesh examples. (a, b) Cortical surface mesh and shaded rendering. (c, d) Facial surface mesh and shaded rendering.

labeled with face IDs from f1 to f8. The vertex data structure shown in Fig. 11.4(c) specifies a 1×3 vector for each vertex, indicating its spatial coordinates x, y, and z. The face data structure shown in Fig. 11.4(b) specifies a 1×3 vector for each face, indicating the indices of its three associated vertices. Fig. 11.5 shows the mesh structure and shaded rendering of two example surfaces: (a and b) a cortical surface [10] and (c and d) a facial surface [11].

Besides capturing geometric surface information, additional signals can be computed from the mesh representation. For example, when the medial curve or axis is calculated for a closed surface, the thickness can be defined as the shortest distance from a specific vertex to the medial curve. This can also be used as surface signals when performing group comparisons. Another example is surface curvature, which can be evaluated on each surface vertex. With this valuable information, many mathematical tools such as differential geometry can be applied.

11.3.2 Surface parameterization

Surface parameterization is a special type of surface registration and aims to find a one-to-one mapping from a suitable domain to a given surface. The goal of surface parameterization is to introduce a meaningful common space for two or more surfaces so that the surface correspondence among them can be established. After such a one-to-one mapping, different surfaces can be projected onto a certain domain such as a disk or a sphere.

Surface parameterizations almost always introduce distortion in either angles or areas. A good mapping in applications is the one which minimizes these distortions as much as possible. Below we discuss two examples of surface parameterization: (1) one parameterizes an open surface to a disk using conformal mapping, and (2) the other performs spherical parameterization for a closed surface using area-preserving mapping.

Of note, surface parameterization establishes the correspondence between an individual surface and a common domain in the parameter space. It is often used as a pre-processing step for creating shape descriptors that register all the surfaces in both parameter and object spaces. We will discuss this type of surface registration methods in Section 11.3.3.1.

11.3.2.1 Conformal open surface parameterization

Here we describe how conformal mapping (i.e., angle-preserving mapping) is implemented to parameterize an open surface onto a 2D domain. The Riemann theorem states that conformal mapping of a smooth surface into a plane exists for any simply connected plane domain [12,13]. Since a mesh representation of a smooth surface can be viewed as an approximation of the surface, it is possible to map a surface to a plane with very little angular distortion.

Suppose a surface $M_1 \subset \mathbb{R}^3$ has the parametric representation

$$\mathbf{x}(u^1, u^2) = (x_1(u^1, u^2), x_2(u^1, u^2), x_3(u^1, u^2))$$

for points (u^1, u^2) in some domain in $\mathbb{R}^2$. The first fundamental form of M_1 is

$$ds_1^2 = \sum_{ij} g_{ij} du^i du^j, \quad \text{where} \quad g_{ij} = \frac{\partial \mathbf{x}}{\partial u^i} \cdot \frac{\partial \mathbf{x}}{\partial u^j} \quad (i, j = 1, 2). \tag{11.15}$$

Another plane $M_2 \subset \mathbb{R}^2$ is similarly represented by $\widetilde{\mathbf{x}}(\widetilde{u}^1, \widetilde{u}^2)$. Let us define a mapping $f : M_1 \mapsto M_2$ between two surfaces. If f is a conformal mapping, then there is some scalar function $\eta \neq 0$ such that $ds_1^2 = \eta(\widetilde{u}^1, \widetilde{u}^2)((d\widetilde{u}^1)^2 + (d\widetilde{u}^2)^2)$. Two Laplace equations are obtained as

$$\Delta_s \widetilde{u}^1 = 0, \quad \Delta_s \widetilde{u}^2 = 0, \tag{11.16}$$

where Δ_s is the Laplace–Beltrami operator, which can be written as $\Delta_s = \text{div}_s \text{grad}_s$. To find the solution to equation (11.16), the conformal mapping f can be viewed as minimizing the Dirichlet energy:

$$E_0(f) = \frac{1}{2}\int_S \|\text{grad}_s f\|^2. \tag{11.17}$$

To compute f, Eck et al. [14] proposed an approach, called discrete harmonic map, which extends the graph embedding method of Tutte [15]. In their method, the boundary vertices of the meshes are first mapped to the boundary of the unit disk. Then the positions of the remaining vertices can be computed by solving equations:

$$L\widetilde{u}^1 = 0, \qquad L\widetilde{u}^2 = 0, \tag{11.18}$$

$$L_{ij} = \begin{cases} -\sum_{k\neq i} L_{ik}, & i = j, \\ w_{ij}, & (i, j) \in E, \\ 0, & \text{otherwise}, \end{cases} \tag{11.19}$$

$$w_{ij} = \cot\alpha_{ij} + \cot\beta_{ij}, \tag{11.20}$$

where α_{ij} and β_{ij} are the opposite angles in the two triangles sharing an edge (i, j).

To illustrate the process of conformal mapping for a surface mesh, Fig. 11.6 shows an example result using the above method, where a facial surface is conformally mapped onto a 2D disk.

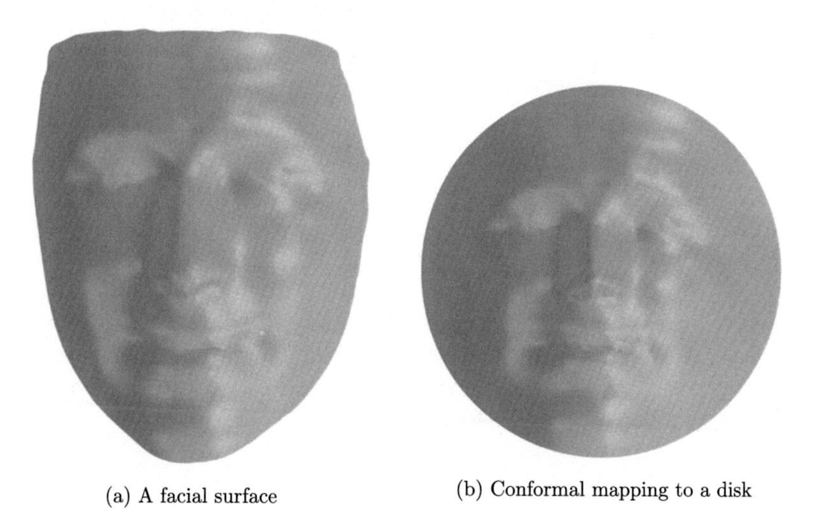

(a) A facial surface (b) Conformal mapping to a disk

FIGURE 11.6

An example of conformal surface mapping to a 2D domain [16]. The facial surface was downloaded from https://github.com/Juyong/3DFace.

11.3.2.2 Area-preserving spherical parameterization

In this section, we introduce a spherical parameterization method used in the spherical harmonics (SPHARM) shape description [17] that will be discussed later. Here

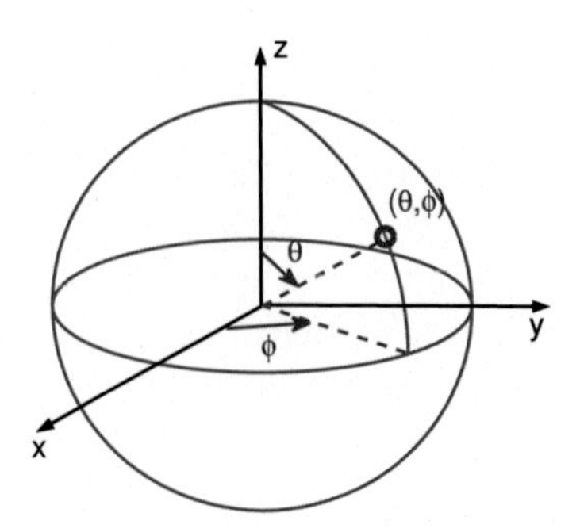

FIGURE 11.7

Spherical coordinates: θ is defined as the polar (or colatitudinal) coordinate with $\theta \in [0, \pi]$, and ϕ is taken as the azimuthal (or longitudinal) coordinate with $\phi \in [0, 2\pi)$.

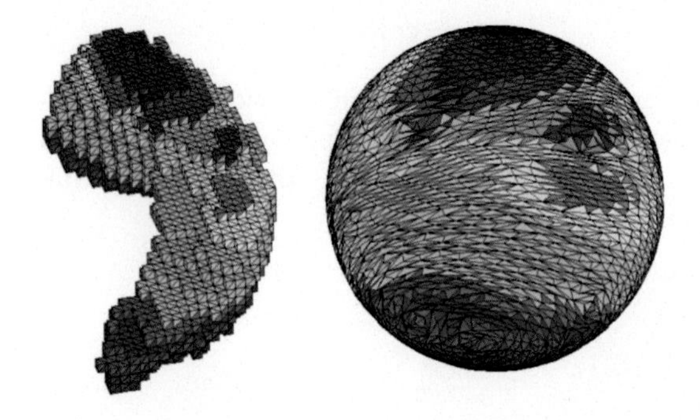

FIGURE 11.8

Hippocampal surface (left) and its spherical parameterization (right). Color indicates the correspondence between the surface and parameterization.

spherical parameterization aims to create a continuous and uniform mapping from the object surface to the surface of a unit sphere, and it results in a bijective mapping between each point v on a surface and a pair of spherical coordinates θ and ϕ that matches the definition of spherical harmonics [17]:

$$\mathbf{v}(\theta, \phi) = (x(\theta, \phi), y(\theta, \phi), z(\theta, \phi))^T, \tag{11.21}$$

where θ is defined as the polar (or colatitudinal) coordinate with $\theta \in [0, \pi]$ and ϕ is taken as the azimuthal (or longitudinal) coordinate with $\phi \in [0, 2\pi)$, as shown in Fig. 11.7. Thus, the north pole is defined as $\theta = 0$ and the south pole as $\theta = \pi$. Fig. 11.8 shows an example spherical parameterization. This parameterization maps the surface of a simply connected 3D object onto the surface of the unit sphere through an area-preserving procedure proposed by Brechbühler et al. [18].

In the initialization step, an initial parameterization is constructed by creating a harmonic map from the object surface to the parameter surface. For colatitude

θ, two poles are selected in the surface mesh by finding the two vertices with the maximum and minimum z-coordinate in object space. Then, a Laplace equation (equation (11.22)) with Dirichlet conditions (equations (11.23) and (11.24)) is solved for colatitude θ:

$$\nabla^2\theta = 0 \quad \text{(except at the poles)}, \tag{11.22}$$

$$\theta_{north} = 0, \tag{11.23}$$

$$\theta_{south} = \pi. \tag{11.24}$$

Since our case is discrete, we can approximate equation (11.22) by assuming that each vertex's colatitude (except at the poles) equals the average of its neighbors' colatitudes. Thus, after assigning $\theta_{north} = 0$ to the north pole and $\theta_{north} = \pi$ to the south pole, we can form a system of linear equations by considering all the vertices and obtain the solution by solving this linear system. For longitude ϕ, the same approach can be employed except that longitude is a cyclic parameter. To overcome this problem, a "date line" is introduced. When crossing the date line, longitude is incremented or decremented by 2π, depending on the crossing direction. After slightly modifying the linear system according to the date line, the solution for longitude ϕ can also be achieved.

In the optimization step, the initial parameterization is refined to obtain an area-preserving mapping. Brechbühler et al. [18] formulate this refinement process as a constrained optimization problem. They establish a few constraints for topology and area preservation and formulate an objective function for minimizing angular distortion. To solve this constrained optimization problem, an iterative procedure is developed to perform the following two operations alternately: (1) satisfying the constraints using the Newton–Raphson method and (2) optimizing the objective function using a conjugate gradient method.

In sum, the key idea in the surface parameterization step is to achieve a homogeneous distribution of parameter space so that the surface correspondence across subjects can be obtained in a later step. Note that equal area parameterization implies such a surface correspondence that the corresponding parts of two surfaces by design occupy the same amount of surface area.

11.3.3 Surface registration strategies

In this section, we discuss a few surface registration strategies. First, we show three methods for registering 3D closed surfaces with spherical topology: (1) SPHARM registration using purely geometric information, (2) landmark-guided surface registration, and (3) signal-guided surface registration. Finally, we discuss a landmark-guided surface registration method for open surfaces.

11.3.3.1 SPHARM surface registration

In this section, we describe an approach that registers a group of 3D surfaces with spherical topology using SPHARM shape description [17]. Given a 3D surface with

FIGURE 11.9

Shown from left to right are the original surface, its spherical parameterization, and its SPHARM reconstruction.

spherical topology, we can use the method introduced in Section 11.3.2.2 to create its spherical parameterization as follows:

$$\mathbf{v}(\theta,\phi) = (x(\theta,\phi), y(\theta,\phi), z(\theta,\phi))^T,$$

which defines a one-to-one mapping between each surface point (x, y, z) and each location (θ, ϕ) on the surface of the unit sphere.

In other words, the surface can be described by three spherical functions, which can be expanded using spherical harmonic basis functions $Y_l^m(\theta, \phi)$ as follows:

$$v(\theta,\phi) = \begin{pmatrix} x(\theta,\phi) \\ y(\theta,\phi) \\ z(\theta,\phi) \end{pmatrix} = \sum_{l=0}^{\infty}\sum_{m=-l}^{l} \begin{pmatrix} c_{xl}^m \\ c_{yl}^m \\ c_{zl}^m \end{pmatrix} Y_l^m(\theta,\phi) = \sum_{l=0}^{\infty}\sum_{m=-l}^{l} c_l^m Y_l^m(\theta,\phi). \tag{11.25}$$

Here Y_l^m denotes the spherical harmonic of degree l and order m and spherical harmonics form a Fourier basis defined on the sphere. The Fourier coefficients c_l^m can be determined by solving a linear system using standard least squares estimation. The object surface can then be reconstructed using these coefficients. Using more coefficients leads to a more detailed reconstruction; see Fig. 11.10(a) for the degree 1 reconstruction and the degree 15 reconstruction for two examples. Fig. 11.9 shows a hippocampal surface, its spherical parameterization, and its SPHARM expansion using coefficients up to degree 15.

Now we describe how to register two SPHARM surfaces together in both parameter and object spaces (see Fig. 11.10 for an example). Of note, the degree 1 reconstruction of a SPHARM model is always an ellipsoid, typically called first-

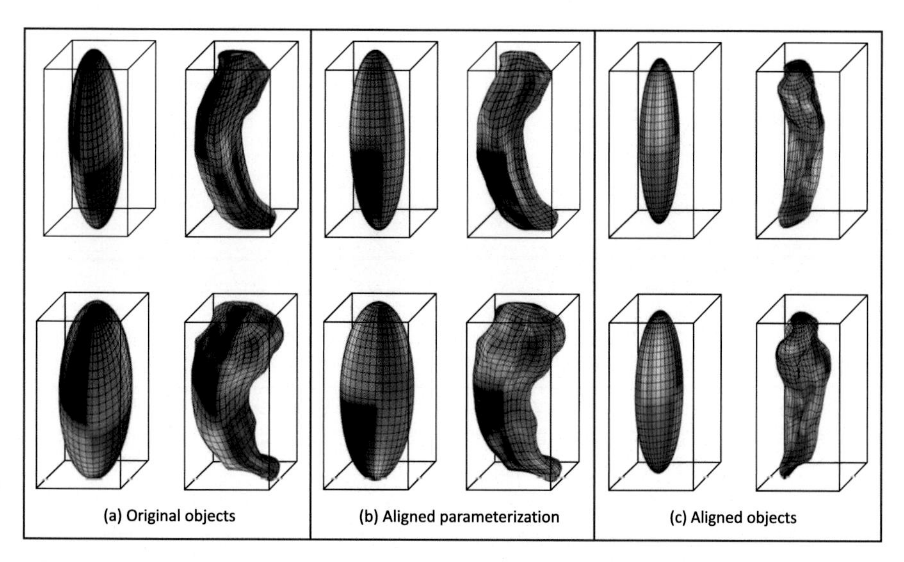

FIGURE 11.10

SPHARM registration using first-order ellipsoids (FOEs). Each of (a–c) shows the FOE on the left and degree 15 reconstruction on the right. Parameterization is indicated by the mesh and color on the surface.

order ellipsoid (FOE) in the literature. We can use this FOE to align two SPHARM models together in both parameter and object spaces [17].

Using two hippocampal surfaces as an example, Fig. 11.10 demonstrates the registration of two SPHARM models by aligning their FOEs. Each of (a–c) shows the FOEs at the left and degree 15 reconstructions at the right. In (a), the original pose and parameterization are shown. Note that the correspondence between two SPHARM models is implied by the underlying parameterization: two points with the same parameter pair (θ, ϕ) on two surfaces are defined to be a corresponding pair. Thus, in (b), the FOE is used to align the parameterization in the parameter space and to establish the surface correspondence while the object pose stays the same. Here the parameter net on each FOE is rotated to a canonical position such that the north pole is at one end of the longest main axis, and the crossing point of the zero meridian and the equator are at one end of the shortest main axis. In (c), the FOE is used to adjust the object pose in the object space: the FOE is rotated to make its main axes coincide with the coordinate axes, putting the shortest axis along x and the longest along z. As a result, these two hippocampal surfaces are aligned to a canonical position in both parameter space and object space.

Optionally, in the object space, one can further remove the effects of scaling and translation as follows. The scaling-invariance can be achieved by adjusting the SPHARM coefficients so that the object volume is normalized. Translation-invariance can be achieved by ignoring the degree 0 coefficient, which essentially centers the object.

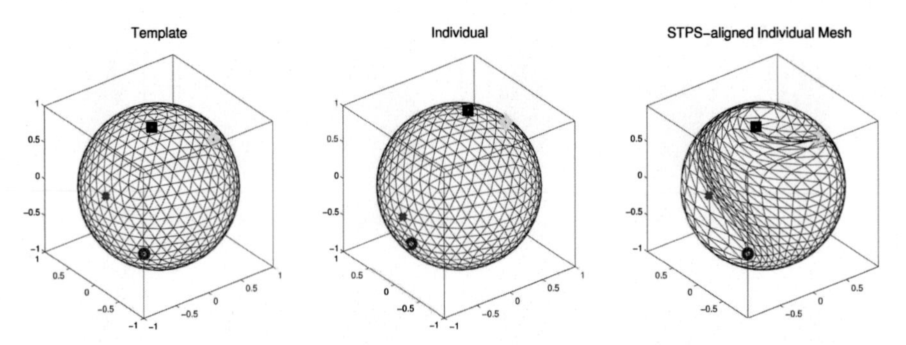

FIGURE 11.11

Application of the basic STPS method. When one or more landmarks are closely located to the north or south pole, the parametric mesh of the individual can be severely distorted. Reprinted by permission from Elsevier: Shen et al. [19].

11.3.3.2 Landmark-guided SPHARM surface registration

The FOE method for registering SPHARM models considers only the geometric information of the surfaces. In many biomedical applications, the studied structures may be marked with landmarks containing important anatomical or biological information, which should not be ignored in surface registration. In this section, we discuss a landmark-guided strategy for SPHARM surface registration.

The strategy is inspired by the spline-based landmark alignment methods such as the 2D TPS method discussed in Section 11.2.4. First, we still use the FOE method to align an individual SPHARM model to the template model. After that, we aim to complete a registration that deforms the individual's spherical parameterization to that of the template so that their corresponding landmarks are perfectly aligned on the surface of the unit sphere. In other words, all pairs of the corresponding landmarks between the individual and the template share the exactly same spherical coordinates. Fig. 11.11 shows such an example: from left to right, we have the spherical parameter nets of the template and the individual before and after registration.

This can be done by applying spherical TPS (STPS), defined in [18]. It is an extension of 2D TPS to a spherical domain and is used here to deform spherical parametric domains. STPS on a spherical domain S^2 minimizes a bending energy $J(u)$, subject to $u(P_i) = z_i$, $i = 1, 2, \cdots, n$, where $P_i \in S^2$ and z_i is the fixed displacement at P_i. This bending energy is formulated as

$$J(u) = \int_0^{2\pi} \int_0^{\pi} (\Delta u(\theta, \phi))^2 \sin\phi d\theta d\phi, \tag{11.26}$$

where $\theta \in [0, \pi]$ is latitude, $\phi \in [0, 2\pi]$ is longitude, and Δ is the Laplace–Beltrami operator.

The solution on the sphere is given by

$$u_n(P) = \sum_{i=1}^{n} c_i K(P, P_i) + d, \tag{11.27}$$

where $\mathbf{c}$ and d are determined by

$$\mathbf{K}_n\mathbf{c} + d\mathbf{T} = \mathbf{z}, \quad \mathbf{T}'\mathbf{c} = 0,$$

$\mathbf{K}_n$ is the $n \times n$ matrix with (i, j)th entry $K(P_i, P_j)$, $\mathbf{T} = (1, ..., 1)'$, and $\mathbf{z} = (z_1, ..., z_n)'$. $K(X, Y)$ between two arbitrary points $X, Y \in S^2$ is defined as follows:

$$K(X, Y) = \frac{1}{4\pi}\int_0^1 (\log h)(1 - \frac{1}{h})(\frac{1}{\sqrt{1 - 2ha + h^2}} - 1)dh,$$

where $a = \cos(\gamma(X, Y))$ and $\gamma(X, Y)$ is the angle between X and Y. However, this is not a computable closed-form expression. Therefore, thin plate pseudospline on the sphere $R(X, Y)$ [18] is used in this study, which is formulated as

$$R(X, Y) = \frac{1}{2\pi}\left[\frac{1}{(2m-2)!}q_{2m-2}(a) - \frac{1}{(2m-1)!}\right] \tag{11.28}$$

and

$$q_m(a) = \int_0^1 (1 - h)^m (1 - 2ha + h^2)^{-1/2} dh, \quad m = 0, 1, \cdots.$$

In our case, we have $m = 2$, giving

$$q_2(a) = \frac{1}{2}\left\{\ln\left(1 + \sqrt{\frac{1}{W}}\right)\left[12W^2 - 4W\right] - 12W^{3/2} + 6W + 1\right\}, \tag{11.29}$$

where $W = (1 - a)/2$. Therefore, equation (11.27) becomes

$$u_n(P) = \sum_{i=1}^{n} c_i R(P, P_i) + d, \tag{11.30}$$

and $\mathbf{c}$ and d are determined by

$$\mathbf{R}_n\mathbf{c} + d\mathbf{T} = \mathbf{z}, \quad \mathbf{T}'\mathbf{c} = 0,$$

where $\mathbf{R}_n$ is the $n \times n$ matrix with (i, j)th entry $R(P_i, P_j)$, $\mathbf{T} = (1, ..., 1)'$, and $\mathbf{z} = (z_1, ..., z_n)'$.

Given n pairs of two corresponding points $\{P_i, P_i^{new}\}$, $i = 1, \cdots, n$, on the sphere, the displacements $(\Delta\theta_i, \Delta\phi_i)$ at a set of points $\{P_i, i = 1, \cdots, n\}$ are calculated. With these displacements, equations (11.28), (11.29), and (11.30) are used to compute displacement $(\Delta\theta_k, \Delta\phi_k)$ of any point $P_k \in S^2$ and the point P_k, located at $p(\theta, \phi)$, is moved to a new point P_k^{new} at $p(\theta + \Delta\theta_k, \phi + \Delta\phi_k)$. For convenience, we call this the *STPS-based displacement scheme*.

The basic method is to directly apply the STPS procedure to two sets of landmarks with known correspondence between the sets, defined in a parametric mesh. However, this naive approach can severely distort the parametric mesh of an individual object and result in a distorted reconstruction, especially when one or more landmarks are located near the north or south pole (see Fig. 11.11).

To avoid large distortion introduced by STPS, we adopt a sampling-based strategy that rotates the individual's landmarks on the unit sphere using Euler angles $(\alpha\beta\gamma)$ to find the best-oriented landmarks for applying STPS. The rotation space can be sampled almost uniformly using icosahedral subdivisions. This assigns rotation angles to α and β. Practically, it is common to set $\gamma = 0$, since the rotation along the z-axis does not affect the distortion level of an STPS result.

For each sampling point, the hierarchical method moves it to the north pole, and the entire parametric mesh is rotated accordingly. Then, STPS is applied to each rotated parametric mesh and the best K rotation angles are selected for minimizing the distortion costs, which will be discussed in later sections. This manner is repeated for a higher level of sampling mesh until a certain criterion is satisfied. More details about this strategy are available in [19].

Fig. 11.12 shows the landmark-guided registration results of one pair of cortical models. The cortical surfaces and their spherical parameterization are shown in the top and bottom panels, respectively. The first column shows the template model, while the second and third columns show individual models before and after the landmark-guided registration, respectively. Comparison between the original individual and the template shows that their landmarks are aligned only in the object space but not in the parameter space. Comparison between the registered individual and the corresponding template shows that their landmarks are aligned not only in the object space but also in the parameter space.

11.3.3.3 Landmark-guided open surface registration

The SPHARM methods discussed above are designed for analyzing closed surfaces such as the surface of the brain cortex or hippocampus. In some biomedical applications, the focus of the study could be open surfaces. Below, we discuss one registration method developed for analyzing open surfaces. Specifically, we present a computational framework for aligning human faces using example data from a facial dysmorphology study in fetal alcohol syndrome (FAS) [20]. We are given two groups (healthy controls vs. FAS patients) of 3D facial surfaces represented by triangular meshes. Each surface mesh is assumed to have a disk topology and does not contain any hole. A set of landmarks are available on each mesh to predefine a coarse correspondence between surfaces. Our goal is to align these surfaces together to enable subsequent statistical shape analysis.

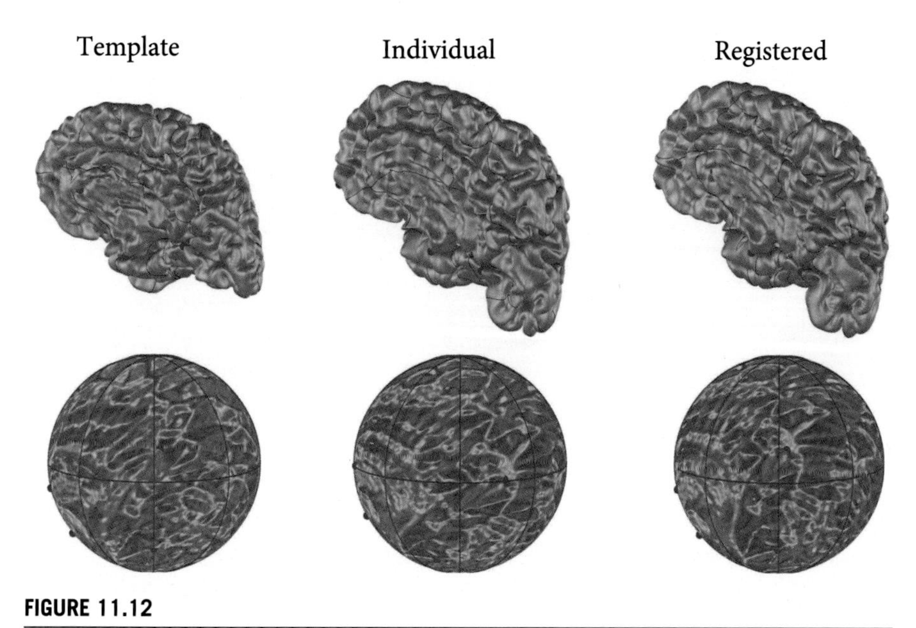

FIGURE 11.12

Landmark-guided registration results between two cortical models. Shown from left to right are the template, the individual before registration, and the individual after registration. A mesh and a bump map are used to illustrate the correspondence between each surface and its parameterization. Blue dots are the landmarks used to guide the registration. Reprinted by permission from Elsevier: Shen et al. [19].

Our computational framework includes two major components: surface alignment in the object space and surface alignment in the parameter space. We first register all the surfaces in the object space. The ICP algorithm discussed in Section 11.2.3 is used to register each surface to a reference surface, which could be pre-selected as a surface from a healthy control. This rigid transformation normalizes the orientation and location of each surface. After that, surface correspondence is established via 2D parameterization, where conformal mapping and 2D TPS are employed to align surfaces in the parameter space. Our goal is to achieve a smoothing mapping between any surface and the 2D parameter domain and assign the same landmark of different surface models with the same location in the parameter space.

We employ Eck's method (see Section 11.3.2.1) to perform conformal mapping from 3D facial meshes to 2D meshes in the unit disk. The goal is to map each vertex of a 3D mesh to the corresponding 2D position in a unit disk with fixed boundary. Suppose an individual is represented as a set of vertices $V = \{(x_i, y_i, z_i) \mid i \in [1, n]\}$. After applying conformal mapping Φ, which is bijective, each individual gets new coordinates $\Phi(V)$ in the 2D disk domain. After that, we pick a healthy individual as the reference and then register all the subjects to the reference using the landmark-based TPS warp Ψ in the 2D domain (see Section 11.2.4). Now the landmarks of each individual are exactly aligned to those of the reference. The remaining parts of

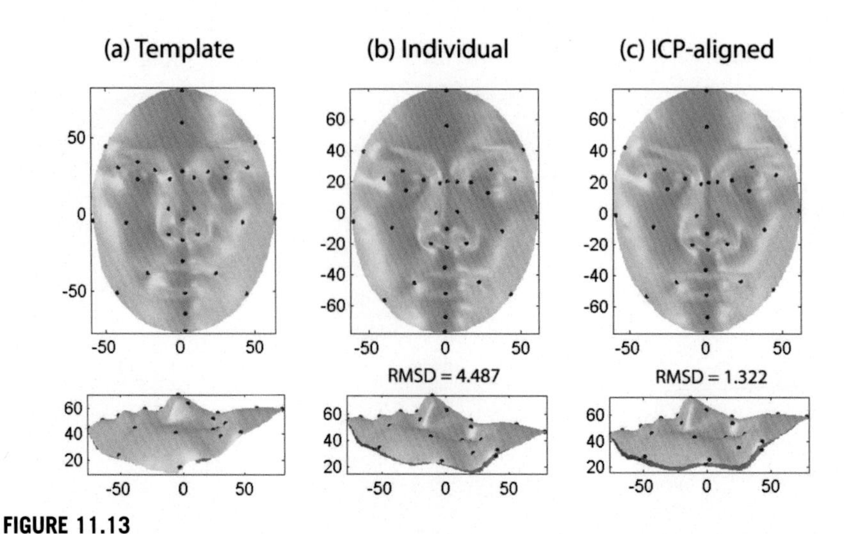

FIGURE 11.13

Registration in the object space by aligning an individual to the reference using ICP: (1) reference, (2) individual, and (3) ICP-aligned individual. Reprinted by permission from Springer Nature: Wan et al. [20].

the individuals are interpolated according to the movement of their landmarks. An individual can then be represented as $\Psi(\Phi(V))$, where Ψ denotes the TPS registration function. Because every single mesh has a different number of vertices and triangles, it is necessary to resample them with a regular mesh grid defined in the disk. After this resampling process, all the individuals in the dataset have the same mesh topology and can be compared with each other.

Fig. 11.13 shows an example registration result in the object space by aligning an individual to the reference using ICP, which normalizes the orientation and location of the initial configuration. The root mean square distance (RMSD) between the individual and the reference before ICP registration is 4.487, and it is reduced to 1.322 after ICP registration.

Fig. 11.14 shows a sample registration procedure in the parameter space by aligning an individual's conformal map (d) to the reference conformal map (c) using TPS. While landmarks are aligned perfectly between the individual and the reference, a smooth mapping from the individual to the reference is obtained for establishing the surface correspondence. To quantify the registration quality, we consider two factors: (1) the area distortion cost (ADC) [21] from the object surface to the parameter domain (i.e., 2D disk) and (2) RMSD between landmarks of the individual and the reference in the parameter domain. Our goal is to achieve RMSD $= 0$ while controlling the ADC. If we just use conformal mapping, we have an ADC of 1.1508 ± 0.0231 (mean $\pm$ std) and an RMSD of 0.6841 ± 0.1124 for all the subjects in our data. If we combine conformal mapping with TPS, we have an ADC of 1.3723 ± 0.0987 and an RMSD of 0 ± 0. In this case, although our ADC gets slightly increased, the fact that RMSD $= 0$ guarantees that all landmarks are perfectly aligned across all subjects.

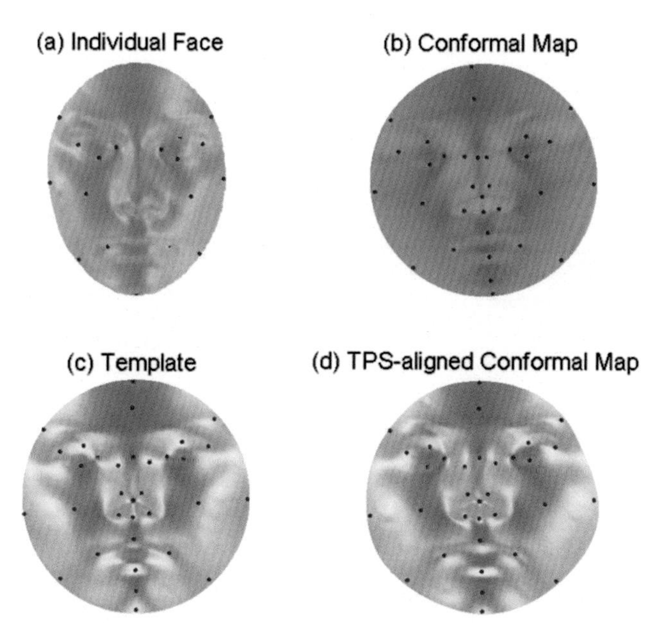

FIGURE 11.14

Registration in the parameter space by aligning an individual's conformal map to the reference conformal map using TPS: (1) individual face, (2) individual conformal map, (3) reference conformal map, and (4) TPS-aligned individual conformal map. Reprinted by permission from Springer Nature: Wan et al. [20].

11.4 Summary

In this chapter, we presented several methods for points and surface registration. We discussed four point registration methods, including: (1) Procrustes analysis for linear registration of multiple corresponding point sets, (2) the quaternion algorithm for rigid registration of two corresponding point sets, (3) the ICP algorithm for rigid points registration in general cases, and (4) TPS for non-rigid alignment of two corresponding point sets. For surface registration, we started our discussion with surface mesh representation. Next we presented surface modeling and parameterization methods for mapping surface meshes onto parameter spaces such as a 2D disk or the surface of a unit sphere. After that, we discussed three 3D surface registration methods, including SPHARM methods for registering 3D closed surfaces, landmark-guided SPHARM surface registration, and landmark-guided open surface registration. Both points and surface registration strategies are valuable tools for many clinical applications, longitudinal health monitoring, as well as statistical shape analyses in biomedical computing.

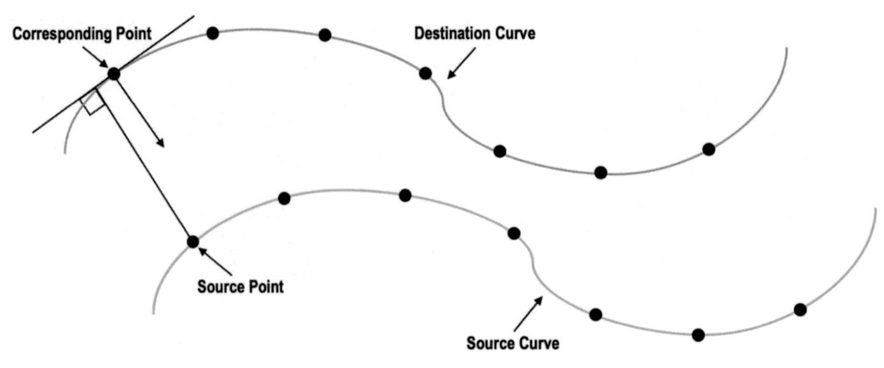

FIGURE 11.15

Applying the ICP algorithm in registering a source curve (green) to a destination curve (blue).

11.5 Exercises

1. Similar to Fig. 11.15, generate two curves and equally sample 10 landmark points on each curve. Then apply the ICP algorithm to register these two curves.
2. As shown in Fig. 11.2, generate two point sets and apply the ICP algorithm in aligning red dots to blue circles.
3. As shown in Fig. 11.3, write a program and apply the TPS algorithm in aligning 2D landmark data.
4. Take a picture of your own face and then apply the conformal mapping algorithm in projecting the face picture onto a circle disk.
5. Given an example hippocampus dataset, apply the spherical parameterization algorithm in projecting the 3D hippocampus onto a unit sphere.
6. Obtain the FOEs with the results from Question 5.
7. Apply the STPS algorithm to the results from Question 5.
8. Reconstruct the hippocampus surfaces with orders of 5, 10, 12, and 15.
9. Apply the landmark-guided open surface registration discussed in Section 11.3.3.3. You can use the data and results from Question 4.

References

[1] Bookstein F.L., Shape and the information in medical images: a decade of the morphometric synthesis, Computer Vision and Image Understanding (ISSN 1077-3142) 66 (2) (1997) 97–118, https://doi.org/10.1006/cviu.1997.0607, http://www.sciencedirect.com/science/article/pii/S107731429790607X.

[2] Klingenberg C.P., Size, shape, and form: concepts of allometry in geometric morphometrics, Development, Genes and Evolution 226 (3) (2016) 113–137.

[3] Besl P.J., McKay N.D., A method for registration of 3-D shapes, IEEE Transactions on Pattern Analysis and Machine Intelligence 14 (2) (1992) 239–256.

[4] Cariow A., Cariowa G., Majorkowska-Mech D., An algorithm for quaternion-based 3D rotation, International Journal of Applied Mathematics and Computer Science 30 (1) (2020).
[5] Horn B.K.P., Closed-form solution of absolute orientation using unit quaternions, Journal of the Optical Society of America A 4 (4) (1987) 629–642, https://doi.org/10.1364/JOSAA.4.000629.
[6] Akyalcin S., Dyer D.J., English J.D., Sar C., Comparison of 3-dimensional dental models from different sources: diagnostic accuracy and surface registration analysis, American Journal of Orthodontics and Dentofacial Orthopedics 144 (6) (2013) 831–837.
[7] Lyu I., Kang H., Woodward N.D., Styner M.A., Landman B.A., Hierarchical spherical deformation for cortical surface registration, Medical Image Analysis 57 (2019) 72–88.
[8] Kim Y., Li R., Na Y.H., Lee R., Xing L., Accuracy of surface registration compared to conventional volumetric registration in patient positioning for head-and-neck radiotherapy: A simulation study using patient data, Medical Physics 41 (12) (2014) 121701.
[9] Choi A., Chae S., Kim T.-H., Jung H., Lee S.-S., Lee K.-Y., Mun J.-H., A novel patient-to-image surface registration technique for ENT- and neuro-navigation systems: proper point set in patient space, Applied Sciences 11 (12) (2021) 5464.
[10] Chung M.K., Bubenik P., Kim P.T., Persistence diagrams of cortical surface data, in: International Conference on Information Processing in Medical Imaging, Springer, 2009, pp. 386–397.
[11] Yueh M.-H., Lin W.-W., Wu C.-T., Yau S.-T., An efficient energy minimization for conformal parameterizations, Journal of Scientific Computing 73 (1) (2017) 203–227.
[12] Ahlfors L.V., Conformal Invariants: Topics in Geometric Function Theory, vol. 371, American Mathematical Soc., 2010.
[13] Krantz S.G., Geometric Function Theory: Explorations in Complex Analysis, Springer, 2006.
[14] Eck M., DeRose T., Duchamp T., Hoppe H., Lounsbery M., Stuetzle W., Multiresolution analysis of arbitrary meshes, in: Proceedings of the 22nd Annual Conference on Computer Graphics and Interactive Techniques, 1995, pp. 173–182.
[15] Tutte W.T., How to draw a graph, Proceedings of the London Mathematical Society 3 (1) (1963) 743–767.
[16] Choi P.T., Lui L.M., Fast disk conformal parameterization of simply-connected open surfaces, Journal of Scientific Computing 65 (3) (2015) 1065–1090.
[17] Shen L., Cong S., Inlow M., Statistical shape analysis for brain structures, in: Statistical Shape and Deformation Analysis, Elsevier, 2017, pp. 351–378.
[18] Shen L., Kim S., Wan J., West J.D., Saykin A.J., Fourier methods for 3D surface modeling and analysis, in: Emerging Topics in Computer Vision and Its Applications, World Scientific, 2012, pp. 175–196.
[19] Shen L., Kim S., Saykin A.J., Fourier method for large scale surface modeling and registration, Computer Graphics (ISSN 0097-8493) 33 (3) (2009) 299–311, https://doi.org/10.1016/j.cag.2009.03.002, https://www.ncbi.nlm.nih.gov/pubmed/20161536.
[20] Wan J., Shen L., Fang S., McLaughlin J., Autti-Rämö I., Fagerlund Å., Riley E., Hoyme H.E., Moore E.S., Foroud T., A framework for 3D analysis of facial morphology in fetal alcohol syndrome, in: MIAR 2010: the 5th Int. Workshop on Medical Imaging and Augmented Reality, in: LNCS, vol. 6326, Springer, 2010, pp. 118–127.
[21] Shen L., Makedon F., Spherical mapping for processing of 3D closed surfaces, Image and Vision Computing 24 (7) (2006) 743–761.

CHAPTER

Graph matching and registration 12

Aristeidis Sotiras[a,f], Mattias Heinrich[b,f], Julia Schnabel[c], and Nikos Paragios[d,e]

[a]*Department of Radiology, and Institute for Informatics, School of Medicine, Washington University in St. Louis, St. Louis, MO, United States*

[b]*Institute of Medical Informatics, Universität zu Lübeck, Lübeck, Germany*

[c]*School of Biomedical Engineering and Imaging Sciences, King's College London, London, United Kingdom*

[d]*Digital Vision Center (CVN), CentraleSupélec, Université Paris-Saclay, Paris, France*

[e]*TheraPanacea, Paris, France*

Learning points

- Probabilistic graphical models
- Markov random fields
- Inference tools for discrete optimization models
- Graph-based modeling of image registration

12.1 Introduction

Image registration is a process of establishing correspondences in space or in time between different sets of imaging data [1–5]. It is a predominant task in medical image analysis, having applications that span population modeling, longitudinal studies, and computer-assisted diagnosis and interventions.

The number of clinical applications involving image registration is phenomenal. Applications of deformable image registration in medical imaging research are also discussed in Chapter 14. Towards keeping the chapter self-contained, and without attempting to present an exhaustive list, let us provide a few indicative examples. Image registration is a core component within the field of brain mapping, where it allows brain researchers to combine data from large numbers of subjects and analyze them to detect subtle group effects or creating atlases that capture the population statistics [6,7]. Importantly, image registration is critical for estimating correspondences in time, allowing the analysis of spatiotemporal acquisitions of deformable moving organs such as a beating heart [8] or breathing lungs [9,10]. Last but not least, image registration is necessary for fusing preoperative and interventional data.

[f] These authors contributed equally.

Medical Image Analysis. https://doi.org/10.1016/B978-0-12-813657-7.00027-3

Specifically, image registration is critical for analyzing possible physiological and anatomical changes during radiation therapy [11], thus ensuring proper dose delivery.

Image registration typically involves three main components: (i) the definition of an objective function based on an image matching criterion, (ii) the adoption of a deformation model (detailed information regarding volumetric affine registration is provided in Chapter 13, while parametric and non-parametric deformable registration is covered in Chapter 14), and (iii) the employment of an optimization strategy to estimate the optimal transformation. The registration objective function is typically comprised of two terms: (i) a regularization term that ensures the solution is well behaved and (ii) a matching criterion that determines how the available imaging information is taken into account in driving the alignment process. The selection of the matching criterion is an important design choice, which is particularly challenging when images from different sources need to be registered. Depending on the choice made, we can distinguish three classes of methods.

The first such class, termed geometric, opts to establish correspondences between geometric primitives, such as points of interest, curves, and surfaces, which have been extracted from the images. The extracted primitives are assumed to be placed in salient image locations, which in turn are assumed to correspond to meaningful anatomical locations. The optimization process seeks for a transformation, which optimally maps corresponding landmarks onto each other [12–14]. This can be simply achieved for example by minimizing the (Euclidean) distance between corresponding point sets. Geometric registration methods enjoy increased robustness to initial conditions and large deformations. However, extracting reliable geometric primitives is not straightforward. Additionally, the extracted primitives typically cover only a small part of the image. As a consequence, correspondence is established only for a sparse set of points between the images, and interpolation strategies such as thin plate splines [15] are required to estimate dense correspondences. For additional information regarding this class of methods, we refer the interested reader to Chapter 11.

The second class of methods is termed iconic. Methods that fall within this category are also referred to as either voxel-based or intensity-based because they quantify the alignment of the images by evaluating an intensity-based criterion over the whole image domain. Different criteria are employed depending on the assumptions about the intensity relationship between the images. Typically, the sum of squared differences (SSD) or the sum of absolute differences is used in intra-modal scenarios, where similar intensity values are assumed to correspond to the same anatomical structures [2]. In the case of inter-modal registration, statistical criteria such as the correlation ratio [16], mutual information [17,18], or Kullback–Leibler divergence [19] are commonly used to determine the similarity between images acquired with different sensors. Intensity-based methods have the potential to better quantify and represent the accuracy of the estimated dense deformation field compared to the geometric approaches. However, this comes at a higher computational cost as all voxels are taken into account when evaluating the matching criterion instead of a few salient points. Unfortunately, since all voxels are usually treated equally, the important information that the salient points carry is not fully exploited for driving the registration.

Additionally, iconic methods tend to be more sensitive to the initial conditions and are often unable to deal with large deformations. For additional information regarding this class of methods, we refer the interested reader to Chapter 13.

The third class of methods, termed hybrid, aims to bridge the two main classes of approaches by taking advantage of both complementing types of information. This is typically done in a decoupled way, where either geometric information is used to provide a coarse registration that is subsequently refined by using the intensity information or, more often, geometric correspondences are used as constraints to guide the estimation of the dense deformation field.

Having defined the objective function, an optimization strategy is used to infer the optimal transformation. Common choices include continuous optimization performed through variants of gradient descent [20,21] (see also Chapter 14 and Chapter 13) or Markov random field (MRF) modeling and discrete optimization techniques [10,22,23]. Continuous optimization methods are constrained to problems where the variables take real values and the objective function is differentiable. The optimal solution is typically estimated by performing a local search in the parameter space. As a consequence, the result depends often greatly on the initialization. Discrete optimization techniques solve problems where the variables take values from a discrete set of solutions. The optimal solution is typically estimated by performing a global search in the solution space. As a result, discrete methods are less sensitive to the initial conditions. However, they are constrained by limited precision due to the quantization of the solution space.

In this chapter, we will detail how MRF theory can be used in conjunction with discrete optimization techniques to model and solve image registration. The cases of geometric, iconic, and hybrid registration will be examined separately. We will present an example application on motion compensation in lung respiration. We will conclude the chapter with a critical presentation of limitations of discrete models.

12.2 Graph-based image registration

Problem description

In this chapter, we focus on pairwise registration, where the goal is to align two n-dimensional intensity images $I, J : \Omega \subset \mathbb{R}^n \mapsto \mathbb{R}$, where J denotes the source or moving image and I denotes the target or fixed image. The two images are defined in the image domain Ω and are related by a transformation T. The transformation is a mapping function of the domain Ω to itself, which maps point locations to other locations, and is defined as the addition of an identity transformation with the displacement field $\boldsymbol{u}$ at every location $\boldsymbol{x} \in \Omega$, or $T(\boldsymbol{x}) = \boldsymbol{x} + \boldsymbol{u}(\boldsymbol{x})$.

The goal of registration is to estimate the optimal transformation by solving a problem of the form

$$\hat{T} = \arg\min_{T} \mu(I, J \circ T) + \rho(T), \tag{12.1}$$

where μ denotes the matching term, which quantifies the level of alignment between a target image I and a source image J, and ρ denotes the regularization term, which aims to instill user-specified, desirable properties in the solution and account for the ill-posedness of the problem.

We should note that, in practice, most registration method implementations estimate the transformation that maps points in I to corresponding points in J. This choice eases the warping of the source image. Suppose that we have estimated the transformation that maps points in J to corresponding points in I. In order to warp the image J, one would need to iterate over each pixel (or voxel) of the input image J, compute new coordinates for it, and then copy its value to the new location. However, this procedure typically generates unevenly spaced points. As a consequence, one needs to solve a challenging interpolation problem to generate the required regularly spaced pixels (or voxels). Suppose now that we have estimated the transformation that maps points in I to corresponding points in J. In this case, in order to warp the image J, one would iterate over each pixel (or voxel) of the output image, estimate their corresponding position in the input image J, and sample the value that needs to be copied. The key difference between the two approaches is that we now iterate over a grid of regularly spaced points and we only need to interpolate the source image J to estimate the values to copy, which is straightforward.

Markov random fields

Image registration along with a number of tasks in medical image analysis can be formulated as a discrete labeling problem. In this setting, one seeks to optimize an objective function related to the quality of the labeling. Such problems are commonly formulated using MRF theory, lending themselves to a representation by a graph $G = (V, E)$.

The graph G consists of a set of vertices, or nodes, V that encode the latent random variables and a set of edges E that encode the interactions between the variables. The graph G is associated with an energy that consists of unary potentials $\mathbf{g} = \{g(\cdot)\}$ and pairwise potentials $\mathbf{f} = \{f(\cdot, \cdot)\}$:

$$\text{MRF}(\mathbf{g}, \mathbf{f}) \equiv \min_{\mathbf{l}} \sum_{p \in V} g_p(l_p) + \sum_{(p,q) \in E} f_{pq}(l_p, l_q), \tag{12.2}$$

where each random variable p models the displacement of a discrete deformation element and takes values in a discrete label set L. The unary potentials $g_p(l_p)$ are typically used to encode some sort of a data likelihood that measures the cost of assigning a value indexed by label l_p to the variable p (data term). The pairwise potentials $f_{pq}(l_p, l_q)$ typically act as regularizers that determine the cost of assigning different values indexed by labels l_p and l_q to the variables p and q, respectively.

Pairwise potentials frequently take the following form:

$$f_{pq}(l_p, l_q) = \alpha_{pq}\, d(l_p, l_q), \tag{12.3}$$

where α_{pq} is a per-edge weight representing the strength of relationship between adjacent vertices p and q, while $d(l_p, l_q)$ is a distance function measuring the dissimilarity between labels. The intuition behind equation (12.3) is that closely related variables encoded by vertices p and q should take similar values.

The construction of the graph (i.e., the definition of the nodes as well as the construction of the edge system), the choice of the unary and pairwise potentials, and the construction of the label set are all important design choices one needs to make when modeling image registration as a discrete MRF. In the continuation of this chapter, we will study a number of available options for designing registration frameworks.

12.2.1 Graphical model construction

Defining vertices

The first step in constructing the graph that models the registration process is to define the nodes of the graph. The nodes of the graph encode the random variables whose value our goal is to infer. Their definition is closely related to the parameterization of the deformation field and the nature (sparse or dense) of the correspondences we opt to estimate.

Let us consider first the case of geometric registration (see also Chapter 11). In this case, points that delineate reliable features are identified manually or automatically in both images. These points are typically referred to as landmarks, and there might be a different number of landmarks identified in each image. The goal is to estimate the displacement that will bring each landmark in one image into correspondence with the best candidate landmark in the other image. Accordingly, the nodes of the graph correspond to the landmarks of the one image, giving rise to an irregular spatial layout (see Fig. 12.1(a)), and the associated random variables model their displacement [24]. This approach allows for estimating displacements for a subset of image locations. As a consequence, interpolation is required to estimate displacements over the entire image.

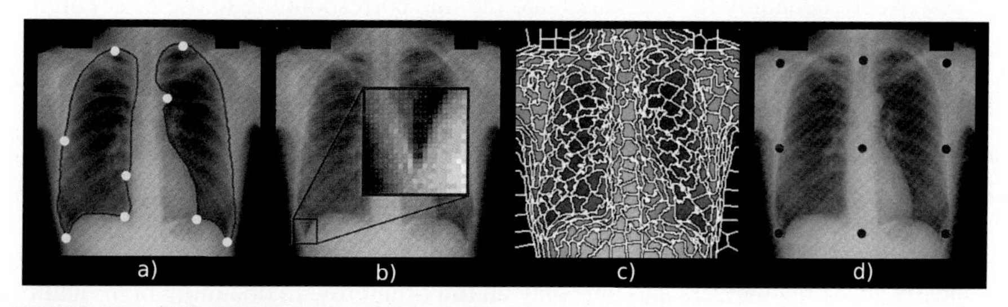

FIGURE 12.1 Defining the nodes of the graph.

Examples of different node systems. (a) Nodes (yellow circles) correspond to landmarks identified in the lung boundary (red contour). (b) Nodes (yellow dots) correspond to each pixel of the image – a close-up view is also provided for better visualization (note that additional zooming might be required). (c) Nodes correspond to supervoxels – an example partitioning of the image to superpixels is shown. (d) Nodes correspond to the control points (red circles) of the mesh that is deforming the image.

Estimating directly dense correspondences based on the intensity information (iconic registration) is often desirable. In this case, the goal is to estimate the optimal displacement for every image element (see also non-parametric methods in Chapter 14). Accordingly, the nodes of the graph correspond to image pixels or voxels (see Fig. 12.1(b)), giving rise to a regular grid [25]. However, the direct voxelwise estimation of displacements comes with the cost of high computational effort due to the high dimensionality of medical imaging data.

To alleviate this challenge, an option is to derive a coarser representation of the image domain by performing supervoxel clustering. Supervoxels remove the largely redundant intensity information of voxels within homogeneous areas, which are likely to belong to the same object, thus providing a more compact image representation with little loss of detail. Importantly, they provide sufficient spatial context and are robust to noise in low-contrast areas. In such a setting, the nodes of the graph correspond to supervoxels (see Fig. 12.1(c)), resulting in an irregular spatial layout [26].

Another possibility to address the challenges that arise due to the high dimensionality of medical imaging data is by making use of free-form deformation (FFD) models, which are typically based on B-splines (see also parametric methods in Chapter 13). The basic idea of FFDs is to deform the image by manipulating an underlying mesh of control points, whose number is much smaller than the dimensionality of the image. In this case, the control points of the mesh constitute the discrete deformation elements whose displacement our goal is to infer. Thus, the nodes of the graph correspond to control points (see Fig. 12.1(d)), resulting in a layout that reflects the form of the mesh [10,22,23].

Lastly, more elaborate node systems may arise by combining the ones previously described to increase our modeling capabilities and derive more sophisticated registration frameworks. Such a case is for example hybrid registration, where one seeks to infer simultaneously landmark correspondences and a dense displacement field parameterized by FFDs. Accordingly, the node system is defined as the union of the set of nodes corresponding to landmarks and the one corresponding to the FFD control points [23].

Edge system definition

Another important decision in constructing the graph that models the registration process is to define the edges that connect the nodes of the graph. The edge system of the graph encodes the relationship between the random variables represented by the nodes of the graph. This relationship is expressed through pairwise potentials, which typically act as regularizers and thus play an important role in obtaining high-quality results.

The assignment of certain labels to nodes could be performed individually and independently for every node based on the corresponding unary potential. This means that even though two pixels are located in very close proximity, they could be assigned completely different displacement vectors. Only when we start to consider pairwise edges that connect two nodes (or even three, or more, using higher-order cliques) within a graph, the MRF model is able to provide spatial regularization of motion and deformation estimates. The existence and the strength of an edge, together with a pairwise potential, determine the probability that a certain joint assignment of two labels to two nodes will be chosen. We can draw a close relation of these pairwise potentials in MRFs to the discretized graph Laplacian used in classical, continuously defined energy optimization on image (grid) graphs. It is well known (see [27] for further reading) that common diffusion regularizers, which enable us to estimate smoothly varying solutions for different kinds of inverse problems for images, can be approximated for discrete images using finite difference operators. This means the interactions of connected nodes within the image graph are modeled by pairwise terms that usually constrain neighboring locations to observe similar motions, and thus result in a spatially smooth displacement field. This concept is of course not unique to image registration and finds use in many processing tasks, including denoising and segmentation.

Depending on the chosen node model, either regular grids or other forms of nearest neighbor graphs can be used to represent the graph. In the case of geometric registration, the use of landmarks (or geometric key points) to define the discrete deformation elements will typically result in an irregular spatial layout of nodes that cannot be connected using a grid. This is also the case when supervoxels are used to derive a more compact representation of the sought deformation. In the aforementioned cases, a k-nearest neighbor (k-NN) graph is typically built, which connects each node to the k closest remaining nodes in a Euclidean space (see Fig. 12.2(a)). An alternative approach is to define a certain threshold distance and connect all nodes whose distance from each other is less than the threshold. However, this may lead to disjoint partial graphs. This is undesirable because it may result in largely diverging solutions between the so-formed separate domains.

In the case of iconic registration, the use of image voxels to define the discrete deformation elements will typically result in a grid graph that connects each node to four spatial neighbors in 2D images (see Fig. 12.2(b)) or six neighbors in volumetric 3D scans. Similarly, a grid graph arises when FFD models are used to provide a low-dimensional representation of the deformation (see Fig. 12.2(c)). We should note that the regular structure has certain advantages for some optimization algorithms discussed in Section 12.2.2. For example, the dependencies of edges in one dimension can be computed in parallel for all nodes orthogonal to the edge direction.

In the case of hybrid registration, geometric and iconic information is used to drive the registration. Accordingly, both landmarks and FFD control points are used to define the discrete deformation elements, which results in a complex graph structure. As previously, nodes corresponding to landmarks are connected through edges based on their distance from each other, forming a k-NN graph. Additionally, nodes that correspond to FFD control points are connected through edges, giving rise to a grid graph. The edges that are part of the k-NN graph promote the smoothness of the landmark displacements. Similarly, the edges that are part of the grid graph ensure the smoothness of the inferred displacements of the FFD control points. However, if there are no edges connecting the nodes that correspond to landmarks with the nodes that correspond to the FFD control points, the estimation of landmark corre-

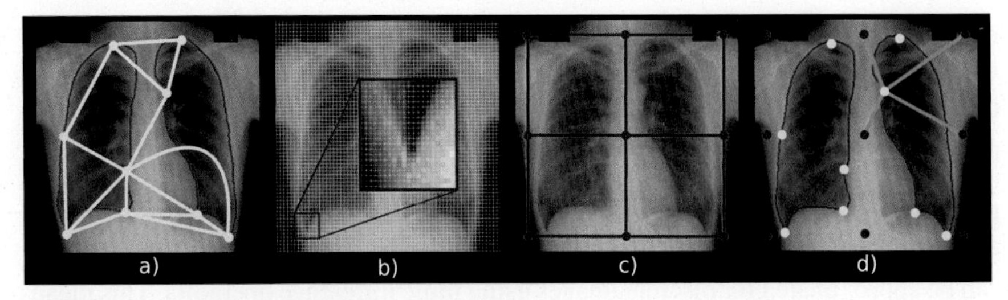

FIGURE 12.2 Defining the edges of the graph.

Examples of different edge systems. (a) Edges connecting nodes (yellow circles) that correspond to landmarks identified in the lung boundary (red contour) – each node is connected to the three nodes closest to it. (b) Example of a grid graph that is constructed by connecting each node (yellow dot) that corresponds to a pixel of the image to its four spatial neighbors – a close-up view is also provided for better visualization (note that additional zooming might be required). (c) Regular grid arising by connecting nodes that correspond to control points of spline-based FFD models (red circles). (d) Edges (green lines) connecting nodes that correspond to landmarks (yellow circles) with nodes that correspond to control points of spline-based FFD models (red circles) – not all possible edges are shown to facilitate visualization.

spondences is not informed by the displacements of the FFD control points that are driven by the iconic information. In other words, the two graphs are disjoint and the two registration problems (i.e., geometric and iconic) are decoupled. To allow information to flow between the two graphs, one needs to connect the two graphs. This can be achieved by connecting with edges the nodes that correspond to landmarks with the k closest in Euclidean distance nodes that correspond to the FFD control points. This results in a complex edge system that is comprised of edges connecting nodes that correspond to landmarks, edges connecting nodes that correspond to FFD control points, and coupling edges that connect the two different types of nodes (see Fig. 12.2(d)).

Lastly, we would like point the interested reader to the existence of more elaborate edge models that are no longer restricted to few localized connections, but consider instead non-local connections between every pair of distinct vertices [28]. These models are in particular beneficial for discrete optimization problems with strong discontinuities, which are either found in segmentation problems or 3D scene analysis with 2D camera images (stereo depth estimation or structure-from-motion). However, the large number of edges makes the MRF energy challenging to optimize and researchers have reverted to approximate conditional random field solutions [28,29], which are beyond the scope of this chapter.

Label definition

The definition of the label set is one of the most important aspects in the discrete formulation of image registration. In the discrete labeling setting, one can explicitly control the definition of the solution space, which allows for important flexibility

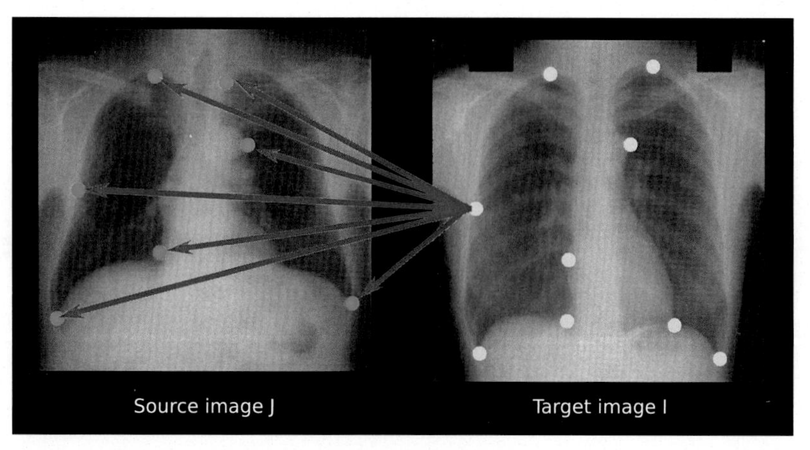

FIGURE 12.3 Label set definition in geometric registration.

Landmarks identified in the target image are shown as yellow circles. Each landmark in the target image may be matched to candidate points identified in the source image (green circles).

in model design. One can directly impact the quality of the obtained solution by appropriately constructing the label set. However, the finite nature of the solution space presents a competitive interplay between solution quality and computational efficiency. A higher number of labels typically leads to more accurate solutions, but at an increased computational cost. Therefore, the label definition should be tailored to each application in order to strike the best balance between solution quality and speed.

Depending on the application, different label sets may be defined. As previously discussed, in the case of geometric registration, one aims to establish correspondences between a set of landmarks and a set of candidate points. The discrete nature of the problem allows for a straightforward label definition. A label may simply index a candidate point that can be matched to a landmark (see Fig. 12.3). In this case, a large set of candidate points gives rise to a large label set, which may significantly increase computational complexity. However, it is important to note that not all nodes need to take values from the same label set. Keeping this in mind, we may define node-specific label sets. Specifically, we can prune the label set for each node (i.e., landmark) to contain only the k closest candidate points to the landmark. As a consequence, the label set for each node contains the same number of labels, but these index different displacements that allow the landmark to match to different candidate points. Thus, it is possible to keep the computations tractable without sacrificing any accuracy. The underlying assumption of the above label sets is that there exists a correspondence between landmarks and candidate points. However, this may not always be the case, and one may need to take into account the existence of outliers. This can be modeled in a relatively straightforward manner by introducing dummy labels encoding the non-correspondence case [30].

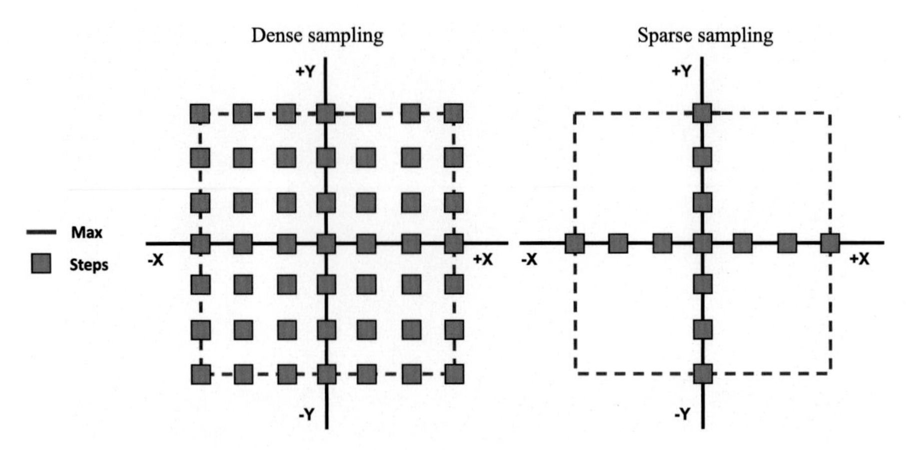

FIGURE 12.4 Label set definition in iconic registration.

Comparison of dense and sparse label sampling in the 2D space.

Contrary to geometric registration, which is inherently of a discrete nature, iconic registration is originally a highly continuous problem. Discretizing the solution space leads to an important trade-off between registration accuracy and computational efficiency. Specifically, as a consequence of the finite nature of the search space, the registration accuracy that can be obtained through labeling is bounded by the range of deformations covered in the set of labels. However, a naive label set definition that aims to approximate the whole range of possible displacements by sampling densely the solution space might lead to an intractable problem. Even when bounding the maximum possible displacement, a dense sampling results in $(2s + 1)^n$ labels (including the zero-displacement vector), where s is the sampling rate and n is the number of dimensions (see Fig. 12.4). Thus, it is important to strike a good balance between computational efficiency and registration accuracy. This can be achieved by combining iterative labeling with a search space refinement strategy.

In iterative labeling, each label corresponds to an update on the transformation and the final registration solution is generated by combining multiple intermediate solutions, which were obtained in previous iterations. Specifically, in each iteration, the optimal MRF labeling with respect to a certain label space is computed. This computation yields an update on transformation, which is applied to the current estimate. In the subsequent iteration, the registration continues on the basis of the updated transformation. As a consequence, incorrect matches can be corrected in subsequent iterations. Importantly, the overall domain of possible transformations is bounded by the number of iterations rather than the set of finite labels. Overall, the iterative labeling allows us to keep the label set small, which is important for efficient optimization. Toward achieving subpixel registration precision, a refinement on the search space is required. In this case, we start with a set of labels that correspond to large deformations. However, during the iterative labeling procedure, we gradually refine the set to include labels that correspond to smaller deformations. In practice, we typically adopt sparse sampling, which selects a fixed number s of displacements along the main axes up to a maximum displacement magnitude, resulting in $4s + 1$ labels in 2D and $6s + 1$ labels in 3D. If the energy decreases sufficiently, we keep the current set of labels for the next iteration. If the energy does not decrease, the label space is rescaled by a factor f, where $0 < f < 1$. The next iteration is then performed on this refined search space. Taken together, the iterative labeling combined with the search space refinement strategy strikes a good compromise by allowing for a small label set, and thus for efficient optimization, while achieving subpixel precision on the solution space by starting with a set of larger displacements and gradually decreasing the maximum allowed displacement to recover finer details.

In the case of hybrid registration, the node system consists of two different types of nodes. There is one set of nodes that corresponds to landmarks and one that corresponds to the FFD control points. The first set of nodes models the matching of the landmarks to respective candidate points, whereas the second set of nodes models the displacement updates of the FFD control points. As discussed in the previous paragraphs, the definition of the label set is different for each task. However, this does not pose a significant modeling challenge. The MRF framework is flexible enough to allow us to associate different label sets to distinct sets of nodes [23] and thus model hybrid registration effectively. In fact, the modeling flexibility is such that the label set need not be the same for all nodes. At the extreme, each node may be associated to its own set of labels, which index a distinct set of displacements. The only requirement that needs to be satisfied when employing different label sets for different nodes is to be able to define pairwise potentials that can account for this fact.

The explicit control over the quantization of the solution space along with the ability to adapt the label set definition per node results in increased modeling flexibility, which enables the creation of versatile registration tools. Describing such registration frameworks in detail is beyond the scope of this chapter. However, we would like to point the interested reader to some examples that illustrate how one may endow image registration with desired properties by appropriately defining the label set. For example, one may achieve diffeomorphic transformations (see Chapter 14 for the definition of diffeomorphisms) in the case of iconic FFD image registration by ensuring that the maximum displacement indexed by any label is set to 0.4 times the grid spacing [22,23]. Additionally, one may extend the definition of labels to represent more complex transformations. For example, one may define labels to represent not one but two displacements of equal magnitude from a common domain toward both image domains, thus achieving registration that is symmetric with respect to image inputs [31]. Similarly, labels may represent combinations of translations and rotations [32] or combinations of displacement and local intensity variations [10].

Markov random field energy

Having described the construction of the graph (i.e., the definition of vertices, edges, and labels), what remains to be detailed is the definition of the energy terms in equation (12.2). The MRF energy typically consists of the unary and the pairwise potentials, encoding the matching or data term and the regularization term of the registration energy (equation (12.1)), respectively.

Data term. The unary potentials are commonly used to construct the matching term of equation (12.1). In other words, the unary potentials quantify the level of the alignment between two images as this is achieved by assigning to a node the value indexed by a label. Depending on the application, the unary potentials may take different forms. Note that in deriving the equations below, we assume that the goal is to estimate the transformation that maps points in the target image I to corresponding points in the source image J, and thus simplifies the warping of the source image.

In the case of geometric registration, the unary potentials may be defined as

$$g_p(l_p) = \int_{\Omega_p} \varrho(J \circ T_{l_p}(\boldsymbol{x}), I(\boldsymbol{x}))d\boldsymbol{x}, \tag{12.4}$$

where Ω_p denotes a patch around the landmark point in the target image I that corresponds to the node p and T_{l_p} denotes the transformation that is implicitly defined by mapping the landmark in the target image J to the candidate point in the source image J that is indexed by label l_p. The unary potentials determine the likelihood of two points being matched by evaluating the dissimilarity function ϱ over patches centered around them (i.e., by integrating the dissimilarity function ϱ for $\boldsymbol{x} \in \Omega_p$). Without loss of generality, the dissimilarity function may be evaluated over a rich set of features extracted from the two points (e.g., [33]). When modeling occlusions, dummy labels may be associated to a fixed cost, which in practice determines the threshold beyond which we believe two points are too dissimilar to correspond to each other.

In the case of iconic registration, the definition of the unary potentials depends on the node system construction. In the case that the vertices model voxel displacements, the unary potentials typically encode pointwise measures:

$$g_p(l_p) = \vartheta(J \circ T_{l_p}(\boldsymbol{x}), I(\boldsymbol{x})), \tag{12.5}$$

where T_{l_p} denotes the transformation that maps the point $\boldsymbol{x}$ of the target image I that corresponds to node p to a point in the source image J according to the value indexed by the label l_p and ϑ is a measure evaluating pointwise dissimilarity (e.g., squared intensity difference). In the case that the nodes correspond to either supervoxels or control points in spline-based deformation models, the unary potentials may encode region-based measures, which allows for increased modeling options:

$$g_p(l_p) = \int_{\Omega_p} \hat{\omega}_p(\boldsymbol{x})\varphi(J \circ T_{l_p}(\boldsymbol{x}), I(\boldsymbol{x}))d\boldsymbol{x}, \tag{12.6}$$

where T_{l_p} denotes the transformation where a discrete transformation element encoded by the node p has been displaced according to the label l_p. Similarly to equation (12.4), Ω_p denotes a region. Depending on whether the node p corresponds to a supervoxel or a control point, this region can be either defined by the supervoxel itself or as a patch centered around the control point. Accordingly, φ denotes a region-based measure that can be defined either by integrating a pointwise measure over the region (i.e., for $\boldsymbol{x} \in \Omega_p$) or by evaluating locally a statistical measure. Different pointwise or statistical measures can be easily adopted. This is a direct result of the discrete nature of the framework that does not require the differentiation of any similarity measure. Lastly, additional flexibility can be achieved by adaptively weighting the contributions of image points onto the local matching term through the weighting function $\hat{\omega}_p$. For example, in the case of FFD models, this function may assign higher weights to image points closer to the control point.

We should note that as more complex spline-based deformation models are employed, the definition of the matching criterion with the use of the unary potentials becomes only an approximation to the real matching energy. This is because we assume conditional independence between the random variables when evaluating the unary potentials. This is in contrast to the fact that the image deformation and thus also the local similarity measure depend on multiple control points. For example, let us examine the case of 2D cubic B-splines FFD, where the transformation is controlled by a $c_x \times c_y$ mesh of control points with uniform spacing δ. The transformation can be written as the tensor product of 1D cubic B-splines:

$$T(\boldsymbol{x}) = \boldsymbol{x} + \sum_{a=0}^{3}\sum_{b=0}^{3} B_a(v_x)B_b(v_y)\boldsymbol{\theta}_{i+a,j+b}, \tag{12.7}$$

where $i = \lfloor x/\delta \rfloor - 1$, $j = \lfloor y/\delta \rfloor - 1$, $v_x = x/\delta - \lfloor x/\delta \rfloor$, $v_y = y/\delta - \lfloor y/\delta \rfloor$, and B_a represents the ath basis function of the B-spline. It is clear from the above equation that the transformation at a location $\boldsymbol{x}$ depends on the coefficients $\boldsymbol{\theta}$ of 4×4 control points. However, when evaluating the unary potentials, only one control point is taken into account. Nonetheless, the above approximation yields very accurate registration, while allowing for computational efficiency.

Regularization term. Potentials encoding interactions between the variables are typically used to construct the regularization term of equation (12.1). This is frequently achieved by modeling interactions between pairs of random variables through pairwise potentials. The definition of pairwise potentials implements a smoothness constraint that penalizes dissimilarities between pairs of labels (see equation (12.3)). Without such a regularization effect of the graph edges, our output displacement field would be highly noisy and unrealistic as visualized in Fig. 12.5.

This may take the form of an isometric constraint in the case of geometric registration:

$$f_{pq}(l_p, l_q) = \alpha_{pq}\|(T_{l_p}(\boldsymbol{x}_p) - T_{l_q}(\boldsymbol{x}_q)) - (\boldsymbol{x}_p - \boldsymbol{x}_q)\|^2, \tag{12.8}$$

where $\boldsymbol{x}_p$ and $\boldsymbol{x}_q$ denote the position of the landmarks modeled by nodes p and q, respectively. As previously, $T_{l_p}(\boldsymbol{x}_p)$ and $T_{l_q}(\boldsymbol{x}_q)$ map the landmarks to candidate points according to labels l_p and l_q. The assumption behind this potential is that the distance between the landmark points in one image will not be very different from the distance between their homologous points in the other image. Moreover, by considering the vector differences, the flipping of the point positions is penalized.

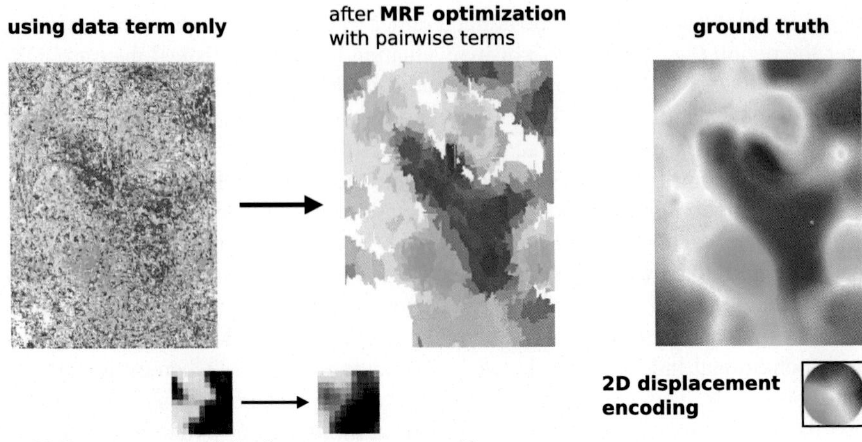

FIGURE 12.5 MRF regularization.

Visual overview of MRF-optimized discrete registration. Estimated displacement fields are shown with and without regularization.

Similarly, in the case of iconic registration, we expect the deformation field to vary smoothly with spatial location. As a consequence, for each edge in the graph that connects nodes corresponding to spatially close deformation elements, a penalty is incurred for differences of pairwise displacements:

$$f_{pq}(l_p, l_q) = \alpha_{pq} \frac{\|\boldsymbol{u}^{l_p} - \boldsymbol{u}^{l_q}\|^2}{\|\boldsymbol{x}_p - \boldsymbol{x}_q\|^2}. \tag{12.9}$$

The above function takes into account the absolute difference between the displacements $\boldsymbol{u}^{l_p}$ and $\boldsymbol{u}^{l_q}$ applied to neighboring nodes p and q through the assignment of the labels l_p and l_q, respectively. The difference is normalized by the spatial distance between positions $\boldsymbol{x}_p$ and $\boldsymbol{x}_p$ of the deformation elements that are represented by the nodes p and q, respectively. Essentially, this penalizes an approximation of the magnitude of the first-order derivatives of the transformation. Penalizing the magnitude of the second-order derivatives of the transformation is also possible through higher-order potentials [34].

12.2.2 Optimization

We previously discussed how one can define image registration as an MRF-based problem and familiarized ourselves with different choices of data terms (unary potentials) and regularization terms (pairwise potentials). In this section, we will focus on how one can solve the optimization problem. There are multiple different and sometimes complex approaches to solve the MRF energy (equation (12.2)). The

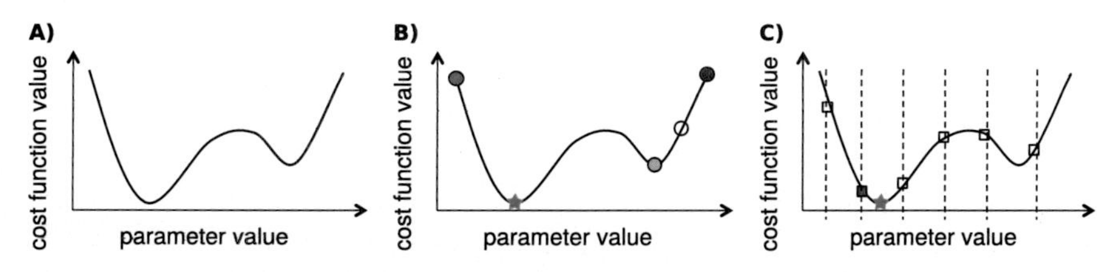

FIGURE 12.6 Continuous vs. discrete optimization.

Comparison of continuous and discrete optimization using 1D parameter space as a toy example.

appropriate choice of MRF solver strongly depends on the employed model. For simplicity, we will focus on so-called belief propagation strategies in this chapter and refer the reader to [29] for a survey on optimization techniques for MRFs and to Chapter 10 for more details on graph cut optimization techniques.

Discrete vs. continuous (intuition – high-level motivation)

As previously discussed, MRF-based image registration strategies will optimize over a discretized distribution of displacement probabilities for each node rather than directly searching for a continuous scalar estimate. To illustrate the difference between discrete and continuous optimization techniques, let us consider a 1D parameter x and its potential cost function value $g_p(\mathrm{x})$ (Fig. 12.6(a)). A continuous optimization starts from a certain initialization x_0 (Fig. 12.6(b), red circle) and calculates the local gradient of the cost function $\frac{dg_p(\mathrm{x})}{d\mathrm{x}}|_{\mathrm{x}_0}$. As discussed in other chapters (e.g., Chapter 13), a gradient descent algorithm will then be used to determine a suitable step size and update the parameter estimate in the (descending) direction of the local gradient. A stopping criterion is applied once the local gradient (nearly) vanishes. In this example, this corresponds to the yellow circle. When considering a more global view on the cost function, we quickly see that this solution is only locally optimal and a better (smaller) value (Fig. 12.6(b), green star) could have been achieved with a different initialization (e.g., green circle). This means that our optimization has become stuck in a bad optimum. Discrete optimization aims to resolve these problems by considering multiple discretized parameter values as potential solutions (Fig. 12.6(c), squares). If we could simply evaluate all these quantized values, then it would be possible to choose the optimal solution (Fig. 12.6(c), blue square) without an iterative optimization and regardless of initialization. The optimization becomes an assignment problem. Given a fixed number of quantized parameter values, we aim to find the most likely one, which should in turn minimize the overall cost function. Note that the quality of the solution depends on the quantization of the solution space. For example, while the optimal solution achieved by the discrete optimization (Fig. 12.6(c), blue square) is better than the local minimum that was achieved by the continuous optimization (Fig. 12.6(b), yellow circle); it does not correspond to the global minimum of the cost function (Fig. 12.6(c), green star).

Broadly speaking, there are two important difficulties to overcome. Firstly, when considering a common registration problem, we search for thousands of node parameters (control points) with thousands of quantized values (displacement vectors). Solving this problem naively leads to an exponentially increasing computation time due to the interdependencies of nodes introduced through our edge system. However, these interdependencies are necessary to regularize the estimated displacement field (Fig. 12.5). In the next subsection, we will discuss solvers that facilitate optimization while achieving near-optimal results.

Secondly, the quantization of the parameter space inherently reduces the accuracy of our solution. In practice, we have to find a compromise between the size of the parameter space, which is usually cubic in the number of discretized displacement steps for volumetric problems, and the desirable (sub)voxel accuracy (depending on the application). Potential solutions include the combination of discrete and continuous optimization steps [35] and the exploitation of the probabilistic nature of the discrete optimization outcome.

In the next sections, we will discuss two commonly used MRF solvers that are relatively easy to follow and implement. Despite their simplicity (there are many more complex alternatives available [29]), the results for medical image registration are good, often outperforming continuous optimization strategies by a margin (see also Section 12.2.3).

Belief propagation

Belief propagation is an established algorithm to solve MRF problems that can be formulated on sequences or graphs without loops. The ideas date back to the Viterbi algorithm developed in the 1960s to calculate error codes for decoding transmitted signals. The main idea behind this approach is to consider the solution to our optimization as a recursive problem, which is possible as long as the sequence has a fixed end and beginning. Two advantages of belief propagation are: (1) its solution is linear with respect to the number of nodes and labels, and (2) the global optimum is found by computing an optimization for each node or edge at most twice. Once the optimal costs have been computed for all nodes, the minimal solution can be selected. This solution corresponds to the optimal displacement field with respect to the employed similarity (data) term and regularization penalty.

For simplicity, let us assume that the parameter space is equally defined for each node and has a certain size λ. This is for example the case when each node can be moved by the same set of possible λ displacements. Let us also consider a simple chain as graphical model that consists of κ nodes and each node is connected to the directly preceding and following node, respectively. It can be assumed that the unary data term g has been precomputed and is stored in an array of size $\lambda \times \kappa$. The total MRF cost depends on the chosen unary term in each node plus the regularization (pairwise term) f that occurs when selecting a somewhat irregular sequence of displacement labels. An example for this would be assigning a leftward displacement to the first node and a rightward motion to the second node. The regularization cost that penalizes a solution where two connected nodes are assigned to different displacements is represented by a $\lambda \times \lambda$ square matrix and can, for example, be formed by a diffusion regularization term. In this case, the squared difference in displacements along each dimension is assigned as a penalty, yielding a smooth transformation.

The recursive solution is obtained by calculating message vectors $m^t_{p\to q}$ between node p and node q at iteration t. Intuitively speaking, these messages pass on information about the likelihood of all potential solutions from one child node p to its respective parent node q. Each message may be computed as follows:

$$m^t_{p\to q}(l_q) = \min_{l_p}\left(g_p(l_p) + \left(\sum_{r\in N(p)\backslash q} m^{t-1}_{r\to p}(l_p)\right) + f_{pq}(l_p, l_q)\right). \tag{12.10}$$

Here, we describe a certain label assignment for a node p as l_p and the neighborhood of a node p is denoted as $N(p)$ and is comprised of the set of nodes that are connected to p with an edge. Belief propagation on directed chains, or trees, is in general performed sequentially, starting from any node without a predecessor and continuing from there in a hierarchical order. After reaching the root node, the so-called forward path ends and the selection of the optimal displacement label is possible. From there, a selection for all nodes in the graph can be obtained by traversing it backwards. In this case, the algorithm converges after just two passes/iterations. An algorithmic description is presented later in this chapter in Algorithm 12.1. Note that the simplified graph model, which is necessary to represent graphs without loops, may yield undesired irregularities between nodes that are spatially close but not connected by an edge.

Loopy belief propagation

Equation (12.10) only holds for graphs without loops. Solving discrete labeling problems for arbitrary graphs exactly is in general not possible. When working with graphs that contain loops, one may first simplify the graphical model to remove any potential loops (e.g., a minimum spanning tree is used in the application detailed in Section 12.2.3), and then apply equation (12.10) to estimate an optimal solution with respect to the simplified graph. An alternative strategy, which has been used with much success in computer vision and other fields, is the use of loopy belief propagation [36]. The basic idea is to employ equation (12.10) for graphs with loops even though it is not formally suited for this. Let us assume a grid graph of 3D images that is constructed by connecting each node to six immediate neighbors. In this case, in order to apply loopy belief propagation, we have to compute six messages using equation (12.10), each of them with a different set of incoming messages $r \in N(p)\backslash q$. Since none of the nodes is a leaf or root node, there is no natural ordering on how to traverse the graph anymore. Therefore, a parallel message scheduling is employed. In each iteration of the algorithm, all nodes compute messages to their neighbors. These are taken into consideration in the next iteration. This has the obvious disadvantage that the propagation of information from one part of the image to another requires many iterations, making the algorithms comparatively slow. However, using a transformation model with fewer nodes (e.g., supervoxel approaches discussed in Section 12.2.1) can alleviate this problem.

Another closely related concept that was popularized in computing disparity (1D displacement) maps between left and right stereo images for depth estimation is semi-global matching (SGM) [37]. Here, multiple complementary graphs are constructed with each graph providing a simplified solution. In the case of 3D image registration, SGM may decompose a six-connected regular grid graph into three graph models: horizontal chains, vertical chains and chains along the depth dimension. Each of these graph models can be optimally solved with (non-iterative) belief iteration and benefit from efficient sequential propagation of information through messages. However, the results of each model may be suboptimal in terms of regularity due to missing links in other directions. A heuristic solution to obtaining globally smooth deformation fields may be achieved by adding (separately for each node) the marginal costs that have been calculated for each node and label when solving each graph. The interested reader is referred to [38] for an application of SGM to lung registration.

An alternative message passing algorithm called mean field inference has been proposed for large image graphs in [28]. The key idea here is to split the propagation of messages and the enforcement of a regularized solution into two alternating steps. Specifically, the message passing is performed using very efficient convolutional filter operations that act on intermediate solutions and replace the slow iterative process of passing information from node to node. The intermediate solution can be thought of as a tensor that has one additional *label* dimension beyond the spatial image dimensions. This tensor represents the data term in the first iteration and is updated with the filtered information as well as with a compatibility function of labels that aims to ensure a regularized solution (i.e., a smooth displacement field).

Uncertainty estimation and its benefits

In most commonly used registration algorithms, only the optimal displacement vector is computed. This can be considered as the mode of the underlying probability distribution or the maximum a posteriori solution. While this is an obvious choice for finding one-to-one correspondences, having knowledge about the certainty of the displacements has several important benefits in medical image analysis. Many different strategies can be used for estimating the solution uncertainty, e.g., variational Bayes methods or stochastic perturbation of the transformation parameters. Interestingly, in discrete graphical optimization, this information may be directly available within the belief propagation framework.

Algorithm 12.1 describes an implementation (based on [26]) that enables the computation of a min-marginal distribution of all probable displacement vectors after inferring regularity of the deformations using belief propagation. In this distribution, displacements with high probability are represented by low cost values. The algorithm employs belief propagation on graphs without loops. It relies on first computing the message passing from leaves to root and then computing it a second time in the opposite direction. Each `min-sum` step requires the solution of equation (12.10). It is useful to introduce a second array, which stores the intermediate costs (marginal

```
1. Initialize marginals and messages:
foreach node do marginals [ node ] ← g(node);
message [ node ] ← 0;
2. Forward-pass of messages:
for node = leaves to root-1 do
|   cost ← marginals [ node ];
|   message [ node ] ← min-sum(cost);
|   marginals [ parent ] ← marginals [ parent ] +
|     message [ node ];
end
3. Backward-pass of messages:
for node = root-1 to leaves do
|   cost ← marginals [ parent ] - message [ node ] +
|     message [ parent ];
|   message [ node ] ← min-sum(cost);
end
4. Add messages to marginals:
foreach node do marginals [ node ] ← marginals [ node ] +
  message [ node ];
```

Algorithm 12.1: Calculation of min-marginals using belief propagation.

distributions). When evaluating a message that is sent from the parent node to one of its children during the backward pass, the incoming messages from all other nodes have to be subtracted (see step 3 in Algorithm 12.1).

There are numerous clinically relevant applications that rely on uncertainty estimation. For example, the user (medical practitioner or researcher) may benefit from the visual presentation of regions where the registration is deemed (locally) unreliable. Another important benefit of quantifying uncertainty is its use in registration-based segmentation propagation [26] and similarly the transfer of other secondary information (e.g., radiation dose planning) to another scan. Here, a one-to-one mapping is not necessarily required and fusing the transformed auxiliary information (i.e., segmentation labels) from multiple possible registration pathways often leads to an improved accuracy. Furthermore, multiple registrations can be combined using uncertainty estimation, e.g., for improved symmetry of the obtained transformation when considering both directions in a pairwise alignment (see the next section for more details). A technical advantage of using message passing for uncertainty estimation, as compared to graph cut approaches, is that marginal distributions for each node are obtained directly.

12.2.3 Application to lung registration

Estimating large non-linear deformations for intra-patient lung motion can be seen as an excellent example for the benefits of MRF-based methods in medical image registration. Let us consider the task of aligning a computed tomography (CT) lung scan at full inspiration with a corresponding expiration CT scan within the same session of the same patient. Even though the movements of the body may have been minimal in between these two scans, the volume of the lungs can change by more than 100% with respect to the volume after exhalation. Finding an accurate local transformation between inspirative and expirative lung scans has many important medical applications, including the diagnosis and analysis of chronic obstructive pulmonary disease (COPD) [39] and the estimation of tumor motion and lung functionality for the planning of radiotherapy [40]. A characterization of the local lung ventilation can be obtained by accurate pulmonary registration and calculations based on the Jacobian determinant of the resulting non-linear displacement fields [41].

In order to establish an accurate deformation field, we need to estimate reliable correspondences of small inner-lung structures (e.g., vessels and airways) that have diameters of a few millimeters and are displaced by more than 3 centimeters. Even for a human expert, it is not easy to discriminate between certain vessels by considering only the appearance from the local neighborhood. Hence, a global optimization strategy is required. In addition, the sliding motion that causes a relatively sharp discontinuity of the displacement field at the border between lung lobes and at the pleura is problematic for conventional iterative optimization schemes that aim to subdivide the registration in many small steps. This means that for a robust computation of lung motion, we should evaluate a set of potential displacements that cover the largest observable motion. The discrete graphical optimization strategies introduced in this chapter offer an elegant solution to this challenging registration problem.

The COPD CT lung registration dataset presented in [42] is comprised of breath-hold inspiration and expiration scans and thereby poses a significantly more complex problem than images that are acquired at more similar respiration levels (e.g., repeated inhale scans). When comparing the best-performing methods for this dataset, it becomes obvious that discrete graphical optimization strategies are consistently among the most accurate algorithms.

The basis for a successful registration of large lung motion can be implemented using a *block-matching strategy* (e.g., as presented in [43]). This strategy typically involves sampling a large number of points within the fixed image. This can be either achieved by following a random spatial distribution within the lungs or by extracting a sparse set of *distinct landmarks* using interest point detectors, such as the Förstner operator, which can extract junction and circular points (see Fig. 12.7 for a visual example). Each point may move according to a large set of displacements that cover the whole spectrum of small and large deformations. For example, the set of displacements was defined as $\boldsymbol{u} \in L = \{0, \pm 2, \pm 4, \dots, \pm 32\}^3$ mm with $|L| = 35{,}937$ in [35]. Given the set of points and the set of possible displacements, one may define the unary data term by selecting a similarity metric that can cope with intensity shifts of the lung parenchyma due to the large volume changes. Potential choices include employing high-dimensional Gabor features (e.g., [44]), modality-independent neighborhood descriptors, or the inner product of normalized gradient fields [4]. These metrics are typically evaluated over a small patch (or block) of voxels (and their gradients) to stabilize the computation of the unary data term. Having computed the unary data term for each point and possible displacement, one may select as optimal displacement for each node the one for which the greatest similarity is achieved.

However, a strategy that is based entirely on similarity (i.e., unary potentials) may yield many outliers (unreliable correspondences) that could lead to implausible volume change estimates. One solution would be to filter out the outliers using a moving least squares fitting as a post-processing step. An alternative solution would be to leverage the discrete optimization tools presented in this chapter and derive a solution for an energy functional that is comprised of both unary and pairwise potentials, thus promoting spatial regularization for smoother displacement fields.

We detail below a potential graphical model formulation for robust and accurate geometric mapping of relevant landmarks within lungs using discrete optimization [45]. We should note that the main goal here is not to match pairs of geometric landmarks, which was detailed in equation (12.8) and is used, for example, in multiview scene reconstruction. Instead, the objective here is to extract landmarks only in the fixed scan and search correspondences in a densely sampled 3D search volume (see Fig. 12.7).

Accordingly, approximately 3500 landmark locations are sampled in the fixed scan (see Fig. 12.7(a)). These correspond to the nodes of the graph, and each node is connected to its k nearest neighbors forming the edge system of the graph. The resulting (k-NN) graph is subsequently converted into a minimum spanning tree with fewer edges that contains no cycles. While this approximation may reduce the regularization of the displacement field, it enables us to find a *globally optimal solution using belief propagation* (see Section 12.2.2) with a relatively simple implementation. Having constructed the graph, the next step is to define the associated MRF energy.

As previously discussed, the MRF energy is comprised of two parts: (1) a dissimilarity metric (data term) and (2) a regularization penalty that discourages unrealistic motion fields. The dissimilarity measure is implemented by calculating the SSD over a patch around a landmark p at location $\boldsymbol{x}_p$ in the fixed scan I and a patch that is shifted in the moving scan J by a relative displacement $\boldsymbol{u}^{l_p}$ indexed by label l_p:

$$g_p(l_p) = \int_{\Omega_p} (J(\boldsymbol{x} + \boldsymbol{u}^{l_p}) - I(\boldsymbol{x}))^2 d\boldsymbol{x}, \tag{12.11}$$

where Ω_p denotes the patch that is centered around the landmark p at location $\boldsymbol{x}_p$.

As far as the regularization penalty is concerned, ternary (higher-order) potentials could be used to obtain several excellent choices of regularization terms. However, using higher-order potentials that model dependencies between three or more random variables would result in higher computational cost. A good compromise may be achieved by using pairwise potentials to promote diffusion-like regularization between nodes p and q that are connected with an edge in the graph:

$$f_{pq}(l_p, l_q) = \alpha_{pq} \frac{\|\boldsymbol{u}^{l_p} - \boldsymbol{u}^{l_q}\|^2}{\|\boldsymbol{x}_p - \boldsymbol{x}_q\|^2 + |I(\boldsymbol{x}_p) - I(\boldsymbol{x}_q)|/\sigma_I}, \tag{12.12}$$

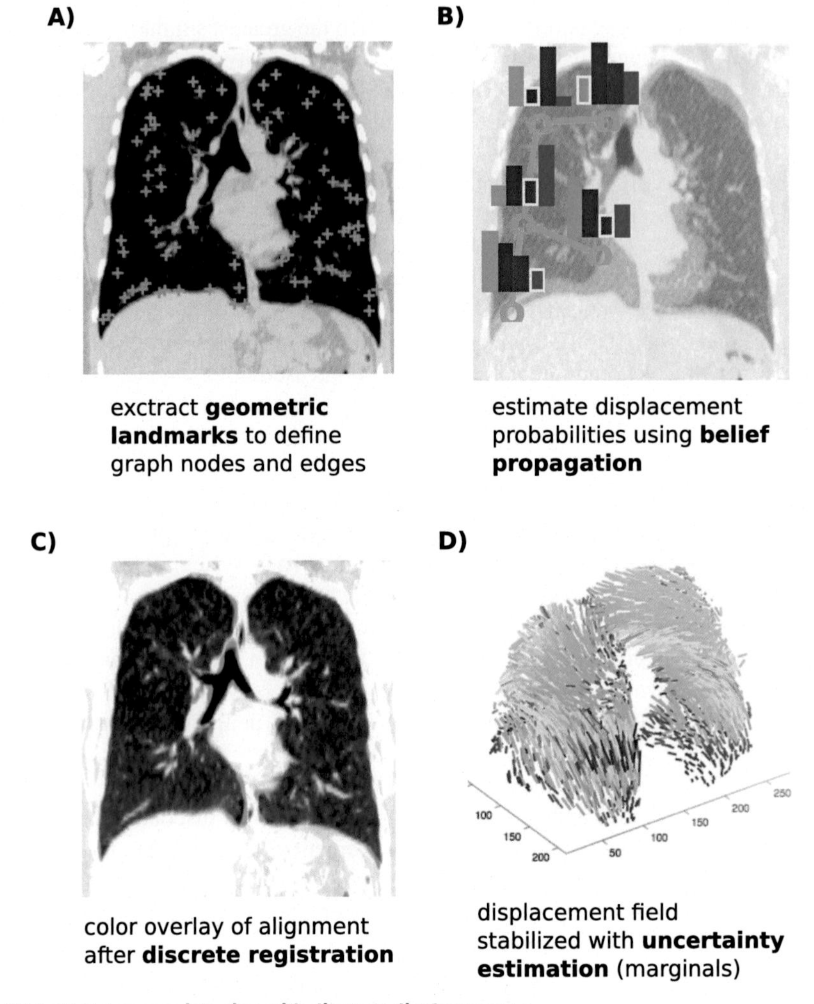
A)
exctract **geometric landmarks** to define graph nodes and edges
B)
estimate displacement probabilities using **belief propagation**
C)
color overlay of alignment after **discrete registration**
D)
100
150
200
50
100
150
200
250
displacement field stabilized with **uncertainty estimation** (marginals)

FIGURE 12.7 Lung registration with discrete displacements.

Deformable registration of a COPD scan pair using the graphical registration method based on geometric landmarks [45]. (a) A subset of landmarks that form the nodes of the graphical model is shown. (b) Computation of the unary data terms and subsequent belief propagation yields a probabilistic estimate of discrete displacements. (c) The inspiration phase is shown in green with an overlay of the registered expiration phase in magenta. A visually very good alignment of inner-lung structures (such as vessels and airways) can be seen. (d) The correspondence field can be further stabilized by taking into account uncertainty estimation of a registration in the reverse direction.

where α_{pq} is an application-specific weighting term. The above pairwise potential penalizes the square root of the squared difference of displacements of two neighboring landmarks p and q, encouraging smoothly varying displacements. The denominator accounts for different lengths of edges, which take into account both the spatial Euclidean distance of the nodes and the absolute difference of their intensity, normalized by $\sigma_I = 150$.

Having defined the graph and the energy, belief propagation (see Section 12.2.2) can be used to calculate the regularized cost for assigning each label to each landmark after two passes of messages. Afterwards, the optimal value and corresponding displacement vector can be chosen for each landmark (see Fig. 12.7(b)). Thin plate spline interpolation can then be used to produce a dense displacement field given the estimated sparse correspondences. This straightforward approach yields accurate displacement estimations. A reference implementation and the option to run the algorithm on any provided lung CT can be found at https://grand-challenge.org/algorithms/corrfield/.

Nevertheless, further improvements are possible by incorporating additional steps. Specifically, the estimated displacement field may contain irregularities as a result of the limited connectivity of the graphical model. Registration uncertainties can be used to improve the quality of the estimated correspondence. The registration uncertainty may be quantified via the marginal energies that can be directly estimated by belief propagation (see Section 12.2.2 and [26]). One can extend the matching from a single direction (i.e., mapping from fixed to moving scan) to a symmetric setting, which additionally includes the optimization from moving to fixed scan (i.e., two directions are estimated). In such a symmetric setting, one can compute registration uncertainties for both directions. Subsequently, a mutual consistency check can be performed that reduces the asymmetry and outliers within the registration. The different steps in defining a graph, evaluating the data term, and optimizing spatial regularity of the correspondences are shown in Fig. 12.7. This algorithm achieves a very low target registration error of 1.08 ± 1.21 mm evaluated at 300 expert-identified anatomical landmarks for each of the 10 publicly available scans of [42]. Finally, the results can be further improved by combining discrete optimization for geometric landmark correspondence with continuous optimization for dense iconic registration [35].

12.2.4 Conclusion

In this chapter, we detailed the use of discrete optimization techniques and MRF modeling for the problem of image registration. We discussed how to construct graphical models for the problems of (1) geometric registration, (2) iconic registration, and (3) hybrid registration. Additionally, we detailed optimization approaches that can be used to perform inference and discussed how these compare to continuous optimization techniques, such as the ones detailed in Chapter 13. Finally, we presented an application of discrete graph-based registration to the problem of intra-patient CT lung registration.

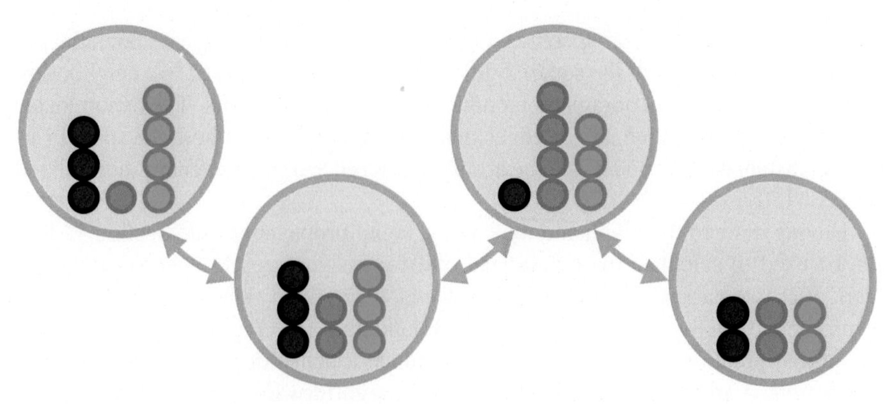
Hint the solution will be blue-blue-black-black with costs of 1-3-5-7

FIGURE 12.8 Exercise 3.

Compute the optimal registration energy (costs) for the given MRF model that consists of four nodes. Each node can be assigned one of three labels. The Pott model is used as pairwise potential.

12.3 Exercises

1. **Regularization energy**
   Sketch out how to penalize a registration cost function in graph-based models using the bending energy term. Which equation would be suitable and how many neighbors have to be considered for each node?
2. **MRF solver**
   The given MRF contains four nodes (yellow circles) connected through three edges that form a chain without loops (see Fig. 12.8). Each node can be assigned one of three labels: black, blue, or green. The unary cost of assigning a label to a node is represented by the number of circles (see Fig. 12.8). We also introduce a pairwise cost, which incurs a penalty of 1 when assigning different labels to two connected nodes. Note that this regularization term is formally known as Pott's model. We aim to find the optimal solution that minimizes the total cost. To solve this optimization problem, we start from the second one and traverse all nodes from left to right. To compute the intermediate optimal costs, we sequentially explore for each label whether it is best to simply add the unary cost of the current label or make a switch. A switch will incur the lowest cost of the preceding nodes plus a constant penalty of 1. To compute the optimal registration energy (costs), one should repeat the previous steps for all remaining nodes, each time for all three labels.
3. **Adapting the energy**
   What is the influence of the regularization model and its weight on the optimal labeling solution? What difference does doubling the pairwise penalty make? What difference does halving the pairwise penalty make? Would you find a different

solution if, e.g., the switch between black and green labels would incur a smaller penalty than the switch between black and blue?

References

[1] Maintz J.B.A., Viergever M.A., A survey of medical image registration, Medical Image Analysis 2 (1) (1998) 1–36.

[2] Hajnal J.V., Hill D.L.G., Hawkes D.J. (Eds.), Medical Image Registration, CRC Press, Baton Rouge, Florida, 2001.

[3] Zitova B., Flusser J., Image registration methods: a survey, Image and Vision Computing 21 (11) (2003) 977–1000.

[4] Modersitzki J., Numerical Methods for Image Registration, Oxford University Press, New York, 2004.

[5] Sotiras A., Davatzikos C., Paragios N., Deformable medical image registration: A survey, IEEE Transactions on Medical Imaging 32 (7) (2013) 1153–1190.

[6] Van Essen D.C., A Population-Average, Landmark- and Surface-based (PALS) atlas of human cerebral cortex, NeuroImage 28 (3) (2005) 635–662.

[7] Evans A.C., Janke A.L., Collins D.L., Baillet S., Brain templates and atlases, NeuroImage 62 (2) (2012) 911–922.

[8] Makela T., Clarysse P., Sipila O., Pauna N., Quoc Cuong Pham, Katila T., Magnin I.E., A review of cardiac image registration methods, IEEE Transactions on Medical Imaging 21 (9) (2002) 1011–1021.

[9] Castillo R., Castillo E., Guerra R., Johnson V.E., McPhail T., Garg A.K., Guerrero T., A framework for evaluation of deformable image registration spatial accuracy using large landmark point sets, Physics in Medicine and Biology 54 (7) (2009) 1849–1870.

[10] Heinrich M.P., Jenkinson M., Brady M., Schnabel J.A., MRF-based deformable registration and ventilation estimation of lung CT, IEEE Transactions on Medical Imaging 32 (7) (2013) 1239–1248.

[11] Kessler M.L., Image registration and data fusion in radiation therapy, British Journal of Radiology 79 (Special Issue 1) (2006) S99–S108.

[12] Joshi S.C., Miller M.I., Landmark matching via large deformation diffeomorphisms, IEEE Transactions on Image Processing 9 (8) (2000) 1357–1370.

[13] Rohr K., Stiehl H.S., Sprengel R., Buzug T.M., Weese J., Kuhn M.H., Landmark-based elastic registration using approximating thin-plate splines, IEEE Transactions on Medical Imaging 20 (6) (2001) 526–534.

[14] Chui H., Rangarajan A., A new point matching algorithm for non-rigid registration, Computer Vision and Image Understanding 89 (2–3) (2003) 114–141.

[15] Bookstein F.L., Principal warps: Thin-plate splines and the decomposition of deformations, IEEE Transactions on Pattern Analysis and Machine Intelligence 11 (6) (1989) 567–585.

[16] Roche A., Malandain G., Pennec X., Ayache N., The correlation ratio as a new similarity measure for multimodal image registration, in: International Conference on Medical Image Computing and Computer-Assisted Intervention, 1998, pp. 1115–1124.

[17] Maes F., Collignon A., Vandermeulen D., Marchal G., Suetens P., Multimodality image registration by maximization of mutual information, IEEE Transactions on Medical Imaging 16 (2) (1997) 187–198.

[18] Viola P., Wells W.M. III, Alignment by maximization of mutual information, International Journal of Computer Vision 24 (2) (1997) 137–154.
[19] Chung A.C.S., Wells W.M., Norbash A., Grimson W.E.L., Multi-modal image registration by minimizing Kullback–Leibler distance, in: International Conference on Medical Image Computing and Computer-Assisted Intervention, 2002, pp. 525–532.
[20] Klein S., Staring M., Pluim J.P.W., Evaluation of optimization methods for nonrigid medical image registration using mutual information and B-splines, IEEE Transactions on Image Processing 16 (12) (2007) 2879–2890.
[21] Klein S., Pluim J.P.W., Staring M., Viergever M.A., Adaptive stochastic gradient descent optimisation for image registration, International Journal of Computer Vision 81 (3) (2009) 227–239.
[22] Glocker B., Komodakis N., Tziritas G., Navab N., Paragios N., Dense image registration through MRFs and efficient linear programming, Medical Image Analysis 12 (6) (2008) 731–741.
[23] Glocker B., Sotiras A., Komodakis N., Paragios N., Deformable medical image registration: setting the state of the art with discrete methods, Annual Review of Biomedical Engineering 13 (2011) 219–244.
[24] Torresani L., Kolmogorov V., Rother C., Feature correspondence via graph matching: models and global optimization, in: European Conference on Computer Vision, 2008, pp. 596–609.
[25] So R.W.K., Tang T.W.H., Chung A.C.S., Non-rigid image registration of brain magnetic resonance images using graph-cuts, Pattern Recognition 44 (10–11) (2011) 2450–2467.
[26] Heinrich M.P., Simpson I.J., Papież B.W., Brady S.M., Schnabel J.A., Deformable image registration by combining uncertainty estimates from supervoxel belief propagation, Medical Image Analysis 27 (Jan. 2016) 57–71.
[27] Gilboa G., Osher S., Nonlocal linear image regularization and supervised segmentation, Multiscale Modeling & Simulation 6 (2) (2007) 595–630.
[28] Krähenbühl P., Koltun V., Efficient inference in fully connected CRFs with Gaussian edge potentials, in: Advances in Neural Information Processing Systems, 2011, pp. 109–117.
[29] Wang C., Komodakis N., Paragios N., Markov Random Field modeling, inference & learning in computer vision & image understanding: A survey, Computer Vision and Image Understanding 117 (11) (2013) 1610–1627.
[30] Kolmogorov V., Zabih R., Computing visual correspondence with occlusions using graph cuts, in: Proceedings Eighth IEEE International Conference on Computer Vision, ICCV 2001, vol. 2, IEEE, 2001, pp. 508–515.
[31] Sotiras A., Paragios N., Discrete symmetric image registration, in: IEEE International Symposium on Biomedical Imaging, 2012, pp. 342–345.
[32] Sotiras A., Neji R., Deux J.-F., Komodakis N., Fleury G., Paragios N., A Kernel-based graphical model for diffusion tensor registration, in: IEEE International Symposium on Biomedical Imaging, 2010, pp. 524–527.
[33] Heinrich M.P., Jenkinson M., Bhushan M., Matin T., Gleeson F.V., Brady S.M., Schnabel J.A., MIND: Modality independent neighbourhood descriptor for multi-modal deformable registration, Medical Image Analysis 16 (7) (Oct. 2012) 1423–1435.
[34] Kwon D., Lee K.J., Yun I.D., Lee S.U., Nonrigid image registration using dynamic higher-order MRF model, in: European Conference on Computer Vision, 2008, pp. 373–386.
[35] Rühaak J., Polzin T., Heldmann S., Simpson I.J.A., Handels H., Modersitzki J., Heinrich M.P., Estimation of large motion in lung CT by integrating regularized keypoint corre-

spondences into dense deformable registration, IEEE Transactions on Medical Imaging 36 (8) (2017) 1746–1757.

[36] Felzenszwalb P.F., Huttenlocher D.P., Efficient belief propagation for early vision, International Journal of Computer Vision 70 (1) (2006) 41–54.

[37] Hirschmuller H., Accurate and efficient stereo processing by semi-global matching and mutual information, in: IEEE Computer Society Conference on Computer Vision and Pattern Recognition, vol. 2, IEEE, 2005, pp. 807–814.

[38] Hermann S., Evaluation of scan-line optimization for 3D medical image registration, in: Proc. IEEE Confer. CVPR, 2014, pp. 3073–3080.

[39] Galbán C.J., Han M.K., Boes J.L., Chughtai K.A., Meyer C.R., Johnson T.D., Galbán S., Rehemtulla A., Kazerooni E.A., Martinez F.J., Ross B.D., Computed tomography-based biomarker provides unique signature for diagnosis of COPD phenotypes and disease progression, Nature Medicine 18 (11) (2012) 1711–1715.

[40] Weiss E., Wijesooriya K., Dill S.V., Keall P.J., Tumor and normal tissue motion in the thorax during respiration: Analysis of volumetric and positional variations using 4D CT, International Journal of Radiation Oncology, Biology, Physics 67 (1) (2007) 296–307.

[41] Reinhardt J.M., Ding K., Cao K., Christensen G.E., Hoffman E.A., Bodas S.V., Registration-based estimates of local lung tissue expansion compared to xenon CT measures of specific ventilation, Medical Image Analysis 12 (6) (2008) 752–763.

[42] Castillo R., Castillo E., Fuentes D., Ahmad M., Wood A.M., Ludwig M.S., Guerrero T., A reference dataset for deformable image registration spatial accuracy evaluation using the COPDgene study archive, Physics in Medicine and Biology 58 (9) (2013) 2861–2877.

[43] Castillo E., Castillo R., Fuentes D., Guerrero T., Computing global minimizers to a constrained B-spline image registration problem from optimal l1 perturbations to block match data, Medical Physics 41 (4) (2014).

[44] Ou Y., Sotiras A., Paragios N., Davatzikos C., DRAMMS: Deformable registration via attribute matching and mutual-saliency weighting, Medical Image Analysis 15 (4) (2011) 622–639.

[45] Heinrich M.P., Handels H., Simpson I.J.A., Estimating large lung motion in COPD patients by symmetric regularised correspondence fields, in: International Conference on Medical Image Computing and Computer-Assisted Intervention, Springer, 2015, pp. 338–345.

CHAPTER

Parametric volumetric registration

13

Paul A. Yushkevich[a], Miaomiao Zhang[b], and Jon Sporring[c]

[a]*Department of Radiology, University of Pennsylvania, Philadelphia, PA, United States*

[b]*Departments of Electrical and Computer Engineering and Computer Science, University of Virginia, Charlottesville, VA, United States*

[c]*Department of Computer Science, University of Copenhagen, Copenhagen, Denmark*

Learning points

- Volumetric registration
- Transformations and displacements
- Jacobian matrix and determinant
- Registration as energy minimization
- Parametric transformations
- Affine, radial basis functions and free-form transformations

13.1 Introduction to volumetric registration

Volumetric registration (VR) is one of the most studied problems in the medical image analysis field. It has led to elegant mathematical solutions that tie together elements of group theory, differential geometry, and probability theory. This chapter firstly describes the basic building blocks of VR and secondly describes *parametric VR* (PVR) by providing the theoretical outlines of some widely used algorithms for PVR. In the following chapter, Chapter 14, the theory and algorithms for *non-parametric VR* (NVR) will be given.

13.1.1 Definition and applications

The process of VR is illustrated in Fig. 13.1. In the example, we seek the translation, rotation, and scaling, which makes the pixel-by-pixel mapping between the two images optimal.

In this chapter, we will use the following definition of VR:

> *VR is the process of transforming the space in which images are defined, in order to match corresponding locations between images.*

There are several key concepts in this definition:

Medical Image Analysis. https://doi.org/10.1016/B978-0-12-813657-7.00028-5

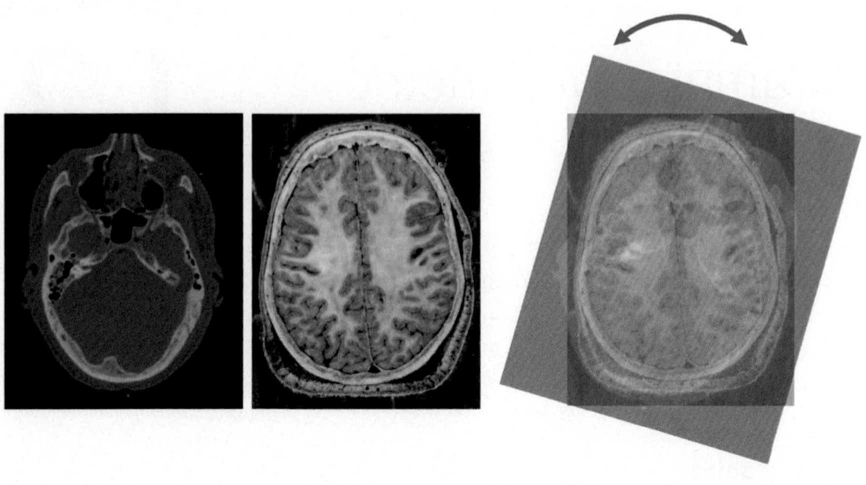

FIGURE 13.1

Two example images of different modalities. A section of a 3D CT image of a human head (left), a cryosection also of a human head (middle), and an illustration of the process of registration (right) using scaling, translation, and rotation. The images were obtained from the Visible Human Project at the U.S. National Library of Medicine.

- VR is about transforming the space in which images "live," rather than images themselves. This is a critical aspect of VR that can be confusing to beginners. Mathematically, VR will be described in terms of functions $\boldsymbol{\phi} : \Omega \to \Omega$ that map points in the image domain $\Omega \subset \mathbb{R}^n$ to their corresponding points in Ω. These functions are called *spatial transformations*, *spatial maps*, or simply *warps*. An example of a warp is shown in Fig. 13.2.
- VR is often characterized by its transformation type such as rigid or deformable. A rigid transformation includes translation and rotation and is an example of a *linear* transformation. A deformable registration uses a *non-linear* transformation, with the possibility of infinitely many degrees of freedom to transform images. Hence, a deformable registration may match images more precisely than a rigid registration, but it is computationally more complex, and care has to be taken to ensure that the deformable transformation is well behaved. For example, a lot of research has focused on constraining transformations $\boldsymbol{\phi}$ to be *invertible*, which prevents them from tearing or folding the image domain Ω.
- The goal of VR is to match up *corresponding locations* between images. For example, when performing VR on images of faces, corresponding locations may be the corners of the eyes and lips, the tip of the nose, etc. Clearly, the concept of correspondence is rather tenuous; for instance, it is ill defined when it comes to matching hair between different people or between different images of the same person. A similar situation also occurs in medical images. We might be able to match up certain places in the lateral ventricles and subcortical gray matter structures such as the thalamus between individual brains, but not so for cortical folds,

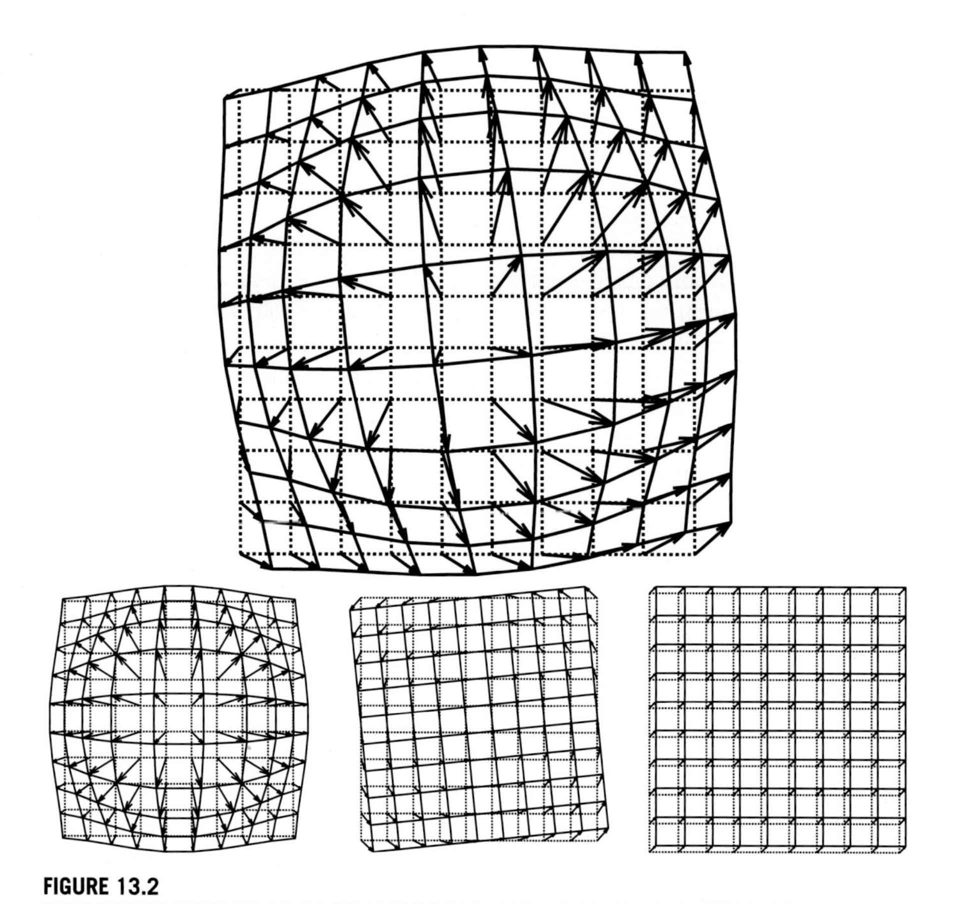

FIGURE 13.2

A 2D vector field: Every point in space is moved from the tail of an arrow to its head. Top: The vector field, which is the result of first a non-linear warp (bottom left), then a rotation (bottom middle), and finally a translation (bottom right). The dotted lines show the original coordinate grid, and the full lines show its transformation.

as the folding pattern of every brain is unique like a fingerprint. Also, the presence of tumors and lesions in some individuals and not in others makes it even more difficult to find anatomical correspondences. However, these challenges have not prevented VR from being one of the most useful techniques in medical imaging research.

VR has many important applications in medical imaging research:

- VR is used for anatomical normalization in population studies that involve imaging. Consider, for example, functional magnetic resonance imaging (fMRI) research, which has transformed the fields of psychology, psychiatry, sociology, linguistics, etc. In a typical study, researchers recruit a cohort of individuals (e.g.,

20 volunteers) and scan their brains using fMRI under various experimental conditions (e.g., reading long sentences vs. reading short sentences). Analyzing the MRI data from each participant results in a "heatmap" that describes experiment-related brain activity at each location of the brain. In order to describe the effects of the experiment at the *population level* rather than the individual level, we transform each individual brain to the space of a "template brain" by using VR, so that the corresponding anatomical regions are matched up [1].

- VR is used to quantify changes in longitudinal imaging. Patients undergoing monitoring or treatment for chronic diseases are often scanned at different times. Changes between these longitudinal scans are used to assess the disease progression or treatment efficacy. For example, in multiple sclerosis, the changes in size and number of white matter lesions are used clinically to determine whether a treatment regimen is effective. VR between images obtained at different time points produces a transformation that matches anatomical locations across time. Examining the geometric properties of this transformation (i.e., whether it shrinks or expands space at different locations in the image domain) can provide accurate measurements of changes in the size of white matter lesions. Such measurements are much more effective than visual examination of longitudinal images, particularly when changes over time are subtle.
- VR is used to account for motion of organs and tissues. In image-guided surgery, it is common to acquire preoperative high-resolution images of the patient's anatomy (e.g., a high-resolution CT scan) for planning purposes and to collect lower-resolution images intraoperatively for surgical guidance (e.g., using 3D ultrasound or 2D fluoroscopy). In many parts of the body, such as the abdomen, organs shift, and performing only rigid or affine registration between the preoperative and intraoperative images could result in misalignment, leading to surgical errors. However, VR is able to account for such non-linear organ motions better.
- VR is a building block of many image analysis algorithms, among them, atlas-based and multiatlas segmentation. When faced with a problem of segmenting a structure in an image, VR can be used to warp existing segmentation of this structure in one or more reference images into the space of the target image. If VR matches up corresponding anatomical locations between the reference images and the target image well, these warped segmentations are themselves good segmentations of the target image. Even when VR is not perfect at matching anatomical locations, it may be possible to obtain good segmentations of the target image by combining the warped segmentation from multiple reference images – this is known as *multiatlas segmentation* [2].

13.1.2 VR as energy minimization

In this subsection, we discuss the general structure of VR algorithms. An example of VR between two chest CT scans of the same individual is shown in Fig. 13.3.

In most algorithms, VR is formulated as an energy minimization problem of the following form:

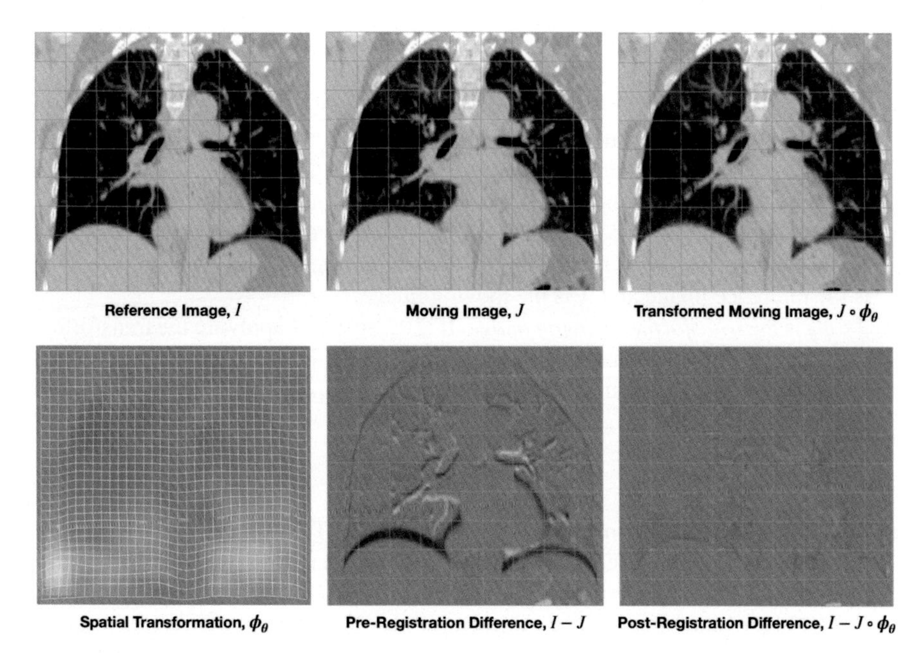

FIGURE 13.3

An example of deformable registration between two 3D chest CT images of the same individual during inhalation/exhalation. The spatial transformation $\boldsymbol{\phi}_\theta$ is visualized by plotting how it deforms a rectangular grid in the image domain Ω.

$$\boldsymbol{\theta}^* = \arg\min_{\boldsymbol{\theta} \in \Theta} E[\boldsymbol{\phi}_\theta; I, J], \quad \text{where} \quad E[\boldsymbol{\phi}_\theta; I, J] = \mu[I, J \circ \boldsymbol{\phi}_\theta] + \lambda\rho[\boldsymbol{\theta}]. \tag{13.1}$$

Let us carefully consider each term in the above expression:

- $\boldsymbol{\phi}_\theta : \Omega \to \Omega$ is the *spatial transformation* that maps the image domain $\Omega \subset \mathbb{R}^n$ onto itself. It is a continuous function of n variables that has n components. For example, in 2D, we may write $\boldsymbol{\phi}_\theta$ as a vector,

$$\boldsymbol{\phi}_\theta(\boldsymbol{x}) = \begin{bmatrix} \phi_\theta^1(x_1, x_2) \\ \phi_\theta^2(x_1, x_2) \end{bmatrix},$$

  where ϕ_θ^1 and ϕ_θ^2 are scalar functions, called the components of $\boldsymbol{\phi}_\theta$. For example, a rigid transformation in 2D may be written as

$$\boldsymbol{\phi}_\theta^{\text{rigid}}(\boldsymbol{x}) = \mathbf{R}_\omega \boldsymbol{x} + \boldsymbol{b} = \begin{bmatrix} x_1 \cos\omega - x_2 \sin\omega + b_1 \\ x_1 \sin\omega + x_2 \cos\omega + b_2 \end{bmatrix}, \tag{13.2}$$

  where $\mathbf{R}_\omega$ is a rotation matrix and $\boldsymbol{b}$ is a translation vector. More examples of transformations are given in Section 13.3.1. A rigid transformation is an example of a linear transformation, since $\boldsymbol{\phi}_\theta^{\text{rigid}}$ is a linear function of $\boldsymbol{x}$. Non-linear transformations have non-linear functions for $\boldsymbol{\phi}_{\boldsymbol{\theta}}$. In both cases, the size of $\boldsymbol{\theta}$ will reflect the complexity of these functions.

- I and J are images. In the theoretical portions of this chapter, we will treat them as continuous functions defined on the image domain, i.e., $I, J : \Omega \to \Gamma$, where for example for gray-valued images $\Gamma \subseteq \mathbb{R}$, and for color images $\Gamma \subseteq \mathbb{R}^3$. We typically distinguish between the *reference image* (alternatively called *fixed image*), which remains unchanged over the course of the energy minimization, and the *moving image*, which is affected by the transformation $\boldsymbol{\phi}_{\boldsymbol{\theta}}$. In expression (13.1), I is the reference image and J is the moving image.
- $J \circ \boldsymbol{\phi}_{\boldsymbol{\theta}}$ is the *transformed moving image*. It is the result of applying the transformation $\boldsymbol{\phi}_{\boldsymbol{\theta}}$ to the moving image J. The mathematical symbol $\circ$ denotes composition of two functions and is used extensively in this chapter. $J \circ \boldsymbol{\phi}_{\boldsymbol{\theta}}$ is a continuous image on the domain Ω, given by

$$(J \circ \boldsymbol{\phi}_{\boldsymbol{\theta}})(\boldsymbol{x}) = J(\boldsymbol{\phi}_{\boldsymbol{\theta}}(\boldsymbol{x})).$$

- $\boldsymbol{\theta}$ represents the set of *transformation parameters*. These parameters determine the function $\boldsymbol{\phi}_{\boldsymbol{\theta}}$, i.e., different values of $\boldsymbol{\theta}$ correspond to different spatial transformations $\boldsymbol{\phi}_{\boldsymbol{\theta}}$. The objective of VR is to determine the transformation parameters that yield the optimal spatial transformation matching the input images I and J.
  - The class of VR methods that we will study in this chapter is called *PVR*, in which case $\boldsymbol{\theta}$ will be a finite vector of unknown values. For example, the rigid transformation in (13.2) has $\boldsymbol{\theta} = [\omega, b_1, b_2]$. The coefficients $\boldsymbol{\theta}$ are used to define a spatial transformation $\boldsymbol{\phi}_{\boldsymbol{\theta}}$ using a system of basis functions. For example, $\boldsymbol{\theta}$ may consist of displacement vectors placed on a grid of control points, from which the continuous transformation $\boldsymbol{\phi}_{\boldsymbol{\theta}}$ is derived using interpolation. When the set of transformation parameters is finite, the minimization problem (13.1) can be solved using multivariate optimization methods, e.g., gradient descent or Newton's method.
  - The second class of VR methods, called *NVR*, which we will study in Chapter 14, has infinitely many transformation parameters, that is to say, $\boldsymbol{\theta}$ is modeled as an unknown function, and solving the minimization problem (13.1) requires tools of variational calculus. In some non-parametric methods one does not distinguish between $\boldsymbol{\theta}$ and $\boldsymbol{\phi}_{\boldsymbol{\theta}}$, i.e., the unknown parameters and the spatial transformation are the same. In this case we rewrite (13.1) omitting $\boldsymbol{\theta}$ altogether:

$$\boldsymbol{\phi}^* = \underset{\boldsymbol{\phi}:\Omega\to\Omega}{\arg\min}\, E[\boldsymbol{\phi}; I, J].$$

- Θ represents the set of all possible values that the transformation parameters $\boldsymbol{\theta}$ may assume.
- $\mu[\bullet, \bullet]$ is the *image dissimilarity measure*. In this chapter, we mainly use the *sum of squared differences* (SSD) dissimilarity measure, which has the simple form

$$\mu_{\text{SSD}}[I, J] = \int_\Omega \|I(\boldsymbol{x}) - J(\boldsymbol{x})\|^2\, d\boldsymbol{x}.$$

However, other dissimilarity measures are also commonly used in VR often with better results than SSD. An often used dissimilarity measure is *mutual information* (MI),

$$\mu_{\mathrm{MI}}[I, J] = \int_{\Gamma}\int_{\Gamma} p(v, w) \log \frac{p(v, w)}{p(v)p(w)}\, dv\, dw,$$

where $p(v, w) = p\,(I(\boldsymbol{x}) = v \wedge J(\boldsymbol{x}) = w)$ is the joint distribution between intensities in I and J at the same locations and $p(v) = p\,(I(\boldsymbol{x}) = v)$ and $p(w) = p\,(J(\boldsymbol{x}) = w)$ are the distributions of intensities of I and J, respectively, or equivalently the marginal distributions of $p(v, w)$. Another often used dissimilarity measure is the *normalized cross-correlation* (NCC) dissimilarity measure,

$$\mu_{\mathrm{NCC}}[I, J] = \frac{\int_{\Omega} I(\boldsymbol{x}) J(\boldsymbol{x})\, d\boldsymbol{x}}{\sigma_I \sigma_J},$$

where σ_I and σ_J are the standard deviation of image I and J, respectively.
SSD is most often used for images which are from the same image modality, e.g., for registering two CT images, while MI is often used across image modalities such as for registering a CT and an MRI image. For between-subject brain MRI registration, the NCC [3,4] achieves greater accuracy than SSD or MI measures [5].

- $\rho[\boldsymbol{\theta}]$ is the regularization term. It can be used to impose soft constraints on the transformation $\boldsymbol{\phi}_{\boldsymbol{\theta}}$. Particularly non-linear PVR and NVR can have a very large number of degrees of freedom in specifying $\boldsymbol{\phi}_{\boldsymbol{\theta}}$, in which case some form of regularization is necessary to make the optimization problem (13.1) well posed and to prevent unrealistic transformations. A common regularization term is

$$\rho[\boldsymbol{\theta}] = \int_{\Omega} |D\boldsymbol{\phi}_{\boldsymbol{\theta}}(\boldsymbol{x})|_p^p\, d\boldsymbol{x},$$

where $p > 0$, $|\mathbf{A}|_p^p = \left(\sum_i \sum_j |a_{ij}|^p\right)^{1/p}$, and $D\boldsymbol{\phi}$ is the Jacobian of $\boldsymbol{\phi}$ with respect to $\boldsymbol{\theta}$ (see Section 13.2.5 for a definition of the Jacobian). For $p = 2$, $|\mathbf{A}|_p^p$ is the Frobenius norm, and ρ is known as a Tikhonov regularization term.
In general, the term $\rho[\boldsymbol{\theta}]$ is designed by the algorithm developer in such a way that when the transformation $\boldsymbol{\phi}_{\boldsymbol{\theta}}$ becomes unrealistic, for example when it folds the space Ω onto itself, then the value of $\rho[\boldsymbol{\theta}]$ becomes large, thus penalizing this particular choice of parameters $\boldsymbol{\theta}$. By minimizing the weighted sum of the dissimilarity and regularization terms (with weighting provided by the constant λ), registration algorithms seek a trade-off between good matching of anatomy that is achieved when dissimilarity is minimized and avoid unrealistic transformations. Regularization terms can vary in different algorithms, but typically involve terms that encourage solutions $\boldsymbol{\phi}_{\boldsymbol{\theta}}$ to be smooth and bounded. In some applications, regularization is used to imbue $\boldsymbol{\phi}_{\boldsymbol{\theta}}$ with specific physical or geometric properties, such as local preservation of area or volume.

13.2 Mathematical concepts

This section introduces several mathematical concepts that will be used throughout the chapter. These concepts primarily involve spatial transformations.

13.2.1 Transformation

In VR we transform the space of the moving image, and thus the choice of the transformation to optimize over is an important modeling decision. We call the transformation $\boldsymbol{\phi}_{\boldsymbol{\theta}} : \Omega \to \Omega$ a *spatial transformation* since it maps the image domain $\Omega \subset \mathbb{R}^n$ onto itself. Continuity and smoothness are key concepts in transformations for registration. Loosely speaking, a continuous transformation Φ has no abrupt changes, i.e., if $\boldsymbol{x}$ and $\boldsymbol{y}$ are close to each other, then $\Phi(\boldsymbol{x})$ and $\Phi(\boldsymbol{y})$ must be as well. A smooth transformation is a transformation which is continuous and where for $\boldsymbol{x} = (x_1, x_2, \ldots)$ all derivatives $\frac{\partial^{i+j+\ldots}}{\partial x_1^i \partial x_2^j \ldots}\Phi(\boldsymbol{x})$ of the transformations exist. The existence of all derivatives is at times too restrictive, and often only C^k-smoothness is required, which means the transformation is continuous and all derivatives up to and including order k exist and are continuous. If a transformation Φ is bijective and smooth and its inverse Φ^{-1} exists and is smooth, then it is called a diffeomorphic transformation, and when only the first k derivatives exist and are smooth it is called a C^k-diffeomorphism. Diffeomorphic transformations are particularly important for NVR to be discussed in Chapter 14.

A parametric transformation is a continuous function of n variables that has n parametric coordinate functions. For example, in 2D, we may write $\boldsymbol{\phi}_{\boldsymbol{\theta}}$ as a vector

$$\boldsymbol{\phi}_{\boldsymbol{\theta}}(\boldsymbol{x}) = \begin{bmatrix} \phi_{\boldsymbol{\theta}}^1(x_1, x_2) \\ \phi_{\boldsymbol{\theta}}^2(x_1, x_2) \end{bmatrix} = \begin{bmatrix} \phi_{\boldsymbol{\theta}}^1(\boldsymbol{x}) \\ \phi_{\boldsymbol{\theta}}^2(\boldsymbol{x}) \end{bmatrix},$$

where $\phi_{\boldsymbol{\theta}}^1$ and $\phi_{\boldsymbol{\theta}}^2$ are scalar functions, called the components of $\boldsymbol{\phi}_{\boldsymbol{\theta}}$, and $\boldsymbol{\theta}$ is a vector of parameters. The components may be both linear and non-linear; examples of both will be given in Section 13.3.1.

13.2.2 Transformation vs. displacement

As noted above, a spatial transformation $\boldsymbol{\phi} : \Omega \to \Omega$ is a mapping of the image domain Ω onto itself. It is sometimes convenient to express spatial transformations as a *displacement* or *displacement field*:

$$\boldsymbol{\phi}(\boldsymbol{x}) = \boldsymbol{x} + \boldsymbol{u}(\boldsymbol{x}). \tag{13.3}$$

We will use Greek letters $\boldsymbol{\phi}$, $\boldsymbol{\psi}$ to denote transformations and boldface Roman letters $\boldsymbol{u}$, $\boldsymbol{v}$ to denote displacements. The distinction between transformations and displacements is simple, but important. The function $\boldsymbol{\phi}(\boldsymbol{x})$ describes the *new position* of point $\boldsymbol{x}$ after the transformation. The displacement $\boldsymbol{u}(\boldsymbol{x})$ describes the *change* in the position of a point $\boldsymbol{x}$ under the transformation $\boldsymbol{\phi}$. For example, in 2D, $\boldsymbol{\phi}$ might map the point with coordinates (2.0, 2.1) to new coordinates (2.4, 1.9). The corresponding displacement is the vector (0.4, −0.2).

The relation between transformations of the form of (13.2) and (13.3) may not be obvious. However, writing the rigid transformations (13.2) as a weighted sum,

$$\boldsymbol{\phi}_{\boldsymbol{\theta}}(\boldsymbol{x}) = \boldsymbol{x} + \sum_{k=1}^{K} \boldsymbol{\theta}_k f_k(\boldsymbol{x}) = \mathbf{R}_{\omega}\boldsymbol{x} + \boldsymbol{b}, \tag{13.4}$$

implies that

$$\begin{aligned}
\boldsymbol{u}_{\boldsymbol{\theta}}(\boldsymbol{x}) = \boldsymbol{\phi}_{\boldsymbol{\theta}}(\boldsymbol{x}) - \boldsymbol{x} &= \sum_{k=1}^{K} \boldsymbol{\theta}_k f_k(\boldsymbol{x}) && (13.5)\\
&= (\mathbf{R}_{\omega} - \mathbf{I})\boldsymbol{x} + \boldsymbol{b} && (13.6)\\
&= \begin{bmatrix} x_1(\cos\omega - 1) - x_2 \sin\omega + b_1 \\ x_1 \sin\omega + x_2(\cos\omega - 1) + b_2 \end{bmatrix} && (13.7)\\
&= \begin{bmatrix} 1 \\ 0 \end{bmatrix} b_1 + \begin{bmatrix} 0 \\ 1 \end{bmatrix} b_2 + \begin{bmatrix} (\cos\omega - 1) \\ \sin\omega \end{bmatrix} x_1 + \begin{bmatrix} -\sin\omega \\ (\cos\omega - 1) \end{bmatrix} x_2. && (13.8)
\end{aligned}$$

Thus, we can identify the coefficients and functions of the weighted sum $\boldsymbol{\theta}_1 = [1, 0]^T$, $f_1 = b_1$, $\boldsymbol{\theta}_2 = [0, 1]^T$, $f_2 = b_2$, etc.

13.2.3 Function composition

Given a pair of transformations $\boldsymbol{\phi}, \boldsymbol{\psi} : \Omega \to \Omega$, the composition of $\boldsymbol{\phi}$ and $\boldsymbol{\psi}$ is a new transformation denoted $\boldsymbol{\phi} \circ \boldsymbol{\psi}$, defined as

$$(\boldsymbol{\phi} \circ \boldsymbol{\psi})(\boldsymbol{x}) = \boldsymbol{\phi}(\boldsymbol{\psi}(\boldsymbol{x})).$$

The combined transformation, $\boldsymbol{\phi} \circ \boldsymbol{\psi}$, is the transformation we get by first applying the transformation $\boldsymbol{\psi}$ and then applying the transformation $\boldsymbol{\phi}$. For example, $\boldsymbol{\phi}$ and $\boldsymbol{\psi}$ are affine, e.g., if they can be written in the form

$$\begin{aligned}
\boldsymbol{\phi}(\boldsymbol{x}) &= \mathbf{A}_1\boldsymbol{x} + \boldsymbol{b}_1, && (13.9)\\
\boldsymbol{\psi}(\boldsymbol{x}) &= \mathbf{A}_2\boldsymbol{x} + \boldsymbol{b}_2, && (13.10)
\end{aligned}$$

then the combined transformation is

$$\begin{aligned}
(\boldsymbol{\phi} \circ \boldsymbol{\psi})(\boldsymbol{x}) &= \mathbf{A}_1(\mathbf{A}_2\boldsymbol{x} + \boldsymbol{b}_2) + \boldsymbol{b}_1 && (13.11)\\
&= \mathbf{A}_1\mathbf{A}_2\boldsymbol{x} + \mathbf{A}_1\boldsymbol{b}_2 + \boldsymbol{b}_1 && (13.12)\\
&= \mathbf{A}_3\boldsymbol{x} + \boldsymbol{b}_3, && (13.13)
\end{aligned}$$

which is itself a linear transformation. It is easy (and a good exercise) to show that the displacement $\boldsymbol{w}$ corresponding to the transformation $\boldsymbol{\phi} \circ \boldsymbol{\psi}$ has the form

$$\boldsymbol{w}(\boldsymbol{x}) = \boldsymbol{v}(\boldsymbol{x}) + \boldsymbol{u}(\boldsymbol{x} + \boldsymbol{v}(\boldsymbol{x})),$$

where $\boldsymbol{u}$ and $\boldsymbol{v}$ are the displacements corresponding to the transformations $\boldsymbol{\phi}$ and $\boldsymbol{\psi}$, respectively.

13.2.4 Computer implementation of transformations

When representing transformations using a computer we will almost always represent them as displacements. The n components of the displacement $\boldsymbol{u}$ can be stored as n-dimensional images. The composition of two transformations can be computed using built-in functions for interpolation available in many software libraries. The following MATLAB® listings show functions for applying a spatial transformation to an image and for composing two spatial transformations.

```
function J_warped = warp_image_2D(I, J, u1, u2)

%       I: reference/fixed image
%       J: moving image
%       u1, u2: Components of the displacement vector field u
%     J_warped: the transformed (warped) moving image

% Create a coordinate grid over the reference image domain
[x1,x1] = ndgrid(1:size(I,1), 1:size(I,2));

% Interpolate the moving image
J_warped = interpn(J, x1 + u1, x2 + u2, '*linear', 0)
```

```
function [w1,w2] = compose_transformations_2D(I, u1, u2, v1, v2)

%       I: reference/fixed image
%       u1, u2: Components of the displacement vector field u
%       v1, v2: Components of the displacement vector field v
%       w1, w2: Components of the displacement field w = v(x) + u(v(x))

% Create a coordinate grid over the reference image domain
[x1,x2] = ndgrid(1:size(I,1), 1:size(I,2));

% Perform the composition
w1 = v1 + interpn(u1, x1 + v1, x2 + v2, '*linear', 0)
w2 = v2 + interpn(u2, x1 + v1, x2 + v2, '*linear', 0)
```

Observe that the last parameter of the function `interpn` is zero, which specifies that when looking up values of $\boldsymbol{u}(\boldsymbol{x} + \boldsymbol{v}(\boldsymbol{x}))$ outside of the domain where $\boldsymbol{u}$ is defined, we assume $\boldsymbol{u}$ to be zero. This assumption makes sense when transformations are represented in terms of displacements, and this is one of the main reasons that displacements are used to represent spatial transformations in code.

13.2.5 Jacobian matrix and determinant

An important property of a transformation around a point is whether it expands or compresses space. For example, with aging, gray matter structures in the brain compress (due to loss of neurons and synapses) while the cerebrospinal fluid in the brain expands to fill in the space (Fig. 13.4). The local expansion or compression caused by a spatial transformation $\boldsymbol{\phi}$ is measured by the determinant of the $n \times n$ matrix of

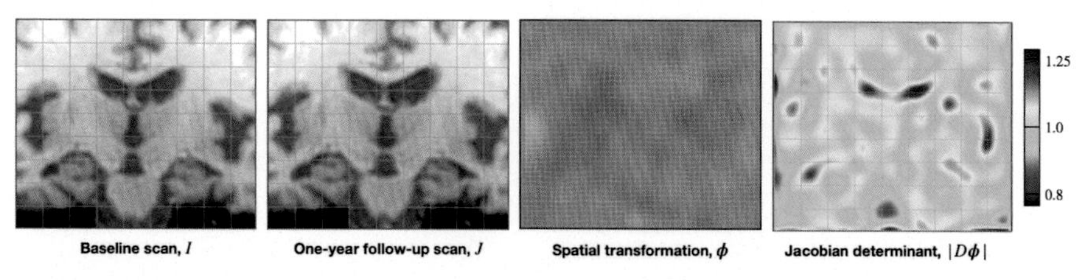

FIGURE 13.4

An example of a Jacobian determinant map derived from deformable registration between two MRI scans of the brain of the same individual taken one year apart. In the presence of neurodegenerative disease, brain gray matter (gray) visibly atrophies and cerebrospinal fluid (dark) expands. The Jacobian map (plotted here on the logarithmic scale) highlights regions of expansion (red) and shrinkage (blue).

partial derivatives of $\boldsymbol{\phi}$, denoted $D\boldsymbol{\phi}(\boldsymbol{x})$ and called the *Jacobian matrix*:

$$D\boldsymbol{\phi}(\boldsymbol{x}) = \begin{bmatrix} \frac{\partial \phi^1}{\partial x_1} & \cdots & \frac{\partial \phi^1}{\partial x_n} \\ \vdots & \ddots & \vdots \\ \frac{\partial \phi^n}{\partial x_1} & \cdots & \frac{\partial \phi^n}{\partial x_n} \end{bmatrix}. \tag{13.14}$$

Note that the Jacobian matrix of a linear function is particularly simple since

$$\boldsymbol{\phi}(\boldsymbol{x}) = \mathbf{A}\boldsymbol{x} + \boldsymbol{b} \Rightarrow D\boldsymbol{\phi}(\boldsymbol{x}) = \mathbf{A}. \tag{13.15}$$

The determinant of the Jacobian matrix, or *Jacobian determinant* for short, is a scalar value, denoted $|D\boldsymbol{\phi}(\boldsymbol{x})|$. The Jacobian determinant is greater than 1 in places where $\boldsymbol{\phi}$ causes a local expansion, less than 1 in places where $\boldsymbol{\phi}$ causes a local contraction, and equal to 1 in places where $\boldsymbol{\phi}$ does not cause a local change in volume. In places where $\boldsymbol{\phi}$ causes the space to fold on itself, the Jacobian determinant has non-positive values.

To interpret the Jacobian determinant geometrically, consider a small equilateral triangle $T_{\boldsymbol{x},\epsilon}$ with sides of length ϵ that is centered around a point $\boldsymbol{x} \in \Omega$ (Fig. 13.5). The transformation $\boldsymbol{\phi}$ will map the vertices of $T_{\boldsymbol{x},\epsilon}$ to new locations, forming a new triangle $\boldsymbol{\phi}T_{\boldsymbol{x},\epsilon}$. The ratio of the areas of $\boldsymbol{\phi}T_{\boldsymbol{x},\epsilon}$ and $T_{\boldsymbol{x},\epsilon}$ describes the local expansion/contraction of space around $\boldsymbol{x}$. The Jacobian determinant $|D\boldsymbol{\phi}(\boldsymbol{x})|$ is the limit as $\epsilon \to 0$ of the ratio of the signed area of $\boldsymbol{\phi}T_{\boldsymbol{x},\epsilon}$ to the signed area of $T_{\boldsymbol{x},\epsilon}$. This interpretation generalizes to higher dimensions, e.g., to tetrahedral volumes in 3D.

13.3 Parametric volumetric registration

In this section, we will present *PVR*. Parametric techniques were among the earliest solutions to the problem of PVR [6,7], but they remain widely used today. Sev-

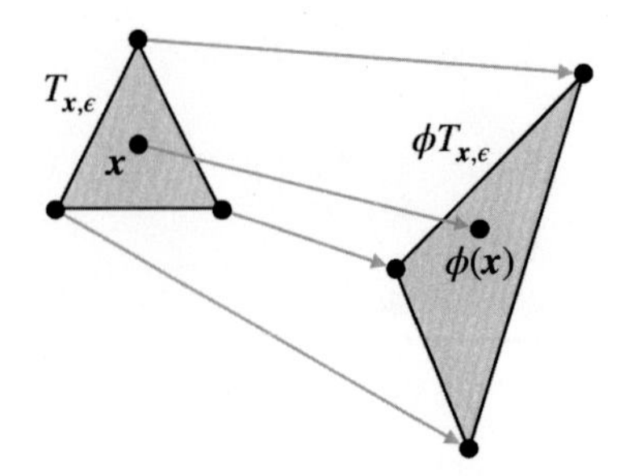

FIGURE 13.5

The Jacobian determinant of the transformation $\boldsymbol{\phi}$ at the point $\boldsymbol{x}$ can be interpreted as the limit of the ratio of the area of the deformed triangle $\boldsymbol{\phi} T_{\boldsymbol{x},\epsilon}$ to the area of triangle $T_{\boldsymbol{x},\epsilon}$ centered on $\boldsymbol{x}$, as the size of the triangle, ϵ, goes to zero.

eral advanced software packages implementing PVR exist, such as `elastix` [8] and `IRTK` [7]. PVR methods are among the most accessible because they do not require advanced mathematics, such as calculus of variations, to describe.

13.3.1 Transformations

The central idea in PVR methods is to model spatial transformations $\boldsymbol{\phi_\theta}$ using families of continuous functions weighted by scalar coefficients. Let $f_1 : \Omega \to \mathbb{R}, \ldots, f_K : \Omega \to \mathbb{R}$ be some "family" of smooth functions such as monomials. These functions are an integral part of the transformation model, but once they are chosen, we will not vary any possible parameters they might have. In PVR, we model $\boldsymbol{\phi_\theta}$ as a weighted sum of these functions with vector-valued coefficients $\boldsymbol{\theta}_1 \in \mathbb{R}^n, \ldots, \boldsymbol{\theta}_K \in \mathbb{R}^n$:

$$\boldsymbol{\phi_\theta}(\boldsymbol{x}) = \boldsymbol{x} + \sum_{k=1}^{K} \boldsymbol{\theta}_k f_k(\boldsymbol{x}). \tag{13.16}$$

Let us note some properties of this transformation:

- When all the coefficients $\boldsymbol{\theta}_k$ are zeroes, $\boldsymbol{\phi_\theta}$ is the identity transform $\boldsymbol{\phi_\theta}(\boldsymbol{x}) = \boldsymbol{x}$.
- The transformation $\boldsymbol{\phi_\theta}$ is a smooth (infinitely differentiable) function as long as each f_k is infinitely differentiable.
- The transformation $\boldsymbol{\phi_\theta}$ is completely determined by the set of coefficients $\boldsymbol{\theta}_k$.
- With the appropriate choice of functions f_k and a sufficiently large number of functions/coefficients K, any smooth and bounded spatial transformation on domain Ω can be approximated by the weighted sum (13.16) with arbitrary accuracy.

In the following, we will present common transformations.

13.3.1.1 Affine transformations

Even for non-linear registrations, affine registration is often performed as an initial registration step to simplify more complicated registration methods. The affine transformation can be deconstructed into the following simpler transformations, where examples are given for transformations in 2D:

$$\text{identity:} \quad \boldsymbol{\phi}_{\boldsymbol{\theta}}^{\text{id}}(\boldsymbol{x}) = \boldsymbol{x} = \begin{bmatrix} x_1 \\ x_2 \end{bmatrix}, \tag{13.17}$$

$$\text{translation:} \quad \boldsymbol{\phi}_{\boldsymbol{\theta}}^{\text{trans}}(\boldsymbol{x}) = \boldsymbol{x} + \boldsymbol{b} = \begin{bmatrix} x_1 + b_1 \\ x_2 + b_2 \end{bmatrix}, \tag{13.18}$$

$$\text{scaling:} \quad \boldsymbol{\phi}_{\boldsymbol{\theta}}^{\text{scale}}(\boldsymbol{x}) = s\boldsymbol{x} = \begin{bmatrix} sx_1 \\ sx_2 \end{bmatrix}, \tag{13.19}$$

$$\text{rotation:} \quad \boldsymbol{\phi}_{\boldsymbol{\theta}}^{\text{rot}}(\boldsymbol{x}) = \mathbf{R}_{\omega}\boldsymbol{x} = \begin{bmatrix} x_1 \cos\omega - x_2 \sin\omega \\ x_1 \sin\omega + x_2 \cos\omega \end{bmatrix}, \tag{13.20}$$

$$\text{general linear:} \quad \boldsymbol{\phi}_{\boldsymbol{\theta}}^{\text{lin}}(\boldsymbol{x}) = \mathbf{A}\boldsymbol{x} = \begin{bmatrix} a_{11}x_1 + a_{12}x_2 \\ a_{21}x_1 + a_{22}x_2 \end{bmatrix}, \tag{13.21}$$

where $\boldsymbol{b}$ is a translation vector, s is a scaling scalar, $\mathbf{R}_{\omega}$ is a rotation matrix, and $\mathbf{A}$ is a general matrix where $\det(\mathbf{A}) \neq 0$. These may be combined by function composition (see Section 13.2.3), where a rigid transformation (13.2) consists of only the translation and rotation transformations and an affine transformation is the combination of the general linear and translation transformations:

$$\text{General affine:} \quad \boldsymbol{\phi}_{\boldsymbol{\theta}}^{\text{affine}}(\boldsymbol{x}) = \boldsymbol{\phi}_{\boldsymbol{\theta}}^{\text{trans}}\left(\boldsymbol{\phi}_{\boldsymbol{\theta}}^{\text{lin}}(\boldsymbol{x})\right) \tag{13.22}$$

$$= \mathbf{A}\boldsymbol{x} + \boldsymbol{b} = \begin{bmatrix} a_{11}x_1 + a_{12}x_2 + b_1 \\ a_{21}x_1 + a_{22}x_2 + b_2 \end{bmatrix}. \tag{13.23}$$

Note that care must be taken since $\boldsymbol{\phi}_{\boldsymbol{\theta}}^{\text{trans}}\left(\boldsymbol{\phi}_{\boldsymbol{\theta}}^{\text{lin}}(\boldsymbol{x})\right) \neq \boldsymbol{\phi}_{\boldsymbol{\theta}}^{\text{lin}}\left(\boldsymbol{\phi}_{\boldsymbol{\theta}}^{\text{trans}}(\boldsymbol{x})\right)$, and for such situations we say that these transformations do not *commute*. In n-dimensional space, affine registrations have at most $n^2 + n$ free parameters, but often this space of transformations is reduced, e.g., to positive definite transformations where $\boldsymbol{x}^T\mathbf{A}\boldsymbol{x} > 0$ and which have $(n+1)n/2 + n$ degrees of freedom or to rigid transformations (13.2) which have only three degrees of freedom in 2D and six in 3D.

13.3.1.2 Rotation in 3D

Rotation in 3D is a particularly troublesome transformation and requires special attention. Rotation in 2D commutes, i.e., $\mathbf{R}_{\omega_1}^{2D}\mathbf{R}_{\omega_2}^{2D} = \mathbf{R}_{\omega_2}^{2D}\mathbf{R}_{\omega_1}^{2D} = \mathbf{R}_{\omega_1+\omega_2}^{2D}$; however, this is not the case in 3D.

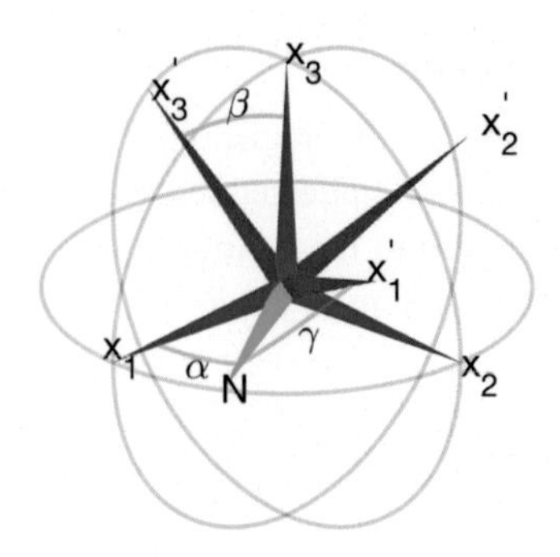

FIGURE 13.6

Rotating the coordinate vectors in blue by $\mathbf{R}_\gamma^{\boldsymbol{x}_3}\mathbf{R}_\beta^{\boldsymbol{x}_1}\mathbf{R}_\alpha^{\boldsymbol{x}_3}$ gives the red frame. $\boldsymbol{N} \parallel \boldsymbol{x}_3 \times \boldsymbol{x}_3'$.

A 3D rotation can be built from a series of rotations around the coordinate axes denoted here as $\boldsymbol{x}_1$, $\boldsymbol{x}_2$, and $\boldsymbol{x}_3$,

$$\mathbf{R}_\omega^{\boldsymbol{x}_1} = \begin{bmatrix} 1 & 0 & 0 \\ 0 & \cos\omega & -\sin\omega \\ 0 & \sin\omega & \cos\omega \end{bmatrix}, \tag{13.24}$$

$$\mathbf{R}_\omega^{\boldsymbol{x}_2} = \begin{bmatrix} \cos\omega & 0 & \sin\omega \\ 0 & 1 & 0 \\ -\sin\omega & 0 & \cos\omega \end{bmatrix}, \tag{13.25}$$

$$\mathbf{R}_\omega^{\boldsymbol{x}_3} = \begin{bmatrix} \cos\omega & -\sin\omega & 0 \\ \sin\omega & \cos\omega & 0 \\ 0 & 0 & 1 \end{bmatrix}. \tag{13.26}$$

A common combination is $\mathbf{R}_\gamma^{\boldsymbol{x}_3}\mathbf{R}_\beta^{\boldsymbol{x}_1}\mathbf{R}_\alpha^{\boldsymbol{x}_3}$, that is, rotate first around the $\boldsymbol{x}_3$-axis, then around the $\boldsymbol{x}_1$-axis, and finally again around the $\boldsymbol{x}_3$-axis. In this case, the angles are called the extrinsic Euler angles. Define the rotation of the coordinate axes $\boldsymbol{x}_i$ as $\boldsymbol{x}_i' = \mathbf{R}_\gamma^{\boldsymbol{x}_3}\mathbf{R}_\beta^{\boldsymbol{x}_1}\mathbf{R}_\alpha^{\boldsymbol{x}_3}\boldsymbol{x}_i$. When $\boldsymbol{x}_3$ and $\boldsymbol{x}_3'$ are non-parallel, then $\boldsymbol{N} = \boldsymbol{x}_3 \times \boldsymbol{x}_3'$ exists, and α is the angle between $\boldsymbol{x}_1$ and $\boldsymbol{N}$, γ is the angle between $\boldsymbol{N}$ and $\boldsymbol{x}_1'$, and β is the angle between $\boldsymbol{x}_3$ and $\boldsymbol{x}_3'$, as illustrated in Fig. 13.6. Euler rotations have several problems; for example, they do not commute, i.e., rotating around $\mathbf{R}_{\omega_1}^{\boldsymbol{x}_1}\mathbf{R}_{\omega_2}^{\boldsymbol{x}_2} \neq \mathbf{R}_{\omega_2}^{\boldsymbol{x}_2}\mathbf{R}_{\omega_1}^{\boldsymbol{x}_1}$. Further, a given rotation result x_1', x_2', and x_3' is non-unique since it can be reached by many different combinations of the axial rotation matrices. Lastly, Euler rotations suffer from gimbal locks, which means that for certain combinations of angles, the three angles of rotation degenerate to two. Gimbal locks are particular troublesome for animating rotating motions and can cause optimization problems in registration. An alternative system for rotation in 3D is the quaternion method.

Quaternions are an extension of complex numbers. A complex number is often written as $a + bi$, where $a, b \in \mathbb{R}$ are called the real and imaginary parts and i is the complex constant, which has the property that $i^2 = -1$. Quaternions have three basic quaternions, i, j, and k, and are often written as

$$q = a + bi + cj + dk, \tag{13.27}$$

where $a, b, c, d \in \mathbb{R}$. The basic quaternions have the property that

$$i^2 = j^2 = k^2 = ijk = -1 \tag{13.28}$$

and permutation of neighboring indices changes the sign in $ijk = -1$, for example, $ikj = jik = 1$. As a consequence, $ijkk = -1k \Rightarrow ij = k$, $ijkj = -j \Rightarrow ikjj = j \Rightarrow ik = -j$, etc. Quaternions have an extension of Euler's formula for complex numbers,

$$e^q = e^a\left(\cos\lambda + \frac{bi + cj + dk}{\lambda}\sin\lambda\right), \tag{13.29}$$

$$e^{-q} = e^a\left(\cos\lambda - \frac{bi + cj + dk}{\lambda}\sin\lambda\right), \tag{13.30}$$

where $\lambda = \sqrt{b^2 + c^2 + d^2}$. With the above, we are now able to use quaternions to rotate a point $\boldsymbol{u} = [u_1, u_2, u_3]$ at an angle θ around the axis $\boldsymbol{v} = [v_1, v_2, v_3]$. Let

$$p = u_1 i + u_2 j + u_3 k, \tag{13.31}$$

$$q = \frac{\theta}{2}(v_1 i + v_2 j + v_3 k) \tag{13.32}$$

and compute

$$p' = e^q p e^{-q}. \tag{13.33}$$

Then $p' = u_1' i + u_2' j + u_3' k$, and the rotated vector is $\boldsymbol{u}' = [u_1', u_2', u_3']$. When using the above system of conversion between 3D vectors and quaternions, the quaternion equivalents of (13.24)–(13.26) are $q_\omega^{\boldsymbol{x}_1} = \frac{\omega i}{2}$, $q_\omega^{\boldsymbol{x}_1} = \frac{\omega j}{2}$, and $q_\omega^{\boldsymbol{x}_3} = \frac{\omega k}{2}$. Compared to Euler rotations, quaternions are more compact, they are simpler to compose, and they are unique up to a rotation of 2π.

13.3.1.3 B-spline transformations

In contrast to affine transformations discussed in Section 13.3.1.1, non-linear transformations have non-linear components in $\boldsymbol{\phi}_{\boldsymbol{\theta}}$. A common system for representing non-linear transformations is based on B-splines [9]. A B-spline of degree m is a piecewise curve of polynomials of degree m. The locations where the polynomials meet are called knots, which we in the following will denote $T = [t_1, \ldots t_K]$, where $0 \le t_i \le 1$ and where the knot-sequence is a non-decreasing sequence, i.e., for all $i < j$, $t_i \le t_j$. Knots may be repeated, and a B-spline has continuous derivative up to order $m - p - 1$ at the knot, where p is the number of times a knot is repeated. Any B-spline of degree m may be written as a linear combination of basis functions

$B^m_{T,i}(t) : [0, 1) \to \mathbb{R}$,

$$\boldsymbol{\phi}_{T,m,\Theta}(t) = \sum_{k=1}^{K} \boldsymbol{\theta}_k B^m_{T,k}(t), \tag{13.34}$$

where $\boldsymbol{\theta} = [\boldsymbol{\theta}_1, \ldots, \boldsymbol{\theta}_K]$ is the sequence of coefficients and $\boldsymbol{\phi}_{T,m,\Theta}, \boldsymbol{\theta}_i \in \mathbb{R}^n$ such that $\boldsymbol{\phi}_{T,m,\Theta}$ is a curve in $\mathbb{R}^n$. In the following, we will write $B^m_k = B^m_{T,k}$ and $\boldsymbol{\phi}_m = \boldsymbol{\phi}_{T,m,\Theta}$ for brevity. The kth basis function is non-zero on the interval $[t_k, t_{k+m+1})$ and the basis functions overlap such that in the interval $t \in [0, 1)$, they sum to unity, i.e.,

$$\sum_k B^m_k(t) = 1. \tag{13.35}$$

They are defined recursively as

$$B^m_k(t) = v_{k,m}(t) B^{m-1}_k(t) + w_{k,m}(t) B^{m-1}_{k+1}(t), \tag{13.36}$$

$$v_{k,m}(t) = \begin{cases} \frac{t-t_k}{t_{k+m}-t_k} & \text{if } t_k < t_{k+m}, \\ 0 & \text{otherwise,} \end{cases} \tag{13.37}$$

$$w_{k,m}(t) = \begin{cases} \frac{t_{k+m+1}-t}{t_{k+m+1}-t_{k+1}} & \text{if } t_{k+1} < t_{k+m+1}, \\ 0 & \text{otherwise,} \end{cases} \tag{13.38}$$

$$B^0_k(t) = \begin{cases} 1 & \text{if } t_k \le t < t_{k+1}, \\ 0 & \text{otherwise.} \end{cases} \tag{13.39}$$

Hence, by construction, B^m_k is a polynomial of degree m since by recursion the factor t is multiplied with itself m times. Examples of basis functions for a uniformly sampled sequence of knots are shown in Fig. 13.7, where repeated knots are added in the beginning and the end of the knot-sequence to ensure fulfillment of (13.35). Since the kth basis function spans the interval $[t_k, t_{k+m+1})$, the number of non-zero basis functions cannot be larger than $K - m - 1$.

The coefficients $\boldsymbol{\theta}_k$ are most often found by minimizing a loss function. In registration, the loss function is related to how well two images match, but for simplicity consider here the problem of finding the B-spline which best fits a target function $f : \mathbb{R} \to \mathbb{R}$ in the sense of quadratic loss. Since f is a 1D function, $\theta_i \in \mathbb{R}$, and we write the optimization as

$$\theta_i^* = \arg\min_{\theta_n} \int_{[0,1)} \|f(t) - y_n(t)\|^2 \, dt. \tag{13.40}$$

In Fig. 13.8 examples of fitting a second degree B-spline to two simple functions are given. As can be seen, the choice of knots strongly influences the resulting fit. The B-spline with repeated inner knots in the bottom row fits the Gaussian function well, but not the sinus function. The opposite is true for the uniformly spaces knots in the top row.

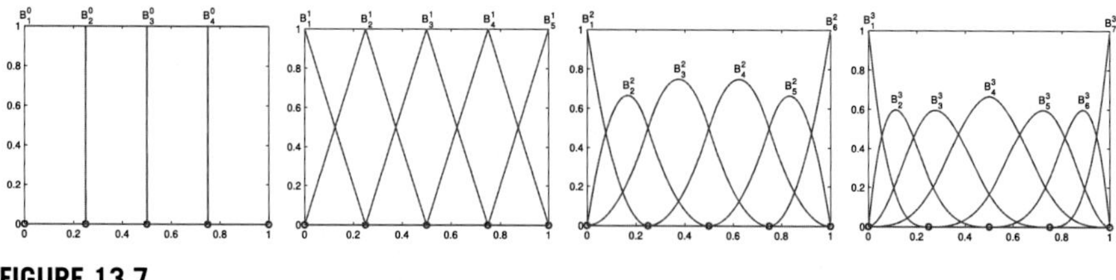

FIGURE 13.7

Example of B_k^m for $m = 0, \ldots, 3$ and where $T = [\overbrace{0, \ldots, 0}^{m}, 0, \frac{1}{4}, \frac{2}{4}, \frac{3}{4}, 1, \overbrace{1, \ldots, 1}^{m}]$. When $m = 0$, the basis functions are a sequence of non-overlapping box functions. When $m = 1$, the basis functions are overlapping triangle functions. For $m > 1$, as a consequence of the padded knots, the basis functions vary in shape.

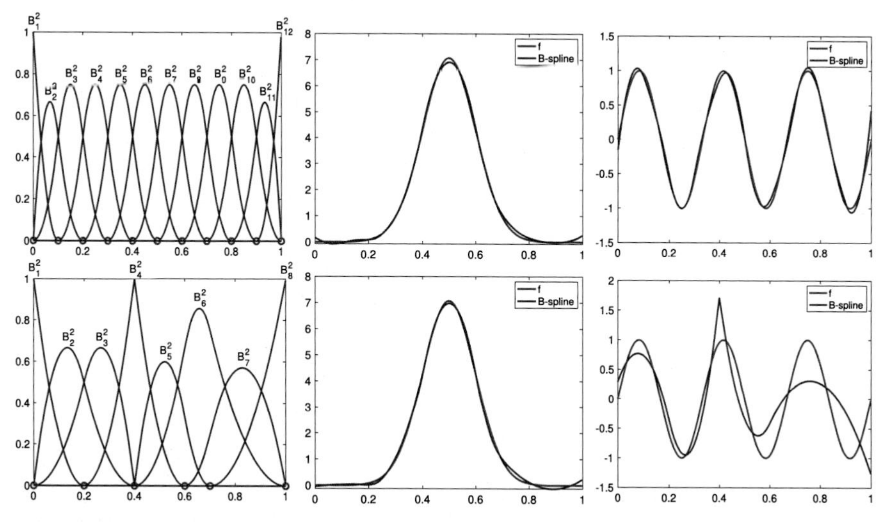

FIGURE 13.8

Example of second-degree B-spline functions with $\theta_k \in \mathbb{R}$. The left column shows the basis functions, and the middle and right columns show the optimal parameters for fitting a Gaussian and a sinus function, where red shows the target and blue the corresponding B-spline. In the top and bottom rows the knots are $[0, 0, 0, 0.1, 0.2, \ldots, 1, 1, 1]$ and $[0, 0, 0, 0.2, 0.4, 0.4, 0.6, 0.7, 1, 1, 1]$, respectively.

B-splines may be generalized to higher dimensions by multiplication of the basis functions in (13.36). For example, in 2D we choose the set of knots T and S, parameterize the plane with the coordinates $t, s \in [0, 1)$, choose a table of coefficients $\boldsymbol{\theta}_{i,j}$, and write

$$\boldsymbol{\phi}_{m_s,m_t}(s,t) = \sum_{i=1}^{|S|} \sum_{j=1}^{|T|} \boldsymbol{\theta}_{i,j} B_{S,i}^{m_s}(s) B_{T,j}^{m_t}(t). \tag{13.41}$$

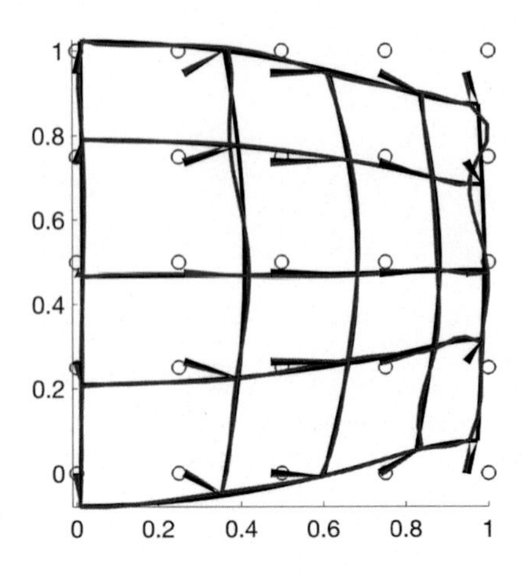

FIGURE 13.9

A third-degree B-spline fitted to a non-linear displacement field. The arrows show the target displacement field in black and the resulting B-spline displacement field in red. In the same colors are also shown examples of grid lines as a result of adding the displacement field to the original, squared, Cartesian grid lines.

In this case, $K = |S|\,|T|$. Often, $m_s = m_t$. The coefficients $\boldsymbol{\theta}_{i,j}$ may again be found by straightforward generalization of (13.40). An example of a fit is shown in Fig. 13.9. Here we fitted a third-degree B-spline to a displacement function as the gradient of the derivative of a Gaussian function, that is, $\boldsymbol{u}(s,t) = \left(D g_{0.5,0.5}(s,t)\right)^T$, where $g_{\mu,\sigma}(s,t) = \frac{s}{\sigma^2\sqrt{2\pi}} \exp\left(-\frac{(s-\mu)^2+(t-\mu)^2}{2\sigma^2}\right)$ for the interval $s,t \in [0,1)$. The figure also shows grid lines, which were generated using (13.16), that is, for the case of the target function, lines in black are drawn as $[s_0, t]^T + \boldsymbol{u}(s,t)$ and $[s, t_0]^T + \boldsymbol{u}(s,t)$ for fixed values of s_0 and t_0, and similarly for the B-spline in red. As the example shows, the fit of the displacement vectors is fairly good for this set of knots and coefficients, but similarly to the 1D example in Fig. 13.8, the grid lines between the knots behave undesirably. Ways to improve this fit include increasing the number of knots and decreasing the B-spline degree.

13.3.1.4 Radial basis functions

A common alternative to B-splines is the family of *radial basis functions*, which are defined as follows. Let Ω be a rectangular domain in $\mathbb{R}^n$ and let $\boldsymbol{c}_1 \in \Omega, \ldots, \boldsymbol{c}_K \in \Omega$ be a set of points in this domain, called *control points*. For example, control points may lie on a uniform grid, as in Fig. 13.10. Associate each control point $\boldsymbol{c}_k$ with a continuous function $f_k(\boldsymbol{x})$,

$$f_k(\boldsymbol{x}) = \eta(|\boldsymbol{x} - \boldsymbol{c}_k|),$$

where $\eta : \mathbb{R} \to \mathbb{R}$ is some continuous function called the *kernel function*. Plugging this form $f_k(\boldsymbol{x})$ into (13.16), we get the following expression for the spatial transfor-

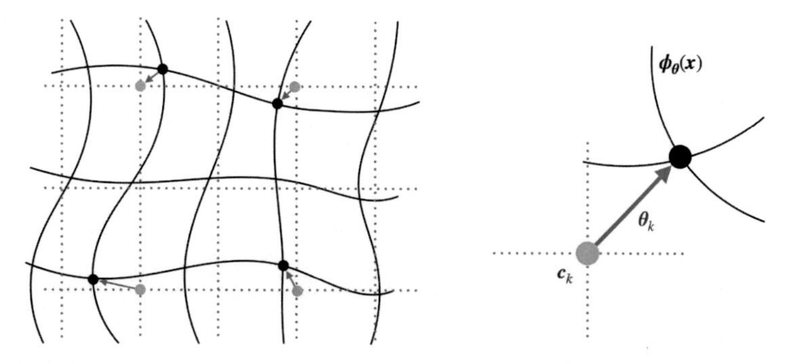

FIGURE 13.10

Construction of a spatial transformation ϕ_θ via interpolation. A grid of control points $\{c_k\}$ is defined over the image domain Ω, and each control point is associated with a coefficient vector θ_k. The spatial transformation ϕ_θ at each point in Ω is computed by interpolating the coefficient vectors, e.g., using radial basis function interpolation or B-spline interpolation.

mation ϕ_θ:

$$\phi_\theta(x) = x + \sum_{k=1}^{K} \theta_k \eta(|x - c_k|).$$

As illustrated in Fig. 13.10, the transformation $\phi_\theta(x)$ is a continuous function on the image domain that is fully determined by the coefficients θ_k.

A common choice of kernel function is the un-normalized Gaussian $\eta(r) = e^{-0.5r^2}$. Note that f_k above only depends on the distance from x to the control point c_k, i.e., it is radially symmetric around c_k. This is why they are called *radial basis functions*. They were used in some of the earliest PVR papers [6], and they are just one example of a suitable family of functions for PVR. Other choices are discussed later in this section.

13.3.2 Optimization

So far, what we have done was to model spatial transformations on the domain Ω using an nK-element vector of coefficients $\theta = \{\theta_1 \ldots, \theta_K\}$. We can now formulate the PVR problem as the problem of finding the set of values θ that minimizes the dissimilarity between image I and the transformed image $J \circ \phi_\theta$ as (13.1):

$$\theta^* = \arg\min_{\theta \in \mathbb{R}^{nK}} E[\phi_\theta; I, J] = \arg\min_{\theta \in \mathbb{R}^{nK}} \mu[I, J \circ \phi_\theta] + \lambda \rho[\theta].$$

For simplicity, consider the SSD dissimilarity measure and no regularization. Then the energy can be written in complete form as

$$\begin{aligned}\boldsymbol{\theta}^* &= \underset{\boldsymbol{\theta}\in\mathbb{R}^{nK}}{\arg\min} \int_\Omega \left[I(\boldsymbol{x}) - J(\boldsymbol{\phi}_{\boldsymbol{\theta}}(\boldsymbol{x}))\right]^2 d\boldsymbol{x} \\ &= \underset{\boldsymbol{\theta}\in\mathbb{R}^{nK}}{\arg\min} \int_\Omega \left[I(\boldsymbol{x}) - J(\boldsymbol{x} + \sum_{k=1}^{K} \boldsymbol{\theta}_k f_k(\boldsymbol{x}))\right]^2 d\boldsymbol{x}. \end{aligned} \tag{13.42}$$

This is an unconstrained energy minimization problem involving nK unknowns that can be solved using gradient-based numerical optimization methods. For completeness, let us derive the partial derivative of the energy function E with respect to the kth coefficient vector $\boldsymbol{\theta}_k$:

$$\frac{\partial E}{\partial \boldsymbol{\theta}_k} = 2 \int_\Omega \left[I(\boldsymbol{x}) - J(\boldsymbol{\phi}_{\boldsymbol{\theta}}(\boldsymbol{x}))\right]\left[-DJ(\boldsymbol{\phi}_{\boldsymbol{\theta}}(\boldsymbol{x}))\, f_k(\boldsymbol{x})\right] d\boldsymbol{x}. \tag{13.43}$$

Implementing the above PVR strategy on real discrete images is not difficult. First, let us consider how to discretize the problem. Images I and J will be discrete images (n-dimensional arrays of intensity values). The spatial transformation $\boldsymbol{\phi}_{\boldsymbol{\theta}}$ may be represented as a set of n separate n-dimensional images, one for each of the components of $\boldsymbol{\phi}_{\boldsymbol{\theta}}$ (i.e., $\boldsymbol{\phi}^1_{\boldsymbol{\theta}}, \boldsymbol{\phi}^2_{\boldsymbol{\theta}}, \ldots$). The main operations in (13.42) and (13.43) above are the computation of $\boldsymbol{\phi}_{\boldsymbol{\theta}}(\boldsymbol{x})$ from the coefficients $\boldsymbol{\theta}$, the computation of $J \circ \boldsymbol{\phi}_{\boldsymbol{\theta}}$, the computation of $DJ \circ \boldsymbol{\phi}_{\boldsymbol{\theta}}$, and integration over the image domain. Computing $J \circ \boldsymbol{\phi}_{\boldsymbol{\theta}}$ in the discrete setting involves sampling the values of J at positions $\boldsymbol{\phi}_{\boldsymbol{\theta}}$. In MATLAB, this is accomplished using the command `interpn`, which interpolates the values of an array at arbitrary sample points.

13.3.3 Real-world approaches

While the approach sketched out above is relatively simple, the same basic principle is followed by the state-of-the-art PVR techniques. Modern techniques employ more efficient ways to parameterize spatial transformations and more advanced image dissimilarity metrics, incorporate multiresolution schemes, and provide options for regularizing the transformation $\boldsymbol{\phi}_{\boldsymbol{\theta}}$.

Choice of basis functions

A significant limitation of the simple approach is that $\boldsymbol{\phi}_{\boldsymbol{\theta}}$ is expensive to compute, requiring $O(KN)$ operations for an image with N pixels or voxels. This is because each of the K basis functions f_k is evaluated at each voxel in (13.16). In fact, after discretization, the computation of the dth component of $\boldsymbol{\phi}_{\boldsymbol{\theta}}$ (for $d \in 1, \ldots, n$) can be expressed in matrix form as follows:

$$\Phi^d_{\boldsymbol{\theta}} = \mathrm{Id}^d + F\boldsymbol{\theta}^d,$$

where $\boldsymbol{\theta}^d$ is a $K \times 1$ vector containing the dth component of each coefficient, F is an $N \times K$ matrix whose columns are the basis functions f_k evaluated at all voxels, Id^d is an $N \times 1$ vector containing the dth coordinate of each voxel, and $\Phi^d_{\boldsymbol{\theta}}$ is an $N \times 1$ vector containing the dth component of $\boldsymbol{\phi}_{\boldsymbol{\theta}}$.

For radial basis functions, the computation can be made more efficient by using basis functions $\{f_k\}$ that have *finite support,* meaning that each f_k is non-zero on only a small subset Ω_k of the image domain Ω. In this case, the computational cost of computing $\boldsymbol{\phi}_\theta$ is $O(Km)$, where $m \ll N$ is the average size of the domains $\{\Omega_k\}$. This is equivalent to having F be a sparse matrix.

Rueckert et al. [7] popularized the use of cubic *B-splines* as the basis for PVR. These bi-cubic (or in 3D, tri-cubic) polynomials are parameterized in terms of a uniform grid of control points and have finite support, such that each control point influences a 4×4 grid cell region around it and has no effect on $\boldsymbol{\phi}_\theta$ outside of this region. Techniques like `IRTK` [7], `NiftyReg` [10], and `elastix` [8] use *B-splines* to represent spatial transformations.

Support for advanced metrics

An important practical consideration in the development of PVR tools is support for multiple image dissimilarity measures, such as MI [11–13] (for across-modality registration) and NCC [3,4]. When using more advanced dissimilarity measures, the computation of the partial derivatives $\frac{\partial E}{\partial \theta_m}$ becomes more complex. Specifically, the term $\left[I(\boldsymbol{x}) - J(\boldsymbol{\phi}_\theta(\boldsymbol{x}))\right]$ in (13.43) should be replaced by the partial derivative of $\mu[I, J \circ \boldsymbol{\phi}_\theta]$ with respect to $J \circ \boldsymbol{\phi}_\theta$. Efficient schemes exist for computation of this partial derivative for the MI and cross-correlation dissimilarity measures, but they are outside of the scope of this book.

Multiresolution schemes

Multiresolution is one of the most effective tools in real-world PVR. A multiresolution pyramid of images is generated from I and J, typically by subsampling by the factor 2 in each dimension several times. For example, if I has dimensions 256×256, the pyramid may contain images of sizes 32×32, 64×64, 128×128, and 256×256. PVR is performed at the lowest resolution first, and the resulting low-resolution transformation $\boldsymbol{\phi}_\theta$ is interpolated by the factor of 2 to initialize PVR at the next resolution level, and so on. The number of control points K typically also scales with the resolution. Multiresolution not only helps PVR converge faster but also helps avoid local minima at early stages of registration by matching more prominent features in the images that are visible at low resolution. When subsampling images for multiresolution schemes, it is important to low-pass filter them appropriately to avoid excessive aliasing due to sampling under the Nyquist frequency (Chapter 5). An advantage of low-pass filtering is that numerical approximation of DJ is more accurate, and hence DE is more accurate as well, leading to better optimization convergence.

Regularization

Additional terms can be added to the PVR objective function to impose desired properties on $\boldsymbol{\phi}_\theta$. For example, if it is known a priori that the deformation is expected to be small or smooth, a penalty on the magnitude or gradient magnitude of $\boldsymbol{\phi}_\theta$ can be imposed.

In some applications, it may be desirable to obtain transformations that approximately preserve local area or volume of structures. Recall that the amount by which the transformation $\boldsymbol{\phi}_\theta$ expands or compresses area/volume at point $\boldsymbol{x} \in \Omega$ is given by the Jacobian determinant of $\boldsymbol{\phi}_\theta$. A regularization term can be formulated to penalize the deviation of the Jacobian determinant from 1 (or the deviation of the logarithm of the Jacobian determinant from 0), thus favoring transformations that do not compress or expand space too much.

13.4 Exercises

These exercises start with the simplest possible registration problem and extend it gradually, incorporating more complex transformations and image similarity metrics. They can be done in MATLAB or Python. However, for the best experience, we recommend using a software framework capable of representing arbitrary-dimensional tensors and performing automatic differentiation, such as PyTorch or TensorFlow. Automatic differentiation allows you to define the objective functions you want to minimize/maximize, without worrying about computing the derivatives of these objective functions with respect to parameters.

1. Implement the simplest possible image registration problem: affine registration ($\phi(x) = a \cdot x + b$) between one-dimensional (1D) images I and J using the sum of squared differences (SSD) metric. To generate input "images" for this problem, you can take a 2D medical image and sample it along two lines at small angles to each other. This will generate 1D datasets that have similar features but with small misalignments. To implement 1D registration, you will need to take the following steps:
   - **a.** Define the coordinate grid on which your fixed image is defined. The grid can be represented as a one-dimensional tensor (array) of coordinates $1, 2, \ldots, n$.
   - **b.** Pick some initial set of transformation parameters, e.g. $a = 1$ and $b = 0$.
   - **c.** Apply the affine transformation $\phi(x) = a \cdot x + b$ to your coordinate grid.
   - **d.** Resample the moving image using the affine transformation - this requires using an interpolation function such as **interp1** in MATLAB or **grid_sample** in PyTorch.
   - **e.** Compute the SSD metric between the fixed image and the resampled moving image.
   - **f.** Compute the gradient of the SSD metric with respect to unknown parameters a and b. Perform a gradient descent step. This step can benefit from auto-differentiation.
   - **g.** Plot the current state of the registration by plotting the fixed image, moving image and resampled moving image as functions of x.
   - **h.** Iterate steps c-f until the metric converges.

2. Improve the implementation above by using a numerical optimizer such as Conjugate gradient descent or limited-memory BFGS (available in MATLAB, PyTorch, etc.).
3. Extend your implementation above to affine registration of 2D or 3D medical images. The main difference will be in the dimensionality of the grid and in the form of the transformation, which will now have the form $\phi(\boldsymbol{x}) = \boldsymbol{A}\boldsymbol{x} + \boldsymbol{b}$, where $\boldsymbol{A}$ is a matrix and $\boldsymbol{b}$ a vector.
4. Extend your implementation to 2D parametric registration using radial basis functions. Define a $10x10$ grid of control points $\boldsymbol{c}_k$ with a displacement vector $\boldsymbol{\theta}_k$ stored at each control point, and use radial basis function interpolation to compute a continuous displacement field over the entire coordinate grid. Minimize the SSD metric with respect to the parameters $\{\boldsymbol{\theta}_k\}$.
5. Extend your implementation by using the NCC metric instead of the SSD metric. To compute the NCC metric between the fixed image I and the resampled moving image $J \circ \boldsymbol{\phi}$, you need to compute correlation coefficients between the intensities of these images in the neighborhood of each pixel. You can do this efficiently by using convolution or average pooling operators to compute the running window averages of I, $J \circ \boldsymbol{\phi}$, I^2, $(J \circ \boldsymbol{\phi})^2$ and $I \cdot (J \circ \boldsymbol{\phi})$, from which the correlation coefficient can be computed at each pixel.
6. Extend your implementation to support a multi-resolution scheme.

References

[1] Evans A.C., Janke A.L., Collins D.L., Baillet S., Brain templates and atlases, NeuroImage 62 (2) (2012) 911–922.

[2] Iglesias J.E., Sabuncu M.R., Multi-atlas segmentation of biomedical images: a survey, Medical Image Analysis 24 (1) (2015) 205–219.

[3] Hill D.L.G., Studholme C., Hawkes D.J., Voxel similarity measures for automated image registration, in: Visualization in Biomedical Computing 1994, vol. 2359, International Society for Optics and Photonics, 1994, pp. 205–216.

[4] Avants B.B., Epstein C.L., Grossman M., Gee J.C., Symmetric diffeomorphic image registration with cross-correlation: evaluating automated labeling of elderly and neurode-generative brain, Medical Image Analysis 12 (1) (Feb. 2008) 26–41, https://doi.org/10.1016/j.media.2007.06.004.

[5] Avants B.B., Tustison N.J., Song G., Cook P.A., Klein A., Gee J.C., A reproducible evaluation of ANTs similarity metric performance in brain image registration, NeuroImage 54 (3) (2011) 2033–2044.

[6] Bookstein F., Principal warps: Thin-plate splines and the decomposition of deformations, IEEE Transactions on Pattern Analysis and Machine Intelligence 11 (1989) 567–585.

[7] Rueckert D., Sonoda L.I., Hayes C., Hill D.L.G., Leach M.O., Hawkes D.J., Nonrigid registration using free-form deformations: application to breast MR images, IEEE Transactions on Medical Imaging 18 (8) (Aug. 1999) 712–721, https://doi.org/10.1109/42.796284.

[8] Klein S., Staring M., Murphy K., Viergever M.A., Pluim J.P.W., Elastix: a toolbox for intensity-based medical image registration, IEEE Transactions on Medical Imaging 29 (1) (2009) 196–205.
[9] de Boor C., A Practical Guide to Splines, Springer, 1978.
[10] Modat M., Ridgway G.R., Taylor Z.A., Lehmann M., Barnes J., Hawkes D.J., Fox N.C., Ourselin S., Fast free-form deformation using graphics processing units, Computer Methods and Programs in Biomedicine 98 (3) (June 2010) 278–284, https://doi.org/10.1016/j.cmpb.2009.09.002.
[11] Wells W.M. III, Viola P., Atsumi H., Nakajima S., Kikinis R., Multi-modal volume registration by maximization of mutual information, Medical Image Analysis 1 (1) (Mar. 1996) 35–51.
[12] Maes F., Collignon A., Vandermeulen D., Marchal G., Suetens P., Multi-modality image registration by maximization of mutual information, IEEE Transactions on Medical Imaging 16 (2) (Apr. 1997) 187–198.
[13] Studholme C., Hill D.L.G., Hawkes D.J., An overlap invariant entropy measure of 3D medical image alignment, Pattern Recognition 32 (1) (1999) 71–86.

CHAPTER

Non-parametric volumetric registration

14

Paul A. Yushkevich[a] **and Miaomiao Zhang**[b]

[a]*Department of Radiology, University of Pennsylvania, Philadelphia, PA, United States*

[b]*Departments of Electrical and Computer Engineering and Computer Science, University of Virginia, Charlottesville, VA, United States*

Learning points

- Diffeomorphisms: motivation and mathematics
- Optical flow and non-parametric registration
- Log-domain diffeomorphic Demons
- Large deformation diffeomorphic metric mapping
- Computational anatomy

14.1 Introduction

Volumetric registration (VR) is one of the most studied problems in the medical image analysis field. It has led to elegant mathematical solutions that tie together elements of group theory, differential geometry, and probability theory. In Chapter 13 the basic building blocks of VR were presented together with *parametric VR* (PVR). In this chapter, we continue with the presentation of *non-parametric VR* (NVR). For basic definitions and applications, the reader should consult Chapter 13.

In this chapter we will explore several representative NVR techniques.

- *Key mathematical concepts* will be reviewed in Section 14.2. Specifically, we will focus on the concept of diffeomorphisms, a subset of spatial transformations that are smooth and have a smooth inverse.
- *Optical flow and related NVR methods* will be discussed in Section 14.3. Optical flow is not strictly a registration technique, but a way of computing velocities of moving objects in pairs of images taken a short time apart. However, the concept of optical flow is very useful to understanding popular non-parametric registration techniques like the Demons algorithm [1] and its more recent variants [2,3].
- *Large deformation diffeomorphic metric mapping* (LDDMM), which will be discussed in Section 14.4, is a category of non-parametric methods that allow NVR to be formulated as an energy minimization problem while constraining the solutions $\boldsymbol{\phi}$ to be diffeomorphic. LDDMM also provides a mathematically sound way to measure distances and perform statistics in the infinite-dimensional space

Medical Image Analysis. https://doi.org/10.1016/B978-0-12-813657-7.00029-7

of images, and for this reason it has found extensive applications in studies of anatomical variability.

14.2 Mathematical concepts

This section introduces several mathematical concepts that will be used throughout the rest of the chapter. These concepts primarily involve spatial transformations.

14.2.1 Diffeomorphisms

Diffeomorphisms are a special class of spatial transformations that have been studied extensively in the NVR literature. A transformation $\boldsymbol{\phi} : \Omega \to \Omega$ is called a *diffeomorphism* if it is differentiable, it is a bijection (i.e., is one-to-one and onto, and thus has an inverse), and its inverse $\boldsymbol{\phi}^{-1} : \Omega \to \Omega$ is also differentiable.

In many applications of VR, we prefer the solutions to be diffeomorphisms. Reasons for this include:

- VR is often used to map pixelwise measurements from the moving image to the fixed image (these data may be in the form of segmentations, statistical maps such as those derived from functional magnetic resonance imaging (MRI) experiments, etc.). In the course of analysis it is often necessary to also be able to map the information in the opposite direction, from fixed to moving. For example, in functional MRI analysis, we may map information from individual brains to a template brain in order to perform group analysis and identify brain regions associated with a particular cognitive function, such as finger tapping. We then want to be able to map these regions back into individual subjects' brains for secondary analyses (e.g., brain network analyses). When VR produces diffeomorphic transformations, we are guaranteed that information can be mapped in both directions.
- In many research studies, the transformations derived by VR are used as the source of features for statistical analysis. In *deformation-based morphometry* (DBM) [4], VR is performed between a pair of scans of the same individual taken some time apart. The Jacobian determinant of the transformation $\boldsymbol{\phi}$ is then analyzed to identify anatomical regions where tissue has shrunk or expanded over time. Such measures of tissue loss and expansion can be sensitive to subtle atrophy that occurs in the brain in some neurodegenerative diseases, such as Alzheimer's. However, when a transformation $\boldsymbol{\phi}$ is not invertible, there exist points in Ω where $|D\boldsymbol{\phi}| \leq 0$, which corresponds to the folding or self-intersection of the space, and cannot be interpreted in an anatomically meaningful way. Therefore, in DBM studies, we aim to restrict solutions of VR to diffeomorphisms.
- When modeling the motion of organs such as the beating heart, the underlying anatomy is known a priori to deform in a way that is diffeomorphic, so it is only natural to require the solutions of VR to be diffeomorphisms.

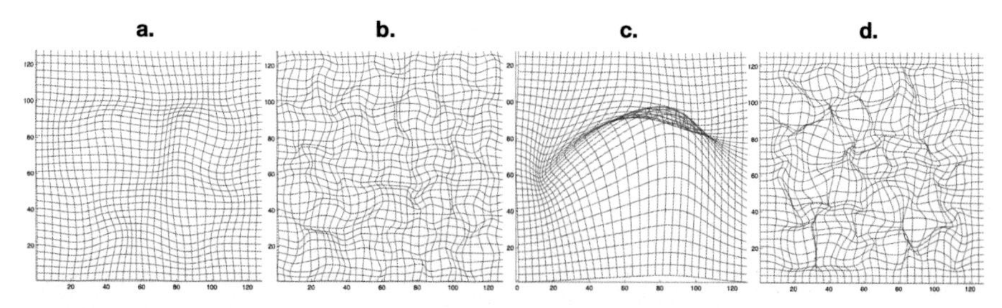

FIGURE 14.1

Examples of diffeomorphic (a, b) and not diffeomorphic (c, d) transformations. Even though transformation (c) is very smooth, it is not invertible.

- However, diffeomorphisms cannot handle VR problems where a certain object or structure is present in one image and absent in another, e.g., a pair of longitudinal MRI scans where lesion appears or disappears over time.

Examples of diffeomorphic and non-diffeomorphic transformations in 2D are shown in Fig. 14.1. An important observation is that *smoothness is not a sufficient condition for a diffeomorphism*: it is possible to have a very smooth transformation that does not have an inverse, just like the function $f(x) = (2x - 1)^2$ does not have an inverse on the interval $x \in [0, 1]$.

14.2.1.1 Group of diffeomorphisms

Diffeomorphisms *form a group under the operation of composition*, called Diff. Specifically, the four conditions of a set being a group are met for diffeomorphisms under composition:

- **Closure**: $\boldsymbol{\phi} \circ \boldsymbol{\psi} \in \text{Diff}$ for all $\boldsymbol{\phi}, \boldsymbol{\psi} \in \text{Diff}$.
- **Identity**: There exists an element $\text{Id} \in \text{Diff}$ that satisfies $\boldsymbol{\phi} \circ \text{Id} = \text{Id} \circ \boldsymbol{\phi} = \boldsymbol{\phi}$ for all $\boldsymbol{\phi} \in \text{Diff}$.
- **Associativity**: For all $\boldsymbol{\phi}, \boldsymbol{\psi}, \boldsymbol{\chi} \in \text{Diff}$, $(\boldsymbol{\phi} \circ \boldsymbol{\psi}) \circ \boldsymbol{\chi} = \boldsymbol{\phi} \circ (\boldsymbol{\psi} \circ \boldsymbol{\chi})$.
- **Inverse**: For all $\boldsymbol{\phi} \in \text{Diff}$, there exists $\boldsymbol{\phi}^{-1} \in \text{Diff}$ such that $\boldsymbol{\phi} \circ \boldsymbol{\phi}^{-1} = \boldsymbol{\phi}^{-1} \circ \boldsymbol{\phi} = \text{Id}$.

One important property of diffeomorphisms is that the Jacobian determinant of $\boldsymbol{\phi} \in \text{Diff}$ is non-negative for all $\boldsymbol{x} \in \Omega$. Conversely, if $\boldsymbol{\phi}$ is a differentiable vector-valued function that satisfies $|D\boldsymbol{\phi}| > 0$ for all $\boldsymbol{x} \in \Omega$, then $\boldsymbol{\phi}$ is a diffeomorphism.

14.2.1.2 Small transformations

While it is easy to define what a diffeomorphism is, it is less obvious how one would go about generating diffeomorphisms. As a starting point, let us consider *small transformations*: transformations that move points by just a small amount. Let $\boldsymbol{u} : \Omega \to \mathbb{R}^n$ be a vector-valued function that is differentiable and bounded and has a bounded derivative, i.e., $|\boldsymbol{u}| + |D\boldsymbol{u}| \leq M$ for all $\boldsymbol{x} \in \Omega$ and some $M > 0$. Let us denote the space of all such functions $\mathcal{U}$.

Then for a sufficiently small $\epsilon > 0$, the transformation

$$\boldsymbol{\psi}(\boldsymbol{x}) = \boldsymbol{x} + \epsilon \boldsymbol{u}(\boldsymbol{x})$$

has a positive Jacobian determinant for every $\boldsymbol{x} \in \Omega$, and therefore is a diffeomorphism. We call $\boldsymbol{\psi}$ a small transformation because it only slightly deviates from Id.

Now, since a composition of two diffeomorphisms is a diffeomorphism, we can use small transformations to build up larger transformations. Given a set of functions $\boldsymbol{u}_1, \ldots, \boldsymbol{u}_n$ in $\mathcal{U}$, we can compose the corresponding small transformations $\boldsymbol{\psi}_i(\boldsymbol{x}) = \boldsymbol{x} + \epsilon \boldsymbol{u}_i(\boldsymbol{x})$, $i \in \{1, \ldots, n\}$, yielding a "large" transformation $\boldsymbol{\phi}$ that is diffeomorphic:

$$\boldsymbol{\phi} = \boldsymbol{\psi}_n \circ \boldsymbol{\psi}_{n-1} \circ \ldots \circ \boldsymbol{\psi}_1. \tag{14.1}$$

The key observation here is that it is relatively easy to generate vector-valued functions that are differentiable and bounded (e.g., this can be done by interpolating a discrete set of vectors placed in Ω). The recipe above provides a way to use such functions to generate diffeomorphic transformations, which form a highly non-linear subspace of the space of all transformations. The general idea of generating diffeomorphic transformations by composing small transformations is used throughout the later portion of this chapter.

14.2.1.3 Flow ordinary differential equation

Another closely related way to generate diffeomorphisms in $\mathbb{R}^n$ is by means of the flow ordinary differential equation (ODE) [5]. Instead of defining a discrete set of functions $\boldsymbol{u}_1, \ldots, \boldsymbol{u}_n$ in $\mathcal{U}$ as we did above, let us consider a function $\boldsymbol{v}(\boldsymbol{x}, t) : \Omega \times [0, 1] \to \mathbb{R}^n$ that is a function of both space and time. Let $\boldsymbol{v}$ be differentiable in t and let $\boldsymbol{v}(\bullet, t) \in \mathcal{U}$ for all t. We will call this function a *time-varying velocity field.* Consider a system of particles that move through space such that the velocity of a particle at position $\boldsymbol{x}$ at time t is $\boldsymbol{v}(\boldsymbol{x}, t)$. The state of such a system of particles is described by the *flow ODE*:

$$\frac{\mathrm{d}\boldsymbol{\phi}}{\mathrm{d}t}(\boldsymbol{x}, t) = \boldsymbol{v}(\boldsymbol{\phi}(\boldsymbol{x}, t), t), \quad \text{subject to } \boldsymbol{\phi}(\boldsymbol{x}, 0) = \boldsymbol{x}, \tag{14.2}$$

where $\boldsymbol{\phi}(\boldsymbol{x}, t)$ describes the position at time t of a particle that at time $t = 0$ was located at position $\boldsymbol{x}$. An example of such a system is dust particles that are being carried around by the wind. The time-varying velocity field $\boldsymbol{v}$ describes the velocity of wind at every point in space and time, and $\phi(\boldsymbol{x}, t)$ describes the trajectories of the dust particles.

The existence and uniqueness theorem for first-order ODEs [6] implies that the solutions of this equation are diffeomorphisms for all $t \in [0, 1]$. In our wind analogy, this is equivalent to saying that two dust particles cannot occupy the same position at the same time. As long as $\boldsymbol{v}$ satisfies the conditions stated above, the solution of the flow ODE, $\boldsymbol{\phi}(\boldsymbol{x}, 1)$, is a bijective and differentiable function with a differentiable inverse, i.e., a diffeomorphism. The flow ODE provides us with another recipe

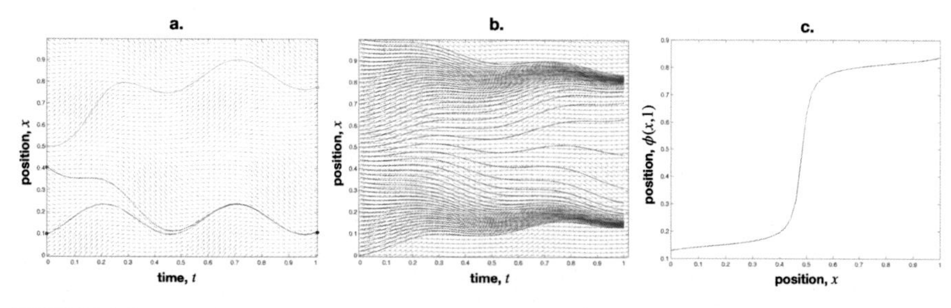

FIGURE 14.2

Illustration of the flow ODE for one spatial dimension. In (a), the gray arrows represent the time-varying velocity field $v(x,t)$ and the blue, red, and green curves represent the solution of the flow ODE, $\boldsymbol{\phi}(x,t)$, for $x=0.1$, $x=0.4$, and $x=0.5$, respectively. The tangent lines to the curves coincide with the velocity field (as prescribed by the flow ODE). The red and blue curves come very close to each other after $t>0.5$, but never cross, so that the order of the curves is maintained through time. In (b), the solution of the flow ODE is plotted for regularly spaced values of x, again illustrating that the paths never cross and the order of the curves is maintained over time. In (c), the end-point of the flow, $\boldsymbol{\phi}(x,1)$, is plotted as a function of x. The function $\boldsymbol{\phi}(x,1)$ is monotonic in x and smooth, and hence it is a diffeomorphism.

to generate diffeomorphisms given only moderately constrained vector-valued functions. We can think of the flow ODE as a "black box" that takes as its input a smooth and bounded time-varying velocity field $\boldsymbol{v}$ and outputs a diffeomorphic transformation $\boldsymbol{\phi}(\boldsymbol{x},1)$. Fig. 14.2 illustrates the flow ODE in one spatial dimension.

There is a close relationship between the flow ODE and our previous recipe of generating diffeomorphisms by composing a sequence of small deformations. Let us approximate the solution of the flow ODE using Euler's method. Euler's method is based on the approximation $\frac{\mathrm{d}\boldsymbol{\phi}}{\mathrm{d}t}(\boldsymbol{x},t)\simeq\frac{\boldsymbol{\phi}(x,t+\Delta t)-\boldsymbol{\phi}(x,t)}{\Delta t}$, where $\Delta t>0$ is a small step size. Plugging this approximation into the ODE leads to the following recurrence:

$$\boldsymbol{\phi}(\boldsymbol{x},t+\Delta t)=\boldsymbol{\phi}(\boldsymbol{x},t)+\Delta t\cdot\boldsymbol{v}(\boldsymbol{\phi}(\boldsymbol{x},t),t),$$

which can be rewritten as follows:

$$\boldsymbol{\phi}(\boldsymbol{x},t+\Delta t)=\boldsymbol{\psi}(\boldsymbol{\phi}(\boldsymbol{x},t)),\quad \text{where}\quad \boldsymbol{\psi}(\boldsymbol{x})=\boldsymbol{x}+\Delta t\cdot\boldsymbol{v}(\boldsymbol{x},t).$$

The function $\boldsymbol{\psi}$ is a small transformation (since $\boldsymbol{v}$ is smooth and bounded and Δt is a small value), and thus for every t, the function $\boldsymbol{\phi}(\boldsymbol{x},t)$ is simply a sequence of compositions of small transformations, just as in (14.1) above.

14.2.1.4 Scaling and squaring algorithm

This section introduces an efficient algorithm to generate diffeomorphic transformations. It is based on generalizing the concept of the natural exponent function e^x to higher-dimensional non-linear spaces. This section assumes knowledge of elementary concepts of group theory, i.e., the definition of a group.

Consider the natural exponential function e^x from calculus. It has a property that it maps the entire real line $\mathbb{R}$ onto the group of positive real numbers under the operation of multiplication. One way to define the natural exponential function is as follows:

$$e^x = \lim_{m \to \infty} \left(1 + \frac{x}{m}\right)^m .$$

This definition of exponentiation can be generalized to other groups, such as the group of all rotation matrices in $\mathbb{R}^3$ under the operation of matrix multiplication. Recall from Chapter 2 that a rotation matrix $\mathbf{R}$ is orthonormal, i.e., satisfies the constraint $\mathbf{R}\mathbf{R}^T = \mathbf{I}_{3\times 3}$, where $\mathbf{I}_{3\times 3}$ is the identity matrix. Rotation matrices in $\mathbb{R}^3$ form a group under the operation of matrix multiplication, called the special orthogonal group (denoted $SO(3)$). The identity element of this group is the identity matrix $\mathbf{I}_{3\times 3}$. Similar to how the natural exponential function maps all real numbers to all positive real numbers, there exists a mapping from the 3D Euclidean space to the space of all rotation matrices in $\mathbb{R}^3$. This mapping, called the *exponential map* for $SO(3)$, has the form

$$\exp(\boldsymbol{w}) = \lim_{m \to \infty} \left\{ \mathbf{I}_{3\times 3} + \frac{1}{m} \begin{bmatrix} 0 & -w_3 & w_2 \\ w_3 & 0 & -w_1 \\ -w_2 & w_1 & 0 \end{bmatrix} \right\}^m ,$$

where $\boldsymbol{w}$ is a point in $\mathbb{R}^3$ and $\{\ldots\}^m$ means multiplying a matrix by itself m times. Observe the similarity between the natural exponent and the matrix exponent map. They both involve taking an element around the identity element of the group and repeatedly applying the group operation. They both provide a way to map points in the Euclidean space into a space of mathematical objects that satisfy a certain constraint (i.e., positivity or orthonormality).

The equivalent notion of an exponential map can also be defined for the group of diffeomorphisms under the operation of composition. This map is given by the similar expression to the ones above:

$$\exp(\boldsymbol{u}) = \lim_{m \to \infty} \underbrace{(\mathrm{Id} + \frac{1}{m}\boldsymbol{u}) \circ \ldots \circ (\mathrm{Id} + \frac{1}{m}\boldsymbol{u})}_{m \text{ times}}, \tag{14.3}$$

where $\boldsymbol{u} \in \mathcal{U}$. Observe that computing $\exp(\boldsymbol{u})$ involves the composition of many small transformations, and therefore is a diffeomorphism. Similar to the natural exponent function and the exponential map for $SO(3)$, this exponential map allows us to map elements of moderately constrained space $\mathcal{U}$ (the space of differentiable and bounded vector fields) to a highly constrained space of diffeomorphisms. One of the attractive properties of the exponential map is that if $\boldsymbol{\phi} = \exp(\boldsymbol{u})$, then the inverse transformation is given by $\boldsymbol{\phi}^{-1} = \exp(-\boldsymbol{u})$, as is also the case for the natural exponential function and for exponentiation in the group of rotation matrices.

In practice, computing the exponential map as in (14.3) appears prohibitive. However, an approximation of the exponential map is provided by the recursive *scaling and squaring algorithm* [7]. Observe that when m in (14.3) is even, due to the associative property of composition, $\exp(\boldsymbol{u})$ can be computed as the composition of two identical terms, each involving $m/2$ compositions. If we let m be a power of 2, i.e., $m = 2^K$ for some integer K, then the term involving $m/2$ compositions can be computed as the composition of two identical terms each involving $m/4$ compositions, and so on. This leads to the following recursive algorithm:

$$\begin{aligned}
\boldsymbol{\psi}^0 &= \mathrm{Id} + \frac{1}{2^K}\boldsymbol{u}, \\
\boldsymbol{\psi}^k &= \boldsymbol{\psi}^{k-1} \circ \boldsymbol{\psi}^{k-1}, \quad \text{for } k = 1, \ldots, K, \\
\exp(\boldsymbol{u}) &\simeq \boldsymbol{\psi}^K.
\end{aligned}$$

With this simple algorithm, an excellent approximation of $\exp(\boldsymbol{u})$ can be obtained in very few iterations (practical values of K are between 5 and 8).

In the literature, the input to the scaling and squaring algorithm, $\boldsymbol{u}$, is often referred to as the *stationary velocity field*. The word "stationary" emphasizes the fact that in scaling and squaring, the diffeomorphism $\boldsymbol{\phi}$ is obtained from a function that does not depend on time. This is in contrast to the flow ODE, where the diffeomorphism $\boldsymbol{\phi}$ is obtained from a time-varying velocity field $\boldsymbol{v}(\boldsymbol{x}, t)$.

14.3 Optical flow and related non-parametric methods

Optical flow is a class of techniques used to measure the velocity of moving objects in video sequences. For example, in a self-driving car application, the continuous video feed from the car's camera could be used to determine the velocity of surrounding vehicles and pedestrians. In medical imaging, optical flow can be used to characterize the dynamics of moving organs, such as the beating heart in cine MRI. Optical flow also serves as a great way to introduce non-parametric NVR algorithms.

14.3.1 Conventional optical flow approach

Let $\tilde{I}(\boldsymbol{x}, t) : \mathbb{R}^n \times [0, t_{\max}] \to \mathbb{R}$ denote an image changing over time, i.e., a video sequence, where time t spans the interval $[0, t_{\max}]$. The objective of optical flow is to determine at each point in space and time a vector $\boldsymbol{v}(\boldsymbol{x}, t)$ that describes the instantaneous velocity of the object located at point $\boldsymbol{x}$ at time t.

14.3.1.1 Preservation of intensity assumption

The optical flow approach is based on the following assumption: *as objects move in the video sequence, their intensity remains nearly constant* (Fig. 14.3). In general computer vision problems, this is a strong assumption since the shading of objects changes as they move through the scene, or objects may become occluded by other

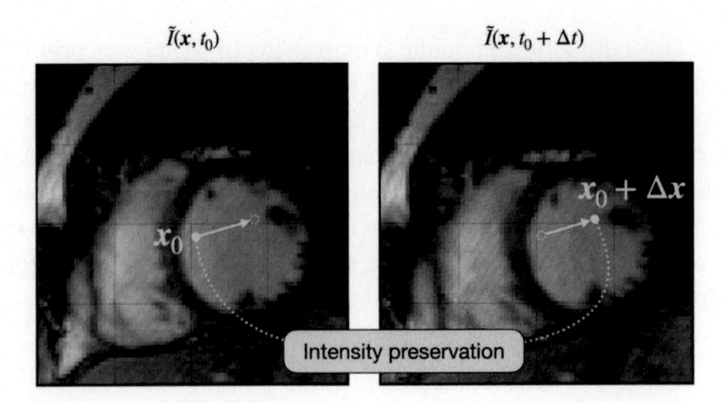

FIGURE 14.3

Intensity preservation in optical flow. The anatomical location at location $\boldsymbol{x}_0$ at time t_0 has moved to position $\boldsymbol{x}_0 + \Delta\boldsymbol{x}$ at time $t_0 + \Delta t$. We assume that the image intensity associated with this anatomical location is the same at time t_0 and at time $t_0 + \Delta t$.

objects. In 3D medical image sequences this assumption is more reasonable, although noise and partial volume artifacts introduce small changes in the intensity of objects as they move.

To express the assumption of intensity preservation mathematically, let us consider two images taken a short time apart, $\tilde{I}(\boldsymbol{x}, t_0)$ and $\tilde{I}(\boldsymbol{x}, t_0 + \Delta t)$. Suppose that a point at location $\boldsymbol{x}_0$ at time t_0 has moved to a new location $\boldsymbol{x}_0 + \Delta\boldsymbol{x}$ at time $t_0 + \Delta t$. Since, under our assumption, the intensity of the point $\boldsymbol{x}_0$ is preserved at its new location $\boldsymbol{x}_0 + \Delta\boldsymbol{x}$, we have

$$\tilde{I}(\boldsymbol{x}_0 + \Delta\boldsymbol{x}, t_0 + \Delta t) = \tilde{I}(\boldsymbol{x}_0, t_0).$$

The expression on the left of the equality can be expanded using Taylor's series as follows:

$$\tilde{I}(\boldsymbol{x}_0 + \Delta\boldsymbol{x}, t_0 + \Delta t) = \tilde{I}(\boldsymbol{x}_0, t_0) + (\Delta\boldsymbol{x})^T \tilde{I}_{\boldsymbol{x}}(\boldsymbol{x}_0, t_0) + (\Delta t)\tilde{I}_t(\boldsymbol{x}_0, t_0) + \mathcal{O}(\Delta^2),$$

where $\tilde{I}_{\boldsymbol{x}}$ and $\tilde{I}_t$ denote the partial derivatives of $\tilde{I}$ with respect to $\boldsymbol{x}$ and t, respectively, and $\mathcal{O}(\Delta^2)$ represents second- and higher-order terms involving $\Delta\boldsymbol{x}$ and Δt. Subtracting the first of the two equations above from the second, we obtain

$$0 = (\Delta\boldsymbol{x})^T \tilde{I}_{\boldsymbol{x}}(\boldsymbol{x}_0, t_0) + (\Delta t)\tilde{I}_t(\boldsymbol{x}_0, t_0) + \mathcal{O}(\Delta^2),$$

and dividing through by Δt, we have

$$\left(\frac{\Delta\boldsymbol{x}}{\Delta t}\right)^T \tilde{I}_{\boldsymbol{x}}(\boldsymbol{x}_0, t_0) = -\tilde{I}_t(\boldsymbol{x}_0, t_0) + \mathcal{O}(\Delta).$$

Now let us consider the limit of this expression as $\Delta t \to 0$. The expression $\frac{\Delta\boldsymbol{x}}{\Delta t}$ is the ratio between the distance an object travels over a short time interval Δt and the

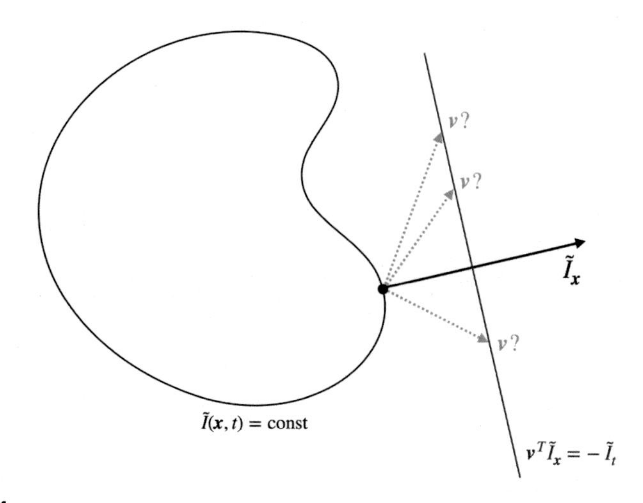

FIGURE 14.4

The intensity preservation constraint implies that the velocity vector v is constrained to lie on a line (plane in 3D), but does not uniquely determine it.

said time interval, and as $\Delta t \to 0$, it converges to the instantaneous velocity of the object, $\boldsymbol{v}(\boldsymbol{x}_0, t_0)$, which is exactly the quantity that we seek to estimate in optical flow. Also, the limit of $\mathcal{O}(\Delta)$ as $\Delta t \to 0$ is zero. Hence, in the limit, the expression above simplifies to

$$\boldsymbol{v}(\boldsymbol{x}_0, t_0)^T \tilde{I}_{\boldsymbol{x}}(\boldsymbol{x}_0, t_0) = -\tilde{I}_t(\boldsymbol{x}_0, t_0),$$

or just

$$\boldsymbol{v}^T \tilde{I}_{\boldsymbol{x}} = -\tilde{I}_t. \tag{14.4}$$

This linear relationship between velocity $\boldsymbol{v}$, the image gradient $\tilde{I}_{\boldsymbol{x}}$, and the image time derivative $\tilde{I}_t$ is called the *optical flow constraint*. Geometrically, this constraint specifies that at every point $\boldsymbol{x}_0$, the tip of the velocity vector $\boldsymbol{v}$ must lie on a certain line (in 2D), plane (in 3D), or hyperplane (in nD) that is orthogonal to the image gradient at $\boldsymbol{x}_0$, as illustrated in Fig. 14.4. Importantly, the restriction of $\boldsymbol{v}$ onto a certain line/plane is not enough to determine $\boldsymbol{v}$ uniquely (except if $n = 1$). At each point $(\boldsymbol{x}_0, t_0)$ the velocity vector $\boldsymbol{v}(\boldsymbol{x}_0, t_0)$ may take as many different values as there are different points in $\mathbb{R}^{n-1}$. Additional assumptions are required in order to uniquely determine $\boldsymbol{v}(\boldsymbol{x}, t)$.

14.3.1.2 Smoothness assumption and formulation as energy minimization

Given that the intensity preservation assumption is not enough to determine $\boldsymbol{v}(\boldsymbol{x}, t)$, Horn and Schunck [8] proposed an additional assumption: that the velocity field

$\boldsymbol{v}(\boldsymbol{x}, t)$ is smooth across the image domain. Rather than formulating this assumption as a hard constraint, they reformulate the optical flow problem as an energy minimization problem in which violations of the optical flow assumption and smoothness assumption are both penalized.

Going forward, let us focus on a particular moment in time, t_0, and let $\boldsymbol{v}(\boldsymbol{x}) = \boldsymbol{v}(\boldsymbol{x}, t_0)$. Horn and Schunck's [8] optical flow minimization is formulated as

$$\boldsymbol{v}^* = \underset{\hat{v} \in \mathcal{V}^n}{\arg\min}\, E[\boldsymbol{v}],$$

$$\text{where} \quad E[\boldsymbol{v}] = \int_{\Omega} \left(\boldsymbol{v}^T \tilde{I}_{\boldsymbol{x}} + \tilde{I}_t\right)^2 + \lambda^2 \sum_{i=1}^{n} \sum_{j=1}^{n} \left(\frac{\partial v^i}{\partial x_j}\right)^2 \mathrm{d}\boldsymbol{x}. \tag{14.5}$$

The left part of the integrand drives $\boldsymbol{v}$ to satisfy the intensity preservation assumption (i.e., the more the assumption is violated, the greater the value of the energy E becomes). The right-hand side of the integrand drives $\boldsymbol{v}$ to be smooth, since discontinuities in $\boldsymbol{v}$ are associated with large values of the partial derivatives $\frac{\partial v_i}{\partial x_j}$. The positive scalar λ is a coefficient of the method, and is used to balance between the two terms; larger values of λ result in smoother solutions.

The energy functional (14.5) is minimized using calculus of variations (Section 2.5.2). The Euler–Lagrange equation for E is

$$\mathcal{L}[\boldsymbol{v}] = \left(\boldsymbol{v}^T \tilde{I}_{\boldsymbol{x}} + \tilde{I}_t\right) \tilde{I}_{\boldsymbol{x}} - \lambda^2 \Delta \boldsymbol{v} = \boldsymbol{0},$$

where Δ is the Laplacian operator. Like in most variational problems involving images, we solve the problem (13.1) iteratively by taking small steps in the direction of $-\mathcal{L}[\boldsymbol{v}]$.

14.3.1.3 Basic implementation

To implement the optical flow algorithm, we must discretize over time. Instead of a continuous image $\tilde{I}(\boldsymbol{x}, t)$, we consider a pair of n-dimensional images $I(\boldsymbol{x}) = \tilde{I}(\boldsymbol{x}, t_0)$ and $J(\boldsymbol{x}) = \tilde{I}(\boldsymbol{x}, t_0 + \Delta t)$. The first-order approximation of the time derivative of $\tilde{I}(\boldsymbol{x}, t)$ is

$$\tilde{I}_t \simeq \frac{J - I}{\Delta t},$$

and $\tilde{I}_{\boldsymbol{x}} = \nabla I$. Hence,

$$\mathcal{L}[\boldsymbol{v}] \simeq \left(\boldsymbol{v}^T \nabla I + \frac{J - I}{\Delta t}\right) \nabla I - \lambda^2 \Delta \boldsymbol{v}.$$

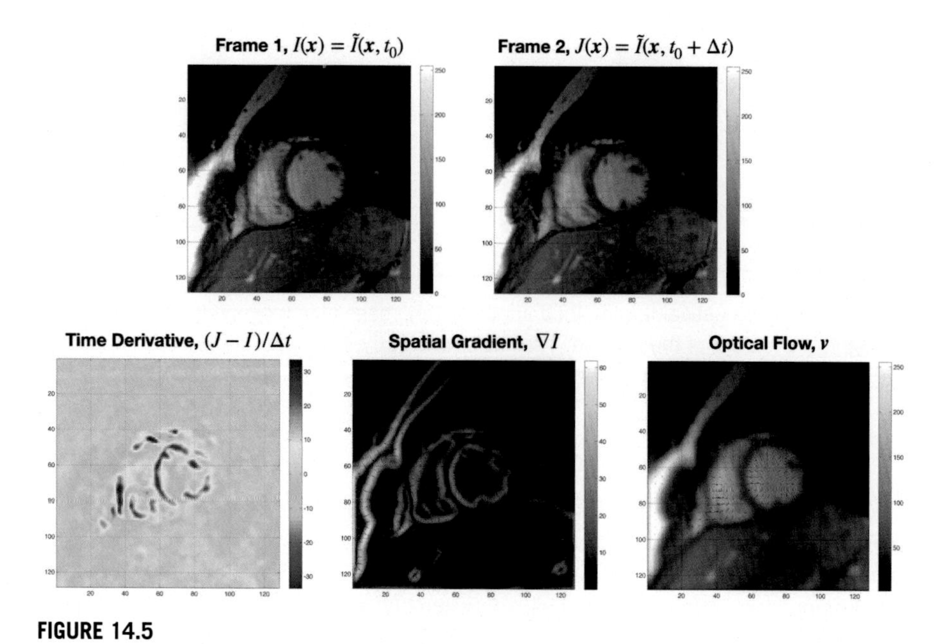

FIGURE 14.5

Horn and Schunck optical flow method applied to a pair of images. The red arrows in the last panel show the estimated velocity field.

We can then perform minimization of $E[\boldsymbol{v}]$ via the iterative scheme $\boldsymbol{v}^{k+1} = \boldsymbol{v}^k - \epsilon \mathcal{L}[\boldsymbol{v}^k]$, i.e., gradient descent. Computing $\mathcal{L}[\boldsymbol{v}^k]$ involves numerical approximation of the second partial derivatives of $\boldsymbol{v}$. Fig. 14.5 gives examples of velocity estimation from a pair of cardiac cine MRI frames. Note how the velocity field seems to capture the compression of the left ventricle and even some of the heart's twisting motion.

14.3.1.4 Limitations of optical flow

A major limitation of optical flow is that the optical flow constraint loses its influence over homogeneous regions in the image. When a large region $R \subset \Omega$ of the image has nearly constant intensity, both the image gradient $\tilde{I}_x$ and the image time derivative $\tilde{I}_t$ vanish on the interior of R. As the result, for any choice of $\boldsymbol{v}$ over the interior of R, the first term in (14.5), $\left(\boldsymbol{v}^T \tilde{I}_x + \tilde{I}_t\right)^2$, will be nearly zero, and hence, the minimization of E is almost exclusively driven by the regularization term over the interior of R. This results in surprising, and almost certainly wrong, estimates of velocity over homogeneous regions, with greater velocity at the boundary of the constant intensity regions than on the interior. This is demonstrated in Fig. 14.6, where several homogeneous objects are displaced rigidly in the image plane. The recovered velocity field is not at all consistent with rigid motion.

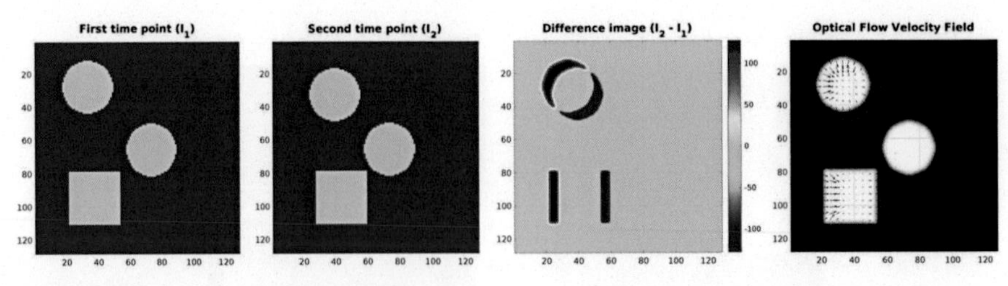

FIGURE 14.6

Horn and Schunck optical flow method applied to an image of homogeneous shapes highlights a limitation of the method. The red arrows in the last panel show the estimated velocity field, which varies across the shapes, suggesting a non-intuitive pattern of motion.

14.3.2 Iterative optical flow

14.3.2.1 Conceptual NVR algorithm based on optical flow

While optical flow can be useful in medical imaging applications for describing the dynamics of moving organs, it is not considered a NVR technique per se. Optical flow only computes an infinitesimal velocity between two time frames in a video sequence, whereas NVR seeks a deformation of the domain Ω that would match up corresponding anatomical locations between two images. These corresponding locations may be quite far apart from each other, requiring a potentially "large" deformation to match them.

However, it turns out that optical flow can be used as a building block for an effective NVR algorithm. Consider the following iterative algorithm:

1. Let I be the fixed image and J the moving image. Let $J^0 = J$.
2. Iterate for $i = 0, 1, \ldots$:
   a. Compute optical flow $\boldsymbol{v}^i$ between images I and J^i
   b. Let $J^{i+1}(\boldsymbol{x}) = J^i(\boldsymbol{x} + \epsilon \boldsymbol{v}^i)$ where ϵ is a (small) constant
   c. Terminate if the difference between J^{i+1} and J^i is below threshold

Algorithm 14.1: A skeleton of a NVR algorithm based on optical flow.

The basic idea of this algorithm is as follows. At every iteration, the velocity field $\boldsymbol{v}^i$ obtained by performing optical flow between images I and J^i can be interpreted as the direction in which objects in I would move if I and J^i were frames from a video. Even if I and J^i are not frames from a video but have similar content (e.g., both are images of brains), the vectors $\boldsymbol{v}^i$ may be expected to point from locations in I toward similarly appearing locations in J^i. This is, of course, a strong assumption. If this assumption is true, then moving the locations in I a small amount along the vectors $\boldsymbol{v}^i$ would place them closer to their corresponding anatomical counterparts in J^i. Conversely, deforming the image J^i by the "small" transformation $J^{i+1}(\boldsymbol{x}) = J^i(\boldsymbol{x} + \epsilon \boldsymbol{v}^i)$ would result in the new image J^{i+1} being better anatomically matched up with I than J^i. Repeating this process iteratively until convergence would result in an image J^M that is most similar anatomically to I.

In practice it turns out that the strong assumption above is actually not unreasonable, and the general algorithm outlined above works quite well for NVR. However, the algorithm above is not quite a practical NVR algorithm yet. One crucial limitation is that at each iteration the image J^i is interpolated, which over just a few iterations would completely degrade the image due to aliasing (Section 5.4.2). Also, the algorithm does not actually yield any spatial transformation, just a deformed image J^M. These two limitations can be easily addressed by a small modification to the algorithm.

1. Let I be the fixed image and J the moving image. Let $J^0 = J$ and $\boldsymbol{\phi}^0 = \text{Id}$.
2. Iterate for $i = 0, \ldots, M$:
   a. Compute optical flow $\boldsymbol{v}^i$ between images I and J^i
   b. Let $\boldsymbol{\psi}^i = \text{Id} + \epsilon \boldsymbol{v}^i$, i.e., $\boldsymbol{\psi}^i(\boldsymbol{x}) = \boldsymbol{x} + \epsilon \boldsymbol{v}^i(\boldsymbol{x})$
   c. Let $\boldsymbol{\phi}^{i+1} = \boldsymbol{\phi}^i \circ \boldsymbol{\psi}^i$, i.e., $\boldsymbol{\phi}^{i+1}(\boldsymbol{x}) = \boldsymbol{\phi}^i(\boldsymbol{x} + \epsilon \boldsymbol{v}^i(\boldsymbol{x}))$
   d. Let $J^{i+1} = J^0 \circ \boldsymbol{\phi}^{i+1}$
   e. Terminate if the difference between J^{i+1} and J^i is below a threshold.

Algorithm 14.2: Conceptual NVR algorithm based on optical flow.

Note that the two algorithms are identical in terms of the outputs J^i when the input images are continuous functions. However, when it comes to practical implementation, the difference is significant! The first algorithm interpolates the moving image repeatedly, whereas in the second algorithm, the moving image is only interpolated once, and the spatial transformations $\boldsymbol{\phi}^i$ are composed repeatedly, which involves interpolation. It turns out that the latter is less problematic, particularly because in practice additional smoothing will be applied to these spatial transformations to limit aliasing.

14.3.2.2 The Demons algorithm

The idea of using optical flow as a step in a NVR algorithm was pioneered by J.P. Thirion [1] in an approach called "Demons," in reference to a famous thought experiment in Maxwell's theory of thermodynamics. One obvious limitation that Thirion's approach overcame is the computational inefficiency of the algorithm above: *at each iteration it requires us to solve the optical flow problem, which itself requires an iterative optimization!* This double iteration would be impractical for real-world registration problems.

Thirion proposed to estimate optical flow in a simpler and more efficient way than in the Horn and Schunck method [8]. The optical flow intensity preservation condition (14.4) can be interpreted as saying that points on any isointensity contour of the image at time t (i.e., points that satisfy $\tilde{I}(\boldsymbol{x}, t) = \alpha$ for some scalar α) should map to points somewhere on the corresponding contour in the image at time $t + \Delta t$ (Fig. 14.7(a)). Thirion proposed to strengthen this constraint, by requiring

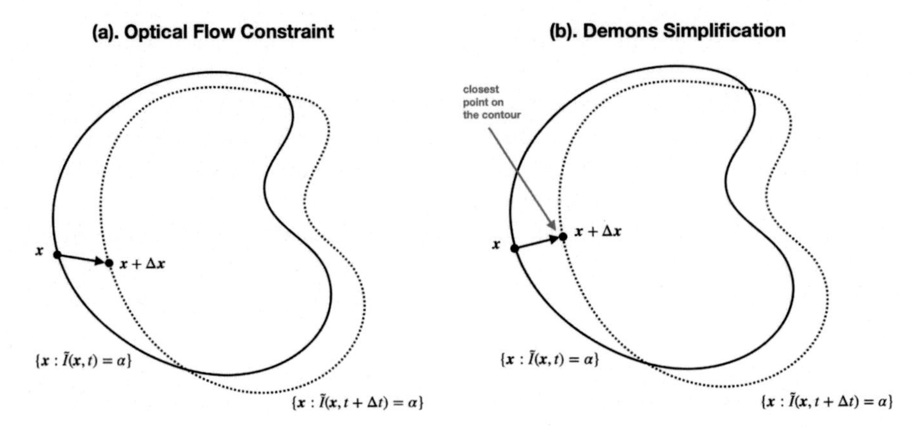

FIGURE 14.7

Additional optical constraint proposed in the Demons method [1]. The optical flow condition dictates that points on a given isointensity contour in I at time t (bold curve, i.e., a set of all points that have intensity α at time t) have to map to the contour of the same intensity in the image at time $t + \Delta t$. The Demons method introduces an additional constraint, that points at time t should map onto the *closest point* on the corresponding contour at time $t + \Delta t$.

these points on the contour at time t to map to the *closest point* on the contour at time $t + \Delta t$ (Fig. 14.7(b)). This is equivalent to requiring $\boldsymbol{v}$ to be perpendicular to the contour, or parallel to $\tilde{I}_{\boldsymbol{x}}$, since the gradient of a function is orthogonal to its isointensity contours. Introducing the additional constraint $\boldsymbol{v} \parallel \tilde{I}_{\boldsymbol{x}}$ makes it possible to solve directly for $\boldsymbol{v}$:

$$\begin{cases} \boldsymbol{v}^T \tilde{I}_{\boldsymbol{x}} = -\tilde{I}_t, \\ \boldsymbol{v} \parallel \tilde{I}_{\boldsymbol{x}} \end{cases} \quad \Longrightarrow \quad \boldsymbol{v} = \frac{-\tilde{I}_t}{|\tilde{I}_{\boldsymbol{x}}|^2} \tilde{I}_{\boldsymbol{x}}.$$

However, the expression on the right is asymptotic at locations with nearly constant intensity, where $|\tilde{I}_{\boldsymbol{x}}| \simeq 0$. It also results in very noisy velocity fields. To overcome this, Thirion proposed modifying the expression for $\boldsymbol{v}$ as follows:

$$\boldsymbol{v} = G_\sigma * \begin{cases} \frac{-\tilde{I}_t}{|\tilde{I}_{\boldsymbol{x}}|^2 + |\tilde{I}_t|^2} \tilde{I}_{\boldsymbol{x}} & \text{if } |\tilde{I}_{\boldsymbol{x}}|^2 + |\tilde{I}_t|^2 > \epsilon, \\ \mathbf{0} & \text{otherwise.} \end{cases} \tag{14.6}$$

This formulation reduces the number of points where $\boldsymbol{v}$ is unstable and results in zero velocity over regions of constant intensity. The convolution operation $*$ with the Gaussian kernel G_σ regularizes the velocity field, somewhat similar to the Horn and Schunck approach, but without optimization (Fig. 14.8). This results in a very fast

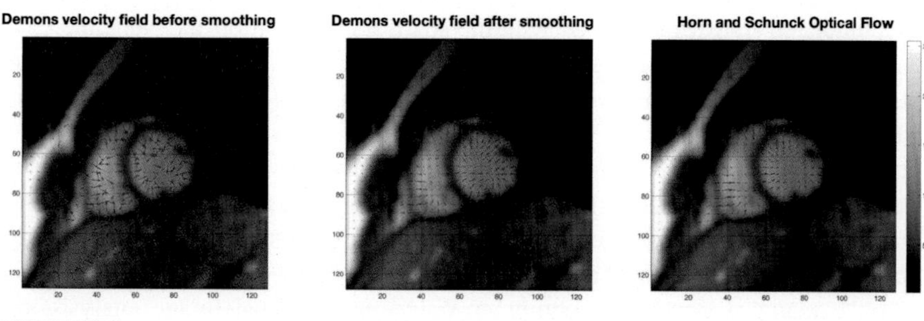

FIGURE 14.8

Illustration of Demons velocity computation [1], contrasted with Horn and Schunck [8] optical flow. The velocity field is computed between cine MRI frames from Fig. 14.5. The Demons velocity field is shown before and after smoothing in (14.6).

computation of optical flow. The following algorithm is approximately equivalent to the original Demons formulation.

1. Let I be the fixed image and let J be the moving image. Let $\boldsymbol{\phi}^0 = \text{Id}$.
2. Iterate for $i = 0, \ldots, M$:
   a. Let $J^i = J \circ \boldsymbol{\phi}^i$
   b. Compute optical flow $\boldsymbol{v}^i$ between images I and J^i as

$$\boldsymbol{v}^i = G_\sigma * \begin{cases} -\frac{J^i - I}{|\nabla J^i|^2 + |J^i - I|^2} \nabla J^i & \text{if } |\nabla J^i|^2 + |J^i - I|^2 > \epsilon, \\ \mathbf{0} & \text{otherwise.} \end{cases}$$

   c. Let $\boldsymbol{\psi}^i = \text{Id} + \epsilon \boldsymbol{v}^i$, i.e., $\boldsymbol{\psi}^i(\boldsymbol{x}) = \boldsymbol{x} + \epsilon \boldsymbol{v}^i(\boldsymbol{x})$
   d. Let $\boldsymbol{\phi}^{i+1} = G_\tau * (\boldsymbol{\phi}^i \circ \boldsymbol{\psi}^i)$
   e. Terminate if $|J \circ \boldsymbol{\phi}^{i+1} - J^i|$ is below a threshold.

Algorithm 14.3: The compositional Demons algorithm [1].

There are several observations to be made about the algorithm:

- The expression in 2(b) is derived from (14.6) by approximating $\tilde{I}_t$ with $J^i - I$ and approximating $\tilde{I}_{\boldsymbol{x}}$ with ∇J^i. It is also possible to approximate $\tilde{I}_{\boldsymbol{x}}$ with ∇I or with $\frac{1}{2}(\nabla J^i + \nabla I)$. How these different approximations affect the behavior of the Demons algorithm is explored in [2].
- In practice, the scaling factor ϵ in step 2(c) can be chosen in different ways. For example, it can be adapted at each iteration so that $\sup_{\boldsymbol{x}\in\Omega} |\boldsymbol{v}^i| \leq \epsilon'$, where ϵ' is fixed across iterations (typically on the order of 1/4 to 1/2 of the voxel dimension).
- Observe that convolution with a Gaussian filter is performed in two places: in step 2(b), when regularizing the optical flow, and in step 2(d), when combining the

accumulated transformation $\boldsymbol{\phi}^i$ with the small transformation $\boldsymbol{\psi}^i$. Vercauteren et al. [2] describe these as "fluid-like" and "diffusion-like" regularization. In practice, both types of regularization are needed to obtain a good quality registration.

- The parameters σ and τ are among the most important in practical image registration. Large values of σ result in smoother deformation fields and large values of τ result in smaller deformations. Experimentation in the context of specific registration problems is often needed. Practical starting values for these parameters are $\sigma = 3$ and $\tau = 0.5$, in units of voxel dimensions.
- Unlike many algorithms in computer vision and medical image analysis, the Demons algorithm is not formulated in a variational fashion, i.e., in terms of some energy functional $E[\boldsymbol{\phi}]$ that is minimized. However, Vercauteren et al. [2] show that it is possible to interpret Demons as energy minimization by introducing an additional concept of a correspondence map.[1]
- A limitation of this algorithm is that it only provides the forward transformation $\boldsymbol{\phi}^i$ that maps points in I to corresponding points in J, but the inverse transformation from J to I is not generated. However, it is possible to estimate the inverse of a diffeomorphic transformation post hoc. Doing so without significant numerical error is non-trivial and computationally expensive. A related limitation is that registration is not symmetric: performing registration with the fixed and moving images swapped is not guaranteed to yield a transformation that is the inverse of the one obtained with the imaged not swapped.

The Demons algorithm is a very fast non-parametric registration algorithm. It only involves very basic image processing operations such as composition of transformation fields (interpolation), computation of image gradients, and convolution with a Gaussian kernel. It is also highly parallelizable.

14.3.2.3 Log-domain diffeomorphic Demons algorithm

The transformations produced by the compositional Demons algorithm are not guaranteed to be diffeomorphic. In practice, they often are because at each iteration the current transformation $\boldsymbol{\phi}^i$ is composed with a small transformation $\boldsymbol{\psi}^i$. However, the convolution with the Gaussian kernel G_τ in step 2(d) may break the diffeomorphic

---

[1] The correspondence map $c : \Omega \to \Omega$ can be interpreted as a more "raw," unregularized mapping of pixels in the fixed image I to the pixels in the moving image J. Vercauteren et al. [2] define an energy functional that involves both the unknown transformation and the unknown correspondence map:

$$E[c, \boldsymbol{\phi}] = \lambda_1 \|I - J \circ c\|^2 + \lambda_2 \|\boldsymbol{\phi} - c\| + \lambda_3 \|\nabla \boldsymbol{\phi}\|^2,$$

where $\lambda_1, \lambda_2, \lambda_3$ are coefficients corresponding to the three terms in the energy: the intensity similarity (how well the correspondence map matches pixels in the two images), the fidelity of the transformation (how well the transformation $\boldsymbol{\phi}$ agrees with the correspondence map c), and the regularization term (how noisy is the transformation $\boldsymbol{\phi}$). The authors then show that the Demons iterative algorithm corresponds to alternatively minimizing $E[c, \boldsymbol{\phi}]$ with respect to c and $\boldsymbol{\phi}$.

property, and also the numerical implementation may eventually lead to solutions that are in fact not diffeomorphic. Also, as noted above, compositional Demons produces forward transformations (from fixed image I to moving image J but not vice versa).

Vercauteren et al. (2008) [9] proposed a modification to the compositional Demons algorithm that guarantees diffeomorphic solutions. This method also outputs both forward and inverse transformations (from fixed to moving and from moving to fixed), overcoming one of the limitations of compositional Demons noted above. This approach is called *log-domain Demons*. It takes advantage of the fact that diffeomorphic transformations can be generated from smooth and bounded vector-valued functions using the exponential map (see Section 14.2.1). The main modification from compositional Demons is that instead of directly updating the spatial transformation $\boldsymbol{\phi}^i$ at each iteration of the algorithm, Vercauteren et al.'s method performs updates in the "logarithmic domain," i.e., it updates the vector-valued function $\boldsymbol{u}^i$, called the *stationary velocity field*, from which the desired spatial transformation is obtained using the exponential map (see Section 14.2.1):

$$\boldsymbol{\phi}^i = \exp(\boldsymbol{u}^i),$$

which in turn is approximated using the scaling and squaring algorithm.

The crux of the algorithm is to figure out how to update the stationary velocity field at each iteration i. Given the optical flow velocity field $\boldsymbol{v}^i$ computed at iteration i, the updated stationary velocity field $\boldsymbol{u}^{i+1}$ should satisfy

$$\exp(\boldsymbol{u}^{i+1}) = \exp(\boldsymbol{u}^i) \circ (\mathrm{Id} + \epsilon \boldsymbol{v}).$$

At the first glance, solving this expression for $\boldsymbol{u}^{i+1}$ requires the ability to take the inverse of the exponential map, which is not well defined for the infinite-dimensional group of diffeomorphisms. However, Vercauteren et al. show that the following approximation, based on the Baker–Campbell–Hausdorff formula from the theory of Lie groups [10], is sufficiently accurate in practice:

$$\boldsymbol{u}^{i+1} \simeq \boldsymbol{u}^i + \epsilon\, \boldsymbol{v}^i + \frac{\epsilon}{2}[\boldsymbol{u}^i, \boldsymbol{v}^i],$$

where the expression $[\boldsymbol{u}, \boldsymbol{v}]$ is called the *Lie bracket* and has the following form:

$$[\boldsymbol{u}, \boldsymbol{v}] = (D\boldsymbol{u})\boldsymbol{v} - (D\boldsymbol{v})\boldsymbol{u},$$

where $D\boldsymbol{u}$ denotes the Jacobian matrix of $\boldsymbol{u}$. This update takes place directly in the domain of stationary velocity fields and is relatively inexpensive computationally compared to the update in step 2(c) and step 2(d) in Algorithm 14.3. Another change in Vercauteren et al.'s algorithm is that regularization with the diffusion kernel G_τ is performed directly in the domain of stationary velocity fields. The final algorithm is as follows:

1. Let I be the fixed image and let J be the moving image. Let $\boldsymbol{u}^0(\boldsymbol{x}) = \boldsymbol{0}$.
2. Iterate for $i = 0, \ldots, M$:
   a. Let $J^i = J \circ \boldsymbol{\phi}^i$
   b. Compute optical flow $\boldsymbol{v}^i$ between images I and J^i as in Algorithm 14.3, step 2(b)
   c. Let $\boldsymbol{u}^{i+1} = G_\tau * \left(\boldsymbol{u}^i + \epsilon\, \boldsymbol{v}^i + \frac{\epsilon}{2}[\boldsymbol{u}^i, \boldsymbol{v}^i]\right)$
   d. Let $\boldsymbol{\phi}^{i+1} = \exp(\boldsymbol{u}^{i+1})$ and $(\boldsymbol{\phi}^{i+1})^{-1} = \exp(-\boldsymbol{u}^{i+1})$
   e. Terminate if $|J \circ \boldsymbol{\phi}^{i+1} - J^i|$ is below a threshold

Algorithm 14.4: The log-domain Demons algorithm [2,9].

The log-domain Demons algorithm is computationally efficient. The added computational cost compared to compositional Demons is the computation of the Lie bracket (requires computing first derivatives of $\boldsymbol{u}^i$ and $\boldsymbol{v}^i$) and the handful of additional compositions involved in the scaling and squaring algorithm. Because it is fast, easy to implement, it guarantees diffeomorphic transformations, and it is very effective in practice, the log-domain Demons algorithm has become one of the most widely used non-parametric registration approaches, and variations of this method are implemented in various registration software packages. A number of extensions and improvements to the baseline log-Demons algorithm have been made, including:

- The algorithm can be extended to work with other intensity dissimilarity measures, such as normalized cross-correlation (NCC) and mutual information (Chapter 13). For an arbitrary dissimilarity measure $\mu[I, J]$, the optical flow computation in step 2(b) is replaced with the vector field

  $$\boldsymbol{v}^i = -G_\sigma * \nabla_J \mu(I, J_i),$$

  which is similar to the optical flow field corresponding to a small transformation that pulls J_i in the direction to lower the dissimilarity measure. In practice, Demons-like registration using the NCC dissimilarity measure [3,11] is particularly effective for many types of medical images.
- The algorithm can be made symmetric, such that the result of the registration with I as the fixed image and J as the moving image is the inverse of the result of the registration with J as the fixed image and I as the moving image. A symmetric extension of log-Demons is described in the original paper [9] and requires only a minor modification to the algorithm, based on computing the optical flow in both directions (fixed to moving and moving to fixed), and a minor increase in computational cost.
- Mansi et al. (2011) [12] proposed a modification of log-domain Demons that constrains transformations to be incompressible ($|D\boldsymbol{\phi}^i| = 1$) over regions of the fixed

image. This constraint is useful when performing registration in anatomical structures with known biomechanical properties, such as deforming cardiac tissue.

14.4 Large deformation diffeomorphic metric mapping

14.4.1 LDDMM formulation

We now turn our attention to another highly influential approach to non-parametric registration with diffeomorphic constraints, called *LDDMM* [13,14]. In LDDMM, the registration problem is formulated explicitly as a problem of minimizing an energy functional over the space of diffeomorphic transformations (Diff). The energy functional is a weighted sum of the dissimilarity measure between the fixed and moving images and a regularization term on the spatial transformation:

$$\boldsymbol{\phi}^* = \underset{\boldsymbol{\phi} \in \mathrm{Diff}}{\arg\min}\, E[\boldsymbol{\phi}], \quad \text{where} \quad E[\boldsymbol{\phi}] = \frac{1}{\sigma^2}\, \mu(\boldsymbol{\phi}; I, J) + \mathrm{Reg}(\boldsymbol{\phi}), \tag{14.7}$$

where σ controls the relative weight of the two terms. This explicit formulation of registration as energy minimization is in contrast to the Demons-like methods, where the energy functional is not explicitly formulated.

The problem formulated in (14.7) requires us to search for solutions in the space Diff. Because of the non-linear nature of Diff, doing so is very difficult when representing the transformations $\boldsymbol{\phi}$ directly, i.e., as vector fields. For example, if $\boldsymbol{\phi}_1$ and $\boldsymbol{\phi}_2$ are in Diff, there is no guarantee that transformations in the form $w_1\boldsymbol{\phi}_1 + w_2\boldsymbol{\phi}_2$ (where w_1 and w_2 are scalars) will be in Diff. Constraining solutions $\boldsymbol{\phi}$ to stay in Diff during the minimization of $E[\boldsymbol{\phi}]$ in (14.7) would be difficult and computationally expensive.

However, as we have seen in Section 14.2.1.3, diffeomorphic transformations can be generated from less constrained representations (i.e., differentiable and bounded vector fields) in various ways. LDDMM leverages the fact that a family of diffeomorphic transformations $\boldsymbol{\phi}(\boldsymbol{x}, t)$ parameterized by time $t \in [0, 1]$ can be generated from a time-varying differentiable and bounded *velocity field* $\boldsymbol{v}(\boldsymbol{x}, t) : \Omega \times [0, 1] \to \mathbb{R}^n$ using the flow ODE (14.2). For a given point $\boldsymbol{x}$ in the space of the fixed image I, the function $\boldsymbol{\phi}(\boldsymbol{x}, t)$ describes a non-linear trajectory of that point as it moves tangentially to $\boldsymbol{v}(\boldsymbol{x}, t)$, like a dust particle in the wind. Let the diffeomorphic transformation $\boldsymbol{\phi}_1(\boldsymbol{x}) = \boldsymbol{\phi}(\boldsymbol{x}, 1)$ denote the final position of $\boldsymbol{x}$. LDDMM reformulates the registration problem (14.7) as the problem of finding a velocity field $\boldsymbol{v}$ for which the final transformation $\boldsymbol{\phi}_1(\boldsymbol{x})$ best matches corresponding points between fixed and moving images, while also subject to a regularization term:

$$\boldsymbol{v}^* = \underset{\boldsymbol{v} \in \mathcal{V}}{\arg\min}\, E[\boldsymbol{v}], \quad \text{where} \quad E[\boldsymbol{v}] = \frac{1}{\sigma^2}\, \mu(\boldsymbol{\phi}_1; I, J) + \mathrm{Reg}(\boldsymbol{v}),$$

$$\text{and} \quad \frac{\mathrm{d}\boldsymbol{\phi}}{\mathrm{d}t}(\boldsymbol{x}, t) = \boldsymbol{v}(\boldsymbol{\phi}(\boldsymbol{x}, t), t), \quad \text{subject to } \boldsymbol{\phi}(\boldsymbol{x}0) = \boldsymbol{x}. \tag{14.8}$$

Unlike (14.7), the search space $\mathcal{V}$ for this energy minimization problem consists of smooth and bounded vector-valued velocity fields, and is much simpler in structure than Diff. For example, given two candidate solutions $\boldsymbol{v}_1, \boldsymbol{v}_2 \in \mathcal{V}$, their weighted sums $w_1\boldsymbol{v}_1 + w_2\boldsymbol{v}_2$ are also members of the search space $\mathcal{V}$.

Kinetic energy of diffeomorphic transformations

The reformulation of the energy functional E in terms of $\boldsymbol{v}(\boldsymbol{x}, t)$ in (14.8) also gives an opportunity to formulate the regularization term Reg in terms of $\boldsymbol{v}(\boldsymbol{x}, t)$. The purpose of the regularization term is to encourage the optimal solution $\boldsymbol{\phi}_1^*$ to have certain desirable geometric properties such as smoothness. However, geometric properties of the transformation $\boldsymbol{\phi}_1$ are directly linked to the geometric properties of the velocity field $\boldsymbol{v}(\boldsymbol{x}, t)$, from which it is derived via the flow ODE. Following our wind/dust analogy, the faster the wind is, the farther the particles are likely to travel, and the smoother the wind field is in space and time, the more simple the trajectories of the particles are likely to be. Particles caught in the middle of a tornado will have much more complex trajectories than particles flying in a light breeze. Formulating regularization in terms of $\boldsymbol{v}(\boldsymbol{x}, t)$ also turns out to be simpler computationally than formulating it in terms of $\boldsymbol{\phi}_1$.

Let us consider for a moment the squared Euclidean or L^2-norm of $\boldsymbol{v}$ as a potential regularization term. It is defined as follows:

$$\|\boldsymbol{v}\|_{L^2}^2 = \int_0^1 \int_\Omega \|\boldsymbol{v}(\boldsymbol{x}, t)\|^2 \, d\boldsymbol{x} \, dt.$$

This norm integrates the magnitude of the velocity across space and time dimensions. The faster the wind in our analogy, the greater $\|\boldsymbol{v}\|_{L^2}^2$ will be. Consequently, the particles whose trajectories are described by $\boldsymbol{\phi}(\boldsymbol{x}, t)$ will travel farther. If we use $\text{Reg}(\boldsymbol{v}) = \|\boldsymbol{v}\|_{L^2}^2$ as the regularization term in (14.8), then we would favor solutions that have shorter trajectories, which is reasonable. However, $\|\boldsymbol{v}\|_{L^2}^2$ does not penalize the noisiness of the velocity field, so we are likely to get noisy solutions.

In LDDMM [13], the authors associate the space of time-varying velocity fields $\mathcal{V}$ with a different non-Euclidean norm called the *Sobolev norm*. The Sobolev norm of a vector field measures both its magnitude and its regularity, and is defined as

$$\|\boldsymbol{v}\|_{\mathcal{V}}^2 = \|\mathcal{L}\boldsymbol{v}\|_{L^2}^2 = \int_0^1 \int_\Omega \|\mathcal{L}\boldsymbol{v}(\boldsymbol{x}, t)\|^2 \, d\boldsymbol{x} \, dt,$$

where $\mathcal{L}$ is a differential operator. In LDDMM, this operator is given the following form:

$$\mathcal{L}\boldsymbol{v} = \gamma\boldsymbol{v} - \alpha\Delta\boldsymbol{v},$$

where Δ is the Laplacian operator, applied separately to each component of $\boldsymbol{v}$, i.e.,

$$\Delta \boldsymbol{v} = \begin{bmatrix} < \frac{\partial^2 v^1}{\partial x_1^2} + \ldots + \frac{\partial^2 v^1}{\partial x_n^2} \\ \vdots \\ \frac{\partial^2 v^n}{\partial x_1^2} + \ldots + \frac{\partial^2 v^n}{\partial x_n^2} \end{bmatrix},$$

where $\gamma > 0$ and $\alpha > 0$ are scalar weights that control the extent to which the magnitude and regularity components contribute to $\|\boldsymbol{v}\|_{\mathcal{V}}^2$. In LDDMM literature, it is common to refer to $\|\boldsymbol{v}\|_{\mathcal{V}}^2$ as *kinetic energy*. This is similar to the familiar concept of kinetic energy from physics in that it involves the squared norm of the velocity, but different in that there is no mass involved in the formulation. In particular, the term "kinetic energy of a diffeomorphic flow $\boldsymbol{\phi}_t$" refers to the Sobolev norm of the velocity field $\boldsymbol{v}$ from which $\boldsymbol{\phi}_t$ is generated by the flow ODE.

Dissimilarity term

In the original formulation of LDDMM [13], the authors used the sum of squared differences (SSD) similarity measure (Section 13.1.2). Other dissimilarity measures can also be used in conjunction with LDDMM, but, due to its simplicity, SSD is best suited for understanding the approach. In the original LDDMM paper [13], the authors compute the dissimilarity measure by warping the fixed image I into the space of the moving image J using the inverse of the final transformation $\boldsymbol{\phi}_1$:

$$\mu(\boldsymbol{\phi}_1; I, J) = \int_{\Omega} \left\| I \circ \boldsymbol{\phi}_1^{-1} - J \right\|^2 \mathrm{d}\boldsymbol{x}.$$

This is different from the parametric and non-parametric approaches described earlier in the paper, where dissimilarity was computed by warping the moving image into the fixed image space (Chapter 13 and Section 14.3). Since it does not matter whether the similarity is computed in the space of the fixed or moving image, we will follow the convention of the original LDDMM paper.

The complete formulation of the LDDMM registration algorithm is the following energy minimization problem:

$$\boldsymbol{v}^* = \arg\min_{\boldsymbol{v} \in \mathcal{V}} E[\boldsymbol{v}; I, J], \quad \text{where}$$

$$E[\boldsymbol{v}; I, J] = \|\boldsymbol{v}\|_{\mathcal{V}}^2 + \frac{1}{\sigma^2} \int_{\Omega} \left\| I \circ \boldsymbol{\phi}_1^{-1} - J \right\|^2 \mathrm{d}\boldsymbol{x},$$

$$\text{and} \quad \frac{\mathrm{d}\boldsymbol{\phi}}{\mathrm{d}t}(\boldsymbol{x}, t) = \boldsymbol{v}(\boldsymbol{\phi}(\boldsymbol{x}, t), t), \quad \text{subject to } \boldsymbol{\phi}(\boldsymbol{x}, 0) = \boldsymbol{x}. \tag{14.9}$$

14.4.2 Approaches to solving the LDDMM problem

14.4.2.1 Direct minimization over the time-varying velocity field

The minimization of the energy functional $E[\boldsymbol{v}; I, J]$ in (14.9) can be solved using standard calculus of variations techniques, i.e., by writing down the Euler–Lagrange equation for E. The derivation is quite involved (largely due to the presence of the ODE in the formulation) and we direct interested readers to the original paper [13]. For a given value of $t \in [0, 1]$, the Euler–Lagrange equation is

$$2\boldsymbol{v}_t - \mathcal{K}\left(\frac{2}{\sigma^2}\left|D\boldsymbol{\phi}_{t,1}\right| \nabla I_t (I_t - J_t)\right) = 0, \tag{14.10}$$

where

$\boldsymbol{v}_t(\boldsymbol{x}) = \boldsymbol{v}(\boldsymbol{x}, t)$	is the velocity field at a fixed time t (note that this notation is used in order to be consistent with [13] and is different from earlier parts of the chapter, where the subscript was used to denote partial derivatives);
$\boldsymbol{\phi}_t(\boldsymbol{x}) = \boldsymbol{\phi}(\boldsymbol{x}, t)$	is the spatial transformation at a fixed time t;
$\boldsymbol{\phi}_{t,1} = \boldsymbol{\phi}_1 \circ \boldsymbol{\phi}_t^{-1}$	is the transformation between points at time point t and the corresponding points at time point 1;
$\boldsymbol{\phi}_{t,0} = \boldsymbol{\phi}_0 \circ \boldsymbol{\phi}_t^{-1} = \boldsymbol{\phi}_t^{-1}$	is the transformation between points at time point t and the corresponding points at time point 0;
$I_t = I \circ \boldsymbol{\phi}_{t,0}$	is the fixed image warped into time point t;
$J_t = J \circ \boldsymbol{\phi}_{t,1}$	is the moving image warped into time point t;
$\left\|D\boldsymbol{\phi}_{t,1}\right\|$	is the determinant of the Jacobian of $\boldsymbol{\phi}_{t,1}$;
$\mathcal{K}$	is the operator satisfying the relation $\mathcal{K}(\mathcal{L}^T\mathcal{L})\boldsymbol{u} = \boldsymbol{u}$ for all vector fields $\boldsymbol{u} \in \mathcal{U}$ ($\mathcal{K}$ is the inverse of the operator $\mathcal{L}^T\mathcal{L}$ and does not have a closed form); in practice it is computed by numerically solving the partial differential equation $\mathcal{L}^T\mathcal{L}\boldsymbol{w} = \boldsymbol{u}$ for $\boldsymbol{w} \in \mathcal{U}$.

To solve the LDDMM optimization problem numerically, we perform gradient descent, i.e., we take small steps in the direction opposite to the left-hand side of the Euler–Lagrange equation (14.10). The following considerations are involved in implementing the LDDMM approach computationally:

- The time dimension is discretized into a fixed number of steps (e.g., 40 or 100).
- During optimization, it is necessary to compute and maintain in memory the transformations $\boldsymbol{\phi}_t(\boldsymbol{x})$ and $\boldsymbol{\phi}_t(\boldsymbol{x})^{-1}$ for each discrete time step t. These transformations are computed using numerical schemes for approximating the flow ODE (14.2). While Euler's method can be used for this, Beg et al. [13] use a more accurate two-step semi-Lagrangian scheme [15].
- The partial differential equation corresponding to the application of the operator $\mathcal{K}$ in (14.10) is solved numerically. It is also possible to define $\mathcal{L}$ in an implicit manner, such that $\mathcal{K}$ corresponds to smoothing with a Gaussian kernel [16].

- Beg et al. [13] show that, theoretically, the velocity field that minimizes (14.9) has constant magnitude (i.e., speed) with respect to time. In practice, the sequential scheme for minimizing (14.9) does not yield solutions with constant speed, and Beg et al. [13] recommend reparameterizing the solution $\boldsymbol{v}(\boldsymbol{x}, t)$ with respect to t to maintain constant speed after every few update iterations.

Overall, the LDDMM approach yields itself well to computational implementation and is highly parallelizable. However, it is quite memory-intensive because of the need to maintain several data arrays of the size $\mathcal{O}(NT)$, where N is the size of I in pixels or voxels and T is the number of time steps.

14.4.2.2 Geodesic shooting

It turns out that we can minimize the LDDMM energy $E[\boldsymbol{v}; I, J]$ without having to optimize in the high-dimensional (space+time) space $\mathcal{V}$. The velocity field $\boldsymbol{v} \in \mathcal{V}$ that minimizes $E[\boldsymbol{v}; I, J]$ is entirely determined by the initial velocity $\boldsymbol{v}_0$. To understand this, consider the following analogy:

- Suppose you want to walk between two points $\boldsymbol{x}_A$ and $\boldsymbol{x}_B$ located in a mountainous terrain. A path that you would follow can be described by its coordinates $\boldsymbol{\phi}(t)$, with $\boldsymbol{\phi}(0) = \boldsymbol{x}_A$ and $\boldsymbol{\phi}(1) = \boldsymbol{x}_A$. It can also be described by your velocity vector at every time, $\boldsymbol{v}(t)$, with the path itself given by the ODE

$$\boldsymbol{\phi}(0) = \boldsymbol{x}_A \quad \text{and} \quad \frac{\mathrm{d}\boldsymbol{\phi}}{\mathrm{d}t}(t) = \boldsymbol{v}(t).$$

  The length of this path is given simply by

$$L[\boldsymbol{\phi}] = \int_0^1 \|\boldsymbol{v}(t)\| \, \mathrm{d}t.$$

- Now suppose you want to walk from $\boldsymbol{x}_A$ to $\boldsymbol{x}_B$ covering the shortest possible distance. This shortest path is called the *minimizing geodesic*. On a plane it is a straight line, but on a curved surface, it is a curve that locally "looks like" a straight line. We can write the problem of finding the minimizing geodesic in terms of $\boldsymbol{v}$ as follows:

$$\boldsymbol{v}^* = \arg\min_{\boldsymbol{\phi}} L[\boldsymbol{\phi}] \quad \text{subject to } \boldsymbol{\phi}(1) = \boldsymbol{x}_B,$$

  or equivalently,

$$\boldsymbol{v}^* = \arg\min_{\boldsymbol{v}} L[\boldsymbol{\phi}] + \frac{1}{\sigma^2} \|\boldsymbol{\phi}(1) - \boldsymbol{x}_B\|^2. \tag{14.11}$$

  This is analogous to the LDDMM energy minimization described above, and similarly requires us to search for a time-parameterized velocity function $\boldsymbol{v}(t)$.
- Let us now elaborate on the notion that a geodesic "looks like" a straight line locally. Let X be some point through which the curve $\boldsymbol{\phi}(t)$ passes, and let T_X

be the tangent plane to the Earth's surface at X. The curvature of the projection of curve $\boldsymbol{\phi}(t)$ onto T_X at X is called *geodesic curvature*. A geodesic is a curve whose geodesic curvature is zero at every point. Mathematically, this property is expressed by the second-order ODE

$$\frac{\mathrm{d}^2\boldsymbol{\phi}_k}{\mathrm{d}t^2} + \sum_{i,j} \Gamma_{ij}^k \frac{\mathrm{d}\boldsymbol{\phi}_i}{\mathrm{d}t}\frac{\mathrm{d}\boldsymbol{\phi}_j}{\mathrm{d}t} = 0, \quad \text{for } i, j, k \in \{1, 2, 3\}, \tag{14.12}$$

or equivalently

$$\frac{\mathrm{d}v_k}{\mathrm{d}t} + \sum_{i,j} \Gamma_{ij}^k v_i v_j = 0, \quad \text{for } i, j, k \in \{1, 2, 3\}, \tag{14.13}$$

where x_k and v_k denote the kth component of $\boldsymbol{\phi}$ and $\boldsymbol{v}$, respectively, and Γ_{ij}^k are the *Christoffel symbols* (second-order properties of the Earth's surface at each point that you might study in a differential geometry course). It is not necessary to know what they are for the purposes of our discussion. What is important is to know that given the initial conditions $\boldsymbol{\phi}(0) = \boldsymbol{x}_A$ and $\boldsymbol{v}(0) = \boldsymbol{v}_0$ this second-order differential equation *determines the rest of the geodesic*. Following a geodesic means setting out from $\boldsymbol{x}_A$ with some initial velocity vector $\boldsymbol{v}_0$ and walking in such a way that our path is always locally straight, having zero geodesic curvature. For some choice of initial velocity vector $\boldsymbol{v}_0$, we will end up reaching village $\boldsymbol{x}_B$ at $t = 1$.

- This general approach of starting out from a given point along a given velocity and following a geodesic is called *geodesic shooting*. It allows us to reformulate the shortest path problem in terms of a much simpler unknown, $\boldsymbol{v}_0$:

$$\begin{aligned} \boldsymbol{v}_0^* = \underset{\boldsymbol{v}_0 \in \mathbb{R}^3}{\arg\min} \, &\|\boldsymbol{\phi}(1) - \boldsymbol{x}_B\|^2 \\ \text{subject to } &\frac{\mathrm{d}v_k}{\mathrm{d}t} + \sum_{i,j} \Gamma_{ij}^k v_i v_j = 0 \quad \text{for } i, j, k \in \{1, 2, 3\}, \\ &\boldsymbol{\phi}(0) = \boldsymbol{x}_A, \quad \text{and} \quad \frac{\mathrm{d}\boldsymbol{\phi}}{\mathrm{d}t}(t) = \boldsymbol{v}(t). \end{aligned} \tag{14.14}$$

This problem is similar to rotating a tank to shoot a target, but with the difference that the projectiles move along geodesics. The advantage of this geodesic shooting formulation compared to the previous formulation (14.11) is that it reduces the search space by a whole dimension. Instead of optimizing over the path of the projectile, we optimize over the initial velocity. Another advantage is that even if we do not find the minimum of (14.14), our solutions $\boldsymbol{\phi}(t)$ obtained by solving the ODE (14.12) are always going to be geodesics (just ones that do not quite get us to $\boldsymbol{x}_B$). This is not the case for the previous formulation (14.11), where solutions short of the minimum are not guaranteed to be geodesics.

This idea of geodesic shooting extends beautifully to diffeomorphisms. Diffeomorphic flows $\boldsymbol{\phi}(\boldsymbol{x}, t)$ with corresponding velocity fields $\boldsymbol{v}(\boldsymbol{x}, t)$ that minimize the LDDMM energy (14.8) are indeed geodesics on the space Diff, with their length given by

$$L[\boldsymbol{\phi}]^2 = \int_0^1 \|\boldsymbol{v}\|_V^2 \, dt.$$

Consequently, these geodesics satisfy the "locally straight" differential equation equivalent to (14.13). This equation is called the Euler–Poincaré differential equation (EPDiff) and has the following form:

$$\frac{d\boldsymbol{m}}{dt} + (D\boldsymbol{m})\boldsymbol{v} + (D\boldsymbol{v})^T \boldsymbol{m} + \boldsymbol{m}\,(\nabla \cdot \boldsymbol{v}) = \mathbf{0}, \tag{14.15}$$

where D denotes the Jacobian matrix, $\nabla \cdot \boldsymbol{v}$ denotes the divergence operator, and $\boldsymbol{m}$ is called the *momentum*, given by

$$\boldsymbol{m} = \mathcal{L}^T \mathcal{L} \boldsymbol{v} \quad \text{(equivalently, } \boldsymbol{v} = \mathcal{K}\boldsymbol{m}\text{)}.$$

The momentum is a time-varying vector field just like $\boldsymbol{v}$, with $\boldsymbol{v}$ essentially being a smoothed version of $\boldsymbol{m}$ due to the form of the operator $\mathcal{L}$. This momentum has some analogy with the concept of momentum in particle physics.

Like the ODE (14.13), the EPDiff equation is a differential equation involving $\boldsymbol{v}$. Consequently, solutions of (14.13) are determined by the initial conditions, i.e., the initial velocity field $\boldsymbol{v}(0) = \boldsymbol{v}(\boldsymbol{x}, 0)$, or equivalently, the initial momentum field $\boldsymbol{m}_0$ (since velocity and momentum are related to each other by operator $\mathcal{L}$). Therefore, we can reformulate diffeomorphic image registration as the problem of geodesic shooting: finding the initial velocity $\boldsymbol{v}_0$ (or initial momentum $\boldsymbol{m}_0$) for which the geodesic $\boldsymbol{\phi}$ optimally matches images I and J at time 1. In complete form, this problem is formulated as follows [14,17]:

$$\boldsymbol{m}_0^* = \arg\min_{\boldsymbol{m}_0} \int_\Omega \left\| I \circ \boldsymbol{\phi}_1^{-1} - J \right\|^2, \quad \text{where}$$

$$\begin{cases} \boldsymbol{v}_t = \mathcal{K}\boldsymbol{m}_t, \\ \frac{d\boldsymbol{m}_t}{dt} + (D\boldsymbol{m}_t)\boldsymbol{v}_t + (D\boldsymbol{v}_t)^T \boldsymbol{m}_t + \boldsymbol{m}_t\,(\nabla \cdot \boldsymbol{v}_t) = \mathbf{0}, \\ \frac{d\boldsymbol{\phi}_t}{dt} = \boldsymbol{v}_t \circ \boldsymbol{\phi}_t. \end{cases} \tag{14.16}$$

Solving this problem via gradient descent is computationally feasible. It involves pulling back the gradient of the image dissimilarity measure μ with respect to $\boldsymbol{\phi}_1$ back through the ODEs for $\boldsymbol{\phi}_t$ and $\boldsymbol{m}_t$. While the math for this is rather involved, the approach can be efficiently implemented computationally. As in the simple example introduced above, geodesic shooting offers the following advantages over the variational formulation of LDDMM:

- The optimization is over a single vector field $\boldsymbol{m}_0$, as opposed to the time-varying velocity field $\boldsymbol{v}$. This reduces the memory requirements of the algorithm and can lead to faster convergence. A recent work on Fourier-approximated Lie Algebras for Shooting (FLASH) has further reduced the computational complexity of geodesic shooting algorithm by developing a low-dimensional representation of such initial vector field $\boldsymbol{v}_0$ in a band-limited space [18,19].
- Even if the minimum of the objective is not reached, solutions that are geodesics in Diff are obtained.

14.4.3 LDDMM and computational anatomy

14.4.3.1 Defining distance on the space of diffeomorphic transformations

In addition to being a formidable image registration method, LDDMM gives rise to a family of medical image statistical analysis approaches known as *computational anatomy* [14,17,20]. The Sobolev norm $\|\boldsymbol{v}\|^2_{\mathcal{V}}$ can be used to extend the concept of "distance" to the space of diffeomorphic transformations Diff, and even to the space of medical images. Let $\boldsymbol{\psi} \in \text{Diff}(\Omega)$ be a diffeomorphic transformation on Ω. We define the distance between the identity transformation (i.e., the origin of the space Diff) and $\boldsymbol{\psi}$ as follows:

$$d(\text{Id}, \boldsymbol{\psi})^2 = \min_{\boldsymbol{v} \in \mathcal{V}: \boldsymbol{\phi}_1[\boldsymbol{v}] = \boldsymbol{\psi}} \|\boldsymbol{v}\|^2_{\mathcal{V}},$$

where $\boldsymbol{\phi}_1[\boldsymbol{v}]$ denotes the solution of the flow ODE (14.2) with input $\boldsymbol{v}$ at time 1. In other words, $d(\text{Id}, \boldsymbol{\psi})^2$ is equal to the kinetic energy of the least energetic diffeomorphic flow that has as its end-point the diffeomorphism $\boldsymbol{\psi}$. For a pair of diffeomorphic transformations $\boldsymbol{\psi}, \boldsymbol{\chi}$, the distance is defined as

$$d(\boldsymbol{\chi}, \boldsymbol{\psi}) = d(\text{Id}, \boldsymbol{\psi} \circ \boldsymbol{\chi}^{-1}).$$

It quantifies the kinetic energy of the least energetic diffeomorphic flow that maps the end-point of $\boldsymbol{\chi}$ to the end-point of $\boldsymbol{\psi}$. Although this distance function is not symmetric, it satisfies the properties of a proper distance function on the space Diff: $d(\boldsymbol{\chi}, \boldsymbol{\psi}) \geq 0$ for all $\boldsymbol{\psi}, \boldsymbol{\chi} \in \text{Diff}$; $d(\boldsymbol{\chi}, \boldsymbol{\psi}) = 0$ if and only if $\boldsymbol{\psi} = \boldsymbol{\chi}$; and the triangle inequality $d(\boldsymbol{\chi}, \boldsymbol{\psi}) \leq d(\boldsymbol{\chi}, \eta) + d(\eta, \boldsymbol{\psi})$ is satisfied for all $\boldsymbol{\psi}, \boldsymbol{\chi}, \eta \in \text{Diff}$ [20].

14.4.3.2 Defining distance on the space of images

While being able to measure distance between elements of Diff is useful, it would be even more useful to measure distance between images. At first, let us make a rather strong assumption that in our collection of images, any two images can be matched perfectly using some diffeomorphic transformation. This assumption makes sense if all images in the collection are deformations of a template image. It also makes sense for simple images, for example, images of fried eggs. In this setting, we can define the distance between images I and J as follows:

$$d(I, J)^2 = \min_{\boldsymbol{v} \in \mathcal{V}: I \circ \boldsymbol{\phi}_1^{-1}[\boldsymbol{v}] = J} \|\boldsymbol{v}\|^2_{\mathcal{V}},$$

in other words, the kinetic energy of the least energetic diffeomorphic flow that warps I to look like J at time 1. This distance function on images also behaves like a proper distance function (which is non-negative, is zero only if $I = J$, and satisfies the triangle inequality).

In real-world applications, the assumption that J can be generated from I by a diffeomorphic transformation is too strong, as images have different intensity characteristics and noise, and often the underlying anatomical content is not one-to-one. The concept of distance between images must be relaxed to allow for inexact matching of I and J. A practical approach is to define the inexact distance between I and J as

$$d_{\text{inexact}}(I, J)^2 = \min_{\boldsymbol{v} \in \mathcal{V}} E[\boldsymbol{v}; I, J],$$

where $E[\boldsymbol{v}; I, J]$ is the LDDMM energy from (14.9), i.e., the sum of the kinetic energy of the diffeomorphic flow that best matches I and J and a residual intensity difference term. While this definition of distance no longer has the rigorous mathematical properties of the exact distance $d(I, J)$, it allows for many practical algorithms where distance between real-world images needs to be computed.[2]

14.4.3.3 Applications to statistical analysis of images

Medical images are data, and we often want to draw inferences from medical images using statistical analysis. Often we derive some summary features from images (e.g., volume of the hippocampus) and perform statistical analysis on these derived measures (What is the average hippocampus volume in a given group? Does the mean hippocampal volume differ between patients and healthy individuals? Does the hippocampal volume change with age?). But summary features do not capture the entirety of information in the image. It would be nice to be able to perform the same kinds of statistical analyses on the images themselves, treating each image not as a source of one or more summary measures, but as a high-dimensional multivariate observation.

Of course, images are collections of voxels, and we can treat the vector of all intensities in an image as a multivariate observation. However, since voxels with the same coordinates across different images come from different anatomical locations, any such analysis will be almost meaningless. For example, averaging a collection of images voxel-by-voxel results in a blurry image that looks almost nothing like the source images and tells us nothing about the average anatomy in the collection.

By defining a distance function on the space of images that captures anatomical differences, rather than intensity differences, the LDDMM approach makes it possible to extend various concepts from statistics to the image domain. Many methods

[2] The theory of *image metamorphosis* [21,22] models transformations between images as pairs consisting of a diffeomorphic geometric component and an intensity variation component, and allows for a mathematically robust definition of distance between images that cannot be matched exactly.

in statistics (finding the mean, linear and non-linear regression, dimensionality reduction) are formulated as least square fitting problems. For example, the problem of finding the mean of a sample of multivariate observations $\boldsymbol{x}_1, \ldots, \boldsymbol{x}_n$ can be formulated in terms of minimizing the sum of squared distances:

$$\overline{\boldsymbol{x}}_{\mathrm{Fr}} = \arg\min_{\boldsymbol{x}\in\mathcal{X}} \sum_{i=1}^{n} d(\boldsymbol{x}, \boldsymbol{x}_i)^2. \tag{14.17}$$

This definition of the mean is called the *Fréchet mean*, and unlike the arithmetic mean given by $\frac{1}{n}\sum_{i=1}^{n}\boldsymbol{x}_i$, it extends to non-Euclidean spaces. As long as the space $\mathcal{X}$ in which the observations $\boldsymbol{x}_1, \ldots, \boldsymbol{x}_n$ lie has a distance function, the Fréchet mean is well defined. It is easy to verify that when $\mathcal{X}$ is a Euclidean space, the Fréchet mean is equal to the arithmetic mean. The Fréchet mean of a set of images $I_1, \ldots, I_n$ defined on the domain Ω can thus be constructed by solving the following minimization problem:

$$\overline{I}_{\mathrm{Fr}} = \arg\min_{I\in\tilde{I}_\Omega} \sum_{i=1}^{n} d_{\mathrm{inexact}}(I, I_i)^2 = \arg\min_{I\in\tilde{I}_\Omega, \boldsymbol{v}_1\in\mathcal{V},\ldots,\boldsymbol{v}_n\in\mathcal{V}} \sum_{i=1}^{n} E[\boldsymbol{v}_i; I, I_i].$$

This involves simultaneously searching for a "synthetic" mean image I and a set of velocity fields $\boldsymbol{v}_1, \ldots, \boldsymbol{v}_n$ that optimally warp I to match the input images. An efficient way to solve this minimization, proposed by Joshi et al. [23], is to alternate minimizing over I and $\boldsymbol{v}_1, \ldots, \boldsymbol{v}_n$. The resulting algorithm alternates between n independent LDDMM registrations and arithmetic averaging of image intensities, and yields excellent mean images. These images are commonly used to create *population-specific templates* for groupwise analysis of medical imaging data [24].

Many other problems in statistics have been extended to the space of images. For example, the equivalent of linear regression (i.e., fitting a slope and an intercept of a line that best describes the relationship between a dependent variable, like hippocampal volume, and an independent variable, like age) is *geodesic regression* [25]. In geodesic regression with a scalar independent variable, we have a set of observations ($t_i \in [0, 1], I_i \in \tilde{I}_\Omega$) and we seek to find a time-parameterized sequence of images $I_t = I_0 \circ \boldsymbol{\phi}_t$ that forms a geodesic in the space of images, such that the sum of squared distances from the observed images to I_t is minimized. Specifically, one solves a minimization problem

$$(I_0^*, \boldsymbol{v}^*) = \arg\min_{I_0\in\tilde{I}_\Omega, \boldsymbol{v}\in\mathcal{V}} \sum_{i=1}^{n} d(I_0 \circ \boldsymbol{\phi}_{t_i}^{\boldsymbol{v}}, I_i)^2$$

that is conceptually similar to the Fréchet mean problem in that it requires nested minimization. The concept of non-linear regression (i.e., fitting a quadratic or cubic curve to data in a least squares sense) has also been extended to the manifold of images [26,27]. So have been the concepts of mixed modeling for longitudinal image analysis [28,29], principal component analysis for data dimensionality reduction

[30–32], and numerous other data analysis techniques that can be framed in terms of distance minimization.

14.5 Exercises

These programming exercises can be done in MATLAB or Python. However, for the best experience, we recommend using a software framework capable of representing arbitrary-dimensional tensors and performing automatic differentiation, such as PyTorch or TensorFlow. Automatic differentiation allows you to define the objective functions you want to minimize/maximize, without worrying about computing the derivatives of these objective functions with respect to parameters.

1. Implement the Horn and Schunck optical flow algorithm (14.5) in 2D or 3D. Apply the algorithm to different pairs of images: frames from a cardiac cine MRI sequence, longitudinal brain MRI scans, MRI scans of different individuals's brain aligned by affine registration, or synthetic images.
2. Implement the Demons approximation of optical flow (14.6) and compare the velocity fields it generates to those from the Horn and Schunck algorithm.
3. Implement the compositional Demons registration algorithm and apply to pairs of brain MRI scans. When implementing this algorithm, it is helpful to represent spatial transformations $\boldsymbol{\phi}(\boldsymbol{x})$ in terms of displacements (see Section 13.2.2 for the explanation of why this is helpful and how the composition of two spatial transformations can be expressed in terms of the corresponding displacements).
4. Extend the compositional Demons registration algorithm with a multi-resolution scheme where registration is first performed on images subsampled by a factor of 8, then by factor of 4, then 2, then at full resolution. Compare the results to the baseline implementation.
5. Write a function that implement the scaling and squaring algorithm. As before, spatial transformations should be expressed in terms of displacements.
6. Implement the Log domain diffeomorphic Demons algorithm. Optionally, implement the symmetric version of this algorithm. Compare registration results and computation time between compositional Demons and Log domain diffeomorphic Demons.

References

[1] Thirion J.-P., Non-rigid matching using demons, in: CVPR '96: Proceedings of the 1996 Conference on Computer Vision and Pattern Recognition (CVPR '96), IEEE Computer Society, ISBN 0-8186-7258-7, 1996, p. 245.

[2] Vercauteren T., Pennec X., Perchant A., Ayache N., Diffeomorphic demons: efficient non-parametric image registration, NeuroImage 45 (1 Suppl) (Mar. 2009) S61–S72, https://doi.org/10.1016/j.neuroimage.2008.10.040.

[3] Avants B.B., Epstein C.L., Grossman M., Gee J.C., Symmetric diffeomorphic image registration with cross-correlation: evaluating automated labeling of elderly and neurodegenerative brain, Medical Image Analysis 12 (1) (Feb. 2008) 26–41, https://doi.org/10.1016/j.media.2007.06.004.
[4] Hua X., Lee S., Yanovsky I., Leow A.D., Chou Y.Y., Ho A.J., Gutman B., Toga A.W., Jack C.R. Jr., Bernstein M.A., Reiman E.M., Harvey D.J., Kornak J., Schuff N., Alexander G.E., Weiner M.W., Thompson P.M., Optimizing power to track brain degeneration in Alzheimer's disease and mild cognitive impairment with tensor-based morphometry: an ADNI study of 515 subjects, NeuroImage 48 (4) (Dec. 2009) 668–681.
[5] Arnold V.I., Sur la géométrie différentielle des groupes de Lie de dimension infinie et ses applicationsa l'hydrodynamique des fluides parfaits, Annales de L'Institut Fourier 16 (1) (1966) 319–361.
[6] Arnold V.I., Ordinary Differential Equations, 1998, translated and edited by Richard A. Silverman.
[7] Arsigny V., Commowick O., Pennec X., Ayache N., A log-Euclidean framework for statistics on diffeomorphisms, in: International Conference on Medical Image Computing and Computer-Assisted Intervention, Springer, 2006, pp. 924–931.
[8] Horn B.K.P., Schunck B.G., Determining optical flow, Artificial Intelligence 17 (1–3) (1981) 185–203.
[9] Vercauteren T., Pennec X., Perchant A., Ayache N., Symmetric log-domain diffeomorphic registration: a demons-based approach, in: Medical Image Computing and Computer-Assisted Intervention - MICCAI 2008, 11th International Conference, Proceedings, Part I, 2008, pp. 754–761, https://doi.org/10.1007/978-3-540-85988-8_90.
[10] Rossmann W., Lie Groups: An Introduction Through Linear Groups, vol. 5, Oxford University Press on Demand, 2006.
[11] Hill D.L.G., Studholme C., Hawkes D.J., Voxel similarity measures for automated image registration, in: Visualization in Biomedical Computing 1994, vol. 2359, International Society for Optics, 1994, pp. 205–216.
[12] Mansi T., Pennec X., Sermesant M., Delingette H., Ayache N., iLogDemons: A demons-based registration algorithm for tracking incompressible elastic biological tissues, International Journal of Computer Vision 92 (1) (2011) 92–111.
[13] Beg M.F., Miller M.I., Trouvé A., Younes L., Computing large deformation metric mappings via geodesic flows of diffeomorphisms, International Journal of Computer Vision (ISSN 0920-5691) 61 (2) (2005) 139–157, https://doi.org/10.1023/B:VISI.0000043755.93987.aa.
[14] Miller M.I., Trouvé A., Younes L., Geodesic shooting for computational anatomy, Journal of Mathematical Imaging and Vision 24 (2) (Jan. 2006) 209–228, https://doi.org/10.1007/s10851-005-3624-0.
[15] Staniforth A., Côté J., Semi-Lagrangian integration schemes for atmospheric models—A review, Monthly Weather Review 119 (9) (1991) 2206–2223.
[16] Sommer S., Nielsen M., Darkner S., Pennec X., Higher order kernels and locally affine LDDMM registration, arXiv preprint, arXiv:1112.3166, 2012.
[17] Younes L., Arrate F., Miller M.I., Evolutions equations in computational anatomy, in: Mathematics in Brain Imaging, NeuroImage (ISSN 1053-8119) 45 (1, Supplement 1) (2009) S40–S50, https://doi.org/10.1016/j.neuroimage.2008.10.050, http://www.sciencedirect.com/science/article/pii/S105381190801166X.
[18] Zhang M., Wells W. III, Golland P., Probabilistic modeling of anatomical variability using a low dimensional parameterization of diffeomorphisms, Medical Image Analysis 41 (2017) 55–62.

[19] Zhang M., Fletcher P.T., Fast diffeomorphic image registration via Fourier-approximated Lie algebras, International Journal of Computer Vision 127 (1) (2019) 61–73.
[20] Miller M.I., Trouve A., Younes L., On the metrics and Euler–Lagrange equations of computational anatomy, Annual Review of Biomedical Engineering 4 (2002) 375–405, https://doi.org/10.1146/annurev.bioeng.4.092101.125733.
[21] Holm D., Trouvé A., Younes L., The Euler–Poincaré theory of metamorphosis, Quarterly of Applied Mathematics 67 (4) (2009) 661–685.
[22] Richardson C.L., Younes L., Metamorphosis of images in reproducing kernel Hilbert spaces, Advances in Computational Mathematics 42 (3) (2016) 573–603.
[23] Joshi S., Davis B., Jomier M., Gerig G., Unbiased diffeomorphic atlas construction for computational anatomy, NeuroImage 23 (Suppl 1) (2004) S151–S160, https://doi.org/10.1016/j.neuroimage.2004.07.068.
[24] Evans A.C., Janke A.L., Collins D.L., Baillet S., Brain templates and atlases, NeuroImage 62 (2) (2012) 911–922.
[25] Fletcher P.T., Geodesic regression and the theory of least squares on Riemannian manifolds, International Journal of Computer Vision 105 (2) (2013) 171–185.
[26] Davis B.C., Fletcher P.T., Bullitt E., Joshi S., Population shape regression from random design data, International Journal of Computer Vision 90 (2) (2010) 255–266.
[27] Singh N., Niethammer M., Splines for diffeomorphic image regression, in: International Conference on Medical Image Computing and Computer-Assisted Intervention, Springer, 2014, pp. 121–129.
[28] Singh N., Hinkle J., Joshi S., Fletcher P.T., A hierarchical geodesic model for diffeomorphic longitudinal shape analysis, in: International Conference on Information Processing in Medical Imaging, Springer, 2013, pp. 560–571.
[29] Schiratti J.-B., Allassonnière S., Colliot O., Durrleman S., A Bayesian mixed-effects model to learn trajectories of changes from repeated manifold-valued observations, Journal of Machine Learning Research 18 (1) (2017) 4840–4872.
[30] Fletcher P.T., Lu C., Pizer S.M., Joshi S., Principal geodesic analysis for the study of nonlinear statistics of shape, IEEE Transactions on Medical Imaging 23 (8) (Aug. 2004) 995–1005.
[31] Sommer S., Lauze F., Hauberg S., Nielsen M., Manifold valued statistics, exact principal geodesic analysis and the effect of linear approximations, in: European Conference on Computer Vision, Springer, 2010, pp. 43–56.
[32] Zhang M., Fletcher T., Probabilistic principal geodesic analysis, in: Advances in Neural Information Processing Systems, 2013, pp. 1178–1186.

CHAPTER

Image mosaicking

15

Sophia Bano and Danail Stoyanov

Wellcome/EPSRC Centre for Interventional and Surgical Sciences (WEISS), Department of Computer Science, University College London, London, United Kingdom

Learning points

- Motion models for image transformations
- Feature based methods for image mosaicking
- Direct methods for image mosaicking
- Deep learning-based methods for image mosaicking
- Image reprojection and blending
- Clinical applications of image mosaicking

15.1 Introduction

Image mosaicking is the process of geometrically aligning multiple images having an overlapping field-of-view (FoV) to generate a high-resolution panoramic image with an expanded FoV. The images to be aligned could be still overlapping images of a scene or a video sequence captured from the same device or different devices. Most existing image mosaicking methods assume that the captured scene is static and also nearly planar, and as a result there is either minimal or no parallax effect, or the images are captured by pivoting the camera around its optical center. The overlapping images are geometrically aligned and reprojected on a global reference plane, also termed mosaic plane, to form an expanded FoV image (as illustrated in Fig. 15.1). Image mosaicking techniques have become common in consumer computer vision and computer graphics devices like mobile phones and in applications such as video stabilization, video summarization, motion detection and tracking, panorama construction (in smartphones and 360 cameras), photomosaics in aerial and satellite imaging [1], and street view [2]. In medicine, however, image mosaicking techniques despite feasibility demonstrations have yet to find widespread use.

Mosaicking for biomedical image stitching is gaining interest as it can provide better understanding and improved imaging of the anatomical site, assist surgeons during minimally invasive surgery, support diagnosis, indexing of surgical procedures and treatment planning, and reduce procedural time and surgical complications. Radiographic image stitching [3,4], whole slide image (WSI) stitching in histopathol-

Medical Image Analysis. https://doi.org/10.1016/B978-0-12-813657-7.00030-3

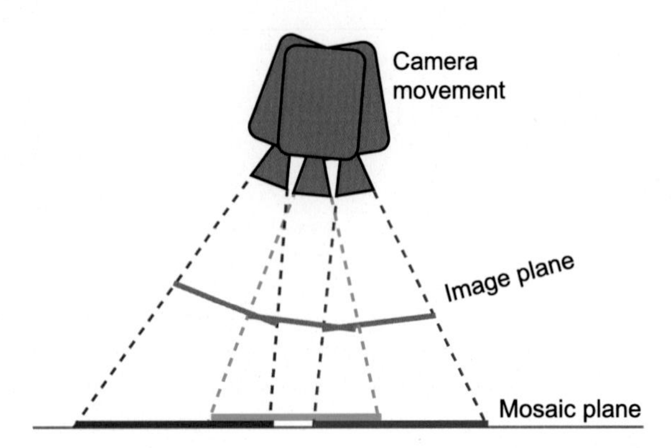

FIGURE 15.1 Image reprojection.

Illustration of camera movement to capture overlapping images on the image plane and their reprojection onto the mosaic plane.

ogy [5,6], endoscopic view expansion in minimally invasive procedures such as fetoscopy [7] and cystoscopy [8,9], and ophthalmologic slit-lamp video mosaicking [10] are some biomedical imaging application areas being actively investigated for enhancing clinical imaging (Fig. 15.2). The requirements of different clinical imaging settings can vary and as such so do the challenges, for example, real-time needs in intraoperative settings or accuracy in histopathological imaging.

Common challenges during minimally invasive surgery mosaicking include limited FoV, constrained maneuverability of the endoscope, and poor visibility due to low resolution and reflection artifacts. This leads to difficulty in inspecting the anatomical site for treatment planning and may result in increased procedural time and surgical complications in some cases. In radiographic imaging and WSI [4,6], high-resolution images are captured with focus on a specific area of interest following radiographic or microscopic camera translation to capture images of the complete region of interest. Through image stitching, an enlarged image is generated, providing the complete high-resolution view that enables better support for diagnosis and treatment. During endoscopic surgeries [7,8], the clinician first visually explores the anatomical sites by maneuvering the endoscopic camera to identify/localize the regions of interest and to build a mental map and treatment plan. Common challenges faced by clinicians during endoscopy include limited FoV, constrained maneuverability of the endoscope, and poor visibility due to low resolution and reflection artifacts. Mosaicking can increase the FoV in endoscopy to ease the localization of different anatomical sites for improved treatment and planning.

The generic pipeline of an image mosaicking approach is shown in Fig. 15.3. Given a set of overlapping images of a scene, the correspondences between pairs of images are first computed by pre-calibrating camera parameters, dense (pixel-by-pixel) feature matching, or keypoint detection and matching. These correspondences are then used for the estimation of the geometric transformation for aligning the pairs

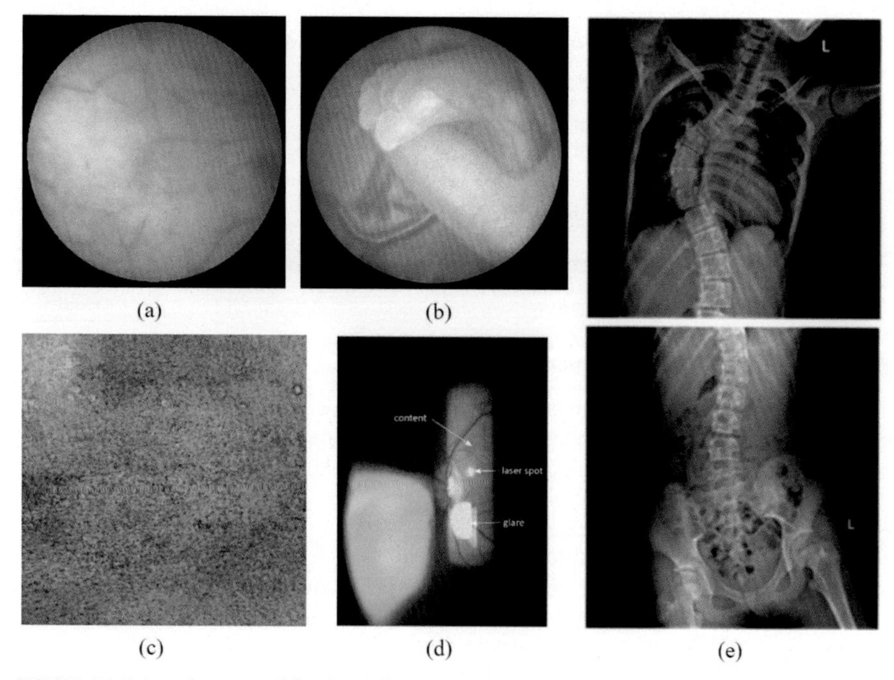

FIGURE 15.2 Imaging modalities investigated for mosaicking.

Example images from imaging modalities that are used in mosaicking. (a) Urinary bladder (cystoscopy). (b) Placenta and fetus (fetoscopy). (c) Whole slide imaging. (d) Slit-lamp imaging. (e) X-ray imaging.

of images. The most common geometric transformations are affine and projective (homography) (discussed in Section 15.2.1). The images are then reprojected onto a common mosaic plane using a chain of geometric transformations. Image mosaicking requires alignment and seamless stitching of the overlapping regions; however, discontinuities and seams are often visible due to photometric errors from illumination variation and geometric misalignment in transformation estimation [11]. To overcome this problem, image blending is applied which merges the overlapping regions such that the discontinuities are minimized, and an overall smooth seam-free image mosaic is obtained.

15.2 Motion models

15.2.1 Image transformations

Image transformation refers to the geometric transformation of the image coordinate system in order to align partially overlapping images taken at different times or with different devices (cameras). Image transformations, similar to image registra-

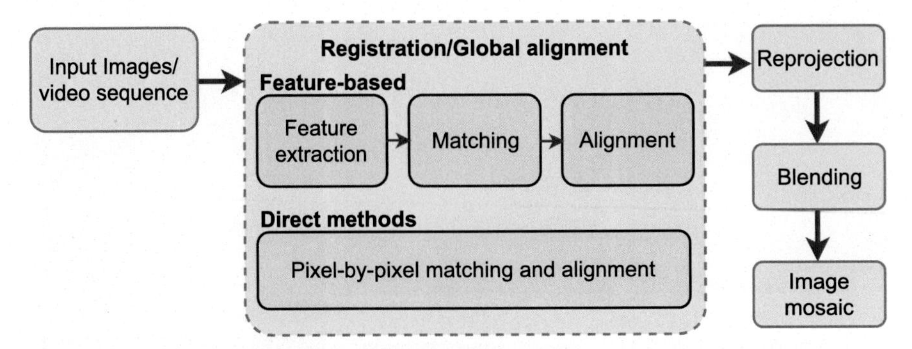

FIGURE 15.3 General image mosaicking framework.

Two commonly used image alignment methods are feature-based and direct, used to compute transformation between image pairs. A complete mosaic is generated after reprojecting the transformed images onto the mosaic place and applying image blending.

tion pipelines, are widely used for image stitching [11], camera calibration and lens distortion correction [12], and image morphing [13]. Given a point $\mathbf{x} = (x, y, 1)^T$ in the homogeneous image coordinate, an image transformation matrix, $\mathbf{H}$, maps $\mathbf{x}$ to the reference coordinate system as

$$\mathbf{x}' = \mathbf{H}\mathbf{x}, \tag{15.1}$$

where $\mathbf{H}$ can only be computed up to a scale factor and $\mathbf{H}$ denotes the transformation matrix which is given by

$$\left[\begin{array}{c|c} \mathbf{R}_{2\times 2} & \mathbf{T}_{2\times 1} \\ \hline \mathbf{P}_{1\times 2} & 1 \end{array}\right] = \left[\begin{array}{cc|c} r_{11} & r_{12} & t_x \\ r_{21} & r_{22} & t_y \\ \hline p_1 & p_2 & 1 \end{array}\right], \tag{15.2}$$

where $\mathbf{R}$ is the rotation matrix responsible for transformations such as scaling, rotation, and shear, $\mathbf{T}$ is the translation vector, and $\mathbf{P}$ is the projection vector. $\mathbf{H}$ is same for any point $\mathbf{x}$ in the image and can be described by a set number of parameters (degrees of freedom [DOFs]), and hence $\mathbf{H}$ is global. Image transformations can be classified into two main types, namely, affine and projective. Fig. 15.4 shows the different types of image transformations.

15.2.2 Affine transformation

An affine transformation is a linear combination of translation, scaling, rotation, and/or shear such that it preserves the collinearity of points, parallelism of lines, and ratios of distances. It is a special case of projective transformation in which the vector $P = [p_1, p_2]$ is always equal to zero. The matrix R is the rotation matrix and is responsible for introducing scaling $\mathbf{R}_s$, rotation $\mathbf{R}_\theta$, and shear $\mathbf{R}_{sh}$ transformations

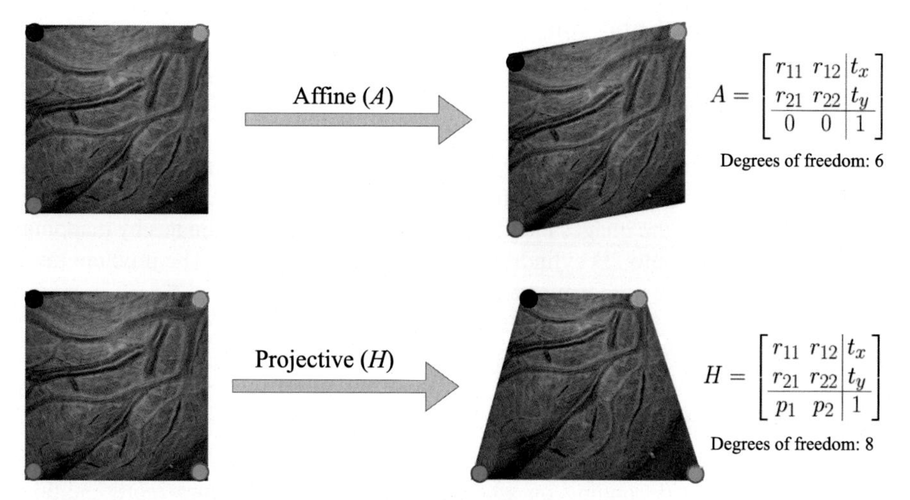

FIGURE 15.4 Illustration of geometric transformations.

An ex vivo human placenta image is used to demonstrate geometric transformations. Affine, having six DOFs, needs at least three-point correspondences, and projective, having eight DOFs, needs at least four-point correspondences for estimating the transformation matrix.

given by

$$\mathbf{R}_s = \begin{bmatrix} s_x & 0 \\ 0 & s_y \end{bmatrix}, \quad \mathbf{R}_\theta = \begin{bmatrix} \cos\theta & -\sin\theta \\ \sin\theta & \cos\theta \end{bmatrix}, \quad \mathbf{R}_{sh} = \begin{bmatrix} 1 & sh_x \\ sh_y & 1 \end{bmatrix}, \tag{15.3}$$

where s_x and s_y define the scaling factor along the x- and y-axes, respectively, θ defines the angle of rotation, and sh_x and sh_y define the shear along the x- and y-axes, respectively. An affine transformation has six DOFs, and hence it requires a minimum of three-point correspondences between a pair of images for solving equation (15.1).

A further simplified form of transformation is called similarity transformation, which is a combination of scaling, rotation, and translation with only four DOFs and can be estimated with two match pairs.

15.2.3 Projective transformation or homography

When the viewpoint of the observer (camera location) changes, the perceived object also changes. This effect is modeled with a projective transformation. A projective transformation or homography is a non-linear transformation that maps image points between two camera views, under the assumption that they belong to a planar scene. The projective transformation preserves collinearity, but does not preserve parallelism, lengths, and angles of lines. It shows the change in the perceived object. A projective transformation is given by equation (15.2) and has eight DOFs.

15.2.4 Cylindrical and spherical modeling

An alternative approach to homography is using a predefined model such as a cylinder or sphere for warping the images; as a result, the problem simplifies to estimating a pure translation motion for aligning the images [11]. A cylinder is used for modeling a camera having panning motion only, and a sphere is used for modeling a camera having both panning and tilting motions. These methods require the focal length to be known and transform the images into cylindrical or spherical coordinates by mapping the 3D image points onto 2D cylindrical or spherical coordinates. The problem then simplifies to estimating pure pan or/and tilt angles only.

15.3 Matching

Image mosaicking algorithms can be grouped into two types, namely, feature-based and direct methods, depending on whether they use sparse feature representation or pixel-by-pixel representation for matching [11,14,15]. Deep learning methods have also been developed recently for estimating homography and geometric matching [16–18].

15.3.1 Feature-based methods

Feature-based methods estimate the geometric transformation between overlapping image pairs using a sparse set of low-level feature points. These methods do not need large overlapping between image pairs to generate mosaics but mostly require image pairs to be of high resolution without motion blur since they rely on detecting edges, corners, and object contours. The extracted features must be discriminative, large in number, and detectable in both images with high repeatability. Moreover, they must be invariant to image noise, scaling, rotation, and translation. The most commonly used features for image mosaicking are Scale-Invariant Feature Transform (SIFT) [19], Principal Component Analysis-based SIFT (PCA-SIFT) [20], Affine SIFT (ASIFT) [21], Speeded Up Robust Features (SURF) [22], and Oriented FAST and rotated BRIEF (ORB) [23]. Additionally, convolutional neural network (CNN)-based features are descriptors that have also been proposed recently for image registration [24,25].

SIFT descriptor is computed by first estimating scale-space extrema by approximating the Laplacian of Gaussian (LoG) with the Difference of Gaussian (DoG), followed by keypoint localization. Keypoint orientation assignment using a local image gradient and descriptor generation for each keypoint is based on a histogram of image gradient magnitude and orientation. PCA is used widely for dimensionality reduction. In PCA-SIFT, instead of a histogram with a normalized gradient, PCA is used, which results in more compact descriptors. ASIFT is an affine-invariant extension of SIFT.

SURF is a time-efficient version of SIFT. SURF estimates the LoG with box filter, since the convolution with square is faster to compute with an integral image and can be done in parallel for different scales. SURF uses a BLOB detector based on a Hessian matrix to find the keypoints. Orientation assignment in SURF is done using the wavelet response in horizontal and vertical directions, followed by feature descriptor generation using the wavelet response.

ORB integrates the Features from Accelerated Segment Test (FAST) [26] keypoint detector and the Binary Robust Independent Elementary Features (BRIEF) [27] descriptor to form a robust and efficient keypoint descriptor. First, FAST is used to obtain keypoints, followed by a Harris corner measure to find high-performing keypoints. FAST is rotation-varying and does not compute orientation, but instead computes the intensity-weighted centroid of a patch with the detected corner at the center. Orientation is given by a directional vector from the corner point to the centroid. Rotation-invariance is improved by computing moments with x and y in a circular region defined by the size of the patch. Since BRIEF performs poorly with rotation, ORB computes a rotation matrix using the patch orientation and steers the BRIEF descriptor based on this orientation.

Convolutional feature descriptors [25] are obtained from pre-trained CNNs (such as the VGG16 network pre-trained using the ImageNet dataset[1]). The input image (224×224) is divided into a 28×28 grid creating image patches, and the output of different pooling layers of the VGG network forms the feature set. A descriptor is generated in the grid where the center of each patch is regarded as a feature point. The generated feature descriptors are built by retaining both convolutional and receptive field location information. These feature descriptors are shown to outperform SIFT in scenarios where SIFT fails to obtain sufficient match pairs.

Once features are computed for each image, the next step is matching features between pairs of images. Two types of methods are commonly used for feature matching, namely, exhaustive (brute-force) and approximate matching. Exhaustive matching methods compute pairwise distances between all feature vectors in an image pair, while approximate matching methods use an approximate search for the nearest neighbor [28] using the smallest distance or distance ratio metrics. A matching threshold is then applied such that if the distance or distance ratio between two feature vectors is less than the threshold, they are considered as a match. This threshold is selected empirically based on the minimum and maximum distance values. Setting a smaller threshold results in fewer but stronger matches. Some false matches are also detected using the abovementioned methods. These outliers (false matches) are eliminated using Random Sample Consensus (RANSAC) [29] or generalized RANSAC [30] algorithms during geometric transformation estimation. Feature extraction and matching results for ex vivo placental overlapping images are shown in Fig. 15.5.

Given the match feature vectors in image pairs, the next step is to find the geometric transformation $\mathbf{H}$ (equation (15.1)) between them. The minimum number of match feature vectors required for estimating homography varies depending on the

---

[1] ImageNet: http://www.image-net.org/ for image classification tasks.

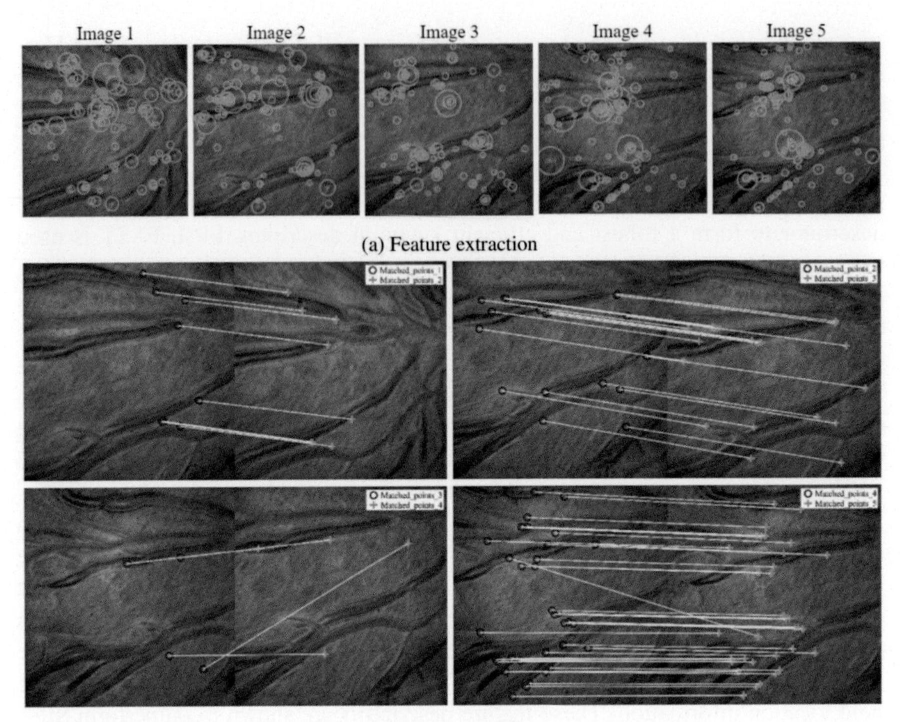

FIGURE 15.5 Feature extraction and matching.

Example overlapping images of an ex vivo human placenta. (a) Extracted SURF features. (b) Matched points between image pairs. Note that in match pair 3–4, only four match pairs are detected, of which two are false matches.

geometric transformation type to be estimated. RANSAC [29] is a robust estimation method that uses a small set of randomly sampled matched feature correspondences to estimate geometric transformation parameters using direct linear transformation. This process is repeated for a large number of trails, and the most consistent $\mathbf{H}$ is selected as the geometric transformation.

Feature-based methods do not provide accurate homography estimates when the image pairs are less textured and structured or contain artifacts due to reflection and motion blur. To overcome these limitations, direct methods can be used since they use global feature alignment and are independent of local features.

15.3.2 Direct methods

Direct (pixel-based) methods match and align the entire information (pixel intensities) of an image with the other image by directly minimizing pixel-to-pixel dissimilarity. This is done by minimizing the cost function such as the photometric error, absolute differences, normalized cross-correlation (NCC), or mutual information (MI) between the overlapping pixels. These methods make full use of the available information and do not require keypoint or feature extraction a priori; however, they are computationally expensive and slower than feature-based methods. Moreover, unlike feature-based methods, direct methods are variant to image scaling, rotation, and pixel intensity changes and have a limited range of convergence as they require a high level of overlap.

Area-based direct methods divide the pair of overlapping images $I_1(x, y)$ and $I_2(x, y)$ into windows and compute the similarity based on pixel intensities between the windows of the two images for each shift using the NCC or MI [11,31]. The NCC is one of the most commonly used similarity metrics and is given by

$$\mathrm{NCC}(i, j) = \frac{\sum_{x,y}(I_1(x, y) - \bar{I}_1)(I_2(x - i, y - j) - \bar{I}_2)}{\sqrt{\sum_{x,y}(I_1(x, y) - \bar{I}_1)^2(I_2(x - i, y - i) - \bar{I}_2)^2}}, \tag{15.4}$$

where $\bar{I}_1$ and $\bar{I}_2$ are the mean of the windowed images, x and y are the pixel coordinates, and i and j represent the horizontal and vertical shifts at which the NCC is computed. The NCC ranges from -1 to $+1$. The shifts i, j corresponding to the maximum NCC represent the geometric transformation between the pair of images. MI is also a widely used similarity metric for image registration; it measures the similarity between two images based on the quality of information shared between them [31]. MI between $I_1(x, y)$ and $I_2(x, y)$ is defined as

$$MI(I_1, I_2) = E(I_1) + E(I_2) - E(I_1, I_2), \tag{15.5}$$

where $E(I_1)$ and $E(I_2)$ are the entropies of $I_1(x, y)$ and $I_2(x, y)$, respectively, and $E(I_1, I_2)$ represents the joint entropy. $E(I_1)$ is given by

$$E(I_1) = \sum_k p_{I_1}(k) \log(p_{I_1}(k)), \tag{15.6}$$

where k is the possible gray level value of $I_1(x, y)$ and $p_{I_1}(k)$ is the probability density function of k. $E(I_1, I_2)$ is given by

$$E(I_1, I_2) = \sum_{k,l} p_{I_1,I_2}(k, l) \log(p_{I_1}(k, l)), \tag{15.7}$$

where l is the possible gray level value of $I_2(x, y)$ and $p_{I_1,I_2}(k, l)$ is the joint probability density function of k and l. The higher the MI between two images, the better is their alignment. The relative position and orientation between I_1 and I_2 are adjusted until the MI is maximized. This gives the geometric transformation (homography) between the pair of images.

Homographies are capable of aligning flat images due to the planar scene assumption. However, the scenes are not always flat, and as a result error accumulation can occur when aligning several overlapping images of the same scene. Such error is referred to as drift. Optical flow establishes a pixelwise mapping between the pair of images to overcome drift error and correct for non-planarity situations. Optical flow methods compute the motion field (gradient) between two overlapping consecutive images [32,33]. Optical flow assumes that the projection of a particular pixel looks the same in every frame. This is referred to as the brightness constancy constraint and is represented as

$$I(x, y, t) = I(x + \Delta x, y + \Delta y, t + \Delta t), \tag{15.8}$$

where $I(x, y, t)$ is the intensity of a pixel at a location (x, y, t) which is displaced by Δx, Δy, and Δt. Solving equation (15.8) using Taylor series approximation gives

$$I_x V_x + I_y V_y = -I_t, \tag{15.9}$$

or

$$\begin{bmatrix} I_x & I_y \end{bmatrix} \begin{bmatrix} V_x \\ V_y \end{bmatrix} = -I_t, \tag{15.10}$$

where V_x and V_y are the x- and y-components of the optical flow (velocity) of $I(x, y, t)$, respectively, and I_x, I_y, and I_t are the derivatives (gradients) of the image in the x-, y-, and t-directions. Equation (15.10) is solved using least square regression by assuming neighboring pixels to have the same optical flow.

There also exist hybrid methods which first align a pair of images using either feature-based or direct methods. Then, within local regions of the homography-aligned images, gradients are computed, and a spline representation of the displacement field is used to refine the alignments and eliminate the ghosting effects [34]. Though this approach of alignment refinement is not new [34], this is still actively used commercially for seamless image stitching in street view mapping.[2,3]

15.3.3 Deep learning-based methods

Deep learning-based methods for homography estimation are gaining interest [16–18]. One such method is the Deep Image Homography (DIH) model [16], which estimates the relative homography between pairs of image patches extracted from a single image. This model uses the four-point homography parameterization ${}^{4p}\mathbf{H}$ [35], instead of the 3×3 parameterization $\mathbf{H}$, as the rotation and shear components in $\mathbf{H}$ have smaller magnitude compared to the translation, and thus have a small effect on the training loss.

Let (u_i, v_i), $i = 1, 2, 3, 4$, denote the four corners of an image patch P_A and let (u_i', v_i') denote the four corners in an overlapping image patch P_B. The displacement

[2] https://ai.googleblog.com/2017/11/seamless-google-street-view-panoramas.html.

[3] http://blog.insta360.com/what-is-optical-flow/.

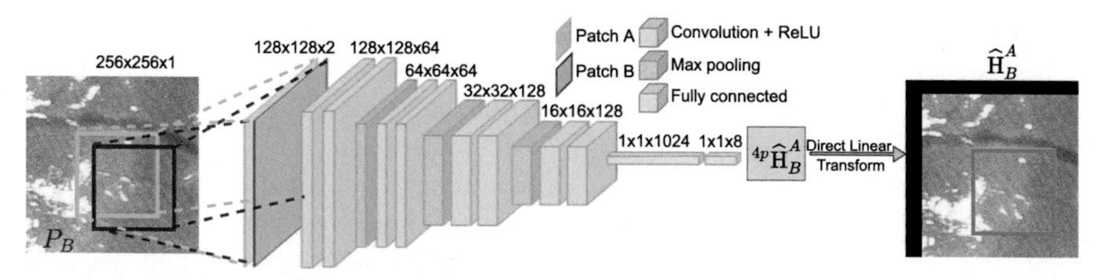

FIGURE 15.6 Deep learning-based image homography.

Network architecture of the deep image homography network with controlled data augmentation [7].

of the ith corner point is given by

$$\begin{aligned} \Delta u_i &= u_i' - u_i, \\ \Delta v_i &= v_i' - v_i. \end{aligned} \tag{15.11}$$

The four-point homography parameterization ${}^{4p}\mathbf{H}$ is given by

$${}^{4p}\mathbf{H} = \begin{bmatrix} \Delta u_1 & \Delta u_2 & \Delta u_3 & \Delta u_4 \\ \Delta v_1 & \Delta v_2 & \Delta v_3 & \Delta v_4 \end{bmatrix}^T. \tag{15.12}$$

DIH [16] uses a VGG-like architecture, with eight convolutional and two fully connected layers (as shown in Fig. 15.6). The input of the network is two image patches P_A and P_B, and the output is the relative homography ${}^{4p}\widehat{\mathbf{H}}_{k+1}^{k}$ between them. This model uses the Euclidean (L_2) loss for gradient back-propagation during the training process:

$$L_2 = \frac{1}{2}\left\| {}^{4p}\mathbf{H} - {}^{4p}\widehat{\mathbf{H}} \right\|^2, \tag{15.13}$$

where ${}^{4p}\mathbf{H}$ and ${}^{4p}\widehat{\mathbf{H}}$ are the ground truth and predicted four-point homographies, respectively. A one-to-one mapping exists between the four-point homography ${}^{4p}\mathbf{H}$ and the 3×3 homography $\mathbf{H}$. The obtained ${}^{4p}\mathbf{H}$ is then converted to $\mathbf{H}$ by applying Direct Linear Transform (DLT) [36]. A limitation of this method is that it is used for still images from the MS-COCO dataset [37] for training, where pairs of patches were extracted from a single real image, free of artifacts (e.g., specular highlights, motion blur, illumination changes) that appear in sequential images or images captured from different devices.

15.3.4 Computing homography – image mosaicking

Given a set of matching features or point correspondences $(x_n, y_n) \rightarrow (x_n', y_n')$ (where $n \in [1, N]$ and N is the total number of matches) between images F_A and F_B, the eight unknowns in the homography matrix $\mathbf{H}$ in equation (15.1) are obtained through least squares solution [11]. Homography assumes that the scene points lie on a plane or are distant (plane at infinity) or captured from the same pivoting point. In panorama imaging of landscapes, this assumption holds true; however, in medical imaging (specially endoscopic vision), the scene is only piecewise planar.

The problem of generating a mosaic from an image sequence is to find the pairwise homographies between consecutive frames followed by computing the relative homographies with respect to a fixed reference frame, which is generally the first frame of the sequence and is also termed the mosaic plane. As a result, all image frames can be mapped to a single planar representation of the scene. Given a sequence of images, the relative homography between frames F_k and F_{k+l} (where k and l are not necessarily consecutive or overlapping frames) is represented by left-hand matrix multiplication of pairwise homographies of all intermediate images:

$$\mathbf{H}_{k+l}^{k} = \prod_{i=k}^{k+l-1} \mathbf{H}_{i+1}^{i}. \tag{15.14}$$

15.3.5 Reprojection and blending

Geometric transformation is followed by reprojection, stitching, and blending to generate the final seamless mosaics (as shown in Fig. 15.3). The process of stitching and blending is often referred to as image compositing.

Once the geometric transformation is obtained between all input images with respect to each other, the next step is to reproject them on a common surface, called mosaic (reference) plane, to produce the stitched image (as shown in Fig. 15.1). This is commonly done by selecting one of the input images as the reference, and then warping all other images using their respective geometric transformation onto this reference plane. Warping pixel values from an image location to the mosaic plane location might result in decimal-valued pixel locations that can create holes. To overcome this problem, linear, nearest neighbor or cubic interpolation is used.

The final step in the image mosaicking pipeline is blending. Directly averaging the values of overlapping pixels results in ghosting artifacts due to moving objects and errors in registration. Moreover, artifacts may also appear due to illumination variation, light source (camera flash) focus, and discontinuities due to camera distortion between the images. To overcome these problems, image blending is used to create a smoother and seamless transition from one image to another image. Two most commonly used methods for image blending are Alpha Feathering, Gradient Domain, and Pyramid Blending [11,14]. Alpha Feathering Blending takes the weighted average from the overlapping region of the two images. This approach is fast but does not give appropriate results if the exposure of images is different. Gradient domain blending uses image gradients instead of image intensities for blending by reconstructing the image that matches the gradients best. This approach creates a seamless and visually appealing mosaic compared to other blending approaches, but is resource-inefficient and only performs well when the reprojection error is small. Pyramid Blending combines the images at different frequency bands and gives a hierarchical representation

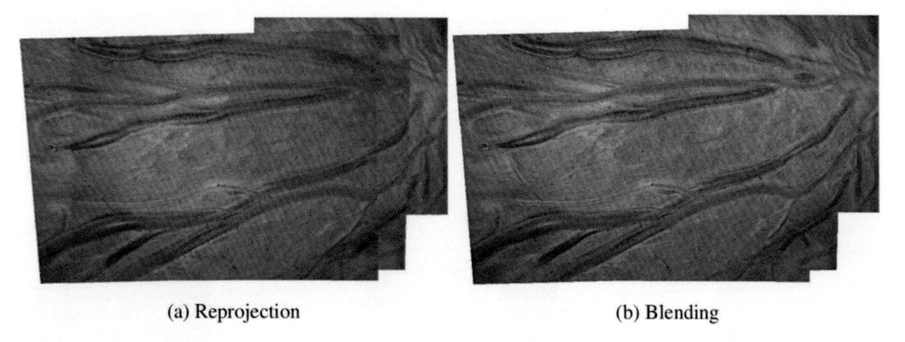

(a) Reprojection (b) Blending

FIGURE 15.7 Reprojection and blending.

Ex vivo placenta images. (a) Reprojection on the mosaic plane. (b) Final seam-free panorama after blending.

of images at different resolution. Pyramid-based approaches prevent blurring but result in ghosting effects when the reprojection error is not small. Fig. 15.7 shows reprojection of placental images onto the mosaic plane using computed geometric affine transformation and the final panorama after blending using the Enblend[4] tool.

15.4 Clinical applications

Image mosaicking has been applied in a number of clinical applications. Some of them are detailed below.

15.4.1 Panoramic X-ray in radiography

Traditional digital radiography systems can only partially capture large anatomical structures in a single scan. Mosaicking is needed to create panoramic X-ray images to observe complete bone structures such as the complete spine, a lower limb, etc., to improve diagnosis. An example of automatic stitching of X-ray images is shown in Fig. 15.8. Most commercial systems perform image alignment by using external marker-based or ruler-based techniques, which are inaccurate, often result in misalignment, and can mislead the clinician in diagnosis. Automatic image mosaicking techniques have been developed to overcome these limitations [3,4], in which several overlapping radiographs are stitched together using either feature-based or phase correlation-based alignment approaches to form a panoramic image with an improved view. However, complex images with significant exposure differences and patient movement result in misalignment in these approaches.

[4] Enblend: https://wiki.panotools.org/Enblend.

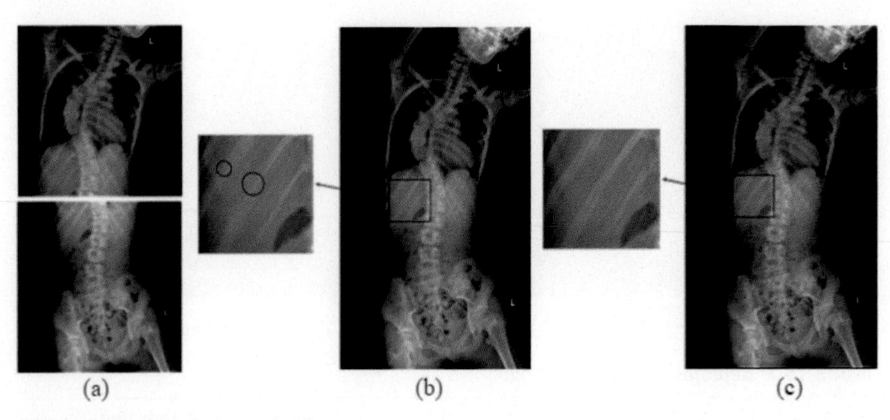

FIGURE 15.8 X-ray image stitching.

X-ray image stitching example from [3] that used a direct alignment (phase correlation) method. (a) Unregistered X-ray image pair with overlap. (b) Stitching results from a clinical software program. (c) Stitching results obtained using the method from [3]. A slight misalignment is visible in the magnified view in (b). On the other hand, the magnified view in (c) shows seam-free stitching. Reproduced with permission from Yang et al. [3].

15.4.2 Whole slide mosaicking in histopathology

Digital light microscopes are taking over non-automatic light microscopes since they are computer-controlled with a digital camera to capture and store images and are capable of imaging specimens larger than the FoV of the microscope by stitching overlapping tiles using integrated positional feedback [6]. Nevertheless, non-automatic light microscopes are still widely available in biological laboratories all over the world, and techniques have been developed for automatic tile stitching of images captured using non-automatic microscopes [5]. Though digital microscopes provide tile locations for image positioning, actuator backlash and stage repeatability errors can occur, causing positional errors in stitching. The Microscopy Image Stitching Tool (MIST) [6] computes geometric transformation parameters using a phase correlation method, followed by estimating the microscope's mechanical stage parameters and jointly optimizing these parameters to minimize the stitching error. Software packages have been developed for stitching microscopic tiles and are available as an integrated plugin for ImageJ[5] application. ImageJ is widely used for microscopic image processing. Commonly used microscopic tile stitching methods include MIST [6], FijiIS [38], TeraStitcher [39], and iStitch [40], which are available as the ImageJ plugin. They are compared in Fig. 15.9.

[5] ImageJ: https://imagej.nih.gov/ij/.

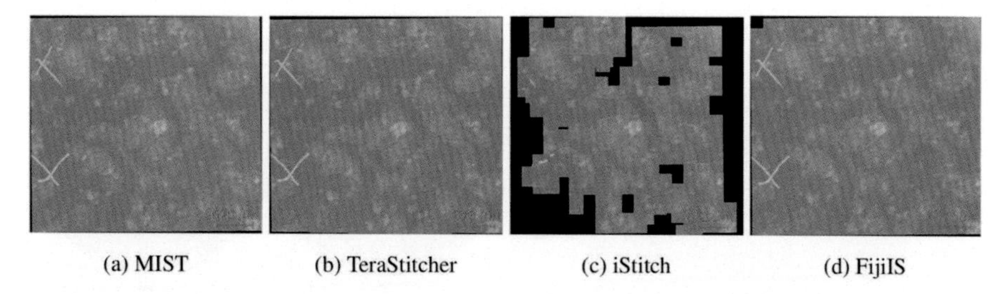

(a) MIST (b) TeraStitcher (c) iStitch (d) FijiIS

FIGURE 15.9 Whole slide image stitching.

Microscope Image Stitching Tool (MIST) [6] and a comparison with other tile stitching methods are shown using a grid with 55 × 55 phase contrast images of stem cell colonies obtained using a 10× microscope with 10% image overlap. The stitched image resolution is 50,000 × 70,000. The average centroid distance error for (a) MIST is 28.56, (b) for TeraStitcher [39] it is 55.50, (c) for iStitch [40] it is 1161.44, and (d) for FijiIS [38] with a regression threshold of 0.01 it is 32.80. Reproduced with permission from Chalfoun et al. [6].

15.4.3 Cystoscopy in urology

A rigid or flexible endoscope is used during cystoscopy for imaging of the urinary bladder for the diagnosis and treatment of cancer. A major challenge faced by surgeons during cystoscopy is the limited FoV, which hinders searching and localization of tumors or potential lesion sites. Therefore, methods have been developed for enlarging the FoV of the bladder to improve diagnosis, navigation, and photodocumentation [8,9,41]. Similar to placental images in fetoscopy, bladder walls are sparsely textured, making it challenging for feature-based methods to detect discriminative features. Therefore, both feature-based and direct alignment methods have been investigated, which use either planar or spherical modeling. Planar modeling tends to result in misalignment and distortion, since these methods project a 3D spherical bladder onto a 2D plane. Spherical modeling is best suited for this application as it represents the shape of the bladder, resulting in minimal alignment errors. Fig. 15.10 shows an example of ex vivo bladder reconstruction using a spherical model. Since the texture of a pig bladder is similar to that of a human bladder, ex vivo videos are generally captured using an excised pig bladder for validating mosaicking algorithms.

15.4.4 Slit-lamp image stitching for ophthalmology

Slit-lamp biomicroscopy or laser indirect ophthalmoscopy is used to perform laser pan retinal photocoagulation for the treatment of retinal ischemia and neovascularization, conditions in which new abnormal blood vessels are formed in the retina or within the eyeball [42]. The FoV in slit-lamp imaging is very narrow compared to fundus imaging, but slit lamps offer higher magnification and control, making it the most common system in clinical settings. Mosaicking facilitates preoperative planning, intraoperative navigation, and photodocumentation [10,43,44]. For mosaicking, the eyeball is considered as a rigid object and planar or spherical models are used for

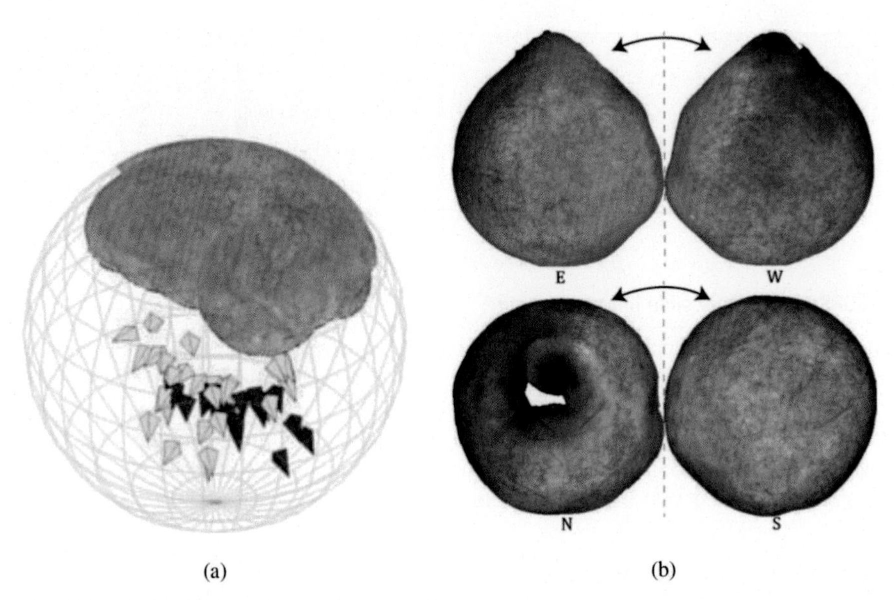

FIGURE 15.10 Bladder reconstruction example.

An ex vivo bladder surface reconstruction example [41]. A flexible fiber endoscope was used to capture a video of an excised pig bladder. (a) An intermediate pass of the spherically constrained incremental bundle adjustment is shown. Following feature extraction and matching, spherically constrained incremental bundle adjustment was imposed for the reconstruction of the 3D surface features and endoscope (camera) motion (yellow). (b) The unfolded view by hemisphere of the 3D reconstructed model of the excised pig bladder. Reproduced with permission from Soper et al. [41].

image alignment and mapping. Both feature-based and direct methods have been used for image alignment and visual tracking. Non-uniform illumination and glare have been the biggest challenge in mosaicking from slit-lamp imaging, and efforts have been made to cater for local illumination variations to overcome these issues. An example of mosaic generation from slit-lamp imaging is shown in Fig. 15.11 for illustration.

15.4.5 Fetal interventions and fetoscopy

Twin-to-twin transfusion syndrome (TTTS) is a placental disease affecting identical twins sharing a common monochorionic placenta. In TTTS, the shared placenta develops abnormal vascular anastomoses, resulting in uneven blood flow between the fetuses [45]. Fetoscopic laser photocoagulation is the most effective treatment for regulating the blood flow in TTTS. However, limited FoV and maneuverability of the fetoscope and poor visibility due to amniotic fluid particles hinder intraoperative planning and treatment. Mosaicking can be employed to generate an image with in-

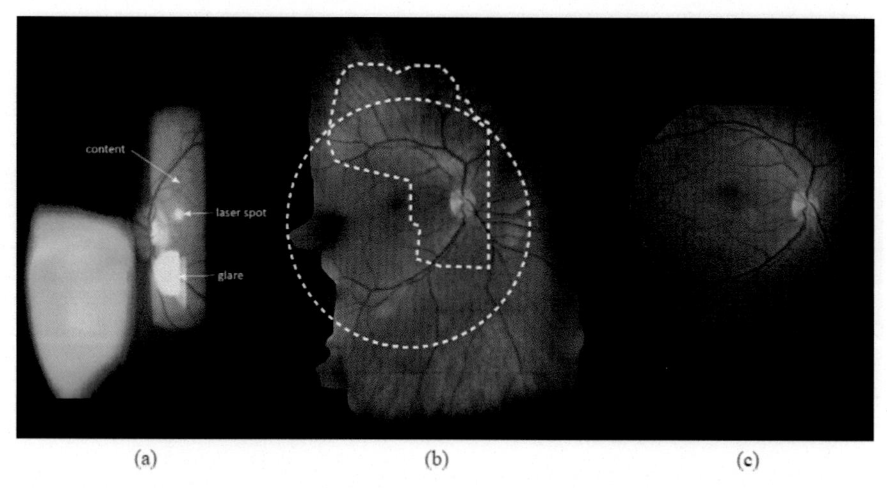

FIGURE 15.11 Slit-lamp image mosaicking.

Slit-lamp image mosaicking example from [43]. (a) A slit-lamp image which is non-uniformly illuminated and contains glare and laser spots occluding the retina. (b) Generated mosaic [43] from a slit-lamp image sequence of 1000 frames covering a much larger area compared to the fundus image shown in (c), whose area is represented by a white dotted circle in (b). The other delineated white region represents the FoV extent of the mosaic generated using the method from [10]. Reproduced with permission from De Zanet et al. [43].

creased FoV and improve localization of the vascular anastomosis sites. Therefore, recently, mosaicking has gained attention to increase the FoV in fetoscopy.

Feature-based [47,48], direct [49], and deep learning-based [7,46,50,51] methods have been proposed, which are tested using synthetic, ex vivo, phantom, and/or in vivo fetoscopic sequences. In an experimental setting, efforts have also been made to integrate an electromagnetic tracker within the fetoscope to minimize any drifting errors [52,53]. However, such integration in clinical settings is still an open challenge due to the limited form factor of the fetoscope and clinical regulations. Though promising results have been achieved, mosaicking of fetoscopy images remains a challenging problem due to monocular cameras, varying visual quality, occlusions, specular highlights, a lack of visual texture, turbid amniotic fluid, and non-planar views. Placental vessel segmentation using deep neural networks followed by direct image alignment using only the segmented vasculature maps has been shown to help in overcoming visibility-related challenges in fetoscopic videos and facilitate building reliable and long-range mosaics [46]. Following up on this work, the first large-scale multicenter dataset of a fetoscopy laser photocoagulation procedure has been presented, capturing intra-procedure [54], inter-procedure, and inter-center variabilities. The dataset is used for the development of semantic segmentation and video mosaicking algorithms for the fetal environment, with a focus on overcoming video mosaicking challenges creating drift-free RGB placental mosaics and placen-

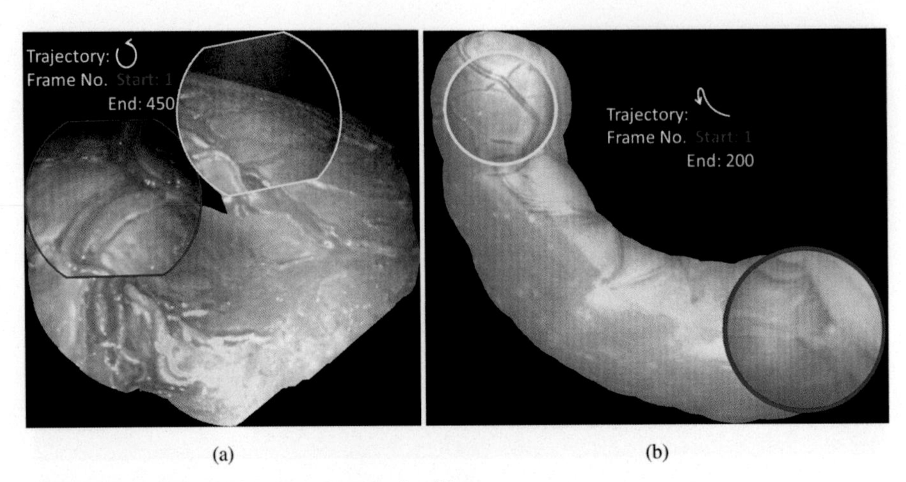

FIGURE 15.12 Mosaicking from fetoscopic videos.

Examples of mosaic generation from fetoscopy videos [7] using a deep learning-based method for sequential frame stitching. (a) A mosaic generated from a phantom placenta sequence that followed a freehand circular trajectory. (b) In vivo sequence containing highly non-planar views with occlusion due to the fetus, low resolution, texture paucity, and amniotic fluid particles posing live intraoperative imaging challenges compared to phantom imaging. Reproduced with permission from Bano et al. [7].

tal vasculature mosaics. Figs. 15.12 and 15.13 show examples of fetoscopic image stitching using two recently developed deep learning-based methods [7,46].

15.4.6 General surgery view expansion

In addition to the abovementioned clinical applications, image mosaicking and surface reconstruction are also applied in other minimally invasive surgical techniques such as laparoscopy [55–57], laryngoscopy (larynx and throat examination) [58], and endonasal procedures [59]. Binocular (stereo vision) endoscopes are also used in laparoscopy for providing a sense of depth to surgeons during the procedures and for surface reconstruction by matching features between left and right camera images [55,56]. The common aims of these methods are to provide dense 3D reconstruction and feature tracking to support simultaneous localization and mapping during surgical procedures.

15.5 Recapitulation

In this chapter, we have covered the main principles of biomedical image mosaicking and then reported and discussed some of the currently used biomedical imaging applications of mosaicking. While image mosaicking is an active research area for various

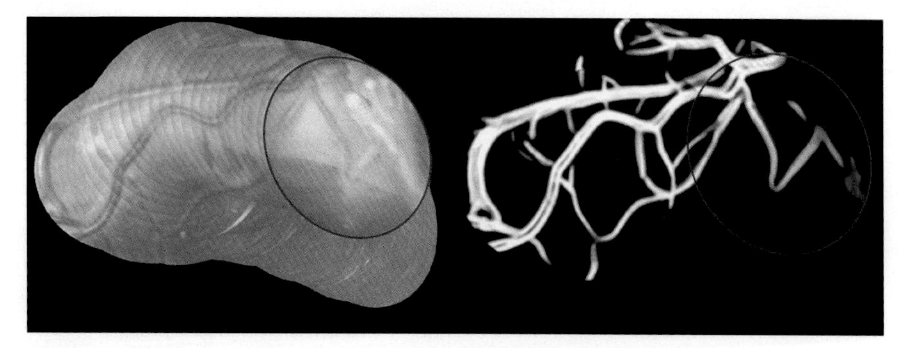

FIGURE 15.13 Video mosaicking of segmented placental vessel maps generated by fetoscopy.

Examples of mosaic generation from fetoscopy videos [46] that were used to train a deep neural network for segmenting placental vessels, followed by applying direct image registration to the placental vessel maps of consecutive frames. (a) RGB mosaic for which the homography transformations are computed through vessel map registration. (b) Registered placental vessels, where the duration of the fetoscopic video clip was 400 frames. Reproduced with permission from Bano et al. [46].

potential clinical specializations, more work is necessary to realize biomedical image mosaicking as a clinically valuable tool, especially in surgery and interventions. To achieve this, a number of technical and clinical challenges need to be addressed.

Technical challenges include (1) matching difficulties either due to paucity of textures or repetitive patterns, making it difficult to reliably estimate the image transformation; (2) non-planar scenes where the transformation model cannot be easily estimated and may incorporate discontinuities, vanishing points, or occlusions; and (3) latency and real-time processing to allow visualization of the generated mosaic during operations.

Clinical challenges include (1) clear requirements and specifications for the necessary alignment needs (for example, in some applications perhaps just an indication of the enhanced FoV is sufficient for better localization); (2) workflow and user interface implementation to display any mosaicked images; and (3) a regulatory pathway for showing augmented information to the clinical team alongside analysis of the potential risk or benefit this augmentation brings.

Progress in both the technical and clinical challenges is needed to advance interventional and surgical imaging and take advantage of computational image mosaicking. Complementary solutions in hardware, for example, low-cost wide-angle lenses, could be important for overcoming some of the current limitations. Advancements are needed towards designing hybrid methods involving global registration followed by fine-grain registration for biomedical imaging mosaicking (in particular for surgical interventional image registration). This could potentially help in overcoming texture paucity and non-planar scene issues, leading to the minimization of the drifting error. Additionally, integration of external tracking sensors [52] could facilitate image location and registration.

15.6 Exercises

1. **Computing homography.** In this problem, you will mathematically derive the system of equations for computing homography and obtaining the least squares solution presented in Section 15.3.4.
   a. Given a set of matching features/points ($(x_n, y_n) \rightarrow (x'_n, y'_n)$, where $n \in [1, N]$ and N is the total number of matches) between images F_A and F_B, derive the system of equations to solve for the homography matrix $\mathbf{H}$.
   b. The system of equation obtained from [a.] can be represented as

$$A\mathbf{h} = 0, \tag{15.15}$$

   where A is known and $\mathbf{h}$ contains the homography parameters which are unknown. A minimum of four pairs of matching points are needed to solve this equation. But, in general, $n > 4$, which results in an overdetermined system of equations. Mathematically, derive the solution for equation (15.15) using the least squares method.
2. **Feature-based image mosaicking.** In this problem, you will perform image stitching using a feature-based method as presented in Section 15.3.1.
   You may use any existing Computer Vision Toolbox (like the OpenCV Toolbox[6] or the MATLAB® Computer Vision Toolbox[7]) for this task. The developed method should be tested on:
   [i] A real-world dataset: Use sequences "goldengate" and "office" from the Adobe Panoramas Dataset.[8]
   [ii] An endoscopic dataset: Use video clip "video01" from the Fetoscopy Placenta Dataset.[9]
   a. Write a code using Python/C++/MATLAB to detect keypoints (SIFT, SURF, ORB, or FAST) and extract local descriptors from a pair of input images from the real-world dataset. Perform feature matching using either an exhaustive or an approximate matching algorithm. Estimate homography by using (a) a least squares solution and (b) RANSAC. Apply a warping transformation using the homography matrix obtained from the (a) and (b) estimators. Visualize the obtained results and comment on the performance of the two estimators.

[6] OpenCV: https://opencv.org/.
[7] MATLAB Computer Vision Toolbox: https://uk.mathworks.com/help/vision/.
[8] Adobe Panoramas Dataset: https://sourceforge.net/adobe/adobedatasets/panoramas/home/Home/.
[9] Fetoscopy Placenta Dataset: https://www.ucl.ac.uk/interventional-surgical-sciences/fetoscopy-placenta-data.

b. Compute the pairwise homography for all images in the two real-world dataset sequences and compute the relative homographies using equation (15.14). Perform reprojection to generate the image mosaic and visually compute the performance on the two sequences when using the (a) and (b) estimators.
Hint: Least squares solution is sensitive to outliers.

c. Repeat [2a.] and [2b.] using the RGB frames endoscopic dataset. Use only the RGB frames "anon001_00901.png" to "anon001_00910.png" from "video01" for this exercise. Comment on the performance, including any issues that you encounter.
Hint: Lowering the feature detector and matching thresholds may help in detecting more features and matches, but with added outliers.

3. **Direct registration-based image mosaicking.** In this problem, you will perform image stitching using a direct method as presented in Section 15.3.2. Similar to Exercise 2, you may use any existing Computer Vision Toolbox for this task and test the developed method on the two real-world dataset (see footnote 8) sequences and the endoscopic dataset (see footnote 9) sequence.

 a. Write a code using Python/C++/MATLAB to compute dense optical flow between a pair of input images from the real-world dataset and estimate homography by using RANSAC. Apply a warping transformation using the homography matrix. Visualize the obtained results and comment on its performance compared to Exercise 2a.

 b. Compute the pairwise homography for all images in the two real-world dataset sequences and compute relative homographies using equation (15.14). Perform reprojection to generate the image mosaic and visually compute the performance on the two sequences with the results obtained in Exercise 2b.

 c. Repeat [2a.] and [2b.] using the RGB frames in the endoscopic dataset. Use only RGB frames "anon001_00901.png" to "anon001_00910.png" from "video01" for this exercise. Comment on the performance and compare the results with Exercise 2c. Are there any improvements?
 Hint: Crop the endoscopic image into a square inscribed in the circular FoV to remove the black regions outside the FoV.

 d. Repeat [2a.] and [2b.] using the placental vessel maps in the endoscopic dataset. Use only the vessel map frames "anon001_00901.png" to "anon001_00910.png" from "video01" for this exercise. Comment on the performance and compare the results with Exercises 2c and 3c. Are there any improvements?

4. **Deep learning-based image mosaicking.** In this problem, you will train a deep neural network for estimating homography between a pair of transformed patches extracted from the same image, as presented in Section 15.3.3. You will use the 483 sparsely sampled RGB frames from the Fetoscopy Placenta Dataset (see footnote 9) for this exercise.

 a. Write a function using Python/C++/MATLAB to generate training data (as presented in Section 15.3.3 and [16]) by extracting the source patch P_A of size 256 × 256 pixels at a random location in an image, randomly perturbing its four corners to get the destination patch P_B, and warping P_A on P_B to compute the homography. The function should take an image as input and give as output the two patches and their homography transformation.

b. Build the Deep Image Homography Model (as presented in Section 15.3.3 and [16]) and use the data generation function to train the model. Use "video002" to "video006" as the training data and "video001" as the testing data.
Hint: If local GPU resources are not available, this model can be easily trained on Google Colab.[10]

c. For the images in the testing data, generate 500 pairs of patches using Exercise 4a and use them for testing the trained model from Exercise 4b. Report the average photometric error and root mean square error (RMSE) for the generated test patches. Comment on the performance of this network. How can you extend this approach to register sequences of images?

References

[1] Kekec T., Yildirim A., Unel M., A new approach to real-time mosaicing of aerial images, Robotics and Autonomous Systems 62 (12) (2014) 1755–1767.
[2] Zhu Z., Martin R.R., Hu S.-M., Panorama completion for street views, Computational Visual Media 1 (1) (2015) 49–57.
[3] Yang F., He Y., Deng Z.S., Yan A., Improvement of automated image stitching system for DR X-ray images, Computers in Biology and Medicine 71 (2016) 108–114.
[4] PyarWin K., Kitjaidure Y., Biomedical images stitching using ORB feature based approach, in: 2018 International Conference on Intelligent Informatics and Biomedical Sciences (ICIIBMS), vol. 3, IEEE, 2018, pp. 221–225.
[5] Piccinini F., Bevilacqua A., Lucarelli E., Automated image mosaics by non-automated light microscopes: the MicroMos software tool, Journal of Microscopy 252 (3) (2013) 226–250.
[6] Chalfoun J., Majurski M., Blattner T., Bhadriraju K., Keyrouz W., Bajcsy P., Brady M., Mist: accurate and scalable microscopy image stitching tool with stage modeling and error minimization, Scientific Reports 7 (1) (2017) 4988.
[7] Bano S., Vasconcelos F., Amo M.T., Dwyer G., Gruijthuijsen C., Deprest J., Ourselin S., Vander Poorten E., Vercauteren T., Stoyanov D., Deep sequential mosaicking of fetoscopic videos, in: International Conference on Medical Image Computing and Computer-Assisted Intervention, Springer, 2019, pp. 311–319.
[8] Lopez A., Liao J.C., Emerging endoscopic imaging technologies for bladder cancer detection, Current Urology Reports 15 (5) (2014) 406.
[9] Behrens A., Bommes M., Stehle T., Gross S., Leonhardt S., Aach T., Real-time image composition of bladder mosaics in fluorescence endoscopy, Computer Science - Research and Development 26 (1–2) (2011) 51–64.

[10] Google Colab: https://research.google.com/colaboratory/.

[10] Richa R., Linhares R., Comunello E., Wangenheim A.W., Schnitz J.Y., Fundus image mosaicking for information augmentation in computer-assisted slit-lamp imaging, IEEE Transactions on Medical Imaging 33 (6) (2014) 1304–1312.

[11] Szeliski R., Image alignment and stitching: A tutorial, Foundations and Trends in Computer Graphics and Vision 2 (1) (2007) 1–104.

[12] Zhang Z., A flexible new technique for camera calibration, IEEE Transactions on Pattern Analysis and Machine Intelligence 22 (2000).

[13] Wolberg G., Image morphing: a survey, The Visual Computer 14 (8–9) (1998) 360–372.

[14] Ghosh D., Kaabouch N., A survey on image mosaicing techniques, Journal of Visual Communication and Image Representation 34 (2016) 1–11.

[15] Pravenaa S., Menaka R., A methodical review on image stitching and video stitching techniques, International Journal of Applied Engineering Research 11 (5) (2016) 3442–3448.

[16] DeTone D., Malisiewicz T., Rabinovich A., Deep image homography estimation, in: RSS Workshop on Limits and Potentials of Deep Learning in Robotics, 2016.

[17] Nguyen T., Chen S.W., Shivakumar S.S., Taylor C.J., Vijay Kumar V., Unsupervised deep homography: A fast and robust homography estimation model, IEEE Robotics and Automation Letters 3 (3) (2018) 2346–2353.

[18] Rocco I., Arandjelovic R., Sivic J., Convolutional neural network architecture for geometric matching, in: Proceedings of the IEEE Conference on Computer Vision and Pattern Recognition, 2017, pp. 6148–6157.

[19] Lowe D.G., Distinctive image features from scale-invariant key-points, International Journal of Computer Vision 60 (2) (2004) 91–110.

[20] Ke Y., Sukthankar R., PCA-SIFT: A more distinctive representation for local image descriptors, in: IEEE/CVF Conference on Computer Vision and Pattern Recognition (2) 4, 2004, pp. 506–513.

[21] Morel J.-M., Yu G., ASIFT: A new framework for fully affine invariant image comparison, SIAM Journal on Imaging Sciences 2 (2) (2009) 438–469.

[22] Bay H., Ess A., Tuytelaars T., Gool L.V., Speeded-up robust features (SURF), Computer Vision and Image Understanding 110 (3) (2008) 346–359.

[23] Rublee E., Rabaud V., Konolige K., Bradski G., ORB: An efficient alternative to SIFT or SURF, in: IEEE/CVF International Conference on Computer Vision, vol. 11.1, Citeseer, 2011, p. 2.

[24] Dosovitskiy A., Fischer P., Springenberg J.T., Riedmiller M., Thomas B., Discriminative unsupervised feature learning with exemplar convolutional neural networks, IEEE Transactions on Pattern Analysis and Machine Intelligence 38 (9) (2015) 1734–1747.

[25] Yang Z., Dan T., Yang Y., Multi-temporal remote sensing image registration using deep convolutional features, IEEE Access 6 (2018) 38544–38555.

[26] Rosten E., Drummond T., Machine learning for high-speed corner detection, in: European Conference on Computer Vision, Springer, 2006, pp. 430–443.

[27] Calonder M., Lepetit V., Strecha C., Fua P., BRIEF: Binary robust independent elementary features, in: European Conference on Computer Vision, Springer, 2010, pp. 778–792.

[28] Muja M., Lowe D.G., Fast approximate nearest neighbors with automatic algorithm configuration, in: International Joint Conference on Computer Vision, Imaging and Computer Graphics Theory and Applications (1) 2.331–340, 2009, p. 2.

[29] Fischler M.A., Bolles R.C., Random sample consensus: a paradigm for model fitting with applications to image analysis and automated cartography, Communications of the ACM 24 (6) (1981) 381–395.

[30] Torr P.H.S., Zisserman A., MLESAC: A new robust estimator with application to estimating image geometry, Computer Vision and Image Understanding 78 (1) (2000) 138–156.
[31] Ghannam S., Lynn Abbott A., Cross correlation versus mutual information for image mosaicing, International Journal of Advanced Computer Science and Applications (IJACSA) 4 (11) (2013).
[32] Levin A., Zomet A., Peleg S., Weiss Y., Seamless image stitching in the gradient domain, in: European Conference on Computer Vision, Springer, 2004, pp. 377–389.
[33] Fortun D., Bouthemy P., Kervrann C., Optical flow modeling and computation: a survey, Computer Vision and Image Understanding 134 (2015) 1–21.
[34] Szeliski R., Coughlan J., Spline-based image registration, International Journal of Computer Vision 22 (3) (1997) 199–218.
[35] Baker S., Datta A., Kanade T., Parameterizing homographies, Technical Report CMU-RI-TR-06-11, 2006.
[36] Hartley R., Zisserman A., Multiple View Geometry in Computer Vision, Cambridge University Press, 2003.
[37] Lin T.-Y., Maire M., Belongie S., Hays J., Perona P., Ramanan D., Dollár P., Zitnick C.L., Microsoft COCO: Common objects in context, in: European Conference on Computer Vision, Springer, 2014, pp. 740–755.
[38] Preibisch S., Saalfeld S., Tomancak P., Globally optimal stitching of tiled 3D microscopic image acquisitions, Bioinformatics 25 (11) (2009) 1463–1465.
[39] Bria A., Iannello G., TeraStitcher–a tool for fast automatic 3D-stitching of teravoxel-sized microscopy images, BMC Bioinformatics 13 (1) (2012) 316.
[40] Yu Y., Peng H., Automated high speed stitching of large 3D microscopic images, in: 2011 IEEE International Symposium on Biomedical Imaging: From Nano to Macro, IEEE, 2011, pp. 238–241.
[41] Soper T.D., Porter M.P., Seibel E.J., Surface mosaics of the bladder reconstructed from endoscopic video for automated surveillance, IEEE Transactions on Biomedical Engineering 59 (6) (2012) 1670–1680.
[42] Evans J.R., Michelessi M., Virgili G., Laser photocoagulation for proliferative diabetic retinopathy, Cochrane Database of Systematic Reviews 11 (2014).
[43] De Zanet S., Rudolph T., Richa R., Tappeiner C., Sznitman R., Retinal slit lamp video mosaicking, International Journal of Computer Assisted Radiology and Surgery 11 (6) (2016) 1035–1041.
[44] Prokopetc K., Bartoli A., SLIM (slit lamp image mosaicing): handling reflection artifacts, International Journal of Computer Assisted Radiology and Surgery 12 (6) (2017) 911–920.
[45] Baschat A., Chmait R.H., Deprest J., Gratacós E., Hecher K., Kontopoulos E., Quintero R., Skupski D.W., Valsky D.V., Ville Y., Twin-to-twin transfusion syndrome (TTTS), Journal of Perinatal Medicine 39 (2) (2011) 107–112.
[46] Bano S., Vasconcelos F., Shepherd L.M., Poorten E.V., Vercauteren T., Ourselin S., David A.L., Deprest J., Stoyanov D., Deep placental vessel segmentation for fetoscopic mosaicking, in: International Conference on Medical Image Computing and Computer-Assisted Intervention, Springer, 2020, pp. 763–773.
[47] Reeff M., Gerhard F., Cattin P., Székely G., Mosaicing of endoscopic placenta images, in: INFORMATIK 2006–Informatik für Menschen, Band 1, 2006.
[48] Daga P., Chadebecq F., Shakir D., Herrera L., Tella M., Dwyer G., David A.L., Deprest J., Stoyanov D., Vercauteren T., Ourselin S., Real-time mosaicing of fetoscopic videos using SIFT, in: Medical Imaging: Image-Guided Procedures, 2016.

[49] Peter L., Tella-Amo M., Shakir D.I., Attilakos G., Wimalasundera R., Deprest J., Ourselin S., Vercauteren T., Retrieval and registration of long-range overlapping frames for scalable mosaicking of in vivo fetoscopy, International Journal of Computer Assisted Radiology and Surgery 13 (5) (2018) 713–720.

[50] Gaisser F., Peeters S.H.P., Lenseigne B.A.J., Jonker P.P., Oepkes D., Stable image registration for in-vivo fetoscopic panorama reconstruction, Journal of Imaging 4 (1) (2018) 24.

[51] Bano S., Vasconcelos F., Tella-Amo M., Dwyer G., Gruijthuijsen C., Vander Poorten E., Vercauteren T., Ourselin S., Deprest J., Stoyanov D., Deep learning-based fetoscopic mosaicking for field-of-view expansion, International Journal of Computer Assisted Radiology and Surgery (2020) 1–10.

[52] Tella-Amo M., Peter L., Shakir D.I., Deprest J., Stoyanov D., Iglesias J.E., Vercauteren T., Ourselin S., Probabilistic visual and electromagnetic data fusion for robust drift-free sequential mosaicking: application to fetoscopy, Journal of Medical Imaging 5 (2) (2018) 021217.

[53] Tella-Amo M., Peter L., Shakir D.I., Deprest J., Stoyanov D., Vercauteren T., Ourselin S., Pruning strategies for efficient online globally-consistent mosaicking in fetoscopy, Journal of Medical Imaging (2019).

[54] Bano S., Casella A., Vasconcelos F., Moccia S., Attilakos G., Wimalasundera R., David A.L., Paladini D., Deprest J., De Momi E., Mattos L.S., Stoyanov D., FetReg: placental vessel segmentation and registration in fetoscopy challenge dataset, arXiv preprint, arXiv:2106.05923, 2021.

[55] Totz J., Mountney P., Stoyanov D., Yang G.Z., Dense surface reconstruction for enhanced navigation in MIS, in: MICCAI, Springer, 2011, pp. 89–96.

[56] Stoyanov D., Scarzanella M.V., Pratt P., Yang G.Z., Real-time stereo reconstruction in robotically assisted minimally invasive surgery, in: International Conference on Medical Image Computing and Computer-Assisted Intervention, Springer, 2010, pp. 275–282.

[57] Puerto-Souza G.A., Mariottini G.-L., A fast and accurate feature-matching algorithm for minimally-invasive endoscopic images, IEEE Transactions on Medical Imaging 32 (7) (2013) 1201–1214.

[58] Sinha A., Ishii M., Hager G.D., Taylor R.H., Endoscopic navigation in the clinic: registration in the absence of preoperative imaging, International Journal of Computer Assisted Radiology and Surgery (2019) 1–12.

[59] Jiang R.S., Liang K.L., Image stitching of sphenoid sinuses from monocular endoscopic views, in: CURAC, 2013, pp. 226–229.

PART

Machine learning in medical image analysis

V

CHAPTER

Deep learning fundamentals

16

Nishant Ravikumar[a], Arezoo Zakeri[b], Yan Xia[b], and Alejandro F. Frangi[b,c,d,e]

[a] *Centre for Computational Imaging and Simulation Technologies in Biomedicine (CISTIB), Schools of Computing and Medicine, University of Leeds, Leeds, United Kingdom*

[b] *Division of Informatics, Imaging, and Data Sciences, School of Health Sciences, Faculty of Biology, Medicine, and Health, The University of Manchester, Manchester, United Kingdom*

[c] *Department of Computer Science, School of Engineering, Faculty of Science and Engineering, The University of Manchester, Manchester, United Kingdom*

[d] *Department of Electrical Engineering (ESAT), KU Leuven, Leuven, Belgium*

[e] *Department of Cardiovascular Sciences, KU Leuven, Leuven, Belgium*

Learning points

- Multilayer perceptron and error back-propagation
- Building blocks of deep neural networks
- Activation functions and Normalization
- Loss functions, optimization, and regularization
- Inductive bias, invariance, and equivariance

16.1 Introduction

Deep learning and so-called deep neural networks led to a step change in the complexity of the task and challenges solved by the medical imaging and image computing communities. The term *deep neural network* loosely refers to neural networks with three or more stacked hidden layers. Neural networks with one or more hidden layers of arbitrary width or depth have been proven to be *universal function approximators*, that is, given a sufficiently large number of nodes/units in the hidden layer(s), a neural network can approximate any (fairly general) function. This result, first attributed to Cybenko [1] and Hornik et al. [2] and known as the *universal approximation theorem*, was since demonstrated for a fairly diverse set of conditions, architectures, and functions [3]. Deep neural networks have been used extensively across several domains in recent years, including healthcare, to address real-world problems involving tasks in signal processing, computer vision, and natural language processing, amongst others. Medical image analysis has benefited greatly with the advent of deep learning too, allowing researchers in the community to tackle increasingly complex and clinically interesting and meaningful tasks.

Medical Image Analysis. https://doi.org/10.1016/B978-0-12-813657-7.00041-8

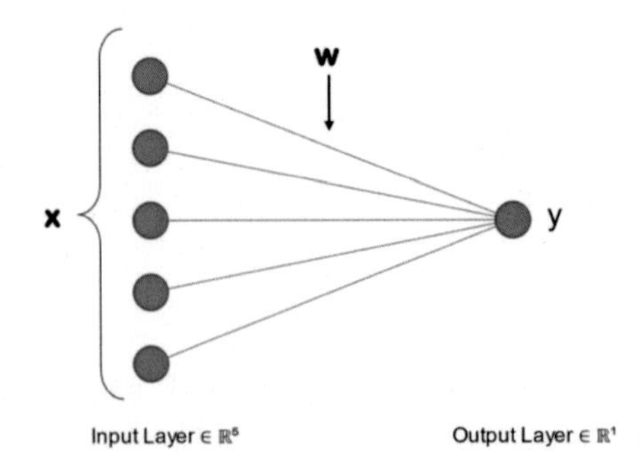

FIGURE 16.1

Single-neuron neural network with linear activation function, which is equivalent to a univariate linear regression model.

The simplest neural network can be interpreted as nothing but a linear regression model applied to a task of predicting a 1D target variable y; put another way, the latter can be expressed as a neural network with a single neuron and a linear activation function. This is visualized in Fig. 16.1, where we see that given an input vector $\mathbf{x} \in \mathbb{R}^D$, the output y of a single neuron with a linear activation function is a weighted linear combination of the input $\mathbf{x} = \{x_i\}$, with weights denoted $\mathbf{w} = \{w_i\}$, where subscript $i = [1, ..., D]$ denotes the dimensionality of the inputs, in this instance with $D = 5$. The resulting function is approximated by the single-neuron network and its associated linear regression model is given by

$$f(\mathbf{x}; \boldsymbol{\theta}) = \mathrm{y} = \mathbf{w}\mathbf{x} + w_0 = w_0 + \sum_{i=1}^{D} w_i x_i, \tag{16.1}$$

where $\boldsymbol{\theta} = \{\mathbf{w}, w_0\}$ represents the set of unknown parameters (i.e., trainable network weights or linear regression coefficients) and w_0 represents the bias term in the network, which plays the same role as the intercept in a linear regression model. A linear regression model and a single-neuron neural network are linear in the weights and assume a linear mapping between the inputs and outputs. However, to capture non-linear input–output relationships, linear regression models can be endowed with feature transformations, specifically by applying so-called basis functions (e.g., polynomial functions, Gaussian or multiquadric radial basis functions, etc.) to the input features in each sample. Deep neural networks, on the other hand, achieve this by introducing non-linear activation functions (discussed in Section 16.2.2) that are applied to outputs of individual neurons and recursively combining several such functions into a chain, expressed as

$$f(\mathbf{x}; \boldsymbol{\Theta}) = f_N(f_{N-1}(\cdots(f_1(\mathbf{x}, \boldsymbol{\theta}_1)\cdots))), \tag{16.2}$$

where $f(\mathbf{x}; \boldsymbol{\theta}_n)$ represents the function of the nth layer. In other words, deep neural networks stack several "hidden layers" of neurons together to recursively compose functions and approximate a non-linear mapping between the inputs and outputs. The term deep neural networks actually encompasses a large family of models, with the example provided above in Eq. (16.2) representing just one type, namely, a feedforward neural network. Additionally, the network described in Eq. (16.2) represents a specific type of feedforward neural network: a multilayer perceptron (MLP). In general, composing several differentiable functions into any kind of directed acyclic graph that maps inputs to outputs constitutes a deep neural network.

16.2 Learning as optimization

Learning with neural networks can be described as an optimization problem. The goal is to estimate suitable values for the network weights, which (at least locally) minimize a predefined *loss/error* function. When training, appropriate loss functions are chosen based on the predictive modeling task of interest. The loss function measures the difference between the network's predictions and the true target values. The optimization goal is to iteratively refine estimates for the network weights to minimize the loss function.

Modern neural networks are trained using gradient-based optimization techniques. This is enabled by a key contribution, viz., *error back-propagation* [4], which is an efficient way of applying the chain rule of differentiation to differentiable networks to update their constituent node weights/parameters. Error back-propagation has profoundly impacted the development and success of deep learning. Before its introduction, training deep neural networks was challenging as no systematic approach existed to updating the network weights. Correspondingly, it was difficult to calculate the gradients of the loss function with respect to the network's weights, especially in deep networks with several layers. Error back-propagation is used to train deep neural networks by calculating the gradients of the loss function with respect to the weights at all layers of the network using the "chain rule" of differentiation and, subsequently, by updating or changing the weights of the connections between the constituent neurons to minimize the loss function, and therefore the prediction error. In this way, the gradients with respect to the loss are back-propagated through the network and used to update the weights iteratively through standard (typically) first-order gradient-based optimization algorithms (e.g., gradient descent) until a minimum value for the loss or a suitable stopping criterion is met.

Deep neural networks with several hidden layers and/or the inclusion of non-linear activation functions (discussed in Section 16.2.2) typically have loss landscapes (i.e., a surface in some high-dimensional space defined by the loss function) that are highly non-convex. Hence, training such networks is a non-convex optimization problem where multiple local minima can exist. The optimization algorithm may yield a suboptimal solution if it converges to a local minimum. To mitigate this issue, various techniques, such as weight initialization, regularization, and normalization, can be used to increase the chances of finding the global minimum. Commonly used gradient-based optimization algorithms, weight regularization strategies, and layer normalization approaches are discussed in subsequent sections (Sections 16.2.4–16.2.6) of this chapter.

16.2.1 Multilayer perceptron

The preceding section (Section 16.1) discussed the equivalence between a linear regression model with a scalar or 1D output/target variable and a single-neuron neural network. Given a multivariate linear regression task where the targets are no longer 1D but an M-dimensional vector, the corresponding neural network interpretation requires a two-layer network comprising an input and output layer (with several neurons, equal to the dimensionality of the target variable), as shown in Fig. 16.2. Such neural networks with two or more layers are called MLPs, and the constituent layers are referred to as fully connected layers. As the term suggests, each neuron in a fully connected layer has connections to all neurons in the preceding layer (where each connection is defined by a trainable weight as indicated in Fig. 16.2), or in other words, the inputs to each neuron in a fully connected layer are the outputs of all neurons in the preceding layer. The inputs to an MLP must be fixed-dimensional vectors (denoted $\mathbf{x} \in \mathbb{R}^D$ in Eq. (16.3)). Hence, MLPs are used in applications that involve "structured" or "tabular" data (i.e., where the data are stored or can be expressed as a matrix). Other types of deep neural networks, such as convolutional neural networks (CNNs) and transformers, are better equipped to learn with/analyze structured data such as images and text. CNNs are discussed in greater detail in a later section (Section 17.2).

MLPs with more than two layers have layers in between the input and output layers known as "hidden layers." An MLP with two layers, representing a multivariate

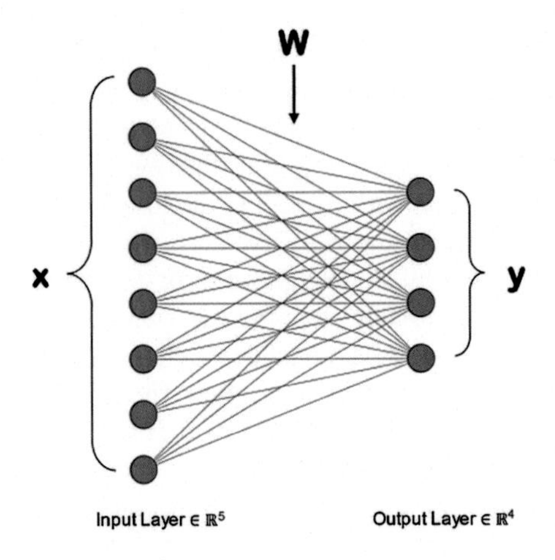

FIGURE 16.2

Two-layer MLP with linear activation functions for each output unit, which is equivalent to a multivariate linear regression model.

linear regression function, is mathematically expressed as

$$f(\mathbf{x}, \mathbf{\Theta}) = \mathbf{y} = \mathbf{W}\mathbf{x} + \mathbf{w}_0, \tag{16.3}$$

where $\mathbf{\Theta} = \{\mathbf{W}, \mathbf{w}_0\}$ are the set of trainable regression coefficients/network weights, $\mathbf{W}$ and $\mathbf{w}_0$ are an $M \times D$ matrix and an M-dimensional column vector, respectively, $\mathbf{x} \in \mathbb{R}^D$ is a D-dimensional column vector comprising features/predictors (often referred to as independent variables in regression) that represent one sample, and $\mathbf{y} \in \mathbb{R}^M$ denotes the corresponding sample's M-dimensional target vector (also referred to as dependent variables in regression). Eq. (16.3) can be written as

$$\begin{bmatrix} y_1 \\ y_2 \\ \cdots \\ y_m \end{bmatrix} = \begin{bmatrix} w_{11} & w_{12} & \cdots & w_{1D} \\ w_{21} & w_{22} & \cdots & w_{2d} \\ \cdots \\ w_{m1} & w_{m2} & \cdots & w_{md} \end{bmatrix} \times \begin{bmatrix} x_1 \\ x_2 \\ \cdots \\ x_d \end{bmatrix} + \begin{bmatrix} w_{10} \\ w_{20} \\ \cdots \\ w_{m0} \end{bmatrix}, \tag{16.4}$$

where subscripts $m = \{1, \cdots, M\}$ and $d = \{1, \cdots, D\}$ represent terms relevant to components of the M-dimensional target vector $\mathbf{y}$ and the D-dimensional features/predictors input vector $\mathbf{x}$, respectively. Eq. (16.4) can be further expressed in a more compact form by including the bias vector $\boldsymbol{w}_0$ as an additional column in the weights matrix $\mathbf{W}$ as follows:

$$\begin{bmatrix} y_1 \\ y_2 \\ \cdots \\ y_m \end{bmatrix} = \begin{bmatrix} w_{10} & w_{11} & \cdots & w_{1D} \\ w_{20} & w_{21} & \cdots & w_{2d} \\ \cdots \\ w_{m0} & w_{m1} & \cdots & w_{md} \end{bmatrix} \times \begin{bmatrix} 1 \\ x_1 \\ x_2 \\ \cdots \\ x_d \end{bmatrix}, \tag{16.5}$$

where the weights matrix $\mathbf{W}$ is now $M + 1 \times D$ in size and the input features/predictors vector $\mathbf{x}$ is $(D + 1)$-dimensional, with the inclusion of a dummy term represented by a "1" to enable the addition of the bias terms $\boldsymbol{w}_0$ as part of the matrix vector multiplication. This allows Eq. (16.3) to be expressed as $\mathbf{y} = \mathbf{W}\mathbf{x}$. This matrix–vector multiplication represents the "forward pass" of a single sample through a two-layer MLP. The forward pass refers to the sequential mathematical operations defined by the layers of the network (and the activation functions applied to outputs of neurons, if included) applied to each input sample. The output of a forward pass is the network's prediction $\hat{\mathbf{y}}$ for the target vector $\mathbf{y}$ corresponding to the input sample $\mathbf{x}$, based on the current estimate for the network's weights $\mathbf{W}$. For batch processing inputs, multiple samples' feature vectors can be stacked as columns in a matrix $\mathbf{X}$, and the resulting forward pass of the MLP for all samples in the batch may be expressed as

$$\hat{\mathbf{Y}} = \mathbf{W}\mathbf{X}, \tag{16.6}$$

where $\hat{\mathbf{Y}}$ denotes a matrix of predictions in the form of column vectors $\hat{\mathbf{y}}$ output by the MLP, corresponding to each sample in the input data matrix $\mathbf{X}$.

Neural networks are trained iteratively through an alternating two-step process involving the "forward pass" and subsequent "backward pass" (also referred to as back-propagation) steps. Given some input training samples, the network outputs predictions for their corresponding targets in the former. It computes the errors or *loss* incurred by the predictions with respect to the actual targets of the input training data. The loss is computed based on some predefined loss function (e.g., mean squared error [MSE] for regression or cross-entropy for classification tasks, discussed in Section 16.2.3), which measures the difference between the network's predictions and actual targets. The chosen loss function and the loss incurred by the network on the input training samples are used to guide the training of the network through optimization and iterative updates to its constituent weights. The network's weights are updated in the subsequent "backward pass" step using the error back-propagation algorithm, which propagates the incurred errors/loss back through the network by calculating the gradient of the loss function with respect to each of the weights in the network. The computed gradients are used to update the network's weights using a suitable optimization algorithm (e.g., stochastic gradient descent (SGD), adaptive moment estimation (ADAM), etc., discussed in greater detail in Section 16.2.4). Changes to the network's weights based on their gradients are in the direction that minimizes the loss function and improves the network's predictions (i.e., pushes them closer to the real targets). Neural networks are trained in this way, iteratively alternating between the forward and backward passes, to update or *learn* their weights, gradually improving the network's performance for the predictive task of interest. Each training iteration for a neural network is typically called an "epoch," and networks are generally trained for many epochs until some suitable termination criterion is reached.

The backward pass for the MLP described above in Eq. (16.6), and for any MLP in general, involves calculating the gradients of the loss function $\boldsymbol{l}$ with respect to the trainable weights $\mathbf{W}$ in the network and using the gradients to update the weights in the direction that minimizes the loss. To update the weights of a specific layer, say the output layer for the MLP in Fig. 16.2 (and defined in Eq. (16.6)), we also need to calculate the gradient of the loss with respect to the outputs $\hat{\mathbf{Y}}$ of the layer (i.e., the error $\mathbf{E}_{\hat{\mathbf{Y}}}$ of the layer), denoted $\frac{\partial \boldsymbol{l}}{\partial \hat{\mathbf{Y}}} = \mathbf{E}_{\hat{\mathbf{Y}}}$. This is because the loss is dependent on the outputs of the layer $\hat{\mathbf{Y}}$, and the outputs are, in turn, dependent on the weights $\mathbf{W}$. Hence, the gradients of the loss with respect to the weights, denoted $\frac{\partial \boldsymbol{l}}{\partial \mathbf{W}}$, are computed using the chain rule as follows:

$$\frac{\partial \boldsymbol{l}}{\partial \mathbf{W}} = \frac{\partial \boldsymbol{l}}{\partial \hat{\mathbf{Y}}} \frac{\partial \hat{\mathbf{Y}}}{\partial \mathbf{W}} = \mathbf{E}_{\hat{\mathbf{Y}}} \mathbf{X}^T. \tag{16.7}$$

Using the gradients of the loss with respect to the weights of a specific layer (say the output layer of the MLP in Fig. 16.2), computed as shown in Eq. (16.7), the weights $\mathbf{W}$ of the layer are updated using a suitable optimization algorithm. The simplest of such optimization algorithms is known as *gradient descent*, using which the weights are updated as

$$\mathbf{W}^{t+1} = \mathbf{W}^{t} - \eta \frac{\partial l}{\partial \mathbf{W}^{t}}, \tag{16.8}$$

where t denotes the current training iteration and η represents the learning rate, which is a hyperparameter that controls the magnitude by which weights are changed in the directions specified by their respective gradients. The learning rate may be tuned empirically or estimated adaptively during training.

If the gradients are calculated using the entire training dataset, the same optimization algorithm is known as *steepest gradient descent*. In practice, however, this may be neither feasible due to memory constraints of computational hardware nor the most efficient way to train a neural network. This is because the computed gradients only provide information on what direction to change the weights *locally* in the loss landscape. Hence, for highly non-convex loss landscapes with several local minima, local gradients computed by the steepest gradient descent may point in the wrong direction, leading to very slow convergence or entrapment in local minima. To circumvent these issues, it is common to group training data into so-called *mini-batches* and to update the weights iteratively within each epoch by conducting forward and backward passes of the network using each mini-batch. One epoch of training thus represents several forward and backward passes of the network until all mini-batches, i.e., all training samples, have been passed through the network and used to update the network's weights. The gradients calculated per sample in each mini-batch are averaged and used to update the weights within each epoch, and this process is repeated for all mini-batches. The resulting optimization algorithm used to update network weights based on the update rule shown in Eq. (16.8) and using the averaged gradients computed over mini-batches is known as the *stochastic gradient descent* (SGD) optimizer (discussed in detail in Section 16.2.4). The update of the weights for a specific layer in an MLP using mini-batch training with an SGD optimizer is given by

$$\mathbf{W}^{t+1} = \mathbf{W}^{t} - \eta \frac{1}{N_t} \sum_{i \in B_t} \frac{\partial f(\mathbf{x}_i, \mathbf{W})}{\partial \mathbf{W}}, \tag{16.9}$$

where B_t and N_t denote the tth mini-batch and the number of samples in the mini-batch, respectively, and subscript i denotes the training samples $\mathbf{x}_i$ in each mini-batch.

Additionally, to back-propagate the error to preceding layers in MLPs with more than two layers, the backward pass also involves calculating the gradients of the loss with respect to the inputs $\mathbf{X}$ of the current layer, denoted $\frac{\partial l}{\partial \mathbf{X}}$, such that the error of the preceding layer may be back-propagated and used to update the weights of the preceding layer. This process generalizes using the chain rule of differentiation to neural networks of any size/depth and is called *error back-propagation*. The error to be back-propagated to the preceding layers is computed as follows:

$$\frac{\partial l}{\partial \mathbf{X}} = \frac{\partial l}{\partial \hat{\mathbf{Y}}} \frac{\partial \hat{\mathbf{Y}}}{\partial \mathbf{X}} = \mathbf{W}^{T} \mathbf{E}_{\hat{\mathbf{Y}}}. \tag{16.10}$$

When all mini-batches of training data have been used to update the weights in all network layers, one *epoch* is said to have been completed. Typically, deep neural networks are trained for several hundreds of epochs until a suitable terminal criterion is reached (e.g., the loss computed for the validation set stops improving for a user-specified number of epochs).

An important aspect to remember when using MLPs is that they assume fixed-dimensional input vectors, and their constituent layers are fully connected. Naive use of such networks for analyzing medical images (which typically have high dimensionality) can result in a network with a very large number of weights, leading to severe *overfitting* of the network to the training data. This can result in very low training loss values but failure to generalize to unseen data and hence a high validation loss (i.e., large errors are incurred on unseen data in the validation set). Therefore, when applying MLPs to medical images or other types of high-dimensional data, care must be taken. It is thus typical to first reduce data dimensionality using suitable representation learning (discussed in Chapter 17) or feature extraction approaches before passing the reduced representation to an MLP for training to solve any given predictive task.

16.2.2 Activation functions

Activation functions have played a significant role in the success of deep neural networks as they introduce non-linearities to the outputs of individual units/neurons in each layer of neural networks. Their inclusion in a network, alongside the sequential combination of several layers with non-linear activations, allows the deep neural network to approximate complex, non-linear mappings between the inputs and outputs. Numerous activation functions have been introduced so far, each with specific properties that make them suitable for particular applications.

The `Sigmoid` or logistic function and hyperbolic tangent (`tanh`) were traditionally widely used as non-linear activation functions for shallow neural networks. Using an exponential function expressed as

$$\texttt{Sigmoid}(x) = \frac{1}{1 + e^{-x}}. \tag{16.11}$$

`Sigmoid` transforms the input values into a value between 0 and 1 (see Fig. 16.3). The `tanh` function is shaped similarly to the `Sigmoid` but outputs zero-centered values between -1 and 1, and is expressed as

$$\texttt{tanh}(x) = \frac{e^{2x} - 1}{e^{2x} + 1}. \tag{16.12}$$

However, an issue with the `Sigmoid` and `tanh` functions is that they saturate the outputs within a bounded range, such as 0 to 1 for the former and -1 to 1 for the latter. As a result, their gradients with respect to the inputs (i.e., the error to be back-propagated to the preceding layer in the network) will be small values close to zero.

Hence, when training a deep neural network with several layers, gradient signals from the higher layers will be small values close to zero that are multiplied successively during back-propagation (due to the chain rule), resulting in gradients not being propagated back to earlier layers. This is known as a *vanishing gradient* [5], which makes it difficult and sometimes impossible to train deep neural networks with such saturating activation functions.

The rectified linear unit (ReLU) [6] is the most widely used activation function in deep neural networks. As shown in Fig. 16.3, ReLU uses a max function and is computationally simple compared to the Sigmoid, which requires computing an exponential function. ReLU is a non-linear activation function, or specifically, it is piecewise-linear, which outputs a 0 for all negative inputs and returns the input values themselves for positive inputs. Thus, the ReLU behaves the same as a linear activation function for positive inputs, but its derivative is the Heaviside or unit step function. The ReLU activation function is expressed as

$$\text{ReLU}(x) = \max(x, 0) = \begin{cases} 0, & x < 0, \\ x, & x \geq 0. \end{cases} \tag{16.13}$$

As the ReLU activation function outputs a 0 for all inputs less than 0, not all neurons are simultaneously activated in a neural network. Hence, ReLU is computationally more efficient and accelerates the training and convergence of deep neural networks relative to Sigmoid or tanh activation functions [7]. A key limitation of non-linear activation functions such as Sigmoid or tanh is the *vanishing gradient* problem in the context of their use in deep neural networks. ReLU helps address this limitation as its output is unbounded in the positive domain. However, the gradient flowing through a ReLU neuron can lead to weight updates in a way that the neuron will never activate because of the zero signal in the negative domain (e.g., $w > 0$, $x < 0$, so, $f(wx) = 0$, always). This situation is called *dead* ReLU because the gradient flowing through the unit will always be zero from that point on. This problem is addressed in some variants of ReLU. Leaky ReLU (LReLU) considers a fixed small positive gradient for negative inputs [8] to prevent saturation at zero (Fig. 16.3). In parametric ReLU (PReLU) [9], the same function as LReLU is utilized, but the slope of the negative part (α) is a learnable parameter. This helps address one of the limitations of LReLU: it cannot adapt to sudden changes given negative inputs due to its fixed slope. With PReLU, on the other hand, the slope parameter for negative inputs can be adaptively learned using back-propagation during training. The PReLU activation function is defined as

$$\text{PReLU}(\alpha, x) = \begin{cases} x, & x \geq 0, \\ \alpha x, & x < 0. \end{cases} \tag{16.14}$$

In randomized leaky ReLU [10], the slope of the negative domain is randomly selected during training. Still, for the test phase, a constant slope computed from the average of all the slopes used in training is utilized. Recently, a paired ReLU [11]

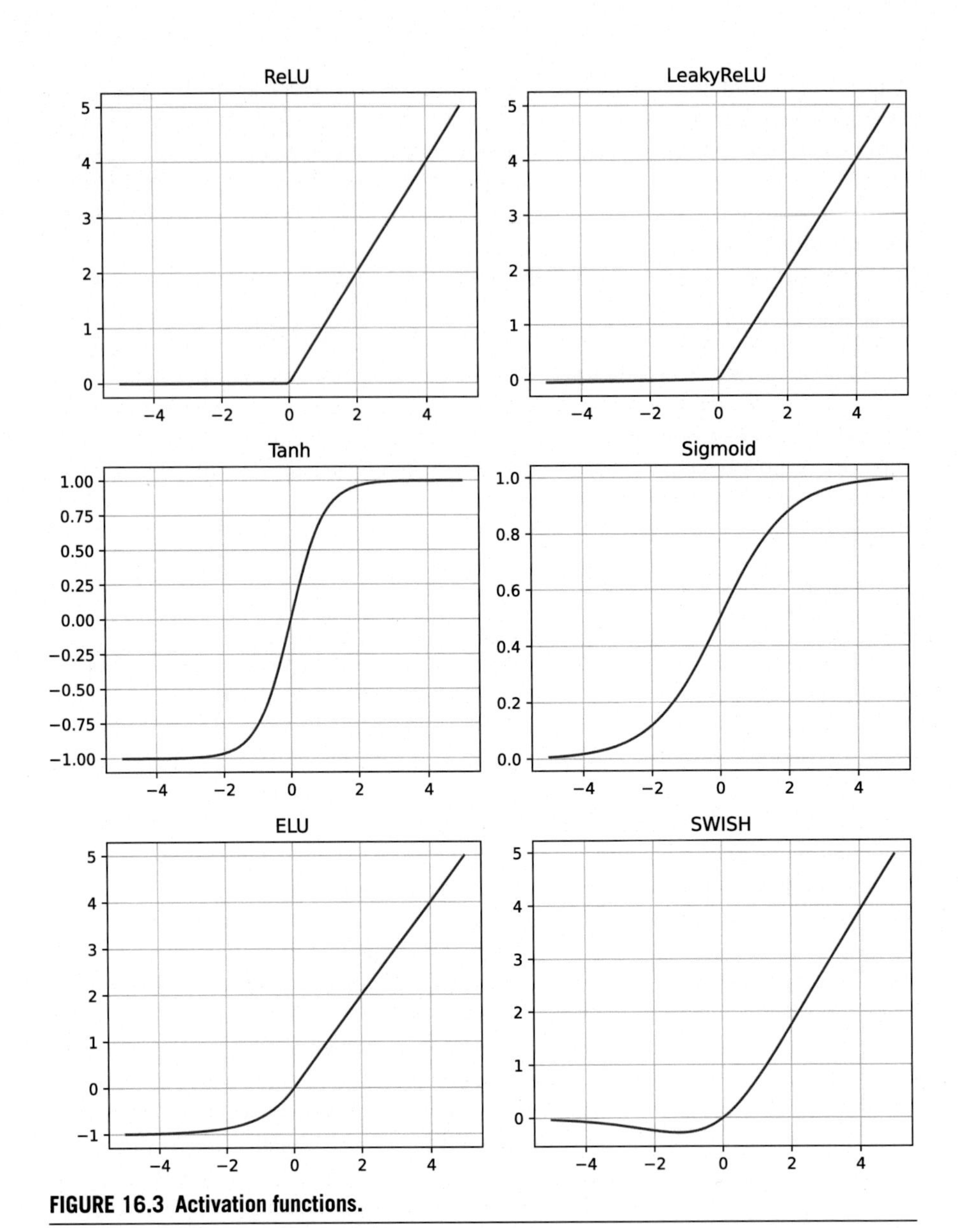

FIGURE 16.3 Activation functions.

Different non-linear activation functions are frequently used in deep neural networks.

activation scheme has been proposed, composed of two parts, one for activating the positive part and the other for the negative phase of the neuron. Hence, applying this activation function leads to double feature maps, and consequently a more efficient local structure can be produced compared to the standard `ReLU`. The paired `ReLU`

(PaReLU) is defined as

$$\text{PaReLU}(x) = (\max(sx - \theta), \max(s_p x - \theta_p, 0)), \tag{16.15}$$

where s and s_p are a pair of scale parameters that are initialized to 0.5 and −0.5, respectively, and θ and θ_p are two trainable thresholds.

Exponential linear unit (ELU) [12] is another alternative to ReLU with improved characteristics. The ELU function is defined as

$$\text{ELU}(x) = \begin{cases} x, & x \geq 0, \\ \alpha(e^x - 1), & x < 0, \end{cases} \tag{16.16}$$

where $\alpha > 0$ is a fixed scalar parameter. A generalization of ReLU activation functions has also been proposed, which is called maxout [13]. In contrast to ReLU, ELU has negative values that become smooth slowly until they are equal to $-\alpha$, using an exponential function. This allows for the mean of the activations to be closer to zero. Because the bias shift effect is reduced, the normal gradient is brought closer to the unit natural gradient, allowing the network to converge faster during training. ELU saturates to a negative value with smaller inputs and, due to small derivatives, decreases the forward propagated variation and information, resulting in noise-robust representation.

An improvement to the ELU activation function was proposed by [14], called the scaled exponential linear unit (SELU). The SELU activation function has a self-normalizing property that ensures the output activations of a deep neural network remain normalized. This can help prevent vanishing and exploding gradient problems and speed up the convergence of the network. Layer normalization techniques (e.g., batch normalization [BN], discussed in Section 16.2.5) in general are used to improve network stability during training and the convergence rate by reducing the internal covariance shift, i.e., the change in the distribution of activations within a hidden layer as the network is trained. Layer normalization techniques typically transform the activations of a hidden layer by subtracting the mean and dividing by the standard deviation, computed across the activations of the layer or across the input batch of training samples (e.g., BN), to ensure that the activations throughout all layers in the network remain close to zero mean and unit standard deviation, effectively tackling the problem of exploding or vanishing gradients during training. The SELU activation function naturally converges to zero mean and unit variance. The SELU activation function is given by

$$\text{SELU}(\alpha, \lambda, x) = \lambda \begin{cases} x, & x > 0, \\ \alpha(e^x - 1), & x \leq 0. \end{cases} \tag{16.17}$$

The Softplus activation function [15] is a smoothed version of ReLU that is defined as

$$\text{Softplus}(x) = \ln(1 + e^x). \tag{16.18}$$

In contrast to ReLU, the Softplus function has non-zero gradients for the negative domain. The derivative of Softplus is the Sigmoid function, meaning that when the input increases, the gradient limits to one, potentially reducing the effects of the vanishing gradients problem. Recently, different variants of Softplus have been proposed. Soft++ [16] is a parametric version of Softplus, which parameterizes the slope in the negative domain and the exponent. A linear combination of parameterized Softplus and ELU is introduced in [17], showing improved performance compared to the most commonly used activation functions.

In addition to the various activation functions mentioned above, Ramachandran et al. [18] utilized automatic search strategies combining exhaustive and reinforcement-learning techniques to discover novel activation functions, including the Swish function. The Swish activation function is defined as

$$\mathrm{Swish}(x) = x \cdot \sigma(\beta x) = \frac{x}{1 - e^{-\beta x}}, \tag{16.19}$$

where β is a constant or a trainable parameter whose scale controls the rate at which the first derivative of the Swish function asymptotes to 0 and 1. In [18], the authors demonstrate that the Swish activation function improves the predictive performance of deep neural networks developed originally using ReLU on standard image classification benchmarks such as ImageNet [19].

In summary, many activation functions have been proposed in recent years. Each new formulation and improvement attempts to address the fundamental limitations of saturating activation functions such as the Sigmoid and tanh and/or those of the piecewise linear ReLU activation function. A comprehensive review of activation functions, with experimental benchmarks evaluating their relative merits in terms of predictive power afforded to deep neural networks, is presented in [20]. Recent activation functions proposed to improve on the Sigmoid and tanh functions are designed to tackle the problems of non-zero means in layer activation and vanishing gradients during back-propagation. Variants of the ReLU activation function, on the other hand, tackle the three main limitations of ReLU, namely, underutilization of negative values, limited non-linearity, and unbounded outputs. Although several new activation functions have been proposed, the ReLU remains among the most commonly used activation functions (alongside LReLU and PReLU) used to train deep neural networks.

16.2.3 Loss functions

Neural networks are trained to approximate a target function representing a mapping from some given inputs to their respective outputs/target variables. The parameters or weights $\mathbf{W}$ of the network are learned, given some training data, by minimizing a suitable loss function $\boldsymbol{l}$ with respect to the network weights. Loss functions are chosen based on the nature of the learning/predictive task of interest, the characteristics of the training data available, the manner in which the target variables are represented/encoded, and whether it is necessary/desirable to constrain the optimization process in some way (i.e., regularization, discussed in Section 16.2.6). Estimation of the (at least locally) optimal parameters ($\mathbf{W}^*$) of a neural network is formulated as a minimization problem as follows:

$$\mathbf{W}^* = \arg\min_{\mathbf{W}} \boldsymbol{l}(\mathbf{W}). \tag{16.20}$$

A brief overview of the most common types of loss functions used to train deep neural networks is given next, and they are grouped according to the types of learning tasks.

Loss functions for classification: Classification with deep neural networks refers to the process of learning a mapping from some input data (e.g., medical images) to one (i.e., binary or multiclass classification) or several co-occurring (i.e., multilabel classification) discrete targets or labels. Classification networks are predominantly trained by mapping the inputs to probabilities (i.e., values between 0 and 1) using either the `Sigmoid` (binary classification) or the `Softmax` (multiclass classification) activation function in the final layer of the network and minimizing the *cross-entropy* loss given by

$$\boldsymbol{l} = -\sum_{c=1}^{C} \mathrm{y}_c \log \mathrm{p}_c, \tag{16.21}$$

where y_c denotes the ground truth label for class $c = \{1 \cdots C\}$, which is binary-valued, i.e., it is either 1 or 0, to indicate whether the sample belongs to a particular class c or not, and p_c denotes the probability of a given sample predicted by the neural network to belong to class c. Target labels for multiclass classification are typically one-hot encoded and used to train classification networks. One-hot encoding refers to representing categorical/discrete class labels as binary vectors, where each element in the vector is a zero except the index that corresponds to the class label for a specific sample. For example, a class label of 2 for a three-class classification problem would be one-hot encoded as $[0, 1, 0]$. For the binary classification case, the cross-entropy loss function shown in Eq. (16.21) reduces to

$$\boldsymbol{l} = -(\mathrm{y}_c \log \mathrm{p}_c + (1 - \mathrm{y}_c) \log(1 - \mathrm{p}_c)). \tag{16.22}$$

The cross-entropy loss measures the divergence of the predicted probabilities from the ground truth class labels and is derived from information theory. Specifically, the information content of a signal or data source can be measured using Shannon's entropy (as bits or nats), and according to Shannon's source coding theorem, cross-entropy measures the number of bits required to compress data drawn from some distribution p using a code based on some other distribution q. This is mathematically expressed as

$$h(\mathrm{p}, \mathrm{q}) = -\sum_{c=1}^{C} \mathrm{p}_c \log \mathrm{q}_c, \tag{16.23}$$

where subscript $c = \{1 \cdots C\}$ denotes the different states that discrete random variables drawn from p and q can take on, and can be verified to be identical to the cross-entropy loss function defined above in Eq. (16.21).

An alternative to the cross-entropy loss function for binary classification (refer to Eq. (16.22)) is the *hinge* loss. It was originally developed for soft-margin support vector machines but it has been used to train classification networks as well. The hinge loss function is expressed as

$$l = \max(0, 1 - \mathrm{y}_c \mathrm{p}_c), \tag{16.24}$$

where y_c denotes the ground truth class labels represented as either $+1$ or -1 for each class, respectively, and p_c is the predicted output from the network which is often mapped to a value in the range $\{-1, +1\}$ using a suitable squashing activation function, such as `tanh`. Although the hinge loss function was originally developed for binary classification, multiclass classification variants have been proposed since, using a one-versus-all strategy, for example. The hinge loss encourages the network to maximize the margin around the decision boundary separating the two classes, which can lead to better generalization performance than using cross-entropy. Additionally, the hinge loss has sparse gradients, which can be useful for training large models with limited memory (unlike cross-entropy with dense gradients).

A frequently used variant of the hinge loss is the squared hinge loss, given by

$$l = \max(0, 1 - \mathrm{y}_c \mathrm{p}_c)^2, \tag{16.25}$$

which is differentiable, unlike the hinge loss. This enables the squared hinge loss to be used with higher-order optimization algorithms (e.g., Levenberg–Marquardt) to train deep neural networks. Higher-order optimization algorithms have been shown to improve the convergence rate and generalizability of deep neural networks [21,22], relative to conventional first-order gradient-based optimizers (e.g., SGD).

Loss functions for classification with noisy labels: Label noise is a common challenge when developing learning-based medical image classification approaches. Label noise in the context of image classification specifically refers to mislabeling of imagewise class labels. This may arise due to incorrect interpretation of the images by clinical experts, errors in the extraction and association of labels with images through the mining of associated radiology reports (e.g., processes used to label chest X-ray images in the CheXpert database [23]), or as a result of errors in the data extraction, preparation, and curation processes. Robust loss functions, therefore, are desirable for training classification networks to mitigate the effects of inherent label noise in the training data.

Cross-entropy has been shown to be sensitive to label noise in several previous studies [24,25], highlighting the need for alternative loss functions to train classification networks. Some robust loss functions of particular note include generalized cross-entropy (GCE) [26], mean absolute error (MAE) [24], focal loss [27], and symmetric cross-entropy (SCE) [28]. The authors in [24] showed theoretically that MAE is robust to label noise under certain assumptions; however, they also noted that training deep neural networks by minimizing the MAE loss function for classification is difficult as MAE treats every sample equally, unlike cross-entropy, which implicitly weights samples according to how well the network's predictions match the ground truth class labels, during training. This leads to suboptimal solutions and decreased performance when training classification networks using the MAE loss function. The MAE loss function for classification is given by

$$l = \sum_{c=1}^{C} \|\mathrm{y}_c - \mathrm{p}_c\|_1, \tag{16.26}$$

where, as before, y_c and p_c denote the ground truth class labels and the predicted output from the classification network, respectively, for a single sample.

To leverage the advantages of the MAE loss function and to address its limitations, the GCE loss was proposed [26] under the assumption of class-dependent label noise, which is given by

$$l = \sum_{c=1}^{C} \frac{(1 - \mathrm{p}_c^{\mathrm{q}})}{\mathrm{q}}, \tag{16.27}$$

where $\mathrm{q} \in (0, 1]$. Using L'Hôpital's rule, Eq. (16.27) can be shown to be equivalent to the cross-entropy loss when $\mathrm{q} \to 0$ and equal to the MAE loss when $\mathrm{q} = 1$. This GCE loss can be viewed as a generalization of the cross-entropy and MAE loss functions. Similarly, the focal loss, which is a popular loss function for object detection, is a generalization of the cross-entropy loss and can also be used as a robust loss function for image classification in the presence of noisy labels. The focal loss [27] is given by

$$l = \sum_{c=1}^{C} \mathrm{y}_c (1 - \mathrm{p}_c)^{\mathrm{q}} \log \mathrm{p}_c, \tag{16.28}$$

where $\mathrm{q} \geq 0$ is a tunable parameter. The focal loss reduces to the cross-entropy loss when $\mathrm{q} = 0$ and acts as a robust loss function for $\mathrm{q} > 0$.

Loss functions for regression: Regression with deep neural networks refers to learning a function that maps the inputs (e.g., medical images or features derived thereof) to some continuous-valued target variables. As with classification, regression is a form of supervised learning where, given a training dataset of predictors/features (defining each input sample) and their corresponding target values, a regression network (or model more generally) is trained by minimizing a suitable loss function. The target values in many regression tasks may be unbounded. Consequently, regression networks trained on such tasks utilize a linear activation function as the output/final layer to make predictions. Given bounded continuous-valued targets, suitable alternative activation functions (e.g., `Sigmoid`) may also be applied to the output layer of regression networks.

The most commonly used regression loss functions are the MAE, MSE, and root mean squared error (RMSE). MAE (also referred to as the L_1 loss) is expressed similarly to Eq. (16.26), except now the target and predicted values are continuous. Given $n = \{1 \cdots N\}$ samples in a mini-batch when training a regression network, the MAE over the mini-batch is calculated as

$$l = \frac{1}{N}\sum_{n=1}^{N} |\mathbf{y}_n - \boldsymbol{f}(\mathbf{x}_n)|, \tag{16.29}$$

where $\mathbf{y}_n$ represents the target values for the nth sample in the mini-batch and $\boldsymbol{f}(\mathbf{x}_n)$ represents the regression network's predictions, given the nth sample's features/predictors $\mathbf{x}_n$ as inputs. As before, as the gradient of the MAE loss function with respect to the network's predictions is not dependent on the magnitude of error incurred, each input sample is treated equally. This leads to slow and, at times, suboptimal convergence when training regression networks using MAE.

The MSE loss (also referred to as the L_2 loss) is expressed as the average of the squared errors incurred by a regression network/model with respect to the ground truth target values of the inputs. Considering a mini-batch of inputs of size N again, the MSE loss is expressed as

$$l = \frac{1}{N}\sum_{n=1}^{N} (\mathbf{y}_n - \boldsymbol{f}(\mathbf{x}_n))^2, \tag{16.30}$$

where, once again, $\mathbf{y}_n$ and $\boldsymbol{f}(\mathbf{x}_n)$ represent the targets and network's predictions for the nth sample, respectively. Due to the squared term in the MSE loss function, it does not suffer from the same type of convergence issues as MAE, as now the calculated gradients of the loss with respect to the network's predictions are weighted by the magnitude of the error incurred for each sample. However, the squared term also amplifies the influence of outliers in the data. Large errors incurred on outlier target values have a larger impact on the computed gradients during network training. This can lead to incorrect/suboptimal network weight updates when the observed targets in the training data contain outliers. The MSE loss is derived based on the assumption that the errors/residuals are Gaussian random variables. Hence, while they are suitable for applications where this assumption is valid, violation of the same can lead to suboptimal results when training regression networks due to the sensitivity of the MSE loss to outliers. The RMSE loss is simply the square root of the MSE loss, and regression networks trained using the former typically converge to the same solution as the latter.

Loss functions for regression with noisy targets: As with medical image classification tasks, regression of target quantities directly from medical images is often challenging due to the presence of noise/outliers in the ground truth targets (e.g., measurements characterizing anatomical structure or function) associated with the images. The source of these outliers may be attributed to errors in the image interpretation and quantitative analyses conducted to extract the ground truth measurements that are used as targets or dependent variables or to issues in the image formation and reconstruction process itself, which subsequently leads to the propagation of errors in downstream quantitative analyses. As discussed above, the MAE, MSE, and RMSE loss functions all have certain limitations, with both MSE and RMSE being sensitive to outliers in the observed targets. While MAE is a robust loss function in this regard, it leads to slow/suboptimal convergence issues.

To address these limitations and leverage the advantages of the MAE and MSE loss functions, the Huber loss function was proposed [29]. The Huber loss is a variant of the MAE and MSE loss that is equal to the MSE loss when the errors/residuals drop below a user-specified value δ; it is given by

$$l = \begin{cases} \frac{1}{2}(\mathbf{y}_n - \boldsymbol{f}(\mathbf{x}_n))^2, & |\mathbf{y}_n - \boldsymbol{f}(\mathbf{x}_n)| \leq \delta, \\ \delta(|\mathbf{y}_n - \boldsymbol{f}(\mathbf{x}_n)| - \frac{1}{2}\delta), & |\mathbf{y}_n - \boldsymbol{f}(\mathbf{x}_n)| > \delta, \end{cases} \tag{16.31}$$

where $\mathbf{y}_n$ and $\boldsymbol{f}(\mathbf{x}_n)$ represent the targets and network's predictions for the nth sample in the mini-batch, respectively. From Eq. (16.31) it is clear that when the errors/residuals incurred by the network's predictions are below a user-specified threshold δ, the Huber loss is similar to the MSE loss. Otherwise, it behaves similarly to the MAE loss. In this way, the Huber loss combines the advantages of both the MAE and MSE losses, increasing robustness during network training to large errors resulting from outliers in the observed targets.

Another popular robust loss function used to train regression networks when the observed targets are noisy/contain outliers is the Log-cosh loss function, which is the logarithm of the hyperbolic cosine of the residuals/errors between the observed targets $\mathbf{y}_n$ and the network's predictions $\boldsymbol{f}(\mathbf{x}_n)$. The Log-cosh loss retains all the benefits of the Huber loss function, and does not require tuning of a hyperparameter (δ in Eq. (16.31) is a hyperparameter that must be tuned). The Log-cosh loss function is given by

$$l = \frac{1}{N}\sum_{n=1}^{N} \log(\cosh(\boldsymbol{f}(\mathbf{x}_n - \mathbf{y}_n))), \tag{16.32}$$

where cosh is the hyperbolic cosine function. An added benefit of the Log-cosh loss function is that it is twice differentiable everywhere, making it suitable for use with higher-order optimization algorithms to train regression networks.

Loss functions for self-supervised learning in generative models: Deep generative networks have found widespread use in recent years in several domains including medical imaging. They are trained to generate synthetic data that resemble real data by approximating the real data distribution of interest (either explicitly or implicitly) and correspondingly to understand the complexities of the input data distributions themselves. In turn, synthetic data generated by such models have their own diverse uses (e.g., augmenting real data for supervised learning tasks). The process of training deep generative networks falls under the category of *self-supervised learning* as the loss functions optimized to train such networks are formulated as functions of the input data/predictors (rather than target variables, as in the case of supervised learning). While a comprehensive review of all loss functions used to train deep generative networks is beyond the scope of this textbook, the most common loss functions used to train popular generative models, such as generative adversarial networks (GANs) and variational autoencoders (VAEs), are discussed here and in Chapter 17, respectively.

Generative adversarial networks (GANs): GANs were proposed by Goodfellow et al. [30,31] in 2014 and have since become one of the most widely used generative models across several application domains, including medical imaging [32]. GANs comprise two subnetworks or network branches: the *generator* and the *discriminator*. They are trained to capture the real data distribution implicitly by generating synthetic/fake data, which in turn are used as information to guide the training of the network and improve the *realism* of the synthesized data. Specifically, the generator learns to generate plausible data and the generated synthetic/fake data instances become negative training examples for the discriminator. The discriminator uses the synthesized fake data instances output by the generator and is trained as a *discriminative model* to distinguish the fake data instances from the real ones. In other words, the generator is trained to try and fool the discriminator so that it can no longer distinguish real from fake data instances. In contrast, the discriminator is simultaneously trained to distinguish real from fake data instances correctly and penalize the generator for synthesizing data that are not realistic.

Training GANs, therefore, involves optimizing both the generator and discriminator networks such that they compete against each other in a two-player zero-sum game. Therefore, the typical loss functions used to train GANs have competing terms optimized to improve the generator and discriminator's performance. Two common loss functions used to train GANs include the minimax loss [30] and the Wasserstein loss [33]. The minimax loss function is given by

$$\min_{\boldsymbol{g}} \max_{\boldsymbol{d}} \boldsymbol{l}(\boldsymbol{d}, \boldsymbol{g}) = \mathbb{E}_{\mathbf{x} \sim p_{data}(\mathbf{x})}[\log \boldsymbol{d}(\mathbf{x})] + \mathbb{E}_{\mathbf{z} \sim p_{\mathbf{z}}(\mathbf{z})}[\log(1 - \boldsymbol{d}(\boldsymbol{g}(\mathbf{z})))], \quad (16.33)$$

where $\boldsymbol{g}$ and $\boldsymbol{d}$ represent the generator and discriminator networks, respectively, and the loss function $\boldsymbol{l}(\boldsymbol{g}, \boldsymbol{d})$ represents a two-player minimax game between $\boldsymbol{g}$ and $\boldsymbol{d}$ (and is derived from the cross-entropy loss discussed previously), where $\boldsymbol{g}$ is trained to synthesize more realistic samples to fool $\boldsymbol{d}$. In contrast, $\boldsymbol{d}$, conversely, is trained to correctly classify real ($\mathbf{x}$) and synthesized/fake ($\hat{\mathbf{x}}$) samples; $\boldsymbol{g}$, with trainable parameters $\boldsymbol{\Theta}_g$, is trained to map latent random variables sampled from some assumed prior distribution, i.e., $\mathbf{z} \sim p(\mathbf{z})$, to synthetic/fake data $\hat{\mathbf{x}} = \boldsymbol{g}(\mathbf{z}; \boldsymbol{\Theta}_g)$, where the aim is to ensure that the distribution of synthesized data resembles the distribution of the real training data, i.e., $\boldsymbol{p}_{\boldsymbol{\Theta}_g}(\hat{\mathbf{x}}|\mathbf{z}) \sim \boldsymbol{p}_{data}$, as closely as possible. Training of $\boldsymbol{g}$ is assisted by training the discriminator network $\boldsymbol{d}$ in tandem, to classify real from synthetic/fake data generated by $\boldsymbol{g}$. Therefore, $\boldsymbol{d}$ is a binary classifier trained to predict $\boldsymbol{d}(\mathbf{x}) = 1$ for real data samples and $\boldsymbol{d}(\hat{\mathbf{x}}) = 0$ for fake data samples, when a minimax loss function is used for training. To optimize the minimax loss function in Eq. (16.33) and train the GAN, $\boldsymbol{d}$ is trained by maximizing $\boldsymbol{l}(\boldsymbol{d}, \boldsymbol{g})$, while $\boldsymbol{g}$ is trained by minimizing the second term in Eq. (16.33), i.e., $\log(1 - \boldsymbol{d}(\boldsymbol{g}(\mathbf{z})))$. GAN training thus involves two alternating steps, where $\boldsymbol{d}$ is trained for one or more epochs to maximize the probability of correctly predicting real from fake data samples and $\boldsymbol{g}$ is also trained for one or more epochs to trick $\boldsymbol{d}$ into classifying a synthetic/fake data sample as being real.

The Wasserstein loss function, proposed for the Wasserstein GAN (WGAN) [33], was introduced to address the issue of vanishing gradients encountered when training GANs using the original minimax loss. As shown by the authors of [30], optimization of Eq. (16.33) minimizes the Jensen–Shannon (JS) divergence between the real and synthesized data distributions, and correspondingly, the global minimum is reached when $\boldsymbol{g}$ perfectly captures the *real* data generating process. Optimizing the JS divergence, however, results in vanishing gradients in some cases, leading to mode collapse when training GANs. Here, mode collapse refers to the phenomenon where $\boldsymbol{g}$ reproduces the same output (or a small set of outputs) repetitively because $\boldsymbol{d}$ is stuck in a local minimum and cannot distinguish $\boldsymbol{g}$ outputs from real data. The Wasserstein loss was introduced in WGAN to address this issue of vanishing gradients and mode collapse when training GANs. WGANs use the Earth Mover or Wasserstein-1 distance in place of the JS divergence in the loss function for training. Specifically, a WGAN does not discriminate or classify real from fake data generated by $\boldsymbol{g}$, as in the case of GANs trained using the minimax loss. Hence, $\boldsymbol{d}$ in WGANs is strictly speaking not referred to as a *discriminator* network but as a *critic* network which scores each real and fake instance with a real-valued score instead of predicting the probability that it is fake. Therefore, the loss function to train WGANs tries to maximize the distance between the scores computed for real and fake data. The two terms in the loss function used to train WGANs are given by

$$l_{critic} = \boldsymbol{d}(\mathbf{x}) - \boldsymbol{d}(\boldsymbol{g}(\mathbf{z})), \tag{16.34a}$$

$$l_{generator} = \boldsymbol{d}(\boldsymbol{g}(\mathbf{z})), \tag{16.34b}$$

where the discriminator/critic network is trained to maximize Eq. (16.34a) and the generator network is trained to maximize Eq. (16.34b). In other words, $\boldsymbol{d}$ tries to maximize the difference between its scores computed for real and fake data and $\boldsymbol{g}$ tries to maximize the scores computed by $\boldsymbol{d}$ for the fake data it generates. The benefits of the Wasserstein loss are that it is continuous and differentiable almost everywhere, allowing for stable (less vulnerable to mode collapse) and continued training of $\boldsymbol{g}$ and $\boldsymbol{d}$.

16.2.4 Weight optimization

Training deep neural networks is an optimization problem, wherein, based on a chosen loss function appropriate for the task of interest, an optimization algorithm (typically, first-order gradient-based) is used to iteratively minimize the loss function with respect to the trainable parameters of the network and correspondingly to update the network's parameters based on the computed gradients via error back-propagation (as discussed in Section 16.2.1). However, it is important to distinguish between the goals of optimizing and training a neural network, as they are not the same. While the former is concerned with minimizing the loss function (training error) given some training data, the objective of training a neural network on a specific task is to find the most suitable set of parameters that helps the network *generalize* to *unseen* data, i.e., the goal of training neural networks, or any machine learning model for that matter, is to reduce the generalization error. Therefore, it is insufficient to reduce the training loss (i.e., error incurred on the training set) alone. To train networks that generalize well to unseen data, controlling for overfitting to the training data is essential, which is a key issue with deep neural networks due to their large number of parameters. This is typically done by also monitoring the network's loss at each training epoch, on an unseen dataset, referred to as the *validation set*. The validation set is only used to evaluate the predictive performance of the network (i.e., only used for network *inference*) at each training step and is not used to update the parameters of the network itself. Monitoring the validation loss and making certain design choices regarding the network and how it is trained is custom to prevent significant divergence between the validation loss and the training loss. These design choices may include tuning the network's architecture and associated hyperparameters (user-specified parameters that are not trainable) and incorporating regularization strategies (discussed in Section 16.2.6) during optimization to prevent overfitting. The chosen configuration of the network is then optimized and encouraged to converge to a suitable model that generalizes to unseen data.

As discussed briefly in Section 16.2.1, deep neural networks are trained using mini-batches, i.e., where the training data are grouped into mini-batches, and each mini-batch is used to iteratively update the parameters of the network through successive forward and backward passes, within each training epoch. We restrict our attention here to the mini-batch training of deep neural networks and to common first-order gradient-based optimization algorithms used for learning with mini-batches.

Mini-batch stochastic gradient descent (SGD): Gradient descent optimization is the simplest type of first-order gradient-based optimization. In the context of neural networks, optimization of the network's parameters using gradient descent (also referred to as steepest gradient descent) means that all training samples are used to update the network parameters at each training epoch. On the other extreme is SGD (also referred to as online learning), where one training sample at a time is used to update the network's parameters within each training epoch. Steepest gradient descent and online learning have their drawbacks, as the former is not data/memory-efficient and the latter is not computationally efficient. However, the middle ground gives rise to mini-batch SGD, which balances data/memory and computational efficiency. Given some loss function denoted $f(\mathbf{X}, \boldsymbol{\Theta})$, where $\boldsymbol{\Theta}$ represents the trainable parameters of a neural network and $\mathbf{X}$ denotes the training data, the mini-batch SGD algorithm iteratively updates the network's parameters at each epoch of training as

$$\boldsymbol{\Theta}^{(t+1)} \leftarrow \boldsymbol{\Theta}^{t} - \eta \frac{1}{N_t} \sum_{i \in B_t} \frac{\partial f(\mathbf{x}_i, \boldsymbol{\Theta})}{\partial \boldsymbol{\Theta}}, \tag{16.35}$$

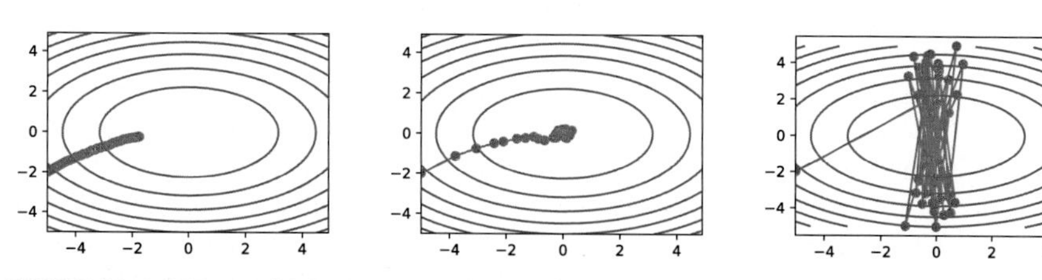

FIGURE 16.4 SGD sensitivity to the learning rate.

From left to right: Trajectory of updates to weights estimated using SGD for increasing learning rates from 0.01 (left), to 0.1 (middle), to 0.5 (right). We see that increasing the learning rate leads to oscillations in the optimizer's updates around the optimal solution.

where subscript t denotes the tth update iteration in a training epoch, B_t denotes the tth mini-batch that is used to update the network's parameters for the next forward pass at iteration $(t+1)$, η denotes the learning rate, and N_t denotes the total number of training samples $\mathbf{x}_{i=1 \ldots N_t}$ in each mini-batch. During mini-batch training, each training sample within each mini-batch is sampled uniformly at random from the training without repetitions. The mini-batch gradient used to update the network's parameters is computed as the average of the individual gradients resulting from each training sample in the mini-batch. This averaging reduces the variance in computed gradients and, correspondingly, in updates to the network's parameters, resulting in quicker and more stable convergence to (at least local) minima than afforded by online learning. Some of the limitations of the mini-batch SGD optimizer are its sensitivity to batch size and learning rate. If the batch size is too small, the resulting gradients computed to update the network's parameters may be noisy, slowing down convergence and leading to suboptimal solutions. Conversely, if the batch size is too large, this can lead to overfitting of the network and convergence to a suboptimal solution. Similarly, if the learning rate is too low, the convergence of the network will be slow and there is a risk of getting trapped in suboptimal solutions. If the learning rate is too high, then the steps taken in the directions of the estimated gradients to update the network's parameters are large. This can lead to oscillations around local/global minima or divergence of the optimization process. Effects of varying the learning rate on the convergence properties of SGD are illustrated in Fig. 16.4.

As a rule of thumb, it is necessary to tune both the batch size and the learning rate to each dataset in a manner that is compatible with each other. Large batch sizes can typically be used with higher learning rates as the resulting gradients are less likely to be noisy, whereas smaller batch sizes should typically be used with lower learning rates to prevent large steps in the direction of noisy gradients when updating the network's parameters.

SGD with Momentum: In order to reduce the sensitivity of SGD to the learning rate in the presence of noisy gradients, SGD with Momentum or just the momentum optimizer was proposed. The momentum optimizer belongs to a family of optimizers referred to as accelerated gradient-based optimization methods. The key motivation for such methods was to use gradient averaging to further benefit from the effect of variance reduction beyond what is afforded by averaging across instances in a mini-batch, in mini-batch SGD. In the case of the momentum optimizer, this is achieved by adding a fraction of the gradient from the previous iteration within an epoch (i.e., the gradient computed using the previous mini-batch) to the current gradient estimate. The resulting quantity can be thought of as a weighted averaging of multiple past gradients and is referred to as the *velocity* ($\mathbf{V}$). The new gradient is given by the velocity $\mathbf{V}$ and is not oriented towards the average steepest descent direction of the instances in any given mini-batch (as with mini-batch SGD), but instead is oriented in the direction of a weighted average of all past gradients (based on all observed mini-batches). The velocity is used to update the network's parameters and is computed as follows:

$$\mathbf{V}^t \leftarrow \beta \mathbf{V}^{(t-1)} + \frac{1}{N_t} \sum_{i \in B_t} \frac{\partial f(\mathbf{x}_i, \mathbf{\Theta}^{(t-1)})}{\partial \mathbf{\Theta}}, \tag{16.36a}$$

$$\mathbf{\Theta}^t \leftarrow \mathbf{\Theta}^{(t-1)} - \eta \mathbf{V}^t, \tag{16.36b}$$

where in Eq. (16.36a), $\beta \in (0, 1)$ is an empirically tuned hyperparameter that controls the relative influence of past gradients, referred to as *momentum*, on the updates to the network's parameters. Large values for β amount to a greater influence of past gradients, i.e., a long-range average, and conversely, small values limit past gradients' influence on updates to the network's parameters. If $\beta = 0$ the resulting optimizer is nothing but the mini-batch SGD optimizer. In Eq. (16.36b), as defined previously, η represents the learning rate which needs to be tuned empirically and $\mathbf{\Theta}$ denotes the trainable parameters of the network.

Another class of optimization methods that try to address the limitations of SGD and SGD with Momentum is adaptive gradient methods, which dynamically update the learning rate during training. Adaptive gradient methods remove the need for fixing an appropriate learning rate before training a network, as is the case with SGD and SGD with Momentum. Additionally, SGD and SGD with Momentum use the same learning rate to update all network parameters, while adaptive gradient methods relax this constraint and allow different learning rates to be dynamically estimated for updating each network parameter. Dynamic estimates for the learning rates are guided by the training process and the calculated gradients for the network's parameters. This provides more precise control of the step sizes used to update each parameter. This makes intuitive sense as each parameter's relationship to the loss function differs during different network training phases. Commonly used adaptive gradient optimizers include AdaGrad [34], Adadelta [35], and ADAM [36]. We restrict our attention to the ADAM optimizer in this chapter as, like SGD/SGD with Momentum, it is one of the most commonly used optimization methods for training deep neural networks.

Adaptive moment estimation (ADAM): ADAM is one of the most robust and effective optimization algorithms currently used to train deep neural networks and has been demonstrated to improve the convergence rate relative to SGD and SGD with Momentum when training deep neural networks in several applications. It dynamically estimates individual learning rates for different network parameters based on the first (i.e., the mean) and second (i.e., the uncentered variance) moments of their gradients concerning the loss function. Specifically, ADAM stores exponentially weighted moving averages of past gradients (first moment, also known as *momentum*) and past squared gradients (second moment) computed as

$$\mathbf{V}^t \leftarrow \beta_1 \mathbf{V}^{(t-1)} + (1-\beta_1)\mathbf{G}_t, \tag{16.37a}$$

$$\mathbf{U}^t \leftarrow \beta_2 \mathbf{U}^{(t-1)} + (1-\beta_2)\mathbf{G}_t^2, \tag{16.37b}$$

where $\mathbf{G}_t = \frac{1}{N_t}\sum_{i\in B_t} \frac{\partial f(\mathbf{x}_i, \boldsymbol{\Theta}^{(t-1)})}{\partial \boldsymbol{\Theta}}$ are mini-batch averaged gradients of the network's parameters and β_1 and β_2 are non-negative hyperparameters that control the decaying rates of these moving averages and the relative influence of the first and second moments of the gradients on the dynamic updates to the learning rates for each parameter. As the moving averages are initially zero, i.e., $\mathbf{V}^0 = \mathbf{U}^0 = 0$, the first and second moments are biased towards zero during the initial training phase. Hence, ADAM utilizes correction terms to counteract this bias, computed as

$$\hat{\mathbf{V}}^t = \frac{\mathbf{V}^t}{1-\beta_1}, \tag{16.38a}$$

$$\hat{\mathbf{U}}^t = \frac{\mathbf{U}^t}{1-\beta_2}. \tag{16.38b}$$

Finally, using these corrected terms for the moving averages, ADAM updates the network's parameters with dynamic learning rates using

$$\boldsymbol{\Theta}^t \leftarrow \boldsymbol{\Theta}^{(t-1)} - \frac{\eta \hat{\mathbf{V}}^t}{\sqrt{\hat{\mathbf{U}}^t} + \epsilon}, \tag{16.39}$$

where ϵ is a small value ($\sim 10^{-6}$ is typically used) included to provide numerical stability and as before η is the learning rate that is set initially but is now scaled/updated dynamically using $\hat{\mathbf{V}}^t$ and $\hat{\mathbf{U}}^t$ as training progresses to update the network's parameters $\boldsymbol{\Theta}$. Despite its success in various deep learning applications, ADAM is not without issues and there are scenarios where optimization can diverge (e.g., when the training set size is small) due to poor variance control (i.e., caused by the moving average of the squared gradients) [37]. These convergence issues have been attributed to the "short-term" history of past gradients used to guide parameter updates, and correspondingly extensions to ADAM to deal with such issues have been proposed (e.g., AMSGrad [37], Rectified ADAM [38], etc.), but these are beyond the scope of this textbook.

16.2.5 Normalization in deep learning

In addition to choosing appropriate loss functions and optimizers, another fundamental concept in deep learning, necessary for training deep neural networks effectively, is using *normalization* techniques. Deep neural networks are composed of several stacked layers of units that represent piecewise linear/non-linear transformations of the inputs (i.e., units refer to neurons in the case of MLPs and convolutions in the case of CNNs, discussed in Section 17.2). A sequential combination of piecewise linear/non-linear functions allows complex non-linear mappings between inputs/predictors and outputs/targets to be approximated by learning a hierarchy of features/representations of the input data. However, this powerful representation learning capacity of deep neural networks comes at the cost of increasing difficulty in training such networks due to the highly non-convex nature of the optimization involved. Therefore, strategies that can help improve/accelerate the convergence rate of training/optimizing deep neural networks are extremely useful and widely used. One such strategy is through the use of *normalization* techniques (others include improved optimization algorithms, as discussed in the previous sections, or the use of regularization strategies, which will be discussed in the next section), which may be loosely described as transformations that ensure data are propagated through a deep neural network and preserve certain desirable statistical properties that are beneficial for training the network. Normalization techniques commonly used to stabilize the training of deep neural networks and enhance their generalization capacity can be broadly categorized as methods that operate on the input data space, the *layer activation* space, or the *network weights/parameters* space. The most common types of normalization techniques used to train deep neural networks include BN, layer normalization, instance normalization, conditional BN, and conditional instance normalization.

Batch normalization (BN): BN is the most widely used normalization technique to train deep neural networks and has been demonstrated to improve training stability, optimization efficiency, and the generalization ability of trained networks. When training deep neural networks (e.g., MLP or CNN) the inputs being forward propagated through the network undergo several sequential piecewise linear/non-linear transformations, which can result in widely varying magnitudes for the resulting features, both within a layer (i.e., across units) and across layers, for the same inputs. BN was proposed [39] specifically to deal with this *drift*, also known as *internal covariance shift* in feature distributions, based on the hypothesis that left unmitigated, it could hamper the stability of network training and convergence.

BN is applied to a layer or all layers of a network and is typically applied before applying the activation function to the outputs of a layer. At each training iteration, BN normalizes the outputs of the units of a layer by subtracting their mean and dividing by their standard deviation, where both are estimated based on the current mini-batch (within each training epoch). This transformation to a mean-centered and unit variance representation of the input features at each layer is then scaled and shifted by learnable parameters to recover the degrees of freedom lost from the preceding transformation. BN for any given layer may be expressed as a function of the inputs $f(\mathbf{x})$ as

$$f(\mathbf{x}) = \gamma \cdot \frac{\mathbf{x} - \boldsymbol{\mu}_B}{\boldsymbol{\sigma}_B + \epsilon} + \beta, \tag{16.40}$$

where $\boldsymbol{\mu}_B$ represents the mean of the input feature representations to a specific layer, computed across the mini-batch B, $\boldsymbol{\sigma}_B$ represents the mini-batch variance, and γ and

β represent the learnable scaling and shift parameters, respectively. BN thus actively centers and scales the feature representations of each layer during training, preventing them from diverging and accelerating the convergence of deep neural networks. While other normalization techniques mentioned previously are also frequently used to train deep neural networks, their in-depth discussion is beyond the scope of this textbook. Interested readers are referred to [40] for a comprehensive survey on normalization techniques used in deep learning.

16.2.6 Regularization in deep learning

Regularization is an umbrella term used for several groups of techniques designed to improve the *generalization* of deep neural networks (and more broadly, all machine learning models) by reducing the tendency of models to overfit to the training data and improving their performance on unseen test data. In traditional machine learning literature, regularization was used to refer to a constraint or penalty term in the loss function optimized to train models. In the context of deep learning, however, regularization has taken on a much broader meaning. Regularization techniques for deep learning may be broadly categorized as follows: (i) loss-based regularization, (ii) data-based regularization, and (iii) network architecture-based regularization. The most common types of regularizers employed to train deep neural networks from each of these categories is discussed below. Readers are referred to the review presented by Kukacka et al. [41] for a comprehensive overview of regularization techniques employed to date in deep learning.

Loss-based regularization: As discussed previously, deep neural networks are trained by optimizing a suitable loss function with respect to the trainable parameters of the network, to update and learn the latter iteratively. The chosen loss function contains the prediction error (E) evaluated between the target variables and the network's predictions in the case of supervised learning or between the inputs/predictors themselves and the outputs of the network in the case of self-supervised learning. Typical loss functions used for any task (e.g., classification, regression, segmentation, etc.) in either learning paradigm (i.e., supervised or self-supervised learning) can be designed to include an additional term that penalizes the network's trainable parameters ($\mathbf{\Theta}$) in some way to control model complexity and encourage learning of simpler models. This, in turn, helps constrain the learning process to achieve a suitable trade-off between model bias and variance. In other words, achieving a balance between model overfitting and underfitting leads to improved model *generalization*. Two common such penalties/constraints used in loss functions when training deep neural networks are the L_1 and L_2 norm regularizers, also referred to as sparsity and weight decay regularizers, respectively. These forms of regularization are common in traditional machine learning models; for example, in the context of linear regression, they are employed for similar reasons to achieve a suitable bias–variance trade-off and referred to as *lasso* (L_1) and *ridge* (L_2) regression. L_1 norm regularization promotes sparsity by driving network weights to zero, effectively reducing overall network complexity. It is formulated as

$$l = e(f(\mathbf{X}), \mathbf{Y}) + \lambda \|\boldsymbol{\Theta}\|_1, \tag{16.41}$$

where $f(\mathbf{X})$ represents the network's predictions given some input training data $\mathbf{X}$, $\mathbf{Y}$ represents the corresponding target variables, and λ represents the regularization coefficient that controls the relative importance of the L_1 norm regularization term and the error function e on the updates to the network's parameters $\boldsymbol{\Theta}$ (i.e., λ is a hyperparameter that controls the bias–variance trade-off).

L_2 norm regularization, also known as weight decay, shrinks the magnitude of the network's parameters to control for overfitting and is formulated as

$$l = e(f(\mathbf{X}), \mathbf{Y}) + \lambda \|\boldsymbol{\Theta}\|_2, \tag{16.42}$$

where minimizing l promotes minimization of the magnitude of the network's parameters given by the second term, i.e., $\|\boldsymbol{\Theta}\|_2$. Both L_1 and L_2 norm regularizers are a form of inductive bias (discussed in Section 16.3) used to encode certain desirable properties into the training of the network to improve *generalization*. From a Bayesian perspective, these regularizers can be viewed as the *maximum a posteriori* solution to the network's parameters, based on some assumed *prior* belief/distribution over the network's parameters. Specifically, L_2 norm regularization corresponds to assuming a symmetric multivariate Gaussian prior distribution for the network's parameters, while L_1 norm regularization corresponds to assuming a Laplace prior distribution.

Data-based regularization: A key ingredient in the success of deep learning, especially for vision problems, has been the incorporation of extensive, often domain-specific data *augmentation* strategies. The performance of machine learning models in general, and correspondingly of deep neural networks, for any specific task, depends on the training data used. In addition to utilizing appropriate, high-quality training data, *augmenting* the data by using stochastic (random) transformations to alter the original training samples and create new versions of the same helps significantly expand the volume and intrinsic variability of data available to train deep neural networks. In other words, it is possible to beneficially *augment* the training data using suitable stochastic transformations. Training models using such augmented data effectively acts as a form of data-based regularization. As shown by Bishop et al. [42], transforming inputs/training data with the simple additive injection of Gaussian noise is equivalent to (for small noise amplitudes) employing an L_2 norm regularization term in the loss function used to train neural networks, and correspondingly, it can help improve the generalization of neural networks. More recently, in the context of deep learning, stochastic transformations of several other types have been used to transform the inputs/training samples (both predictors and targets, but we will focus here on the former) and to augment the training data. Commonly used stochastic transformation/augmentation techniques in the medical image analysis domain include rigid and/or elastic spatial transformations; intensity transformations, e.g., Gaussian smoothing, histogram normalization, intensity shifting/scaling, etc.;

cropping and padding transformations, e.g., "Cutout" [43], where fixed size regions of interest are cropped out at random from the input images, "CutMix" [44], where cropped out regions from one part of the image are swapped with other cropped out regions, or "RandomErasing" [45], where cropped regions of the images are replaced by noise, etc.; and several others. In practice, a typical augmentation pipeline utilizes several combinations of the aforementioned transformations, selected at random, to perform either online (during training) augmentation of each mini-batch or offline augmentation where several versions of each training sample (transformed in different ways) are generated. An effective method for transforming and augmenting both inputs and associated targets is the "MixUp" strategy [46], which essentially constructs synthetic training samples as linear interpolations of pairs of input samples in the training data and their corresponding targets. Overall, data-based regularization techniques are an effective means for improving the performance of deep neural networks, and they are especially popular in the computer vision and medical image analysis domains.

Network architecture-based regularization: Another category of regularization techniques often employed when training deep neural networks is to imbue the network architecture itself with certain properties that have a regularizing effect. We loosely refer to network architecture here as both specific layers/mathematical operations within networks and specific criteria used in the training process/algorithm. In this sense, the two most commonly adopted types of network architecture-based regularization are "Dropout" [47] and "early stopping." In addition to these, other network architectural features such as parameter sharing between layers or training paradigms such as multitask learning also facilitate regularized training of deep neural networks.

Dropout [47] is a simple yet effective regularizer used frequently to train deep neural networks. This regularizer is also stochastic in nature and works by randomly switching off neurons/units in a network during each forward pass whilst training the network. This may be viewed as a way of reducing model complexity by artificially, and at random, setting the outputs of individual network units to zero. This prevents network units from coadapting too much and overfitting the training data. The probability of dropping each unit in a network is estimated using the Bernoulli distribution at each training step, making the process stochastic. This process may also be viewed as training an exponential number of "thinned" networks, each with a distinct and random configuration of network connections arising from random dropout of units during training. The effect of training an exponential number of thinned networks in this way is that, following training, the effect of averaging predictions from all these thinned networks may be approximated during inference by simply using a single "unthinned"/original network to make predictions for any input test samples. In other words, Dropout leverages the well-established strategy of reducing model variance (and hence overfitting) through averaging several models' predictions (a strategy commonly employed in ensemble approaches by "bagging," such as random forests). Given an MLP with $l = [1 \cdots L]$ hidden layers, where the input to each lth layer is denoted $\mathbf{X}_l$, the outputs of the layer before and after activation may be expressed as

$$y_l^i = \mathbf{w}_l^i \mathbf{x}_l + b_l^i, \tag{16.43a}$$

$$x_{l+1}^i = f(y_l^i), \tag{16.43b}$$

where y_l^i represents the linear transformation of the inputs $\mathbf{x}_l$ based on the trainable weights ($\mathbf{w}_l^i$) and bias (b_l^i) associated with hidden unit i, and the output of the hidden unit (x_{l+1}^i), which forms a part of the input to the subsequent $(l+1)$-th layer in the MLP, is given by the application of any activation function (e.g., `Sigmoid`) to the pre-activation output y_l^i. With Dropout, the same feedforward operation in the MLP may be expressed as

$$\mathbf{r}_l \sim Bernoulli(\mathrm{p}), \tag{16.44a}$$

$$\hat{\mathbf{x}}_l = \mathbf{r}_l \odot \mathbf{x}_l, \tag{16.44b}$$

$$\hat{y}_l^i = \mathbf{w}_l^i \hat{\mathbf{x}}_l + b_l^i, \tag{16.44c}$$

$$\hat{x}_{l+1}^i = f(\hat{y}_l^i), \tag{16.44d}$$

where $\mathbf{r}_l$ is a vector of independent Bernoulli random variables, each having a probability p of being 1, which, when multiplied elementwise (denoted $\odot$) with the outputs of the $(l-1)$th layer (denoted as $\mathbf{x}_l$ and referred to here as the inputs to the lth layer) results in “thinned” inputs $\hat{\mathbf{x}}_l$. The thinned outputs $\hat{x}_{l+1}^i$ are then used as inputs to the subsequent $(l+1)$th layer. Dropout can be added to any number of layers within a neural network and is often applied to every layer. Application of Dropout results in sampling of a subnetwork from the original network architecture during each training step and back-propagating gradients through the subnetwork. The probability p for the Bernoulli random variables is a user-specified hyperparameter that needs to be tuned empirically (e.g., through several cross-validation experiments and grid search optimization to identify the most suitable value). During inference, the parameters of each lth layer are scaled as $\mathbf{W}_l^{test} = \mathrm{p}\mathbf{W}_l$, resulting in approximately averaging predictions from the various trained subnetworks.

Another commonly used form of regularizing deep neural networks is “early stopping.” Unlike other approaches discussed thus far, this technique is a training strategy rather than a mathematical operation that directly constrains the parameters of the network. Specifically, early stopping constrains the number of training epochs based on a suitable termination criterion for training, which is typically defined by monitoring the validation loss/error across all training epochs. Often, a “patience” hyperparameter is defined empirically to specify the number of epochs to monitor the validation loss and see whether it decreases by at least some predefined amount (also a hyperparameter set empirically) before terminating training. Recent work [48] has revealed that in the presence of significant label noise (i.e., corrupted or mislabeled targets) neural networks tend to represent the cleanly labeled data first and subsequently fit to the mislabeled data. Therefore, effective use of early stopping helps prevent overfitting to training data, which is especially useful when training with noisy labels.

16.3 Inductive bias, invariance, and equivariance

The success of deep neural networks in specific supervised learning tasks and their excellent in-distribution generalization performance may at least in part be attributed to their intrinsic *inductive biases*. *Inductive bias* refers to a set of preferences, rules, priors or assumptions that are defined over the space of all functions in order to constrain the search space for learning algorithms and enable the latter to generalize from the training data, to unseen data, for a specific task. The no-free-lunch theorem in machine learning [49] states that no completely general-purpose learning algorithm exists and for any learning algorithm to generalize well to unseen data, a set of constraints over the space of all functions is required. In other words, *inductive biases* encourage learning algorithms to prioritize solutions that preserve certain desirable properties, for any given task.

Deep learning algorithms have been designed to exploit several *inductive biases* to enable efficient learning of representations and facilitate generalization to unseen data. The current types of *inductive biases* used in deep learning are encoded into the learning process in several ways, such as network architectural constraints (e.g., parameter sharing in convolution operations in CNNs, discussed in more detail in Chapter 17), choice of loss functions and associated regularization constraints (this refers to loss-based regularization approaches, refer to Section 16.2.6 for more details), choice of optimization algorithm and its implicit effects/constraints (refer to Section 16.2.4), choice of prior distributions over weights/parameters in Bayesian neural networks, and the use of transfer learning for fully supervised learning tasks through self-supervised or fully supervised pre-training (discussed in more detail in Chapter 17), to name a few. The aforementioned *inductive biases* imbue their corresponding neural networks with certain favorable properties that improve generalization to unseen data, for example, convolution operations used in CNNs for automatic feature learning/extraction from the input data involve processing each input sample with the same convolution kernel/filter (i.e., weights/parameters are shared across the convolution kernels that parse each entire input sample). This yields the property of group equivariance in a spatial sense, for example, in the case of image data, convolution operations enable learning of translation-equivariant features. Equivariance is the property of predicting the output of a function given some transformation of the input data. For example, the feature map resulting from convolving an input image with a kernel/filter in a CNN is translation-equivariant. This means that if the input image is translated in some way, the corresponding feature map resulting from the convolution operation is also translated by the same amount. Equivariance differs from invariance in that the latter is a property of a function that means its outputs are unchanged under certain transformations of the input data. For example, a desirable property for object recognition systems in medical imaging might be invariance to changes in intensity distributions resulting from variability in imaging system/scanner characteristics and acquisition protocols. Equivariance is a particularly powerful property of CNNs as it enables learning of generalizable features that may be useful for various input data that exhibit variable spatial positions for objects/features of interest for a specific task. For example, convolutional feature maps learned regarding the specific characteristics of a brain tumor in one patient's image are likely to be useful for recognizing brain tumors in another patient's images regardless of the position of the tumor.

There are several lines of thought surrounding the future of artificial intelligence and deep learning, particularly of developing artificial general intelligence that closes the gap to human cognitive abilities. Although deep learning has yielded remarkable progress in a variety of highly specific tasks (e.g., semantic segmentation or object recognition in images) by exploiting certain *inductive biases* (such as those discussed above) for learning, in order to take the next step towards strong out-of-distribution generalization and transfer learning to new tasks with limited sample complexity (i.e., few training samples required to generalize well to new tasks), growth in diversity and volume of data and computational resources alone is unlikely to be sufficient without formulation of new and instructive *inductive biases* [50].

16.4 Recapitulation

The take-home messages from this chapter are:

- Univariate and multivariate linear regression models can be interpreted as the simplest type of neural networks. The former is equivalent to a single-neuron network with a linear activation function, while the latter is equivalent to an MLP with just two layers, an input layer and an output layer also with linear activation functions applied to each of the output neurons.
- Efficient and scalable training of all modern neural networks is enabled by the error back-propagation algorithm. Training progresses iteratively alternating between the forward and backward pass (also referred to as back-propagation) and may be implemented using various gradient-based optimization techniques. Error back-propagation involves calculating the gradients of the loss function concerning the weights in all layers of the network using the chain rule of differentiation and subsequently updating the weights in the direction that minimizes the loss function.
- Neural networks with one or more hidden layers, given a sufficient number of neurons/units in the hidden layer(s), can approximate any function, i.e., neural networks with non-linear activation functions are *universal function approximators*.
- Saturating non-linear activation functions such as `Sigmoid` and `tanh` lead to vanishing gradients when used in deep neural networks that are trained by back-propagation. This makes it difficult and at times impossible to train deep neural networks effectively with such activation functions. The `ReLU` activation function, which is piecewise linear, is non-saturating for positive-valued inputs and helps address the vanishing gradient problem when used in deep neural networks instead of saturating activation functions. `ReLU` remains the most widely used type of activation function for deep neural networks.

- A limitation of the `ReLU` activation function is that it can lead to dead neurons/units in neural networks, wherein neurons/units never activate (i.e., their outputs are always zero). Leaky `ReLU` and parametric `ReLU` activation functions are suitable alternatives to address this limitation of `ReLU`, as they permit negative inputs without saturating the outputs to zero.
- Neural networks are trained to approximate a mapping between some given inputs and their respective targets/outputs, by optimizing a suitable loss function with respect to the trainable parameters of the network. Loss functions are chosen based on the nature of the learning task, characteristics of the training data, the manner in which target variables are represented and based on desired constraints for the optimization process.
- Label noise or noisy targets are common problems encountered in medical image classification and regression problems. They arise due to incorrect interpretation of images, errors in mining radiology reports to extract labels associated with patients' images, or errors in the image formation and reconstruction process itself leading to propagation of errors in downstream tasks. Commonly used loss functions for classification and regression with deep neural networks such as cross-entropy and MSE are sensitive to noise/outliers. Alternative robust loss functions should be considered under such conditions.
- GANs are a type of deep generative model comprising two competing subnetworks referred to as the generator and discriminator. The generator iteratively synthesizes data that are as close to the real observed data as possible, in an attempt to fool the discriminator. The discriminator meanwhile tries to distinguish between the generated and real data. The two subnetworks are trained to compete against each other in a two-player zero-sum game.
- Optimization of deep neural networks is a highly non-convex problem as the loss function typically has several local minima. Mini-batch training leads to quicker and more stable convergence than online learning (i.e., training with a batch size of one) due to the averaging of gradients across randomly selected training samples in each mini-batch, which reduces the variance in the gradients computed to update network weights. The selection of batch size is an important hyperparameter that must be tuned to each problem/dataset as too small a batch can lead to network weights being trapped in local minima, whereas too large a batch size can lead to overfitting of the training data. While mini-batch SGD has been used effectively to train deep neural networks in several studies, its limiting factor is its sensitivity to the chosen learning rate. Consequently, optimization algorithms such as SGD with Momentum or ADAM, which utilize the history of the weight gradients from previous training iterations to inform weight updates, are often more desirable.
- Regularization in deep learning refers to any technique that helps improve the generalization of deep neural networks by reducing their tendency to overfit to training data. Regularization methods may be broadly grouped into loss-based, data-based, or network architecture-based approaches. The underlying common

thread across approaches in all three groups is the injection of noise/stochasticity to the training process.

16.5 Further reading

1. Interactive online book on deep learning (https://d2l.ai/index.html).
2. Review of loss functions for deep learning [51].
3. Review of GANs for medical image analysis [32].
4. Reviews of regularization approaches in deep learning [41,52].

16.6 Exercises

1. How do you determine the optimal number of hidden layers and neurons in an MLP for a given problem/training dataset?
   Answer: The numbers of hidden layers and neurons to use in each layer are hyperparameters that must be tuned by trial-and-error. This is typically done by conducting several experiments using a combination of grid search and cross-validation.
2. How do you train an MLP when your training dataset is large and high-dimensional (e.g., hundreds of 3D MRI volumes)?
   Answer: MLPs require fixed-size vectors as inputs and their constituent layers are fully connected (i.e., there is no weight sharing between neurons or layers). Naive use of MLPs for analyzing medical images can result in a very large number of network weights, which may be computationally intractable due to memory constraints and can lead to severe overfitting. Similarly, given a very large dataset, full-batch training of MLPs may be infeasible, also due to memory constraints. Hence, it is often necessary to reduce the dimensionality of the training data through automatic or handcrafted feature extraction and to use mini-batch learning to train MLPs with large and high-dimensional data.
3. Why is it difficult and at times infeasible to train deep neural networks with `Sigmoid` or `tanh` activation functions applied to every layer in the network?
   The bounded nature of the outputs of such activation functions results in gradients that are very close to zero. Hence, as multiple small gradients are multiplied successively during back-propagation, the errors from higher layers are decreased significantly in magnitude and may not reach lower layers of the network as the gradient vanishes (i.e., becomes too small). This effect is known as vanishing gradients and prevents effective training of deep neural networks using such saturating activation functions.
4. How does SGD with Momentum help reduce the sensitivity of the SGD optimizer to the presence of noisy gradients?
   Answer: SGD with Momentum exploits the benefit of further gradient averaging to reduce the variance resulting from noisy gradients, beyond the averaging

of gradients across training samples in a mini-batch employed by SGD. Specifically, SGD with Momentum computes a weighted average of the gradient from the previous training iterations (i.e., calculated for previous mini-batches) with the current iteration's average gradient estimate (i.e., averaged across samples of the current mini-batch). This weighted average, which is used to update the network's parameters, thus takes into account the past gradients computed for previous mini-batches, thereby reducing the sensitivity of the optimizer to noisy gradients that may be calculated for any given mini-batch during training.

5. How do you choose an appropriate learning rate for optimization algorithms that train deep neural networks?
Answer: Choosing appropriate learning rates for optimizers when training deep neural networks is critical to ensure stable training and convergence to a reasonable local minimum. The learning rate is a hyperparameter which requires tuning through trial-and-error by conducting several experiments, typically using a combination of grid search and cross-validation, which can be time consuming. An alternative approach is to use adaptive learning rate optimizers such as ADAM, which automatically adjust the learning rate based on the history of past gradients.

References

[1] Cybenko G., Approximation by superpositions of a sigmoidal function, Mathematics of Control, Signals and Systems 2 (4) (1989) 303–314.
[2] Hornik K., Stinchcombe M.B., White H.L., Multilayer feedforward networks are universal approximators, Neural Networks 2 (1989) 359–366.
[3] Calin O., Deep Learning Architectures: A Mathematical Approach, Springer Series in the Data Sciences, Springer, 2020.
[4] Rumelhart D.E., Hinton G.E., Williams R.J., Learning representations by back-propagating errors, Nature 323 (6088) (1986) 533–536.
[5] Basodi S., Ji C., Zhang H., Pan Y., Gradient amplification: an efficient way to train deep neural networks, Big Data Mining and Analytics 3 (3) (2020) 196–207, https://doi.org/10.26599/BDMA.2020.9020004.
[6] Nair V., Hinton G.E., Rectified linear units improve restricted Boltzmann machines, in: ICML, 2010.
[7] Glorot X., Bordes A., Bengio Y., Deep sparse rectifier neural networks, in: Proceedings of the Fourteenth International Conference on Artificial Intelligence and Statistics. JMLR Workshop and Conference Proceedings, 2011, pp. 315–323.
[8] Maas A.L., Hannun A.Y., Ng A.Y., Rectifier nonlinearities improve neural network acoustic models, in: Proc. ICML, vol. 30. 1, Atlanta, Georgia, USA, 2013, p. 3.
[9] He K., Zhang X., Ren S., Sun J., Delving deep into rectifiers: surpassing human-level performance on ImageNet classification, in: Proceedings of the IEEE International Conference on Computer Vision, 2015, pp. 1026–1034.
[10] Xu B., Wang N., Chen T., Li M., Empirical evaluation of rectified activations in convolutional network, arXiv preprint, arXiv:1505.00853, 2015.
[11] Tang Z., Luo L., Peng H., Li S., A joint residual network with paired ReLUs activation for image super-resolution, Neurocomputing 273 (2018) 37–46.

[12] Clevert D.-A., Unterthiner T., Hochreiter S., Fast and accurate deep network learning by exponential linear units (ELUs), arXiv preprint, arXiv:1511.07289, 2015.
[13] Goodfellow I., Warde-Farley D., Mirza M., Courville A., Bengio Y., Maxout networks, in: International Conference on Machine Learning, in: PMLR, 2013, pp. 1319–1327.
[14] Klambauer G., Unterthiner T., Mayr A., Hochreiter S., Self-normalizing neural networks, Advances in Neural Information Processing Systems 30 (2017).
[15] Dugas C., Bengio Y., Belisle F., Nadeau C., Garcia R., Incorporating second-order functional knowledge for better option pricing, Advances in Neural Information Processing Systems 13 (2000).
[16] Ciuparu A., Nagy-Dăbâcan A., Mureşan R.C., Soft++, a multi-parametric non-saturating non-linearity that improves convergence in deep neural architectures, Neurocomputing 384 (2020) 376–388.
[17] Vargas V.M., Gutierrez P.A., Barbero-Gomez J., Hervas-Martinez C., Activation functions for convolutional neural networks: proposals and experimental study, IEEE Transactions on Neural Networks and Learning Systems 34 (2021) 1478–1488.
[18] Ramachandran P., Zoph B., Le Q.V., Searching for activation functions, arXiv preprint, arXiv:1710.05941, 2017.
[19] Deng J., Dong W., Socher R., Li L., Li K., Fei-Fei L., ImageNet: a large-scale hierarchical image database, in: 2009 IEEE Conference on Computer Vision and Pattern Recognition, IEEE, 2009, pp. 248–255.
[20] Dubey S.R., Singh S.K., Chaudhuri B.B., Activation functions in deep learning: a comprehensive survey and benchmark, Neurocomputing 503 (2022) 92–108.
[21] Anil R., Gupta V., Koren T., Regan K., Singer Y., Scalable second order optimization for deep learning, arXiv preprint, arXiv:2002.09018, 2020.
[22] Osawa K., Tsuji Y., Ueno Y., Naruse A., Yokota R., Matsuoka S., Large-scale distributed second-order optimization using Kronecker-factored approximate curvature for deep convolutional neural networks, in: Proceedings of the IEEE/CVF Conference on Computer Vision and Pattern Recognition, 2019, pp. 12359–12367.
[23] Irvin J., Rajpurkar P., Ko M., Yu Y., Ciurea-Ilcus S., Chute C., Marklund H., Haghgoo B., Ball R., Shpanskaya K., Seekins J., Mong D.A., Halabi S.S., Sandberg J.K., Jones R., Larson D.B., Langlotz C.P., Patel B.N., Lungren M.P., Ng A.Y., CheXpert: a large chest radiograph dataset with uncertainty labels and expert comparison, Proceedings of the AAAI Conference on Artificial Intelligence 33 (01) (2019) 590–597.
[24] Ghosh A., Kumar H., Sastry P.S., Robust loss functions under label noise for deep neural networks, Proceedings of the AAAI Conference on Artificial Intelligence 31 (1) (2017).
[25] Karimi D., Dou H., Warfield S.K., Gholipour A., Deep learning with noisy labels: exploring techniques and remedies in medical image analysis, Medical Image Analysis 65 (2020) 101759.
[26] Zhang Z., Sabuncu M., Generalized cross entropy loss for training deep neural networks with noisy labels, Advances in Neural Information Processing Systems 31 (2018).
[27] Lin T.-Y., Goyal P., Girshick R., He K., Dollár P., Focal loss for dense object detection, in: Proceedings of the IEEE International Conference on Computer Vision, 2017, pp. 2980–2988.
[28] Wang Y., Ma X., Chen Z., Luo Y., Yi J., Bailey J., Symmetric cross entropy for robust learning with noisy labels, in: Proceedings of the IEEE/CVF International Conference on Computer Vision, 2019, pp. 322–330.
[29] Huber P.J., A robust version of the probability ratio test, in: The Annals of Mathematical Statistics, 1965, pp. 1753–1758.

[30] Goodfellow I.J., Abadie J.P., Mirza M., Xu B., Farley D.W., Ozair S., Courville A., Bengio Y., Generative adversarial networks, arXiv preprint, arXiv:1406.2661, 2014.

[31] Goodfellow I.J., Abadie J.P., Mirza M., Xu B., Farley D.W., Ozair S., Courville A., Bengio Y., Generative adversarial networks, Communications of the ACM 63 (11) (2020) 139–144.

[32] Kazeminia S., Baur C., Kuijper A., van Ginneken B., Navab N., Albarqouni S., Mukhopadhyay A., GANs for medical image analysis, Artificial Intelligence in Medicine 109 (2020) 101938.

[33] Arjovsky M., Chintala S., Bottou L., Wasserstein generative adversarial networks, in: International Conference on Machine Learning, in: PMLR, 2017, pp. 214–223.

[34] Duchi J., Hazan E., Singer Y., Adaptive subgradient methods for online learning and stochastic optimization, Journal of Machine Learning Research 12 (7) (2011).

[35] Zeiler M.D., Adadelta: an adaptive learning rate method, arXiv preprint, arXiv:1212.5701, 2012.

[36] Kingma D.P., Ba J., Adam: a method for stochastic optimization, arXiv preprint, arXiv:1412.6980, 2014.

[37] Reddi S.J., Kale S., Kumar S., On the convergence of Adam and beyond, arXiv preprint, arXiv:1904.09237, 2019.

[38] Liu L., Jiang H., He P., Chen W., Liu X., Gao J., Han J., On the variance of the adaptive learning rate and beyond, arXiv preprint, arXiv:1908.03265, 2019.

[39] Ioffe S., Szegedy C., Batch normalization: accelerating deep network training by reducing internal covariate shift, in: International Conference on Machine Learning, in: PMLR, 2015, pp. 448–456.

[40] Huang L., Qin J., Zhou Y., Zhu F., Liu L., Shao L., Normalization techniques in training DNNs: methodology, analysis and application, in: IEEE Transactions on Pattern Analysis and Machine Intelligence, 2023.

[41] Kukačka J., Golkov V., Cremers D., Regularization for deep learning: a taxonomy, arXiv preprint, arXiv:1710.10686, 2017.

[42] Bishop C.M., Training with noise is equivalent to Tikhonov regularization, Neural Computation 7 (1) (1995) 108–116.

[43] DeVries T., Taylor G.W., Improved regularization of convolutional neural networks with cutout, arXiv preprint, arXiv:1708.04552, 2017.

[44] Yun S., Han D., Oh S.J., Chun S., Choe J., Yoo Y., CutMix: regularization strategy to train strong classifiers with localizable features, in: Proceedings of the IEEE/CVF International Conference on Computer Vision, 2019, pp. 6023–6032.

[45] Zhong Z., Zheng L., Kang G., Li S., Yang Y., Random erasing data augmentation, Proceedings of the AAAI Conference on Artificial Intelligence 34 (07) (2020) 13001–13008.

[46] Zhang H., Cisse M., Dauphin Y.N., Paz D.L., mixup: beyond empirical risk minimization, arXiv preprint, arXiv:1710.09412, 2017.

[47] Srivastava N., Hinton G., Krizhevsky A., Sutskever I., Salakhutdinov R., Dropout: a simple way to prevent neural networks from overfitting, Journal of Machine Learning Research 15 (1) (2014) 1929–1958.

[48] Rolnick D., Veit A., Belongie S., Shavit N., Deep learning is robust to massive label noise, arXiv preprint, arXiv:1705.10694, 2017.

[49] Wolpert D.H., Macready W.G., No free lunch theorems for search, Tech. Rep. Citeseer, 1995.

[50] Goyal A., Bengio Y., Inductive biases for deep learning of higher-level cognition, Proceedings of the Royal Society A 478 (2266) (2022) 20210068.

[51] Ciampiconi L., Elwood A., Leonardi M., Mohamed A., Rozza A., A survey and taxonomy of loss functions in machine learning, arXiv preprint, arXiv:2301.05579, 2023.

[52] Dos Santos C.F.G., Papa J.P., Avoiding overfitting: a survey on regularization methods for convolutional neural networks, ACM Computing Surveys (CSUR) 54 (10s) (2022) 1–25.

CHAPTER

17 Deep learning for vision and representation learning

Arezoo Zakeri[a], Yan Xia[a], Nishant Ravikumar[b], and Alejandro F. Frangi[a,c,d,e]

[a]*Division of Informatics, Imaging, and Data Sciences, School of Health Sciences, Faculty of Biology, Medicine, and Health, The University of Manchester, Manchester, United Kingdom*

[b]*Centre for Computational Imaging and Simulation Technologies in Biomedicine (CISTIB), Schools of Computing and Medicine, University of Leeds, Leeds, United Kingdom*

[c]*Department of Computer Science, School of Engineering, Faculty of Science and Engineering, The University of Manchester, Manchester, United Kingdom*

[d]*Department of Electrical Engineering (ESAT), KU Leuven, Leuven, Belgium*

[e]*Department of Cardiovascular Sciences, KU Leuven, Leuven, Belgium*

Learning points

- Convolutional neural networks
- Convolutional arithmetic. Forward and backwards passes
- Pooling layers and dilated convolutions
- Skip connections
- Deep representation learning
- Autoencoder. Denoising and variational autoencoders
- Contrastive learning

17.1 Introduction

Machine learning (ML) for vision problems (e.g., object detection or semantic segmentation) traditionally relied on domain expertise and handcrafted features to transform the raw data (e.g., images or video frames/sequences) into a condensed set of representations to train models of interest to solve specific tasks (e.g., classification or regression). This significantly limited the generalizability of developed ML systems, especially in the medical image analysis domain, as handcrafted features were seldom easily transferable across imaging applications. Deep learning models are inherently representation learning methods that automatically learn features from raw input data, so that the learned features are tailored to solve the task(s) of interest. By constructing a sequence of non-linear transformations, deep neural networks can learn multiple levels or hierarchies of representations of the raw input data, driven by

Medical Image Analysis. https://doi.org/10.1016/B978-0-12-813657-7.00044-3

optimizing a suitable *loss* function to solve the task of interest. While several types of deep neural networks have been developed in recent years, the tremendous success of deep learning in solving vision problems may be attributed to a specific type, namely, deep convolutional neural networks (CNNs). Deep CNNs were at the heart of the deep learning revolution, particularly starting with Alexnet [1] winning the ImageNet challenge in 2012. Since then, CNNs have dominated the fields of image classification, object detection, image segmentation, and many other tasks involving images and videos.

17.2 Convolutional neural networks

The convolution operation effectively describes transformations that apply the same linear transformation of a small, local region throughout the whole input. Unlike fully connected layers, where each neuron has a separate weight vector, CNNs share weights. Rather than learning a different set of parameters for every location, they learn only one set. This reduction in the overall number of trainable weights introduces sparse interactions or connectivity in CNNs (only a small number of input units contribute to a particular output unit). Thus, convolution is vastly superior to dense matrix multiplication regarding memory requirements and statistical efficiency. Discrete convolutions perform a linear operation by sliding a kernel across the input feature map. Fig. 17.1 presents a simple example of a 2D discrete convolution [2]. The light blue grid is called the input feature map. For simplicity, a single input feature map is shown here, while in practical applications, multiple feature maps are stacked one onto another, showing different channels of the input image. The shaded area presents the kernel that moves over the input. The output is calculated at each location by computing the product between each kernel element and its overlapping input value. Then, the results are added up to obtain the output at the current place. The kernel can only move to locations where it entirely fits within the image. This example depicts a unit stride convolution in which the kernel slides in one-pixel sliding steps. However, in strided convolutions, the strides are chosen to be more than one. This results in downsampling the output of the convolution compared to the case of unit stride convolution. The reason is that the kernel skips some locations in the image, reducing the computational cost at the expense of coarser extraction of feature maps. One can also define a different stride for each direction of sliding.

Zero-padding is a technique that allows the management of the output size in a convolutional layer and prevents the output feature maps from being shrunk if it is not required. It performs by padding the input's outer edges with zeroes. Fig. 17.2 illustrates the convolution of a 3×3 kernel over a 5×5 input using padding 1×1 and 2×2 strides. The output has the same size as in Fig. 17.1, while a non-unit stride is utilized.

A distinct kernel must be convolved with each input feature map in practical applications with multiple parallel input channels. Finally, the resulting output feature maps should be summed up to form the output corresponding to the utilized kernel.

3 3 2 1 0
0 0 1 3 1
3 1 2 2 3
2 0 0 2 2
2 0 0 0 1

12.0 12.0 17.0
10.0 17.0 19.0
9.0 6.0 14.0

3 3 2 1 0
0 0 1 3 1
3 1 2 2 3
2 0 0 2 2
2 0 0 0 1

12.0 12.0 17.0
10.0 17.0 19.0
9.0 6.0 14.0

3 3 2 1 0
0 0 1 3 1
3 1 2 2 3
2 0 0 2 2
2 0 0 0 1

12.0 12.0 17.0
10.0 17.0 19.0
9.0 6.0 14.0

3 3 2 1 0
0 0 1 3 1
3 1 2 2 3
2 0 0 2 2
2 0 0 0 1

12.0 12.0 17.0
10.0 17.0 19.0
9.0 6.0 14.0

3 3 2 1 0
0 0 1 3 1
3 1 2 2 3
2 0 0 2 2
2 0 0 0 1

12.0 12.0 17.0
10.0 17.0 19.0
9.0 6.0 14.0

3 3 2 1 0
0 0 1 3 1
3 1 2 2 3
2 0 0 2 2
2 0 0 0 1

12.0 12.0 17.0
10.0 17.0 19.0
9.0 6.0 14.0

3 3 2 1 0
0 0 1 3 1
3 1 2 2 3
2 0 0 2 2
2 0 0 0 1

12.0 12.0 17.0
10.0 17.0 19.0
9.0 6.0 14.0

3 3 2 1 0
0 0 1 3 1
3 1 2 2 3
2 0 0 2 2
2 0 0 0 1

12.0 12.0 17.0
10.0 17.0 19.0
9.0 6.0 14.0

3 3 2 1 0
0 0 1 3 1
3 1 2 2 3
2 0 0 2 2
2 0 0 0 1

12.0 12.0 17.0
10.0 17.0 19.0
9.0 6.0 14.0

FIGURE 17.1 Discrete convolution.

Computing a 2D discrete convolution. A 3×3 kernel (shaded area) is applied to a 5×5 input image (in blue), resulting in a 3×3 output feature map (in green) [2].

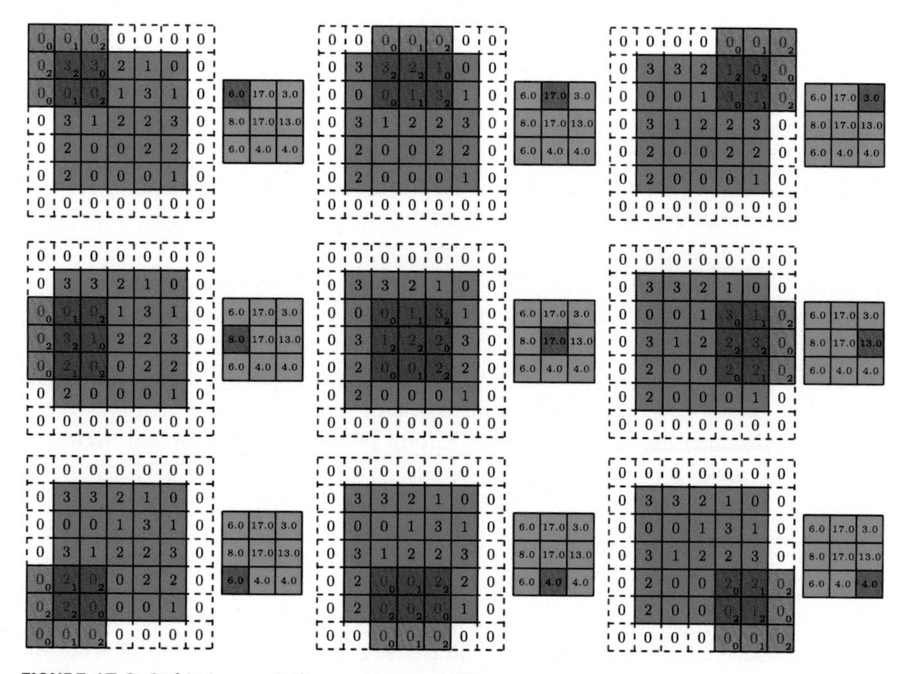

FIGURE 17.2 Strided convolution and zero-padding.

Computing convolution when a 3×3 kernel is applied to a 5×5 input image padded with a 1×1 border of zeroes and 2×2 strides, resulting in a 3×3 output feature map [2].

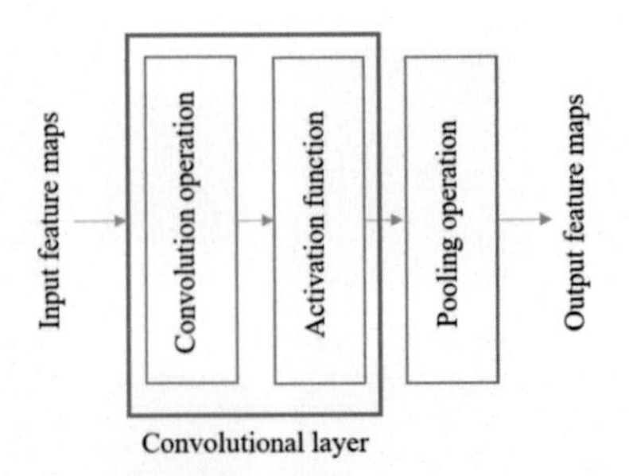

FIGURE 17.3 Components of a typical CNN layer.

The main stage in a typical CNN layer is performing the convolution operation, followed by a non-linear process using an activation function and a pooling stage to make the output representation approximately invariant to small translations of the input and provide down-sampled output feature maps.

This is because when we use a single kernel, we extract one kind of feature in different spatial locations. However, we desire each layer of the convolution network to extract many kinds of features at many locations from the input. Hence, for image data types, the input and output feature maps of a convolution layer are defined by 3D tensors; one index describes the different channels, and two indices are for the spatial coordinates of each channel. A typical convolutional layer of a CNN is shown in Fig. 17.3. First, the layer runs several convolutions in parallel to create a set of linear activations. An activation function is included in the second stage to give the neural network non-linear capabilities. Various activation functions affect the model's ability to fit non-linear functions differently (see Section 16.2.2). A spatial pooling operation usually follows a convolutional layer over a local neighborhood of outputs to create equivariance to small translations of the input. In addition, the pooling operation results in more compact representations and improved robustness to noise and irrelevant details (Section 17.2.3). A deep CNN consists of many convolutional layers to increase the network's capacity to learn a large training dataset and efficiently extract complex structures from inputs. Depending on the application, e.g., for classification purposes, the convolutional layers are followed by dense or fully connected layers that classify each pattern according to the previously extracted features. Very deep networks such as Visual Geometry Group (VGG) [3] and residual networks (ResNets) [4] are of this kind.

17.2.1 Convolution arithmetic

This section explores some special cases of convolution operation settings for controlling the output size and determining the relation between the convolutional layer parameters. The output width o in a convolution along a particular axis j depends on the input size i, kernel size k, stride selection s, and zero padding p, only along axis j.

- Valid convolution: A particular case of having no zero-padding is called *valid* convolution in deep learning literature [5]. In this case, the size of the output decreases at each layer. One can determine the output width by counting the number of possible positions for the kernel sliding with unit strides ($s = 1$) or non-unit strides ($s > 1$) across the input from the initial location on the top left to the right side. The number of kernel positions along one axis (output width) will be $o = i - k + 1$ if strides are unit and $o = \lfloor \frac{i-k}{s} \rfloor + 1$ for $s > 1$.

- Same convolution: A special case of having $o = i$ is obtained by padding enough zeroes to the input. If k is odd ($k = 2n + 1$ and $n \in \mathbb{N}$), using $s = 1$ and $p = \lfloor k/2 \rfloor = n$, it can be proved that $o = i$. That is because by padding p zeroes, the effective size of the input will be $i + 2p$, and therefore $o = (i + 2p) - k + 1 = i + 2n - 2n - 1 + 1 = i$.
- Full convolution: In this case, an appropriate zero-padding is applied to ensure that the output size will be larger than the input size. In a particular case where each input pixel is considered k times in each direction, full convolution occurs by $p = k - 1$ and $s = 1$, which results in $o = (i + 2p) - k + 1 = i + 2k - 2 - k + 1 = i + k - 1$.
- Having zero-padding and $s > 1$: In this general case of having padding p and $s > 1$, the output size can be determined by counting the number of possible positions for the kernel while sliding with strides $s > 1$ over an input with the effective size of $i + 2p$, which is

$$o = \lfloor \frac{(i + 2p) - k}{s} \rfloor + 1. \tag{17.1}$$

17.2.2 Forward and backward passes

In the convolution operation, the forward pass entails applying overlapping filter (kernel) weights to the input, multiplying them, and adding the results to obtain the output. The convolution between an image $\mathbf{I}$ of an arbitrary size and C channels with the kernel $\mathbf{K}$ of size $(K1 \times K2)$ is defined by

$$[\mathbf{I} * \mathbf{K}]_{i,j} = \sum_{c=0}^{C-1} \sum_{m=0}^{K1-1} \sum_{n=0}^{K2-1} [\mathbf{K}]_{m,n,c} . [\mathbf{I}]_{i-m,j-n,c} . \tag{17.2}$$

Many machine learning libraries implement the *cross-correlation* function and still call it convolution [5], as defined by

$$[\mathbf{I} \otimes \mathbf{K}]_{i,j} = \sum_{c=0}^{C-1} \sum_{m=0}^{K1-1} \sum_{n=0}^{K2-1} [\mathbf{K}]_{m,n,c} . [\mathbf{I}]_{i+m,j+n,c} . \tag{17.3}$$

Comparison of Eqs. (17.2) and (17.3) shows that the convolution is equivalent to the cross-correlation function but with a flipped kernel. Here, we consider the flipped kernel and use Eq. (17.3) for the convolution operation to explain forward and backward equations in the convolutional layers.

The convolutional layer transforms the input feature maps, which are the output of the previous layer $\mathbf{O}^{l-1}$ with size $H \times W \times C$, into the output $\mathbf{O}^{l}$ with size

$H' \times W' \times F$. Hence, this layer can be defined by having F filters with weights $[\mathbf{W}]^f_{i,j,c}$ and biases b^f. We generally add the bias term to the output of the linear convolution operation before applying the activation function. It is usual to have one bias per output channel and to share it among all locations on each convolution map. In the forward pass, considering the kernel sizes of $K1 \times K2$, the convolutional layer forwards the $\mathbf{O}^{l-1}$ first into the variable $\mathbf{Z}$, and then applying the activation function $f(.)$ results in $\mathbf{O}^l$ as

$$[\mathbf{Z}]_{i,j,f} = \sum_{c=0}^{C-1} \sum_{m=0}^{K1-1} \sum_{n=0}^{K2-1} ([\mathbf{W}]^f_{m,n,c}.[\mathbf{O}]^{l-1}_{i+m,j+n,c}) + \mathrm{b}^f, \tag{17.4}$$

$$[\mathbf{O}]^l_{i,j,f} = f([\mathbf{Z}]_{i,j,f}). \tag{17.5}$$

After each forward pass through a network, the model's prediction error ($\boldsymbol{e}$) back-propagates in the backward pass and adjusts the weights and biases to minimize the loss function. Hence, the derivative of the error with respect to the weights and biases should be computed. We perform the backward pass of a layer by knowing $\frac{\partial \boldsymbol{e}}{\partial [\mathbf{O}]^l_{i,j,f}}$. Given Eq. (17.5), $d[\mathbf{Z}]_{i,j,f}$ is computed as

$$d[\mathbf{Z}]_{i,j,f} = \frac{\partial \boldsymbol{e}}{\partial [\mathbf{Z}]_{i,j,f}} = \frac{\partial \boldsymbol{e}}{\partial [\mathbf{O}]^l_{i,j,f}}.f'([\mathbf{Z}]_{i,j,f}). \tag{17.6}$$

Using the chain rule, we can compute $\frac{\partial \boldsymbol{e}}{\partial [\mathbf{W}]^f_{i,j,c}}$ as

$$\frac{\partial \boldsymbol{e}}{\partial [\mathbf{W}]^f_{i,j,c}} = \sum_{k=0}^{F-1} \sum_{m=0}^{H'-1} \sum_{n=0}^{W'-1} \frac{\partial \boldsymbol{e}}{\partial [\mathbf{Z}]_{m,n,k}} \frac{\partial [\mathbf{Z}]_{m,n,k}}{\partial [\mathbf{W}]^f_{i,j,c}}. \tag{17.7}$$

Given Eq. (17.4), $[\mathbf{Z}]_{m,n,k}$ is only related to the kth output filter. In other words, the weights of the fth kernel are only linked to the fth channel of $d\mathbf{Z}$. Hence, Eq. (17.7) reduces to

$$\frac{\partial \boldsymbol{e}}{\partial [\mathbf{W}]^f_{i,j,c}} = \sum_{m=0}^{H'-1} \sum_{n=0}^{W'-1} d[\mathbf{Z}]_{m,n,f} \frac{\partial [\mathbf{Z}]_{m,n,f}}{\partial [\mathbf{W}]^f_{i,j,c}} = \sum_{m=0}^{H'-1} \sum_{n=0}^{W'-1} d[\mathbf{Z}]_{m,n,f} [\mathbf{O}]^{l-1}_{i+m,j+n,c}. \tag{17.8}$$

Note that Eq. (17.8) presents the cross-correlation of $[\mathbf{O}]^{l-1}_{i+m,j+n,c}$ with the kernel $d\mathbf{Z}$. In the same manner, $\frac{\partial \boldsymbol{e}}{\partial \mathrm{b}^f}$ is computed by

$$\frac{\partial \boldsymbol{e}}{\partial \mathrm{b}^f} = \sum_{m=0}^{H'-1} \sum_{n=0}^{W'-1} d[\mathbf{Z}]_{m,n,f} \frac{\partial [\mathbf{Z}]_{m,n,f}}{\partial \mathrm{b}^f} = \sum_{m=0}^{H'-1} \sum_{n=0}^{W'-1} d[\mathbf{Z}]_{m,n,f}. \tag{17.9}$$

The back-propagation of the error or determining the relation between $\frac{\partial \boldsymbol{e}}{\partial [\mathbf{O}]^{l-1}_{i,j,c}}$ and $d\mathbf{Z}$ is the final task to complete. Using the chain rule, we have

$$\frac{\partial e}{\partial[\mathbf{O}]^{l-1}_{i,j,c}} = \sum_{f=0}^{F-1}\sum_{m=0}^{H'-1}\sum_{n=0}^{W'-1} \frac{\partial e}{\partial[\mathbf{Z}]_{m,n,f}} \frac{\partial[\mathbf{Z}]_{m,n,f}}{\partial[\mathbf{O}]^{l-1}_{i,j,c}}$$
$$= \sum_{f=0}^{F-1}\sum_{m=0}^{H'-1}\sum_{n=0}^{W'-1} d[\mathbf{Z}]_{m,n,f} \frac{\partial[\mathbf{Z}]_{m,n,f}}{\partial[\mathbf{O}]^{l-1}_{i,j,c}}. \tag{17.10}$$

Given Eq. (17.4), the second term can be rewritten as

$$\frac{\partial[\mathbf{Z}]_{m,n,f}}{\partial[\mathbf{O}]^{l-1}_{i,j,c}} = \frac{\partial}{\partial[\mathbf{O}]^{l-1}_{i,j,c}}\left(\sum_{c'=0}^{C-1}\sum_{m'=0}^{K1-1}\sum_{n'=0}^{K2-1}([\mathbf{W}]^{f}_{m',n',c'}.[\mathbf{O}]^{l-1}_{m+m',n+n',c'}) + \mathrm{b}^{f}\right), \tag{17.11}$$

which is not equal to zero if and only if $m+m'=i$, $n+n'=j$, and $c'=c$. Therefore,

$$\frac{\partial[\mathbf{Z}]_{m,n,f}}{\partial[\mathbf{O}]^{l-1}_{i,j,c}} = [\mathbf{W}]^{f}_{i-m,j-n,c} \tag{17.12}$$

and

$$\frac{\partial e}{\partial[\mathbf{O}]^{l-1}_{i,j,c}} = \sum_{f=0}^{F-1}\sum_{m=0}^{H'-1}\sum_{n=0}^{W'-1} d[\mathbf{Z}]_{m,n,f}.[\mathbf{W}]^{f}_{i-m,j-n,c}. \tag{17.13}$$

Assuming that $d\mathbf{Z}$ is the kernel and $[\mathbf{W}]^{f}_{i,j,c}$ constitutes the image, Eq. (17.13) represents a convolution between $d\mathbf{Z}$ and the layer's filters.

In this way, it is possible to achieve backward propagation for a convolutional layer with any number of input channels and filters.

17.2.3 Pooling operations

As explained in Section 17.2.2, convolutional layers in the CNNs systematically apply learned kernels to input images to extract feature maps. However, these feature maps are highly sensitive to the location of features in the input images. A spatial pooling operation reduces this sensitivity and makes them invariant to small translations in the input. The pooling operations downsample the output of several nearby feature detectors to local or global features so that the resulting downsampled feature maps are more robust to changes in the position of the feature in the image. In general terms, the main goals of a pooling operation in the convolutional layers are to preserve important information from the feature detectors, discard irrelevant details, and achieve equivariance to changes in position or lighting conditions in the input images, robustness to clutter, and compactness of the output feature maps [6].

Average pooling and max pooling are two commonly used pooling techniques that summarize the joint distribution of the features over some region of interest with the average or max activation in several neighboring feature detectors, respectively. Fig. 17.4 presents an example of average pooling. The pooling operation is performed by sliding a window over the input and applying the pooling function to the window

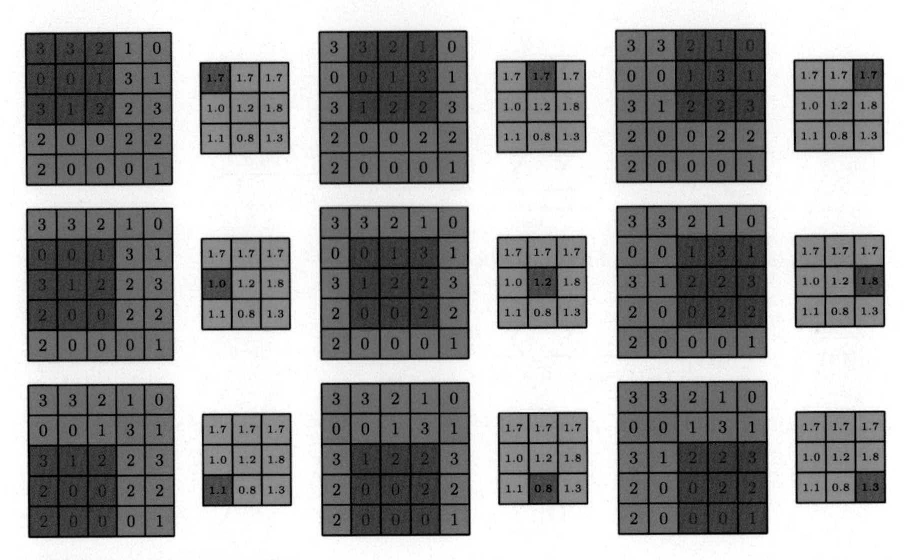

FIGURE 17.4 Average pooling.

Computing the output values of a 3 × 3 average pooling operation on a 5 × 5 input image using unit strides, resulting in a 3 × 3 downsampled output feature map [2].

content. The output size for any square input with size i, pooling size k, and stride s is obtained in the same manner as the convolution operation as

$$o = \lfloor \frac{i-k}{s} \rfloor + 1. \tag{17.14}$$

17.2.4 Dilated convolutions

The *algorithme à trous* [7], which applies a filter at various scales for wavelet decomposition, provided the inspiration for the dilated convolution operator. However, no *dilated filter* is constructed or represented in the dilated convolution introduced and frequently used in the deep learning literature [8]. Instead, it expands the kernel by inserting holes between its consecutive elements. Hence, at the same computational cost as a regular convolution operation, this technique offers a broader field of view of input by involving pixel skipping. The dilated convolution can apply the same kernel at different dilation factors l. The hyperparameter l defines the number of spaces between kernel parameters, which is usually $l-1$. Hence, $l=1$ represents a regular convolution. The effective size of a square kernel of size k that has been expanded by a factor l is $k' = k + (k-1)(l-1)$. Similar to Eq. (17.1) for the convolution operation, the output size considering the effective width of the kernel is as follows:

$$o = \lfloor \frac{i+2p-k-(k-1)(l-1)}{s} \rfloor + 1. \tag{17.15}$$

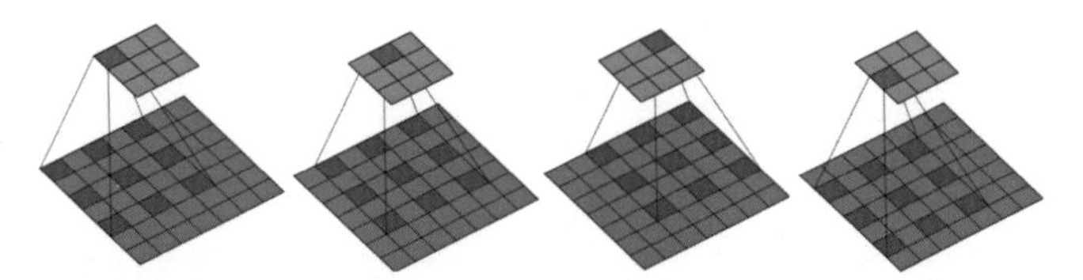

FIGURE 17.5 Dilated convolution.

Dilated convolution with kernel size 3×3 over a 7×7 input with dilation factor 2 and strides of 1 without zero padding, resulting in a 3×3 output [2].

Fig. 17.5 shows an example of the dilated convolution with kernel size 3×3 over a 7×7 input with dilation factor 2 and strides of 1 without zero padding.

17.2.5 Transposed convolutions

The CNN layers introduced so far, such as convolutional and pooling layers, typically reduce the input dimensions (height and width) or keep them unchanged. However, for applications such as superresolution or semantic segmentation and object localization, which process classification at the pixel level, we require the spatial dimensions of the output to be greater than or equal to the input. Another type of CNN layer called transposed convolutional layer, fractionally strided convolutional layer, or deconvolution layer [2] is useful for these purposes, especially when the spatial dimensions are reduced by CNN layers. Transposed convolutions reverse downsampling operations performed by the convolutional layers and upsample the intermediate input feature map to a desired output feature map using some learnable parameters. Transposed convolutional layers are the constructive building blocks in the U-net architecture proposed for image segmentation, explained in Chapter 18.

Ignoring channels and assuming a given input feature map of size $H \times W$ and kernel $K1 \times K2$ with unit strides and no padding, the transposed convolution is performed by sliding the kernel window on each element of the input, resulting in HW intermediate tensors with the size of $(H + K1 - 1) \times (W + K2 - 1)$ that are initialized as zeroes. Each intermediate tensor is obtained by multiplying each input element by the kernel such that the resulting weighted kernel replaces a portion in the intermediate tensor corresponding to the position of the input element. Finally, all the intermediate tensors are summed to form the output transposed convolution result. Fig. 17.6 shows the basic transposed convolution operation in a simple example. In the transposed convolution, strides are applied for the intermediate tensors (and hence output), not for input. Fig. 17.7 shows the transposed convolution for the same input and kernel in Fig. 17.6, but using strides of 2, resulting in an upsampled output. The transposed convolution operates similarly to the regular convolution for inputs and outputs with multiple channels.

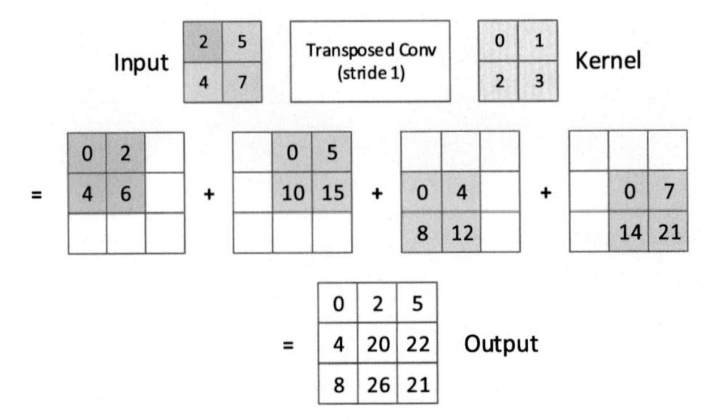

FIGURE 17.6 Transposed convolution.

Transposed convolution with kernel size 2 × 2 over a 2 × 2 input with strides of 1 without zero padding, resulting in a 3 × 3 output.

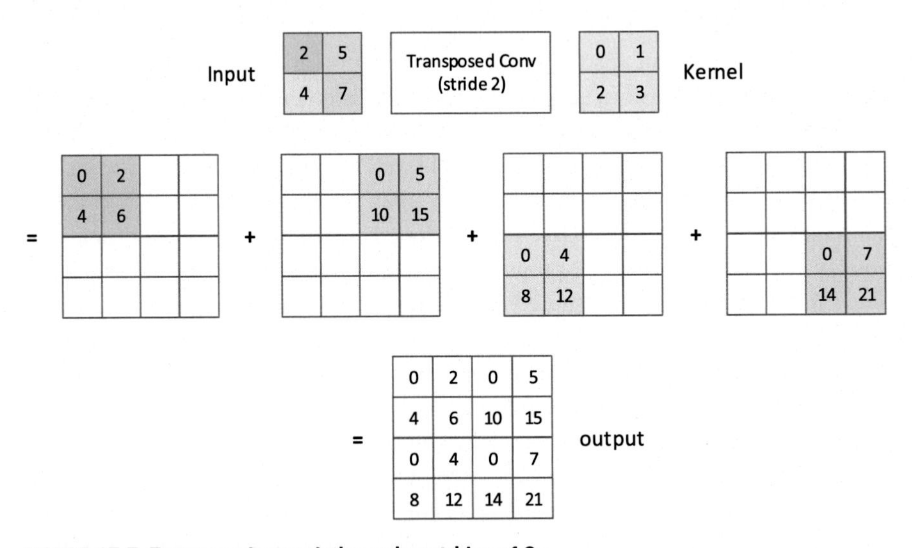

FIGURE 17.7 Transposed convolution using strides of 2.

Transposed convolution with kernel size 2 × 2 over a 2 × 2 input with strides of 2, resulting in a 4 × 4 output. The shaded parts show the input elements and kernel and their multiplication for computing the intermediate tensors.

17.2.6 Residual (skip) connections

In deep neural networks, by increasing the number of layers, the problem of vanishing and exploding gradients becomes increasingly prominent because each additional layer increases the number of multiplications required for computing the gradient in the back-propagation step. Multiplying several values less than one will exponentially decrease the result, while multiplying several values greater than one will exponen-

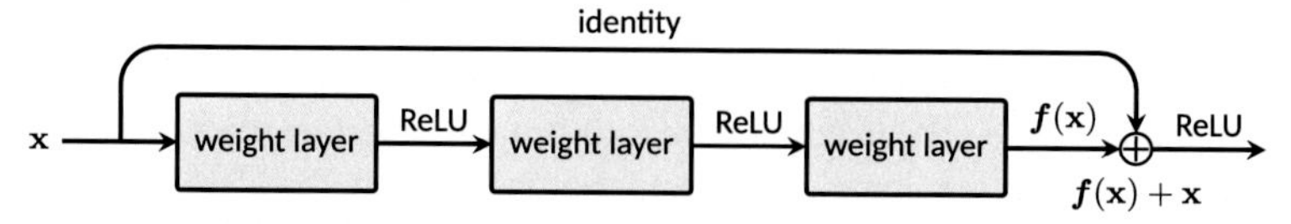

FIGURE 17.8 Residual (skip) connections.

A residual block in a ResNet architecture consists of two or three convolution operations, shown here as weight layers, followed by ReLU activation functions and a skip connection, which adds the input of the block ($\mathbf{x}$) to the last weight layer's output ($\boldsymbol{f}(\mathbf{x})$) before applying the ReLU function (i.e., adding the identity mapping of input to $\boldsymbol{f}(\mathbf{x})$).

tially increase that. Hence, training the traditional deep networks is problematic and leads to degradation of the network performance in learning complex mappings even if the network can start converging. As a remedy, skip connections are used to bypass one or more layers and their corresponding operations. One of the main architectures based on skip connections is residual networks (ResNet) [4], which are made by stacking several residual blocks together. A residual block consisting of three convolutional layers (weight layers + ReLU activations) and an identity skip connection is illustrated in Fig. 17.8. Considering $\boldsymbol{h}(\mathbf{x})$ is an underlying mapping to be fit by a few stacked layers in a residual block, the function $\boldsymbol{f}(\mathbf{x})$ represents all the intermediate operations performed in the block without skip connection, where

$$\boldsymbol{h}(\mathbf{x}) = \boldsymbol{f}(\mathbf{x}) + \mathbf{x}. \tag{17.16}$$

It is more likely in very deep networks that $\boldsymbol{f}(\mathbf{x})$ become very small or alter in a way that would degrade the network's accuracy. However, the elementwise addition of $\mathbf{x}$ to $\boldsymbol{f}(\mathbf{x})$ through the skip connection followed by application of the ReLU activation function causes an identity mapping for the residual block in the worst case since $\text{ReLU}(\mathbf{0} + \mathbf{x}) = \mathbf{x}$, assuming $\mathbf{x}$ is positive. This enables the deep networks to disregard the mapping of particular layers. During back-propagation, we must differentiate through all the intermediate functions represented in the preceding layers to reach the weights in the earlier layers. Assuming the last ReLU function in the residual block gives the final output that is used to calculate the loss function $\boldsymbol{e}$, to back-propagate the gradient to layers before the residual block, we have to compute the gradient of the loss with respect to $\mathbf{x}$ (i.e., $\frac{\partial e}{\partial \mathbf{x}}$). Here, in a network with skip connections, we can differentiate through $\boldsymbol{h}(\mathbf{x})$:

$$\frac{\partial e}{\partial \mathbf{x}} = \frac{\partial e}{\partial \boldsymbol{h}(\mathbf{x})} \frac{\partial \boldsymbol{h}(\mathbf{x})}{\partial \mathbf{x}} = \frac{\partial e}{\partial \boldsymbol{h}(\mathbf{x})} \left(\frac{\partial \boldsymbol{f}(\mathbf{x})}{\partial \mathbf{x}} + 1\right). \tag{17.17}$$

Therefore, even if $\frac{\partial \boldsymbol{f}(\mathbf{x})}{\partial \mathbf{x}}$ vanishes in a deep network due to the multiple multiplications of intermediate convolution layers through the chain rule of the back-propagation algorithm, the second term in Eq. (17.17) (i.e., $\frac{\partial e}{\partial \boldsymbol{h}(\mathbf{x})}$) prevents the total gradient from becoming vanishingly small. Hence, using skip connections, the gradient can be kept from vanishing or expanding by allowing the network to skip some of the gradient calculations during back-propagation [9].

ResNets have achieved outstanding, record-breaking performance in image classification, object detection, localization, and segmentation in the 2015 ImageNet and COCO competitions. Deep ResNets (e.g., with over 100 layers) are easy to optimize and, because of significantly increased depth, yield results that are far superior to those of traditional networks that just stack layers [4]. Stochastic depth [10] proposed later presents variations of ResNets by randomly dropping layers during training to improve information and gradient flow and was successful in training a 1202-layer ResNet.

17.3 Deep representation learning

The ease or difficulty of various information processing tasks is heavily influenced by the way the information is represented. This concept applies to machine learning algorithms as well. The performance of machine learning algorithms is significantly impacted by the quality of the data representations generated during the learning process. Representation learning involves encoding and decoding feature representations from raw data, which can be used in learning tasks. Effective representation learning can simplify subsequent learning tasks and is particularly valuable as it enables unsupervised/self-supervised learning. In this section, we explore different techniques for representation learning that can facilitate the discovery of underlying latent patterns and trends in data in an unsupervised/self-supervised manner, ultimately enhancing the learning of multiple tasks.

17.3.1 Autoencoder architecture

An autoencoder is a type of neural network used to learn data encodings in a self-supervised manner. The aim of an autoencoder is to learn a lower-dimensional representation $\mathbf{z}$ for a higher-dimensional data $\mathbf{x}$, typically for dimensionality reduction, by training the network to capture the most important features of the input image. As shown in Fig. 17.9, the general architecture of an autoencoder consists of two parts: an encoder and a decoder. The encoder q_ϕ compresses the data from a higher-dimensional space to a lower-dimensional space (also called the latent space), while the decoder p_θ does the opposite, i.e., reconstruct the low-dimensional latent representation back to the input data $\hat{\mathbf{x}}$ in the higher-dimensional space. The output $\hat{\mathbf{x}}$ is then compared with a ground truth $\mathbf{x}$ as follows:

$$l\left(\mathbf{x}, \hat{\mathbf{x}}\right) = \left\|\mathbf{x} - \hat{\mathbf{x}}\right\|_2 = \left\|\mathbf{x} - p_\theta\left(\mathbf{z}\right)\right\|_2 = \left\|\mathbf{x} - p_\theta\left(q_\phi\left(\mathbf{x}\right)\right)\right\|_2. \tag{17.18}$$

From an architectural perspective, the encoder is a set of convolutional blocks followed by pooling modules that reduce the input to a strongly compressed encoded form. The bottleneck is designed in such a way that the most relevant and useful features possessed by the data (e.g., an image) are captured and helps to form a knowl-

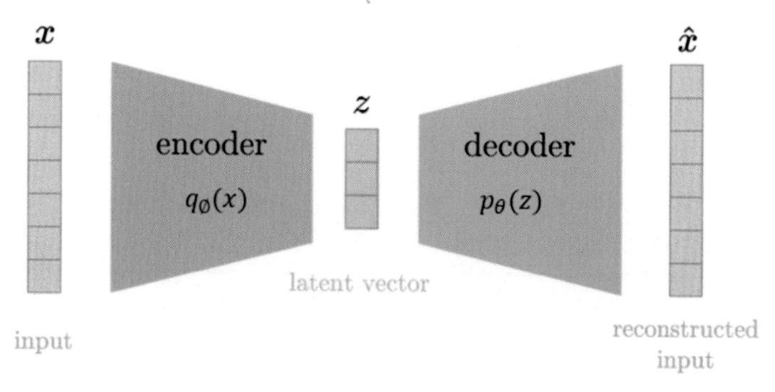

FIGURE 17.9 Autoencoders.

The general architecture of an autoencoder consists of two parts: an encoder and a decoder.

edge representation of the input. The bottleneck further prevents the network from memorizing the input and overfitting the data. The decoder follows the bottleneck, consisting of a series of upsampling and convolutional layers to bring the compressed feature back into its original form. The output is expected to have the same dimensions as the input. Many variants of autoencoders have been proposed in the literature; we introduce a few representative autoencoders in the following subsections.

17.3.2 Denoising autoencoders

Denoising autoencoders (DAEs) [11] corrupt the data by adding stochastic noise and reconstructing it back into intact data. Thus, a denoising autoencoder minimizes

$$\left\| \mathbf{x} - p_\theta\left(q_\phi\left(\tilde{\mathbf{x}}\right)\right)\right\|_2, \tag{17.19}$$

where $\tilde{\mathbf{x}}$ is a copy of $\mathbf{x}$ that is corrupted by some form of noise. DAEs must therefore undo this corruption rather than simply copying their input. As depicted in Fig. 17.10, the added noise to the input is the only difference between this method and the traditional autoencoders. DAEs give another example of how useful properties can emerge as a byproduct of minimizing reconstruction error and result in better feature extraction and generalization in classification tasks. They are also an example of how overcomplete, high-capacity models may be used as autoencoders as long as care is taken to prevent them from learning the identity function.

17.3.3 Variational autoencoders

Although variational autoencoders (VAEs) [12] comprise components similar to the traditional autoencoder (cf. Fig. 17.11), they enable probabilistic generation of data by constraining the learnable latent space/embedding to follow some prior chosen distribution, and correspondingly are trained using variational inference. VAEs address the issue of non-regularized latent space encountered with traditional autoen-

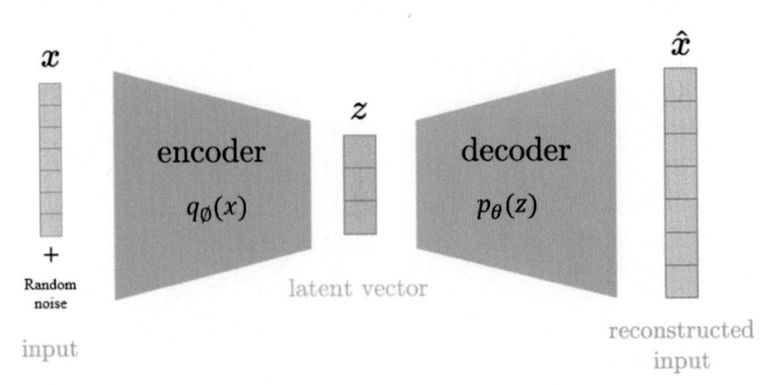

FIGURE 17.10 Denoising autoencoders.

A typical architecture of a denoising autoencoder (DAE) that results in better feature extraction.

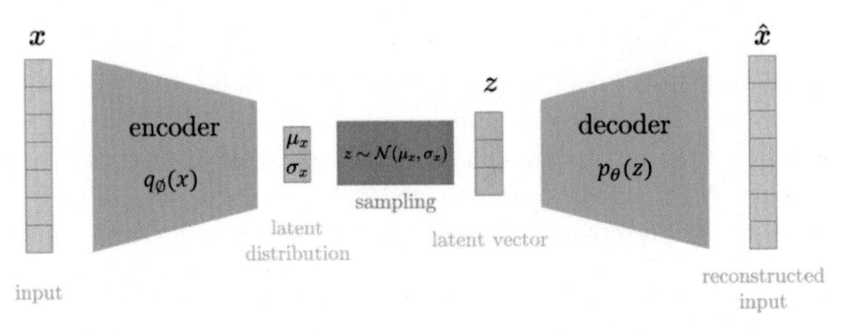

FIGURE 17.11 Variational autoencoders.

A typical architecture of a variational autoencoder (VAE) that is used to model probabilistic generation of data.

coders by imposing a constraint on this latent space, i.e., forcing it to follow a prior distribution typically chosen as a multivariate Gaussian/normal distribution. So instead of outputting the vectors in the latent space, the encoder of VAE outputs the mean and the standard deviation for each latent variable. The latent vector is then sampled from this mean and standard deviation, which is then fed to the decoder to reconstruct the input. On the inference, various samples conforming to the same distribution as the training data can be sampled from the interpolation of the latent space.

VAEs define an intractable density function with latent $\mathbf{z}$:

$$p_\theta(\mathbf{x}) = \int p_\theta(\mathbf{z}) p_\theta(\mathbf{x}|\mathbf{z}) d\mathbf{z}. \tag{17.20}$$

However, we can not optimize Eq. (17.20) directly. Instead, we drive and optimize the lower bound on the likelihood as

$$
\begin{aligned}
\log p_\theta(\mathbf{x}) &= \mathbb{E}_{\mathbf{z}\sim q_\phi(\mathbf{z}|\mathbf{x})}\left[\log p_\theta(\mathbf{x})\right] \\
&= \mathbb{E}_{\mathbf{z}}\left[\log \frac{p_\theta(\mathbf{x}|\mathbf{z})\, p_\theta(\mathbf{z})}{p_\theta(\mathbf{z}|\mathbf{x})}\right] \\
&= \mathbb{E}_{\mathbf{z}}\left[\log \frac{p_\theta(\mathbf{x}|\mathbf{z})\, p_\theta(\mathbf{z})}{p_\theta(\mathbf{z}|\mathbf{x})} \frac{q_\phi(\mathbf{z}|\mathbf{x})}{q_\phi(\mathbf{z}|\mathbf{x})}\right] \\
&= \mathbb{E}_{\mathbf{z}}\left[\log p_\theta(\mathbf{x}|\mathbf{z})\right] - \mathbb{E}_{\mathbf{z}}\left[\log \frac{q_\phi(\mathbf{z}|\mathbf{x})}{p_\theta(\mathbf{z})}\right] + \mathbb{E}_{\mathbf{z}}\left[\log \frac{q_\phi(\mathbf{z}|\mathbf{x})}{p_\theta(\mathbf{z}|\mathbf{x})}\right] \\
&= \mathbb{E}_{\mathbf{z}}\left[\log p_\theta(\mathbf{x}|\mathbf{z})\right] - D_{KL}\left(q_\phi(\mathbf{z}|\mathbf{x}) \,\|\, p_\theta(\mathbf{z})\right) + \\
&\qquad D_{KL}\left(q_\phi(\mathbf{z}|\mathbf{x}) \,\|\, p_\theta(\mathbf{z}|\mathbf{x})\right).
\end{aligned} \tag{17.21}
$$

Therefore, to train a VAE, we need to optimize and maximize this likelihood lower bound, which consists of (i) maximizing the log-likelihood of the original input being reconstructed and (ii) computing the Kullback–Leibler (KL) term to make the approximate posterior distribution close to the prior.

The original VAE formulation, however, cannot synthesize images conditioned on auxiliary information. To facilitate controllable image synthesis, conditional VAE [13] was proposed by incorporating additional conditioning information in the inputs and has been widely applied in the computer vision domain. The early VAEs [12,14] tended to produce blurred images due to the pixelwise loss and log-likelihood loss, which limited their applications in image synthesis. Strategies using the perceptual loss or other feature space loss were suggested to improve the quality of synthesized images [15,16]. To capture a common latent space between the data from different domains and address the multi-input scenarios, multichannel VAEs have also been explored [17–19].

17.3.4 Multichannel variational autoencoders

The multichannel VAE (mcVAE) [17–19] is a multivariate approach that extends VAEs in a multichannel scenario. The model jointly analyzes heterogeneous data and enforces latent representation encoded from each type of the data to match a common target distribution. The model is trained using a variational optimization framework, allowing to effectively generate and reconstruct missing data by decoding from the learned joint latent representation.

Specifically, in mcVAE, an observation $\mathbf{x} = \{\mathbf{x}_1, \ldots, \mathbf{x}_C\}$ represents the heterogeneous data associated with a single subject (i.e., a set of different types/channels of information), where each $\mathbf{x}_c$ has different size and dimension. The l-dimensional latent vector commonly shared by each type of data is denoted as $\mathbf{z}$. Then, the generative process for the observation can be formulated as follows:

$$
\mathbf{z} \sim p(\mathbf{z}), \tag{17.22}
$$

$$
\mathbf{x}_c \sim p(\mathbf{x}_c|\mathbf{z}, \boldsymbol{\theta}_c), \qquad \text{for } c \text{ in } 1 \ldots C, \tag{17.23}
$$

where $\boldsymbol{p}(\mathbf{z})$ denotes a prior distribution for the latent variable $\mathbf{z}$, $\boldsymbol{p}(\mathbf{x}_c|\mathbf{z},\boldsymbol{\theta}_c)$ denotes the likelihood distributions for the observations conditioned on the latent variable $\mathbf{z}$, and Eq. (17.23) indicates decoding the latent representation to data space using the likelihood functions that belong to a distribution family $\mathcal{P}$ and are parameterized by the set $\boldsymbol{\theta}=\{\boldsymbol{\theta}_1,\ldots,\boldsymbol{\theta}_C\}$. Solving the inference allows finding the joint latent representation from which each type or channel of the observed data is generated. As the true posterior $\boldsymbol{p}(\mathbf{z}|\mathbf{x},\boldsymbol{\theta})$ is intractable, the variational inference is used to compute an approximate posterior [20]. Assuming each type of data contributes independent information about the distribution of latent variables, we can approximate the posterior distribution with $\boldsymbol{q}\left(\mathbf{z}|\mathbf{x}_c,\boldsymbol{\phi}_c\right)$ belonging to a distribution family $\mathcal{Q}$ and parameterized by the set $\boldsymbol{\phi}=\left\{\boldsymbol{\phi}_1,\ldots,\boldsymbol{\phi}_C\right\}$, and impose a constraint enforcing each $\boldsymbol{q}\left(\mathbf{z}|\mathbf{x}_c,\boldsymbol{\phi}_c\right)$ to be close to the target posterior distribution. The KL divergence is used to minimize the distribution mismatch as

$$\arg\min_{q\in\mathcal{Q}}\mathbb{E}_c\left[D_{\mathrm{KL}}\left(\boldsymbol{q}\left(\mathbf{z}_c|\mathbf{x}_c,\boldsymbol{\phi}_c\right)||\boldsymbol{p}\left(\mathbf{z}|\mathbf{x}_1,\ldots,\mathbf{x}_C,\boldsymbol{\theta}\right)\right)\right], \tag{17.24}$$

where $\mathbb{E}_c$ is the average over channels computed empirically. Solving Eq. (17.24) allows minimizing the discrepancy between the variational approximations and the target posterior. Antelmi et al. [19] showed that optimizing Eq. (17.24) is equivalent to maximizing the evidence lower bound $\boldsymbol{l}(\boldsymbol{\theta},\boldsymbol{\phi},\mathbf{x})$:

$$\boldsymbol{l}(\boldsymbol{\theta},\boldsymbol{\phi},\mathbf{x})=\mathbb{E}_c\left[\boldsymbol{l}_c-D_{\mathrm{KL}}\left(\boldsymbol{q}\left(\mathbf{z}|\mathbf{x}_c,\boldsymbol{\phi}_c\right)||\boldsymbol{p}(\mathbf{z})\right)\right], \tag{17.25}$$

with

$$\boldsymbol{l}_c=\mathbb{E}_{\boldsymbol{q}\left(\mathbf{z}|\mathbf{x}_c,\boldsymbol{\phi}_c\right)}\sum_{i=1}^{C}\ln\boldsymbol{p}\left(\mathbf{x}_i|\mathbf{z},\boldsymbol{\theta}_i\right), \tag{17.26}$$

where $\boldsymbol{l}_c$ is the expected log-likelihood of decoding each channel from the latent representation of the channel $\mathbf{x}_c$ only. In native VAEs, all the channels are concatenated into a single one and thus cannot handle missing data when inferring the latent space variables. In contrast, Eqs. (17.25) and (17.26) in mcVAE imply that we can reconstruct multichannel data using the encoded information from a single channel only.

17.3.5 Contrastive learning

Contrastive learning falls within the realm of self-supervised learning, aiming to leverage sample comparisons within an embedding space. Samples belonging to the same distribution are encouraged to be closer to each other, while samples from different distributions are pushed apart. In supervised learning, the availability of large-scale labeled datasets is often limited, especially in domains like biomedical imaging that require specialized clinical expertise. The quality and size of labeled data greatly impact the performance of supervised models. Self-supervised learning, particularly in the image domain, has been extensively explored. Self-supervised approaches can be categorized into generative learning and discriminative learning. Generative learning methods, such as VAE and GAN, focus on reconstructing or generating images. These methods require pixel-level reconstruction and detailed embeddings. However, pixel-level generation is computationally intensive and may not be essential for representation learning. On the other hand, discriminative approaches based on contrastive learning in the latent space have emerged as promising techniques, achieving state-of-the-art results [21,22]. The fundamental contrastive learning framework consists of selecting a data sample, namely an "anchor," a data point belonging to the same distribution as the anchor called the "positive" sample, and another data point belonging to a different distribution called the "negative" sample. Positive samples are formed by different data augmentations of the same input sample, while negative samples are formed by different input samples. The model tries to minimize the distance between the anchor and positive samples, i.e., the samples belonging to the same distribution, in the latent space, and simultaneously maximize the distance between the anchor and the negative samples.

17.3.6 Framework, loss functions, and interpretation

A typical framework for contrastive learning of visual representations is shown in Fig. 17.12. The input $\mathbf{x}$ is augmented to $\mathbf{x}_i$ and $\mathbf{x}_j$ with different augmentation operators. Two separate data augmentation operators are sampled from the same family of augmentations ($\boldsymbol{t} \sim T$ and $\boldsymbol{t}' \sim T$) and applied to each data example to obtain two correlated views. A base encoder network $\boldsymbol{f}(\cdot)$ and a projection head $\boldsymbol{g}(\cdot)$ are trained to maximize agreement using a contrastive loss. After training is completed, one can throw away the projection head $\boldsymbol{g}(\cdot)$ and use encoder $\boldsymbol{f}(\cdot)$ and representation $\boldsymbol{h}$ for downstream tasks.

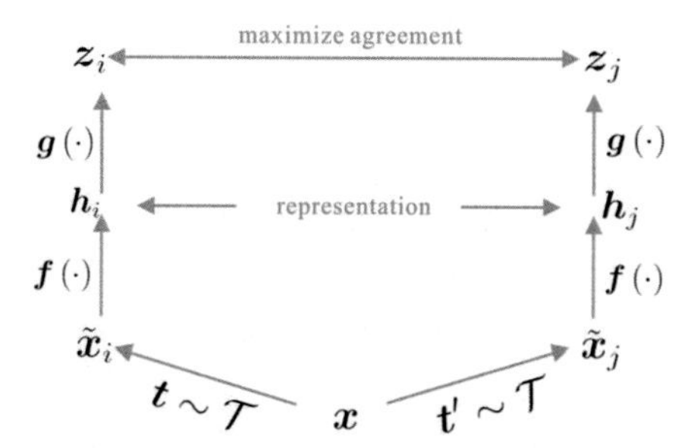

FIGURE 17.12 Framework, loss functions, and interpretation.

A typical framework for contrastive learning of visual representations.

The contrastive learning loss function is modified from the cross-entropy loss, in which the similarity of positive samples is maximized while it is minimized for negative samples. Let $(\mathbf{z}_i, \mathbf{z}_j)$ be a positive pair and let them be normalized so that their magnitudes are 1. The contrastive loss between a pair of positive examples i, j (augmented from the same image) is given as follows:

$$\ell_{i,j}^{\text{NT-Xent}} = -\log \frac{\exp\left(sim\left(\mathbf{z}_i, \mathbf{z}_j\right)/\tau\right)}{\sum_{k=1}^{2N} \mathbb{1}_{[k \neq i]} \exp sim\left(\mathbf{z}_i, \mathbf{z}_k\right)/\tau}, \tag{17.27}$$

where $sim(\cdot, \cdot)$ is the cosine similarity between two vectors, τ is a temperature scalar, NT-Xent indicates the normalized temperature-scaled cross-entropy loss, and $\mathbb{1}_{[k \neq i]}$ is an indicator function evaluating to 1 iff $k \neq i$. The loss function can also be interpreted geometrically. Since $\mathbf{z}_i$ and $\mathbf{z}_j$ are high-dimensional latent vectors and normalized, they can be seen as two points on a hypersphere. The cosine similarity between $\mathbf{z}_i$ and $\mathbf{z}_j$ is the Euclidean distance between them:

$$\text{Similarity} = \cos(\theta) = \frac{\mathrm{A} \cdot \mathrm{B}}{\|\mathrm{A}\| \, \|\mathrm{B}\|}. \tag{17.28}$$

Since vectors A and B can be viewed as two points on a hypersphere, the smaller the angle between them, the nearer they are. That is, if the two vectors form a pair of positive samples, they would be pushed towards each other through learning. Otherwise, if they form a negative pair, they would be pulled against each other through learning.

17.3.7 Contrastive learning with negative samples

A Simple Framework for Contrastive Learning of Visual Representations (SimCLR) [23] is a reference contrastive learning architecture because (i) the model yields a significant improvement compared to previous approaches, (ii) it adopts a symmetrical structure that is relatively straightforward, and (iii) its structure has become a standard component and inspired other contrastive learning models. The SimCLR framework comprises the following major components.

In the data augmentation part, the authors adopted a stochastic data augmentation module that randomly transforms any given data example, resulting in two correlated views of the same example, considered a positive pair. They applied three augmentations: random cropping followed by resizing back to the original size, random color distortions, and random Gaussian blur (cf. Fig. 17.13). They observed that combining random crop and color distortion is crucial to achieving good performance. One main reason is that when using only random cropping, most patches from an image share a similar color distribution. Neural nets may exploit this shortcut to solve the predictive task. Therefore, composing cropping with color distortion is critical to learn generalizable features.

The authors adopted the commonly used ResNet as the encoder $f(\cdot)$ that extracts representation vectors from augmented data examples, although the framework allows various choices of network architecture without any constraints. The small neural network projection head $g(\cdot)$ maps representations to the space where contrastive loss is applied. They used a multilayer perceptron (MLP) with one hidden layer to obtain $\mathbf{z}_i = g(\mathbf{h}_i) = W^{(2)}\sigma(W^{(1)}\mathbf{h}_i)$, where σ is a `ReLU` non-linearity. Moreover, large batch sizes and more training steps benefit contrastive learning more than supervised learning. This is due to a large amount of unlabeled data and rich patterns in contrastive learning, which benefit training large models and prevent overfitting.

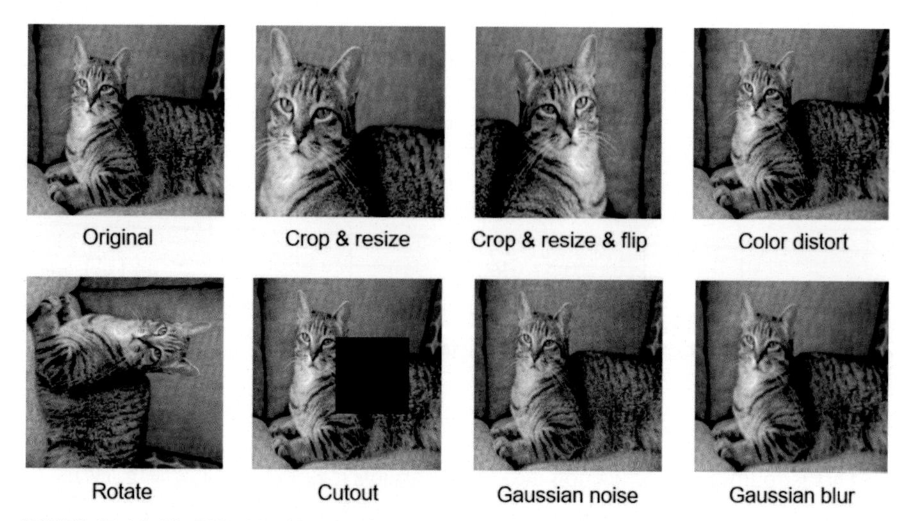

FIGURE 17.13 SimCLR: CL with negative samples.

Illustrations of the studied data augmentation operators. Each augmentation can transform data stochastically with some internal parameters (e.g., rotation degree, noise level).

The purpose of using negative pairs is to prevent mode collapse. However, it is not essential. Some contrastive methods without negative pairs also prevent collapse, as introduced in the following.

17.3.8 Contrastive learning without negative samples

Bootstrap Your Own Latent (BYOL) [24] was the first contrastive learning method that achieved state-of-the-art results without using negative pairs. BYOL uses two neural networks, namely the online and target networks, that interact and learn from each other. As shown in Fig. 17.14, the online network parameterized by a set of weights $\boldsymbol{\theta}$ consists of three stages: an encoder $\boldsymbol{f}_\theta$, a projector $\boldsymbol{g}_\theta$, and a predictor $\boldsymbol{q}_\theta$. The target network parameterized by $\boldsymbol{\xi}$ has the same architecture as the online network. The target network provides the regression targets to train the online network, and its parameters $\boldsymbol{\xi}$ are an exponential moving average of the online parameters $\boldsymbol{\theta}$. More precisely, given a target decay rate $\tau \in [0, 1]$, after each training step, the authors perform the following update:

$$\boldsymbol{\xi} \leftarrow \tau\boldsymbol{\xi} + (1 - \tau)\boldsymbol{\theta}. \tag{17.29}$$

Given a set of images D, an image $\mathbf{x} \sim D$ sampled uniformly from D, and two distributions of image augmentations T and T', the BYOL is constructed as follows:

- Create two augmented views, $\mathbf{v} = \boldsymbol{t}(\mathbf{x})$ and $\mathbf{v}' = \boldsymbol{t}'(\mathbf{x})$, with augmentations sampled $\boldsymbol{t} \sim \mathcal{T}, \boldsymbol{t}' \sim \mathcal{T}'$.
- The two augmented views are encoded into representations: $\mathbf{y}_\theta = \boldsymbol{f}_\theta(\mathbf{v})$, $\mathbf{y}' = \boldsymbol{f}_\xi(\mathbf{v}')$.

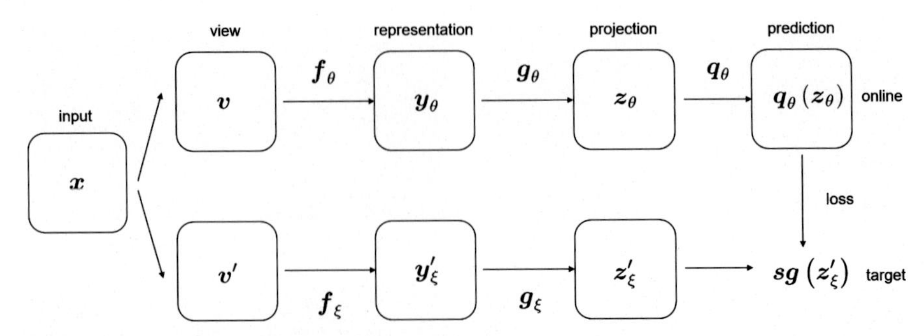

FIGURE 17.14 BYOL: CL without negative samples.

BYOL's architecture.

- This is followed by projection into latent variables: $\mathbf{z}_\theta = g_\theta(\mathbf{y}_\theta)$, $\mathbf{z}' = g_\xi(\mathbf{y}')$.
- The online network predicts a result $q_\theta(\mathbf{z}_\theta)$.
- Both $q_\theta(\mathbf{z}_\theta)$ and $\mathbf{z}'$ are L_2-normalized, resulting in $\bar{q}_\theta(\mathbf{z}_\theta)$ and $\bar{\mathbf{z}}'$.
- The loss l_θ^{BYOL} is the MSE between the L_2-normalized prediction $\bar{q}_\theta(\mathbf{z})$ and $\bar{\mathbf{z}}'$.
- The symmetric loss $\tilde{l}_\theta^{\text{BYOL}}$ can be generated by switching $\mathbf{v}'$ and $\mathbf{v}$, that is, feeding $\mathbf{v}'$ to the online network and $\mathbf{v}$ to the target network;
- The final loss is $l_\theta^{\text{BYOL}} + \tilde{l}_\theta^{\text{BYOL}}$ and only parameters $\boldsymbol{\theta}$ are optimized.

The authors evaluated the performance obtained when fine-tuning BYOL's representation on a classification task with a small subset of ImageNet's training set. Following the semi-supervised protocol and the same fixed splits of respectively 1% and 10% of ImageNet labeled training data, they found that BYOL consistently outperforms previous approaches across various architectures.

17.4 Recapitulation

The take-home messages from this chapter are:

- CNNs are an example of a neural network architecture that exploits inductive bias to learn effective representations from the input data. The inductive bias here refers to the prior knowledge that given input data with a regular grid-like structure (e.g., images or videos), the same feature extractors/filters are likely to be useful to process the entire domain of the input samples. CNNs implement this reuse of filters across the domain of input samples by sharing weights across the convolution filters or kernels used to process all regions of the input samples. This weight sharing across convolution operations ensures translation-equivariance in the learned features and significantly reduces the number of trainable parameters required for effective feature learning from high-dimensional data such as images or videos (relative to MLPs).

- Convolution operations with kernel sizes greater than one naturally downsample the inputs. Hence, using several such operations sequentially combined with additional downsampling operations, such as pooling, leads to effective learning of hierarchical representations of input data. This hierarchical feature learning underpins the tremendous success of CNNs at vision tasks such as object recognition/segmentation, instance or sample classification, etc., from input images or videos.
- A key ingredient to hierarchical feature learning with CNNs is the increasing receptive field size afforded by sequential convolution and pooling layers combinations. However, often, it is desirable to increase the receptive field size of the network without significantly downsampling the inputs and/or intermediate feature maps, as this could lead to loss of relevant information, or utilizing several successive convolution layers, as this would significantly increase the model complexity (leading to overfitting issues) and correspondingly the computational burden. Increasing the network's receptive field size without increasing the downsampling rate or the number of convolution layers is enabled through dilated convolution operations, which provides a greater global context to the learned features without increasing the number of trainable parameters or the computational cost.
- Autoencoders are neural networks that utilize self-supervised learning for dimensionality reduction as they map input data to a low-dimensional embedding through an encoder network and can reconstruct an approximation to the original data through a decoder network. They are typically trained by minimizing a reconstruction loss between the original inputs and the network-reconstructed approximations of the inputs. The learned low-dimensional embedding for input data may be used for downstream tasks, i.e., it serves a similar purpose to how conventional dimensionality reduction techniques are used (e.g., principal component analysis [PCA]).
- VAEs are a probabilistic variant of autoencoders and are generative models analogous to probabilistic PCA. They differ from autoencoders in the constraint placed on the learned latent space, which drives the posterior latent distribution to be similar to some assumed prior distribution over the latent variables (e.g., conventionally, a multivariate Gaussian distribution is used as the prior). The probabilistic constraint in VAEs results in a regularized latent space that is easier to interpret and interpolate to generate meaningful and realistic synthetic data close to the observed real data, a property not afforded by autoencoders. VAEs may be extended joined representations of multiview/multimodal data by formulating them as multichannel VAEs.
- Contrastive learning falls into the category of discriminative self-supervised learning wherein networks are trained to maximize the similarity between representations of two similar data points or views of the same data point and minimize the similarity between representations of two different data points. Contrastive networks are trained with input pairs of similar or different data points by optimizing a contrastive loss function. Contrastive learning can learn useful representations of data without requiring labeled training data.

17.5 Further reading

1. A comprehensive introduction to CNNs is provided in Chapter 9 of the book *Deep Learning* by Goodfellow et al. [5], available at https://www.deeplearningbook.org/.
2. A comprehensive introduction to VAEs is available in [25].
3. Reviews of self-supervised learning approaches employed in computer vision and healthcare applications [26,27].

17.6 Exercises

1. Why are CNNs better suited to vision problems and learning with image/video data than MLPs?
   Answer: Vision problems typically involve high-dimensional image or video data. As a result, MLPs are computationally inefficient for learning features from such data as they require many parameters to process the same. CNNs, on the other hand, use parameter/weight sharing by convolving the inputs with the same convolution kernel/filter across different spatial positions to significantly reduce the number of parameters required to process such data (relative to MLPs), enabling efficient learning of features. This also helps CNNs exploit the spatial structure in image/video data and ensure translation-equivariance in the learned features.
2. How do residual connections help with training deep neural networks?
   Answer: Training deep neural networks often leads to the issue of vanishing or exploding gradients due to the successive multiplication of either very small or very large gradients during back-propagation, respectively. Residual connections can bypass one or more layers and provide a direct path for gradients to flow through the network. This prevents small or large gradients of specific layers from attenuating or amplifying the gradients being back-propagated. This, in turn, helps improve stability when training deep neural networks.
3. What are some challenges and limitations of VAEs?
   Answer: One challenge with VAEs is that the learned latent representation may not capture sufficient meaningful variations observed in the training data distribution, leading to poor generative performance. This may be addressed by using more expressive prior distributions (in place of the conventional multivariate Gaussian distribution) to constrain the latent space or modifying the objective function to encourage learning of better representations. Additionally, VAEs can be combined with other techniques, such as adversarial training or hierarchical modeling, to retain fine-grained details.
4. What is the difference between contrastive learning and other self-supervised learning approaches such as VAE and GAN?
   Answer: Contrastive learning falls under the category of discriminative learning, while VAE and GAN are types of generative self-supervised learning. Specifically, contrastive learning explicitly maximizes the similarity between represen-

tations of similar samples (i.e., samples belonging to the same distribution) while minimizing the similarity between dissimilar samples (i.e., samples belonging to different distributions). Contrastive learning approaches thus learn effective representations of training data by learning to discriminate between samples that belong to different distributions and cannot be used to generate new/synthetic data. In contrast, generative self-supervised approaches such as VAE and GAN learn to reconstruct the input samples and approximate the training data distribution. Correspondingly, they can generate new/synthetic data once trained.

5. Why is self-supervised learning using discriminative or generative approaches relevant/important for supervised learning tasks?
   Answer: Supervised learning with deep neural networks often requires a large volume of labeled training data, which is especially difficult to source in the medical imaging domain. One approach to mitigate this and reduce reliance on a large sample size is to pre-train deep neural networks using unlabeled data from the same domain as the target supervised learning task of interest, i.e., semi-supervised learning. Self-supervised learning using both discriminative and generative approaches has been demonstrated to learn effective representations from unlabeled data that, subsequently, help improve predictive performance in downstream supervised learning tasks.

References

[1] Krizhevsky A., Sutskever I., Hinton G.E., ImageNet classification with deep convolutional neural networks, Communications of the ACM 60 (6) (2017) 84–90.

[2] Dumoulin V., Visin F., A guide to convolution arithmetic for deep learning, ArXiv e-prints, arXiv:1603.07285, Mar. 2016.

[3] Simonyan K., Zisserman A., Very deep convolutional networks for large-scale image recognition, arXiv preprint, arXiv:1409.1556, 2014.

[4] He K., Zhang X., Ren S., Sun J., Deep residual learning for image recognition, in: Proceedings of the IEEE Conference on Computer Vision and Pattern Recognition, 2016, pp. 770–778.

[5] Goodfellow I., Bengio Y., Courville A., Deep Learning, MIT Press, 2016, http://www.deeplearningbook.org.

[6] Boureau Y-L., Ponce J., LeCun Y., A theoretical analysis of feature pooling in visual recognition, in: Proceedings of the 27th International Conference on Machine Learning (ICML-10), 2010, pp. 111–118.

[7] Holschneider M., Martinet R.K., Morlet J., Tchamitchian Ph., A real-time algorithm for signal analysis with the help of the wavelet transform, in: Wavelets, Springer, 1990, pp. 286–297.

[8] Yu F., Koltun V., Multi-scale context aggregation by dilated convolutions, arXiv preprint, arXiv:1511.07122, 2015.

[9] He K., Zhang X., Ren S., Sun J., Identity mappings in deep residual networks, in: European Conference on Computer Vision, Springer, 2016, pp. 630–645.

[10] Huang G., Sun Y., Liu Z., Sedra D., Weinberger K., Deep networks with stochastic depth, in: European Conference on Computer Vision, Springer, 2016, pp. 646–661.

[11] Vincent P., Larochelle H., Bengio Y., Manzagol P.A., Extracting and composing robust features with denoising autoencoders, in: Proceedings of the 25th International Conference on Machine Learning, 2008, pp. 1096–1103.
[12] Kingma D.P., Welling M., Auto-encoding variational Bayes, in: Proceedings 2nd International Conference on Learning Representations (ICLR), 2014.
[13] Sohn K., Lee H., Yan X., Learning structured output representation using deep conditional generative models, Advances in Neural Information Processing Systems 28 (2015) 3483–3491.
[14] Rezende D.J., Mohamed S., Wierstra D., Stochastic backpropagation and approximate inference in deep generative models, in: International Conference on Machine Learning, in: PMLR, 2014, pp. 1278–1286.
[15] Larsen A.B.L., Sønderby S.K., Larochelle H., Winther O., Autoencoding beyond pixels using a learned similarity metric, in: International Conference on Machine Learning, in: PMLR, 2016, pp. 1558–1566.
[16] Hou X., Shen L., Sun K., Qiu G., Deep feature consistent variational autoencoder, in: 2017 IEEE Winter Conference on Applications of Computer Vision (WACV), IEEE, 2017, pp. 1133–1141.
[17] van Tulder G., de Bruijne M., Learning cross-modality representations from multi-modal images, IEEE Transactions on Medical Imaging 38 (2) (2018) 638–648.
[18] Shi Y., Siddharth N., Paige B., Torr P.H.S., Variational mixture-of-experts autoencoders for multi-modal deep generative models, in: Advances in Neural Information Processing Systems 32, Curran Associates Inc, Dec. 2019, pp. 15718–15729.
[19] Antelmi L., Ayache N., Robert P., Lorenzi M., Sparse multi-channel variational autoencoder for the joint analysis of heterogeneous data, in: International Conference on Machine Learning, in: PMLR, 2019, pp. 302–311.
[20] Blei D.M., Kucukelbir A., McAuliffe J.D., Variational inference: a review for statisticians, Journal of the American Statistical Association 112 (518) (2017) 859–877.
[21] van den Oord A., Li Y., Vinyals O., Representation learning with contrastive predictive coding, arXiv preprint, arXiv:1807.03748, 2018.
[22] Bachman P., Devon Hjelm R., Buchwalter W., Learning representations by maximizing mutual information across views, Advances in Neural Information Processing Systems 32 (2019).
[23] Chen T., Kornblith S., Norouzi M., Hinton G., A simple framework for contrastive learning of visual representations, in: International Conference on Machine Learning, in: PMLR, 2020, pp. 1597–1607.
[24] Grill J.B., Strub F., Altché F., Tallec C., Richemond P., Buchatskaya E., Doersch C., Pires B.A., Guo Z., Azar M.G., Piot B., Kavukcuoglu K., Munos R., Valko M., Bootstrap your own latent-a new approach to self-supervised learning, Advances in Neural Information Processing Systems 33 (2020) 21271–21284.
[25] Kingma D.P., Welling M., An introduction to variational autoencoders, Foundations and Trends® in Machine Learning 12 (4) (2019) 307–392.
[26] Jing L., Tian Y., Self-supervised visual feature learning with deep neural networks: a survey, IEEE Transactions on Pattern Analysis and Machine Intelligence 43 (11) (2020) 4037–4058.
[27] Krishnan R., Rajpurkar P., Topol E.J., Self-supervised learning in medicine and healthcare, Nature Biomedical Engineering 6 (2022) 1–7.

CHAPTER

Deep learning medical image segmentation

18

Sean Mullan, Lichun Zhang, Honghai Zhang, and Milan Sonka
Iowa Institute for Biomedical Imaging, The University of Iowa, Iowa City, IA, United States

Learning points

- Segmentation with CNNs, Transformers, Hybrid networks
- Assisted and sparse annotation for training data
- Understanding of DL segmentation through a case study

18.1 Introduction

Human beings come in a massive variety of shapes and sizes, and the diversity of this population must be accounted for if medical image segmentation is to be a useful part of patient care. Fortunately, deep learning (DL) is able to model these complex domains and learn the underlying representations without requiring explicitly crafted features from medical experts. This chapter starts by discussing how DL methods have been applied to medical image segmentation tasks and follows with the improvements and considerations developed to enable the application of DL methods to such a potentially high-stakes domain as medical imaging.

The chapter builds on the fundamentals of DL presented in Chapter 16 and initially examines the most relevant approaches to DL segmentation – the first section covers convolutional neural networks (CNNs) such as the fully convolutional network (FCN) [1] and U-Net [2], which utilize convolution operations to generate pixel-level segmentation predictions. Furthermore, CNNs are extended from 2D slice-by-slice methods to fully 3D volumetric methods that better capture the context of modern medical images. One recently developed CNN-based approach, nnU-Net [3], has been applied to dozens of prominent medical image segmentation tasks and stands as a benchmark for medical image segmentation approaches in general.

The following section looks at an alternative to convolution operations – Transformer networks [4]. Transformers were originally proposed for modeling long-range dependencies in natural language processing. Since then, their unique attention-based mechanisms have become key components in many state-of-the-art DL segmentation architectures.

Medical Image Analysis. https://doi.org/10.1016/B978-0-12-813657-7.00042-X

The third section steps outside the field of pure DL and discusses how non-learning-based methods can be used to overcome the inherent DL limitations. Specifically, this section discusses a hybrid DL and graph optimization approach that not only improves the topological consistency and surface delineation compared to pure DL methods, but also supports highly efficient human–algorithm interaction when manual correction is required.

Following the discussion of these approaches, the final two sections explore important considerations surrounding DL. DL can be an incredibly data-hungry technique, which is a major limitation in medical applications, where well-annotated data come at a premium. Section 18.5 discusses two ways to alleviate the burden of annotating a large, high-quality dataset: (1) assisted annotation to reduce the manual effort required and (2) selective or sparse annotation to reduce the size of the required training set without sacrificing model performance. The latter section in this part addresses the explainability of DL. A drawback associated with the inherent complexity of DL methods is that it is essentially impossible to explain the concepts and relationships encoded in the millions to hundreds of millions of parameters learned by the models. Section 18.6 explores post-hoc and structural approaches allowing better understanding to the processes of DL segmentation and thereby building the trust necessary for patient care.

The final part of the chapter presents a case study to demonstrate the power of the discussed approaches. The performance of convolution-based, Transformer-based, and hybrid DL methods is evaluated on segmentation of calf muscle compartments in 3D magnetic resonance (MR) images. In addition, the study demonstrates how each approach can benefit when their training sets are expanded through the use of assisted annotation.

For approaches discussed in this chapter, links to publicly available implementations have been added to the related references at the end of the chapter should such code be available.

18.2 Convolution-based deep learning segmentation

FCN is a milestone in semantic segmentation [1]. It transformed the fully connected layers in typical recognition CNNs (e.g., LeNet [5], AlexNet [6], etc.) into convolution layers and utilized deconvolution layers for upsampling the coarse outputs to pixel-dense outputs. Doing so, FCN can take input of arbitrary size and produce a correspondingly sized likelihood (probability) map with efficient inference and learning. The final segmentation is achieved via the argmax operation, which selects the category with the highest probability on the likelihood map. It also demonstrated that combining the semantic information from deep layers with appearance information from shallow layers via the skip connection strategy is beneficial for improving semantic segmentation accuracy. FCN dramatically improved the segmentation accuracy of challenges including PASCAL VOC and NYUDv2. The overall architecture of FCN is demonstrated in Fig. 18.1.

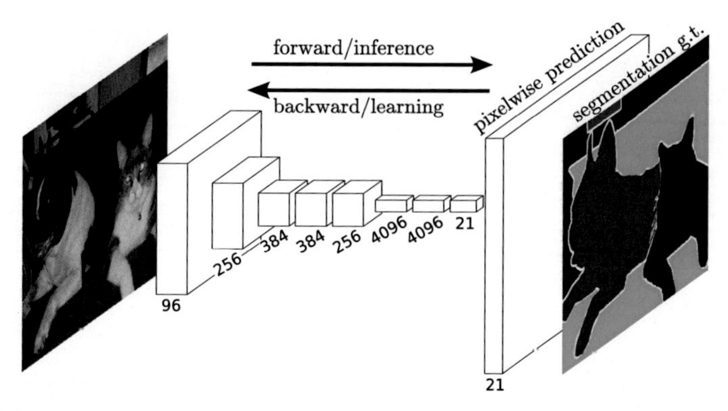

FIGURE 18.1

FCN can efficiently learn to make dense predictions for per-pixel tasks like semantic segmentation. From [1].

The idea of FCN was later extended one step further, yielding an elegant symmetrical encoder–decoder architecture called U-Net [2]. As illustrated in Fig. 18.2, U-Net consists of a contracting path (left side) to capture context features and a symmetric expansive path (right side) that enables precise localization, which makes the architecture U-shaped. Compared to FCN, the main idea of the modification in U-Net is to fully leverage the features by propagating context information to higher-resolution layers. The features from each layer in the contracting path are directly connected to the symmetrical layers in the expanding path via dense skip connections. With excessive data augmentation and a distance-based weighted cross-entropy loss function, U-Net won the IEEE-ISBI tracking challenge in 2015 and the ISBI challenge for segmentation of neuronal structures in electron microscopic stacks.

With its straightforward and successful architecture, U-Net has drawn extensive attention from the medical image analysis community and quickly evolved to a commonly used benchmark in medical image segmentation. Numerous modified versions of U-shaped networks have been applied to medical image segmentation, and the U-Net architecture was extended to 3D U-Net by replacing all 2D operations with their 3D counterparts [7]. A VNet variant followed that incorporates residual blocks and a Dice loss layer to directly minimize this commonly used segmentation error metric [8]. Similarly, a FilterNet+ approach utilizes contextual information in a larger neighborhood by residual connections and embeds edge-aware constraints by edge gate [9,10]. In yet another modification called UNet++, the encoder and decoder subnetworks are connected through a series of nested, dense skip pathways [11]. As shown in Fig. 18.3, the main idea behind UNet++ is to bridge the semantic gap be-

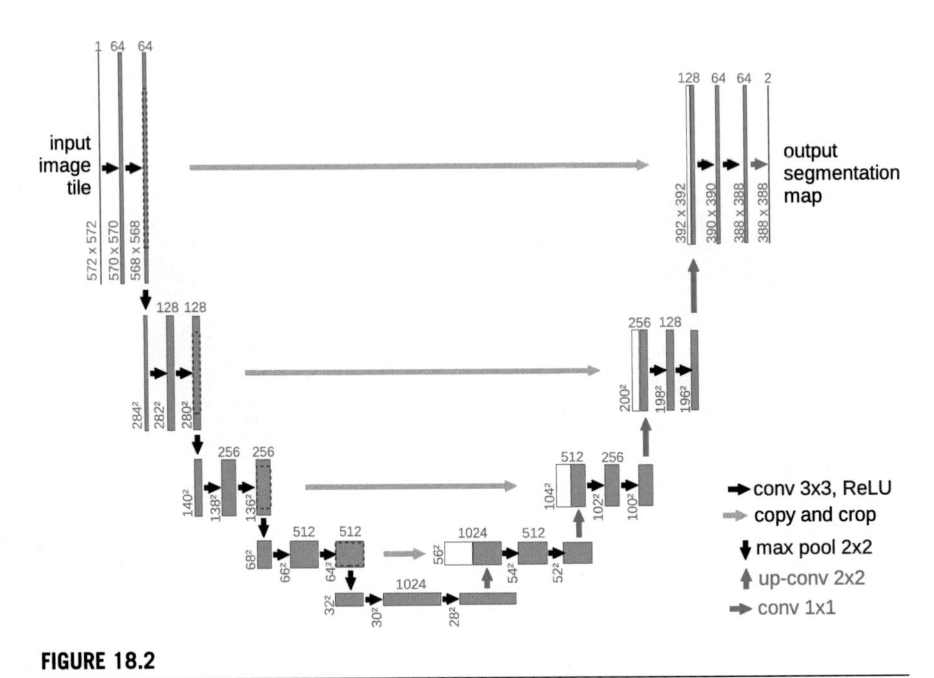

FIGURE 18.2

U-Net architecture (example for 32 × 32 pixels in the lowest resolution) from [2]. Each blue box corresponds to a multichannel feature map. The number of channels is denoted on top of the box. White boxes represent copied feature maps.

tween the feature maps of the encoder and decoder prior to fusion using a dense convolution block.

Another trend of U-Net extensions is combining the basic architecture of U-Net and novel mechanisms such as self-attention and Transformer models. After noticing that computational resources and model parameters are used excessively and redundantly, for instance when similar low-level features are repeatedly extracted in general U-Net models with multistage cascaded CNNs, a novel self-attention gating (AG) module was designed. The resulting Attention U-Net automatically learns to focus on target structures of varying shapes and sizes by integrating AG within a standard U-Net model [13]. As shown in Fig. 18.4, information extracted from coarse scale (the contracting path on the left side) is used in gating to disambiguate irrelevant and noisy responses in skip connections, performed right before the concatenation operation to merge only relevant activations. Attention U-Net can implicitly learn to suppress irrelevant regions in an input image while highlighting salient features useful for a specific task. Inspired by the success of Transformer models in natural language processing (NLP), combinations of Transformer and U-Net, such as TransUNet [14] and UNETR [15], have emerged as strong alternatives for medical image segmentation. Details of Transformer models are given in Section 18.3.

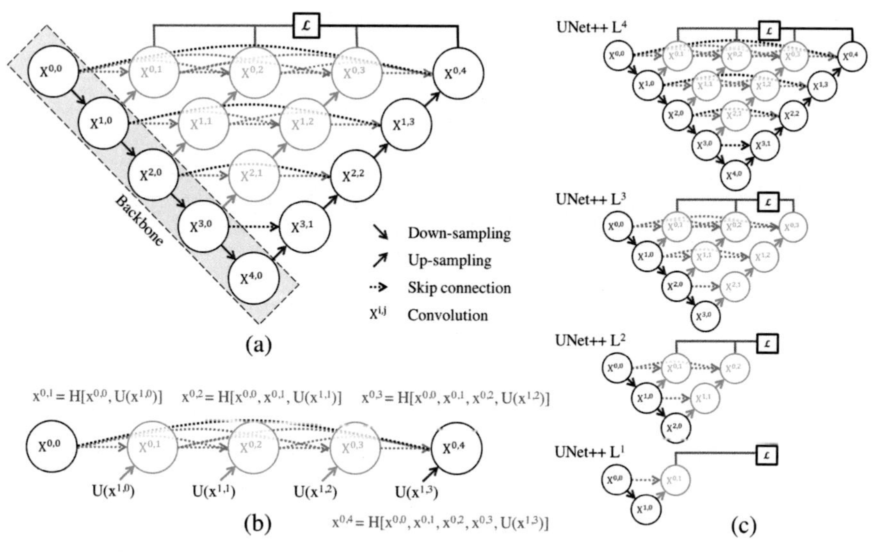

FIGURE 18.3

UNet++ architecture from [11]. (a) The black part indicates the original U-Net, green and blue show dense convolution blocks on the skip pathways, and red indicates deep supervision. The red, green, and blue components distinguish UNet++ from the original U-Net. (b) The detailed computation in the first skip pathway. H denotes convolution operations followed by an activation function, U is an upsampling layer, and [ ] denotes the concatenation layer. (c) UNet++ can be pruned at inference time if trained with deep supervision [12]. UNet++ L^i denotes UNet++ pruned at level i.

Despite the success of variants of U-Net in medical image segmentation tasks, it was noticeable that some of the architectural modifications may be partly overfitted to specific problems and thus fail to generalize beyond the experiments they were demonstrated on [16]. Based on the hypothesis that the influence of non-architectural aspects in segmentation methods is much more impactful, they proposed nnU-Net (no-new-Net) [3], a self-adapting framework for U-Net-based medical image segmentation. For its network architecture, nnU-Net omits fancy extensions such as residual connections and attention mechanism and only resides on three simple U-Net models (2D, 3D, and U-Net cascade) that contain only minor modifications to the original U-Net. Instead, nnU-Net focuses on the automatic configuration for arbitrary datasets and training pipelines. As shown in Fig. 18.5, nnU-Net is holistic in that the configuration covers the entire segmentation pipeline (including essential topological parameters of the network architecture) without any explicitly defined hyperparameters. nnU-Net demonstrated that the non-architectural steps are more impactful than we estimated in segmentation methods and can help gain performance. nnU-Net therefore thoroughly defines all these steps including pre-processing (e.g., resampling

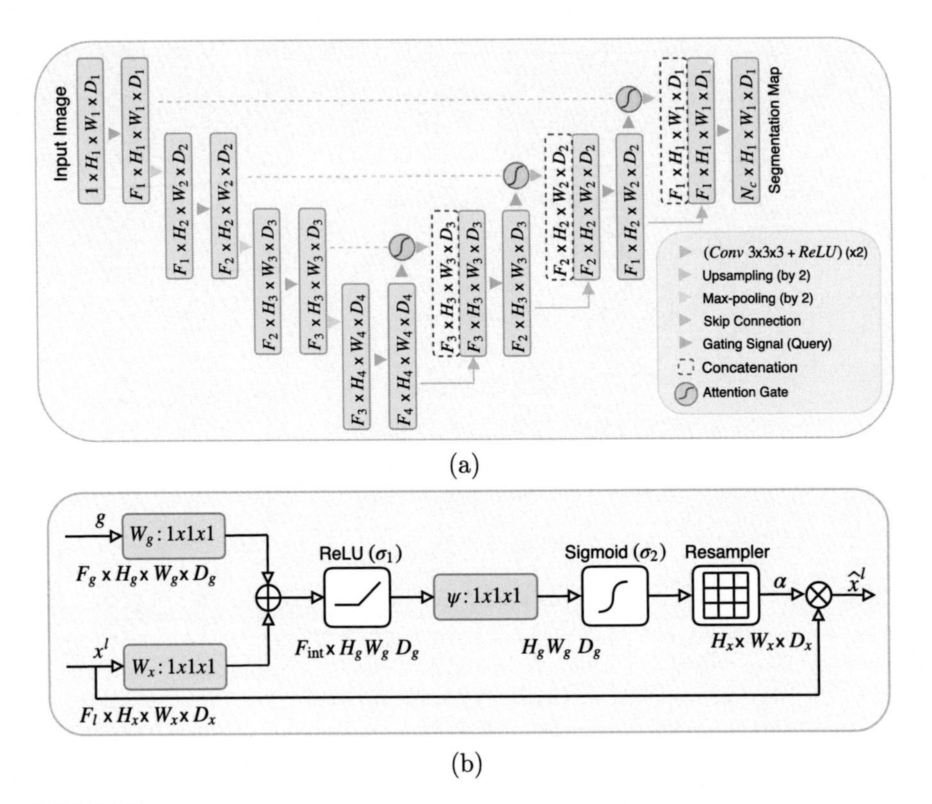

FIGURE 18.4

Attention U-Net and attention gate (AG) from [13]. (a) Attention U-Net. Input image is progressively filtered and downsampled by factor of 2 at each scale in the encoding part of the network. N_c denotes the number of classes. (b) Attention gate. Input features x^l are scaled with attention coefficients α. The spatial regions are selected by analyzing both the activations and contextual information provided by the gating signal g. Grid resampling of attention coefficients is done using trilinear interpolation.

and normalization), training (e.g., loss, optimizer setting, and data augmentation), inference (e.g., patch-based strategy and ensembling across test time augmentations and models), and a potential post-processing step (e.g., enforcing single connected components if applicable) [16]. The value of nnU-Net was further demonstrated by applying it to 11 international biomedical image segmentation challenges comprising 23 different datasets and 53 segmentation tasks (https://cremi.org). Overall, nnU-Net outperformed 33 existing segmentation solutions, even though the latter were specifically optimized for the respective task and otherwise showed performance on par with or close to the top leaderboard entries. nnU-Net thus quickly became a current benchmark for biomedical image segmentation comparisons.

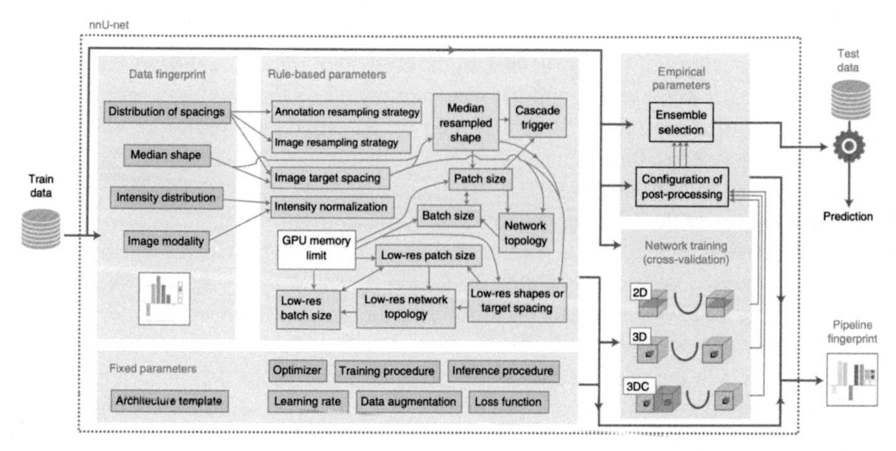

FIGURE 18.5

Automated method configuration of nnU-Net for deep biomedical image segmentation from [3]. Given a new segmentation task, dataset properties are extracted in the form of a *dataset fingerprint* (pink). A set of heuristic rules models parameter interdependencies (shown as thin arrows) and operates on this fingerprint to infer the data-dependent *rule-based parameters* (green) of the pipeline. These are complemented by *fixed parameters* (blue), which are predefined and do not require adaptation. Up to three configurations are trained in a five-fold cross-validation. Finally, nnU-Net automatically performs empirical selection of the optimal ensemble of these models and determines whether post-processing is required (*empirical parameters*, yellow).

18.3 Transformer-based deep learning segmentation

Transformers are a novel form of DL architecture that was originally introduced as a method for effectively processing long sequences of text data. Their self-attention mechanisms allow the model to consider the entire input sequence simultaneously rather than iteratively processing small neighborhoods, as is done with convolutions. Since their introduction in 2017 [17], Transformers have been adapted for 2D and 3D tasks and are considered the state-of-the-art on many major medical imaging benchmarks.

18.3.1 Scaled dot-product attention

One of the defining traits of the Transformer network is the ability to contextualize tokens in a sequence against every other token regardless of the distance between them. While the receptive field, the region of input that may contribute to the processed output, of a CNN must be grown through the iterative stacking of layers, the receptive field of a Transformer can use a single layer to encompass the entire input region.

The innovation that allows Transformers to have this wide field of view is the fully attention-based nature of their structure. In the context of natural language, each word in a sentence is a single token, and the sentence would be the full input sequence. In practice, each of these tokens is represented by an embedded vector $\boldsymbol{x}_i$. These tokens are then passed to the attention unit of a Transformer, the "attention head." The attention head contains three learned weight matrices $\mathbf{W}_Q$, $\mathbf{W}_K$, and $\mathbf{W}_V$ that are optimized as part of the training process, similar to kernel weights in a convolutional layer. Multiplying the input token values by these matrices provides a linear projection of the tokens into a query vector $\boldsymbol{q}_i = \boldsymbol{x}_i \mathbf{W}_Q$, a key vector $\boldsymbol{k}_i = \boldsymbol{x}_i \mathbf{W}_K$, and a value vector $\boldsymbol{v}_i = \boldsymbol{x}_i \mathbf{W}_V$. For each position i in the input sequence, the query vector can be interpreted as a question, "How much attention do you need?" In response, every position j in the sequence responds with their key vector $\boldsymbol{k}_j$, and the pairwise similarity computed using the dot product between the two represents the attention weight that i is applying to j, $a_{(i,j)}$.

These weights are then scaled by dividing by the square of the dimension of the key vectors and normalized using a softmax operation. For large sizes of $\boldsymbol{q}_i$ and $\boldsymbol{k}_j$, the dot product between them can have very large magnitudes. During training, this would push the softmax operation into regions where the gradient becomes small enough to inhibit meaningful back-propagation, so scaling before the softmax helps to stabilize the gradients of the model during optimization.

Once the normalized attention weights have been determined between each position in the input sequence, the output values for each position i are computed as the weighted sum of all value vectors $\boldsymbol{v}_j$ multiplied by their respective attention weight $a_{(i,j)}$. Of important note, the weight matrices $\mathbf{W}_Q$ and $\mathbf{W}_K$ are non-symmetric, meaning that the attention between two positions does not have to be the same in both directions. This allows the model to learn richer attention representations. In practice, this entire process is done simultaneously using a series of highly optimized matrix operations, with all separate token query, key, and value vectors stacked to create matrices **Q**, **K**, and **V**, respectively:

$$\text{Attention}(\mathbf{Q}, \mathbf{K}, \mathbf{V}) = \text{softmax}\left(\frac{\mathbf{Q}\mathbf{K}^T}{\sqrt{d_k}}\right)\mathbf{V}. \tag{18.1}$$

Similar to how each layer of a CNN has multiple filters that are used to learn different representations, Transformer networks use multiple attention heads in each layer which together comprise a Multihead Attention (MHA) module. Each MHA will utilize a number of query-, key-, and value-weight matrices $\mathbf{W}_Q$, $\mathbf{W}_K$, $\mathbf{W}_V$ to jointly learn multiple different representations for each token position. The final values from all of the attention heads are concatenated and multiplied by a final weight matrix $\mathbf{W}_O$ to get the combined result.

In Fig. 18.6, a full Transformer Encoder layer is presented. The embedded tokens are normalized and passed through the MHA module, and a recurrent connection adds them back to the input values. Following a normalization step, they pass through a multilayer perceptron (MLP), and a second recurrent connection performs another addition to combine the outputs of the first and second stages.

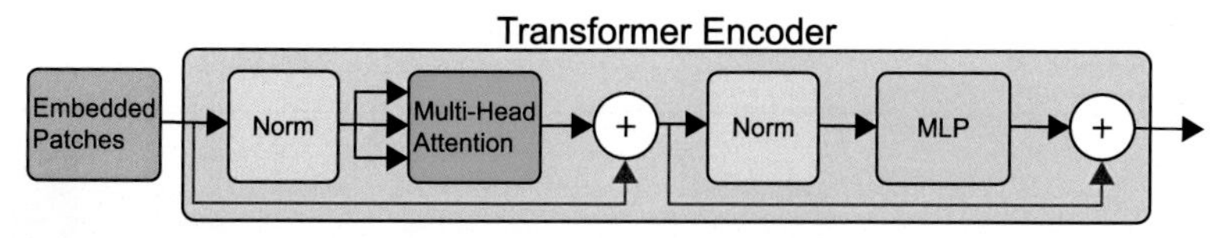

FIGURE 18.6

A standard Vision Transformer encoder as used by Dosovitskiy et al. (2021) [4]. Red arrows represent recurrent connections and white circles represent addition operations.

18.3.2 Positional embedding

The previously described attention mechanism allows a Transformer model to simultaneously process all tokens in an input sequence; however, it loses the implicit spatial information of other architectures. The order of tokens in a sequence often carries important contextual information, so position information is added to the input tokens using positional embedding (PE). Types of PE can be roughly subdivided by two defining features: absolute vs. relative position and fixed vs. learned values.

Absolute PE makes use of a set of vectors with the same length as the input sequence so that the two can be summed before computing the query, key, and value vectors. These vectors are generally defined using sine/cosine wave functions of different frequencies so that each dimension of the input tokens corresponds to a separate sinusoid. The functions used in the original Transformer implementation [17] can be seen in Eq. (18.2), where i is the index in the input sequence and d is the dimension of the token vector at that index:

$$\begin{aligned} PE_{(i,2d)} &= \sin\left(i/(10000^{2d/|input|})\right), \\ PE_{(i,2d+1)} &= \cos\left(i/(10000^{2d/|input|})\right). \end{aligned} \tag{18.2}$$

Rather than encoding a single position value for each location of the input sequence, *relative PE* [18] associates each position in the input sequence with a set of relative position values representing the distance between the token position and all other positions in the sequence. That allows the model to learn more flexible representations of the positional relationships between tokens. As relative PE is specific to the interaction between two tokens, it is generally added to the attention weight after the query and key have been combined but before the softmax operation:

$$\text{Attention}(\mathbf{Q}, \mathbf{K}, \mathbf{V}) = \text{softmax}\left(\frac{\mathbf{QK}^T}{\sqrt{d_k}} + PE_{rel}\right)\mathbf{V}. \tag{18.3}$$

The difference between *fixed PE* and *learned PE* is that the first must be defined by functions that are static while the latter is treated as a bias term that is optimized during the training process. Experimentally, learned absolute PE does not show large improvements over fixed absolute PE, but learned relative PE performs significantly better than other forms of PE [19].

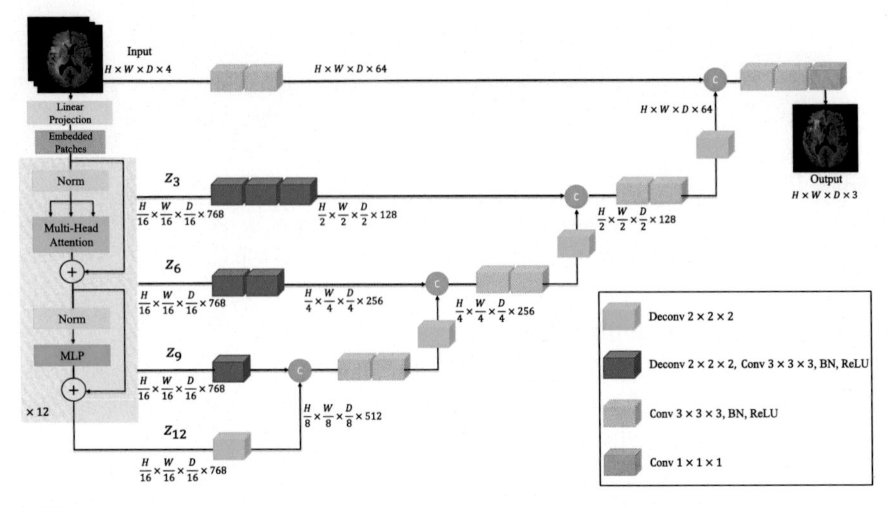

FIGURE 18.7

The UNETR architecture from [15]. Sequence representations are extracted by 12 repeating Transformer Encoder layers and then merged into a U-Net-style decoder using skip connections.

18.3.3 Vision transformers

While the previous discussion has been primarily in the context of sequences of text, the extension from text to images is not as complicated as it may seem. Just as a sentence can be subdivided into words that are embedded to create a sequence of word vectors, an image can be subdivided into patches of pixels that are then reshaped to create a sequence of patch vectors. From this point, the same Transformer architecture can be applied in almost the exact same way. Experiments with PE have even found that there is no significant improvement when going from 1D to 2D PE in the case of image classification [4].

When applying this fully attention-based architecture to computer vision tasks, Transformer models avoid the inductive biases of previous approaches. The CNNs used in previous state-of-the-art approaches have a heavy reliance on local neighborhood structures and have a translational equivariance enforced by the nature of convolution operations that limit their ability to consider the entire input space.

On the other hand, by making full use of the local neighborhood representations, the resolution of a CNN-based segmentation can match that of the input-level resolution. Since a Transformer subdivides an image into patches, the output from a purely Transformer segmentation model will only have a single output for each patch and is insufficient for most medical segmentation tasks even when upsampled to the input sizing. The combination of a Transformer encoding module and a convolutional decoding module in the style of a U-Net architecture, as seen in Fig. 18.7, combines the power of both architectures to generate high-resolution segmentations that encode the long-range dependencies of the input space [15].

While this fixes the issue of output resolution in image segmentation tasks, there is another issue involving input resolution. As a price to pay for the powerful attention mechanism, the computation required for a Transformer to compare every input token to every other token in the input sequence can grow massively, especially in the context of medical imaging. Going from an image size of 256×256 to one of 512×512 means that the Transformer is now comparing $4\times$ the number of patches (or $16\times$ the number of comparisons) if patch size remains the same. This scaling is even more exaggerated when you want to operate over 3D image volumes, as is common in medical tasks.

To handle these issues, novel implementations such as Deformable Attention [20, 21] and Shifted Windows (Swin, [19]) have altered the attention mechanisms within Transformers to learn which tokens need to be compared or to only compare tokens within certain windowed regions of each other.

18.4 Hybrid deep learning segmentation

The DL-based segmentation methods presented above can be trained on segmented examples and – once properly trained with sufficient examples – typically achieve performance that is comparable to or better than that of traditional non-learning-based methods such as those described in Chapters 8–10. In addition, a DL-based method can often be easily adapted to a new application by training it on a fresh set of examples or via transfer learning. Due to these advantages, the DL-based methods have become the most popular segmentation methods in recent years.

However, DL methods are not free of several fundamental limitations. First, most DL segmentation approaches are not topology preserving – in other words, segmentation regions carrying the same label may then be discontinuous even if the respective anatomy requires a single connected region (e.g., one heart rather than several heart regions). Second, DL segmentations typically perform better with respect to regional correctness (finding the target object) than to surface positioning delineation accuracy (accurately identifying object boundaries). Last but not least, if a DL segmentation regionally fails, the correction requires tedious and time-consuming slice-by-slice adjudication by an expert and DL segmentations do not offer inherent smart segmentation correction mechanisms that allow human users to interact with the underlying algorithm to provide additional guidance.

To overcome these limitations, one possible approach is to use a hybrid method that combines DL with a traditional method and is free of the above limitations. One such traditional method is Layered Optimal Graph Image Segmentation of Multiple Objects and Surfaces (LOGISMOS) [22,23], an early version of which is referred to as *optimal surfaces* or *graph search* in Chapter 10, Section 10.7.3.

18.4.1 Deep LOGISMOS

Deep LOGISMOS is an automated n-dimensional image segmentation approach, which introduced a novel combination of DL and optimal graph-based identification of multiple mutually interacting objects and their possibly multiple surfaces.

The LOGISMOS approach is initiated by a deep-learning preliminary segmentation followed by graph-based optimization, guaranteeing topological consistency and global optimality with respect to a (task-specific machine-learned) cost function. The LOGISMOS approach uses a set of surface- and/or region-specific costs, representing the multisurface segmentation task as a graph optimization problem, and finds the optimal solution via s–t cut graph optimization [22–24]. As a result, all target objects and their surfaces are segmented simultaneously in n dimensions in a single optimization process that employs not only information about the surface properties but also utilizes regional information about image appearance of the individual objects and anatomy-based information about mutual object-to-object interactions of the surfaces (e.g., anatomical priors, adjacency, etc.).

Fig. 18.8 (red box) outlines the main steps of automated Deep LOGISMOS, shown on a 3D segmentation task of the prostate and nearby organs: (1) A 3D mesh is generated from DL-based pre-segmentation that gives useful information about the topological structures of the target objects and is considered the approximation to the (unknown) surfaces for target object by DL probability map thresholding and marching cubes algorithm. Any properly trained DL segmentation method can be used for LOGISMOS initialization. The triangulated mesh is computed to specify the structure of the base graph that defines the neighboring relations among voxels on the sought surfaces. (2) A weighted directed graph is built on the vectors of voxels in the image. Each such voxel vector corresponds to a list of nodes in the graph (graph columns) and is typically normal to the mesh surface at mesh vertices. (3) The cost function for LOGISMOS emulates the unlikeness of the graph node residing on the desired surface and/or in a region between surfaces. (4) A single multiobject graph is constructed by linking object-specific subgraphs while incorporating geometric surface constraints and anatomical priors (Chapter 10, Section 10.7.3: hard constraints). (5) The graph construction scheme ensures that the sought optimal surfaces correspond to an optimal closed set in the weighted directed graph [25] and the optimal solution is found by standard s–t cut (or the dual maximal flow) algorithms as the closed set of nodes, one node per column, with a globally optimal total cost. A practical guide to graph construction and cost function design can be found in [26]. More details of the LOGISMOS algorithm can be found in [22,23,27]. Typical examples where LOGISMOS is employed in specific tasks are described in [28,29].

18.4.2 Machine-learned cost function

LOGISMOS can optimize two types of cost functions – on-surface and in-region cost – individually or combined. The on-surface cost, for each node, reflects the unlikeliness that the sought target surface passes through that node. Intensity derivatives and the inverse of image gradient magnitudes are typical choices for on-surface cost.

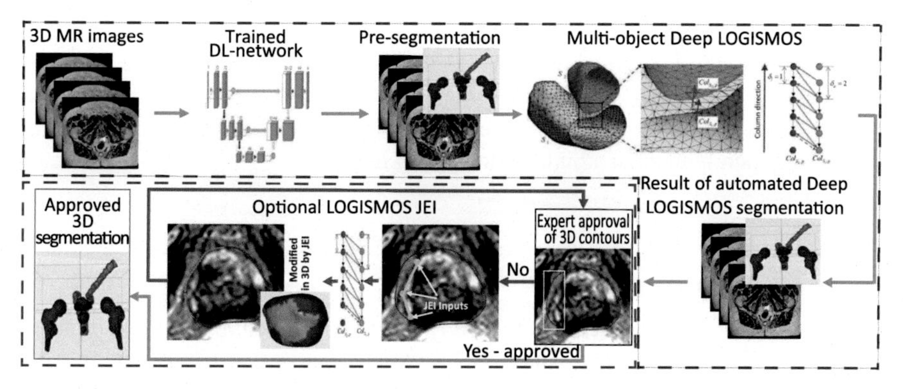

FIGURE 18.8

The main steps of Deep LOGISMOS + JEI are shown using 3D segmentation of prostate and nearby organs as an example. Fully automated Deep LOGISMOS consists of steps outlined in dashed red. Optional JEI steps are outlined in dashed blue.

When multiple layered surfaces of a single object are to be segmented, the in-region cost [27] can be used as unlikeliness of a node to reside in a specific region between a pair of target surfaces. A known Gaussian distribution of the intensity of the target region is often used to model in-region cost.

Designing appropriate cost functions is crucial, and human-designed and machine-learned cost functions can be employed individually or together. Machine-learned cost functions are better suited for segmentation of objects with complex patterns and/or no clean-cut local boundary [29]. DL segmentations of the retina layer [30] also indicate DL-produced probability maps can provide better in-region costs. In addition, as described in Section 18.7, machine learning can be used to identify optimal parameters of human-designed cost functions.

18.4.3 Just-enough interaction

Certain popular maximum flow algorithms [31,32] allow dynamic graph optimization – more flow can be pushed into a optimized graph and dynamically achieve new updated maximum flow without re-starting the optimization from scratch. This dynamic nature enables highly efficient minimally interactive expert guidance (called just-enough interaction [JEI]) for LOGISMOS. JEI utilizes user inputs about correct surface locations to directly interact with the maximum flow algorithm via modifying costs on a small portion of graph nodes and quickly – often in milliseconds [24] – produce updated segmentation that still maintains its global optimality and geometry regulated by constraints. Fig. 18.8 outlines the main steps of this general JEI approach (blue box).

In practice, user interaction on one 2D slice is often enough to correct segmentation errors three-dimensionally in its neighboring 2D slices and thus reduce the amount of human effort. In addition, due to the existence of embedded inter-object constraints, in regions where multiple compartments are close to each other, editing is only needed on one compartment. More details about the design of the JEI architecture are covered in [26]. LOGISMOS-JEI approaches have been successfully applied in [28,29] and are useful in reducing the inter-observer and intra-observer variability compared to purely manual adjudication approaches.

18.5 Training efficiency

Despite the considerable success of DL, its adoption in medical imaging is hampered by the lack of large annotated datasets. Annotating medical images is not only tedious and time consuming but also demands highly specialized and thus costly skills. Assisted and sparse annotation strategies aim at easing this critical issue by dramatically reducing the annotation effort while maintaining the best possible performance. There are two main directions for alleviating the burden of annotation: (1) reduce the annotation time needed for each image dataset and (2) reduce the size of the training dataset for annotation.

An assisted annotation method that employs the iterative loop introduced above (Section 18.4) to efficiently and iteratively add new annotated datasets was reported in [10]. The core idea of this approach is utilizing the automated DL segmentation results as the preliminary labels of unannotated images and employing efficient adjudication (JEI) for label correction by experts. A new larger training set of both the initial images with fully manual ground truth and the new images with expert-adjudicated ground truth are formed and iteratively used to train the increasingly better versions of the automated segmentation model. Benefiting from the advantage of JEI in highly efficient minimally interactive expert guidance and the effectiveness of the iteratively trained segmentation models, the time required for careful 3D annotation can be dramatically reduced. Automatic segmentation quality assessment (SQA) is utilized throughout this process of assisted annotation to reduce the effort of reviewing and editing a large dataset even further. For this purpose, a DL framework can be developed to locate erroneous regions on segmented results without ground truth associated with the to-be-segmented data [33] that works in three main steps: (1) determine difference images between the machine segmentation and ground truth, (2) train a U-Net-based network for volumetric error prediction using the original image and segmentation as two channel inputs and the difference image calculated as target, and (3) visualize the predicted erroneous volume on the surfaces of the segmentation volume to guide the human expert to segmentation regions likely needing corrections, greatly increasing the overall annotation efficiency. Performance-improving approaches exist that use click and scribble corrections for additional efficiency gains [34–36].

Another useful approach to improve training efficiency is to actively select data samples that can best contribute to successful learning and focus the annotation effort accordingly. The key idea is that the learning algorithm can perform well with less training data if it is allowed to choose the *valuable* data. Clearly, the data selection criterion that properly assesses the task-specific value of available data is crucial. Four major approaches were presented so far according to the data selection criteria – methods based on uncertainty, diversity, change, and their hybrid combinations. Uncertainty-based methods [37–39], as the name suggests, select data by defining and measuring the level of data uncertainty. Diversity-based methods [40–42] resort to diversity that represents the whole distribution of the unlabeled dataset. Change-based methods [43,44] identify data that would cause the greatest change to the current model parameters or outputs if their labels are available. Hybrid methods [45–48] combine multiple of the aforementioned strategies. In all cases, the annotation effort is reduced by decreasing the training set size; a 50% training data reduction is common. Different from these approaches, a novel task-agnostic and loss-based method was reported in [49]. It utilized a small parametric *loss prediction module* that is trained by fused multilevel features from the mid-blocks of the main network to predict target losses of unlabeled inputs. By selecting only the data identified by the module as likely producing a wrong prediction by the main model, the method effectively reduces the size of the training set. Similar to assisted annotations, relevant SQA methods can be trained to identify areas of likely segmentation errors, which are then considered as a type of *loss* between the segmentation and unavailable ground truth. Datasets that exhibit more frequent or larger areas of such predicted segmentation errors suggest that the main model is inclined to produce incorrect segmentation and thus the identified datasets are more *valuable* to annotate and subsequently use for training.

Different from the subject level of assisted annotation, sparse annotation identifies a subset of the most effective and representative 2D slices from the 3D (or higher-dimensional) dataset for annotation. The key strategy of sparse annotation is mining and utilizing effective data, and therefore unsupervised learning and self-training are usually involved. A DL CMC-Net framework bridges the gap between full annotation and sparse annotation in 3D medical image segmentation [50] and consists of three main steps: (1) Representative slice selection: An auto-encoder is utilized as the representation extractor for each slice in the unannotated dataset and representative features are learned in an unsupervised manner. Then the similarity between any two slices is determined as the cosine similarity of the corresponding two representative features. An optimal slice subset selection task is converted to a maximum set cover problem and a good approximate solution can be obtained using a greedy method. (2) Pseudolabel generation by training ensemble models: For a 3D segmentation task, three 2D segmentation modules are trained using slices selected and annotated in the previous step along each of the three orthogonal planes and the inference results of the unlabeled slices generated as pseudolabels. A single 3D segmentation module is then trained using the 3D images that contain both the expert annotations and pseudolabels. The final pseudolabels are obtained by averaging the four probability maps from the ensemble models. (3) Segmentation generation by self-training: A single 3D segmentation model is trained using the final pseudolabels by employing a self-training mechanism, in which the segmentation results of the unlabeled slices from the 3D segmentation model are regarded as pseudo-*ground truth* and used to further retrain the four ensemble models. The CMC-Net method achieved good performance with sparsely annotating less than 40% of slices from a 3D MR calf muscle dataset (Section 18.7) and outperformed 3D Attention U-Net that was trained with full annotation. A similar sparse annotation method (KCB-Net [51]) was successful in segmenting 3D MR knee joint datasets, yielding high-quality results when annotating less than 10% of training data slices.

18.6 Explainability

When it comes to medical imaging, understanding how the DL model reaches its decision can be just as important as the decision itself. To develop and deploy an automated process that has impact in a clinical setting requires a significant amount of trust that it is not only reaching the correct conclusion but that it is reaching it in a robust and generalizable way.

Fig. 18.9 shows an attribution map generated for a model trained to classify radiographs based on the presence of pneumonia. We can see that a surprising amount of importance is attributed to the heart rather than the more diagnostically relevant lungs. On closer inspection of the dataset, it turns out that a disproportionate majority of subjects with pneumonia represented in the dataset were imaged using a portable X-Ray device as compared to those without pneumonia. The images were acquired using the erect anteroposterior (AP) chest view as opposed to the posteroanterior (PA) chest view. The AP view is of lesser quality than the PA view, but it can be the only one available in patients unable to leave their bed. By learning the position of the heart on the radiograph, the model was able to learn a feature strongly cor-

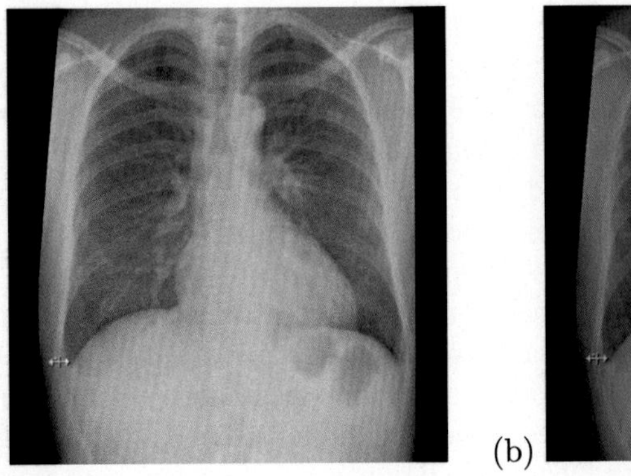

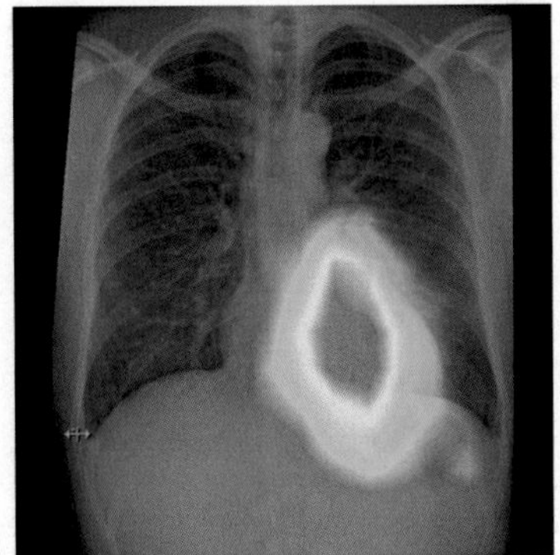

(a) (b)

FIGURE 18.9

An example attribution map for pneumonia classification. (a) A radiograph image from the CheXpert dataset [39] predicted as positive for pneumonia by a DL model. (b) The Grad-CAM [52] attribution map for the prediction, showing a high level of attribution focused on the heart.

related with the target label in the training data without having to actually learn the more difficult features of the actual presentation of the pathology. Due to the unfortunate sparsity of labeled data in many applications, the training and test data used to train DL models often come from the same institution and thus may exhibit the consistently same albeit often subtle biases. Explanatory methods can shine light on the internal processes of these "black boxes" and enable to better understand their strengths and limitations.

18.6.1 Image attribution

The predominant class of methods for explainability in DL classification is post-hoc attribution. Attribution models assign a value to each location in the input image that represents the influence that area has on the output prediction of the model. The post-hoc label attached to these methods indicates that they can be applied to a model after it has already been trained, without any changes to structure or fine-tuning. Class Activation Map (CAM) methods, such as GradCAM [52] or ScoreCAM [53], take the output activations A_k^l from a convolutional layer l in a model f and perform a weighted sum over the features using method-specific weights w_k^l to generate the attribution map for the model's prediction (Eq. (18.4) shows the general form). The output of the layer represents the position of features that activate the convolutional kernels, and the inherent dependence on local neighborhoods means that the positions of the feature map are spatially consistent with positions in the input image. We have

$$M_{f|X}^l = \text{ReLU}\left(\sum_k w_k^l A_k^l\right). \tag{18.4}$$

Unfortunately, structural assumptions were made when CAM methods were developed for use with classification models that lead them to being ineffective with segmentation models. The structure of a segmentation model such as the U-Net in Fig. 18.2 contains multiple skip connections allowing data to cross from early to later parts of the model. This means that the only layer that can be available to the CAM methods to fully capture the model's flow of data is the final convolutional layer. However, all of the surrounding contextual information has already been refined down to generate the final segmentation prediction prior to reaching the final layer. One easy adaptation could be to apply a CAM method to all layers of the model after the bottleneck and average them together. This approach is known to provide a decent approximation, but generally results in multiple false positive regions.

One CAM method specifically designed for segmentation attribution is kernel-weighted contribution (KWC) [54]. Similarly to other CAM methods, KWC uses a heuristic to obtain a weighted sum of activation maps. Unlike other CAM methods, KWC obtains this weighted sum simultaneously across all post-bottleneck layers. For each feature map A_k^l in each layer, KWC computes the dependent contribution (D_k^l) and the independent contribution (I_k^l) of the convolutional kernel that generated the feature map with respect to the output prediction. DC is defined as the percentage of the segmentation predicted by the model that is removed when the corresponding kernel is turned off, and IC is the percentage of the predicted segmentation that remains when all kernels in a given layer are turned off *except* for the corresponding kernel. The product of the two is the contribution weight for the given kernel and is used to perform a weighted sum of all activation maps from all post-bottleneck kernels.

18.6.2 Attention gates

Aside from CAM methods, another approach to explainability in segmentation models is to employ attention gates. The model architecture is supplemented with a number of gated modules [55]. In these, the model learns a function $C_{1\times1}$ that returns a feature map with the same spatial dimensions as the input to the module but with only a single feature channel. This map is then elementwise multiplied against the input to the module, so that the single channel feature map is explicitly controlling the flow of information through the model. Not only do these modules encourage the model to better focus on relevant parts of the image, but as the images exist in the same spatial dimensions as the input, they are just as interpretable as the heatmaps generated by the CAM methods discussed above. In exchange for the addition of architectural constraints, attention gates provide a built-in method of explainability for segmentation models.

18.7 Case study

To demonstrate the power of DL and hybrid segmentations and to show how they can benefit from assisted annotation training, a 3D calf muscle segmentation case study is presented. Anatomically, the calf muscles are composed of five individual compartments shown in Fig. 18.10 (tibialis anterior [TA], tibialis posterior [TP], soleus [Sol], gastrocnemius [Gas], and peroneal longus [PL] [56]). Structural and volumetric changes of these compartments provide valuable information about the diagnosis, severity, and progression of various muscular diseases. MR imaging facilitates noninvasive muscle analysis. The hybrid DL + optimization approach (Section 18.4), benefiting from assisted annotation (Section 18.5), allows highly accurate calf muscle compartment segmentation.

Popular purely DL segmentation and hybrid approaches, trained on assist-annotated datasets of varying sizes, were selected for performance comparison – nnU-Net, FilterNet+, TransUNet, CoTr, and Deep LOGISMOS (Sections 18.2 and 18.3) – all working in a fully automated regime. In total, 350 lower-leg MR images from 93 subjects with annotated calf muscle compartments were available (47 healthy, 46 with different severity of muscular dystrophy DM1 disease, $512 \times 512 \times 30$, voxel size $0.7 \times 0.7 \times 7$ mm). These expert annotations served as an independent standard. All methods were initially trained on a small fully expert-annotated subset of 80 legs from 40 subjects, with the remaining 270 leg MR images added to the second training set following their JEI-assisted annotation

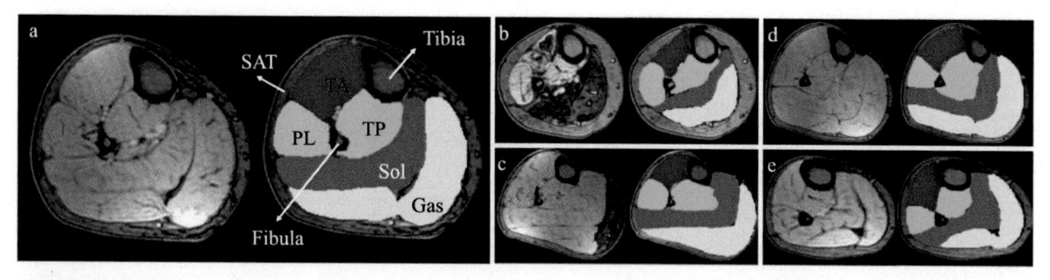

FIGURE 18.10

Examples of T1-weighted MR images of calf muscle cross-sections and corresponding expert segmentations of TA, TP, Sol, Gas, and PL (see text). (a) Normal subject. (b, c) Patients with severe muscular dystrophy DM1. (d) Patient at risk for DM1. (e) Patient with juvenile onset of DM1 (JDM). Note MR appearance variability.

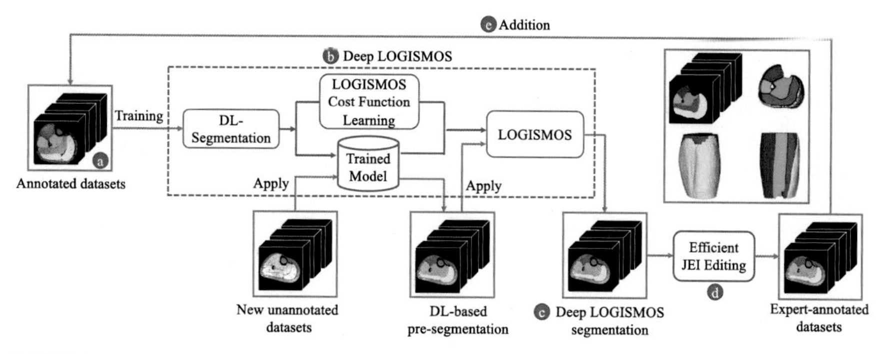

FIGURE 18.11

Workflow of the Deep LOGISMOS segmentation framework in the context of assisted annotation. Processing steps are shown in blue, datasets are shown in orange.

(Section 18.4.3). Once trained on the 80 datasets, individual automated segmentation methods delivered decent segmentations of the five calf compartments in the remaining 270 MR datasets. Segmentations resulting from the automated Deep LOGISMOS, for which JEI adjudication is feasible, were reviewed and – if needed – interactively JEI-corrected, serving as additional training data used for all approaches in the assisted annotation training loop (Fig. 18.11). The average time of reviewing and editing each 3D MR image used in the assisted annotation loop was approximately 25 minutes; expert annotation effort was thus decreased by 95% compared to fully manual annotation; each dataset required 8 hours of expert manual tracing in 3D Slicer on average [57]. Performance of each method was assessed when trained on the initial set of 80 legs and trained/tested using cross-validation on a full set of 350 legs. Deep LOGISMOS was initialized by the FilterNet+ segmentation method.

Performance of five DL or hybrid segmentation approaches trained on two differently sized datasets was compared. Fig. 18.12 shows the achieved results; increasing the training size had a uniformly positive impact on each method's performance.

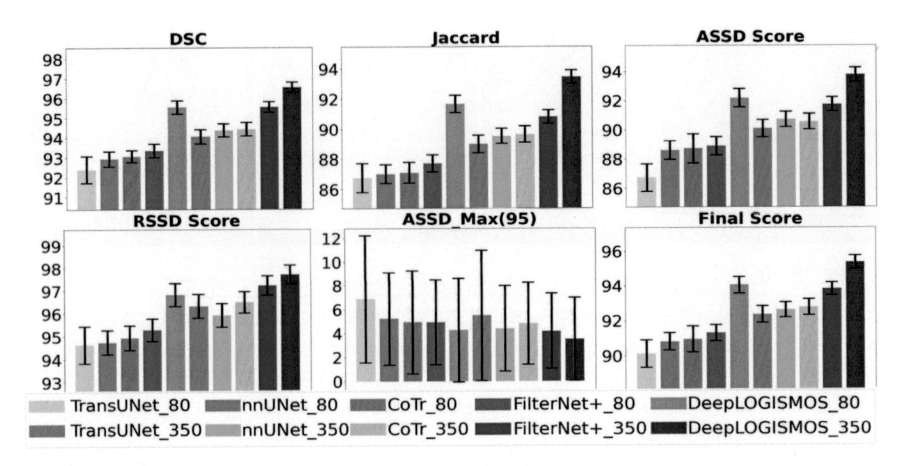

FIGURE 18.12

Performance comparison for the segmentation of five calf muscle compartments by four DL and one hybrid segmentation method. Performance metrics evaluated for each leg segmentation in 3D: DSC (Dice similarity coefficient), the Jaccard coefficient, ASSD (absolute surface-to-surface distance) (unsigned), RSSD (relative surface-to-surface distance) (signed), $\text{ASSD}_{max(95)}$ (the 95th percentile of Hausdorff distances), and the final score (weighted combination of the individual metrics listed). Subscript 80 or 350 indicates cardinality of the employed training set.

Note, however, that the hybrid Deep LOGISMOS approach offered not only the best performance when trained on each of the two differently sized datasets, its variant trained on only 80 legs also outperformed all purely DL segmentation approaches, even those trained on a full set of 350 MR leg images (Fig. 18.13).

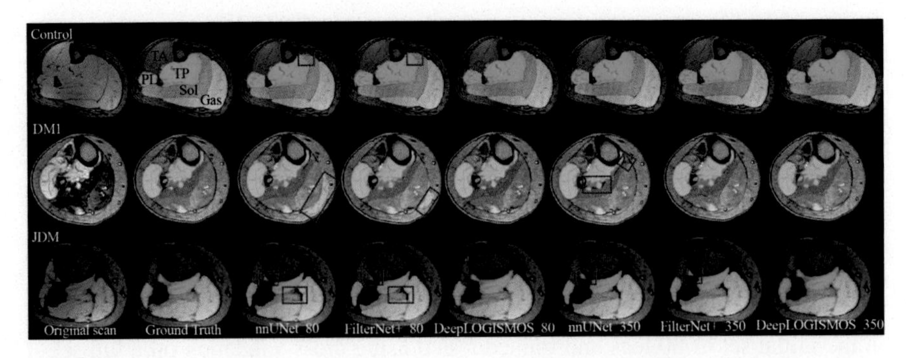

FIGURE 18.13

MR segmentation of calf muscle compartments. Red rectangles highlight regions with segmentation errors. For each presented dataset, segmentation improvements are noticeable for each more advanced method from nnU-Net to Deep LOGISMOS.

This case study demonstrates that a hybrid machine learning framework combining the main advantages of DL-based segmentation with a subsequent optimization approach is well positioned to outperform purely DL segmentation methods. Furthermore, by maximizing the value of an initial small dataset of fully annotated MR images of 80 lower legs and initially training a segmentation method on this small dataset, an efficient assisted annotation strategy can be designed to markedly increase the efficiency of data annotation for segmentation training purposes, offering clinically acceptable performance in an automated regime. The presented hybrid combination additionally offers efficient JEI adjudication tools if such steps would be needed, an approach not directly available for purely DL segmentation methods.

18.8 Recapitulation

This chapter was devoted to medical image segmentation approaches that learn their parameters from sets of annotated segmentation examples. Understanding the learning concepts presented in Chapter 16 is a prerequisite to comprehension of methods, techniques, and approaches concerning learning-based medical image segmentation, the most relevant of which were discussed here:

- FCNs take image inputs of arbitrary size and produce a correspondingly sized likelihood (probability) map based on efficient inference and learning from segmentation examples. The final segmentation is achieved via the argmax operation, which selects the category with the highest probability on the likelihood map.
- U-Net, an extension to the original FCN structure, is based on a symmetrical encoder–decoder architecture that exhibits a contracting path to capture context features and a symmetric expansive path facilitating precise localization. This structure leverages the image features by propagating context information to higher-resolution layers.
- Following its success, numerous modifications of the U-Net architecture were introduced with nnU-Net. This represents a powerful well-generalizing strategy, considered today's state-of-the-art for FCN DL segmentations.
- Transformer networks utilize a fully attention-based structure to contextualize tokens in a sequence against every other token regardless of the distance between them, allowing Transformers to have a wide field of view. For image segmentation, an image is subdivided into patches of pixels that are then reshaped to create a sequence of patch vectors providing a single input and model output for each patch. As such, segmentation resolution of pure Transformers is insufficient for most medical segmentation tasks even when upsampled to the input sizing. Jointly using a Transformer encoding module with a convolutional decoding module combines the power of both architectures to generate desirable high-resolution segmentations.
- Hybrid approaches combine DL and traditional segmentation strategies. Concatenation of DL segmentation (for robust localization) and traditional approaches

(for accurate delineation) overcomes their individual limitations and may enable efficient algorithmic correction currently not available for pure DL segmentations.

- Learning-based image segmentation approaches require large annotated image datasets for training. Annotating medical images requires highly specialized skills and as such is not only tedious and time consuming but also difficult to achieve. Assisted and sparse annotation strategies aim at easing this critical issue by dramatically reducing the annotation effort while maintaining the best possible performance.
- Use of learning-based medical image segmentation for clinical care requires that segmentation results can be understood and thus trusted. This understanding is not immediately available since DL approaches typically function as opaque "black boxes." Explanatory methods shine light on DL internal processes and enable better understanding of their strengths and limitations.

18.9 Further reading

1. Key papers on convolution-based segmentation networks: [1–3]
2. Reading on Vision Transformers: [4,19]
3. An interesting comparison of CNNs and Transformers: [58]
4. Reading on hybrid models: [22,26]
5. Reading on DL segmentation explainability: [53–55]

18.10 Exercises

1. Why is U-Net (and/or its variants) an effective approach to medical image segmentation? Identify U-Net's main architectural features that contribute to its success and explain how these features are linked to the typical properties of medical images.
2. Compared to Transformer-based methods, what are the merits and demerits of CNN-based methods?
3. Evaluation of different hyperparameters has revealed that Transformer networks can often display improved performance with decreased patch sizes. Assume you reduce the size of patches used by a Transformer network from 16×16 to 8×8.
   a. How does this affect the number of parameters needed for the network?
   b. How does this affect the number of computations required during a forward pass through the MHA module?
4. Fig. 18.9 gives an example where a model is using incorrect image features to predict the presence of pneumonia. For what task (different from the detection of pneumonia) would the shown attribution map indicate that the model has correctly learned to use image features?
5. **Hybrid segmentation** Consider a 3D U-Net and a 3D statistical shape model (Chapter 8) that are trained individually.

a. How can we use the statistical shape model to restrict the U-Net segmentation result?
b. How can we design an active shape model segmentation approach that utilizes the probability map produced by U-Net?

References

[1] Long J., Shelhamer E., Darrell T., Fully convolutional networks for semantic segmentation, in: Proceedings of the IEEE Conference on Computer Vision and Pattern Recognition, 2015, pp. 3431–3440, https://github.com/BVLC/caffe/wiki/Model-Zoo#fcn.
[2] Ronneberger O., Fischer P., Brox T., U-Net: Convolutional networks for biomedical image segmentation, in: International Conference on Medical Image Computing and Computer-Assisted Intervention, Springer, 2015, pp. 234–241, https://lmb.informatik.uni-freiburg.de/people/ronneber/u-net.
[3] Isensee F., Jaeger P.F., Kohl S.A.A., Petersen J., Maier-Hein K.H., nnU-Net: a self-configuring method for deep learning-based biomedical image segmentation, Nature Methods 18 (2) (2021) 203–211, https://github.com/MIC-DKFZ/nnunet.
[4] Dosovitskiy A., Beyer L., Kolesnikov A., Weissenborn D., Zhai X., Unterthiner T., Dehghani M., Minderer M., Heigold G., Gelly S., Uszkoreit J., Houlsby N., An image is worth 16x16 words: transformers for image recognition at scale, in: International Conference on Learning Representations, 2021, https://github.com/google-research/vision_transformer.
[5] LeCun Y., Boser B., Denker J.S., Henderson D., Howard R.E., Hubbard W., Jackel L.D., Backpropagation applied to handwritten zip code recognition, Neural Computation 1 (4) (1989) 541–551.
[6] Krizhevsky A., Sutskever I., Hinton G.E., ImageNet classification with deep convolutional neural networks, Communications of the ACM 60 (6) (2017) 84–90.
[7] Çiçek Ö., Abdulkadir A., Lienkamp S.S., Brox T., Ronneberger O., 3D U-Net: learning dense volumetric segmentation from sparse annotation, in: International Conference on Medical Image Computing and Computer-Assisted Intervention, Springer, 2016, pp. 424–432, https://lmb.informatik.uni-freiburg.de/resources/opensource/unet.en.html.
[8] Milletari F., Navab N., Ahmadi S.-A., V-Net: Fully convolutional neural networks for volumetric medical image segmentation, in: 2016 Fourth International Conference on 3D Vision (3DV), IEEE, 2016, pp. 565–571, https://github.com/mattmacy/vnet.pytorch.
[9] Guo Z., Zhang H., Chen Z., van der Plas E., Gutmann L., Thedens D., Nopoulos P., Sonka M., Fully automated 3D segmentation of MR-imaged calf muscle compartments: neighborhood relationship enhanced fully convolutional network, Computerized Medical Imaging and Graphics 87 (2021) 101835.
[10] Zhang L., Guo Z., Zhang H., van der Plas E., Koscik T.R., Nopoulos P.C., Sonka M., Assisted annotation in Deep LOGISMOS: Simultaneous multi-compartment 3D MRI segmentation of calf muscles, Medical Physics (2023), https://doi.org/10.1002/mp.16284.
[11] Zhou Z., Siddiquee M.M.R., Tajbakhsh N., Liang J., UNet++: a nested U-Net architecture for medical image segmentation, in: Deep Learning in Medical Image Analysis and Multimodal Learning for Clinical Decision Support, Springer, 2018, pp. 3–11, https://github.com/MrGiovanni/UNetPlusPlus.
[12] Lee C.Y., Xie S., Gallagher P., Zhang Z., Tu Z., Deeply-supervised nets, in: Artificial Intelligence and Statistics, in: PMLR, 2015, pp. 562–570.

[13] Oktay O., Schlemper J., Folgoc L., Lee M., Heinrich M., Misawa K., Mori K., McDonagh S., Hammerla N.Y., Kainz K., Ben B., Rueckert D., Attention U-Net: Learning where to look for the pancreas, arXiv preprint, arXiv:1804.03999, 2018, https://github.com/ozan-oktay/Attention-Gated-Networks.
[14] Chen J., Lu Y., Yu Q., Luo X., Adeli E., Wang Y., Lu L., Yuille A.L., Zhou Y., TransUNet: transformers make strong encoders for medical image segmentation, arXiv preprint, arXiv:2102.04306, 2021, https://github.com/Beckschen/TransUNet.
[15] Hatamizadeh A., Tang Y., Nath V., Yang D., Myronenko A., Landman B., Roth H., Xu D., UNETR: transformers for 3D medical image segmentation, in: Proceedings of the IEEE/CVF Winter Conference on Applications of Computer Vision, 2022, pp. 574–584, https://github.com/Project-MONAI/research-contributions/tree/main/UNETR.
[16] Isensee F., Petersen J., Klein A., Zimmerer D., Jaeger P.F., Kohl S., Wasserthal J., Koehler G., Norajitra T., Wirkert S., Maier-Hein K.H., nnU-Net: Self-adapting framework for U-Net-based medical image segmentation, arXiv preprint, arXiv:1809.10486, 2018.
[17] Vaswani A., Shazeer N., Parmar N., Uszkoreit J., Jones L., Gomez A.N., Kaiser L., Polosukhin I., Attention is all you need, in: Guyon I., et al. (Eds.), Advances in Neural Information Processing Systems, vol. 30, Curran Associates, Inc., 2017.
[18] Shaw P., Uszkoreit J., Vaswani A., Self-attention with relative position representations, in: Association for Computational Linguistics, Mar. 2018, pp. 464–468, https://doi.org/10.18653/v1/N18-2074.
[19] Liu Z., Lin Y., Cao Y., Hu H., Wei Y., Zhang Z., Lin S., Guo B., Swin transformer: hierarchical vision transformer using shifted windows, https://github.com/microsoft/Swin-Transformer, Mar. 2021.
[20] Xie Y., Zhang J., Shen C., Xia Y., CoTr: efficiently bridging CNN and transformer for 3D medical image segmentation, in: de Bruijne M., et al. (Eds.), Medical Image Computing and Computer Assisted Intervention - MICCAI 2021, Springer International Publishing, Cham, ISBN 978-3-030-87199-4, 2021, pp. 171–180.
[21] Zhu X., Su W., Lu L., Li B., Wang X., Dai J., Deformable DETR: deformable transformers for end-to-end object detection, in: International Conference on Learning Representations, 2021, https://github.com/fundamentalvision/Deformable-DETR.
[22] Li K., Wu X., Chen D.Z., Sonka M., Optimal surface segmentation in volumetric images — a graph-theoretic approach, IEEE Transactions on Pattern Analysis and Machine Intelligence 28 (1) (2006) 119–134.
[23] Yin Y., Zhang X., Williams R., Wu X., Anderson D.D., Sonka M., LOGISMOS — layered optimal graph image segmentation of multiple objects and surfaces: cartilage segmentation in the knee joint, IEEE Transactions on Medical Imaging 29 (12) (2010) 2023–2037.
[24] Sun S., Sonka M., Beichel R.R., Graph-based IVUS segmentation with efficient computer-aided refinement, IEEE Transactions on Medical Imaging 32 (8) (2013) 1536–1549.
[25] Wu X., Chen D.Z., Optimal net surface problems with applications, in: Automata, Languages and Programming, in: Lecture Notes in Computer Science, vol. 2380, Springer, 2002, pp. 1029–1042, https://link.springer.com/chapter/10.1007/3-540-45465-9_88.
[26] Zhang H., Lee K., Chen Z., Kashyap S., Sonka M., Chapter 11 - LOGISMOS-JEI: segmentation using optimal graph search and just-enough interaction, in: Zhou S.K., Rueckert D., Fichtinger] G. (Eds.), Handbook of Medical Image Computing and Computer Assisted Intervention, in: The Elsevier and MICCAI Society Book Series, Academic Press, ISBN 978-0-12-816176-0, 2020, pp. 249–272.

[27] Garvin M.K., Abràmoff M.D., Wu X., Russell S.R., Burns T.L., Sonka M., Automated 3-D intraretinal layer segmentation of macular spectral-domain optical coherence tomography images, IEEE Transactions on Medical Imaging 28 (9) (2009) 1436–1447.
[28] Kashyap S., Zhang H., Sonka M., Just enough interaction for fast minimally interactive correction of 4D segmentation of knee MRI, Osteoarthritis and Cartilage 25 (2017) S224–S225.
[29] Kashyap S., Zhang H., Rao K., Sonka M., Learning-based cost functions for 3-D and 4-D multi-surface multi-object segmentation of knee MRI: data from the osteoarthritis initiative, IEEE Transactions on Medical Imaging 37 (5) (2018) 1103–1113.
[30] Xie H., Xu W., Wu X., A deep learning network with differentiable dynamic programming for retina OCT surface segmentation, arXiv preprint, arXiv:2210.06335, 2022.
[31] Boykov Y., Kolmogorov V., An experimental comparison of min-cut/max-flow algorithms for energy minimization in vision, IEEE Transactions on Pattern Analysis and Machine Intelligence 26 (9) (2004) 1124–1137.
[32] Goldberg A.V., Hed S., Kaplan H., Kohli P., Tarjan R., Werneck R.F., Faster and more dynamic maximum flow by incremental breadth-first search, in: Algorithms-ESA 2015, in: Lecture Notes in Computer Science, vol. 9294, Springer, 2015, pp. 619–630.
[33] Zamana F.A., Zhang L., Zhang H., Sonka M., Wu X., Segmentation quality assessment by automated detection of erroneous surface regions in medical images, TechRxiv preprint, https://doi.org/10.36227/techrxiv.19767661.v2, 2022.
[34] Xu N., Price B., Cohen S., Yang J., Huang T., Deep interactive object selection, in: Proceedings of the IEEE Conference on Computer Vision and Pattern Recognition, 2016, pp. 373–381.
[35] Wang G., Zuluaga M.A., Li W., Pratt R., Patel P.A., Aertsen M., Doel T., David A.L., Deprest J., Ourselin S., Vercauteren T., DeepIGeoS: a deep interactive geodesic framework for medical image segmentation, IEEE Transactions on Pattern Analysis and Machine Intelligence 41 (7) (2018) 1559–1572.
[36] Benenson R., Popov S., Ferrari V., Large-scale interactive object segmentation with human annotators, in: Proceedings of the IEEE/CVF Conference on Computer Vision and Pattern Recognition, 2019, pp. 11700–11709.
[37] Wang K., Zhang D., Li Y., Zhang R., Lin L., Cost-effective active learning for deep image classification, IEEE Transactions on Circuits and Systems for Video Technology 27 (12) (2016) 2591–2600.
[38] Beluch W.H., Genewein T., Nurnberger A., Kohler J.M., The power of ensembles for active learning in image classification, in: Proceedings of the IEEE Conference on Computer Vision and Pattern Recognition, 2018, pp. 9368–9377.
[39] Irvin J., Rajpurkar P., Ko M., Yu Y., Ciurea-Ilcus S., Chute C., Marklund H., Haghgoo B., Ball R., Shpanskaya K., Seekins J., Mong D.A., Halabi S.S., Sandberg J.K., Jones R., Larson D.B., Langlotz C.P., Patel B.N., Lungren M.P., Ng A.Y., CheXpert: a large chest radiograph dataset with uncertainty labels and expert comparison, Proceedings of the AAAI Conference on Artificial Intelligence 33 (01) (2019) 590–597.
[40] Nguyen H.T., Smeulders A., Active learning using preclustering, in: Proceedings of the Twenty-First International Conference on Machine Learning, 2004, p. 79.
[41] Mustafa B., Getoor L., Link-based active learning, in: NIPS Workshop on Analyzing Networks and Learning with Graphs, vol. 4, 2009.
[42] Guo Y., Active instance sampling via matrix partition, in: Advances in Neural Information Processing Systems, vol. 23, 2010.
[43] Settles B., Craven M., Ray S., Multiple-instance active learning, in: Advances in Neural Information Processing Systems, vol. 20, 2007.

[44] Freytag A., Rodner E., Denzler J., Selecting influential examples: active learning with expected model output changes, in: European Conference on Computer Vision, Springer, 2014, pp. 562–577.
[45] Paul S., Bappy J.H., Roy-Chowdhury A.K., Non-uniform subset selection for active learning in structured data, in: Proceedings of the IEEE Conference on Computer Vision and Pattern Recognition, 2017, pp. 6846–6855.
[46] Liu B., Ferrari V., Active learning for human pose estimation, in: Proceedings of the IEEE International Conference on Computer Vision, 2017, pp. 4363–4372.
[47] Yang L., Zhang Y., Chen J., Zhang S., Chen D.Z., Suggestive annotation: a deep active learning framework for biomedical image segmentation, in: International Conference on Medical Image Computing and Computer-Assisted Intervention, Springer, 2017, pp. 399–407.
[48] Zhou Z., Shin J.Y., Gurudu S.R., Gotway M.B., Liang J., Active, continual fine tuning of convolutional neural networks for reducing annotation efforts, Medical Image Analysis 71 (2021) 101997.
[49] Yoo D., Kweon I.S., Learning loss for active learning, in: Proceedings of the IEEE/CVF Conference on Computer Vision and Pattern Recognition, 2019, pp. 93–102.
[50] Peng Y., Zheng H., Zhang L., Sonka M., Chen D.Z., CMC-Net: 3D calf muscle compartment segmentation with sparse annotation, Medical Image Analysis 79 (2022) 102460.
[51] Peng Y., Zheng H., Liang P., Zhang L., Zaman F., Wu X., Sonka M., Chen D.Z., KCB-Net: A 3D knee cartilage and bone segmentation network via sparse annotation, Medical Image Analysis 82 (2022) 102574.
[52] Selvaraju R.R., Cogswell M., Das A., Vedantam R., Parikh D., Batra D., Grad-CAM: visual explanations from deep networks via gradient-based localization, International Journal of Computer Vision (ISSN 0920-5691) 128 (2 Feb. 2020) 336–359, https://doi.org/10.1007/s11263-019-01228-7, https://github.com/jacobgil/pytorch-grad-cam.
[53] Wang H., Wang Z., Du M., Yang F., Zhang Z., Ding S., Mardziel P., Hu X., Score-CAM: score-weighted visual explanations for convolutional neural networks, in: IEEE/CVF Conference on Computer Vision and Pattern Recognition Workshops (CVPRW), IEEE, ISBN 978-1-7281-9360-1, June 2020, pp. 111–119, https://doi.org/10.1109/CVPRW50498.2020.00020, https://github.com/haofanwang/Score-CAM.
[54] Mullan S., Sonka M., Visual attribution for deep learning segmentation in medical imaging, in: Išgum I., Colliot O. (Eds.), Medical Imaging 2022: Image Processing, SPIE, ISBN 9781510649392, Apr. 2022, p. 25, https://doi.org/10.1117/12.2612288, https://github.com/Mullans/KernelWeighted.
[55] Sun J., Darbehani F., Zaidi M., Wang B., SAUNet: shape attentive U-Net for interpretable medical image segmentation, in: LNCS, vol. 12264, Springer Science and Business Media Deutschland GmbH, ISBN 9783030597184, Jan. 2020, pp. 797–806, https://doi.org/10.1007/978-3-030-59719-1_77, https://github.com/sunjesse/shape-attentive-unet.
[56] Yaman A., Ozturk C., Huijing P.A., Yucesoy C.A., Magnetic resonance imaging assessment of mechanical interactions between human lower leg muscles in vivo, Journal of Biomechanical Engineering 135 (9) (2013).
[57] Fedorov A., Beichel R., Kalpathy-Cramer J., Finet J., Fillion-Robin J.C., Pujol S., Bauer C., Jennings D., Fennessy F., Sonka M., Buatti J., Aylward S., Miller J.V., Pieper S., Kikinis R., 3D Slicer as an image computing platform for the Quantitative Imaging Network, Magnetic Resonance Imaging 30 (9) (2012) 1323–1341.
[58] Park N., Kim S., How do vision transformers work?, in: International Conference on Learning Representations, 2021.

CHAPTER

Machine learning in image registration

19

Bob D. de Vos[a], **Hessam Sokooti**[b], **Marius Staring**[b], **and Ivana Išgum**[a,c]

[a]*Department of Biomedical Engineering and Physics, Amsterdam UMC, Amsterdam, the Netherlands*

[b]*Department of Radiology, Leiden University Medical Center, Leiden, the Netherlands*

[c]*Informatics Institute, University of Amsterdam, Amsterdam, the Netherlands*

Learning points

- Deep learning image registration
- CNN architectures for transformation models
- Supervised image registration
- Unsupervised image registration
- Image synthesis for image registration

19.1 Introduction

Image registration, the task of aligning images, has been one of the key methods in medical image analysis. It can be used for subtraction imaging, to improve the analysis of longitudinal images, to transfer information between images, or to analyze motion patterns among images. The goal in image registration is to find a coordinate mapping between a fixed target image and a moving source image as shown in Fig. 19.1. Image registration can be accomplished in several ways, ranging from approaches using (physical) fiducial markers to data-driven approaches. Many data-driven image registration approaches rely on image intensities and can be applied globally or locally on an image. In global image registration, a transformation matrix describes one transformation for every voxel in an image. In local image registration, different transformations are locally applied to each voxel in an image.

Not long ago machine learning was mainly employed for image registration as an aid, for example to predict alignment errors in registered images [1,2]. However, now, in the age of deep learning, the role of machine learning in image registration has become more prominent. Instead of merely aiding image registration, deep learning methods are powerful enough to perform image registration themselves. The benefit is that the spatial statistical relation between images can be learned in an offline

Medical Image Analysis. https://doi.org/10.1016/B978-0-12-813657-7.00031-5

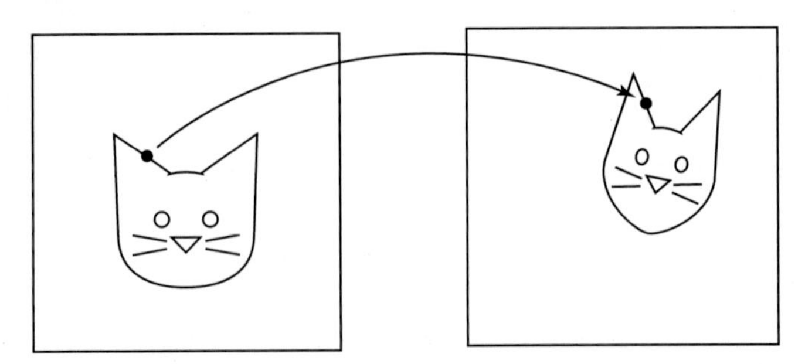

FIGURE 19.1 Image registration.

The goal in image registration is to align images to each other, i.e., the goal is to find a mapping for every point in source and target images.

training phase and that at test time only a single forward pass through the network is required. This makes registration, which is typically a slow procedure, extremely fast, and it opens up new applications in real-time scenarios since it performs image registration in one shot, i.e., non-iteratively.

Deep learning-based image registration can be divided into supervised and unsupervised approaches. Supervised approaches are trained on images that are accompanied by a ground truth that defines a correspondence among images, for example by using corresponding anatomical landmark points. Unsupervised methods do not require predefined correspondences, but use similarity metrics between the fixed and moving images, very similar to conventional intensity-based image registration. In the next section the basic building blocks of image registration are introduced. These will help in understanding the subsequent sections about supervised and unsupervised training. Thereafter, several applications of deep learning-based image registration will be discussed as well as the future directions. The figures in this chapter illustrate concepts in 2D but can be easily extended to 3D.

19.2 Image registration with deep learning

The goal in image registration is to align a source image, the moving image I_M, to a target image, the fixed image I_F. Ultimately, each voxel in the moving image should be mapped to a fixed image by finding a displacement $\boldsymbol{u}(\boldsymbol{x})$, thereby aligning $I_M(\boldsymbol{x} + \boldsymbol{u}(\boldsymbol{x}))$ to $I_F(\boldsymbol{x})$. The resulting displacement vector field (DVF) is used to warp the moving image into alignment with the fixed image. A DVF typically maps $T : \boldsymbol{I_F} \rightarrow \boldsymbol{I_M}$, which is also referred to as backward warping. Mapping from the fixed domain to the moving domain ensures that all voxels in the fixed image will be linked to voxels in the moving image, so the warped image will not contain holes.

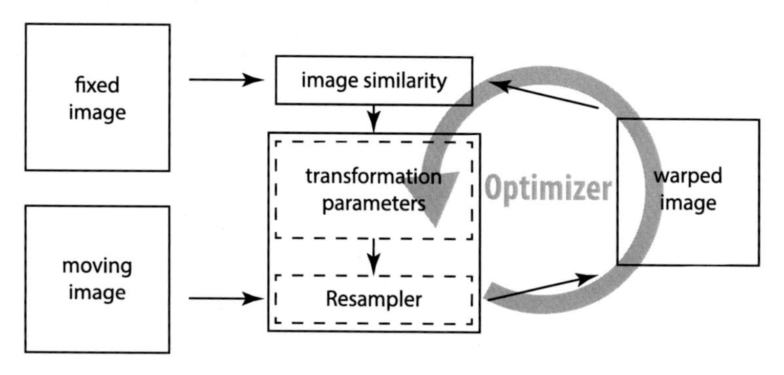

FIGURE 19.2 Conventional image registration.

In conventional intensity-based image registration, iterative optimization is employed to update registration parameters.

Common conventional image registration methods use intensity-based optimization, where an image similarity metric L, defined as a dissimilarity by negating, is optimized by updating the transformation T that describes the DVF:

$$\hat{T} = \arg\min L(T; I_F, I_M) + R(T), \tag{19.1}$$

where R is additional regularization, which ensures well-behaved optimization. Fig. 19.2 shows the schematics of a typical conventional image registration framework. This schematic shows that transformation parameters are updated in multiple iterations using an optimizer. Provided that the image similarity metric is differentiable, gradient descent can be used. Image similarity is often constrained by a regularization term R because optimal similarity does not equal optimal spatial alignment [3]. The regularization term can be used to penalize transformation parameters, e.g., by using an L^2-norm, or it can be used to enforce smoothness of the DVF. Calculation of the similarity metric and the regularization term can be very computationally demanding. This puts a limit on the speed of conventional image registration, since these terms are recalculated every iteration during registration.

Deep learning-based methods for image registration predominantly rely on convolutional neural networks (CNNs) that can predict transformation parameters directly, thus non-iteratively [1,4–7]. A CNN takes a pair of fixed and moving images as its input and it processes the pair through multiple layers of convolutional filters and non-linear functions. Its prediction is a continuous output that constitutes the vectors of a DVF, or parameters for, e.g., a thin plate spline, a B-spline, or an affine transformation model.

Deep learning image registration methods learn from multiple registration cases, and they generally perform single-pass registration. However, an alternative deep learning approach is reinforcement learning (RL). Instead of conventional optimization, an agent will perform registration [8]. Although RL approaches are deep learning-based, they are similar to conventional methods in the sense that they are iterative and hence they may be time consuming.

19.3 Deep neural network architecture

The possibilities for network architecture design are innumerable, but the used transformation model imposes some restrictions on the architecture. The CNN architecture has to be designed such that it outputs the number of parameters the transformation model requires: for direct prediction of a DVF, the CNN should predict a displacement vector for every voxel in the fixed image; for spline registration, the CNN should predict a grid of coefficients; and for affine registration, the CNN should predict a fixed number of output parameters constituting the transformation matrix entries and the translation vector. In short, the transformation model heavily impacts the used CNN architecture.

Global transformations. Global transformations, like rigid and affine transformations, use a transformation matrix that is of size 3×3 for 2D transformations and 4×4 for 3D transformations. The elements of the transformation matrix can be directly predicted by a CNN, or the CNN can output meaningful parameters that compose the transformation matrix, such as translations, rotation angles, scales, and shearing angles. When a fixed number of parameters is required, a vanilla CNN[1] can be used. However, since affine registration is often employed as the first coarse step for pre-alignment, the input images may be of different size. To capture the full extent of images, a Siamese network can be used, as shown in Fig. 19.3. If input images are of different size, the CNN will have different size output maps. Global average pooling can be employed to obtain an average per feature map, thereby ensuring an output with a fixed number of features that are connected to the final MLP layer.

Parameterized deformable registration. For spline-based transformation models, the CNN should predict a grid of coefficients evenly distributed over the fixed image. When a grid of transformation parameters is required, a patch-based CNN can be used that provides an output per image patch. Fig. 19.4 shows a schematic of this task. A CNN takes fixed and moving patch pairs as input channels and outputs a two-channel matrix that constitutes x, y spline coefficients. This two-channel matrix represents the grid points. Adjoining image patches may be taken, but overlapping patches show better results [5] for cubic B-spline registration. It is advisable to let the receptive field of a CNN coincide with the local support range of the employed spline. A vanilla CNN could be used, or, for variable-sized inputs, a patch-based fully convolutional CNN.

[1] A "vanilla-flavored" CNN is a standard architecture of several alternating layers of convolutions and max pooling, and a final multilayer perceptron (MLP). Due to the architecture, the input and output are required to be of fixed size.

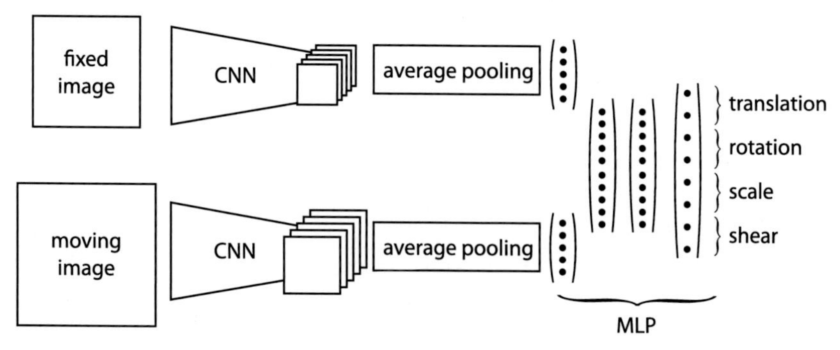

FIGURE 19.3 Affine registration.

Affine image registration is typically used for coarse pre-alignment of images. Input images may be of different size. In this example different-size fixed and moving images are analyzed separately with a CNN, resulting in different output sizes. To ensure fixed-sized output, global average pooling is used, which is connected to a regular multilayer perceptron (MLP) with the fixed number of transformation parameters.

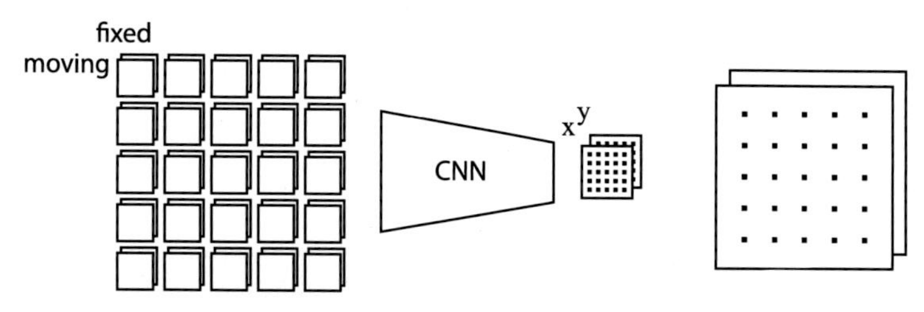

FIGURE 19.4 Patchwise prediction.

Thin plate spline or B-spline registration requires a grid of coefficients. For this a CNN can be used that outputs parameters per patch. For increased performance in image registration, the CNN should analyze overlapping patches [9].

Direct DVF prediction. For direct DVF prediction, the CNN should predict a displacement vector for every voxel position in the fixed image (Fig. 19.5). The fixed and moving images may be stacked as channels for input in a CNN. Any network architecture that is designed for image segmentation can be employed, with the sole requirement that the CNN should predict a displacement vector for each voxel in the fixed image. Thus, the main difference with CNNs for segmentation is that the output layer predicts a DVF instead of a label map. Such networks are typically fully convolutional, which means that all layers in a network use convolutional filters, i.e., dense layers are implemented as convolutional layers. The result is that a CNN can theoretically analyze inputs of any size.

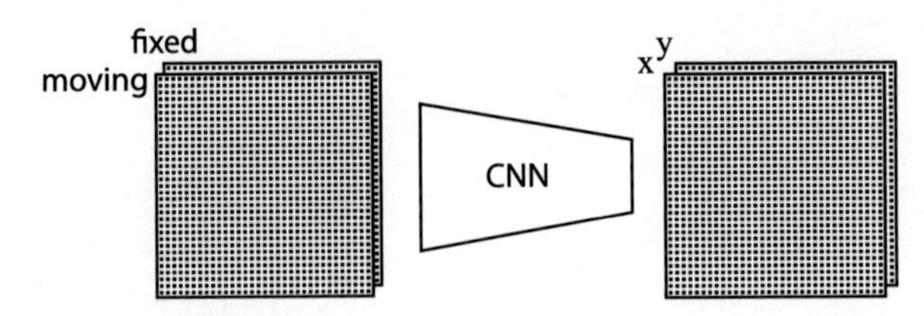

FIGURE 19.5 Direct DVF prediction.

For direct prediction of a DVF, the used CNN should output a displacement vector for each voxel. Most of these networks use a variant of a fully convolutional network [4].

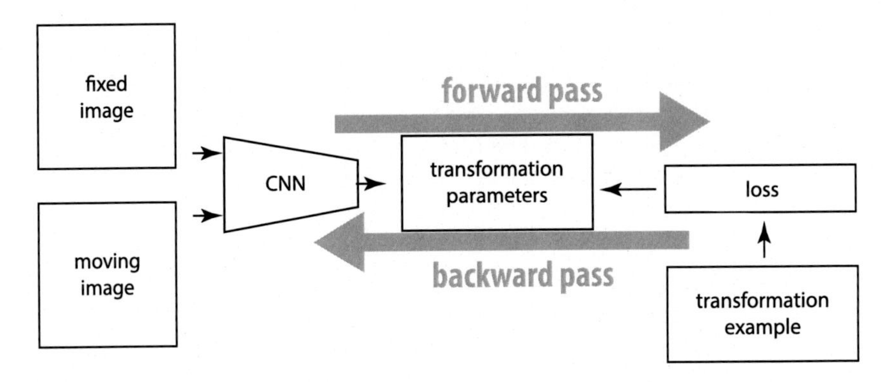

FIGURE 19.6 Supervised image registration.

In supervised deep learning-based image registration, a CNN predicts transformation parameters in the forward pass. A loss is calculated between the known transformation and the predicted transformation, and the error is back-propagated through the CNN.

19.4 Supervised image registration

In supervised training, a CNN learns from known image correspondences (Fig. 19.6). To teach an image registration task, a sufficient number of fixed and moving image pairs with known correspondences are required. Each image pair requires a reference transformation that aligns the images. Reference transformations can be obtained by aligning corresponding segmentations, contours, or landmarks; by registering images with conventional iterative methods; or by generating synthetic transformations. Once references have been obtained for fixed and moving image pairs, a CNN can be trained with them. The CNN aims to minimize the difference between the predicted transformations and the ground-truth transformations. Since the CNN outputs transformation parameters, it can be cast into a regression problem. Common losses are the mean absolute error (MAE), mean square error (MSE), and the Huber loss.

19.4.1 Supervision via conventional image registration

Image pairs may be registered using conventional iterative methods, thus obtaining transformations that can act as a (pseudo) ground truth. The network is then trained to predict these transformations. In a way it learns to mimic conventional image registration.

Note that the transformations obtained by the iterative methods might not be optimal and can be dependent on the hyperparameters of these methods. To increase the performance, a set of well-tuned hyperparameters for the specific database can be used. Moreover, registration assessment can be done to automatically discard uncertain pairs [2,10,11]. Another possibility to increase the accuracy of registration is to incorporate manual annotations like landmarks or segmentation maps to the loss function [12]. For instance, if the segmentation maps are available, the registration can be performed over the distance transform of them, which potentially improves the registration performance. Finally, manual interaction can further improve the quality of the golden standard.

19.4.2 Supervision via synthetic transformations

Training data for supervised approaches can also be obtained by generating synthetic transformations that are applied to a single image. This way, the original image and a synthetically deformed image form the image pair, for which we know the exact corresponding transformation.

A simple way to generate synthetic transformation is to randomly assign values to the transformation parameters. In order to avoid folding or harsh transformation, we can smooth the random values. By checking the Jacobian of the generated transformation, it can be ensured that the smoothness of transformation is reasonable. A negative Jacobian indicates folding and positive values correspond to local volume changes; when the Jacobian equals 1, this indicates no local volume change. An example of random transformation is given in Fig. 19.7. It has been shown in [1,4,13] that learning from this synthetic random transformation generalizes to real medical image pairs.

A more realistic way of generation is to simulate the movement of organs and tissue that is visible in that image. To give an example, the respiratory motion in chest images can be simulated with the prior information. It can be constructed with two components: (1) expansion of the chest in the transversal plane and (2) transition of the diaphragm in the cranio-caudal direction. A sample respiratory transformation is given in Fig. 19.7. Adding respiratory transformation was shown to improve performance in chest computed tomography (CT) registration [13].

Another realistic way to generate training data is to make a statistical model from the real transformations and subsequently use this model to generate transformations. With this approach, the model potentially learns the elasticity and softness of the tissue in the image. Therefore, the generated deformed image can imitate the real transformation of the organs. Uzunova et al. [14] experimented with a model-based augmentation scheme on cardiac and brain magnetic resonance imaging (MRI) data.

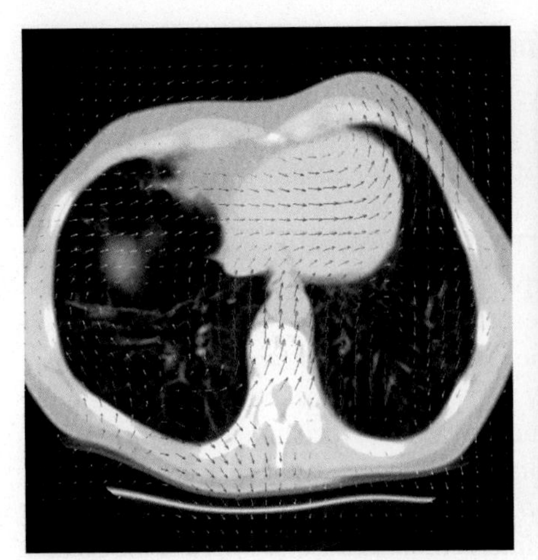
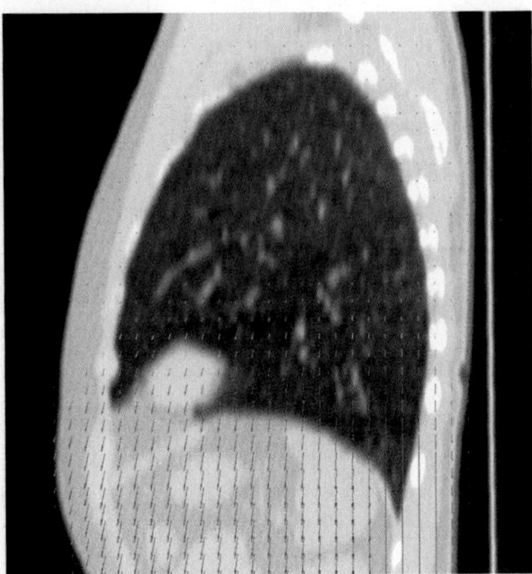

FIGURE 19.7 Examples of synthetic transformations.

Left: Random transformation. Right: Simulated respiratory motion.

How to generate many image pairs. For each single image in the training set, potentially a large number of synthetic transformations can be generated. However, if this image is to be reused for multiple transformations, then for many training pairs we have one image unaltered. Consequently, the trained model can be overfitted to few samples. To tackle this problem, one potential solution is to generate deformed versions of the original image. A schematic design of utilizing artificial image pairs is depicted in Fig. 19.8. In this approach, the original image is only used once to generate the artificial image $\mathrm{I_{F0}}$. Deformed versions of the original image I_{Mi} are generated using gray blocks in Fig. 19.8 and are used afterwards. Training pairs are thus $(I_{M0}, I_{F0}), (I_{M1}, I_{F1}), (I_{M2}, I_{F2}), \ldots$.

The drawback of this approach is that the appearance of deformed generated images might be very similar to the original image. However, in real image pairs, the appearance of fixed and moving images are not completely similar. This issue can be avoided by adding several intensity augmentations to the deformed image.

Intensity noise. To achieve more accurate simulation of real images a Gaussian noise can be added to the transformed moving image. The standard deviation of this noise is specific to the modality of the images. For instance, Sokooti et al. [13] proposed to set the standard deviation to 5 in CT images.

Gamma transform. The gamma transform is a gray-value transformation which can mimic the contrast variation in image pairs. The gamma transform is defined as $\boldsymbol{I}(x) = \left(\boldsymbol{I}^{\mathbf{clean}}(x)\right)^{\gamma}$, where the value of γ can be chosen randomly. Elmahdy et al. [15] proposed a uniform distribution between -0.4 and 0.4 in prostate CT images.

Sponge intensity model. The sponge model is applicable to CT images. In a dry sponge model [13] intensity values are inversely proportional to the local volume

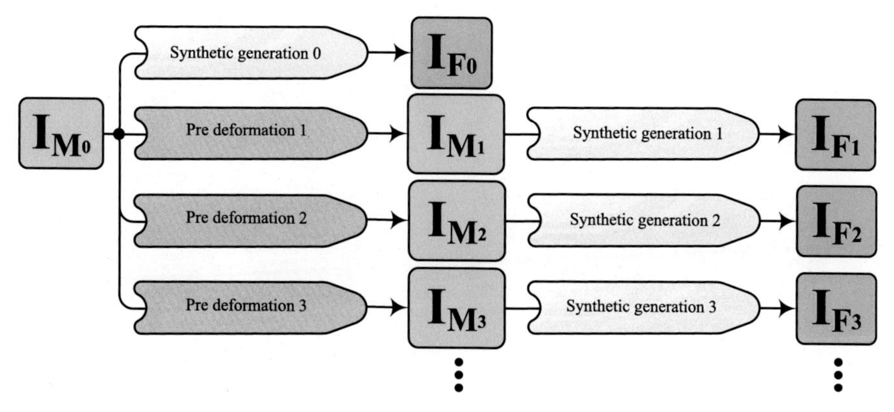

FIGURE 19.8 Extensive pair generation.

The generation of training pairs from a single input image I_{M0}. The input image is deformed slightly using the pre-deformation blocks to generate moving images I_{Mi}. These are then each deformed and post-processed multiple times using all categories to generate fixed images I_{Fi}.

change (see Fig. 19.9):

$$I(x) = I^{\text{clean}}(x)[J_T(x)^{-1}], \tag{19.2}$$

where J denotes the determinant of the Jacobian of the transformation.

Occlusion. In order to simulate the retrospective variation of anatomical tissue, the occlusion technique can be utilized. An example of chest CT scan is given in Fig. 19.9, where several regions inside the lung are occluded to the Hounsfield unit of air.

In some scenarios it might be non-trivial to devise synthetic examples. In such cases it would be better to employ an unsupervised training method, which will be covered in the next section.

19.5 Unsupervised image registration

Unsupervised training of CNNs for image registration is very similar to conventional image registration [9,16,17]. Both methods use stochastic gradient descent and an image similarity metric for optimization. However, unlike conventional image registration, where the similarity metric is used to directly update transformation parameters, in unsupervised deep learning image registration the metric is used to update the weights of a CNN, as shown in Fig. 19.10. In unsupervised training, no predefined training examples are required; hence, all pairs of fixed and moving images can be used to train a CNN for image registration.

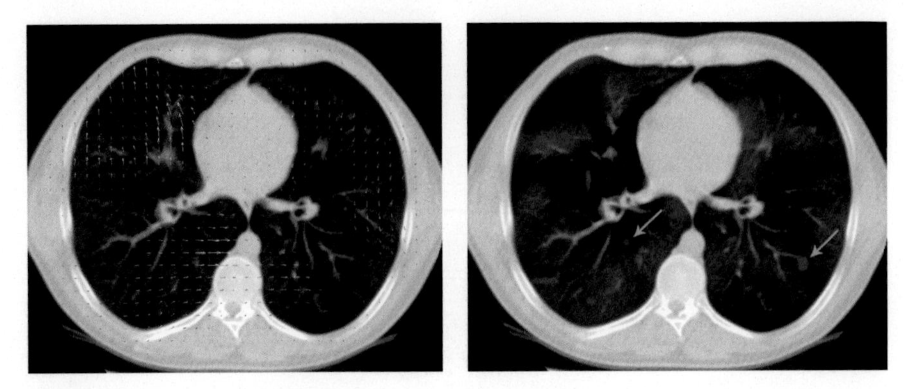

FIGURE 19.9 Intensity augmentation.

Left: Deformed image generated using synthetic transformation. Right: Augmented image where the intensity values vary with the inverse ratio of the Jacobian. The occluded regions are indicated by blue arrows.

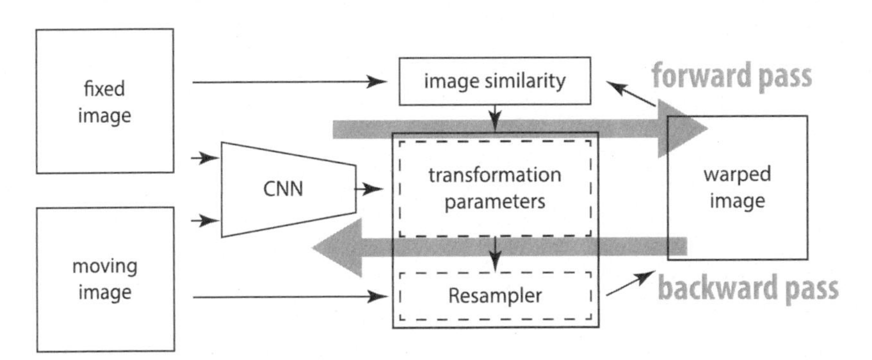

FIGURE 19.10 Unsupervised image registration.

In unsupervised deep learning-based image registration, a CNN predicts transformation parameters that warp the moving image, and a similarity metric is used as a loss for back-propagation.

19.5.1 Image similarity as a loss

Depending on the registration task, several similarity metrics can be used. The sum of squared distances (SSD) loss is especially suited for quantity-based modalities like CT. When slight differences in intensity are present, e.g., caused by contrast agents, normalized cross-correlation (NCC) might be preferred. For inter-modality image registration, a mutual information metric is better suited [5], because it can model a non-linear probabilistic relation between intensities of modalities. Note that for use as a loss in gradient descent optimization, image similarity is negated; hence, it is sometimes called dissimilarity.

19.5.2 Metric learning

An upcoming field where machine learning is applied in image registration is that of metric learning. In metric learning the goal is to learn an optimal descriptor of image alignment. Such methods can focus on finding matching key points, i.e., anatomical landmarks, in images for alignment. When such key points are found a transformation matrix can be easily acquired for affine registration [18]. Metric learning can also learn local regularization. Instead of a global predefined regularizer, a CNN can be trained for localized smoothing, which is essentially localized regularization [19].

19.5.3 Regularization and image folding

Unconstrained optimization might introduce image folding. Therefore, similar to conventional image registration, regularization is required to achieve anatomically plausible deformations. As with conventional image registration, this introduces an extra computational load. However, this load is only applicable during offline training. After training the CNN ought to have learned its registration task, and during online inference, no calculation of regularization is needed.

Image folding might also be prevented using specific transformation models. Global linear transformations, like affine transformations, are parameterized with a transformation matrix, and as such cannot be subject to image folding. However, these transformations can only model linear transformations and are not deformable. On the other hand, thin plate spline transformations or B-spline transformations provide inherently smooth transformations and are therefore less susceptible to image folding. While these strategies might limit image folding, regularization is often applied to further mitigate image folding.

A very common strategy to prevent folding is coarse-to-fine image registration. This optimization technique stems from conventional image registration, where it has been shown that image registration improves by executing several subsequent stages of coarse-to-fine image resolution and detail. Several approaches have been introduced for coarse-to-fine deep learning-based image registration. A straightforward approach is training multiple CNNs, each with their own registration task [7,20]. By subsequently applying CNNs that are specifically trained for coarse or fine correspondences, image registration performance increases.

Another approach is to embed the coarse-to-fine strategy during training of a single CNN architecture [21]. For example, a U-Net-based network can be modified such that it is able to analyze multiple resolution levels of fixed and moving image pairs. By progressively training this CNN, starting with a low resolution and ending with a high resolution, the CNN learns coarse-to-fine image registration during training. During inference it is applied in one shot.

19.5.4 Auxiliary losses

In cases that prior information, e.g., segmentation maps, are available, image registration may be improved by using a metric like the Dice coefficient. For example, Hu et al. [22] proposed a Gaussian-based multiscale Dice for measuring label similarity between MRI and ultrasound images of the prostate. Elmahdy et al. [15] demonstrated that adding Dice loss in addition to NCC can improve unsupervised learning using prostate CT data.

To improve handcrafted similarity metrics, extra implicit metrics can be added to the loss function using a generative adversarial neural network. The discriminator network is utilized to learn an implicit loss function in order to minimize the binary classification of "fake" and "real" labels. Elmahdy et al. [15] generated noisy version of the fixed images as "real" samples and considered the transformed moving images as "fake" samples.

19.6 Recapitulation

Since the introduction of deep learning-based image registration, there is a clear paradigm shift; instead of aiding, image registration is now performed with machine learning. The benefit of deep learning-based image registration is that the objective function, the deformation model, and the optimization strategy can be delegated to offline training. Since dealing with these components can be computationally demanding, this means that most computational burden is only performed offline. There is a limited burden during online inference, i.e., the act of image registration. Since CNNs are highly parallelizable, they are extremely fast when running on a graphics processing unit (GPU) or a tensor processing unit (TPU). This makes deep learning-based image registration particularly interesting for time-dependent applications.

To train a CNN for machine learning, a sufficiently large dataset is required. Obtaining a training set can be an issue. When it is too limited it might be necessary to increase the number of training data. However, especially in medical imaging, obtaining representative data can be a problem by itself. To increase dataset sizes it might be necessary to introduce augmentations. Augmentations artificially increase the training set size. Straightforward augmentations are image cropping, intensity flipping, and rotations. For some tasks this can be sufficient, but for image registration spatial augmentations or intensity-based augmentations need to make sense. When the size of the training set is limited, it might be necessary to increase the training data size, e.g., by creating synthetic transformations.

Acquiring correct registration examples for supervised learning can be a challenging task. It is a very tedious and time-consuming task to manually align images using rigid transformations and it is infeasible to perform deformable registration manually. For supervised deformable image registration it is easier to use synthetic examples. In the case of synthetic examples, random deformations are not guaranteed to be sensible, and training with these might cause suboptimal registration performance. Instead anatomically plausible deformation examples provide better examples for supervised training. However, modeling correct deformations can be a very challenging task, since it requires prior knowledge of anatomical landmarks. For example, for modeling breathing motion it is essential to know the location of the diaphragm and the lung boundaries and lobes. This is particularly difficult when pathology is present. When provided with a sufficiently large and sufficiently varying dataset, it might be preferable to employ unsupervised methods.

Instead of training a CNN for pairwise image registration, a CNN can also be trained for atlas registration. This can easily be achieved by designing a CNN that takes one image as its input and that learns registration to a single atlas image. When training is finished, the CNN has internalized the structure of the atlas image, and it can then be used to align unseen input images to the atlas. This approach can also be employed as a memory-efficient groupwise registration technique. In most conventional groupwise image registration methods, all images have to be analyzed jointly, which impacts scaling of these methods to a group with many images. In contrast, the atlas registration approach can scale easily to very large groups of images.

Very related to image registration are optical flow and spatial transformer networks. Optical flow methods employ image alignment, but the methodology is primarily aimed at describing motion in video sequences. Spatial transformer networks are neural networks embedded as subnetworks in larger deep learning models. The task of a spatial transformer is to reorient, i.e., to transform, an input image such that the main task of the CNN, which can be any task like classification or regression, is simplified. The weights of the spatial transformer are updated by propagating the loss of the main task through the neural network. As a result, images might be aligned as an implicit result, but since image alignment is not enforced image registration is uncertain.

To conclude, deep learning-based image registration bears many similarities with conventional iterative intensity-based image registration. Many methodologies from conventional image registration can be readily implemented in deep learning image registration.

19.7 Exercises

Practical experience with image registration is important. Some details are easily overlooked when reading about image registration instead of practicing with it. For the exercises, we use the Pytorch Image Registration module TorchIR.[2]

1. Get the code from the GitHub repository and run the MIA tutorial notebook. The notebook shows an example using handwritten digits from the MNIST dataset. In this tutorial we register number 9s to each other. Please run the example of 10 epochs and inspect the loss using Tensorboard. The network is now trained and it can be applied to all digits from the test set. If all went well, you will see a noticeable difference between the moving images before and after registration.
2. The images are clearly aligned after registration. The question now is: are the generated fields useful for image registration? Let us check the registration of single

[2] https://github.com/BDdeVos/TorchIR.

images and the DVF. Although registration may seem acceptable when looking at the images, the vector fields are not useful. There is a considerable amount of folding. Think about several solutions to prevent folding from happening. Now implement one, retrain the model, and compare the new DVF to the previous one.

3. What happens when you flip the intensity of the moving images? What do you need to change to make this happen? Implement the solution. Do not modify the input, i.e., by simply flipping the intensity of the fixed image, but try to think of a methodological solution.

References

[1] Eppenhof K.A.J., Pluim J.P.W., Pulmonary CT registration through supervised learning with convolutional neural networks, IEEE Transactions on Medical Imaging 38 (5) (2018) 1097–1105.

[2] Sokooti H., Saygili G., Glocker B., Lelieveldt B.P.F., Staring M., Quantitative error prediction of medical image registration using regression forests, Medical Image Analysis (ISSN 1361-8415) 56 (2019) 110–121, https://doi.org/10.1016/j.media.2019.05.005, http://www.sciencedirect.com/science/article/pii/S1361841518300811.

[3] Rohlfing T., Image similarity and tissue overlaps as surrogates for image registration accuracy: widely used but unreliable, IEEE Transactions on Medical Imaging 31 (2) (Feb. 2012) 153–163, https://doi.org/10.1109/TMI.2011.2163944.

[4] Sokooti H., De Vos B.D., Berendsen F., Lelieveldt B., Išgum I., Staring M., Nonrigid image registration using multi-scale 3D convolutional neural networks, in: International Conference on Medical Image Computing and Computer-Assisted Intervention, Springer, 2017, pp. 232–239.

[5] De Vos B.D., van der Velden B.H.M., Sander J., Gilhuijs K.G.A., Staring M., Išgum I., Mutual information for unsupervised deep learning image registration, in: Išgum I., Landman B.A. (Eds.), Medical Imaging 2020: Image Processing, vol. 11313, International Society for Optics, SPIE, 2020, pp. 155–161, https://doi.org/10.1117/12.2549729.

[6] Balakrishnan G., Zhao A., Sabuncu M.R., Guttag J., Dalca A.V., VoxelMorph: a learning framework for deformable medical image registration, IEEE Transactions on Medical Imaging 38 (8) (Aug. 2019) 1788–1800, https://doi.org/10.1109/TMI.2019.2897538.

[7] Hering A., van Ginneken B., Heldmann S., mlVIRNET: Multilevel Variational Image Registration Network, in: Shen D., et al. (Eds.), Medical Image Computing and Computer Assisted Intervention – MICCAI 2019, Springer International Publishing, Cham, ISBN 978-3-030-32226-7, 2019, pp. 257–265.

[8] Krebs J., Mansi T., Delingette H., Zhang L., Ghesu F.C., Miao S., Maier A.K., Ayache N., Liao R., Kamen A., Robust non-rigid registration through agent-based action learning, in: International Conference on Medical Image Computing and Computer-Assisted Intervention, Springer, 2017, pp. 344–352.

[9] De Vos B.D., Berendsen F.F., Viergever M.A., Staring M., Išgum I., End-to-end unsupervised deformable image registration with a convolutional neural network, in: Deep Learning in Medical Image Analysis and Multimodal Learning for Clinical Decision Support, Springer, Cham, 2017, pp. 204–212.

[10] Murphy K., et al., Evaluation of registration methods on thoracic CT: the EMPIRE10 challenge, IEEE Transactions on Medical Imaging 30 (11) (2011) 1901–1920.

[11] Sokooti H., Saygili G., Glocker B., Lelieveldt B.P.F., Staring M., Accuracy estimation for medical image registration using regression forests, in: International Conference on Medical Image Computing and Computer-Assisted Intervention, Springer, 2016, pp. 107–115.

[12] Rohé M.-M., Datar M., Heimann T., Sermesant M., Pennec X., SVF-Net: Learning deformable image registration using shape matching, in: International Conference on Medical Image Computing and Computer-Assisted Intervention, Springer, 2017, pp. 266–274.

[13] Sokooti H., De Vos B., Berendsen F., Ghafoorian M., Yousefi S., Lelieveldt B.P.F., Išgum I., Staring M., 3D convolutional neural networks image registration based on efficient supervised learning from artificial deformations, arXiv preprint, arXiv:1908.10235, 2019.

[14] Uzunova H., Wilms M., Handels H., Ehrhardt J., Training CNNs for image registration from few samples with model-based data augmentation, in: International Conference on Medical Image Computing and Computer-Assisted Intervention, Springer, 2017, pp. 223–231.

[15] Elmahdy M.S., Wolterink J.M., Sokooti H., Išgum I., Staring M., Adversarial optimization for joint registration and segmentation in prostate CT radiotherapy, in: International Conference on Medical Image Computing and Computer-Assisted Intervention, Springer, 2019, pp. 366–374.

[16] Balakrishnan G., Zhao A., Sabuncu M.R., Dalca A.V., Guttag J., An unsupervised learning model for deformable medical image registration, in: The IEEE Conference on Computer Vision and Pattern Recognition (CVPR), June 2018.

[17] Dalca A.V., Balakrishnan G., Guttag J., Sabuncu M.R., Unsupervised learning for fast probabilistic diffeomorphic registration, in: International Conference on Medical Image Computing and Computer-Assisted Intervention, Springer, 2018, pp. 729–738.

[18] Hu J., Sun S., Yang X., Zhou S., Wang X., Fu Y., Zjou J., Yin Y., Cao K., Song Q., Wu X., Towards accurate and robust multi-modal medical image registration using contrastive metric learning, IEEE Access 7 (2019) 132816–132827, https://doi.org/10.1109/ACCESS.2019.2938858.

[19] Niethammer M., Kwitt R., Vialard F.-X., Metric learning for image registration, in: The IEEE Conference on Computer Vision and Pattern Recognition (CVPR), June 2019.

[20] De Vos B.D., Berendsen F.F., Viergever M.A., Sokooti H., Staring M., Išgum I., A deep learning framework for unsupervised affine and deformable image registration, Medical Image Analysis 52 (2019) 128–143.

[21] Eppenhof K.A.J., Lafarge M.W., Veta M., Pluim J.P.W., Progressively trained convolutional neural networks for deformable image registration, IEEE Transactions on Medical Imaging 39 (5) (2020) 1594–1604, https://doi.org/10.1109/TMI.2019.2953788.

[22] Hu Y., Modat M., Gibson E., Li W., Ghavami N., Bonmati E., Wang G., Bandula S., Moore C.M., Emberton M., Ourselin S., Noble J.A., Barratt D.C., Vercauteren T., Weakly-supervised convolutional neural networks for multimodal image registration, Medical Image Analysis 49 (2018) 1–13.

PART

Advanced topics in medical image analysis

CHAPTER

Motion and deformation recovery and analysis

20

James S. Duncan[a,b] **and Lawrence H. Staib**[a]

[a]*Division of Bioimaging Sciences, Departments of Radiology & Biomedical Imaging, Biomedical Engineering and Electrical Engineering, Yale University, New Haven, CT, United States*

[b]*Department of Statistics & Data Science, Yale University, New Haven, CT, United States*

Learning points

- Dense flow field estimation
- Eulerian and Lagrangian descriptions
- Intensity-based and feature-based methods
- Image-centric and object-centric
- Spatiotemporal modeling
- Data-driven learning

20.1 Introduction

Biologic motion and deformation are essential for the characterization of normal and pathologic behavior in many applications and medical imaging is ideal for the quantitative analysis of these changes. Temporal information from an image sequence is critical for approaching the underlying problem. Depending on the application, the relevant changes may occur over different timescales ranging from milliseconds to months. Example applications include the analysis of tumor growth over months from magnetic resonance imaging (MRI) or computed tomography (CT) longitudinal images, tracking the motion and deformation of a dense set of points on the left ventricle (LV) of the human heart from ultrasound in real-time, or tracking the trajectories of a dense set of subcellular transport vesicles from high-resolution microscopy. In all of these cases, the starting point is typically determining the correspondence of local information between a pair of temporal frames leading to tracking entire temporal trajectories. While the problems in this chapter are related to the image registration methods described in Part IV, they are unique due to the sequential and multiframe nature of the spatiotemporal image information.

Motion and deformation analysis can be categorized in a number of ways. Some methods are *image-based* and estimate flow densely, specifying the motion vector at all points in the image. Others are object-centric and estimate the motion of just the moving object (e.g., organ, cell). *Flow fields* are functions of time and space that

Medical Image Analysis. https://doi.org/10.1016/B978-0-12-813657-7.00033-9

specify the motion. *Eulerian* flow fields specify the motion at each fixed point in space. Alternatively, *Lagrangian* flow fields follow the material as it moves through space. We can directly use the image intensities to track motion or use features extracted either densely or sparsely. More advanced techniques incorporate statistical models, biomechanical models, data-driven methods, and information from multiple frames.

In this chapter, we will discuss motion and deformation analysis in a number of areas. In Section 20.3 we will discuss image-based non-rigid registration mapping techniques as well as image-based frame-to-frame Eulerian motion analysis. The classic method for detecting image-based displacement is via *optical flow* methods. In Section 20.4.1, we will discuss the assembly of local information into dense flow fields using frame-to-frame, Lagrangian feature tracking based on knowledge of the moving object (such as from object segmentation) with methods including shape tracking, speckle tracking, and intensity patch tracking. Then, in Section 20.6, we will discuss the use of models for more complete motion estimation, including multiframe analysis with Kalman filtering, complete trajectory analysis, biomechanical deformation/strain models, and data-driven approaches using deep learning with feedforward artificial neural networks.

At the end of the chapter, in Section 20.7, we will discuss evaluation approaches for characterizing algorithm performance based on synthetic data, implanted markers (typically in pre-clinical experiments), and human perceived motion analysis (including limitations).

20.1.1 Notation: displacement, deformation, and strain

In the approaches discussed below, we will use the following notation. First, we define $u = \frac{dx}{dt}$, $v = \frac{dy}{dt}$, and $w = \frac{dz}{dt}$ as the three components of the velocities between two time frames in 3D. If only 2D image data are being considered, we will limit to two components. When we assume that the flow fields are computed between two time frames, these components, (u, v, w), are equivalent to the corresponding *displacements*. We assemble these components into a single velocity/displacement vector: $\mathbf{u} = (u, v, w)$ (or $\mathbf{u} = (u, v)$ for the 2D case). We can visualize these velocity/displacement vectors as emanating from a single position from a starting time frame t in 3D space $\mathbf{x} = (x, y, z)$ or in 2D space $\mathbf{x} = (x, y)$ such that this velocity/displacement vector can be indexed as $\mathbf{u}[\mathbf{x}(t)]$, where this expression represents the displacement vector u starting at position $\mathbf{x}$ at time t. We can further specify the location that the point moves to in a second (usually sequential) time point by including a second position and time index, together as $\mathbf{x}(t + \Delta t)$, resulting in an overall index for the displacement vector of $\mathbf{u}[\mathbf{x}(t), \mathbf{x}(t + \Delta t)]$.

In addition to displacement estimation from image-derived information, we will also discuss the modeling and estimation of deformation. Biomedical deformation occurs in organs such as the brain during surgery or the heart over the cardiac cycle. In general, it refers to the non-rigid movement in an image from one time frame to the next. Incorporating concepts related to deformation requires the description of some basic concepts in, and notation related to, continuum mechanics [1] that we present here. To begin with, consider an object (e.g., an organ) $\mathcal{B}(0)$ seen in an image at time 0 that moves and deforms to become object $\mathcal{B}(t)$ at time t. Next, suppose that a material particle that is part of the object at some position X on $\mathcal{B}(0)$ moves to a new position x on $\mathcal{B}(t)$. We further assume that material does not appear or disappear between the two time frames and thus there will always be a one-to-one correspondence between X and x. We can define the displacement vector for this particle as $u(t) = x(t) - X$, a relationship that is invertible such that given x and t we can find X. Furthermore, if we consider two neighboring particles at time 0 located at X and $X + dX$ on $\mathcal{B}(0)$, we can write the relationship

$$dx = \frac{\partial x(X,t)}{\partial X} dX. \tag{20.1}$$

The Jacobian matrix, $\frac{\partial x(X,t)}{\partial X}$, of this mapping or transformation is called the *deformation gradient matrix*. By definition, $F(0) = I$ (the identity matrix). The mapping defined by equation (20.1) can be decomposed into two components: a rigid motion component, R, and a non-rigid deformation component that captures the change in the shape of the object, U: $F = RU$. We will discuss the Jacobian in the context of image registration-based motion estimation (see Section 20.3.1). For strategies below which invoke biomechanical models (see Section 20.6.1), we assume that the local deformations and rotations between any two consecutive frames in an image sequence are small (i.e., less than about 5%). We invoke the approximation that $\frac{\partial u}{\partial x} \approx \frac{\partial u}{\partial X}$, recalling that u is the local displacement. If we further express F as $F = RU = (I + \omega)(I + \epsilon)$, where ω is the small rotation tensor and is antisymmetric and ϵ is the small (actually infinitesimal) strain tensor and is symmetric, we can further define these tensors as

$$\omega_{i,j} = \frac{1}{2}\left(\frac{\partial u_i}{\partial x_j} - \frac{\partial u_j}{\partial x_i}\right), \tag{20.2}$$

$$\epsilon_{i,j} = \frac{1}{2}\left(\frac{\partial u_i}{\partial x_j} + \frac{\partial u_j}{\partial x_i}\right). \tag{20.3}$$

We write the strain tensor, e, in terms of its components as

$$e = \begin{bmatrix} \epsilon_{11} & \epsilon_{12} & \epsilon_{13} \\ \epsilon_{21} & \epsilon_{22} & \epsilon_{23} \\ \epsilon_{31} & \epsilon_{32} & \epsilon_{33} \end{bmatrix} = [\epsilon_{11}, \epsilon_{22}, \epsilon_{33}, \epsilon_{12}, \epsilon_{13}, \epsilon_{23}]^T. \tag{20.4}$$

Here we take advantage of symmetry and write the six independent elements in vector form as well. The definition of e is the classical definition for infinitesimal strain under conditions of linear elasticity [1]. Strain can be approximated from the discrete displacements estimated by the techniques described in the chapter.

20.2 The unmet clinical need

There are many biomedical problems that require the analysis of a temporal sequence of image frames. Characterization of the spatiotemporal motion and deformation of the chambers of the heart is important for diagnosis and therapy guidance. The estimation of cardiac or respiratory motion is also needed in order to better characterize other anatomical structures or measure physiological function *without* movement artifacts. Deformations due to surgery need to be accounted for in order to provide accurate interventional guidance, such as brain shift and deformation during neurosurgery. Estimating blood flow through vessels is another application area for motion estimation. Image datasets that can be used to estimate biomedical motion or deformation include video image sequences, real-time echocardiography, cardiac gated MRI sequences, dynamic contrast-enhanced MRI, and contrast-enhanced CT. In most cases, one needs to quantitatively characterize displacement and sometimes strain in local spatial regions and track information over temporal sequences.

20.3 Image-centric flow fields: Eulerian analysis

20.3.1 Motion analysis with non-rigid image registration

The non-rigid image registration approaches described in Chapter 14 can also be applied to motion and deformation estimation. In this case, the registration aims to find the non-rigid transformation T of one frame to the next in a temporal sequence: $T_{t\to t+1}$. We write this transformation as $\mathbf{y} = T_{t\to t+1}\mathbf{x}$. A typical formulation would be to find the optimal transformation given a match metric, M:

$$\hat{T}_t = \arg\min_{T_t} M(I_{t+1}, T(I_t)) + \alpha R(T_t). \tag{20.5}$$

The match metric M in this context is typically squared intensity difference, although other metrics, including those based on derived features, can also be used where appropriate. A spatial regularization term, R, with weighting λ, may also be included to control, for example, the smoothness of the transformation.

The displacement field between the time points registered $\mathbf{u}$ can be simply determined from the transformation: $\mathbf{u} = T_{t\to t+1}\mathbf{x} - \mathbf{x}$. The transformation can be modeled as a combination of a global affine transformation (which includes translation, rotation, scaling, and shearing) with a parameterized local deformation. The local deformation, using, for example, a free-form deformation based on B-splines, constrains the displacements to be spatially and/or temporally smooth. Typically, consecutive time points are registered; this provides Eulerian displacements. The change between consecutive images is less, making the registration relatively easier. Multiple registrations may be computed sequentially or simultaneously. Deformation over multiple time points can then be determined by summing (accumulating) the deformations over the sequence. The deformation between consecutive frames can be used as an initialization for the deformation at the subsequent frame. Alternatively,

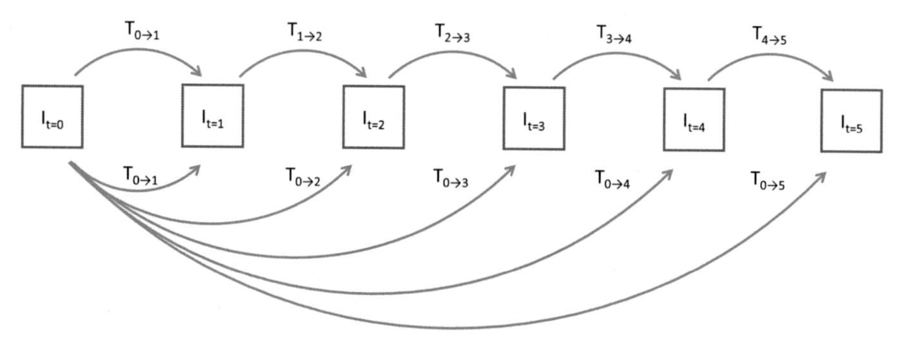

FIGURE 20.1

Transformations. Top: Frame-to-frame Eulerian transformations. Bottom: Lagrangian transformations.

Lagrangian displacements can be estimated by registering the first time point to each subsequent time point. While these registrations are more difficult due to the larger displacements, they provide direct Lagrangian displacements (see Fig. 20.1).

Given the spatiotemporal nature of the problem, it may be desirable to determine the multiple transformations *simultaneously*. Additional regularization may be designed that accounts for the entire temporal sequence. Thus,

$$\left(\hat{T}_0, \ldots \hat{T}_{N-1}\right) = \arg \min_{T_0 \ldots T_{N-1}} \sum_{t=0}^{N-1} M\left(I_{t+1 \bmod N}, T_t(I_t)\right) + \alpha R(T_t) + \sum_{t=0}^{N-1} \beta R'. \tag{20.6}$$

Multiframe smoothness can be constrained, for example, using

$$R' = \sum_{t=1}^{N} (T_t - T_{t+1 \bmod N})^2. \tag{20.7}$$

Note that in the case of cardiac motion and other cyclic phenomena, we would like to include a cyclicity constraint. This is achieved by including a term comparing T_N and T_1 using the modulus (mod) operator.

An alternative method for incorporating multiple time points involves generalizing from a spatial transformation to a *spatiotemporal* transformation, u, that represents the instantaneous velocity and is a function of both $\mathbf{x}$ and t [2]. We can extend the B-spline to 4D to encompass this additional dimension:

$$\mathbf{u}(\mathbf{x}, t) = \sum_{p=0}^{3} \sum_{q=0}^{3} \sum_{r=0}^{3} \sum_{s=0}^{3} \theta_p(x')\theta_q(y')\theta_r(z')\theta_s(t')\phi_{i+p,j+q,k+r,l+s}, \tag{20.8}$$

where

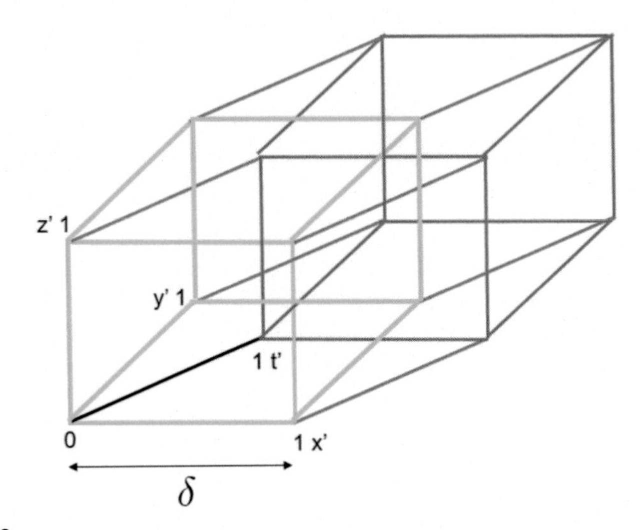

FIGURE 20.2

3D+t B-spline control point cube indicating spacing in x, y, z, and t (from green to purple).

$$
\begin{aligned}
x' &= \frac{x}{\delta_x} - \left\lfloor \frac{x}{\delta_x} \right\rfloor, \quad p = \left\lfloor \frac{x}{\delta_x} \right\rfloor - 1, \\
y' &= \frac{y}{\delta_y} - \left\lfloor \frac{y}{\delta_y} \right\rfloor, \quad q = \left\lfloor \frac{y}{\delta_y} \right\rfloor - 1, \\
z' &= \frac{z}{\delta_z} - \left\lfloor \frac{z}{\delta_z} \right\rfloor, \quad r = \left\lfloor \frac{z}{\delta_z} \right\rfloor - 1, \\
t' &= \frac{t}{\delta_t} - \left\lfloor \frac{t}{\delta_t} \right\rfloor, \quad s = \left\lfloor \frac{t}{\delta_t} \right\rfloor - 1,
\end{aligned} \tag{20.9}
$$

where $(\delta_x, \delta_y, \delta_z, \delta_t)$ are the control point spacings, θ are the B-spline basis functions, and ϕ are the control point parameters and thus the parameters to be optimized (see Fig. 20.2).

The transformation from time 0 to time t_n is given by

$$
\mathbf{x}(t_n) = \mathbf{x_0} + \int_0^{t_n} \mathbf{u}(\mathbf{x}, t)dt, \tag{20.10}
$$

which can be integrated numerically.

This spatiotemporal transformation can then be used to register the entire sequence

$$
\hat{\phi} = \arg\min_{\phi} \sum_{t=0}^{N-1} M(I_{t+1}(x_0), I_t(x_t)) + \alpha R, \tag{20.11}
$$

where R can be used to constrain smoothness or enforce incompressibility.

20.3.2 Optical flow

One of the earliest biomedical image analysis techniques to address motion tracking came out of the computer vision community and was aimed at simply detecting perceived motion from a fixed observer position using a sequential pair of image frames. This approach was developed by Horn and Schunck [3] in the early 1980s and was based on two fundamental constraints: (i) that the observed brightness of any point on a particular object is constant over time and (ii) that nearby points on the same object move in a similar manner. While intended originally for video-based tracking applications, the approach has since been extensively applied in a variety of biomedical applications, including cardiac motion, and is the basis for the "Demons" image registration strategy [4].

The optical flow approach computes Eulerian motion in a field from a pair of images and attempts to capture motion across a 2D image sequence. Importantly, it is desirable to avoid confusion due to intensity variation such as from shadows (in typical videos) or other unwanted intensity changes. Optical flow can also be used as an initial pre-processing step for more complex motion estimation. As shown in Fig. 20.3, we desire to find the motion velocity vector $\mathbf{u} = (u, v)$. If $\mathbf{x} = (x, y)$ is a 2D position vector, then $u = \frac{dx}{dt}$ and $v = \frac{dy}{dt}$. Furthermore, given a discrete image sequence equally sampled in time, $\mathbf{u}$ can also be thought of as a displacement vector.

We start this development by assuming that $f(x, y, t)$ represents an image intensity value located at spatial location (x, y) at time frame t. Furthermore, we can express the information in a dynamic sequence of moving images as changes in both space and time, i.e., $f(x + dx, y + dy, t + dt)$. Next, we expand this dynamic expression in a Taylor Series:

$$f(x + dx, y + dy, t + dt) = \\ f(x, y, t) + f_x(x, y, t)dx + f_y(x, y, t)dy + f_t(x, y, t)dt + \mathcal{O}(\partial^n), \tag{20.12}$$

where $f_x = \frac{\partial f(x,y,t)}{\partial x}$, $f_y = \frac{\partial f(x,y,t)}{\partial y}$, $f_t = \frac{\partial f(x,y,t)}{\partial t}$, and $\mathcal{O}(\partial^n)$ represents higher-order derivative terms. Now we invoke the brightness consistency assumption (i) from above by assuming that the point (x, y) is translated only a small distance in the small time instant dt, i.e.,

$$f(x + dx, y + dy, t + dt) \simeq f(x, y, t). \tag{20.13}$$

If we impose brightness consistency by inserting equation (20.13) into equation (20.12) and ignore the higher-order derivative terms, we end up with the *optical flow* constraint equation:

$$f(x, y, t) = f(x, y, t) + f_x(x, y, t)dx + f_y(x, y, t)dy + f_t(x, y, t)dt, \tag{20.14}$$

which can be rewritten as

$$- f_t(x, y, t) = f_x(x, y, t)\frac{dx}{dt} + f_y(x, y, t)\frac{dy}{dt}. \tag{20.15}$$

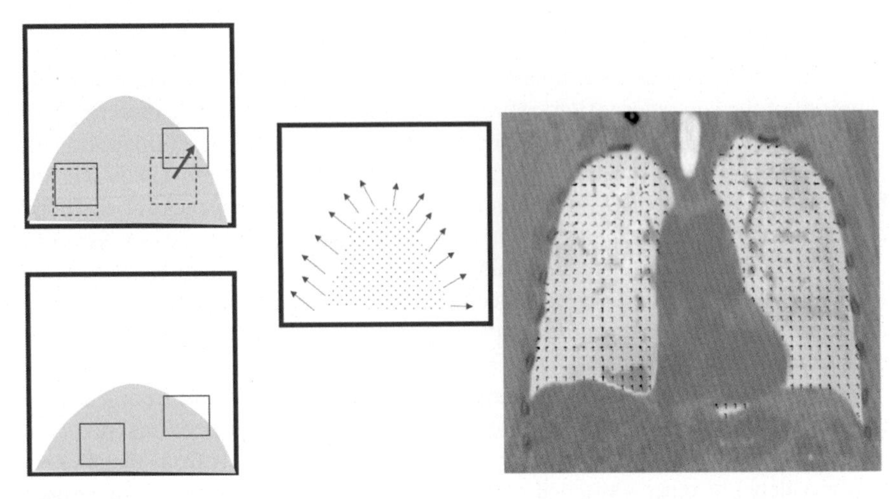

FIGURE 20.3

Optical flow. Left: Brightness consistency information is matched between time frames, but can only reliably estimate displacements perpendicular to brightness gradients. The inability to track information within a window (aperture) along a gradient is known as the "aperture problem." Right: Example optical flow computation in the lungs using the strategy defined by equation (20.16) [5].

This expression can be interpreted as saying that the local change of the intensity values over time is proportional to the local spatial changes in the current image scaled by an optical flow velocity vector (or optical flow displacement vector if discrete), and is sometimes written in an abbreviated form as $-f_t = f_x u + f_y v$, recalling that $u = \frac{dx}{dt}$ and $v = \frac{dy}{dt}$. It is important to realize that the local information about the brightness gradient (f_x and f_y) and the rate of change of brightness with time (f_t) provide just one constraint to solve for the component of the flow velocity (or displacement) vector $\mathbf{u} = (u, v)$ in the direction of the brightness gradient (f_x, f_y). This constraint provides no information about the flow component at right angles to this direction or along isobrightness contours, as can be seen by considering the search windows in Fig. 20.3. This difficulty is known as the *aperture problem* [3], see also Section 14.3.

In most biomedical problems (take for instance motion of the heart or lungs) we are dealing with deformations that are locally smooth. Hence, the most common additional constraint for estimating dense fields of Eulerian optical flow fields is to minimize a departure of the flow vectors from local smoothness, which can be written in continuous form as an error $e_s = (u_x^2 + u_y^2) + (v_x^2 + v_y^2)$, where (u_x, u_y, v_x, v_y) are local velocity/displacement changes in the x- and y-directions between any two frames. Thus, to find a smooth optical flow velocity/displacement field between two (typically consecutive) sets of images indexed by $\mathbf{x}(t)$ and $\mathbf{x}(t + \Delta t)$, namely $\mathbf{u}[\mathbf{x}(t), \mathbf{x}(t + \Delta t)]$, we would find the argument of the expression that minimizes a functional $F(u, v, u_x, v_x, u_y, v_y)$ that combines the optical flow constraint $e_c = f_x u + f_y v + f_t$ from above and a weighted (by λ) version of the smoothness constraint e_s. Thus, this functional can be written as $F(u, v, u_x, v_x, u_y, v_y) = e_c + \lambda e_s$ and the minimization to solve for a field of optical flow velocity/displacement vectors can be written as

$$\mathbf{u}[\mathbf{x}(t), \mathbf{x}(t + \Delta t)] = \arg\min_{\mathbf{u}} \int_x \int_y F(u, v, u_x, v_x, u_y, v_y) dx dy. \quad (20.16)$$

This continuous expression can be solved for the continuous velocity optical flow field using calculus of variations (see [3]). It is key to remember that a functional is defined as a "function of functions" such that the optimal solution is the function representing a continuous field of displacement vectors $\mathbf{u}[\mathbf{x}(t), \mathbf{x}(t + \Delta t)]$ that can be sampled into vectors at discrete positions.

The optical flow approach has been applied to a variety of biomedical problems, including cardiac motion estimation and lung motion quantification (see Fig. 20.3) for the purpose of motion correction or respiratory measurement [5]. Deep learning with neural networks has been used to estimate optical flow in a number of problems as well [6]. See Section 20.6.2 for more on data-driven learning techniques.

20.4 Object-centric, locally derived flow fields: Lagrangian analysis

20.4.1 Feature-based tracking

The non-rigid registration and optical flow techniques described above are primarily useful when time intervals are relatively short and when objects (e.g., organs) in the image are not already segmented. An important class of approaches for motion tracking and deformation analysis in the presence of irregular or longer times between frames and when specific objects (e.g., the heart) are identified for tracking uses feature-based correspondence. Feature matching aims to provide a set of high-confidence, local frame-to-frame correspondences (often sparse). Regularization is included to interpolate the displacements/velocities to derive dense flow fields. Some important biomedical examples are the use of shape-based features from segmented object surfaces [7] or image-acquired embedded intensity tags (e.g., MR tagging) [8], most often in the analysis of cardiac LV motion/deformation, but in other areas as well (e.g., tongue movement [9]).

For these feature-based strategies, the core idea is that some form of a local tracking token must be identified at an initial frame and then located at the next sequential time frame. Often, the information at the second time frame needs to be considered within a search window W from which an optimal position would be selected. Using the notation developed at the beginning of this chapter, we refer to this optimal position as $\mathbf{x}^*(t + \Delta t)$ and solve for it by minimizing the match in tracking tokens at positions $\mathbf{x}(t)$ at the first frame and candidate tracking tokens at surrounding positions

$\mathbf{x}(t+\Delta t)$ at the next frame within the search window W, i.e.,

$$\mathbf{x}^*(t+\Delta t) = \arg \min_{\mathbf{x}(t+\Delta T)\in W} M[\mathbf{x}(t), \mathbf{x}(t+\Delta t)], \tag{20.17}$$

where M is a match metric that compares information in the neighborhood of the point at coordinates $\mathbf{x}(t)$ and $\mathbf{x}(t+\Delta t)$. Once this optimum coordinate is found at the second time frame, it can be used to compute the velocity/displacement vector $\mathbf{u}^*[\mathbf{x}(t), \mathbf{x}^*(t+\Delta t)]$.

The information defined within match metric $M(\cdot,\cdot)$ helps define the different motion tracking methods. As noted above, two interesting examples we consider here are (i) tracking local shape from segmented objects [7] and (ii) tracking image intensity tag lines created by spatially modulating the longitudinal magnetization of a deforming object, as in cardiac MR tagging [8]. First, in (i) the primary application is for tracking the motion of the LV of the heart from MRI, ultrasound, or CT images. Here, positions $\mathbf{x}_b(t)$ in an image that lie on bounding segmented curves or surfaces, $\mathcal{B}(t)$, that have already been segmented are considered, i.e., $\mathbf{x}_b(t) = \mathbf{x}(t) \in \mathcal{B}(t)$. These segmented curves or surfaces could be manually delineated or found automatically using one of the approaches described in Part IV.

Shape features for tracking are derived from these sets of bounding surface points by finding the local bending energy on the curve surrounding each point. For curves that define the LV that were derived from 2D image frames, this is simply the square of the curvature at that point (using a fixed distance on either side of the point) $= \kappa^2$; for surfaces that have been extracted from 3D image frames, this is the squared average of the principal curvatures of a patch (again defined as a fixed distance) surrounding the point $= \frac{\kappa_1^2+\kappa_2^2}{2}$. In the case of surfaces segmented from 3D image frames, it is helpful to note that the principal curvatures are the maximum and minimum curvatures defined at two perpendicular angles when an imaginary cutting plane perpendicular to the tangent plane centered at each point slices through the patch of interest at all possible angles. The match metric M for shape matching of curve sections or surface patches can now be defined by comparing the bending energies of a patch at time t at location $\mathbf{x}(t)$ to that of candidate patches within window W at time $t+\Delta t$ by looking at the squared differences in the principal curvatures as follows (note that one could include a term α that is an overall scaling that could be thought of as an overall stiffness of the surface, although it is often ignored):

$$\begin{aligned} M_{BE}\,[\mathbf{x}_b(t), \mathbf{x}_b(t+\Delta t)] &= \epsilon_{BE}\,[\mathbf{x}_b(t), \mathbf{x}_b(t+\Delta t)] \\ &= \alpha\left[\frac{1}{2}[\kappa_1(\mathbf{x}_b(t)) - \kappa_1(\mathbf{x}_b(t+\Delta t))]^2 + \frac{1}{2}[\kappa_2(\mathbf{x}_b(t)) - \kappa_2(\mathbf{x}_b(t+\Delta t))]^2\right]. \end{aligned} \tag{20.18}$$

See also Fig. 20.4.

In case (ii) for MR tagging, we use Active Geometry matching (see Chapters 9 and [8]). While there are many approaches to tag line identification, this approach is

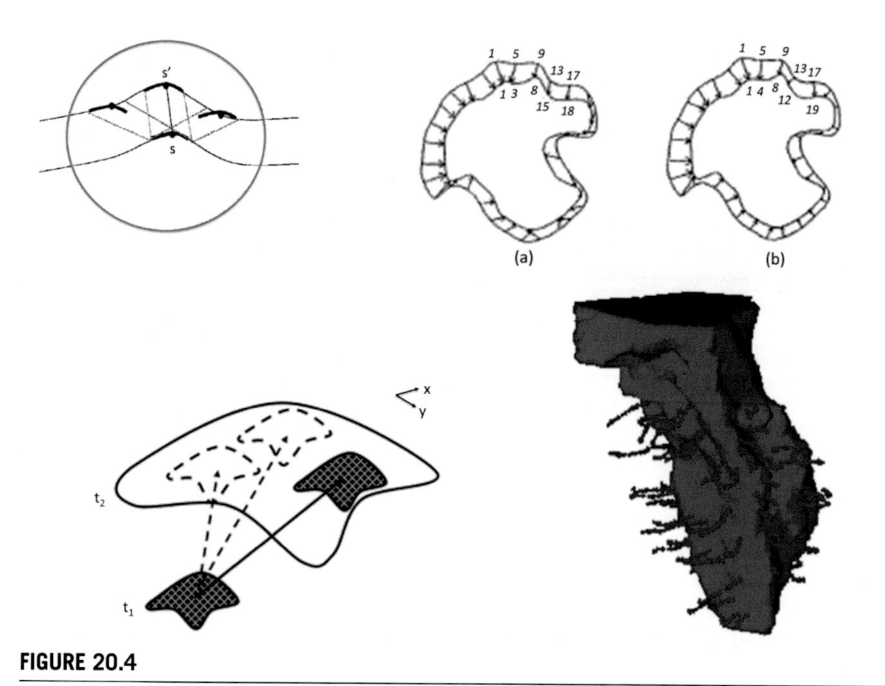

FIGURE 20.4

Segmented curve and surface shape tracking. Upper left: 2D curve segment at time t_1 to be matched to a segment within a search window at time t_2. Upper right: initial shape-matched displacements based on curvature (a) and after regularization (b). Lower left: 3D surface patch segment at time t_1 to be matched to a patch segment within search window at time t_2. Lower right: Spatially regularized shape matches between many pairs of frames connected end-to-end to show the trajectory time sequence.

one of the more basic ideas that also allows us to pose the problem within the local feature-based Lagrangian tracking methods discussed in this section. Thus, we start with the notion that an initial MR tag identification process identifies a deformable curve segment or set of curve segments (termed a "snake" in the segmentation literature) using the methods described in Chapter 9 (see Kerwin et al. [8]). These curve segments are detected so that they adhere to a tag line in the underlying MR tag images (see Fig. 20.5). In practice, each separate curve segment (or snake) within a set or grid of such segments is represented by a set of K points along one tag line. The kth point on the detected curve segment at the next $(t + \Delta t)$ time frame can then be written as $\mathbf{x}(k, t + \Delta t)$. To determine the set of points that best identifies a tag line at time $(t + \Delta t)$, namely $\mathbf{x}^*(1, t + \Delta t), ..., \mathbf{x}^*(K, t + \Delta t)$, we begin with the solution at time t or for the first time point with a user-specified (e.g., manually traced or automatically segmented) set of points for the curve. Next, a set of 2D displacements $(\mathbf{u}(k, t + \Delta t))$ are found that move the entire curve segment/snake at time t to the optimal position(s) at time $(t + \Delta t)$, which is written for each curve/snake point as $\mathbf{x}^*(k, t + \Delta t) = \mathbf{x}(k, t) + \mathbf{u}(k, t + \Delta t)$. Finding these optimal points is performed by

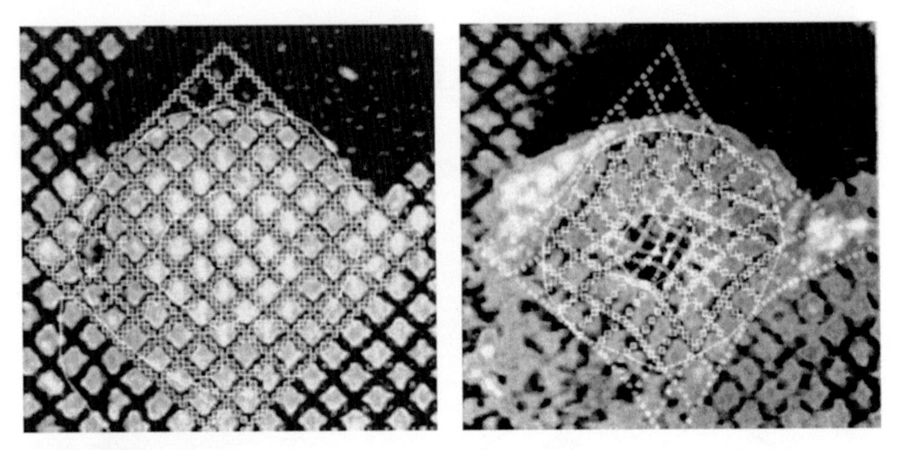

FIGURE 20.5

MR tagging-based tracking using an Active Geometry approach. Left: A 10×10 grid of coupled snakes superimposed on the first MR tagged image in a sequence (following the ideas in [8] and [10]). Right: After minimizing an energy expression that includes regularization (see Section 20.4.2) and equation (20.19), the coupled snakes move to new positions. Positions marked by diamonds are points outside the myocardium that are turned off for further analysis.

looking for signal voids or minimum intensities in the underlying MR tag image that are located within a search window W near the candidate curve/snake at time t but displaced to candidate positions in the next image in the MR tagging time sequence, $I_{MRtag}\left[\mathbf{x}^*(k, t+\Delta t)\right]$. Localization and regularization of the snake tags curve can be performed discretely in one minimization step (see the original development in Young et al. [10]). For our purposes, we separate out the matching to signal voids and perform regularization below. Thus, we refer to equation (20.17) and write down an MR tagging image signal void metric term to be minimized (in other words, the $M_{MRtag}[\mathbf{x}_b(t), \mathbf{x}_b(t+\Delta t)]$ term is optimal when the image intensity values are at their lowest/darkest values) as

$$M_{MRtag}[\mathbf{x}_b(t), \mathbf{x}_b(t+\Delta t)] = I_{MRtag}\left[\mathbf{x}^*(k, t+\Delta t)\right]. \tag{20.19}$$

Now, referring back to the discussion above related to equation (20.17), for our two examples of shape-based tracking using local bending energies from segmented curves or surfaces (e.g., LV endocardial surfaces) using $M_{BE}[\mathbf{x}_b(t), \mathbf{x}_b(t+\Delta t)]$ or matching using curve segments and MR tagged images using $M_{MRtag}[\mathbf{x}_b(t), \mathbf{x}_b(t+\Delta t)]$ we can insert a local feature matching metric into equation (20.17), search within a search window W, arrive at an optimal coordinate or position at a second time frame, and from this find or compute the velocity/displacement vector $\mathbf{u}^*[\mathbf{x}(t), \mathbf{x}^*(t+\Delta t)]$. It is important to note that the search window only permits considering candidate points *on* the segmented surface at the next time point in the shape matching case and only looking at intensity values underneath a coherent displaced tag curve in the MR tag case. The Active Geometry MR tag matching approach is shown in Fig. 20.5.

20.4.2 Displacement regularization

In most applications it is important to derive a set of spatially (and often also temporally) smooth displacement fields. This smoothness constraint will have additional importance as we consider finding strain and/or deformation fields below. The simplest form of constraint is inserting the above displacements found from pointwise tracking into objective functions that include regularization terms aimed at minimizing the distance between neighboring displacements. For more complicated scenarios, data-driven ideas based on neural networks and deep learning can also be used.

First, we consider thinking of the set of feature-matched displacements described above as a continuous set of vector-valued information $\mathbf{u}^*[\mathbf{x}(t), \mathbf{x}^*(t+\Delta t)]$ (abbreviated as $\mathbf{u}^*$). We now want to derive a smoothed set of vector values $\overline{\mathbf{u}}[\mathbf{x}(t), \mathbf{x}^*(t+\Delta t)]$ (abbreviated as $\overline{\mathbf{u}}$). Thus, the desired new set of vectors can be found from

$$\overline{\mathbf{u}}[\mathbf{x}(t), \mathbf{x}^*(t+\Delta t)] = \arg\min_{\overline{\mathbf{u}}} \int \int_{\overline{\mathbf{u}} \in B} \left\{ A(\mathbf{u}^*) \left[\overline{\mathbf{u}} - \mathbf{u}^*\right]^2 + \left[\frac{d\overline{\mathbf{u}}}{d\mathbf{x}^*}\right]^2 \right\} d\mathbf{x}^*, \quad (20.20)$$

where $A(\mathbf{u}^*)$ are confidence values placed on the initial feature-based matches as described above. In this basic setup, only smoothing of only the first derivative of $\overline{\mathbf{u}}$ is included in the functional, which is used to minimize stretching. One could include a second derivative term as well $\left[\frac{d^2\overline{\mathbf{u}}}{d\mathbf{x}^{*2}}\right]^2$ to minimize overall bending, especially related to the MR tagging grids just mentioned.

Confidence values for tracking. It is often helpful when applying the displacement regularization approach, defined in equation (20.20), to include matching confidence-based weights to the data-driven term. In other words, one could weight the leftmost data term in this equation, which rewards agreement with $\mathbf{u}$ values found using the displacements estimated by applying either the bending energy or MR tagging matching metric in equation (20.17), differently than the smoothing or regularizing term. We write this term as $A(\mathbf{u}^*)$ and note that it can vary for different displacements at different positions. We design it such that it takes on higher values when there is higher confidence in the underlying data match used to generate the displacement. More concretely, for the $\mathbf{u}$ displacements found using the curve-based bending energy metric $M_{BE}[\mathbf{x}_b(t), \mathbf{x}_b(t+\Delta t)]$, $A(\mathbf{u}^*)$ is larger when the curvature match between segments at time t and $t+\Delta t$ is both strong and unique. For the $\mathbf{u}$ displacements found using the MR tagging energy metric $M_{MRtag}[\mathbf{x}_b(t), \mathbf{x}_b(t+\Delta t)]$, the confidence value $A(\mathbf{u}^*)$ is larger when the image segment overlays a full set of low-intensity image signal void values (I) more completely.

20.5 Multiframe analysis: Kalman filters, particle tracking

While many techniques in this chapter discuss the estimation of displacements or velocities between pairs of frames, there are some motion models that endeavor to estimate motion using information from a number of (typically consecutive) frames in an image sequence, such as for predicting cellular motion or motion of subcellular particles (e.g., vesicles) or the movement of points on the wall of the heart or lungs. These dynamic models typically stem from a single approach based on the statistical estimation of the states of a time-varying system from a noisy image sequence. More precisely, in general for this type of problem, one wants to estimate the current (or next) state of a dynamic system from a series of noisy measurements. For our purposes here we can think of the *state(s)*, typically referred to as a state vector $\mathbf{x}$, as a set of positions (in (x, y) or even (x, y, z)) of an object or a portion of the image at a particular moment in time which could in turn be used to derive displacements. The state vector is derived from a time sequence of local typically noisy image information (*the measurements*), usually referred to as the measurement vector $\mathbf{z}$. For our purposes here we can think of the measurement vector as consisting of either raw image intensities or image-derived features.

The simplest of these analyses is a linear technique known as Kalman filtering and a set of more complicated, non-linear approaches are referred to as particle filtering. We will briefly discuss both of these methods here in the context of the motion estimation applications mentioned above. The general state space diagram for estimating states in a time-varying system is shown in Fig. 20.6. This system could represent the motion of points on an object such as a heart or lung wall, or a set of independently moving objects such as vesicles or cells, as derived from noisy image sequences, e.g., echocardiograms or sequences of microscopy images. As shown in the figure, there are two key steps: (i) prediction, which contains a dynamic model (A) that describes the evolution of the states, and (ii) update (or correction) of the prediction via an image-derived measurement, which is represented by an observation model (H). In general, the prediction process can be described as $\mathbf{x}_k = \mathbf{A}_k(\mathbf{x}_{k-1}, \boldsymbol{\nu}_{k-1})$, where $\mathbf{x}_k$ is the current state vector (i.e., all object positions at state or time k), $\mathbf{x}_{k-1}$ is the previous state vector, $\boldsymbol{\omega}_{k-1}$ is process noise involved in the prediction, and $\mathbf{A}_k$ is the state evolution model. The update/correction process can be described as $\mathbf{z}_k = \mathbf{H}_k(\mathbf{x}_k, \boldsymbol{\nu}_k)$, where $\mathbf{x}_k$ again is the current state vector, $\mathbf{z}_k$ is now the image-derived measurement vector, $\boldsymbol{\nu}_k$ is the image-based measurement noise, and $\mathbf{H}_k$ is the model that relates the measurements to the states. In most uses of this approach, one tries to find the maximum a posteriori estimate of the next state of the time-evolving system, e.g., positions or displacements at the next point in time k given the image sequence data from the initial frame up to and including data at the current time frame.

In the most readily understandable use of this approach, assumptions are made that both the state transition and the measurement-to-state models are linear and can be written in matrix form ($\mathbf{A}_k$ and $\mathbf{H}_k$), and furthermore that both the process noise (now written as a vector $\mathbf{w}_k$) and the measurement noise (now written as vector $\mathbf{v}_k$) can be assumed to be sampled from probability distributions that are zero mean,

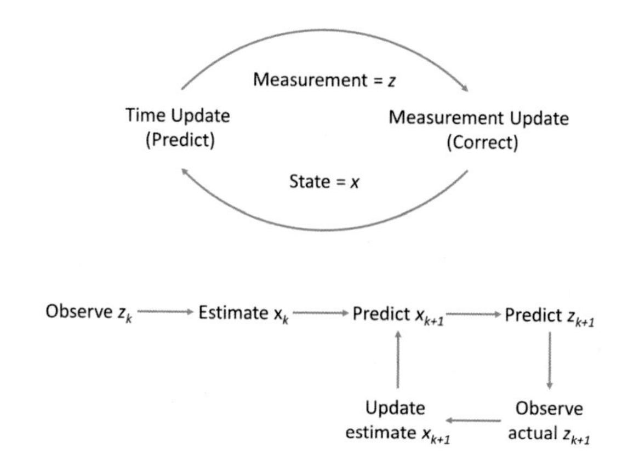

FIGURE 20.6

Kalman filtering. Top: Prediction and update steps. Bottom: More detailed processing sequence.

Gaussian processes with known covariances, $\mathbf{Q}_k = E(\mathbf{w}_k \mathbf{w}_k)^T$ and $\mathbf{R_k} = E(\mathbf{v}_k \mathbf{v}_k)^T$. For image-based motion tracking/estimation, in most cases it is helpful to understand that the measurement noise information comes from the physical characteristics of the type of images being used (e.g., MRI, ultrasound, light microscopy), whereas process noise is more used as something to adjust the influence (e.g., how much smoothing) that the Kalman filter introduces. Thus, the basic Kalman filter model is defined by these two linear equations:

$$\begin{aligned} \mathbf{x}_{k+1} &= \mathbf{A}_k \mathbf{x}_k + \mathbf{w}_k, \\ \mathbf{z}_k &= \mathbf{H}_k \mathbf{x}_k + \mathbf{v}_k. \end{aligned} \tag{20.21}$$

To further define the Kalman filter, it is necessary to define a few more terms. First for a particular state k (which typically implies we are considering a particular time frame in an image sequence), we define the "true" underlying state to be $\mathbf{x}_k$. Now if we are given the actual values (or an estimate) of the previous state $\mathbf{x}_{k-1}$, we can use the top line in equation (20.21) to provide a "pre-observation at k estimate" of the state at k, which we write as $\hat{\mathbf{x}}_{k-1}^-$. Furthermore, if we define $\hat{\mathbf{x}}_{k-1}^+$ as the "post-observation estimate," we can define the following errors in these two estimates as

$$\begin{aligned} \mathbf{e}_k^- &= \mathbf{x}_k - \hat{\mathbf{x}}_k^- \quad \text{with covariance} \quad \mathbf{P}_k^- = E\left\{\mathbf{e}_k^- \mathbf{e}_k^{-T}\right\}, \\ \mathbf{e}_k^+ &= \mathbf{x}_k - \hat{\mathbf{x}}_k^+ \quad \text{with covariance} \quad \mathbf{P}_k^+ = E\left\{\mathbf{e}_k^+ \mathbf{e}_k^{+T}\right\}. \end{aligned} \tag{20.22}$$

Then, using all of the above, it can be shown that the Kalman filter fundamentally operates by looking at the computed *residual* between observed data from an image and the linear estimate in the second line of equation (20.21), namely $(\mathbf{z}_k - \mathbf{H}_k \hat{\mathbf{x}}_k^-)$. Both the measurement noise $\mathbf{v}_k$ and the pre-observation error $\mathbf{e}_k^-$ contribute to this residual (and, it should be noted, if there were no measurement noise and the estimate were perfect, the residual would be zero). Now, one can write down the equation for the updating phase (post-observation) of the Kalman process as

$$\hat{\mathbf{x}}_k^+ = \mathbf{x}_k^- + K_k \left(\mathbf{z}_k - \mathbf{H}_k \hat{\mathbf{x}}_k^- \right), \tag{20.23}$$

where K_k is the Kalman gain matrix, which alters the estimate by filtering/compensating for the residual difference between the actual observed data and the estimate of the same information. Since the best situation is when this residual is zero, we solve for K_k by taking the partial derivative of the trace of the post-observation covariance matrix $\mathbf{P}^+$ (essentially the sum of all of the posterior error variances) with respect to K_k and set it equal to zero, i.e., $\partial \text{trace}(\mathbf{P}_k^+)/\partial K_k = 0$. After a number of steps, recalling all the definitions in the last several paragraphs, it can be shown that the Kalman gain term that will be used in equation (20.23) reduces to

$$K_k = \mathbf{P}_k^- \mathbf{H}_k^T \left(\mathbf{H}_k \mathbf{P}_k^- \mathbf{H}_k^T + \mathbf{R}_k \right)^{-1}. \tag{20.24}$$

As part of this development and manipulation, an expression for the pre-observation error covariance (the current state) falls out as $\mathbf{P}_k^- = \mathbf{A}_k \mathbf{P}_{k-1}^+ \mathbf{A}_k^T + \mathbf{Q}_k$ and then after each estimate, the update for the filter's post-observation error becomes

$$\mathbf{P}_k^+ = (I - \mathbf{K}_k \mathbf{H}_k) \mathbf{P}_k^-. \tag{20.25}$$

The notation used here follows the development in an earlier excellent text by Sonka et al. [11]; refer to that book for additional detail. Note that while the state and hence feature vectors $\mathbf{x}$ are basically thought of as positions in 2D or 3D space within an image, they could also be used to represent other parameters, such as displacement or velocity, relevant to the motion/deformation tracking problem highlighted in this chapter.

Example. Here, we present a simple simulated object tracking example to illustrate the basic Kalman filtering principles. The initial test image is shown in Fig. 20.7, made up of seven white objects, simulating seven individual cells, on a dark background, and the image sequence is made up of 50 frames, illustrating motion that causes occlusion (blocking one object by another), fusion (merging objects), fission (splitting objects), and blinking (some objects disappear in certain frames) over the temporal sequence. Testing is performed by injecting Gaussian noise, realizing ground truth by having a human observer locate the centroids of each object at all frames, and comparing algorithm-estimated and human expert-localized trajectories. In this example, for each object, the state vector at each time point k is made up of both the 2D centroid and the velocity vector, i.e., $\mathbf{x}_k = (s_k, y_k, u_k, v_k)$. Furthermore, both the last position and the last velocity serve as a basis for updating position estimates and the last velocity is carried forward as the estimate to update the next

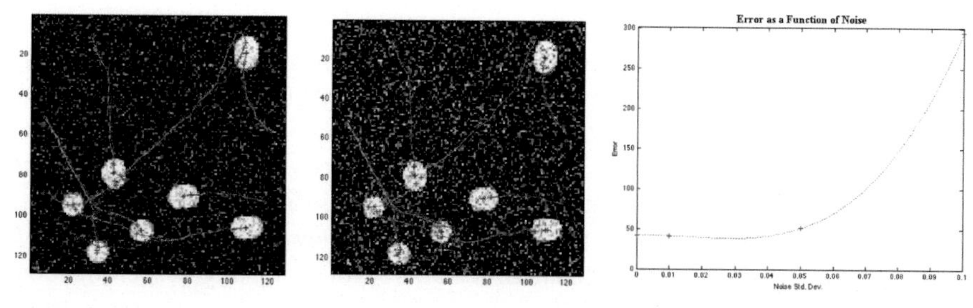

FIGURE 20.7

Kalman tracking results. Left: Tracking of seven cell-like objects with 10 frames per image sequence (low noise). Human expert tracking (blue) and algorithmic results (red, when there is a mismatch) are shown. Middle: Same objects with higher noise. Right: Plot of tracking error vs. noise level in each image sequence.

velocity; thus, the state transition matrix for this basic example is written as

$$\mathbf{A} = \begin{bmatrix} 1 & 0 & 1 & 0 \\ 0 & 1 & 0 & 1 \\ 0 & 0 & 1 & 0 \\ 0 & 0 & 0 & 1 \end{bmatrix}. \tag{20.26}$$

Finally, it is assumed that the states and data are the same, i.e., $\mathbf{H} = I$, and the process and measurement noise and hence covariances are constant everywhere. Tracking results as a function of image noise are shown in Fig. 20.7.

The errors plotted in Fig. 20.7 refer to the pixel-based Euclidean distance between the human expert-marked object centroid and that estimated by the algorithm over all frames, averaged over all objects.

If one relaxes the linear constraints and goes to either an extended linear or fully non-linear model for A and H, the particle filters that can be realized can be used for advanced tracking such as following subcellular objects such as vesicles. Biomedical examples of these approaches have been reported in Chenouard et al. [12] for intracellular object tracking from different types of fluorescent microscopy, including quantifying motion of endocytic processes [13].

20.6 Advanced strategies: model-based analysis and data-driven deep learning

20.6.1 Model-based strategies: biomechanics and deformation analysis

In some biomedical motion and deformation analysis problems, most notably analysis of the ventricular chambers of the heart from MRI or echocardiographic image data, it is both convenient and prudent to employ biomechanical models for the purposes of both obtaining a smooth displacement and/or strain field and computing strain values or material properties that can be used for further analysis. Key examples of this are work based on MR tagging (e.g., see [14]) and on shape/speckle tracking from echocardiography (e.g., see [15]). Similar biomechanical models have been used for estimating brain shift or deformation during neurosurgery as well [16,17].

To give the reader a sense of these methods, we follow a simplified version of the development described in Papademetris et al. [15] for estimating LV deformation from medical images, including MRI images, CT scans, and echocardiograms. Here we assume (i) that the bounding surfaces of the object being studied, i.e., the LV endocardial and epicardial surfaces, have been segmented for all 2D or 3D time frames of interest (either manually or using one of the techniques described in Part IV on segmentation); (ii) that a volumetric mesh made up of hexahedral (cuboid) elements has been generated [15] between the bounding surfaces at one starting frame (typically the end diastolic frame in the cardiac image sequence); (iii) that one or more initial displacement tracking methods have been applied to a pair of frames in the 2D or 3D temporal sequence of images to be analyzed (shape tracking or MR tag tracking are key examples for this initial tracking, as described in Section 20.4.1, without regularization applied yet); and (iv) that the set of initial displacements $\mathbf{u}[\mathbf{x}(t), \mathbf{x}^*(t + \Delta t)]$ have been computed with confidence values $A(\mathbf{u}^*)$.

We desire to find a dense set of displacements at all nodes of the volumetric mesh model of the LV between a pair of (typically consecutive) image frames. We concatenate the set of these dense displacements between any two frames t and $t + \Delta t$ (at all mesh nodes) into a long vector $\mathbf{U}$. Furthermore, although higher-confidence values from initial (typically feature-based) displacement tracking lie in a sparser set of nodes of this same volumetric mesh, we compile and refer to a complete set of initial displacements at all mesh nodes as $\mathbf{U}^*$ with an associated long confidence vector for all nodes $\mathbf{A}^*$. In [15], the authors proceed to follow a finite element method (FEM) formulation to solve for the dense long displacement vector $\mathbf{U}$ over the entire mesh by setting the problem up as estimating a Bayesian posterior probability estimation with the initial displacement tracking $\mathbf{U}^*$ as a data term and smoothing or regularization of final displacements through a use of a strain energy function prior that incorporates a model of the stiffness of the LV material through the specification of an LV material matrix $\mathbf{C}$ that is taken to be the same at all nodes in the mesh. If one assumes Gaussian probability density functions for the data and priors and takes the log of both terms, the estimation of the vector of smoothed displacements that agree with the initial (feature-tracked) $\mathbf{U}^*$ displacement estimates can be written as

$$\hat{\mathbf{U}} = \arg\min_{\mathbf{U}} \left(\mathbf{U}^t \mathbf{K} \mathbf{U} + \left(\mathbf{U}^* - \mathbf{U}\right)^t \mathbf{A}^{-1} \left(\mathbf{U}^* - \mathbf{U}\right) \right). \tag{20.27}$$

Note that the biomechanical model, as incorporated in the strain energy term $\mathbf{U}^t\mathbf{K}\mathbf{U}$ (usually referred to as W) in equation (20.27), performs a regularization that can be thought of as an alternative to the simple regularization terms described above for feature tracking (Section 20.4.1). The design of the biomechanical term is more readily understood by first writing down the more typical form of the strain energy in terms of strain e as defined above in Section 20.1.1, i.e.,

$$W = \mathbf{e}^t \mathbf{C} \mathbf{e}. \tag{20.28}$$

If $\mathbf{e}$ is a 1×6 vector representation of the strain tensor (equation (20.4)) and we employ the simplest useful continuum mechanics model for our purposes, namely one that uses a linear elastic strain energy function W, then $\mathbf{C}$ is a 6×6 material matrix that defines the material properties at each mesh node for our deforming LV. While the most basic model is isotropically linear elastic [15], the LV is more appropriately modeled as a transversely elastic material to account for preferential stiffness along LV fibers as noted in the biomechanics literature [18]. In this case, the inverse of the material matrix $\mathbf{C}$ takes the following form:

$$\mathbf{C}^{-1} = \begin{bmatrix} \frac{1}{E_p} & \frac{-\nu_{pp}}{E_p} & \frac{-\nu_{fp}}{E_f} & 0 & 0 & 0 \\ \frac{-\nu_{pp}}{E_p} & \frac{1}{E_p} & \frac{-\nu_{fp}}{E_f} & 0 & 0 & 0 \\ \frac{-\nu_{fp}E_f}{E_p} & \frac{-\nu_{fp}E_f}{E_p} & \frac{1}{E_f} & 0 & 0 & 0 \\ 0 & 0 & 0 & \frac{2(1+\nu_{pp})}{E_p} & 0 & 0 \\ 0 & 0 & 0 & 0 & \frac{1}{G_f} & 0 \\ 0 & 0 & 0 & 0 & 0 & \frac{1}{G_f} \end{bmatrix}, \tag{20.29}$$

where E_f is the fiber stiffness, E_p is the cross-fiber stiffness, ν_{fp} and ν_p are the corresponding Poisson ratios, and G_f is the shear modulus across fibers, $G_f \approx \frac{E_f}{2(1+\nu_{fp})}$. If $E_f = E_p$ and $\nu_p = \nu_{fp}$, the model reduces to the more common isotropic linear elastic model. A good LV model is to set the fiber stiffness to 3.5 times the cross-fiber stiffness and set the values for ν_{fp} and ν_p to be about 0.4 to enforce the approximate incompressibility that is appropriate for the LV (see Papademetris et al. [15]). Finally, note that to get from the designed material matrix noted here to the stiffness matrix $\mathbf{K}$ embedded in the smoothing term of equation (20.27), one must pre- and post-multiply the matrix $\mathbf{C}$ by the strain displacement Jacobian matrix $\mathbf{B}$.

Once all of the above is taken care of in terms of performing steps (i) to (iii), including specifying the biomechanical modeling parameters, note that equation (20.27) can be differentiated to yield a final analytical form for estimating the dense set of biomechanically smoothed displacements $\mathbf{U}$ that agree with the initial (feature-tracked) displacements $\mathbf{U}^*$ (weighted by their confidences $\mathbf{A}$) by numerically solving the following expression for $\mathbf{U}$:

$$\mathbf{KU} = \mathbf{A}\left(\mathbf{U}^* - \mathbf{U}\right). \tag{20.30}$$

Using this method and shape-tracked features derived from endocardial and epicardial surfaces segmented from cine MRI data (see Fig. 20.8), a set of strains can be

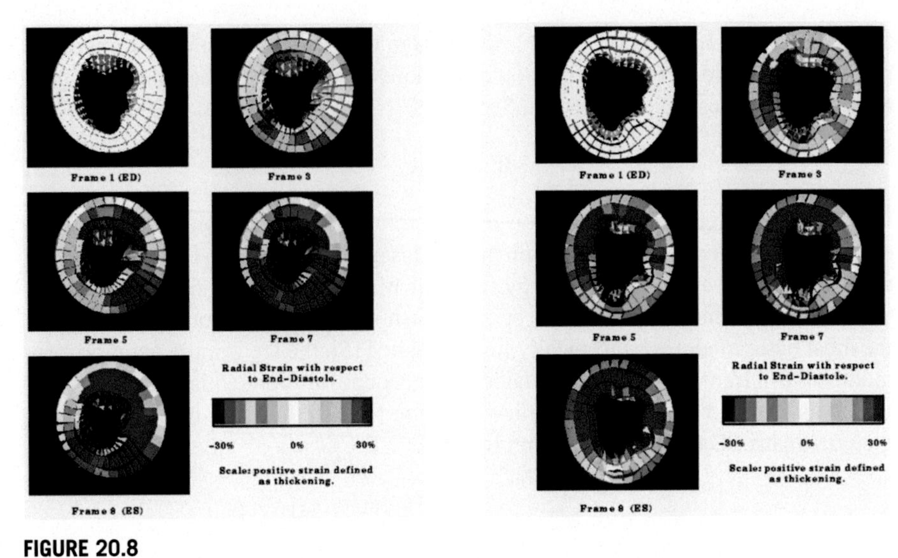

FIGURE 20.8

LV strain computation. Example strains shown before and after LAD occlusion using biomechanical regularization and shape tracking from MRI data [15].

derived from the **U** matrix. One component, radial strain, is shown for an interesting test case, before and after occlusion of the left anterior descending coronary artery (LAD).

20.6.2 Deep learning for displacement field regularization

For many of the techniques already presented in this chapter, different researchers have begun to look at the use of data-driven neural network learning strategies (see Part VIII) to complement or replace the model-based ideas that have been reported. These approaches include work on deep optical flow, deep Kalman, deep learning for MR tagging analysis, and deep learning for displacement integration. Here we describe a basic technique for integrating and regularizing displacement information looking at full spatiotemporal information. This method achieves smooth strains based on learning from synthetic data, using 4D Lagrangian displacement patches and a robust loss function designed for a multilayered neural network.

We generate 4D Lagrangian displacement patches from different input sources as training data and learn the regularization procedure via a multilayered perceptron (MLP) network (see Lu et al. [19]). The learned regularization procedure is applied to initial noisy tracking results. The approach is initiated by applying one or more of a variety of frame-to-frame displacement estimation methods (e.g., intensity block matching or feature tracking as noted above) to get an initial set of displacements. Next, the 3D (or 2D) image data, typically 16 or 32 frames in a periodic (e.g., cardiac) image sequence, such as from echocardiography or cine MRI, are aligned using intensity-based non-rigid registration. Then, using points on the object surface in a reference frame (e.g., end diastole), a sparser set of points are sampled and $m \times m \times m$ spatial patches are defined, each patch extending over all time frames. Earlier processing defines the data as a set of frame-to-frame 3D displacement vectors over time, registered to a reference frame, thus defining a 4D "patch," as shown in Fig. 20.9 (top). For each initial tracking method, the resulting frame-to-frame displacement field is temporally interpolated and propagated to produce Lagrangian displacement fields and sampled into X_{train} and X_{test}, as seen in Fig. 20.9 (bottom). Next, we spatially interpolate a sparse set of synthetic ground truth trajectories (provided using the method of Alessandrini et al. [20]) with radial basis functions (see Parajuli et al. [21]). These frame-to-frame displacement fields are also temporally interpolated to produce the Lagrangian displacement fields, and are used as ground truth trajectory patches Y_{train} and Y_{test} during training and testing.

In the training stage, given initial Lagrangian noisy tracking data, the optimal parameters θ^* are found by solving the following equation (minimization of a loss function):

$$\theta^* = \arg\min_{\theta} \frac{1}{N} \sum_{i=0}^{N-1} \log\cosh\left[Y_{train}^{(i)} - f_\theta(X_{train}^{(i)})\right], \tag{20.31}$$

where $Y_{train}^{(i)}$ is the ground truth trajectory patch and $f_\theta(X_{train}^{(i)})$ is the regularized trajectory patch for sample i over N samples. While the L_2-norm (i.e., sum of squared distances between the patches' pixels) is often used, this approach uses the mean log-cosh error, which is more robust to noise and outliers; f_θ is approximated using an MLP network f with three fully connected hidden layers and parameters θ. Rectified linear units (ReLU) are used as the activation function. To avoid overfitting, a dropout layer is incorporated after each activation layer. Dropout randomly drops the output of each neuron during training in order to avoid coadaptation among neurons [14]. During testing, the neural network with learned parameters θ^* is applied to the noisy trajectory patches X_{test} to produce corresponding regularized displacement trajectories. Finally, a dense displacement field is reconstructed by averaging the overlapping regularized trajectories. This learned network was trained on synthetic data and tested on both synthetic test data and related in vivo test data, all based on 4D echocardiography image sequences. Voxel sizes were 0.5 mm^3 with an image size of $75 \times 75 \times 61$ voxels. To test the method, a leave-one-image-out scheme was employed, training on seven images and testing on the eighth image. Training patches were sampled with a stride of 2 in each direction, and we used $5 \times 5 \times 5 \times 32 \times 3$ (3 for the x-, y-, and z-directions) for normal geometry images and $5 \times 5 \times 5 \times 39 \times 3$ for dilated geometry images (around 100,000 patches). Test patches were sampled with a stride of 1 (around 22,000 patches). For each MLP, we utilized three hidden layers with 1000 neurons each along with dropout with a probability of 0.2. Average test time was around 800 seconds (see Fig. 20.10).

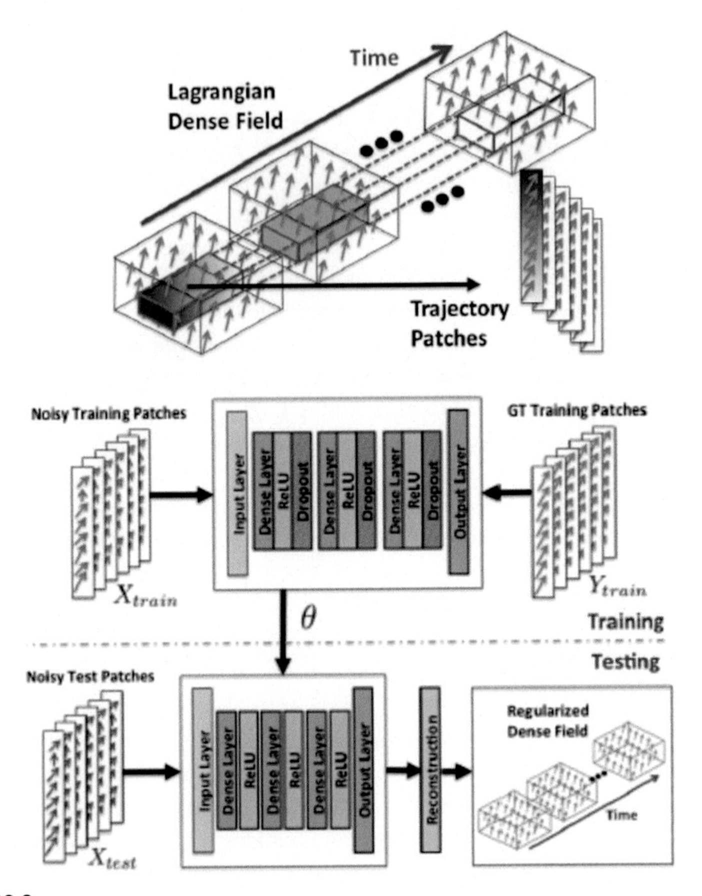

FIGURE 20.9

4D patch-based deep learning for spatiotemporal regularization. Top: Patch representation. Bottom: Neural network architecture.

20.6.3 Deep learning for integrated motion tracking and segmentation

Deep learning methods have shown remarkable performance and fast inference times compared to traditional motion tracking and registration algorithms. However, unlike segmentation or classification problems that often have concrete ground truth labels to train from, motion vector ground truth data are typically difficult to obtain, especially in medical imaging (see Section 20.7). Considering the problem of motion tracking the heart in the previous section, MR tagging has been considered the golden standard for motion vectors, but acquiring large samples of MR tagging data to train a supervised neural network is a difficult task.

To address this issue, many researchers have pursued unsupervised or semi-supervised approaches to track the motion of an object in medical imaging. For

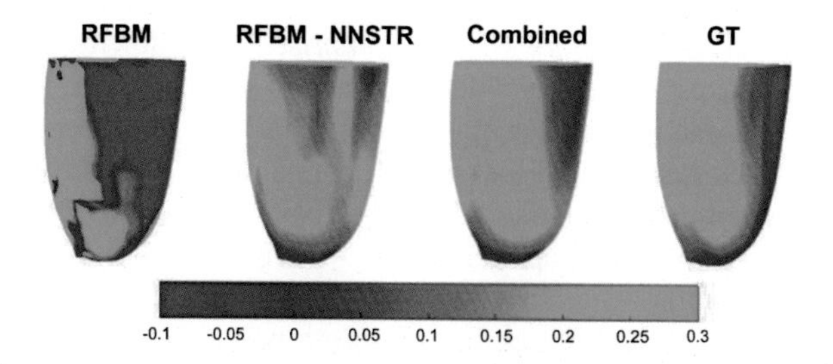

FIGURE 20.10

Results of a neural network for spatiotemporal regularization of 4D displacement patches. Radial strain is shown during end systole computed from a synthetic cardiac image sequence with a simulated left circumflex coronary artery (LCx) infarction. The results of three tracking methods are shown: (i) initial radiofrequency ultrasound block mode (RFBM) speckle tracking, (ii) RFBM plus neural network spatiotemporal regularization (NNSTR), and (iii) combined RFBM using NNSTR in conjunction with flow network tracking, a graph-based approach for further improving the consistency of the motion correspondences. The combined method shows excellent agreement with the synthetically generated ground truth, representing a significant improvement over the other techniques [19].

example, Balakrishnan et al. [22] looked at ways to directly learn the registration vectors between two images in brain MRI images. The neural network takes two images as input which then go through a standard encoder–decoder architecture (see Part VIII). The output is the x-, y-, and z-components of the displacement vectors $\mathbf{u}$ that are used to spatially transform and interpolate one image to the other. The model then uses the mean square error between the transformed image and the original image as the loss function to update the model parameters during the learning process. Similar unsupervised motion tracking has been done in echocardiography as well [23]. Qin et al. [24] address *multimodal* image registration, where the images come from different imaging modalities (domains), using a neural network that extracts features in an unsupervised approach. The extracted features are the so-called disentangled attributes of the images. These disentangled features effectively embed images into a domain-invariant latent space where the registration takes place. The neural network, thus, registers images from different modalities by aligning the latent space image features, thereby reducing a multimodal registration problem into a monomodal problem.

Segmentation and motion tracking tasks can often be complementary, especially in cardiac imaging. The tracked displacement fields and the estimated segmentation masks should be consistent. Thus, the displacements can be propagated to transform the myocardium segmentation masks as well as the raw images to augment the performance of the individual tasks. Here, we describe a deep learning method that combines motion tracking and segmentation in an iterative manner. Following the procedure of Ta et al. [25], and as illustrated in Fig. 20.11, we start with an initial motion tracking network branch to generate rough motion estimates. This branch is a convolutional neural network (CNN) comprised of a downsampling encoder path followed by an upsampling decoder path with skip connections to connect features learned in the encoder path with features learned in the decoder path. Each path is built using convolutional blocks, each containing a 3D convolutional layer, a batch normalization operation, and a ReLU activation function. Downsampling is performed using a 3D maxpool operation and upsampling is performed using nearest neighbor interpolation. The input of the CNN is a pair of sequential volumetric images, I_{source} and I_{target}, representing two time points. The objective is to estimate a displacement field $\mathbf{U}$ that properly maps the source image to the target image. The CNN achieves this by minimizing the voxelwise ($\Omega \subset \mathbb{R}^3$) mean square error between the target image and the source image that has been warped using trilinear interpolation $\mathcal{T}$ using the estimated displacement field:

$$\mathcal{L}_{motion} = \frac{1}{\Omega} \sum_{\Omega} \left(I_{target} - \mathcal{T}(I_{source}, \mathbf{U}) \right)^2 . \tag{20.32}$$

Once fully trained, the estimated motion can be used to warp a manually segmented source image, which will serve as a pseudoground truth for the segmentation branch. The segmentation branch is architecturally the same as the motion tracking branch. The input is a single volumetric image, ideally the target image of the pair used in the motion tracking branch. The objective is to segment the object of interest in the image. Segmentation is achieved by maximizing a combined Dice and binary cross-entropy loss between the predicted segmentation S_{pred} and the pseudoground truth S_{gt}:

$$\mathcal{L}_{seg} = \left(1 - \frac{2|S_{pred} \cap S_{gt}|}{|S_{pred}| + |S_{gt}|} \right) + \mathcal{L}_{bce}(S_{pred}, S_{gt}). \tag{20.33}$$

The training of both the motion tracking and segmentation networks constitutes the first iteration. Once fully trained, the segmentation predictions can be reinserted into the motion tracking branch to constrain the shape of the object of interest as the branch is trained to estimate motion for a second iteration. This is achieved by incorporating an additional loss term, which seeks to minimize the difference between the segmentation branch-predicted target segmentations S_{target} and the manually traced source image segmentation S_{source} warped using the estimated displacement fields:

$$\mathcal{L}_{shape} = \frac{1}{\Omega} \sum_{\Omega} \left(S_{target} - \mathcal{T}(S_{source}, \mathbf{U}) \right)^2 . \tag{20.34}$$

This term may also be formulated as a Dice loss similar to L_{seg}, which provides more accurate tracking along the object's border:

$$\mathcal{L}_{shape} = 1 - \frac{2|\mathcal{T}(S_{source}, \mathbf{U}) \cap S_{target}|}{|\mathcal{T}(S_{source}, \mathbf{U})| + |S_{target}|}. \tag{20.35}$$

The shape constraint assists in localizing the object of interest and improves motion estimation along the edges.

To further encourage more realistic motion patterns and discourage jumps and discontinuities in the motion estimations, an additional term is added which minimizes the voxelwise squared norm of the spatial gradients of the displacement field:

$$\mathcal{L}_{smooth} = \frac{1}{\Omega} \sum_{\Omega} \|\nabla \mathbf{U}\|^2. \tag{20.36}$$

Once fully trained, the shape-constrained motion estimates can be used to produce a second set of pseudoground truth segmentations to train the segmentation branch for a second iteration. This process may be repeated until performance ceases to improve.

When tested on an in vivo canine dataset with implanted sonomicrometer crystals to track sparse motion [25], performance improved for both motion tracking and segmentation when the methods were integrated. For motion tracking, the mean square error reduces by approximately 29.6% across the x-, y-, and z-directions, indicating better alignment with densely interpolated sonomicrometer crystals. For segmentation, Dice coefficient scores improved by approximately 2.9%, while Hausdorff distances simultaneously reduced by approximately 5.9% across the endo- and epicardium, indicating better performance in both mask and contour overlap with manually traced segmentations.

While the process performs well as a single pipeline, practical improvements can be achieved by incorporating inter-processing steps between the training of each branch, such as smoothing operations or manual corrections in the segmentation predictions. Furthermore, the motion estimation and segmentation predictions in the first iteration may be substituted with various motion tracking and segmentation algorithms as the user sees fit. In addition to the iterative training framework described above, an alternative effort to integrate motion tracking and segmentation includes a joint feature encoder framework that uses a Siamese style CNN to concurrently train both segmentation and motion tracking [26]. As both tasks are being optimized together, the joint feature encoder allows complementary information in the latent feature space to enhance the performances of both the motion tracker and the segmenter.

20.7 Evaluation

There are a number of ways to approach the evaluation of motion estimation methods. Some aspects of evaluation can be achieved using general properties that are desirable in a solution. For example, cardiac motion trajectories should be cyclic as well as spatially and temporally smooth. Thus, we can evaluate these aspects of our solutions directly using measures of cyclicity and smoothness.

For a more fine-grained analysis, ground truth data that provide an independent and reliable estimate of motion are valuable for the assessment of motion estimation accuracy. Sparse indicators of ground truth can be determined from the correspondence of anatomic landmarks or implanted markers and evaluated based on the degree

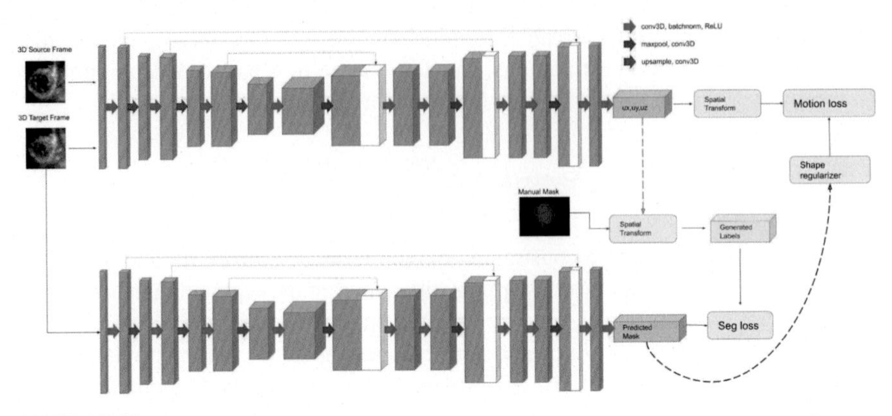

FIGURE 20.11

Deep learning architecture showing iterative integrated motion tracking (top branch) and segmentation (bottom branch) [25].

of agreement with the estimated flow field at those sparse locations. Realistic synthetic data can be used to perform an even more detailed assessment. Synthetic data generated with known dense ground truth motion patterns can be used to assess the error in pixel-by-pixel estimation. Their utility may be limited since they are not real images, but the more accurate synthetic image generators, e.g., ultrasound or MRI [27], provide very realistic images that can be suitable substitutes for the real thing.

20.8 Recapitulation

Motion and deformation analysis continues to be an important problem with significant potential as these techniques translate into practice. The basic elements covered in this chapter, including displacement, deformation, and strain representation/recovery, are fundamental. Moving forward, deep learning methods will continue to proliferate and advance the field in accuracy and reliability. However, models from biomechanics and statistics will likely continue to play an important role.

20.9 Exercises

1. **Optical flow.** In this problem, you will examine a variation on the optical flow formulation presented in Section 20.3.2.
   - **a.** Using calculus of variations, the optical flow constraint e_c, and the smoothness constraint e_s defined in Section 20.3.2, write down the first-order Euler equations that would solve equation (20.15). Furthermore, show that these equations can be manipulated into a coupled pair of elliptic second-order

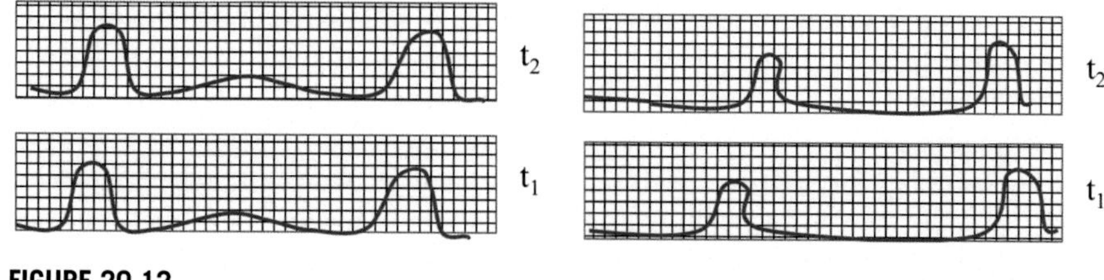

FIGURE 20.12

Shape tracking examples on segmented curves: Left: Shift. Right: Contraction.

partial differential equations that can be solved using iterative methods. The final Euler equations should yield two Laplace equations of the form $\nabla^2 \mathbf{u}$ and $\nabla^2 \mathbf{v}$, where $\nabla^2 = \frac{\partial^2}{\partial x^2} + \frac{\partial^2}{\partial x^2}$.

b. Repeat the first part, only now replace e_s with a second-order regularizer e_{s2}, defined as

$$e_{s2} = \int\int \left(\left(u_{xx} + u_{yy}\right)^2 + \left(v_{xx} + v_{yy}\right)^2 \right) dx dy.$$

c. Sketch the optical flow field of a rigid disk rotating around image point origin (x_0, y_0) with counterclockwise angular velocity $-\omega$. What would you expect the solutions of the Laplace equations to be in the center of the disk if the motion is known at the boundary of the disk? You do not need to solve fully, but use the equations to inform your answer.

2. Shape-based matching of curves. Consider two different sets of curves shown in Fig. 20.12(a and b). Assume that each pair of curves represents the segmented boundaries of a portion of the LV of the heart that were derived from two MRI images at time frames 1 and 2 of an image sequence. Using the grid points to help you estimate the values, do the following:

a. First, compute the curvatures at every third grid sample point along each curve at each time point, and list the values in a table (~20 values should be entered). Next, for every third point at time 1, determine and draw the set of displacements translating them to time 2. Use a search window of five points at time 2 (two on each side of the point being considered) and determine these displacements by imposing equation (20.18) in 1D (only one curvature). Assume the confidence weighting is constant (i.e., $A = 1$).

b. Next, design a confidence term A (could be a function) that weights unique high curvature matches as optimal and low curvature matches as less desirable. Apply this term to the same curvatures computed above on test examples (a) and (b) and draw the new set of displacements.

c. Finally, apply a discrete form of equation (20.20) and show the resulting smoothed displacement vectors. Incorporate your newly designed confidence term as $A(\mathbf{u}^*)$ in this equation. The regularization term $d\mathbf{x}^*$ is now discrete and is applied to adjacent points on the curve.

3. **FEM-based deformation analysis.** Show that differentiation of equation (20.27) yields the analytical form of equation (20.30). Suppose that the confidence matrix $\mathbf{A}$ is the identity matrix. What would the final form be for the smoothed displacement matrix $\mathbf{U}$?
4. **Deep learning for motion analysis.** Given a set of two-frame time sequences of 2D images and corresponding ground truth (obtained from the supplementary materials for this book), design, implement, and test a deep learning approach to motion estimation. Your approach will learn to compute image displacements from pairs of raw intensity data using the log-cosh loss function given in equation (20.31), where the input is a pair of vectors of intensity inputs $\mathbf{X}$ in each image and an output vector of corresponding displacements $\mathbf{Y}$ for each image pair. You may use any combination of fully connected and/or convolutional neural network layers (see Part VIII). Show the architecture you plan to use. Implement the approach using your preferred deep learning framework (e.g., Pytorch, Tensorflow). Augment the training data using random shifts and rotations. Train your model and report the training and validation curves. Test the trained model on the test data and report your results.
5. **Deep learning for integrated motion/segmentation.** Given an input of three-frame time sequences of 64×64 2D images and paired segmentation masks, design a supervised U-Net similar to the one described in Section 20.6.3 to estimate displacement fields between frames 1 and 2, between frames 2 and 3, and between frames 1 and 3. Show the architecture you would use. Describe how you would train and test the network and what data you would need.

References

[1] Spencer A., Continuum Mechanics, Longman, 1980.

[2] De Craene M., Piella G., Camara O., Duchateau N., Silva E., Doltra A., D'hooge J., Brugada J., Sitges M., Frangi A.F., Temporal diffeomorphic free-form deformation: Application to motion and strain estimation from 3D echocardiography, Medical Image Analysis 16 (2) (2012) 427–450.

[3] Horn B., Robot Vision, MIT Press, Cambridge, MA, 1986.

[4] Thirion J.-P., Image matching as a diffusion process: an analogy with Maxwell's demons, Medical Image Analysis 2 (3) (1998) 243–260.

[5] Hermann S., Werner R., High accuracy optical flow for 3D medical image registration using the census cost function, in: Pacific-Rim Symposium on Image and Video Technology (PSIVT 2013), in: LNCS, vol. 8333, 2013, pp. 23–35.

[6] Dosovitskiy A., Fischer P., Ilg E., Häusser P., Hazirbas C., Golkov V., van der Smagt P., Cremers D., Brox T., FlowNet: learning optical flow with convolutional networks, in: Proceedings of the IEEE International Conference on Computer Vision, 2015, pp. 2758–2766.

[7] Shi P., Sinusas A.J., Constable R.T., Ritman E., Duncan J.S., Point-tracked quantitative analysis of left ventricular surface motion from 3-D image sequences, IEEE Transactions on Medical Imaging 19 (1) (2000) 36–50.

[8] Kerwin W.S., Osman N.F., Prince J.L., Image processing and analysis in tagged cardiac MRI, in: Bankman I.N. (Ed.), Handbook of Medical Image Processing and Analysis, Academic Press, 2009, Chap. 26, pp. 435–452.

[9] Parthasarathy V., Prince J.L., Stone M., Murano E.Z., Nessaiver M., Measuring tongue motion from tagged cine-MRI using harmonic phase (HARP) processing, The Journal of the Acoustical Society of America 121 (1) (2007) 491–504.

[10] Young A.A., Kraitchman D.L., Dougherty L., Axel L., Tracking and finite element analysis of stripe deformation in magnetic resonance tagging, IEEE Transactions on Medical Imaging 14 (3) (1995) 413–421.

[11] Sonka M., Hlavac V., Boyle R., Motion analysis, in: Image Processing, Analysis and Machine Vision, Thomson, 2007, pp. 791–798.

[12] Chenouard N., Smal I., de Chaumont F., Maška M., Sbalzarini I.F., Gong Y., Cardinale J., Carthel C., Coraluppi S., Winter M., Cohen A.R., Godinez W.J., Rohr K., Kalaidzidis Y., Liang L., Duncan J., Shen H., Xu Y., Magnusson K.E.G., Jaldén J., Blau H.M., Paul-Gilloteaux P., Roudot P., Kervrann C., Waharte F., Tinevez J.-Y., Shorte S.L., Willemse J., Celler K., van Wezel G.P., Dan H.-W., Tsai Y.-S., de Solórzano C.O., Olivo-Marin J.-C., Meijering E., Objective comparison of particle tracking methods, Nature Methods 11 (3) (2014) 281–290.

[13] Liang L., Shen H., De Camilli P., Duncan J.S., A novel multiple hypothesis based particle tracking method for clathrin mediated endocytosis analysis using fluorescence microscopy, IEEE Transactions on Image Processing 23 (4) (2014) 1844–1857.

[14] Hu Z., Metaxas D., Axel L., In vivo strain and stress estimation of the heart left and right ventricles from MRI images, Medical Image Analysis 7 (2003) 435–444.

[15] Papademetris X., Sinusas A.J., Dione D.P., Constable R.T., Duncan J.S., Estimation of 3-D left ventricular deformation from medical images using biomechanical models, IEEE Transactions on Medical Imaging 21 (7) (2002) 786–800.

[16] DeLorenzo C., Papademetris X., Staib L.H., Vives K.P., Spencer D.D., Duncan J.S., Volumetric intraoperative brain deformation compensation: model development and phantom validation, IEEE Transactions on Medical Imaging 31 (8) (2012) 1607–1619.

[17] Dumpuri P., Thompson R.C., Cao A., Ding S., Garg I., Dawant B.M., Miga M.I., A fast and efficient method to compensate for brain shift for tumor resection therapies measured between pre-operative and post-operative tomograms, IEEE Transactions on Biomedical Engineering 57 (6) (2010) 1285–1296.

[18] Guccione J.M., McCulloch A.D., Finite element modeling of ventricular mechanics, in: Theory of Heart, Springer, 1991, pp. 121–144.

[19] Lu A., Ahn S.S., Ta K., Parajuli N., Stendahl J.C., Liu Z., Boutagy N.E., Jeng G.-S., Staib L.H., O'Donnell M., Sinusas A.J., Duncan J.S., Learning-based regularization for cardiac strain analysis via domain adaptation, IEEE Transactions on Medical Imaging 40 (9) (2021) 2233–2245.

[20] Alessandrini M., De Craene M., Bernard O., Giffard-Roisin S., Allain P., Waechter-Stehle I., Weese J., Saloux E., Delingette H., Sermesant M., D'hooge J., A pipeline for the generation of realistic 3D synthetic echo-cardiographic sequences: methodology and open-access database, IEEE Transactions on Medical Imaging 34 (7) (2015) 1436–1451.

[21] Parajuli N., Compas C.B., Lin B.A., Sampath S., O'Donnell M., Sinusas A.J., Duncan J.S., Sparsity and biomechanics inspired integration of shape and speckle tracking for cardiac deformation analysis, in: Functional Imaging and Modeling of the Heart, in: LNCS, vol. 9126, Springer, 2015, pp. 57–64.

[22] Balakrishnan G., Zhao A., Sabuncu M.R., Dalca A.V., Guttag J., An unsupervised learning model for deformable medical image registration, in: Proceedings of the IEEE Conference on Computer Vision and Pattern Recognition, 2018, pp. 9252–9260.
[23] Ahn S.S., Ta K., Lu A., Stendahl J.C., Sinusas A.J., Duncan J.S., Unsupervised motion tracking of left ventricle in echocardiography, in: Medical Imaging 2020: Ultrasonic Imaging and Tomography, vol. 11319, International Society for Optics and Photonics, 2020.
[24] Qin C., Shi B., Liao R., Mansi T., Rueckert D., Kamen A., Unsupervised deformable registration for multi-modal images via disentangled representations, in: International Conference on Information Processing in Medical Imaging, Springer, 2019, pp. 249–261.
[25] Ta K., Ahn S.S., Stendahl J.C., Sinusas A.J., Duncan J.S., A semi-supervised joint network for simultaneous left ventricular motion tracking and segmentation in 4D echocardiography, in: Martel A.L., et al. (Eds.), Medical Image Computing and Computer Assisted Intervention – MICCAI 2020, Springer International Publishing, Cham, ISBN 978-3-030-59725-2, 2020, pp. 468–477.
[26] Qin C., Bai W., Schlemper J., Petersen S.E., Piechnik S.K., Neubauer S., Rueckert D., Joint learning of motion estimation and segmentation for cardiac MR image sequences, in: Frangi A.F., et al. (Eds.), Medical Image Computing and Computer Assisted Intervention (MICCAI), Springer, Cham, ISBN 978-3-030-00934-2, 2018, pp. 472–480.
[27] Zhou Y., Giffard-Roisin S., De Craene M., Camarasu-Pop S., D'Hooge J., Alessandrini M., Friboulet D., Sermesant M., Bernard O., A framework for the generation of realistic synthetic cardiac ultrasound and magnetic resonance imaging sequences from the same virtual patients, IEEE Transactions on Medical Imaging 37 (3) (2018) 741–754.

CHAPTER

Imaging Genetics

21

Marco Lorenzi[a] and Andre Altmann[b]

[a]*CMIC, University College London, London, United Kingdom*

[b]*Inria Sophia Antipolis, Epione Research Group, UCA, Sophia Antipolis, France*

Learning points

- What is genetic data?
- Genetic data and imaging phenotypes
- Genome-wide association studies
- Multivariate methods in imaging-genetics
- Regularization, stability and validation
- Discovery of genetic markers

21.1 Introduction

Conceptually, the human genome spans about 3 billion characters. The alphabet used to store the genetic information, which determines how the human body is assembled from a single cell and how the entire organism functions, is comprised of only four letters: A, C, G, and T. The physical carrier of this information is *deoxyribonucleic acid* (DNA). Each character of the genetic sequence is a nucleotide in the DNA chain; a nucleotide is composed of a sugar (deoxyribose), a phosphate group, and one of the name-giving nitrogen-containing nucleobases: adenine (A), cytosine (C), guanine (G), and thymine (T). The nucleotides are joined together to form a chain via covalent bonds between the sugar of one nucleotide and the phosphate group of the next, thereby forming the backbone of the DNA molecule. Furthermore, if the sequences of two DNA molecules are compatible according to the base pairing rules (adenine can pair with thymine and cytosine with guanine), then the two separate DNA molecules can bind together via hydrogen bonds between the nucleobases and form the iconic double helix. That is, the two chains have to be complementary: AAAGGTCCAGGA will form a bond with TTTCCAGGTCCT, but not with CCCCCCAAAGGG (see Fig. 21.1).

The genetic information in humans is packaged into separate *chromosomes*: 22 autosomes (numbered simply based on their physical size from 1 through 22) and the sex chromosomes named X and Y. Each chromosome is a long uninterrupted DNA double helix. Commonly, every cell in the body carries exactly two copies of each autosome, where each parent provided one set. The sex chromosomes, typically, come either in XX pairs (female) or in XY pairs (male).

Medical Image Analysis. https://doi.org/10.1016/B978-0-12-813657-7.00034-0

```
AAAGGTCCAGGA    AAAGGTCCAGGA
||||||||||||       ||
TTTCCAGGTCCT    CCCCCCAAAGGG
```

FIGURE 21.1 Base-pairing in two short DNA sequences.

The left pair is complementary and form hydrogen bonds (indicated by |) along the entire string and therefore forms a proper double strand. The pair on the right has many mismatching nucleotides (indicated by a space) and therefore cannot form a double-stranded DNA molecule.

Probably the most renowned part of the genome are the genes. Formally, a gene is a sequence of genetic code that encodes a molecule, typically a protein or *ribonucleic acid* (RNA), that carries out one or more functions in the organism. According to the GENCODE project, the human genome is composed of about 19,901 protein-coding genes and an additional 15,779 non-coding genes; estimates, however, vary based on the applied methodologies [1]. Before a protein can be synthesized from a gene, the genetic code is *transcribed* from the DNA into a working copy, named *messenger RNA* (mRNA), which serves as the template for protein production. The RNA molecule uses ribose instead of deoxyribose as the sugar in the backbone. In eukaryotes, such as the human, many genes contain *exons* and *introns*; introns are removed from the mRNA through a mechanism termed *splicing* and only the sequence in the exons is retained and used to synthesize the protein. A protein is a chain of amino acid residues, and the synthesis or *translation* from the mRNA follows a very precise genetic code: every group of three nucleotides (a *triplet*) encodes one of 20 amino acids or the code to terminate the synthesis. Splicing can lead to different proteins by altering the exon configuration of the mRNA that is produced from the DNA. This phenomenon is known as *alternative splicing* and can occur, e.g., by skipping one or more exons or retaining introns. Alternative splicing is the rule rather than the exception and can occur in a condition- or tissue-specific manner. Thus, one gene can be the source of many different but related proteins.

Although genes are the most prominent part of the genome, they only make up for a minority of genetic sequence in the human genome: only about 1% of the genetic sequence in humans is at some point translated into proteins [2]. The non-coding DNA, i.e., the remaining 99%, which is comprised of introns and non-coding genes, among others, was referred to as "*junk DNA*" in the past. However, non-coding DNA has been found recently to contain important sequences that regulate the expression of genes, i.e., where and in which context a gene is to be transcribed and how much protein should be produced [2].

21.1.1 Heritability

Differences in the appearance of individuals, that is, their *phenotype* (such as hair color and height), are the result of the individual genetic information (*genotype*) and environmental influences. Simply put, people appear different partly because of their difference in the genotype.

A means to quantify the degree by which variation of a trait in a population is due to the genetic variation between subjects, as opposed to environmental influence or chance, is the heritability statistic [3]. Formally, the phenotypic variance σ_P^2 is defined as

$$\sigma_P^2 = \sigma_G^2 + \sigma_E^2 + 2\Sigma(G, E),$$

where σ_G^2 and σ_E^2 are the genetic and environmental variance, respectively, and $\Sigma(G, E)$ is the covariance between genetics and environment (which can be set to 0 in controlled experiments). Thus, broad sense heritability (H^2) is simply defined as

$$H^2 = \frac{\sigma_G^2}{\sigma_P^2}.$$

The most prominent method to estimate heritability is through twin studies that compare the phenotype similarity in monozygotic (MZ; i.e., identical or maternal) twins to the phenotype similarity in dizygotic (DZ; i.e., non-identical or fraternal) twins. MZ twins share nearly 100% of the genetic code, while DZ twins, like non-twin siblings, share on average 50% of the genetic code. It is in general assumed that MZ and DZ twins share the same environment. Thus, one can compute H^2 using the Falconer formula [3]:

$$H^2 = 2(r_{\mathrm{MZ}} - r_{\mathrm{DZ}}),$$

where r_{MZ} and r_{DZ} are the Pearson correlation coefficients of the trait between pairs of MZ twins and pairs of same-sex DZ twins, respectively. There are further methods to obtain heritability estimates outside twin studies, e.g., comparing trait similarities in close relatives including parents and non-twin siblings. Other methods are based on linear mixed models and can estimate heritability on more complex family trees (pedigrees). Furthermore, recent developments in statistical genetics enable us to estimate heritability from large datasets of unrelated subjects [4,5].

21.1.2 Genetic variation

There are many changes that can shape an individual genome. The most drastic ones are gains or losses of entire chromosomes as in trisomy 21 (*Down syndrome*), in which three copies of chromosome 21 are present in cells, or monosomy X (*Turner syndrome*), where only a single X chromosome is present in cells. In some cases parts of chromosomes are deleted, duplicated, inverted, or moved to other chromosomes; these changes are referred to as either *copy number variations* or *structural variations*. For instance, a well-characterized deletion of a small part of chromosome 22 (22q11.2 microdeletion) typically results in the loss of 30–40 genes in that region and causes *DiGeorge syndrome* and a 20–30-fold increased risk to develop schizophrenia [6]. Other variations are on a much smaller scale and concern either the gain or loss of a few nucleotides (*insertion* or *deletion*, respectively), or simply the identity of the nucleotides at a given position in the reference genome, e.g., `A` is replaced by G. Depending on their location in the genome, these single nucleotide exchanges can have a serious effect on the organism. For instance, a nucleotide exchange or a deletion in the non-coding DNA is likely to be of little consequence to the organism, but replacement of a nucleotide in a gene's exon, which also leads to a different triplet in the mRNA and thus potentially to an amino acid replacement in the final protein, may have severe consequences. While some amino acid replacements can be tolerated by the organism, e.g., if their biochemical properties are similar, and are considered benign, others have a deleterious effect and alter the protein function. This is the case for instance with mutations in the *APP* gene causing the familial variant of Alzheimer disease (AD) (a map of different deleterious and benign mutations in *APP* can be found online: https://www.alzforum.org/mutations/app). Luckily, exchanges with such drastic consequences are not shared by many people, i.e., their frequency in the population is low. Conversely, many of those nucleotide exchanges are part of the natural variation in our genome and are shared by many individuals, i.e., such variants show a high frequency (1% or more) in the population and are referred to as *single nucleotide polymorphisms* (SNPs). Interestingly, in most of the cases of such nucleotide exchanges, the reference nucleotide (or *major allele*), which occurs more than 50% in the population, is exchanged for only one of the other three nucleotides (referred to as *minor allele*), that is, for example, we only see a replacement of A by G at a given position in the genome, but never a replacement of A by C or A by T at the same position (though, occasional exceptions do exist). Often the major allele is simply denoted by a capital "A," while the minor allele is denoted by a lowercase "a." Discovered SNPs have been cataloged and can be referred to either by their position in the reference genome and nucleotide substitution (e.g., `19:44908684:T-C`) or by an identifier consisting of an integer number and the prefix `rs` (e.g., `rs429358`). Of note, the *minor allele frequency* (MAF) of some SNPs is close to 50%, and therefore the identity of the minor and major allele in these SNPs may change between datasets due to sampling.

In this chapter we will focus on the analysis of SNP data. The fact that SNPs are shared by many individuals renders them easy and cheap to measure: microarray technology allows the measurement of 100,000s of SNPs in one experiment with more recent iteration of the technology reaching up to 2 million SNPs in one experiment, thus producing (after processing the data) per subject an array of 0, 1, 2 entries quantifying how many copies of the minor allele are present in the individual, where 0, 1, and 2 simply reflect the count of the minor (the non-reference) allele (Fig. 21.2).

The availability of a method to cheaply measure 100,000s of genetic positions in the human genome enabled a type of analysis known as *genome-wide association study* (GWAS). Simply put, this framework univariately tests for every measured SNP whether the frequency of its non-reference allele in cases is higher (or lower) than the frequency in controls, thus indicating SNPs that increase (or decrease) the disease risk. With little change in statistical methodology, GWAS also allows to test the genetic influence of SNPs on quantitative measures, referred to as *traits*, such as height. GWAS conducted over the recent past uncovered many associations (https://www.ebi.ac.uk/gwas/) often identifying single SNPs that affect disease risk.

	SNP$_1$	SNP$_2$	SNP$_3$	$\cdots$	SNP$_{l-2}$	SNP$_{l-1}$	SNP$_l$
1	1	0	0	$\cdots$	2	1	0
2	0	0	1	$\cdots$	0	1	0
3	0	0	0	$\cdots$	1	0	0
$\vdots$	$\vdots$	$\vdots$	$\vdots$	$\ddots$	$\vdots$	$\vdots$	$\vdots$
$n-2$	1	0	1	$\cdots$	1	0	0
$n-1$	1	0	1	$\cdots$	0	2	1
n	2	0	1	$\cdots$	0	1	0

FIGURE 21.2 SNP data matrix.

Example of a SNP data matrix for n subjects organized in rows and l SNPs organized in columns. Each cell $c_{i,j}$ contains the count of the number of minor alleles in subject i for SNP j.

Especially in the realm of brain disorders the question is often how these disease-associated variants affect structure and function of the brain. A typical *imaging-genetics* approach is to collect people with the disease-associated variant (*carriers*) and people without that variant (*non-carriers*) and compare their brain imaging data using standard tools. For instance, the APOE-ϵ4 variant increases the risk to develop AD; thus, one could assess the effect of APOE-ϵ4 on gray matter by conducting a voxel-based morphometry analysis and treating carriers as cases and non-carriers as controls. Similarly, carriers of the 22q11.2 microdeletion can be compared to people without this genetic change.

The previous approach allowed to investigate effects of a single genetic variant on the brain. However, the limitation is that such suitable *candidate variants* must have been previously identified through other genetic (non-imaging) analyses. In brain disorders there is the strong assumption that changes in brain structure and function underly the observed symptoms and altered cognitive performance. Thus, phenotypes derived from brain imaging data can act as *endophenotypes* in genetic studies of brain disorders, because they are closer to the actual biological substrate. For example, in such studies a diagnosis of AD would be the phenotype and hippocampal atrophy or cortical burden of the amyloid protein would be considered the endophenotype. The underlying assumption is that the effect sizes of genetic variations on endophenotypes are greater than those on the disease risk or phenotype itself [7]. Therefore, one expects to increase the chance to identify genetic variants that would go unnoticed when studying just the case control status alone. If there is just one or a few of such endophenotypes of interest, then one can follow the standard pipeline for GWAS for quantitative traits (see Section 21.2) with the imaging-derived phenotypes. For instance, one could analyze the genetic contribution to subcortical volumes [8]. After exploring imaging-genetics approaches based on the standard GWAS framework, we will introduce machine learning-based methods to link genetic variation to variation in brain imaging phenotypes at high resolution (see Section 21.3).

21.2 Genome-wide association studies

The field of genetics is rapidly expanding and recent advances in the technology enable us now to obtain the entire sequence of a human at a reasonable price (i.e., less than $1000). In this chapter, however, we will focus on the type of genetic information that is utilized in GWAS and still constitutes the main genetic information used in imaging-genetics.

21.2.1 Genotyping using microarray technology

Microarray technology is used to assess the identity of a genetic variant at a given position in the genome. The technology is based on the property that a single DNA strand will bind (*hybridize*) to its complementary DNA strand and form a double-stranded DNA molecule. The general principle of such SNP arrays is that they carry two copies of short DNA strands (e.g., 121 letters) to measure the identity of a SNP. Both such *DNA probes* contain the complementary sequence of the reference genome before and after the targeted SNP (e.g., 60 letters in each direction); one probe contains the complement of the reference allele, while the other probe contains the complement of the alternative allele. The binding affinity of a subject DNA sample will be higher to the probe containing the correct complementary SNP. Importantly, these DNA probes are immobilized on the array and their positions (e.g., x- and y-coordinates) are known. For this approach to work, the test subject DNA is fragmented (i.e., chopped into small pieces) and a fluorescent dye is attached to each fragment. The amount of binding strength of a subject genetic sample to both probes is then quantified through fluorescence microscopy. After analyzing the raw imaging data in a number of subjects, one can assess whether a subject is *homozygous* for the reference allele (*AA*, i.e., both versions of the SNP show the reference allele), *heterozygous* (*Aa*, i.e., one copy is the reference and the other is the alternative allele), or *homozygous* for the alternative allele (*aa*), which are encoded for computations as 0, 1, or 2, respectively, indicating the number of non-reference allele counts.

21.2.2 Processing SNP data

The first step in the process is to *call* the SNPs from the fluorescence signal. In some subjects the signal may not be sufficiently strong to make a call, resulting in a missing entry for that SNP in these subjects. The end product of the calling step is essentially a data matrix of dimension $n \times l$, where n represents the number of subjects and l the number of measured SNPs (Fig. 21.2). There are various popular data formats for efficiently storing SNP information along with programs to manipulate the data and conduct association studies, e.g., PLINK is a widely used tool [9]. There are a number of quality control (QC) steps being carried out prior to the actual statistical analysis:

1. Removal of SNPs that show a poor quality across the n subjects, i.e., SNPs that have more than 10% missing entries are typically removed.

2. Removal of subjects with too many missing calls, i.e., subjects missing more than 10% of SNP calls are removed. Many missing SNP calls are an indicator for poor sample quality.
3. Removal of subjects in whom the reported sex and the sex inferred from SNP data are mismatching.
4. Removal of SNPs with a low frequency in the study sample. Depending on the size of the study, SNPs with frequencies lower than 5% (or 1% in large samples) are removed before analysis.
5. Removal of SNPs deviating from the Hardy–Weinberg equilibrium (HWE) in control subjects. Here the allele counts for *AA, Aa*, and *aa* are tested for consistency with population genetics theory. Violation of the HWE among healthy controls may indicate technical problems with measuring the SNP.

These are the main steps in SNP data quality assessment. In practice, there exist further QC steps involving more advanced statistics capturing participants' relatedness, genetic ancestry, and population structure [10], but they are outside of the scope of this chapter.

21.2.3 Imputation

Imputation is the process where missing (or unobserved) data are inferred from available data using statistical methods. When working with SNP data there are generally two types of missing data:

1. Missing SNP calls for a few subjects with low signal intensity preventing a high-quality SNP call, resulting in missing entries within a column of the data matrix.
2. SNPs that were not measured due to lack of probes on the microarray platform, resulting in unobserved (missing) columns.

This latter type of missing data is a major challenge for large collaborative GWAS involving dozens of teams around the world. Teams generate the SNP data using different genotyping microarray platforms (often referred to as *SNP chips*) and chip versions, resulting in different SNPs being queried by different researchers. Moreover, often a team may have generated their data with different chip versions based on the availability at the time of data generation. Thus, simply taking the intersection of genotyped SNPs across all used platforms in a collaborative study often leaves too few SNPs for a meaningful analysis. One way to circumvent this problem is to impute the calls for SNPs that were not measured on these chips. This feat is possible because humans are a relatively young species, resulting in a great deal of correlation between genetic variants that are in close proximity on the genome; this is referred to as *linkage disequilibrium* (LD). Intuitively, this has the same effect as spatial correlation between voxels in imaging data. Thus, knowing a few SNPs and having access to the LD information from a large database, missing SNPs in the vicinity of genotyped SNPs can be accurately imputed. The imputation results in posterior probabilities for each of the three possible genotype calls: *homozygous reference* (*AA*), *heterozygous* (*Aa*), and *homozygous alternative* (*aa*). See the review by Marchini and Howie for additional details [11]. In addition to the posterior probability, a quality score for the imputed SNPs across all subjects in the dataset is provided by most available tools. The imputation result can be converted into standard 0, 1, 2 genotype calls based on a hard cutoff for the genotype posterior probability. For instance, if the cutoff was set to 0.9 and none of the posterior probabilities for *AA*, *Aa*, or *aa* reached 0.9, then the imputed SNP in that subject would be set to missing. Alternatively, the imputed genotypes can be converted into a *dosage*, i.e., sum of the minor allele counts (0, 1, 2) multiplied by their corresponding posterior probabilities (P_0, P_1, P_2), resulting in genotype values in $[0, 2]$:

$$\text{dosage} = 0 \times P_0 + 1 \times P_1 + 2 \times P_2.$$

The advantage of the latter approach is that no missing entries are produced within a column. Post-imputation QC typically involves:

1. Removal of SNPs with a low imputation quality, e.g., a quality score of <0.9, using recent tools.
2. Removal of SNPs with high missingness rate in case of hard-called SNPs.

Overall the imputation quality depends on the used reference set, where larger databases typically lead to better imputation results. A widely used reference database is provided by the Haplotype Reference Consortium and is comprised of data on 32,488 subjects [12]. In fact, this reference dataset requires a lot of storage and it was more economical to be made available through cloud-based free-for-all imputation servers, e.g., the Michigan [13] (http://imputationserver.sph.umich.edu/) and the Sanger (https://imputation.sanger.ac.uk/) imputation server. A very positive side effect of this development is the increased reproducibility of research and that users are less likely to accidentally misuse the software.

21.2.4 Univariate analysis

Once the SNP data have been cleaned, imputed, and cleaned again, one can proceed to the statistical analysis. The predominant approach to GWAS is a mass univariate testing strategy in which every SNP is tested individually for its association with the phenotype, regardless of the identity of the remaining SNPs. The initial method was to compare allele counts between cases and controls using a one-degree of freedom χ^2 test on a 2×2 contingency table of allele counts (Table 21.1).

Of note, each subject contributes two alleles (one inherited from each parent), thus the overall sample size for this test is $2n$. However, the χ^2 test approach does not allow for adjusting the association test for relevant covariates such as age, sex, or genetic ancestry. Thus, the common method of choice is logistic and linear regression for case control and quantitative studies, respectively. Here the output is modeled as Y, x is the minor allele count per subject (SNP), $C_1, \ldots, C_k$ are the covariates, and $\varepsilon \sim N(0, 1)$ is the residual error:

$$Y = a_0 + b_0 \times x + b_1 C_1 + \ldots + b_k C_k + \varepsilon. \tag{21.1}$$

Table 21.1 Allele counts for the $\epsilon4$ allele of the *APOE* gene (*a*) and the reference allele (*A*) in an Alzheimer disease study compared to cognitively normal controls. There is a significant increase in APOE-$\epsilon4$ in participants with Alzheimer disease: $P = 6.03 \times 10^{-22}$ using a χ^2 test with one degree of freedom.

	A	*a*
Alzheimer disease	358	256
Control	471	89

Table 21.2 Different encodings of the SNP variable in GWAS to realize different genetic models.

	AA	*Aa*	*aa*
Additive	0	1	2
Dominant	0	1	1
Recessive	0	0	1
Over-/underdominance	0	1	0

For every SNP, such a linear model is estimated using all n subjects, and the intercept (a_0) and the coefficients for the variables ($b_0, \ldots, b_k$) are estimated. Naturally, in genetic studies the main interest is the statistical significance of b_0, i.e., the influence of the SNP on the phenotype. There are different assumptions on how a SNP can act biologically: the *additive* model assumes that each copy of the minor allele has an independent effect; the *dominant* model assumes that the presence of one allele has the same effect as both alleles; and the *recessive* model assumes that there is only an effect on the phenotype when both copies exhibit the minor allele. To realize these main models in the linear framework, the encoding of the SNP is adjusted as shown in Table 21.2. There are also cases where the effect is only present in heterozygous subjects (*Aa*). Depending on the effect direction this is termed *overdominance* or *underdominance*. In brain science, examples of overdominant effects include the effect of a variant in the nicotine receptor gene *CHRNA4* on cognitive control [14] and the enhancing effect of a variant in the longevity gene *KLOTHO* on cognition [15].

In addition to the classic linear models, there are more advanced methods that increase the computational speed on large datasets with many subjects and increase statistical power [16]. Genome-wide results are typically presented in the form of a *Manhattan plot*, where the $-\log_{10}(P)$ value of b_0 for a SNP is given on the y-axis and the SNPs' locations in the genome are arranged on the x-axis. Strong associations between a lead SNP as well as its correlated neighboring SNPs with a phenotype lead to peaks in the plot, mimicking the name-giving skyline of Manhattan. The mass univariate strategy of testing genetic variants leads to a massive multiple testing problem.

21.2.5 Multiple testing correction

In statistical hypothesis testing, so-called *test statistics* are computed from the given observations, e.g., in the two-sample testing procedure, the t-score is the test statistic when conducting a t-test to compare the means of a measurement of two groups. In this setting, we assume that the test statistic is distributed under the *null hypothesis* (H_0). In case of two-sample testing, the test statistic is the distribution of the difference between the means, when there is actually no difference between two groups. Thus, combining the observed test statistic value and the null distribution, we can compute the likelihood of achieving the observed test statistic (or better) when there is no difference between two groups, i.e., when H_0 is correct. This probability is the definition of the P-value. A widely accepted threshold (α-level) for calling a result *statistically significant* is $P = 0.05$. That is, there is only a 5% chance that a test statistic could be produced from random data where there is no value as or more extreme than the observed difference.

In situations where more than a single hypothesis is tested, one has to reconsider the 5% cutoff. The problem is related to the situation of casting a die. The chance to get a 6 is $1/6 \approx 0.167$. However, when the die is cast 20 times, the chance of getting a single 6 in any of the rolls is $1.0 - (5/6)^{20} \approx 0.974$. So it is almost certain to hit at least one 6 in 20 rolls. The same holds true for a significant finding in statistical testing: if enough hypotheses are tested, then the likelihood increases that at least one of them will get a P-value of 0.05 or lower, even if the data are randomly generated. Therefore, the need to adjust the P-value for multiple testing arises. There are various methods available, but the most widely used ones are the Bonferroni method [17] for controlling the familywise error rate (FWER) and the Benjamini–Hochberg method to control for the false discovery rate (FDR) [18].

The Bonferroni method aims to control the FWER, i.e., the probability to incorrectly reject H_0 (calling something significant when it is not) once across all m alternative hypothesis tests ($H_1, \ldots, H_m$) at a predefined α-level. This is simply achieved by testing each individual alternative hypothesis at α/m. That is, we divide the P-value threshold by the number of tests, or alternatively, we multiply each raw P-value with m to get a multiple testing-adjusted P-value. This method assumes that all the tested hypotheses are statistically independent.

The Benjamini–Hochberg method aims to control the FDR, i.e., the fraction of all hypothesis tests that were called "significant" but where there was no true difference (and H_0 was correct). This is a less stringent correction than the FWER correction, where the α-level is adjusted such that the occurrence of a single false discovery is minimized. Practically, the P-values of the m tests are ordered from smallest to largest: $P_{(1)} < \ldots < P_{(m)}$. Next, the largest k such that $P_{(k)} < \frac{k}{m}\alpha$ is identified and all hypotheses corresponding to the P-values $P_{(1)}, \ldots, P_{(k)}$ are accepted. The Benjamini–Hochberg method and its extensions are also valid when hypotheses are dependent on (or correlated to) each other. FDR correction is therefore often preferred when it is known that hypotheses are correlated, e.g., when testing voxels in the brain.

The widely used approach of correcting P-values using the Bonferroni method for all l tests (SNPs) is overly conservative in GWAS: SNPs are (spatially) correlated – this fact is exploited by the genotype imputation approach described above – reducing the number of effective independent tests below the number of tested SNPs. It has been estimated that in GWAS with common SNPs in the European population the effective number of independent tests, while accounting for the correlation between genetic variants (i.e., the LD structure), is about 1 million [19]. Thus, the Bonferroni-adjusted P-value threshold for genome-wide significance was set to $P = 5 \times 10^{-8}$. However, with recent advances leading to the inclusion of rare SNPs and efforts to include non-European subjects, this cutoff value may need to be revised.

SNPs passing the genome-wide significance threshold are typically followed up with additional research in order to reveal the underlying biology; this often involves costly wet-lab experiments. In order to limit the risk of false positive associations, GWAS typically employ a two-step process where parts of the data are used as the *discovery* dataset (often referred to as stage `I`) and a *validation* dataset (stage `II`), where SNPs that passed the genome-wide threshold in stage `I` will be tested again at a more lenient threshold. This approach resembles the train–test split applied in statistical learning (Section 21.3).

21.2.6 Adjustments for quantitative traits

Initially GWAS were only conducted for dichotomous traits, but minor changes in modeling facilitate the analysis of quantitative traits as well. However, additional precautions have to be taken when working with quantitative traits due to the sensitivity and model assumptions of the underlying linear models:

1. Phenotypic outliers can easily produce a false positive association and outlier removal techniques such as windosorizing may be applied.
2. Phenotypes that are not approximately Gaussian-distributed violate model assumptions, thus the raw phenotype may have to be transformed prior to analysis.

Furthermore, many datasets with brain imaging data are comprised of healthy subjects as well as diseased subjects. Thus, one recurring question is how to best handle the resulting bias and whether it is appropriate to adjust the model for disease status [20].

21.2.7 Statistical power

Before actually collecting data for a GWAS, the question is often whether the sample size will be sufficiently large to discover variants given the many tests (∼1 million). Like in many other disciplines this answer is addressed via a statistical power analysis. Statistical power quantifies the likelihood of rejecting the null hypothesis (H_0) when the alternative hypothesis (H_1) is correct (i.e., $1.0 -$ `type II` error probability). In the context of classification, statistical power would be equivalent to the classifier

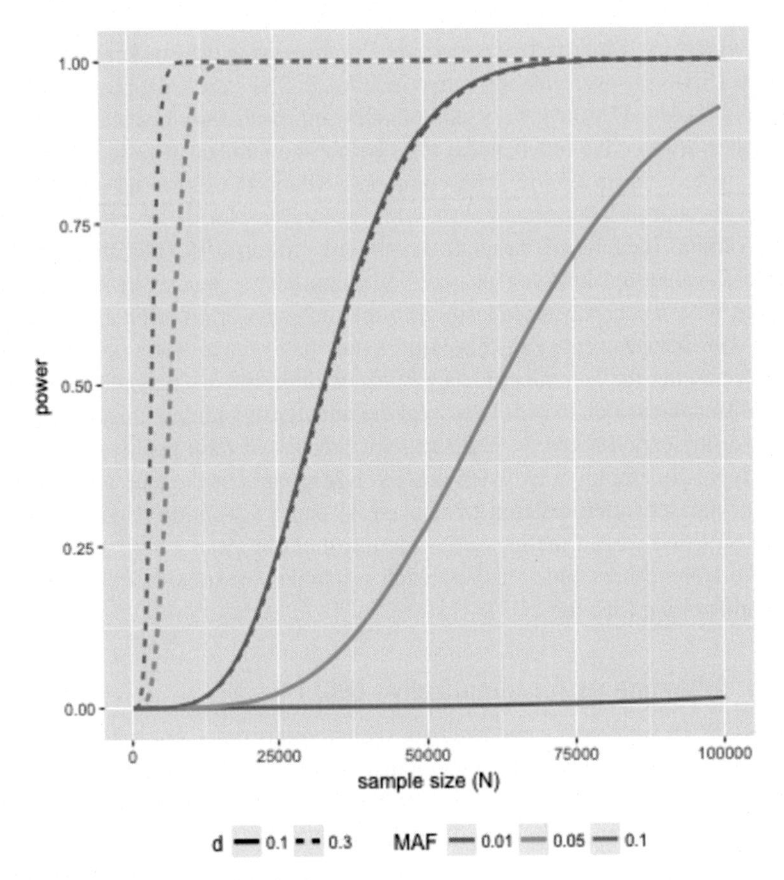

FIGURE 21.3 Power analysis in GWAS.

Power curves generated using a t-test with unequal sample sizes. The significance threshold was set to $\alpha = 5 \times 10^{-8}$, minor allele frequency (MAF) was varied from 0.01 to 0.1 (different colors), effect size (Cohen d) was varied from 0.1 to 0.3 (different line types), and the sample size (n) was varied from 200 to 100, 000 subjects (horizontal axis). Alternatively, for quantitative traits the power can be computed based on a test for the significance of Pearson r, where in practice the correlation r is a function of the MAF and Cohen d [21].

true positive rate for the ideal effect to be identified in the study. In GWAS the power depends on the SNP MAF, its effect size, and the size of the study sample.

As evident from Fig. 21.3, larger samples in GWAS lead to an increased power to detect true associations at genome-wide significance ($\alpha = 5 \times 10^{-8}$) given that frequency and effect size of the variant remain unchanged. This fact led to an ever-increasing size of genetic studies over the last decade, with recent GWAS including more than 1 million subjects [22]. Further, power can be increased by studying a phenotype that is more directly influenced by the genetic variant, leading to the rationale of *endophenotypes*. For instance, a SNP with MAF of 0.1 (blue line) may have an effect of $d = 0.1$ (solid) on disease status achieving only a power of 0.25 with 25,000 subjects; the same SNP shows an effect of $d = 0.3$ (dashed line) on, say, hippocampal volume, thus increasing the power to 0.9 with only 5,700 subjects.

21.2.8 GWAS and imaging phenotypes

The concept of endophenotypes is clearly appealing from the power perspective. However, when there is more than one phenotype to study, the statistical power will be reduced because of the need to adjust for the number of (independent) tested phenotypes in addition to the number of genetic tests (1 million). For instance, in the case of 10 brain imaging biomarkers this would amount to 10 million tests ($= 10 \times 1$ million) leading to a P-value cutoff of 5×10^{-9}. Applying this adjusted α-level to the above power example in Fig. 21.3 ($d = 0.3$, $\text{MAF} = 0.1$, $n = 5700$) would reduce the power to 0.83, which is still acceptable. However, in brain imaging it is common to study thousands of derived phenotypes, reducing the power even further (e.g., to 0.58 when corrected for 1 million SNPs and 1000 imaging phenotypes). Thus, due to the increased multiple testing burden, the classic univariate GWAS approach works with large datasets and a few hundred imaging-derived phenotypes, an approach followed by the ENIGMA consortium [8]. However, methodological advances are spurring ever finer-grained structural [23] and functional [24] brain parcellations, thereby increasing the number of available imaging phenotypes. Moreover, depending on the cohort there are many different imaging modalities that can be studied from gray matter volume, cortical thickness, or measures based on diffusion-weighted imaging, thus easily amounting to a few thousand phenotypes with a rather coarse granularity [25]. In some cases imaging phenotypes are combined to generate novel biomarkers; for example, Scelsi et al. combined hippocampal volume and cortical amyloid burden into a novel multimodal imaging phenotype [26]. However, even despite finer-grained parcellations, it is conceivable that genetic effects are not limited to boundaries of regions of interest, e.g., only a fraction of a candidate region may be affected by a given SNP, thereby weakening the association when the phenotype is defined over the entire region of interest. Taking the resolution spectrum to the extreme end leads to *voxelwise genome-wide association studies* (vGWAS). Though, the power analysis for this approach is not very encouraging: modern MRI whole brain acquisition techniques generate voxels of less than 1 mm^3, which together with an approximate gray matter volume of 0.8 dm^3 results in approximately 1 million relevant voxels to be tested in vGWAS. Fortunately, due to spatial correlation these tests will not constitute independent tests, but the estimate will still be around at least 100,000 independent tests, pushing the threshold for genome-wide and voxelwise significance to at least 5×10^{-13}. In their first vGWAS in AD, Stein et al. analyzed data of 740 subjects and developed a novel method based on FDR to address the multiple comparisons problem [27]. While this first analysis was still limited in statistical power, the availability of large databases with paired imaging and genetics data such as the UK BioBank (http://www.ukbiobank.ac.uk/) or the collaborative efforts such as the ENIGMA consortium (http://enigma.ini.usc.edu/) [28] will ultimately lead to a statistical viability of vGWAS. Last but not least, the brute-force concept of vGWAS induces a massive computational burden: studies with a few hundred phenotypes can easily be parallelized on conventional hardware; running whole genome studies on millions of voxels requires advanced computational methods such as fast vGWAS [29].

GWAS-based approaches to analyze brain imaging phenotypes are by their nature mass-univariate on the genetic part; therefore, they neither explore potential interactions between SNPs (referred to as *epistasis*) nor make explicit use of the spatial correlation within the imaging data. Multivariate approaches, which we will introduce in the next section, are a promising way to make efficient use of the available data.

21.3 Multivariate approaches to imaging genetics

In recent years many domains have seen an increased use of multivariate approaches including neuroscience [30] and GWAS [31]. Also recent methodological advances in the imaging-genetics domain rely on multivariate approaches to capture meaningful genotype–phenotype interactions [32,33]. The appeal of these methods lies in their ability to identify complex relationships between the genome and the brain by simultaneously modeling the joint effect of genetic variants on brain features. The promising potential of multivariate imaging-genetics approaches is to explicitly highlight the underlying biology of macroscopic processes, such as brain atrophy, by identifying sets of genetic variants that are jointly associated with the phenotype.

21.3.1 A sketch on classical multivariate approaches

Typical multivariate approaches for the joint analysis of brain imaging phenotype and large arrays of genetic variants are based on the identification of latent modes of maximal association between imaging and genetics features. The underlying principle of these procedures consists in looking for pairs of feature combinations – one "combination" or mode for each of the two distinct data types – that have maximal association. For example, while *partial least squares* aims at maximizing the covariance between these combinations (or projections on the modes' directions), *canonical correlation analysis* (CCA) maximizes their statistical correlation, and *reduced-rank regression* (RRR) minimizes the error in predicting a target data type from the optimal feature combination of a source one.

In what follows, genetic and imaging data for a given individual k are encoded by arrays $\mathbf{x}_k$ and $\mathbf{y}_k$, respectively, with dimensions $dim(\mathbf{x}_k) = d_g$ and $dim(\mathbf{y}_k) = d_i$. An imaging-genetics data matrix for n individuals is therefore represented by the pair of centered matrices $\mathbf{X}$ and $\mathbf{Y}$, with $dim(\mathbf{X}) = n \times d_g$ and $dim(\mathbf{Y}) = n \times d_i$.

The basic principle of classical multivariate analysis techniques relies on the identification of *linear transformations* of $\mathbf{X}$ and $\mathbf{Y}$ into a lower-dimensional subspace where the projected data exhibit the desired statistical properties of similarity.

21.3.1.1 Canonical correlation analysis

In CCA, this problem is formulated by looking for linear transformations parameterized by the vectors $\mathbf{w}_x$ and $\mathbf{w}_y$ such that $\mathbf{X}\mathbf{w}_x$ and $\mathbf{Y}\mathbf{w}_y$ are maximally correlated. In mathematical terms,

$$\mathbf{w}_x, \mathbf{w}_y = \arg\max_{\mathbf{w}_x, \mathbf{w}_y} \rho(\mathbf{X}\mathbf{w}_x, \mathbf{Y}\mathbf{w}_y), \tag{21.2}$$

where $\rho(\mathbf{a}, \mathbf{b}) = \mathbf{a}^T\mathbf{b}/(\sqrt{\mathbf{a}^T\mathbf{a}}\sqrt{\mathbf{b}^T\mathbf{b}})$. The definition of the correlation as expressed in equation (21.2) is

$$\rho(\mathbf{X}\mathbf{w}_x, \mathbf{Y}\mathbf{w}_y) = \frac{\mathbf{w}_x^T \mathbf{S}_{\mathbf{XY}} \mathbf{w}_y}{\sqrt{\mathbf{w}_x^T \mathbf{S}_{\mathbf{XX}} \mathbf{w}_x}\sqrt{\mathbf{w}_y^T \mathbf{S}_{\mathbf{YY}} \mathbf{w}_y}}, \tag{21.3}$$

where $\mathbf{S}_{\mathbf{XY}}$ is the cross-covariance matrix between $\mathbf{X}$ and $\mathbf{Y}$, while $\mathbf{S}_{\mathbf{XX}}$ and $\mathbf{S}_{\mathbf{YY}}$ are the sample covariances of $\mathbf{X}$ and $\mathbf{Y}$, respectively.

Since the optimization of CCA (21.2) is independent of the norm of the projection vectors, without loss of generality we can restrict the problem by introducing the constraints $\|\mathbf{X}\mathbf{w}_x\|^2 = 1$ and $\|\mathbf{Y}\mathbf{w}_y\|^2 = 1$. This implies that we look for projections into a low-dimensional subspace of unitary variance. The maximization of (21.2) can be tackled through the optimization of the Lagrangian

$$\mathcal{L}(\mathbf{w}_x, \mathbf{w}_y, \lambda_x, \lambda_y) = \mathbf{w}_x^T \mathbf{S}_{\mathbf{XY}} \mathbf{w}_x - \lambda_x(\mathbf{w}_x^T \mathbf{S}_{\mathbf{XX}} \mathbf{w}_x - 1) - \lambda_y(\mathbf{w}_y^T \mathbf{S}_{\mathbf{YY}} \mathbf{w}_y - 1), \tag{21.4}$$

and it can be easily shown that the CCA solution can be obtained through the following generalized eigenvalue problem [34]:

$$\begin{bmatrix} \mathbf{0} & \mathbf{S}_{\mathbf{XY}} \\ \mathbf{S}_{\mathbf{YX}} & \mathbf{0} \end{bmatrix} \begin{bmatrix} \mathbf{w}_x \\ \mathbf{w}_y \end{bmatrix} = \lambda \begin{bmatrix} \mathbf{S}_{\mathbf{XX}} & \mathbf{0} \\ \mathbf{0} & \mathbf{S}_{\mathbf{YY}} \end{bmatrix} \begin{bmatrix} \mathbf{w}_x \\ \mathbf{w}_y \end{bmatrix}, \tag{21.5}$$

or equivalently,

$$\begin{bmatrix} \mathbf{0} & \mathbf{S}_{\mathbf{XX}}^{-1/2} \mathbf{S}_{\mathbf{XY}} \mathbf{S}_{\mathbf{YY}}^{-1/2} \\ \mathbf{S}_{\mathbf{YY}}^{-1/2} \mathbf{S}_{\mathbf{YX}} \mathbf{S}_{\mathbf{XX}}^{-1/2} & \mathbf{0} \end{bmatrix} \begin{bmatrix} \mathbf{v}_x \\ \mathbf{v}_y \end{bmatrix} = \lambda \begin{bmatrix} \mathbf{v}_x \\ \mathbf{v}_y \end{bmatrix}, \tag{21.6}$$

where $\mathbf{v}_x = \mathbf{S}_{\mathbf{XX}}^{1/2T} \mathbf{w}_x$ and $\mathbf{v}_y = \mathbf{S}_{\mathbf{YY}}^{1/2T} \mathbf{w}_y$. This last formula shows that the CCA functional can be rewritten as

$$\rho(\mathbf{X}\mathbf{v}_x, \mathbf{Y}\mathbf{v}_y) = \frac{\mathbf{v}_x^T \mathbf{S}_{\mathbf{XX}}^{-1/2} \mathbf{S}_{\mathbf{XY}} \mathbf{S}_{\mathbf{YY}}^{-1/2} \mathbf{v}_y}{\sqrt{\mathbf{v}_x^T \mathbf{v}_x}\sqrt{\mathbf{v}_y^T \mathbf{v}_y}}. \tag{21.7}$$

Through standard algebraic derivations formula (21.7) provides an alternative formulation for the projection vectors. Introducing the vector $\mathbf{u}^T = \mathbf{v}_x^T \mathbf{S}_{\mathbf{XX}}^{-1/2} \mathbf{S}_{\mathbf{XY}} \mathbf{S}_{\mathbf{YY}}^{-1/2}$, we observe that the product $\mathbf{u}^T\mathbf{v}_y$ at the numerator of formula (21.7) is bounded by $\sqrt{(\mathbf{u}^T\mathbf{u})(\mathbf{v}_y^T\mathbf{v}_y)}$, with the equality holding if and only if $\mathbf{v}_y$ is parallel to $\mathbf{u}$ (Cauchy–Schwarz inequality). In this case, formula (21.7) can be rewritten as the new functional

$$\rho'(\mathbf{X}\mathbf{v}_x, \mathbf{X}\mathbf{v}_x) = \frac{\mathbf{u}^T\mathbf{u}}{\sqrt{\mathbf{v}_x^T\mathbf{v}_x}} = \frac{\mathbf{v}_x^T\mathbf{S}_{\mathbf{XX}}^{-1/2}\mathbf{S}_{\mathbf{XY}}\mathbf{S}_{\mathbf{YY}}^{-1}\mathbf{S}_{\mathbf{YX}}\mathbf{S}_{\mathbf{XX}}^{-1/2}\mathbf{v}_x}{\sqrt{\mathbf{v}_x^T\mathbf{v}_x}}. \tag{21.8}$$

By following an analogous derivation as in (21.4) we conclude that the solution of this functional is the eigenvector of the matrix $\Sigma = \mathbf{S}_{\mathbf{XX}}^{-1/2}\mathbf{S}_{\mathbf{XY}}\mathbf{S}_{\mathbf{YY}}^{-1}\mathbf{S}_{\mathbf{YX}}\mathbf{S}_{\mathbf{XX}}^{-1/2}$. Thus, the CCA solution can be alternatively obtained by computing $\mathbf{w}_x = \mathbf{S}_{\mathbf{XX}}^{-1/2}\mathbf{v}_x$, where $\mathbf{v}_x$ is the eigenvector of Σ, and by defining $\mathbf{w}_y = \mathbf{S}_{\mathbf{YY}}^{-1/2}\mathbf{u}$.

Formula (21.5) highlights the numerical drawbacks of CCA, as its computation depends on the sample covariances $\mathbf{S}_{\mathbf{XX}}$ and $\mathbf{S}_{\mathbf{YY}}$. In particular, these matrices being estimated from sample data, the associated eigenvalues may quickly become small and unavoidably compromise the stability of the estimation. For this reason it is common practice to reformulate the CCA problem with a *regularized* version aimed at improving stability. This is performed by stabilizing the right-hand side of (21.4) by adding a constant diagonal term depending on a regularization parameter δ:

$$\begin{bmatrix} \mathbf{0} & \mathbf{S}_{\mathbf{XY}} \\ \mathbf{S}_{\mathbf{YX}} & \mathbf{0} \end{bmatrix} \begin{bmatrix} \mathbf{w}_x \\ \mathbf{w}_y \end{bmatrix} = \lambda \begin{bmatrix} \mathbf{S}_{\mathbf{XX}} + \delta\mathbf{I} & \mathbf{0} \\ \mathbf{0} & \mathbf{S}_{\mathbf{YY}} + \delta\mathbf{I} \end{bmatrix} \begin{bmatrix} \mathbf{w}_x \\ \mathbf{w}_y \end{bmatrix}. \tag{21.9}$$

In typical applications, the value for the parameter δ is chosen a priori or is estimated through cross-validation techniques.

21.3.1.2 Partial least squares

Analogously to CCA, partial least squares (PLS) is based on the identification of linear projections maximizing the *covariance* between the projected data:

$$\mathbf{w}_x, \mathbf{w}_y = \underset{\mathbf{w}_x, \mathbf{w}_y}{\arg\max}\, \Sigma(\mathbf{X}\mathbf{w}_x, \mathbf{Y}\mathbf{w}_y), \tag{21.10}$$

where

$$\Sigma(\mathbf{X}\mathbf{w}_x, \mathbf{Y}\mathbf{w}_y) = \frac{\mathbf{w}_x^T\mathbf{S}_{\mathbf{XY}}\mathbf{w}_y}{\sqrt{\mathbf{w}_x^T\mathbf{w}_x}\sqrt{\mathbf{w}_y^T\mathbf{w}_y}}. \tag{21.11}$$

As before, the PLS problem can be optimized through the solution of an ordinary eigenvalue problem,

$$\begin{bmatrix} \mathbf{0} & \mathbf{S}_{\mathbf{XY}} \\ \mathbf{S}_{\mathbf{YX}} & \mathbf{0} \end{bmatrix} \begin{bmatrix} \mathbf{w}_x \\ \mathbf{w}_y \end{bmatrix} = \lambda \begin{bmatrix} \mathbf{w}_x \\ \mathbf{w}_y \end{bmatrix}. \tag{21.12}$$

PLS has interesting analogies with CCA. In particular, the PLS solution can be seen as the maximally regularized version of CCA obtained when the regularizing parameter δ of formula (21.9) tends towards infinity. The solution corresponding to the eigenmodes of the problem (21.12) is known as PLS-SVD, and has been popularized in the field of neuroimaging in the seminal works [35,36] for the study of positron emission tomography (PET) and functional magnetic resonance imaging (fMRI) through the analysis of the associated eigenmodes of intensity variation.

21.3.1.3 Iterative numerical schemes for PLS and CCA: NIPALS

In practice, to avoid the numerical instabilities related to the decomposition of potentially large sample covariance matrices, both PLS and CCA can be computed by leveraging on stable numerical schemes. In particular, the *non-linear iterative partial least squares* (NIPALS) is a classical algorithm proposed by H. Wold [37] for the iterative computation of the PLS (and CCA) solution. Within this method, the principal eigenmodes $\mathbf{w}_x^{(0)}$ and $\mathbf{w}_y^{(0)}$ are initially computed from the data matrices $\mathbf{X}^{(0)} = \mathbf{X}$ and $\mathbf{Y}^{(0)} = \mathbf{Y}$ with the iterative scheme detailed in Algorithm 21.1 [38].

After the estimation of the eigenmode of iteration i, the high-order components $\mathbf{w}_x^{(i)}$ and $\mathbf{w}_y^{(i)}$ are subsequently computed by applying NIPALS on the deflated data matrices $\mathbf{X}^{(i)}$ and $\mathbf{Y}^{(i)}$ obtained by regressing out the current projections in the latent space $\boldsymbol{t}^{(i)} = \mathbf{X}^{(i)}\mathbf{w}_x^{(i)}$, $\boldsymbol{u}^{(i)} = \mathbf{Y}^{(i)}\mathbf{w}_y^{(i)}$:

$$\mathbf{X}^{(i+1)} = \mathbf{X}^{(i)} - \boldsymbol{t}^{(i)} \frac{\boldsymbol{t}^{(i)T}\mathbf{X}^{(i)}}{\boldsymbol{t}^{(i)T}\boldsymbol{t}^{(i)}},$$

$$\mathbf{Y}^{(i+1)} = \mathbf{Y}^{(i)} - \boldsymbol{u}^{(i)} \frac{\boldsymbol{u}^{(i)T}\mathbf{Y}^{(i)}}{\boldsymbol{u}^{(i)T}\boldsymbol{u}^{(i)}}.$$

Initialize $\mathbf{w}_y^{(i)}$. While not converged do:

1. $\boldsymbol{u}^{(i)} = \mathbf{Y}^{(i)}\mathbf{w}_y^{(i)}$
2. Estimate the projection for $\mathbf{X}^{(i)}$ from the latent space of $\mathbf{Y}^{(i)}$
   PLS. $\mathbf{w}_x^{(i)} = \mathbf{X}^{(i)T}\boldsymbol{u}^{(i)}/\boldsymbol{u}^{(i)T}\boldsymbol{u}^{(i)}$.
   CCA. $\mathbf{w}_x^{(i)} = \mathbf{X}^{(i)*}\boldsymbol{u}^{(i)}$,
   where $\mathbf{X}^{(i)*}$ is the Moore–Penrose inverse of $\mathbf{X}^{(i)}$.
3. Normalize $\mathbf{w}_x^{(i)}$.
4. $\boldsymbol{t}^{(i)} = \mathbf{X}^{(i)}\mathbf{w}_x^{(i)}$
5. Estimate the projection for $\mathbf{Y}^{(i)}$ from the latent space of $\mathbf{X}^{(i)}$
   PLS. $\mathbf{w}_y^{(i)} = \mathbf{Y}^{(i)T}\boldsymbol{t}^{(i)}/\boldsymbol{t}^{(i)T}\boldsymbol{t}^{(i)}$,
   CCA. $\mathbf{w}_y^{(i)} = \mathbf{Y}^{(i)*}\boldsymbol{t}^{(i)}$,
   where $\mathbf{Y}^{(i)*}$ is the Moore–Penrose inverse of $\mathbf{Y}^{(i)}$.
6. Normalize $\mathbf{w}_y^{(i)}$.

Algorithm 21.1: NIPALS iterative eigenmode computation for component i [38].

The implementation illustrated in Algorithm 21.1 can be found in standard statistical and machine learning packages, such as Scikit-learn[1] [39].

21.3.1.4 Reduced-rank regression

RRR differs from CCA and PLS since the relationship between the data matrices **X** and **Y** is estimated by optimizing a standard regression problem with Gaussian noise $\boldsymbol{\varepsilon}$:

$$\mathbf{Y} = \mathbf{X}\mathbf{C} + \boldsymbol{\varepsilon}. \tag{21.13}$$

The model parameters **C** are a matrix of dimensions $d_i \times d_g$ with rank $r \leq min(d_i, d_g)$ [40]. The decomposition $\mathbf{C} = \mathbf{AB}$, with $dim(A) = d_i \times r$ and $dim(B) = r \times d_g$, provides a way to interpret RRR as the composition of respectively a linear projection of **X** into a latent space of dimension r and a linear reconstruction of **Y** from the latent representation. Thanks to this representation, **A** and **B** can be interpreted to jointly provide a quantification of the relationship between imaging and genetic features [41].

The optimization of RRR is equivalent to the minimization of the loss:

$$f(\mathbf{A}, \mathbf{B}) = tr\{(\mathbf{Y} - \mathbf{XAB})\boldsymbol{\Gamma}(\mathbf{Y} - \mathbf{XAB})^T\}, \tag{21.14}$$

for any given positive definite matrix $\boldsymbol{\Gamma}$ [40]. Within this formulation, the RRR solution is

$$\mathbf{A} = \Gamma^{-1/2}\mathbf{U}, \quad \mathbf{B} = \mathbf{U}^T\Gamma^{1/2}\mathbf{S_{YX}}\mathbf{S_{XX}^{-1}},$$

where **U** are the eigenvectors associated to the matrix $\mathbf{R} = \Gamma^{1/2}\mathbf{S_{YX}}\mathbf{S_{XX}^{-1}}\mathbf{S_{XY}}\Gamma^{1/2}$. The role of Γ is to account for the covariance between the features of **Y** and can be set to identity or any other suitable form. For example, by setting $\Gamma = \mathbf{S_{YY}}$, we obtain the CCA solution derived from formula (21.8).

21.3.1.5 Parallel independent component analysis

Although less standard in the multivariate analysis literature, *parallel independent component analysis* (pICA) is an established method in medical imaging and imaging-genetics [42], building on the classical framework of *independent component analysis* (ICA) [43].

ICA-based methods are conceptually different from the multivariate approaches previously described in this chapter (CCA, PLS, RRR). Indeed, the latter are essentially based on the identification of a low-dimensional data representation in which the statistical association expressed by the covariance, or the correlation, is maximized. In practice, this amounts to assuming a Gaussian distribution of the data and identifying an opportune linear change of coordinates leading to maximum data likelihood.

[1] http://scikit-learn.org/stable/.

On the contrary, when applying ICA to a single modality $\mathbf{X}$, we assume that the observed data are a linear mixture of *non-Gaussian distributed and independent* latent sources (or there is at most a single Gaussian component). The associate generative model is

$$\mathbf{X} = \mathbf{S}\mathbf{W}_x, \tag{21.15}$$

where $\mathbf{S}$ and $\mathbf{W}$ are respectively sources and mixing matrix. Several ICA approaches for the joint optimization of $\mathbf{S}$ and $\mathbf{W}$ have been proposed in the literature [44]. For example, when the measure of non-Gaussianity is given by the negentropy [45], the FastICA algorithm is a very efficient ICA approach based on Newton gradient-based optimization [46].

Given imaging-genetics data matrices, pICA jointly optimizes the ICA problem separately on the two modalities [47],

$$\mathbf{X} = \mathbf{S}_x\mathbf{W}_x,$$
$$\mathbf{Y} = \mathbf{S}_y\mathbf{W}_y,$$

while also maximizing the correlation between the columns of the respective mixing matrices $\mathbf{W}_x[:, i]$ and $\mathbf{W}_y[:, j]$. This approach thus aims at identifying the mixing parameters to reconstruct the data while at the same time highlighting their relationship.

21.3.2 Regularization in multivariate imaging-genetics

When dealing with high-dimensional data, such as the one available in imaging-genetics, multivariate approaches are highly prone to the problem of overfitting and to the identification of spurious associations. This problem ultimately leads to low stability and poor interpretability of the analysis results.

A common strategy for mitigating this problem is through model regularization, achieved by constraining the solution to belong to a suitable parameter space. These constraints are usually specified by introducing a penalization term on the parameters' norm. In the following section we illustrate the main approaches to regularization along with their principal applications in imaging-genetics.

21.3.2.1 Sparsity and smoothness

Sparsity is a classical regularization choice, known as *least absolute shrinkage and selection operator* (LASSO), looking for a limited subset of non-zero parameters' coefficients. In its simplest formulation, given an objective function $f(\mathbf{w})$, sparsity on the solution $\mathbf{w}$ is imposed by solving the associated functional:

$$\arg\min_{\mathbf{w}} f(\mathbf{w}) + \lambda\|\mathbf{w}\|_1, \tag{21.16}$$

which corresponds to the Lagrangian of the optimization of $f(\mathbf{w})$ constrained to $|\mathbf{w}| \leq t$, for an opportune t. The left panel of Fig. 21.4 illustrates the sparsity ef-

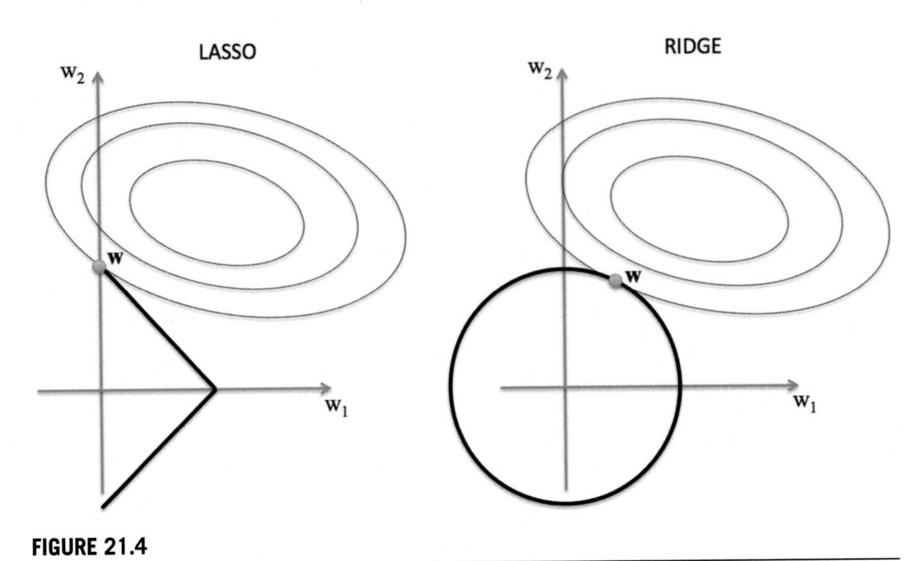

FIGURE 21.4

An illustration of regularized regression with LASSO and ridge penalization. Red: Contours for the solution of the function $f(\mathbf{w})$. Black: Regularization contours. In LASSO, the geometric constraint associated to the ℓ_1 penalization promotes sparse solutions (e.g., $w_1 = 0$).

fect of the LASSO penalty on the identification of the minimum for a given function f (red contours). Increasing the sparsity constraint reduces the size of the regularization contour (black line), enforcing the solution parameters to zero (in this case the parameter w_1).

Another classical form of regularization, known as *ridge* or *Tikhonov* regularization, consists in introducing a penalization of the ℓ_2-norm of the parameters:

$$\arg\min_{\mathbf{w}} f(\mathbf{w}) + \lambda\|\mathbf{w}\|_2^2. \tag{21.17}$$

While this regularization form does not promote sparsity (Fig. 21.4, right), it is used to enhance regularity of the solution and promote the well-posedness of the optimization problem.

In applied studies, approaches to regularized CCA [48,49], PLS [50,51], and RRR [41] have been proposed by introducing an additional parameter penalization term of the ℓ_1-norm (LASSO, [52]), or jointly to ℓ_1 and ℓ_2 (so-called elastic net regularization [53]).

In its simplest formulation, sparsity in CCA and PLS can be implemented by soft-thresholding of the parameters' weights, based on a predefined ratio of desired resulting non-zero coefficients [48,51]. Regularity can also be introduced in NIPALS [50], by estimating regularized projections $\overline{\mathbf{w}_x}$ and $\overline{\mathbf{w}_y}$ through elastic net (Algorithm 21.2). It is interesting to note that the method reduces to univariate LASSO soft-thresholding when the ℓ_2-parameters $\lambda_{y_2}, \lambda_{x_2}$ are set to infinity [53].

Given current estimates of $\mathbf{w}_x$ and $\mathbf{w}_y$.
While not converged do:

1. compute $\mathbf{t} = \mathbf{X}\mathbf{w}_x$,
2. compute $\mathbf{u} = \mathbf{Y}\mathbf{w}_y$,
3. compute $\overline{\mathbf{w}_x}$ by solving the elastic-net regression:
   $\overline{\mathbf{w}_x} = \arg\min_{\mathbf{v}} (\mathbf{t} - \mathbf{X}\mathbf{v})^2 + \lambda_{x_2} \|\mathbf{v}\|_2^2 + \lambda_{x_1} \|\mathbf{v}\|_1,$
4. compute $\overline{\mathbf{w}_y}$ by solving the elastic-net regression:
   $\overline{\mathbf{w}_y} = \arg\min_{\mathbf{v}} (\mathbf{u} - \mathbf{Y}\mathbf{v})^2 + \lambda_{y_2} \|\mathbf{v}\|_2^2 + \lambda_{y_1} \|\mathbf{v}\|_1,$
3. Normalize $\overline{\mathbf{w}_x}$ and $\overline{\mathbf{w}_x}$,
4. Set $\mathbf{w}_x = \overline{\mathbf{w}_x}$, $\mathbf{w}_y = \overline{\mathbf{w}_y}$.

Algorithm 21.2: Regularization of projection parameters $\mathbf{w}_x$ and $\mathbf{w}_y$ in NIPALS.

The experimental investigation proposed in [51] provides a useful comparison of sparse vs. non-sparse implementations of multivariate approaches in imaging-genetics. In particular, although sparsity was shown to generally help in identifying more stable solutions, the choice of regularization parameters, cross-validation, and data pre-processing strategies (especially data filtering and dimensionality reduction) seems to play a critical role for the reliability of the solution.

Moreover, an important issue is still represented by the intrinsic signal correlation characterizing both imaging and genetic data. While the imaging data are characterized by non-trivial spatial correlations, affecting for example the signal of neighboring voxels, SNPs are spatially correlated due to LD. In this case, the solutions provided by LASSO and ridge regression are in opposition and thus not straightforwardly interpretable. Indeed, LASSO tends to isolate a single variable from a set of correlated ones, while ridge identifies the whole set of correlated predictors. However, in the practical context, this behavior is not guaranteed and may be strongly related to data variability and dimensions [51].

For this reason, more complex regularization strategies accounting for known relationships across imaging and genetic features have been proposed in the last years. The next section provides an overview of this more advanced modeling topic.

21.3.2.2 Groupwise penalization

Imaging-genetics data are characterized by non-trivial correlations representing precise biological and anatomical mechanisms.

On the one hand, SNPs are known to act "in concert" through *biological pathways* representing specific biological processes (metabolic, cellular, genetic). These processes can be represented via relation networks, as provided in the KEGG pathway database.[2] Similarly, we can rely on handcrafted ontologies describing gene functions

[2] https://www.genome.jp/kegg/pathway.html.

and relationships (Gene Ontology Consortium[3]). On the other hand, medical imaging data are characterized by both local correlations across neighboring voxels and non-local ones. For example, due to the organization of the brain in anatomical and functional networks, the signal measured in brain imaging data exhibits important network properties representing specific physiological mechanisms.

In this context, the stability and interpretability of the multivariate analysis results can be promoted by introducing an additional regularization constraint on the model parameters. This constraint aims at enforcing the model to provide solutions compatible with our prior information on the relationship between features.

This problem can be addressed by introducing a groupwise penalization [54]. Within this approach, variables known to act together are jointly regularized with respect to the same penalization parameter. In its basic formulation, we assume the d features of a predictor $\mathbf{x} = (x_1, \dots, x_d)$ are grouped in s feature sets taking the form $\mathcal{L}_k = (x_{i_1}, \dots, x_{i_k})$. The groupwise penalized regression of the independent variable $\mathbf{y}$ with respect to the predictors $\mathbf{X}$ is thus expressed as

$$f(\mathbf{a}) = \|\mathbf{y} - \mathbf{X}\mathbf{a}\|_2^2 + \lambda \sum_{k=1}^{s} w_k \|\mathbf{a}_k\|_2,$$

where the vectors $\mathbf{a}_k = (a_{i_1}, \dots, a_{i_k})$ group the elements of $\mathbf{a}$ associated to each feature set $\mathcal{L}_k$. Computational issues may arise when the sets $\mathcal{L}_k$ are not disjoint, and specific optimization strategies need to be defined to address the problems of identifiability and computational efficiency [55].

Building on this methodological framework, an extension of RRR to account for groupwise relationships was introduced in [56]. In particular, by considering a rank-one RRR model, formula (21.14) can be extended to

$$f(\mathbf{a}, \mathbf{b}) = tr\{(\mathbf{Y} - \mathbf{X}\mathbf{a}\mathbf{b})(\mathbf{Y} - \mathbf{X}\mathbf{a}\mathbf{b})^T\} + \lambda \sum_{k=1}^{s} w_k \|\mathbf{a}_k\|_2, \tag{21.18}$$

where the coefficients $\mathbf{a}_k$ account for known relationships on the genetic data. Similarly to the RRR case, other extensions to groupwise penalization in multivariate imaging-genetics models have been proposed, for example applied to CCA [57] or to sparse regression [58].

21.3.3 Stability and validation of multivariate models

The promise of multivariate models to unveil hidden relationship in high-dimensional data is often hindered by the problem of overfitting and lack of stability of the results. In particular, overfitting is a critical issue concerning the drop in the generalization ability of the model when tested on independent data. In the previous sections we

[3] http://geneontology.org/.

have seen that regularization can be used to identify more meaningful and stable solutions. However, the introduction of regularization comes with the problem of tuning the associated regularity parameter (e.g., the LASSO parameter λ of equation (21.16)). Since the model solution often greatly varies depending on the imposed degree of regularization, this problem ultimately affects the stability of the results. For this reason it is common practice to tune the parameters by grid search on a subset of hold-out data, to identify the parameter set leading to the optimal cross-validation results [41,51,56]. Unfortunately this practice may also lead to overfit and selection bias [59], especially when the sample size is low compared to the data dimensionality. Again, in this case the marginal improvement related to the tuning of the parameters does not lead to an improvement in the generalization of the model in unobserved data.

All in all, the validity of multivariate models should be ultimately established via testing on datasets not used to estimate the model parameters. This is a critical aspect related to the availability of independent data compatible with the one used for model fitting. Current efforts in standardization and opening of high-quality datasets are becoming fundamental for the reliable use of multivariate methods in imaging-genetics and, more generally, in biomedical applications.

21.4 Exercises

1. A simplified imaging-genetics analysis example.

- **Univariate Genetic Analysis.** Load the two data sets `simulated_snp.raw` and `simulated_volumes.txt` and conduct a univariate genetic analysis of the region `Lhippo`. Display the $-\log_{10}$(p-value) for all genetic variants in the form of a Manhattan plot. Which SNP shows the strongest association with volume in the left Hippocampus?
- **Univariate Imaging Analysis.** Conduct a univariate imaging analysis for the SNP `rs438811_T`. Which brain region shows the strongest association with that SNP?
- **Multiple Testing.** Under the simplified assumption that all SNPs and all imaging phenotypes are independent. In this dataset, what p-value would be considered statistically significant for a genetic study of a single imaging phenotype, for an imaging study evaluating a single SNP, and for a full imaging-genetics study investigating all SNPs and imaging phenotypes?

2. Understanding multivariate modeling through data generation and analysis.

- Define a 2-dimensional latent space of 100 points where each latent dimension is sampled from a Gaussian distribution $N(0, 1)$.
- Define the following two linear transformations with random integer coefficients. `Transformation_x` maps the 2-dimensional latent space to a 5-dimen-

sional space of modality X:

$$t_x = \begin{pmatrix} -6 & 2 & 7 & -7 & 4 \\ -5 & -8 & -1 & 1 & 2 \end{pmatrix},$$

while `Transformation_y` maps the 2-dimensional latent space to a 10-dimensional space of modality Y:

$$t_y = \begin{pmatrix} 5 & -4 & 7 & 6 & 1 & 5 & -5 & 7 & -4 & 5 \\ 4 & 0 & 0 & 7 & -5 & 1 & -4 & -6 & 1 & -4 \end{pmatrix}$$

- Use the transformations to map each of the 100 latent points into the observed modalities X and Y. The dimensions of X and Y will be respectively 100×5, and 100×20.
- Add some Gaussian random noise $N(0, \sigma^2)$ to the observations X and Y. The Gaussian noise for X will be sampled from $N(0, 4)$, and the Gaussian noise for Y will be sampled from $N(0, 3)$.
- Plot the first 3 dimensions of modality X versus the first 3 dimensions of modality Y. Verify that the two data modalities are correlated across these dimensions.
- Split the data into two non-overlapping folds of 50 observations each.
  - From `sklearn.cross_decomposition` fit a `PLSCanonical` model on the first fold (training fold) by setting the number of `latent_components = 1`.
  - Plot the weights of both X and Y projections estimated by the PLS model against the weights of the transformations t_x andt_y defined above. What do you conclude?
  - Using the second data fold (testing fold) project both modalities X and Y into the latent space. Plot the projections with respect to the two dimensions of the corresponding latent variables defined in point 1. What do you conclude?
  - Using the second data fold (testing fold) predict the modality Y from X, and assess the quality of the prediction (testing predictive accuracy). What do you conclude?
- Repeat the previous analysis by setting the number of latent dimensions of different PLS models from 2. to 5. Compare the different models with respect to the respective testing predictive accuracy.
  - Which model is the best performing one?
  - Is the conclusion compatible with the data generation process of the modalities X and Y?
- Repeat the analysis using CCA. Is there any noticeable difference between the CCA and PLS models for this kind of data?

References

[1] Willyard C., New human gene tally reignites debate, Nature 558 (7710) (2018) 354.

[2] ENCODE Project Consortium, An integrated encyclopedia of DNA elements in the human genome, Nature 489 (7414) (2012) 57.

[3] Falconer D.S., Introduction to Quantitative Genetics, Oliver and Boyd, Edinburgh, London, 1960.

[4] Yang J., Lee S.H., Goddard M.E., Visscher P.M., GCTA: a tool for genome-wide complex trait analysis, American Journal of Human Genetics 88 (1) (2011) 76–82.

[5] Bulik-Sullivan B.K., Loh P.R., Finucane H.K., Ripke S., Yang J., LD Score regression distinguishes confounding from polygenicity in genome-wide association studies, Nature Genetics 47 (3) (2015) 291.

[6] Perez E., Sullivan K.E., Chromosome 22q11. 2 deletion syndrome (DiGeorge and velocardiofacial syndromes), Current Opinion in Pediatrics 14 (6) (2002) 678–683.

[7] Flint J., Munafò M.R., The endophenotype concept in psychiatric genetics, Psychological Medicine 37 (2) (2007) 163–180.

[8] Hibar D.P., et al., Common genetic variants influence human subcortical brain structures, Nature 520 (7546) (2015) 224.

[9] Purcell S., Neale B., Todd-Brown K., Thomas L., Ferreira M.A., Bender D., Maller J., Sklar P., de Bakker P.I., Daly M.J., Sham P.C., PLINK: a tool set for whole-genome association and population-based linkage analyses, American Journal of Human Genetics 81 (3) (2007) 559–575.

[10] Novembre J., Johnson T., Bryc K., Kutalik Z., Boyko A.R., Auton A., Indap A., King K.S., Bergmann S., Nelson M.R., Stephens M., Bustamante C.D., Genes mirror geography within Europe, Nature 456 (7218) (2008) 98.

[11] Marchini J., Howie B., Genotype imputation for genome-wide association studies, Nature Reviews. Genetics 11 (7) (2010) 499.

[12] McCarthy S., et al., A reference panel of 64,976 haplotypes for genotype imputation, Nature Genetics 48 (10) (2016) 1279.

[13] Das S., Forer L., Schönherr S., Sidore C., Locke A.E., Kwong A., Vrieze S.I., Chew E.Y., Levy S., McGue M., Schlessinger D., Stambolian D., Loh P.R., Iacono W.G., Swaroop A., Scott L.J., Cucca F., Kronenberg F., Boehnke M., Abecasis G.R., Fuchsberger C., Next-generation genotype imputation service and methods, Nature Genetics 48 (10) (2016) 1284.

[14] Sadaghiani S., Ng B., Altmann A., Poline J.B., Banaschewski T., Bokde A.L.W., Bromberg U., Büchel C., Burke Quinlan E., Conrod P., Desrivières S., Flor H., Frouin V., Garavan H., Gowland P., Gallinat J., Heinz A., Ittermann B., Martinot J.L., Paillère Martinot M.L., Lemaitre H., Nees F., Papadopoulos Orfanos D., Paus T., Poustka L., Millenet S., Fröhner J.H., Smolka M.N., Walter H., Whelan R., Schumann G., Napolioni V., Greicius M., Overdominant effect of a CHRNA4 polymorphism on cingulo-opercular network activity and cognitive control, The Journal of Neuroscience 37 (40) (2017) 9657–9666.

[15] Dubal D.B., Yokoyama J.S., Zhu L., Broestl L., Worden K., Wang D., Sturm V.E., Kim D., Klein E., Yu G.Q., Ho K., Eilertson K.E., Yu L., Kuro-o M., De Jager P.L., Coppola G., Small G.W., Bennett D.A., Kramer J.H., Abraham C.R., Miller B.L., Mucke L., Life extension factor klotho enhances cognition, Cell Reports 7 (4) (2014) 1065–1076.

[16] Loh P.R., Kichaev G., Gazal S., Schoech A.P., Price A.L., Mixed-model association for biobank-scale datasets, Nature Genetics (2018) 1.

[17] Bonferroni C., Teoria statistica delle classi e calcolo delle probabilita, Pubblicazioni del R Istituto Superiore di Scienze Economiche e Commerciali di Firenze 8 (1936) 3–62.
[18] Benjamini Y., Hochberg Y., Controlling the false discovery rate: a practical and powerful approach to multiple testing, Journal of the Royal Statistical Society, Series B, Methodological (1995) 289–300.
[19] Hoggart C.J., Clark T.G., De Iorio M., Whittaker J.C., Balding D.J., Genome-wide significance for dense SNP and resequencing data, Genetic Epidemiology: The Official Publication of the International Genetic Epidemiology Society 32 (2) (2008) 179–185.
[20] Kim J., Pan W., A cautionary note on using secondary phenotypes in neuroimaging genetic studies, NeuroImage 121 (2015) 136–145.
[21] Aaron B., Kromrey J.D., Ferron J., Equating "r"-based and "d"-based effect size indices: problems with a commonly recommended formula, ERIC Clearinghouse, 1998.
[22] Lee J.J., et al., Gene discovery and polygenic prediction from a genome-wide association study of educational attainment in 1.1 million individuals, Nature Genetics (2018) 1.
[23] Iglesias J.E., Augustinack J.C., Nguyen K., Player C.M., Player A., Wright M., Roy N., Frosch M.P., McKee A.C., Wald L.L., Fischl B., Van Leemput K., A computational atlas of the hippocampal formation using ex vivo, ultra-high resolution MRI: application to adaptive segmentation of in vivo MRI, NeuroImage 115 (2015) 117–137.
[24] Glasser M.F., Coalson T.S., Robinson E.C., Hacker C.D., Harwell J., Yacoub E., Ugurbil K., Andersson J., Beckmann C.F., Jenkinson M., Smith S.M., Van Essen D.C., A multimodal parcellation of human cerebral cortex, Nature 536 (7615) (2016) 171–178.
[25] Elliott L.T., Sharp K., Alfaro-Almagro F., Shi S., Miller K.L., Douaud G., Marchini J., Smith S.M., Genome-wide association studies of brain structure and function in the UK Biobank, Nature 562 (3) (2018) 210–216.
[26] Scelsi M.A., Khan R.R., Lorenzi M., Christopher L., Greicius M.D., Schott J.M., Ourselin S., Altmann A., Genetic study of multimodal imaging Alzheimer's disease progression score implicates novel loci, Brain 141 (7) (2018) 2167–2180.
[27] Stein J.L., Hua X., Lee S., Ho A.J., Leow A.D., Toga A.W., Saykin A.J., Shen L., Foroud T., Pankratz N., Huentelman M.J., Craig D.W., Gerber J.D., Allen A.N., Corneveaux J.J., Dechairo B.M., Potkin S.G., Weiner M.W., Thompson P., Voxelwise genome-wide association study (vGWAS), NeuroImage 53 (3) (2010) 1160–1174.
[28] Thompson P.M., et al., The ENIGMA Consortium: large-scale collaborative analyses of neuroimaging and genetic data, Brain Imaging and Behavior 8 (2) (2014) 153–182.
[29] Huang M., Nichols T., Huang C., Yu Y., Lu Z., Knickmeyer R.C., Feng Q., Zhu H., FVGWAS: Fast voxelwise genome wide association analysis of large-scale imaging genetic data, NeuroImage 118 (2015) 613–627.
[30] Schrouff J., Rosa M.J., Rondina J.M., Marquand A.F., Chu C., Ashburner J., Phillips C., Richiardi J., Mourão-Miranda J., PRoNTo: pattern recognition for neuroimaging toolbox, Neuroinformatics 11 (3) (2013) 319–337.
[31] Szymczak S., Biernacka J.M., Cordell H.J., González-Recio O., König I.R., Zhang H., Sun Y.V., Machine learning in genome-wide association studies, Genetic Epidemiology 33 (S1) (2009) S51–S57.
[32] Liu J., Calhoun V.D., A review of multivariate analyses in imaging genetics, Frontiers in Neuroinformatics 8 (2014) 29.
[33] Lorenzi M., Altmann A., Gutman B., Wray S., Arber C., Hibar D.P., Jahanshad N., Schott J.M., Alexander D.C., Thompson P.M., Ourselin S., Susceptibility of brain atrophy to TRIB3 in Alzheimer's disease, evidence from functional prioritization in imaging genetics, Proceedings of the National Academy of Sciences 115 (12) (2018) 3162–3167, https://doi.org/10.1073/pnas.1706100115.

[34] De Bie T., Cristianini N., Rosipal R., Eigenproblems in pattern recognition, in: Handbook of Geometric Computing, Springer, 2005, pp. 129–167.

[35] McIntosh A.R., Bookstein F.L., Haxby J.V., Grady C.L., Spatial pattern analysis of functional brain images using partial least squares, NeuroImage 3 (3) (1996) 143–157.

[36] Worsley K.J., An overview and some new developments in the statistical analysis of PET and fMRI data, Human Brain Mapping 5 (4) (1997) 254–258.

[37] Wold H., Path models with latent variables: The NIPALS approach, in: Quantitative Sociology, Elsevier, 1975, pp. 307–357.

[38] Tenenhaus M., L'approche PLS, Revue de Statistique Appliquée 47 (2) (1999) 5–40.

[39] Pedregosa F., Varoquaux G., Gramfort A., Michel V., Thirion B., Grisel O., Blondel M., Müller A., Nothman J., Louppe G., Prettenhofer P., Weiss R., Dubourg V., Vanderplas J., Passos A., Cournapeau D., Brucher M., Perrot M., Duchesnay É., Scikit-learn: machine learning in Python, Journal of Machine Learning Research 12 (2011) 2825–2830.

[40] Velu R., Reinsel G.C., Multivariate Reduced-Rank Regression: Theory and Applications, vol. 136, Springer Science & Business Media, 2013.

[41] Vounou M., Nichols T.E., Montana G., Discovering genetic associations with high-dimensional neuroimaging phenotypes: A sparse reduced-rank regression approach, NeuroImage 53 (3) (2010) 1147–1159.

[42] Pearlson G.D., Calhoun V.D., Liu J., An introductory review of parallel independent component analysis (p-ICA) and a guide to applying p-ICA to genetic data and imaging phenotypes to identify disease-associated biological pathways and systems in common complex disorders, Frontiers in Genetics 6 (2015) 276.

[43] Comon P., Independent component analysis, a new concept?, Signal Processing 36 (3) (1994) 287–314.

[44] Hyvärinen A., Oja E., Independent component analysis: algorithms and applications, Neural Networks 13 (4–5) (2000) 411–430.

[45] Cover T.M., Thomas J.A., Elements of Information Theory, John Wiley & Sons, 2012.

[46] Hyvarinen A., Fast and robust fixed-point algorithms for independent component analysis, IEEE Transactions on Neural Networks 10 (3) (1999) 626–634.

[47] Liu J., Pearlson G., Windemuth A., Ruano G., Perrone-Bizzozero N.I., Calhoun V., Combining fMRI and SNP data to investigate connections between brain function and genetics using parallel ICA, Human Brain Mapping 30 (1) (2009) 241–255.

[48] Parkhomenko E., Tritchler D., Beyene J., Genome-wide sparse canonical correlation of gene expression with genotypes, in: BMC Proceedings, vol. 1, BioMed Central, 2007, S119.

[49] Witten D.M., Tibshirani R., Hastie T., A penalized matrix decomposition, with applications to sparse principal components and canonical correlation analysis, Biostatistics 10 (3) (2009) 515–534.

[50] Waaijenborg S., Zwinderman A.H., Penalized canonical correlation analysis to quantify the association between gene expression and DNA markers, in: BMC Proceedings, vol. 1, BioMed Central, 2007, p. S122.

[51] Le Floch E., Guillemot V., Frouin V., Pinel P., Lalanne C., Trinchera L., Tenenhaus A., Moreno A., Zilbovicius M., Bourgeron T., Dehaene S., Thirion B., Poline J.B., Duchesnay E., Significant correlation between a set of genetic polymorphisms and a functional brain network revealed by feature selection and sparse Partial Least Squares, NeuroImage 63 (1) (2012) 11–24.

[52] Tibshirani R., Regression shrinkage and selection via the lasso, Journal of the Royal Statistical Society, Series B, Methodological (1996) 267–288.

[53] Zou H., Hastie T., Regularization and variable selection via the elastic net, Journal of the Royal Statistical Society, Series B, Statistical Methodology 67 (2) (2005) 301–320.

[54] Yuan M., Lin Y., Model selection and estimation in regression with grouped variables, Journal of the Royal Statistical Society, Series B, Statistical Methodology 68 (1) (2006) 49–67.

[55] Silver M., Montana G., Fast identification of biological pathways associated with a quantitative trait using group lasso with overlaps, Statistical Applications in Genetics and Molecular Biology 11 (1) (2012) 1–43.

[56] Silver M., Janousova E., Hua X., Thompson P.M., Montana G., Identification of gene pathways implicated in Alzheimer's disease using longitudinal imaging phenotypes with sparse regression, NeuroImage 63 (3) (2012) 1681–1694.

[57] Lin D., Zhang J., Li J., Calhoun V.D., Deng H.W., Wang Y.P., Group sparse canonical correlation analysis for genomic data integration, BMC Bioinformatics 14 (1) (2013) 245.

[58] Wang H., Nie F., Huang H., Yan J., Kim S., Nho K., Risacher S.L., Saykin A.J., Shen L., From phenotype to genotype: an association study of longitudinal phenotypic markers to Alzheimer's disease relevant SNPs, Bioinformatics 28 (18) (2012) i619–i625.

[59] Mendelson A.F., Zuluaga M.A., Lorenzi M., Hutton B.F., Ourselin S., Selection bias in the reported performances of AD classification pipelines, NeuroImage: Clinical 14 (2017) 400–416.

PART

Large-scale databases

VII

CHAPTER

Detection and quantitative enumeration of objects from large images

22

Cheng Lu[a], Simon Graham[b], Nasir Rajpoot[b], and Anant Madabhushi[c,d]

[a]*Medical Research Institute, Guangdong Provincial People's Hospital (Guangdong Academy of Medical Sciences), Southern Medical University, Guangzhou, China*

[b]*Department of Computer Science, University of Warwick, Coventry, United Kingdom*

[c]*Department of Biomedical Engineering, Emory University and Georgia Institute of Technology, Atlanta, GA, United States*

[d]*Atlanta Veterans Administration Medical Center, Atlanta, GA, United States*

Learning points

- Detection and counting in histology images: classical methods
- Thresholding-based methods. Global, Otsu, local, fuzzy and hysteresis thresholding
- Filtering-based methods. Template matching. Laplacian of Gaussians
- Machine learning and deep learning-based methods
- Bayesian estimation and correlation learning
- Deep learning for detection and counting

22.1 Introduction

The automatic detection of objects in biomedical images using computational techniques is a fundamental component in most medical image analysis pipelines. Methods for localization of regions of interest in biomedical images can help identify areas of abnormality and may also enable subsequent downstream analysis, such as object counting. Generally, the task of detection and counting first involves separating the region of interest from the rest of the image and then the result is analyzed to determine the number of objects. Separating the region of interest from the background can be done using various image processing, machine learning, and advanced deep learning methods, where the chosen approach may reflect the desired accuracy, computational budget, and available data. For example, traditional image processing techniques are often computationally efficient, yet may not reach the performance of deep learning-based approaches. Moreover, image processing techniques are typi-

Medical Image Analysis. https://doi.org/10.1016/B978-0-12-813657-7.00036-4

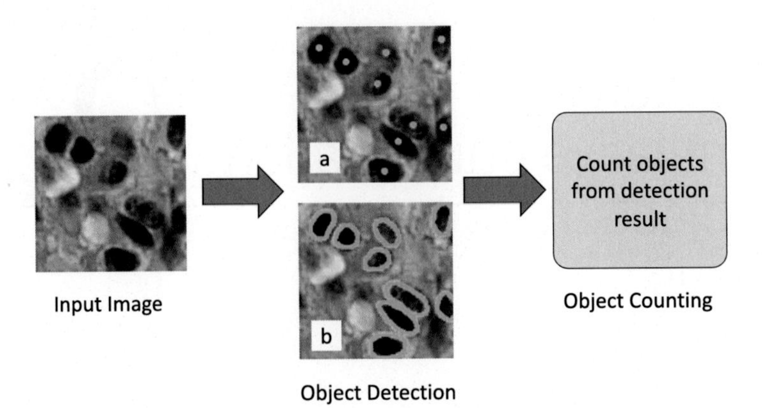

FIGURE 22.1 Overview of the main steps for object detection and counting in biomedical images.

An automated method is used to detect regions/objects in the input image; then the number of objects in the image is counted using the detection result.

cally not *supervised* and therefore do not require lots of labeled data to ensure a good level of performance. For supervised object detection, the labeled data may be in the form of bounding boxes, object centroids, or an instance segmentation map. Therefore, automated methods aim to map the input image to the target output, as described above. An overview of the task of object detection and counting, along with some examples of the labeled data for supervised learning, can be viewed in Fig. 22.1. We observe from this figure that first the objects are detected using a chosen technique and then the count is inferred from the detection result.

Object detection is important in various applications in medical image analysis, including lung nodule detection in thoracic computed tomography (CT) images, breast cancer recognition in mammogram images, and detection of nuclei in histology images. Histology images contain tens of thousands of nuclei, where their count, appearance, and organization can help inform the overall diagnosis of the tissue sample. Therefore, accurate detection and quantification of nuclei can help empower a computational system for cancer diagnosis and abnormality detection. Furthermore, detecting specific types of nuclei, such as those from tumor cells, can further improve the performance of computational systems for histology image analysis. For example, counting the number of mitotic cells (MCs; cells undergoing proliferation) within a region in breast cancer tissue is widely used by pathologists to determine the grade of cancer. In addition to nuclei, there exist other structures in the tissue, such as blood vessels and glands, that may be automatically detected and counted, which may further improve downstream tissue quantification systems.

In this chapter, we introduce various methods for object detection and counting, focusing on histology image analysis. We first introduce a range of traditional image processing techniques, primarily describing a series of thresholding- and filtering-based methods. Then, we go on to describe some machine learning and advanced deep learning techniques for object detection. To conclude the chapter, we include two extended examples: the first describes a classical approach for mitotic figure detection and the second approach describes a deep learning method for detection of different nucleus subtypes.

22.2 Classical image analysis methods

Traditional image analysis methods have been used for detecting objects in medical images. These image analysis methods generally incorporate prior or domain-specific knowledge, which may require understanding of the task at hand. For example, in a template matching (TM) method, one may need to construct a template in order to detect the object of interest accurately. Advantages of using these classical methods include: (1) they are quite easy to interpret and understand; (2) they are easy to implement and computationally cheap; and (3) they work well on datasets with a small number of images.

In this section, we first introduce the concept of basic global thresholding and a widely utilized method, called Otsu thresholding. We then present local thresholding and hysteresis thresholding, which use different thresholds for various regions of an image. Next, we focus on filtering-based methods, which include TM and Laplacian of Gaussian (LoG) filters.

22.2.1 Thresholding

22.2.1.1 Global thresholding

One simple method for object detection is the thresholding-based method, in which the foreground and background are separated by applying a single threshold on the pixel intensity. Denote a pixel in a 2D image as $f(x, y)$, where x and y are the horizontal and vertical coordinates, respectively. A unique threshold of T applied to every image pixel is called a global threshold. An easy way to detect or extract objects from the background is to choose an optimal threshold t that can separate the foreground, i.e., object, and background. The point (x, y) is called an object pixel if $f(x, y) \geq t$; otherwise it is called a background pixel. The segmented image $g(x, y)$ is given as follows:

$$g(x, y) = \begin{cases} 1, & \text{if } f(x, y) < t, \\ 0, & \text{if } f(x, y) \geq t. \end{cases} \tag{22.1}$$

When t is a constant that is applied to the entire image, the process given above is called global thresholding. There are many methods to determine a global threshold for segmenting the object from the background, such as the minimum error threshold method [1] and the Otsu threshold method [2].

22.2.1.2 Otsu thresholding

In Otsu's method, we exhaustively search for a threshold, t, that minimizes the intra-class variance $\sigma_w^2(t)$, defined as a weighted sum of variances of the two classes, i.e., foreground and background:

$$\sigma_w^2(t) = \omega_b \sigma_b^2(t) + \omega_f \sigma_f^2(t), \tag{22.2}$$

where ω_b and ω_f are the portions of the two pixel classes, background and foreground, separated by threshold t; σ_b^2 and σ_f^2 are variances of the background and foreground pixel class, respectively.

An example is shown in Fig. 22.2. Fig. 22.2(a) displays a grayscale histology image containing several nuclei. The intra-class variance $\sigma_w^2(t)$ calculated by equation (22.2) with respect to threshold t is shown in Fig. 22.2(b). The minimum value of intra-class variance $\sigma_w^2(t)$ was achieved with $t = 134$, so the global threshold $t = 134$ was applied on the grayscale image (Fig. 22.2(a)) to obtain the nuclear segmentation result (Fig. 22.2(c)). One may observe that even though it is not perfect, all the nuclei present in the image were segmented by the Otsu threshold method.

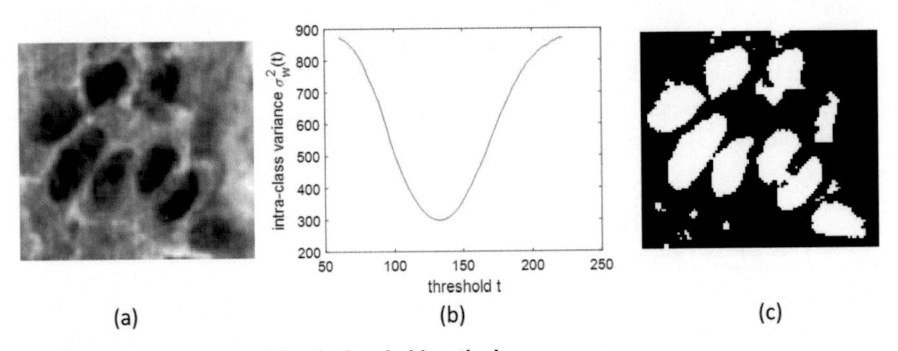

FIGURE 22.2 An example of Otsu's threshold method.

(a) A grayscale histology image containing several nuclei. (b) The intra-class variance $\sigma_w^2(t)$ calculated as shown in equation (22.2). (c) The binary image obtained by applying the threshold based on Otsu's threshold method ($t = 134$).

22.2.1.3 Local threshold

While a global threshold uses a single and fixed threshold for the whole image, the local threshold method determines the threshold value in local regions. For example, we can divide the whole image into small non-overlapping patches/regions and calculate a unique threshold, e.g., the median value, within each patch. This kind of method is able to avoid the intensity variation to some extent; however, it may suffer from stepwise jumps of identified thresholds across borders of non-overlapping regions. One solution could be to constrain the local thresholds of adjacent regions within a certain range, i.e., the changes of the local thresholds of adjacent regions cannot be larger than a predefined threshold.

22.2.1.4 Fuzzy threshold

Instead of using a hard threshold, fuzzy thresholding involves assigning each image pixel a membership value, ranging from 0 to 1, in turn reflecting the implicit likelihood of belonging to a specific tissue class (e.g., foreground and background). The membership function can be modified to also include prior information, e.g., image structure, object distribution, nature of noise, in order to avoid assigning pixels to a wrong class due to the intensity variations. An example can be found in [3], in which the membership of pixels was adaptively changed by local spatial information.

22.2.1.5 Hysteresis threshold

Hysteresis thresholding uses two thresholds in order to avoid the disconnected segmentation result when local variations are present. In other words, a high threshold h and a low threshold l are jointly utilized to distinguish the object pixels. If the pixel value is greater than the high threshold h, it is considered a pixel belonging to the foreground object and denoted as a strong object pixel. If the pixel value is less than the high threshold h and greater than the low threshold l, it is marked as a weak object pixel. Pixels below the low threshold l are suppressed. At this point, strong object pixels can be considered as true pixels that belong to the foreground object. The weak object points may be true foreground objects, or they may be caused by noise or intensity changes. In order to get accurate results, the weak object points caused by the latter should be removed. Hysteresis thresholding makes the assumption that weak object points and strong object points caused by the real foreground object are connected to each other, and weak object points caused by noise are not connected to the real foreground object. Therefore, hysteresis thresholding performs the following steps to achieve foreground segmentation:

- every pixel value below l is set to 0 (background);
- every pixel value above h is set to 1 (foreground);
- the rest of the pixels are set to 1 if they are four-way/eight-way connected to any other 1-valued blob (area), and otherwise they are set to 0.

22.2.2 Filtering-based methods

22.2.2.1 Template matching

TM is a widely used technique for object detection in medical imaging. TM compares portions of a source image against a template image, which may be an image of an object or small pattern and is used as a template to detect similar objects/patterns in the source image. In the case where difference between the template image and the source image is small, TM may be useful.

The matching process moves the template image to all possible positions in the larger source image in a pixel-by-pixel manner and computes a numerical index, e.g., correlation, that indicates how well the template matches the image in that position. An illustration is shown in Fig. 22.3. Correlation is a measure of the degree to which two variables agree. In 2D TM, the two variables are the corresponding pixel values between the template image and source image. Discrete cross-correlation $\gamma(x, y)$

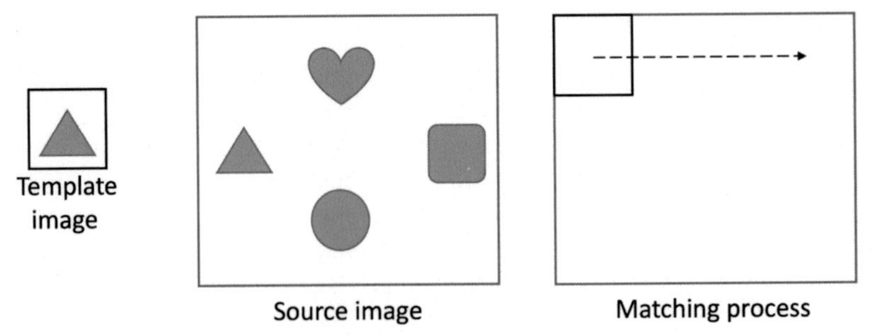

FIGURE 22.3 Illustration of the template matching process.

The template image slides across the source image pixel-by-pixel and computes the similarities between the template image and the overlaid portion of the source image.

between template image f_t and source image f_s at location (x, y) can be defined as follows:

$$\gamma(x,y) = \frac{\sum_m \sum_n (f_s(m+x, n+y) - \bar{f_s}) \cdot (f_t(m,n) - \bar{f_t})}{\sqrt{\sum_m \sum_n (f_s(m+x, n+y) - \bar{f_s})^2 \cdot \sum_m \sum_n (f_t(m,n) - \bar{f_t})^2}}, \quad (22.3)$$

where $\bar{f_t}$ is the average intensity value of the template image and $\bar{f_s}$ is the average intensity value of the source image region overlapped with the template image; m and n are the horizontal and vertical indices of the template, respectively. The value of correlation is between -1 and $+1$, with larger absolute values representing a stronger relationship between the two. In an example shown in Fig. 22.4, one may observe that on the output image (i.e., the correlation map), regions with higher correlation values (brighter region) correspond to the center region of lymphocytes on the source image.

22.2.2.2 Laplacian of Gaussian filter

In the traditional TM method, the template image is a pre-selected image that contains an object sample. The choice of the template image may affect the detection performance. Instead of using a template image to perform the filtering, we could use a parametric kernel to perform the filtering.

An example of such parametric kernels is the LoG filter, which is defined as

$$g(x, y; \sigma) = \frac{\partial^2 h(x, y; \sigma)}{\partial x^2} + \frac{\partial^2 h(x, y; \sigma)}{\partial y^2}, \quad (22.4)$$

where σ is the scale factor to control the size of the LoG filter and $h(x, y; \sigma)$ is a Gaussian kernel with zero mean and scale σ, i.e.,

$$h(x, y; \sigma) = \frac{1}{2\pi\sigma} \exp(-\frac{x^2 + y^2}{2\sigma^2}). \quad (22.5)$$

An LoG filter with $\sigma = 6$ is shown in Fig. 22.5(a). With varying scale factor σ, the multiscale LoG approach accumulates the maximum values over the spatial domain

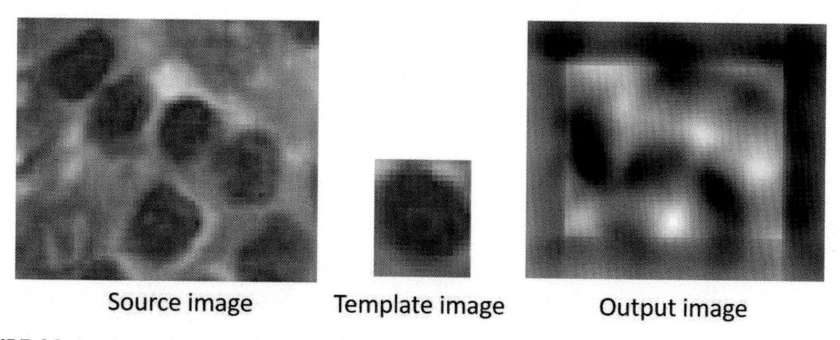

FIGURE 22.4 Illustration of template matching with correlation as measurement.

The source image is an H&E stained lung histology image, whereas the template image is a sample of a lymphocyte. The output image shows the correlation measurement, where each pixel has a value between -1 and $+1$, with larger values (brighter) higher correlations between source image and template image. Note that there exists a boundary effect on the edge of the output image. This boundary effect can be resolved by padding the source image, performing the template matching, and cropping the output image back to the original size of the source image.

to determine the final response surface. After the application of the multiscale LoG to the histopathological image that contains blob-like nuclei, we can obtain higher responses near the center of the nuclei. An example of the multiscale LoG filter for nuclear seed point detection is shown in Fig. 22.5(b and c). Fig. 22.5(b) shows an original image with two nuclei, whereas Fig. 22.5(c) shows the response surface after the application of the LoG filter. Another example is shown in Fig. 22.5(d and e), with an image containing multiple nuclei and the corresponding response surface after applying the LoG filter on the image.

22.3 Learning from data

Traditionally, classical image processing methods and handcrafted feature extraction techniques, like those described in Section 22.2, have been used for object detection and counting in medical images [2,5,6]. These methods are especially advantageous when working with low computational budgets and limited datasets. On the other hand, deep learning techniques automatically learn from the data and therefore a substantially large dataset is imperative to ensure the task at hand is performed well. In the case of object detection, deep learning can be used to locate specific objects and regions in biomedical images (see also Chapters 17 and 18). Traditional machine learning methods for image analysis are limited in their ability to model the 2D structure of images, where they usually treat the input image as a vector of pixel values. On the other hand, convolutional neural networks (CNNs) have shown great promise in the field of computer vision and specifically in medical image analysis. As opposed

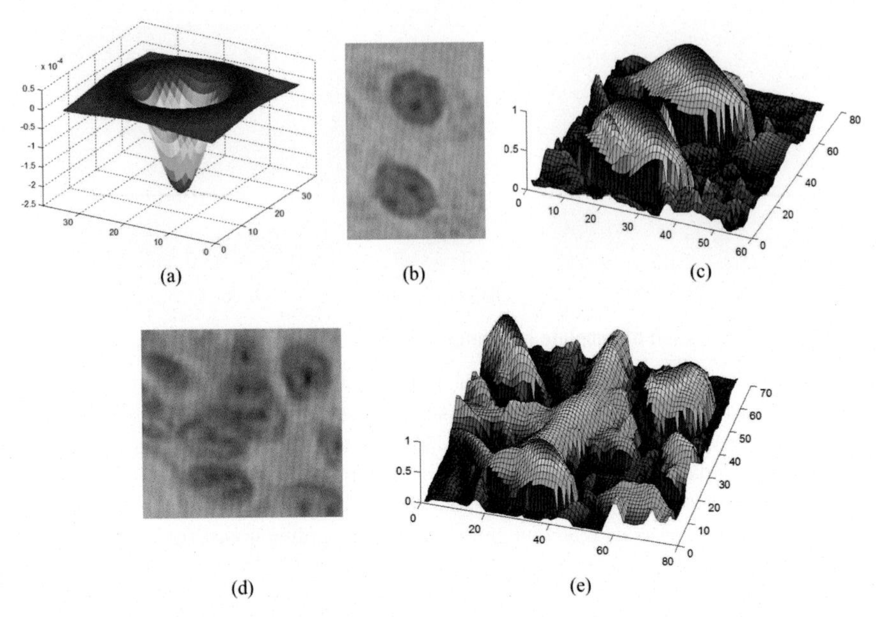

FIGURE 22.5 Example of the LoG filter.

(a) LoG filter with $\sigma = 6$. (b) An image containing two nuclei. (c) The response surface after applying the multiscale LoG filter on the image in (b). (d) An image containing multiple nuclei. (e) The response surface applying the multiscale LoG filter on the image. The reader is referred to [4] for more information on blob detection.

to many traditional machine learning methods, CNNs consider the 2D organization of images and automatically learn a set of 2D filters that are correlated across all spatial locations of the input. This operation is highlighted in Fig. 22.6, where we show the computed output from correlating a 3×3 filter at the top left position of the original image. Reusing 2D filters over the input drastically reduces the number of model parameters compared to its traditional counterparts and leverages the fact that filter responses should be independent of the spatial location of objects in the image. The filter values are automatically learned during optimization of the network and therefore filters do not need to be predefined, as described in Section 22.2.2. Below, we will describe some machine learning and deep learning techniques for object detection and counting.

22.3.1 Classical machine learning methods

22.3.1.1 Bayesian modeling

In thresholding methods described above, each pixel in the image was compared with a hard threshold and pixels were classified as foreground and background. In

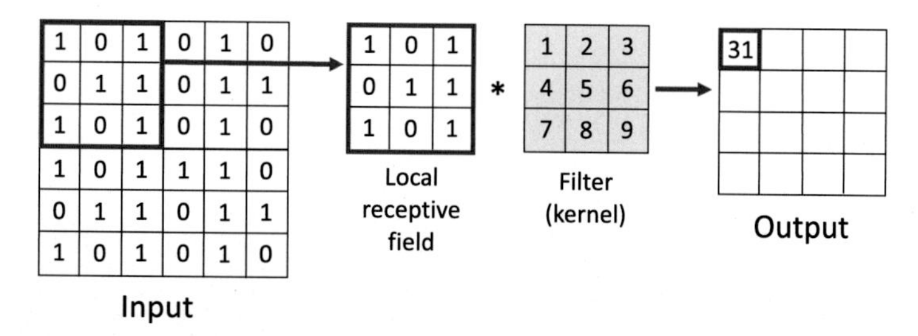

FIGURE 22.6 Illustration of the convolution operation.

Illustration of the convolution operation between an input image and a 3 × 3 filter. Here, we show that the linear operation between the receptive field and the filter gives the output on the right-hand side. This is repeated at all spatial locations of the input.

Bayesian modeling, we estimate the posterior, i.e., the probability a given pixel belongs to the foreground or background, based on a set of training images.

Using the Bayesian theorem, the posterior is calculated as follows:

$$p(\omega_j | f(x, y)) = \frac{p(f(x, y)|\omega_j)P(\omega_j)}{p(f(x, y))}, \tag{22.6}$$

where $\omega_{j, j \in \{1,2\}}$ indicates the class label of the pixel. Class ω_1 corresponds to the foreground and class ω_2 corresponds to the background. The likelihood function $p(f(x, y)|\omega_j)$ indicates the probability of a certain pixel intensity value given a certain class ω_j. The prior $P(\omega_j)$ indicates the proportion of different classes present in the image; $p(f(x, y))$ represents the probability of a specific pixel intensity. The likelihood term $p(f(x, y)|\omega_j)$ can be determined using training data. Fig. 22.7(b) illustrates an example of an estimated probability density function (PDF) in terms of the pixel intensity for foreground and background. These PDFs were estimated from training data, in which the nuclei correspond to foreground (one example is shown in Fig. 22.7(a)).

Given a pixel from the unseen image, the posterior $p(\omega_j | f(x, y))$ that a pixel belongs to class ω_j given intensity value $f(x, y)$ can be calculated using equation (22.6). Fig. 22.7(c) shows an unseen image, and the corresponding posterior map for $p(\omega_1 | f(x, y))$ is shown in Fig. 22.7(d). In Fig. 22.7(d), each pixel has a probability that was calculated using equation (22.6).

22.3.1.2 Learning correlation filters

In comparison to the traditional TM method and LoG filtering method, correlation filters can learn target patterns from annotated training images, which allows for accurate detection of objects with complex structure.

A correlation filter can be conceptually visualized as an image whose correlation output with an input image is as close to the corresponding target image as possible. Given a correlation filter C, the correlation response R for an image F in the Fourier

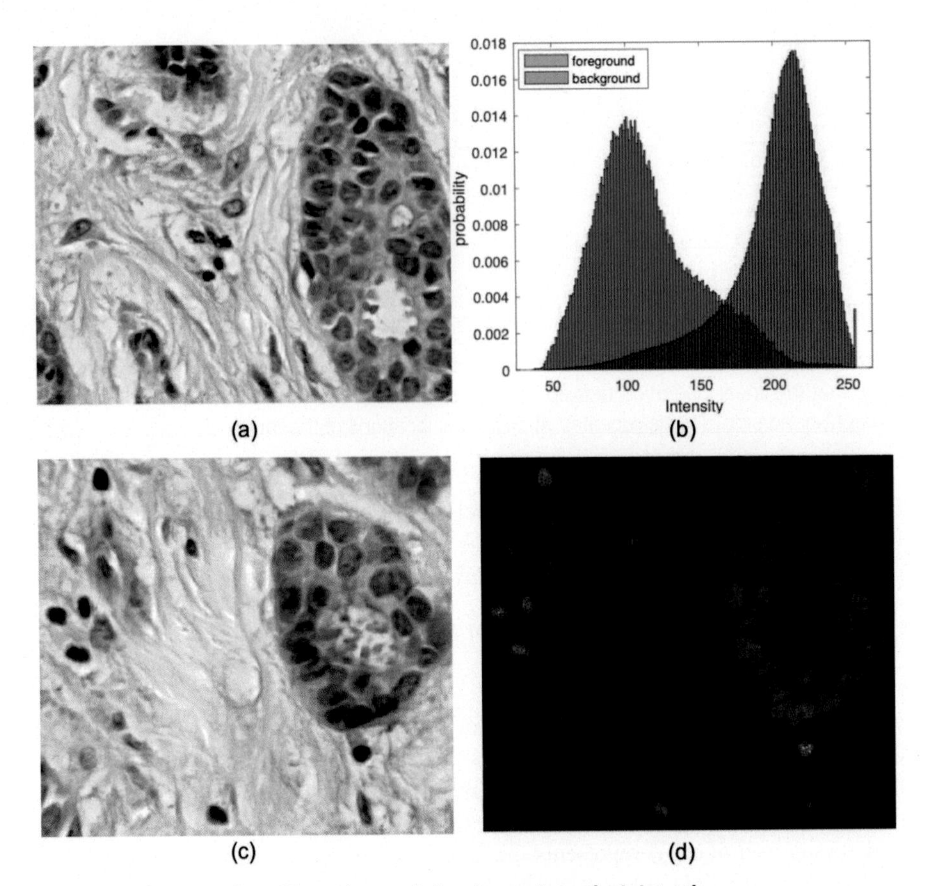

FIGURE 22.7 Illustration of Bayesian modeling for nuclear pixel detection.

(a) An image from training data. (b) Estimated probability density function (PDF) from training data for foreground and background. (c) An unseen image. (d) Posterior map of (c) for foreground.

domain can be written as follows:

$$R = C^* \odot F, \tag{22.7}$$

where the capital letters represent the Fourier domain representation of their corresponding spatial domain image, $\odot$ represents elementwise multiplication, and $*$ indicates the complex conjugate. The filter C can be obtained from training such that the correlation response of the filter for a given training image F is as close to its target image G as possible.

Different correlation filter methods differ, primarily, in how to construct the filter using training data. Minimum Output Sum of Squared Error (MOSSE) is one particular kind of linear filter construction method. Given n training images I_i, $i = 1 \dots n$, and corresponding annotated response images G_i, construction of the correlation filter C can take the following form:

$$C^* = \arg\min_{C^*} \left[\sum_{i=1}^{N} \|I_i \odot C^* - G_i\|^2 + \lambda \|C\|^2 \right], \tag{22.8}$$

where $\lambda > 0$ represents the regularization term's parameter and $\|\cdot\|$ indicates the norm of an argument matrix. The regularization term in equation (22.8) aims to reduce the norm of the filter and it ensures filter stability while the training examples are limited. The regularization parameter λ controls the trade-off between the empirical error and regularization terms. Equation (22.8) can be solved analytically and the closed-form expression for the filter can be written as follows:

$$C^* = \left(\sum_{i=1}^{N} G_i \odot F_i^* \right) \oslash \left(\sum_{i=1}^{N} F_i \odot F_i^* + \lambda \right), \tag{22.9}$$

where $\oslash$ represents the elementwise division operation.

In Fig. 22.8(a and b), a training image for constructing the MOSSE filter and the associated target image obtained by human annotation are shown. Fig. 22.8(c) shows the constructed MOSSE filter, and Fig. 22.8(d) shows the correlation response image obtained by the MOSSE filter.

22.3.2 Deep learning methods

22.3.2.1 Convolutional neural networks

As described in Section 22.3 and covered in detail in Chapters 17 and 18, CNNs perform a linear operation between a set of learned filters and the input at all spatial positions of the image. Here, each local position of the original image considered by the filter is called the receptive field. This operation is repeated throughout the network and is typically accompanied by a non-linear function, such as a rectified linear unit, that enables more complex functions to be learned by the model. CNNs will usually utilize a series of pooling operations to reduce the spatial dimensions of the input. Reducing the spatial dimensions increases the relative field-of-view of the filter, which is important for learning a hierarchical set of features and is most commonly done by returning either the maximum or average value in a rectangular neighborhood. Applying a series of convolution, non-linear, and pooling layers enables a representative set of task-specific features to be extracted. For an in-depth description of CNNs and more broadly deep learning, we recommend reading the recent work by Goodfellow et al. [8].

There exist many deep learning methods in the literature for object detection, which mainly differ in the model architecture and the considered prediction target. Two of the main types of CNN architectures for object detection include region proposal and encoder–decoder methods. In this chapter, we will describe the general

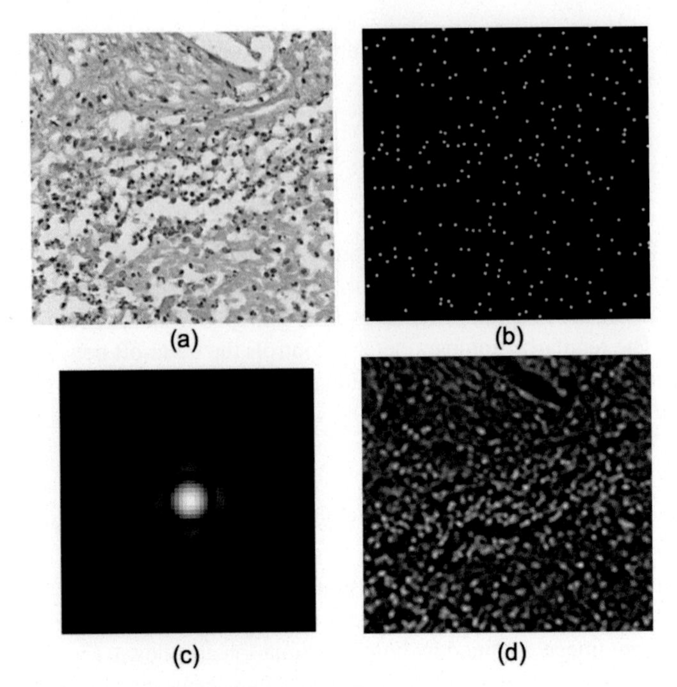

FIGURE 22.8 Example of a MOSSE correlation filter.

(a) A training image. (b) Associated target image/annotated image of (a). (c) Visualization of a MOSSE filter in the spatial domain. Even though the learned filter looks similar to the LoG filter discussed in Section 22.2.2.2, they are different. Note that the MOSSE filter is learned from data, whereas the LoG filter is predefined. (d) Correlation response image obtained by the MOSSE filter for (a). Figure adapted from [7].

framework for an encoder–decoder CNN and highlight different possible prediction targets.

22.3.2.2 Encoder–decoder architecture

Before, we mentioned that CNNs typically consist of multiple convolution, non-linear, and pooling operations which automatically learn features from a set of images. Typically, filters in early layers of a CNN learn simple features, such as edges, whereas deeper layers learn complex and semantically meaningful features. After feature learning, the output can be converted to a vector representation, which can be used for a subsequent classification task. However, this may not be suitable for object detection. Instead, a common strategy is to upsample the learned features to the same dimensions as the original image, which enables a prediction to be made per input pixel. Hypothetically this can be done by applying a single resizing operation, but this will not give good performance. This is because important edge information is typically lost in deeper layers. Instead, a widely used technique involves repeatedly upsampling features by a factor of 2, interspersed with additional convolution operations. Then, each time features are upsampled, features from earlier layers are combined, which increases the model's ability to consider fine details in the input. This technique became popularized with the development of U-Net [9].

During training, the encoder–decoder CNN learns to make a prediction for each input pixel by optimizing a defined loss function $\mathcal{L}$, which takes the model prediction and target label as input. There are various options for the loss function, but often the target label is limited by the ground truth that is available. For example, it may not make sense to utilize the object boundary as a prediction target if we are only provided with a single point per object. In this chapter we will explain some common strategies used for predicting the object centroid and localizing the entire object boundary.

22.3.2.3 CNNs for centroid prediction

Proximity map prediction. It can be difficult for a classifier to locate each *exact* detection point, given that each point is comprised of a single location at the center of each nucleus. Instead, it is possible for a CNN to predict a continuous variable encoded within a 2D target label $\mathbf{y} \in [0, 1]$. Given the original ground truth $\mathbf{g}$, which has the value 1 at the nuclear centers and 0 elsewhere, $\mathbf{y}$ is defined by placing a 2D Gaussian centered at every point in the ground truth $\mathbf{g}$. The pixel j of label $\mathbf{y}$ can be defined as

$$y_j = \begin{cases} \frac{1}{1+(\|\mathbf{z}_j - \mathbf{z}_m^0\|_2^2)/2}, & \text{if } \forall m \neq m', \|\mathbf{z}_j - \mathbf{z}_m^0\|_2 \leq \|\mathbf{z}_j - \mathbf{z}_m'^0\|_2 \leq d, \\ 0, & \text{otherwise}. \end{cases} \tag{22.10}$$

Here, $\mathbf{z}_j$ and $\mathbf{z}_m^0$ are the coordinates of y_j and the coordinates of the mth nucleus, respectively. For every point within $\mathbf{g}$, its corresponding 2D Gaussian is enclosed within a disk with radius d. In other words, the proximity map has high values (close to 1) for pixel values near the center of a nucleus and low values (close to 0) for background areas. With $\mathbf{y}$ in a suitable form, the CNN takes image $\mathbf{x}$ as input and extracts a set of features using a series of convolutional, non-linear, and pooling operations, as explained in Section 22.3. These features are then upsampled to the full resolution of the input image at the output of the network to yield the 2D proximity prediction map.

During training, the weights of the network are typically updated such that either the mean square error (MSE) or the cross-entropy (CE) between the output of the network and the label is minimized. MSE and CE are usually used for regression and classification tasks, respectively. When using MSE, the network learns to minimize the distance between each prediction and the corresponding pixel in the proximity map. If using CE, it is standard to apply a softmax function at the output of the network to convert the prediction to a probability. However, if this is done, then we must ensure that the target is between 0 and 1, in line with the range of the softmax

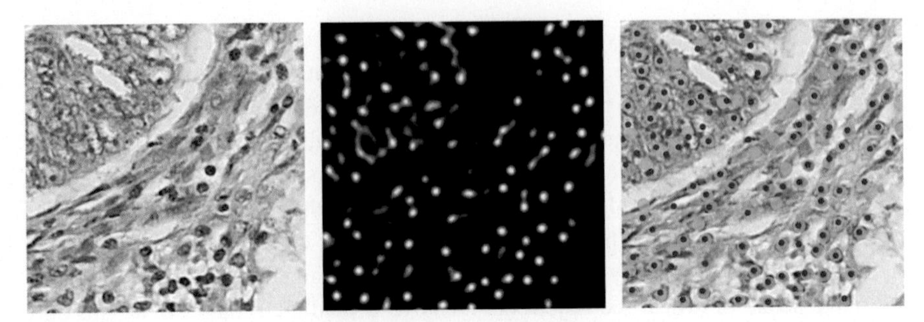

FIGURE 22.9 Illustration of nucleus detection results by proximity map prediction.

Left: Input image. Middle: Proximity map prediction. Right: Post-processed detection results. On the right image, the green disks denote the pathologist annotation and the red points denote the model prediction. Image adapted from [10].

output. Formally, we define the MSE and CE as follows:

$$\mathcal{L}_{MSE} = \frac{1}{N}\sum_{i=1}^{N}(\hat{y}_i - y_i)^2, \tag{22.11}$$

$$\mathcal{L}_{CE} = -\frac{1}{N}\sum_{i=1}^{N} y_i \log \hat{y}_i + (1 - y_i)\log(1 - \hat{y}_i), \tag{22.12}$$

where $\hat{y}_i = f(x_i; \Theta)$ denotes the output of the network for pixel i of image $\mathbf{x}$ with corresponding network parameters Θ, N is the total number of input pixels, and y_i denotes the label of pixel i within the label. Of course, the task of proximity map prediction is not limited to the above two loss functions and others can be explored, such as the mean absolute error (MAE).

Given a processed image $\hat{\mathbf{y}}$, the final detection point is typically extracted by finding the local maxima. A local maximum is defined as a pixel that is surrounded by pixels with a lower value. (See Fig. 22.9.)

Centroid disk prediction. The same encoder–decoder-based approach described above can be used to perform binary classification of the object centroids. For this, rather than representing the target as a 2D proximity map with continuous values between 0 and 1, the target is binary and all pixels within a fixed radius of the nuclear centroids are set to 1. This radius is explicitly set to minimize overlap between disks placed at neighboring nuclei, while also ensuring that there does not exist a significant class imbalance between positive and negative classes. Then, most commonly the network is optimized for binary classification by minimizing the CE loss, as given in equation (22.11), between the prediction and the target for each output pixel. To obtain the centroid predictions, a threshold is applied to the output probability map to indicate which pixels belong to a nuclear disk and the centroid is determined from each binary disk.

22.3.2.4 CNNs for object localization

CNNs for centroid or bounding box prediction subsequently enable counting because the format of the target clearly indicates the location of individual objects. However, sometimes it may be preferable to delineate the entire object boundary to enable downstream morphological analysis. Despite this, performing binary segmentation of the regions of interest will often be infeasible for counting tasks because objects may be clustered together. In Fig. 22.10 we can see that utilizing a binary target will often be inadequate for certain tasks, such as counting nuclei in tumorous regions where cells are usually overlapping. Therefore, if we want to localize the entire object, but still enable counting, we must perform *instance segmentation*, which aims to assign a unique value for each object in the image of a given class.

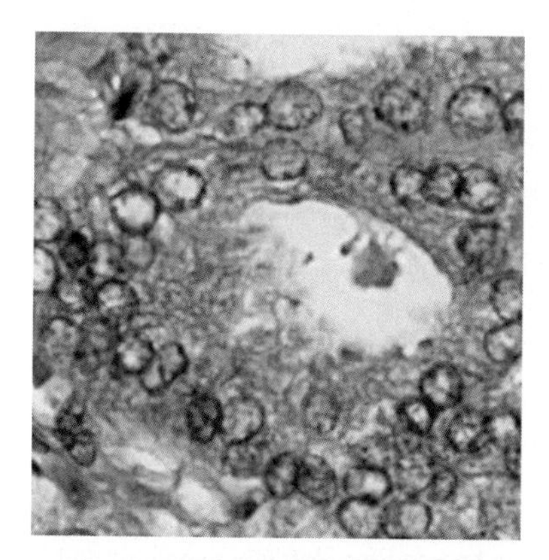

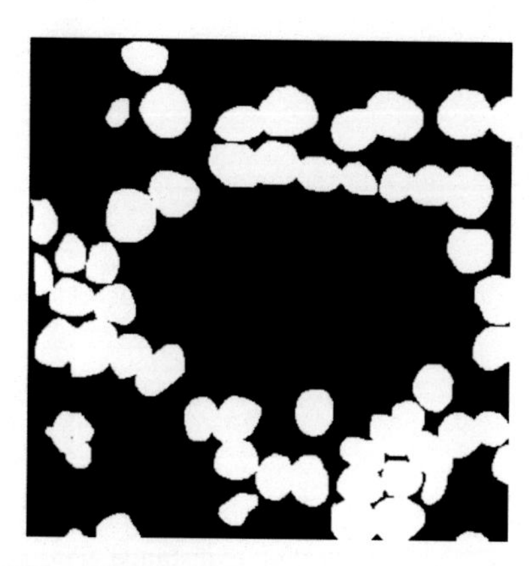

FIGURE 22.10 Illustration of binary target for nuclear segmentation.

It is evident from the figure that using a binary target does not enable differentiation between clustered objects.

To allow instance segmentation, additional information is usually needed to help separate touching objects. For example, the CNN may simultaneously segment both the object and the contour (or boundary), where the additional contour information can be used to separate touching objects and achieve instance segmentation. A recent example of such an approach is the Deep Contour-Aware Network (DCAN) [11] for instance segmentation, which utilizes two upsampling branches after extracting features with a CNN to output the probability of each input pixel belonging to either the object or the contour. During training, the weights of the network are optimized by minimizing the CE loss between the binary targets and the predicted output, as defined in equation (22.12). After, the object and contour probability maps are fused together to generate the final segmentation masks $m(\mathbf{x})$ using

$$m(\mathbf{x}) = \begin{cases} 1, & \text{if } \hat{\mathbf{y}}_o \geq t_o \text{ and } \hat{\mathbf{y}}_c \leq t_c, \\ 0, & \text{otherwise.} \end{cases} \tag{22.13}$$

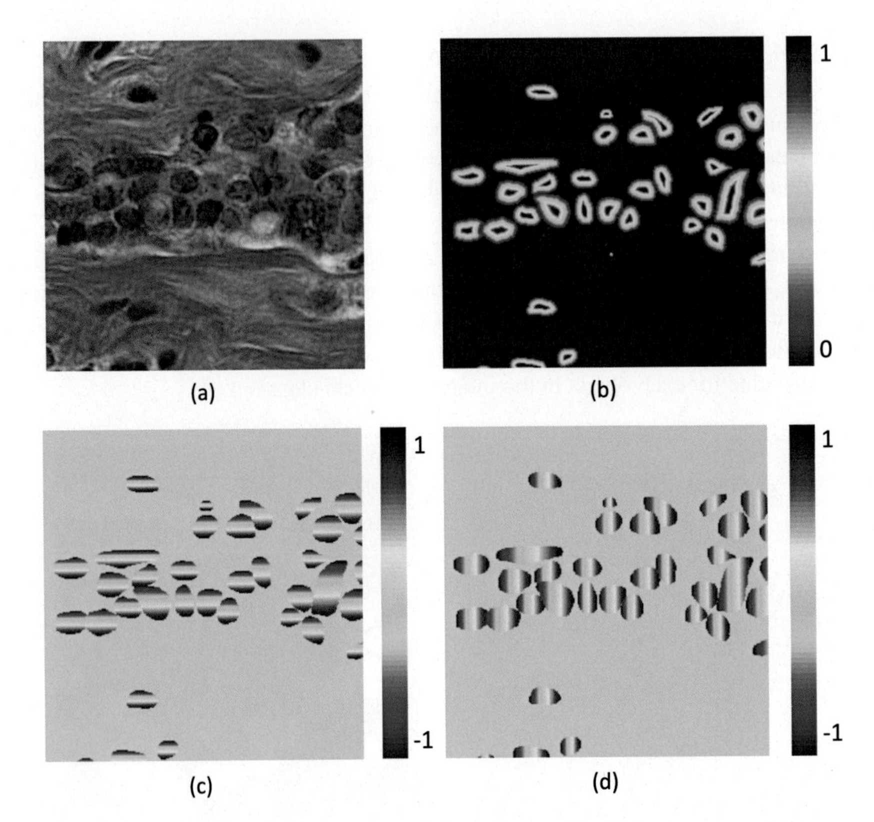

FIGURE 22.11 Illustration of distance maps that can be used for instance segmentation.

(a) Input image. (b) Map of distances of object pixels to their nearest boundary. (c, d) Horizontal and vertical distances of nuclear pixels to their centers of mass.

Here, $\hat{\mathbf{y}}_o$ and $\hat{\mathbf{y}}_c$ are the probability maps of the object and contour, respectively, t_o and t_c are the selected thresholds, and $\mathbf{x}$ is the input image. This method was first applied to the task of gland and nucleus instance segmentation, but can be applied to various other instance segmentation tasks as well, including segmentation of vertebrae in MRI and CT scans.

Instead of using the object contour, distance maps can be used to help separate objects in close proximity and can be accurately predicted using an encoder–decoder approach. In Fig. 22.11 we display a selection of distance maps that can be used during nucleus detection. By inspecting these images, it is clear where individual nuclei are located, as opposed to those in Fig. 22.10. Therefore, if these targets can be learned, then it may be possible to separate touching objects during post-processing. In Fig. 22.11(b) we display distances of nuclear pixels (NPs) to their nearest boundary, whereas in Fig. 22.11(c and d) we show horizontal and vertical distances of NPs to their centers of mass, which has proven to be a particularly successful target for nuclear segmentation [12]. During training, a regression-based loss function is commonly used, such as MSE or MAE. Note that if the distance maps are not normalized between 0 and 1, then a softmax function should not be applied at the output of the network.

Upon prediction of the distance maps, the method for extracting individual nuclei usually differs depending on the type of distance map considered. A widely used approach involves utilizing the predicted distance maps as input to marker-controlled watershed, which considers the input image as a topographic surface and simulates its flooding from specific seed points or markers to determine the object boundaries. Obtaining the marker and energy landscape (described above as a topographic surface) from the predicted distance maps involves a series of post-processing steps. These specific steps may be dependent on the format of the distance map. In Section 22.5 we describe how to use predicted horizontal and vertical distance maps for object detection.

22.4 Detection and counting of mitotic cells using Bayesian modeling and classical image processing

The count of MCs is a critical factor in most cancer grading systems. Extracting each MC from histopathological images is a very challenging task, due to the large variation in the appearance of MCs and the similarity in appearance to certain non-MCs. In this section, we demonstrate how to detect and count the mitotic figures in high-resolution histology images. A flowchart is shown in Fig. 22.12. At first, a set of MC candidate regions is detected and segmented by the Bayesian modeling and local region threshold method. Next, a set of features that quantify texture and shape is extracted from the MC candidates and their surrounding regions. These features are then used in a classifier to differentiate between the real MCs and other regions.

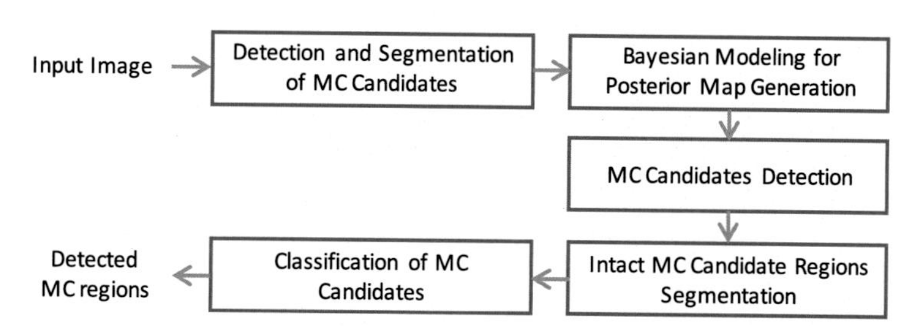

FIGURE 22.12 Flowchart for detection and counting of MCs using Bayesian modeling and classical image processing.

Flowchart of detection and counting of MCs using Bayesian modeling and classical image processing.

22.4.1 Detection and segmentation of MC candidates

The goal of this module is to detect and segment the MC candidate regions. There are three steps that are discussed further in the following sections.

22.4.2 Bayesian modeling for posterior map generation

At first, a training set comprised of the MCs and non-MCs is utilized to build the PDF in terms of the pixel intensity.

For each pixel, the posterior for class ω_1, i.e., the MC region, is calculated as follows:

$$P(\omega_1|f(x,y)) = \frac{p(f(x,y)|\omega_1)P(\omega_1)}{p(f(x,y))}. \tag{22.14}$$

The posterior of each pixel will form a posterior map, denoted as M_p. An example is show in Fig. 22.13. The original image is shown in Fig. 22.13(a). Fig. 22.13(b) shows the PDF of MC (class ω_1) and non-MC regions (class ω_2) in terms of intensity values. Fig. 22.13(c) shows the posterior map.

22.4.3 MC candidate detection

In order to remove noise and achieve a better segmentation result, a hybrid grayscale morphological reconstruction method (HGMR) [13] is applied for Fig. 22.13(c). By comparing Fig. 22.13(c) and Fig. 22.13(d), it is clear that the HGMR is able to let the pixels within nuclear regions have similar intensity levels, which is useful for the subsequent nucleus segmentation, while at the same time suppressing the noise in the posterior map.

To detect the potential MCs, a global threshold $T_p = 0.5$ is then applied to M_h, such that the pixels in M_h whose value is greater than T_p are regarded as the pixels located in the potential MC regions (we denote these pixels as potential pixels [PPs]). This will result in a binary map, M_b, where the white pixels correspond to the PPs and black pixels correspond to other regions. Note that since other cytological components also have a similar intensity value to that of the MC regions, the above operations will result in a large number of PPs corresponding to the MCs. We assume that in the MC regions, all PPs are grouped to form local clusters. Therefore, we calculate the connected components in M_b and remove the components whose size is smaller than a predefined threshold T_r. This threshold is obtained from the training set such that all the true MCs are included in the results and fewer false MC regions are included. So far, we get a binary map M_t, where each white component indicates a potential MC region. An example is shown in Fig. 22.13(e), where the white components represent the candidate MC regions.

22.4.4 Intact MC candidate region segmentation

Due to the non-uniform absorption of the staining dye and the variation in the staining procedure, most of the potential MC regions in M_t are not intact (a close up

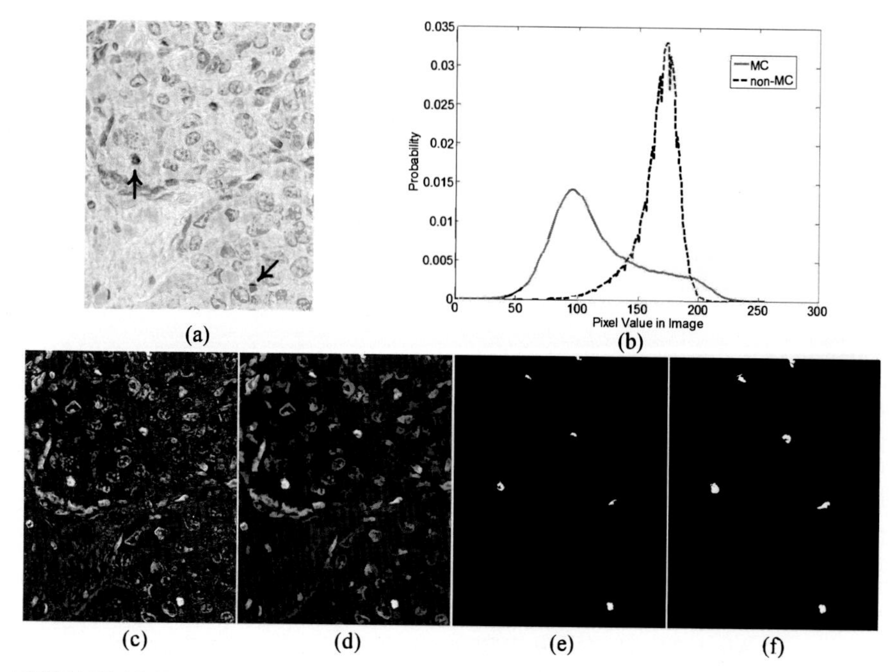

FIGURE 22.13 Example of MC candidate detection and intact region segmentation.

(a) An example image in which two MCs are indicated by arrows. (b) Probability density functions of MC and non-MC regions from the training set. (c) Posterior map M_p. (d) After applying a hybrid grayscale morphological reconstruction method (HGMR) [13] on (a), we got this enhanced map M_h. (e) Binary image M_t which contains potential MC regions. (f) Binary image which contains all intact MC candidate regions. Figure adapted from [14].

example is shown in Fig. 22.14). It is observed that the pre-segmented region is not intact (shown in Fig. 22.14(b)) compared to the ground truth nuclear region shown in Fig. 22.14(c). For each potential MC region, we used an efficient local threshold method to obtain intact nuclear regions which are required in the next module (i.e., the classification module).

The local threshold method is performed as follows:

1) *Best-fitted ellipse calculation*: Since nuclei normally appear as ellipse-shaped objects in the image, we then ellipses to approximate the shape of nuclei. Denote all the candidate regions in the binary map M_t as $R_{j|j=1...Q}$, where Q is the total number of the candidate regions. An ellipse is fitted based on the boundary points of a candidate region R_j using the direct least square fitting algorithm [15].

2) *Local region determination*: In order to obtain a good local threshold for later local region segmentation, we now need to determine the nuclear region by the fitted ellipse. We denote the major axis length of the fitted ellipse as L_j. In order to

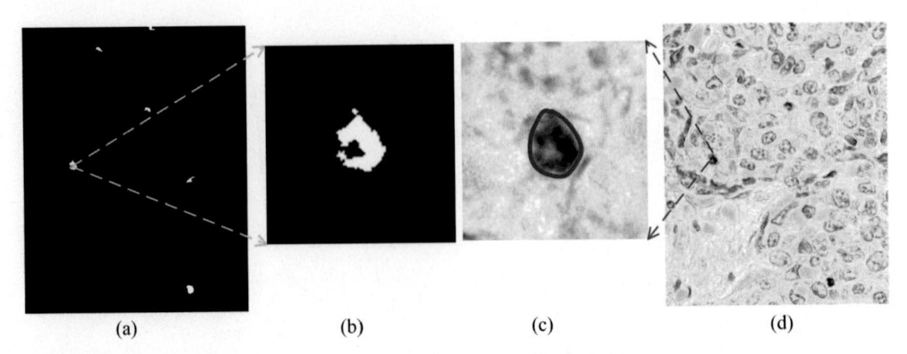

FIGURE 22.14 A close-up example of a non-intact candidate MC region.

(a) A binary map M_t which contains several candidate MC regions. (b) A zoomed in version of an MC region with a non-intact region is shown for illustration. (c) The corresponding image obtained from the original image. The expected intact region is manually drawn as a thick contour. (d) The original image. Figure adapted from [14].

recover the intact nuclei, a local circular area S_j with an enlarged radius $(2 \times L_j)$ is determined.

3) *Local threshold calculation*: Within the local circular area S_j, pixels are retrieved from the original image. A local threshold (denoted as T_j^{local}) for this local circular area is calculated by using the Otsu thresholding method described in Section 22.2.1.1.

4) *Local area segmentation*: The local threshold T_j^{local} is used to segment its corresponding local area S_j into nuclear and background regions. Based on the prior knowledge that the intensity value in the nuclear region is lower than that in the background, the segmentation is calculated as follows:

$$x = \begin{cases} \text{nuclear}, & \text{if } f(x,y) \leq T^{local}, \\ \text{background}, & \text{if } f(x,y) > T^{local}. \end{cases} \tag{22.15}$$

At the end, the main component in the local circular area is retained, whereas other unrelated noisy components are removed by size filtering. The morphological opening operation is then applied to obtain a smooth boundary.

An illustration of the local threshold method is shown in Fig. 22.15. A binary map M_t which contains several candidate MC regions obtained from the last step is shown in Fig. 22.15(a) (the same image shown in Fig. 22.13(e)). An MC region with a non-intact region is selected for the illustration, and is shown in Fig. 22.15(b). The centroid of this region and the circular area are shown as the red point and the blue circle, respectively. This local circular area will map to the original image, and a local threshold is calculated for segmentation. Fig. 22.15(d) shows the final result for the intact region recovery. It shows that the region is more intact compared to the original region in Fig. 22.15(b). The binary map for all the intact regions is shown in

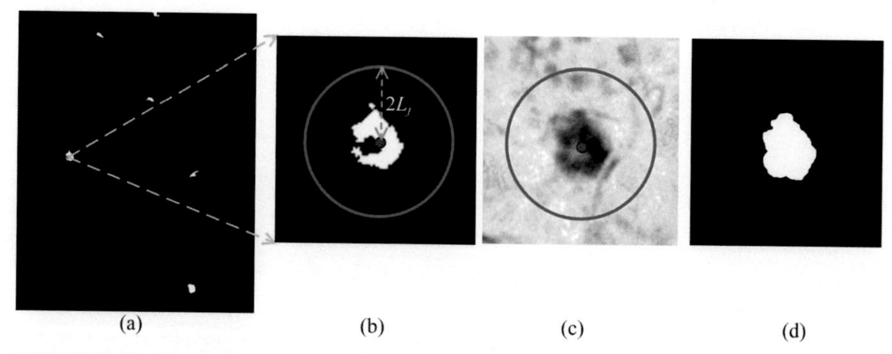

FIGURE 22.15 Illustration of the local threshold method.

(a) Binary map M_t which contains several candidate MC regions. (b) An MC region with non-intact region is selected for the illustration. The centroid of the region is shown as the red point in the middle. The circular area is shown as the blue circle in the image. (c) The local circular area will map to the original image, and a local threshold for this local circular area is calculated. (d) The result after the local thresholding within the local circular area. Figure adapted from [14].

Fig. 22.13(f). The intact regions will help to extract useful features and are required in the following classification module.

Since there exist other cytological components which have a similar appearance to the MCs, the candidate regions that have been obtained so far still will contain many false positives. In the following step, classification is performed to distinguish between MCs and other cytological components.

22.4.5 Classification of MC candidates

Due to the high similarity in the color and morphological appearance between MCs and other cytological components, we have obtained a large number of non-MC regions compared to a small number of MC regions. In this module, all MC candidate regions are classified into MC regions and other regions based on the features extracted from the intact candidate regions. These features include shape-based, intensity-based, and texture-based features.

Fig. 22.16 presents three visual examples for MC detection. Each column in Fig. 22.16 corresponds to one example. The first row shows the original image for all three examples. The manually labeled MCs (i.e., the ground truth) are indicated by solid arrows in the image. As shown in the first row, there are two, one, and five MCs in Example 1, Example 2, and Example 3, respectively. The second row shows the segmentation and detection results obtained by the method that we discussed in Section 22.4.4. For each image, the result is a binary image, where the white regions represent the detected/segmented MC regions. A few dotted-line circles are superimposed onto the binary image to highlight the location of the manually labeled MCs for better comparison. The blue circles indicate the true positive MC locations.

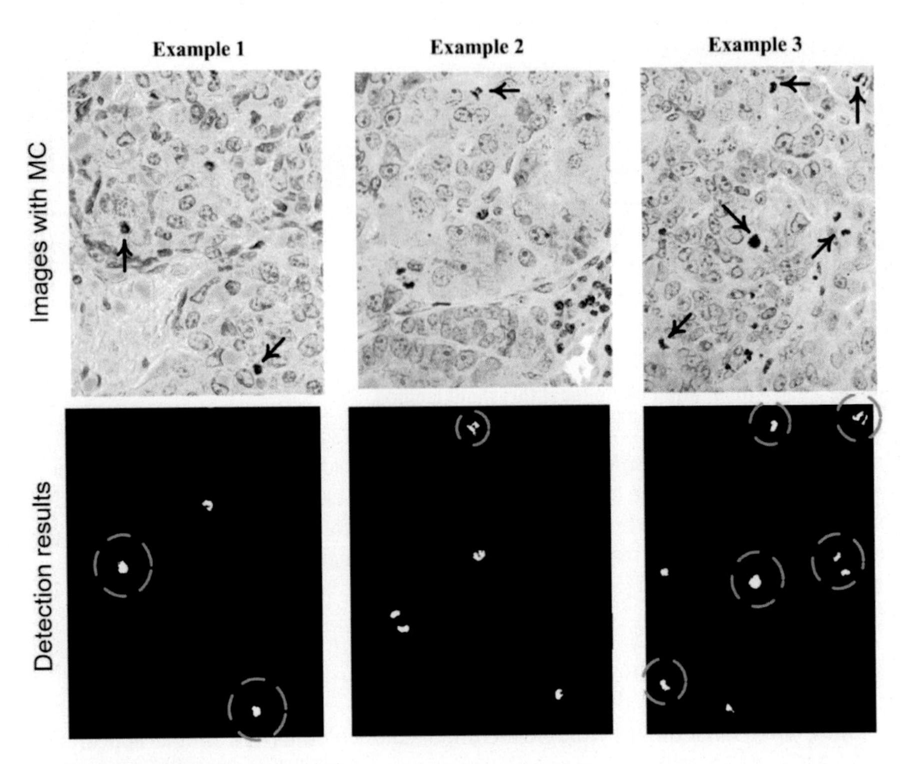

FIGURE 22.16 MC detection result.

The manually labeled (i.e., the ground truth) MCs are indicated by solid arrows in the images on the first row. The detection results are shown in the images on the second row, where the circles indicate the true positive detection results. Figure adapted from [14].

22.5 Detection and counting of nuclei using deep learning

The nuclei within histology images are key indicators of the grade of cancer, and the relationship between locally connected nuclei can provide insight into the tumor microenvironment. Detecting each nucleus within a histology image enables subsequent counting and also enables sophisticated analysis of the spatial relationship between cells. In this section, we demonstrate how to detect and classify nuclei in histology images using HoVer-Net [12], enabling profiling of different cell types in the tissue. We will initially describe how to prepare the data for training the CNN. Then, we will give a brief overview of the method and the optimization strategy for training the network. Finally, we will discuss the post-processing steps and how to infer the object counts from the output of the model.

22.5.1 Patch extraction and data generation

For this example, we use the CoNSeP dataset [12], which provides the segmentation masks of several nucleus types, including epithelial, spindle-shaped, inflammatory, and miscellaneous nuclei.

Given an input image $\mathbf{I}$ of size $n \times n$, we extract non-overlapping patches $\mathbf{x}$ of size $p \times p$ from our chosen histology dataset, where typically $p \leq n$. In our case, we set $p = 270$. Next, we generate the targets that the CNN will learn to map the input image to during training. Target output generation can either be done before (offline) or during training (online) and is left to discretion of the user. For our specific choice of network, for each input image patch, we require (i) a binary nuclei segmentation mask, (ii) horizontal and vertical distance maps, and (iii) binary segmentation masks per class.

22.5.2 Convolutional neural network architecture

The feature extraction component of HoVer-Net, also known as the encoder, is a recent state-of-the-art CNN with 50 layers, known as ResNet50 [16]. This popularized the use of residual units within CNNs that use skip connections to help ease optimization. However, unlike the original ResNet, we downsample features by a factor of 8, rather than 32. A series of consecutive residual units is denoted as a residual block. The number of residual units within each residual block that are applied at downsampling levels 1, 2, 4, and 8 is 3, 4, 6, and 3, respectively. Following feature extraction, three identical upsampling branches based on the DenseNet architecture [17] are used to perform instance segmentation and classification. These branches are named as follows: (i) the NP branch, (ii) the HoVer branch, and (iii) the nuclear classification (NC) branch. The NP branch predicts whether a pixel belongs to the nuclear or background region, whereas the HoVer branch predicts the horizontal and vertical distance maps. Then, the NC branch predicts the nucleus category for each pixel. In particular, the NP and HoVer branches jointly achieve nuclear instance segmentation by first separating NPs from the background (NP branch) and then separating touching nuclei (HoVer branch). The NC branch determines the type of each nucleus by aggregating the pixel-level nuclear type predictions within each instance.

22.5.3 Neural network optimization

During training, the loss between the generated targets and the predictions at the output of each of the branches of the CNN is calculated. The overall loss function is given as

$$\mathcal{L} = \underbrace{\lambda_a \mathcal{L}_a + \lambda_b \mathcal{L}_b}_{\text{HoVer branch}} + \underbrace{\lambda_c \mathcal{L}_c + \lambda_d \mathcal{L}_d}_{\text{NP branch}} + \underbrace{\lambda_e \mathcal{L}_e + \lambda_f \mathcal{L}_f}_{\text{NC branch}}, \tag{22.16}$$

where $\mathcal{L}_a$ and $\mathcal{L}_b$ represent the regression loss with respect to the output of the HoVer branch, $\mathcal{L}_c$ and $\mathcal{L}_d$ represent the loss with respect to the output of the NP branch, and finally, $\mathcal{L}_e$ and $\mathcal{L}_f$ represent the loss with respect to the output at the NC branch;

$\lambda_a ... \lambda_f$ are scalars that give weight to each associated loss function. Specifically, λ_b is set to 2 and all other scalars are set to 1, based on empirical selection.

At the output of the NP and NC branches, the CE and the Dice losses are computed. The CE loss function is given in equation (22.12), but this is specifically for the binary case. In the case of multiclass classification (output of the NC branch), this equation is generalized to sum over all represented classes. The Dice loss for each class is defined as

$$\mathcal{L}_{dice} = 1 - \frac{2 \times \sum_{i=1}^{N} (\hat{y}_i \times y_i) + \epsilon}{\sum_{i=1}^{N} \hat{y}_i + \sum_{i=1}^{N} y_i + \epsilon}, \tag{22.17}$$

where $\hat{y}_i$ and y_i denote the predicted and true output of pixel i for a given class, respectively, and ϵ is a small smoothness constant that is set as 10^{-3}. In the multiclass setting, such as at the output of the NC branch, the dice loss is computed for each individual class and added together.

At the output of the HoVer branch, the first regression loss term is the standard MSE, as described in equation (22.11). The second loss term calculates the MSE between the horizontal and vertical gradients of the horizontal and vertical maps, respectively, and the corresponding gradients of the targets. This is defined as

$$\mathcal{L}_b = \frac{1}{m} \sum_{i \in M} (\nabla_x(\hat{p}_{i,x}) - \nabla_x(p_{i,x}))^2 + \frac{1}{m} \sum_{i \in M} (\nabla_y(\hat{p}_{i,y}) - \nabla_y(p_{i,y}))^2, \tag{22.18}$$

where $\hat{p}_x$, p_x, $\hat{p}_y$, and p_y denote the predicted and true horizontal maps and the predicted and true vertical maps, respectively; ∇_x and ∇_y denote the gradient in the horizontal x- and vertical y-directions; m denotes the total number of NPs within the image; and M denotes the set containing all NPs.

To optimize the parameters of the network, Adaptive Moment Estimation (Adam) optimization [18] is used, which is a variant of the gradient descent algorithm. There are various alternative optimization strategies based on gradient descent, but Adam is chosen in line with the HoVer-Net paper. Gradient descent is the most popular method used to optimize the loss function of a neural network. The loss function $\mathcal{L}(\Theta)$ is parameterized by the network parameters Θ. Therefore, updating the model parameters will have a direct effect on the value of the loss. Specifically, the model parameters update in the opposite direction of the gradient of the loss function $\nabla_\Theta \mathcal{L}(\Theta)$ with respect to the parameters. In particular, the parameters of the network are updated corresponding to the update rule:

$$\Theta_{t+1} = \Theta_t - \eta \nabla_\Theta \mathcal{L}(\Theta), \tag{22.19}$$

where η denotes the learning rate that controls the update step between each iteration. Instead of using a fixed learning rate, Adam uses an *adaptive* learning rate. To achieve this, the decaying averages of past and past squared gradients m_t and v_t are calculated as follows:

$$m_t = \beta_1 m_{t-1} + (1 - \beta_1) g_t, \tag{22.20}$$

$$v_t = \beta_2 v_{t-1} + (1 - \beta_2) g_t^2, \tag{22.21}$$

where m_t and v_t are estimates of the first moment and the second moment of the gradients, respectively. However, because typically m_t and v_t are initialized with a vector of zeroes, the terms are biased towards very small values. Instead, the terms $\hat{m}_t$ and $\hat{v}_t$ are calculated to counteract this bias; $\hat{m}_t$ and $\hat{v}_t$ are defined as

$$\hat{m}_t = \frac{m_t}{1 - \beta_1^t}, \tag{22.22}$$

$$\hat{v}_t = \frac{v_t}{1 - \beta_2^t}, \tag{22.23}$$

where the default values for β_1 and β_2 are 0.9 and 0.999, respectively. Then, the update rule for Adam is defined as

$$\Theta_{t+1} = \Theta_t - \frac{\eta}{\sqrt{\hat{v}_t} + \epsilon} \hat{m}_t, \tag{22.24}$$

where 10^{-8} is set as the default value for ϵ. We set the initial learning rate η within our framework as 10^{-4}. During training, we input a batch of images to the network before performing an update of the parameters. Passing a batch to the network, rather than a single image, reduces the variance of the parameter updates and therefore helps with model convergence. Specifically, we select a batch size of 8.

22.5.4 Post-processing and cell counting

During test time, each patch $\mathbf{x}_i$ within an image $\mathbf{I}$ is processed with the optimal parameters Θ of the network. Here, $1 \leq i \leq n$ and n is the total number of patches within $\mathbf{I}$. All patches are then merged, where an example of the overall output of each branch of the network can be observed in Fig. 22.17. Following this, a series of post-processing steps are applied to the output of the NP and HoVer branches to yield the instance segmentation result, as shown in Fig. 22.17(a). For this, first the gradients of the HoVer output in the horizontal and vertical directions are computed using the Sobel operator [19]. Then, computing the maximum over both of these gradient maps results in a map with high values between neighboring nuclei. Specifically, the map S_m is defined as

$$S_m = max(H_x(\hat{p}_x), H_y(\hat{p}_y)), \tag{22.25}$$

where $\hat{p}_x$ and $\hat{p}_y$ refer to the horizontal and vertical predictions at the output of the HoVer branch and H_x and H_y refer to the horizontal and vertical components of the Sobel operator, respectively. Then, S_m is combined with the NP prediction to give the markers and the energy landscape, which are then used to obtain the nuclear instances

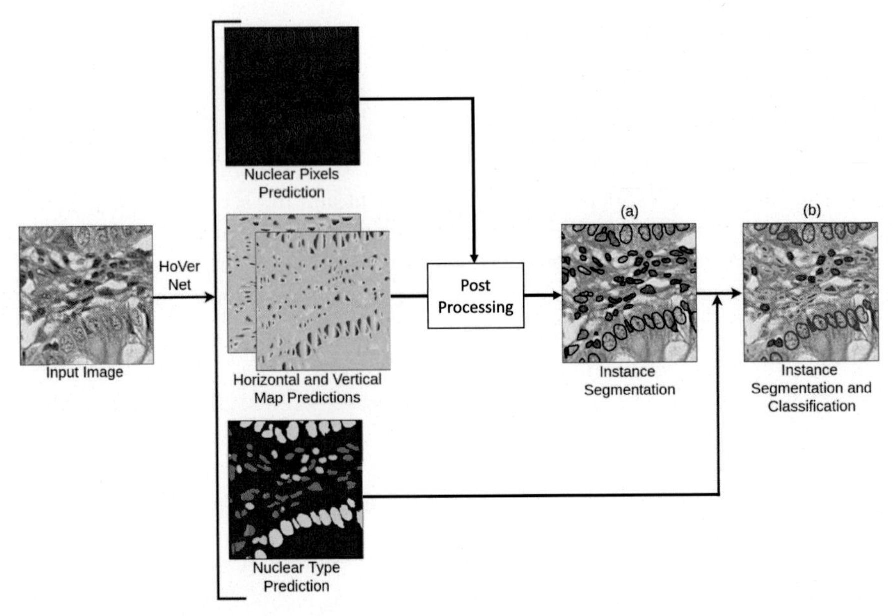

FIGURE 22.17 Illustration of overall HoVer-Net workflow.

The outputs of the three branches in the CNN are combined to give the overall instance segmentation and classification result. Figure adapted from [12].

using marker-controlled watershed. As a final step, the NC branch output is used to label each nuclear instance using a majority voting strategy. The final segmentation and classification results can be seen in Fig. 22.17(b), where the color of the boundary denotes the type of nucleus predicted. Here, blue refers to epithelial, green refers to spindle-shaped, and red refers to inflammatory nuclei. Following post-processing, the number of each type of nucleus can easily be inferred from the result.

22.6 Recapitulation

In this chapter we described a range of methods for object detection and counting, specifically concentrating on histology image analysis. We initially gave examples of classical image analysis techniques, using thresholding- and filter-based methods. Then, we described how objects can be detected by learning from the data using machine learning. In particular, we described Bayesian modeling and correlation filters and then reported various deep learning approaches for object detection that mainly differed in their prediction target. To conclude the chapter, we gave two extended examples of object detection and counting. We first, showed a classical approach for MC detection and then finally included an example of nucleus detection for tumor microenvironment profiling using HoVer-Net.

22.7 Exercises

1. Two-dimensional convolution.
   - **(a)** Download gray-scale image "E1.jpg" from the MIA book website. Read the image in MATLAB® and store the gray values in a 2-D matrix x. Determine the size of the image. Display the image x in subplot (2, 2, 1). Note down the 5 × 5 pixel values of the image in all four corners.
   - **(b)** Consider the 2-D impulse response

$$h(m,n)=\begin{bmatrix}-1 & -1 & -1\\ 0 & 0 & 0\\ 1 & 1 & 1\end{bmatrix},$$

     Calculate the 2-D convolution of x and y. Note that the filter h has a size of 3 × 3. Therefore, in order to do the convolution, first pad one layer of zeros around the image. Note down the 5 × 5 pixel values of the zero-padded image in all four corners and ensure that the zero-padding is correct.
   - **(c)** Use MATLAB (do not use the MATLAB built-in *conv* function), calculate the 2D convolution using the following equation

$$y(m,n)=x*h=\sum_{k=0}^{M-1}\sum_{l=0}^{N-1}x(k,l)h(m-k,n-l).$$

     Display the image $y(m,n)$ in subplot (2, 2, 2). In order to see $y(m,n)$, you may use $imshow(y,[])$ to rescale the image display range.
   - **(d)** Use the built-in MATLAB function *conv* to perform the above convolution. (ii) Display the convolution output in subplot (2, 2, 3).
   - **(e)** Explain the features observed in the output image. What did the filter h do?
   - **(f)** Compare the convolution outputs obtained in (d), and (e). Note down the first 5×5 pixel values of the y matrix obtained in (d), and (e). Compare their values. Comment on the output values corresponding to the first row and column.
2. Thresholding and quantifying.
   - **(a)** Download image "E2.jpg" from the MIA book website. Read the image in MATLAB and store it in a matrix I. Determine the size and the channel of the image. Display the image I. Display the first channel of image I, denote it as I_r.
   - **(b)** Apply thresholding on I_r, display the thresholded result as a binary image, in which the nuclear regions should shown as white pixels, whereas the background as black pixels.
   - **(c)** Apply morphological operation, *imopen()*, to the binary image, to let the boundary of the foreground white regions become smoother.
   - **(d)** Count all foreground regions in the binary image using MATLAB function *bwconcomp()*; Quantify area, centroid and perimeter for all foreground re-

gions using function *regionprops()*. Display these quantified values on top of the image.

References

[1] Kittler J., Illingworth J., Minimum error thresholding, Pattern Recognition (ISSN 0031-3203) 19 (1) (Jan. 1986) 41–47, https://doi.org/10.1016/0031-3203(86)90030-0, https://www.sciencedirect.com/science/article/pii/0031320386900300. (Accessed 6 October 2021).

[2] Otsu N., A threshold selection method from gray-level histograms, IEEE Transactions on Systems, Man and Cybernetics (ISSN 0018-9472) 9 (1) (Jan. 1979) 62–66, https://doi.org/10.1109/TSMC.1979.4310076.

[3] Aja-Fernández S., Curiale A.H., Vegas-Sánchez-Ferrero G., A local fuzzy thresholding methodology for multiregion image segmentation, Knowledge-Based Systems (ISSN 0950-7051) 83 (July 2015) 1–12, https://doi.org/10.1016/j.knosys.2015.02.029, https://linkinghub.elsevier.com/retrieve/pii/S095070511500129X. (Accessed 1 September 2021).

[4] Lindeberg T., Feature detection with automatic scale selection, International Journal of Computer Vision 30 (2) (1998) 77–116.

[5] Raza S.-e-A., Qaisar Marjan M., Arif M., Butt F., Sultan F., Rajpoot N.M., Anisotropic tubular filtering for automatic detection of acid-fast bacilli in Ziehl-Neelsen stained sputum smear samples, in: Gurcan M.N., Madabhushi A. (Eds.), Medical Imaging 2015, Orlando, Florida, United States, Mar. 2015, p. 942005, https://doi.org/10.1117/12.2081835, http://proceedings.spiedigitallibrary.org/proceeding.aspx?doi=10.1117/12.2081835. (Accessed 24 September 2020).

[6] Yao Z., Directional edge and texture representations for image processing, PhD Thesis, University of Warwick, 2007, http://webcat.warwick.ac.uk/record=b2242778~S9.

[7] Ahmad A., Asif A., Rajpoot N., Arif M., Minhas F.u.A.A., Correlation filters for detection of cellular nuclei in histopathology images, Journal of Medical Systems 42 (1) (Jan. 2018) 7, https://doi.org/10.1007/s10916-017-0863-8, http://link.springer.com/10.1007/s10916-017-0863-8. (Accessed 23 September 2020), ISSN: 0148-5598, 1573-689X.

[8] Goodfellow I., Bengio Y., Courville A., Deep Learning, MIT Press, 2016, http://www.deeplearningbook.org.

[9] Ronneberger O., Fischer P., Brox T., U-Net: convolutional networks for biomedical image segmentation, arXiv:1505.04597 [cs.CV], 2015.

[10] Sirinukunwattana K., Raza S.E.A., Tsang Y.-W., Snead D.R.J., Cree I.A., Rajpoot N.M., Locality sensitive deep learning for detection and classification of nuclei in routine colon cancer histology images, IEEE Transactions on Medical Imaging 35 (5) (2016) 1196–1206.

[11] Chen H., Qi X., Yu L., Dou Q., Qin J., Heng P.-A., DCAN: Deep contour-aware networks for object instance segmentation from histology images, Medical Image Analysis 36 (2017) 135–146.

[12] Graham S., Vu Q.D., Raza S.E.A., Azam A., Tsang Y.W., Kwak J.T., Rajpoot N., Hover-Net: Simultaneous segmentation and classification of nuclei in multi-tissue histology images, Medical Image Analysis 58 (2019) 101563.

[13] Lu C., Mahmood M., Jha N., Mandal M., A robust automatic nuclei segmentation technique for quantitative histopathological image analysis, Analytical and Quantitative Cytology and Histopathology 12 (2012) 296–308.
[14] Lu C., Mandal M., Towards automatic mitosis detection and segmenation using multispectral H&E staining histopathological images, IEEE Journal of Biomedical and Health Informatics 18 (2) (2014) 594–605.
[15] Halır R., Flusser J., Numerically stable direct least squares fitting of ellipses, in: Proc. 6th International Conference in Central Europe on Computer Graphics and Visualization, WSCG'98, Citeseer, 1998, pp. 125–132.
[16] He K., Zhang X., Ren S., Sun J., Deep residual learning for image recognition, arXiv preprint, arXiv:1512.03385, 2015.
[17] Huang G., Liu Z., Pleiss G., van der Maaten L., Weinberger K.Q., Convolutional networks with dense connectivity, IEEE Transactions on Pattern Analysis and Machine Intelligence (12) (2019) 8704–8716.
[18] Kingma D.P., Ba J., Adam: A method for stochastic optimization, arXiv preprint, arXiv: 1412.6980, 2014.
[19] Kanopoulos N., Vasanthavada N., Baker R.L., Design of an image edge detection filter using the Sobel operator, IEEE Journal of Solid-State Circuits 23 (2) (1988) 358–367.

CHAPTER 23

Image retrieval in big image data

Sailesh Conjeti[a], Stefanie Demirci[b], and Vincent Christlein[c]

[a]*Siemens Healthineers, Erlangen, Germany*

[b]*Computer Aided Medical Procedures, Technical University of Munich, Garching, Germany*

[c]*Pattern Recognition Lab, Friedrich-Alexander-Universität Erlangen-Nürnberg, Erlangen, Germany*

Learning points

- Image retrieval
- Global image descriptors for image retrieval
- Deep learning-based image retrieval
- Efficient indexing strategies

23.1 Introduction

The evolution of medical imaging technologies as discussed in Chapter 1 and its widespread use over decades has resulted in image data with hundreds of thousands samples being stored within image databases, as well as image archiving and communication systems. Containing images from a diverse range of modalities and dimensions, these repositories are of immeasurable value for evidence-based diagnosis, teaching, and research. In order to be able to take full advantage of these benefits, there is a need for appropriate search methods that take an image as input and deliver all those images within one or a collection of databases that have characteristics similar to the case of interest.

Content-based image retrieval (CBIR) describes an image search technique that bases the indexing of images entirely on visual features, such as color, texture, and shape, as search criteria [1]. This is in contrast to the conventional text-based indexing of images that may employ keywords, subject headings, captions, or natural language text. Due to the increasing and ubiquitous availability of visual data on the internet, CBIR systems have attracted much attention, especially in the fields of computer vision and image processing. Its potential has also been revealed for biomedical purposes [2].

In general, any CBIR system follows the common workflow [3] depicted in Fig. 23.1: Given a reference database, feature extraction methods are employed to represent each image. In case of supervised retrieval, features need to encode the

Medical Image Analysis. https://doi.org/10.1016/B978-0-12-813657-7.00037-6

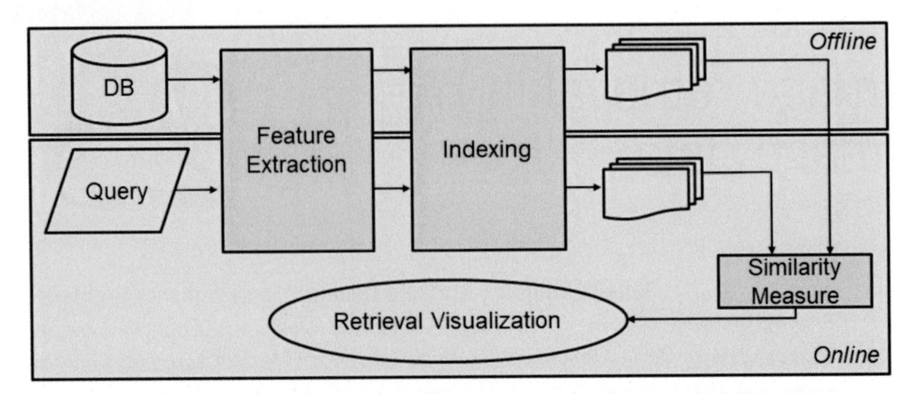

FIGURE 23.1 Schematic illustrating the overall process of content-based image retrieval.

This image visualizes the overall process of content based image retrieval.

defined semantics associated to the special purpose of the database. Unlike traditional image retrieval methods that directly compare the image similarity via original feature vectors, large-scale approaches often first reorganize the feature space into semantically similar neighborhoods using different indexing strategies. In the query phase, the query image is compared only to similar images within respective neighborhoods rather than an exhaustive search of the whole database. Eventually, the retrieval results can be visualized to users for further analysis.

Let us consider a repository of B images, $\mathcal{I}^b$ (with $b \in \{1, \ldots, B\}$), where each image belongs to a particular class. Let us denote as $\mathcal{L}(\mathcal{I}^b) \in \{+1, -1\}$ the labels used to denote the class of $\mathcal{I}^b$. The main objective within CBIR is: given a query image $\mathcal{I}^q$ and a repository image $\mathcal{I}^l$, find an appropriate distance function $\mathcal{D}$ using metric ϕ such that $\mathcal{D}_\phi(\mathcal{I}^q, \mathcal{I}^b)$ yields a small value for those $\mathcal{I}^b$ with $b \in \{1, \ldots, B\}$ for which $\mathcal{L}(\mathcal{I}^q) = \mathcal{L}(\mathcal{I}^b)$ and correspondingly a large value for $\mathcal{L}(\mathcal{I}^q) \neq \mathcal{L}(\mathcal{I}^b)$.

The major challenges for CBIR include the application-specific definition of similarity, extraction of image features that are relevant to this definition of similarity, and organizing these features into indices for fast retrieval from large repositories. Feature extraction techniques have been discussed in detail in Section 15.3.1, and basically any combination of features will serve the purpose if designed in accordance with the intended application.

However, the similarity computation and the indexing can be applied on two different levels of features: (1) on the local feature descriptor level and (2) on the global image descriptor level. In the former, local feature descriptors are computed at specific landmarks (a. k. a. keypoints) in the image, e.g. at corner points. They are then stored in databases or index structures to compute similarities between images using sets of local descriptors. In order to achieve real-time retrieval, efficient indexing schemes are needed that store and partition the database such that the data can be accessed and explored quickly by filtering out irrelevant data. While efficient indexing strategies allow to store millions of local descriptors, this becomes more and more challenging with the increase of dataset sizes. The use of global image descriptors can alleviate this problem. The indexing strategies can then be employed on a global image descriptor level instead of a local descriptor level. The computation of global image descriptors is in focus of the first part of this chapter, while the second part focuses on the efficient organization of indices, which can refer to both local and global descriptors.

23.2 Global image descriptors for image retrieval

In this section, techniques to compute robust image representations are described that are computed using local image feature descriptors. Afterwards, several normalization techniques to improve their robustness are outlined. Finally, this formation process is related to convolutional neural networks (CNNs).

23.2.1 Encoding

The process (and its product) of computing a single representation per image from many local feature descriptors is called encoding or embedding. In computer vision, this is often referred to as bag-of-(visual)-words (BoW), where a visual word means a local feature descriptor. Originally, BoW was introduced in document retrieval and text categorization, where a document is represented by the set of its components, i.e., words. Each word occurrence of a predefined vocabulary is counted and the histogram of the counts serves as the document representation.[1] BoW in its original form can be seen as a special case of vector quantization (VQ) [6] for discrete data.

The goal of the encoding process is the formation of a *global* feature representation given *local* feature descriptors. Generally, the encoding consists of two steps: (1) embedding of local descriptors and (2) aggregation (a.k.a. pooling) in which the embedded local descriptors are combined into a global descriptor of fixed length. Aggregation is usually performed using sum aggregation, i.e., the local embeddings are summed up; alternatively, a weighted sum aggregation can be used [7]. The embedding of the local descriptors can be realized via different techniques. Probably the most popular group of methods computes embeddings using statistics of image descriptors (of the image under study) to a background model. Popular background models are formed by k-means or Gaussian mixture models (GMMs), which are computed based on a representative sample set of local descriptors of the training set.

One of the simplest encoding techniques is VQ [4]. For each cluster center, obtained through a background model, the closest local descriptors are counted per cluster center. Formally, given a background model with K cluster centers $\{\boldsymbol{\mu}_k\}_{k=1}^K$ and associations α_{tk} for each local feature vector $\{\mathbf{x}_t \in \mathbb{R}^D\}_{t=1}^T$ of an image computed

[1] The document retrieval approach was first applied to the image domain by Sivic and Zisserman [4], while the term "bag-of-words" was coined by Fei-Fei and Perona [5].

as

$$\alpha_{tk} := \alpha_k(\mathbf{x}_t) = \begin{cases} 1 & \text{if } k = \arg\min_{k=1,\dots,K} \|\mathbf{x}_t - \boldsymbol{\mu}_k\|_2^2, \\ 0 & \text{else,} \end{cases} \tag{23.1}$$

the VQ follows as

$$\psi_k = \sum_{t=1}^{T} \alpha_{tk}, \tag{23.2}$$

and the final embedding $\boldsymbol{\psi}_{\text{vq}} \in \mathbb{R}^K$ follows as

$$\boldsymbol{\psi}_{\text{vq}} = (\psi_1, \dots, \psi_K)^\top. \tag{23.3}$$

Another popular encoding technique is Vectors of Locally Aggregated Descriptors (VLAD) [8], in which the residuals between cluster centers and the sample's closest local descriptors serve as an embedding:

$$\boldsymbol{v}_k = \sum_{t=1}^{T} a_{tk}(\mathbf{x}_t - \boldsymbol{\mu}_k). \tag{23.4}$$

The full encoding $\boldsymbol{\psi}_{\text{VLAD}} \in \mathbb{R}^{KD}$ follows as

$$\boldsymbol{\psi}_{\text{VLAD}} = (\boldsymbol{v}_1^\top, \dots, \boldsymbol{v}_K^\top)^\top. \tag{23.5}$$

23.2.2 Normalization

In general, the encodings need to be normalized, e.g., by means of an ℓ_2-normalization, i.e., dividing by the vector's ℓ_2-norm, in order to compare two different encodings because the number of local descriptors per image may differ.

In this section, several other factors that can affect the similarity of two vectors and methods to mitigate them are outlined. Typically, those methods can be applied to local descriptors as well as to the global descriptors.

- *Visual burstiness*: The phenomenon that one visual element appears disproportionately more frequently than others is referred to as visual burstiness, e.g., windows in a house facade. Since this contradicts the assumption that local descriptors are independent and identically distributed, it can negatively influence the similarity measure of two feature descriptors. A popular method to counter visual bursts is power normalization [9], where each element x_i of the vector $\mathbf{x}$ under consideration is transformed as follows:

$$x_i = \text{sign}(x_i)|x_i|^p \quad \forall x_i \in \mathbf{x}, \tag{23.6}$$

where $p \in \mathbb{R}^+$ is typically chosen as 0.5.

- *Correlation*: While visual burstiness describes single frequently occurring descriptors, co-occurrence refers to two or more local descriptors that frequently occur together. A standard method to decorrelate features is principal component analysis (PCA), where the dimensionality can additionally be reduced. The problem of co-occurrences can be reduced with *whitening* where the eigenvectors are additionally scaled with their variance during PCA.
- *Inter-session variability*: This phenomenon describes the variance between different records. In contrast to the unsupervised techniques usually applied to address the previous two points, vectors can also be decorrelated in a supervised manner. In this way, the variability between recordings can be significantly reduced. A possible technique would for example be linear discriminant analysis.

23.3 Deep learning-based image retrieval

A CNN can be seen as a representation engine that contains both retrieval steps: (1) local feature extraction through the CNN backbone up to the layer before the penultimate layer and (2) feature aggregation through the global pooling layer (typically the penultimate layer), which results in a global embedding.

The simplest deep learning-based option for image retrieval is the use of a pre-trained network. For example, a network trained on the 1000 classes of the popular ImageNet dataset using cross-entropy can be used to produce local embeddings. For this purpose, the last classification layer, which would be used to compute the association to the 1000 ImageNet classes, is removed. Given an input image, the activations of the penultimate layer would then represent the global embedding. Nearly all modern architectures use a global pooling layer, e.g., average pooling, as penultimate layer, which allows images of arbitrary size to be processed.

Inspired by traditional encoding methods, many of the steps outlined in Section 23.2.1 can also be viewed as layers of a neural network, e.g., NetVLAD for VLAD encoding [10] or deep generalized max-pooling (DGMP) that realizes a weighted aggregation as special pooling layer [11].

Instead of using a network that has been pre-trained by means of cross-entropy on possibly irrelevant classes (in the sense of image similarity), the network can be adapted to classes of the training set by means of fine-tuning or trained from scratch. In the case that these classes do not appear in the test set or there are too few samples per class, the multiclass problem can also be converted to a binary class problem (same identity vs. different identity), or alternatively other surrogate classes can be chosen. The latter is nowadays often denoted as self-supervised learning.

Furthermore, global descriptors which follow a certain metric can be learned explicitly. This process is called *metric learning* and can be enforced by training a CNN with a specific loss function.

A simple yet powerful loss function is the *contrastive* loss. Given a pair of images and their global embeddings $\boldsymbol{\psi}_1$ and $\boldsymbol{\psi}_2$, the goal is to enforce a low distance D (typically Euclidean distance or cosine distance) between them if they have the same class label and otherwise a high distance, i.e., the following loss function is going to be minimized for the whole (or a reasonable subset of) the training set:

$$yD^2(\psi_1, \psi_2) + (1 - y) \max(m - D(\psi_1, \psi_2), 0)^2. \tag{23.7}$$

If both samples have the same class labels, then y becomes 1 and the first part is evaluated. In case of different samples $y = 0$ and the distance needs to become larger than a margin parameter m, which should be cross-validated for the respective dataset.

Another popular loss function is the so-called *triplet* loss [12]. The goal is to minimize the distance between two (global) image descriptors of the same class while maximizing descriptors of different classes. This can be achieved by minimizing the following loss of all N triplets of the training set:

$$\max\left(0, D(\psi_a, \psi_p) - D(\psi_a, \psi_n) + m\right). \tag{23.8}$$

Here $D(\psi_a, \psi_p)$ is the distance of a global embedding of an image ψ_a (anchor) and a positive one ψ_p, i.e., both samples are of the same class, and $D(\psi_a, \psi_n)$ denotes the distance of the anchor embedding ψ_a to an embedding ψ_n where the class is different from that of the anchor. The loss of a triplet is only considered if the difference $D(\psi_a, \psi_n) - D(\psi_a, \psi_p)$ is smaller than a certain margin parameter m.

23.4 Efficient indexing strategies

The following description is based on the reference image repository containing B image entries $\mathcal{I}^b$ (with $b \in \{1, \ldots, B\}$). As indicated by the general CBIR workflow (refer to Fig. 23.1), each database entry I^b is represented by a feature vector $\mathbf{x}^b$ describing m landmarks within image $\mathcal{I}^b$. The nature of these landmarks is mostly application-specific and they are commonly defined by domain experts. In order to yield robust and accurate results, the landmarks are supposed to be scale- and rotation-invariant [13].

Note that the following techniques are described on a local feature basis but can similarly well be applied on global image descriptors.

23.4.1 Vocabulary tree

The vocabulary tree method employs a hierarchical tree and inverted files that can significantly improve the retrieval efficiency. The framework can be divided into two phases. The offline training phase builds the indexing model (hierarchical tree structure) from given image sets and the online query phase returns images that are similar to the query image [13].

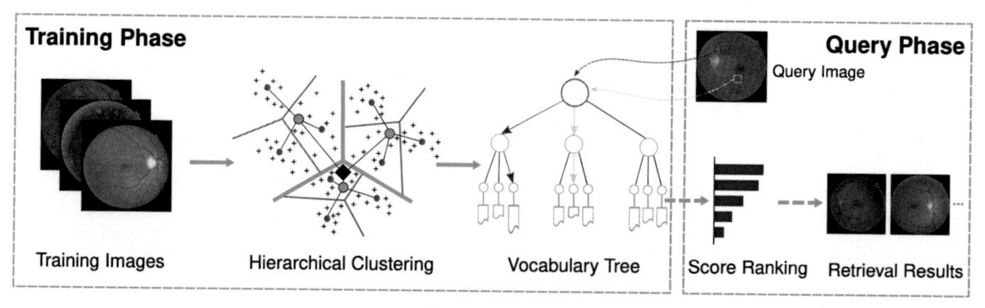

FIGURE 23.2 Vocabulary tree-based image retrieval.

This image shows a pipeline for vocabulary tree-based image retrieval.

-Adapted from [13]

Training phase

In order to yield a hierarchical structure on all our image descriptors

$$\left\{\mathbf{x}^b\right\}_{i=1}^{m}, \quad b \in \{1, \dots, B\},$$

an iterative k-means clustering for L recursions is performed. In contrast to the conventional algorithm, which ends up with k clusters, a hierarchical cluster scheme with k^L subclusters on the smallest level is computed after L recursive iterations. A tree structure of depth L and branch factor k with a total of $\sum_{j=1}^{L} k^j$ nodes can be built. Each tree node corresponds to a cluster center $\boldsymbol{\mu}$ and each leaf node includes several landmark descriptors across all image entries that are enclosed within the same subcluster on the smallest scale. In other terms, landmarks within the same subcluster are more similar than landmarks within two different subclusters. This means that a topological indexing of landmarks that is independent of their original image sources has been created. A display of this tree structure is given in Fig. 23.2. Accordingly, all images in the database are attached to the leaf nodes as inverted file structure with respect to their corresponding landmarks. Afterwards, the vocabulary tree and inverted file structure are used for the indexing of the images.

Query phase

Given a query image I^q together with its set of landmark descriptors

$$\left\{\mathbf{x}^q\right\}_{i=1}^{m},$$

the vocabulary tree can now be traversed for each descriptor separately. At each level l ($l \in \{1, \dots, L\}$) of the tree, the distances from a query landmark descriptor d to each level node representing cluster centers are computed using the (squared) ℓ_2-norm and the one node that yields the smallest distance is chosen, i.e.:

$$\arg\min_{\boldsymbol{\mu} \in l} \|\boldsymbol{\mu} - \mathbf{x}_d^q\|_2^2 . \tag{23.9}$$

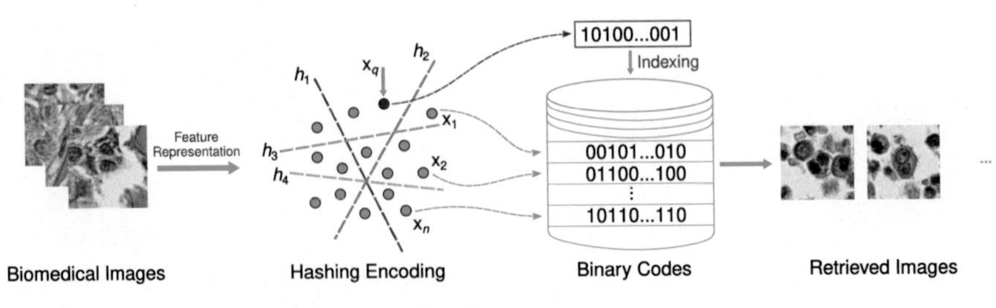

FIGURE 23.3 Hashing-based image retrieval.

This image shows a pipeline for hashing-based image retrieval.

-Adapted from [13]

Repeating this for all tree levels, every query image landmark descriptor reaches a leaf node recording several images relevant to the leaf node. Each of these is then compared to I_q via a dedicated similarity score. By ranking all the similarity scores in descending order, the top-ranked images can be considered as the retrieval results.

Vocabulary trees and their variants have shown great potential for large-scale medical image retrieval. Unlike the brute-force approach that simply compares the similarity of all landmark descriptors in the database to those extracted from the query image, vocabulary tree methods take advantage of the hierarchical tree structure that yields only a total of kL dot products. More importantly, the inverted file strategy can significantly improve the indexing process since it does not need to traverse the whole image database.

Despite this advantage, vocabulary tree-based methods also show several limitations when applied in clinical routine. In particular, the sole usage of local features may not be enough to represent and discriminate specific biomedical findings. Also, when applied to very large image databases such as clinical archives, the creation of a hierarchical vocabulary tree during training is deemed to be very time consuming and heavily depending on the choice of parameters k (number of child clusters) and L (the total number of iterations, respective tree levels).

23.4.2 Hashing

In order to overcome the efficiency problems of vocabulary trees, hashing methods have been intensively investigated for indexing Big Data. Instead of directly searching nearest neighbors from an original dataset, hashing methods first compress the original data into short binary codes based on a defined hashing function. The nearest neighbor search according to equation (23.9) is then more efficiently computed in binary Hamming space rather than in the high-dimensional feature space. The workflow of hashing-based image retrieval is depicted in Fig. 23.3.

Formally, a hash function is defined as

$$y = h(x), \tag{23.10}$$

where y is named the *hash code* and $h(\cdot)$ the *hash function*. When represented with a sequence of N bits, the hash code is

$$\mathbf{y} = \mathcal{H}(x), \tag{23.11}$$

with

$$\mathbf{y} = [y_1\ y_2 \cdots y_N],$$
$$\mathcal{H} = [h_1(x)\ h_2(x) \cdots h_N(x)].$$

For an image I^b with m regions of interest and its corresponding descriptors

$$\left\{\mathbf{x}_i^b\right\}_{i=1}^{m},$$

the feature space of dimension $\mathbb{R}^{d\times m}$ can be split by a set of K hash functions

$$\mathcal{H} = [h_1(x)\ h_2(x) \cdots h_K(x)],$$

each single one encoding one feature vector $\mathbf{x}_i^b$ into one bit of binary hash code $h_k(\mathbf{x}_i^b)$. Hence, the corresponding K bits of binary code of $\mathbf{x}_i^b$ can be denoted as

$$\mathbf{y}_i = \mathcal{H}(\mathbf{x}_i^b) = [h_1(\mathbf{x}_i^b)\ h_2(\mathbf{x}_i^b) \cdots h_K(\mathbf{x}_i^b)].$$

The distance between two such binary encodings $\mathbf{y}_i = \mathcal{H}(\mathbf{x}_i^b)$ and $\mathbf{y}_j = \mathcal{H}(\mathbf{x}_j^b)$ can be computed using the Hamming distance:

$$d_{\mathcal{H}}(\mathbf{y}i, \mathbf{y}j) = |\mathbf{y}i - \mathbf{y}j| = \sum_{k=1}^{K} |h_k(\mathbf{x}_i^b) - h_k(\mathbf{x}_j^b)|. \tag{23.12}$$

Being deemed as an approximation of the true distance between the respective data items, the Hamming distance can be efficiently calculated as the bitwise XOR operation. Hence, conducting an exhaustive search in the Hamming space is significantly faster than doing the same in the original feature space due to its low memory footprint and computational cost [3].

Essential to a good indexing is finding an ideal hash function that provides an optimal binary encoding of the original data while keeping initial similarities and diversity among the data intact. Respective methods are commonly divided into two categories: data-independent and data-dependent hashing, which are presented in detail in the following subsections.

23.4.2.1 Data-independent hashing

Data-independent methods usually design generalized hash functions that can encode any given dataset into binary codes.

Locality-sensitive hashing (LSH) and its variants are the most popular data-independent methods. The idea is that if two points are similar and closely embedded within the original space, then upon randomized linear projections, they will remain close to each other. The hash functions for LSH can be parameterized as

$$h_k(\mathbf{x}) = \text{sgn}(\mathbf{w}_k^T \mathbf{x} + \mathbf{b}_k), \tag{23.13}$$

where $\{\mathbf{w}_k, \mathbf{b}_k\}_{k=1}^K$ are parameters of the hash function representing the projection vector $\mathbf{w}_k$ and the corresponding intercept $\mathbf{b}_k$. LSH is a data-independent method since it randomly generates hash functions regardless of the data distribution. However, this type of method often needs long binary codes and many hash functions to ensure the desired retrieval precision, which dramatically increases the storage costs and the query time.

In summary, since all the previously mentioned hashing methods are developed independently of the training datasets, it is difficult to guarantee the retrieval performance for a particular dataset.

23.4.2.2 Data-dependent hashing

Another category are data-dependent methods (also called learning to hash methods) that learn the hashing functions from a given training dataset. In analogy to machine learning, data-dependent hashing methods are further divided into supervised, unsupervised, and semi-supervised methods based on the existence of labels for the training dataset. Supervised methods employ advanced machine learning techniques such as kernel learning, metric learning, and deep learning to compute the hashing functions from labeled training data. If label information is not available for a given training dataset, unsupervised methods explore the underlying data properties such as distributions and manifold structures to design effective hashing functions.

Another taxonomy of data-dependent methods differentiates between linear and non-linear hashing functions. Linear hashing functions separate and map the original feature space with simple projections (refer to $\{h_1, h_2, \ldots, h_K\}$ in Fig. 23.3). They are computationally efficient and easy to optimize [13]. However, when the training data appear to be linearly inseparable or present only subtle differences among single entities, linear hashing functions will not yield acceptable results. Non-linear hashing methods, in contrast, embed the intrinsic data structure in a high-dimensional space and non-linearly map feature vectors into binary codes. Most non-linear hashing functions are based on learning kernel matrices or manifold structures for the original training dataset. In general, compared with data-independent methods, data-dependent methods can achieve comparable or even better retrieval accuracy with shorter binary codes [13].

Fig. 23.4 shows an overview of a selection of popular hashing methods together with their taxonomy classification. The details of a selection of methods is outlined in the remainder of this section.

Spectral hashing (SH) aims at preserving neighborhoods present in the input space to neighborhoods in the Hamming space. Let $\{\mathbf{y}_n\}_{n=1}^N$ be the hash codes of N data items comprised each of M bits. It is assumed that the inputs are embedded

Taxonomy of data-dependent hashing	supervised	unsupervised
non-linear	Kernelized Hashing	Composite Hashing MIPS Binary Coding Deep Autoencoder
linear		Spectral Hashing Anchor Graph Hashing Hashing Forest

FIGURE 23.4 Taxonomy of data-dependent hashing methods.

Data-dependent hashing methods are divided into supervised and unsupervised methods depending on the availability of training data labels. The underlying hashing functions can be of linear or non-linear nature. For details on these methods, please refer to the text.

in $\mathbb{R}^d$ so that Euclidean distance correlates with similarity. Then s_{ij} denotes the similarity between data item i and j, and $\mathbf{S}$ is the $N \times N$ similarity matrix. Using this notation, the average Hamming distance between similar neighbors can be written as $\sum_{ij} s_{ij} \|\mathbf{y}_i - \mathbf{y}_j\|^2$. This yields the following problem for preserving neighborhoods in Hamming space [14]:

$$\min_{\mathbf{y}_i \in \{-1,1\}^M} \sum_{ij} s_{ij} \|\mathbf{y}_i - \mathbf{y}_j\|^2, \tag{23.14}$$

$$\text{s.t.} \sum_i \mathbf{y}_i = 0, \tag{23.15}$$

$$\frac{1}{n} \sum_i \mathbf{y}_i \mathbf{y}_i^T = \mathbf{I}, \tag{23.16}$$

where the constraint $\sum_i \mathbf{y}_i = 0$ requires each bit to fire 50% of the time and the constraint $\frac{1}{n} \sum_i \mathbf{y}_i \mathbf{y}_i^T = \mathbf{I}$ requires the bits to be uncorrelated. Sadly, solving (23.14) has been proven to be NP-hard [14]. However, by restructuring the hash codes $\{\mathbf{y}_n\}_{n=1}^N$ into an $N \times M$ matrix $\mathbf{Y}$ whose jth row is filled with hash code $\mathbf{y}_j^T$ and introducing a diagonal $N \times N$ matrix $\mathbf{D}$ with entries $d_{nn} = \sum_{i=1}^N s_{ni}$, this can be rewritten as

$$\min_{\mathbf{Y}} \text{Trace}(\mathbf{Y}^T (\mathbf{D} - \mathbf{S}) \mathbf{Y}), \tag{23.17}$$

$$\text{s.t. } \mathbf{Y}^T \mathbf{1} = 0, \tag{23.18}$$

$$\mathbf{Y}^T \mathbf{Y} = \mathbf{I}. \tag{23.19}$$

Note that due to elimination of constraint $\mathbf{Y}(i, j) \in \{-1, 1\}$, a simplified problem is obtained whose solutions are the M eigenvectors of the Laplacian matrix $\mathbf{D} - \mathbf{S}$ with minimal eigenvalue. The final SH algorithm [14] can be seen in Algorithm 23.1.

Result: Spectral hashing
Find the principal components of the N-dimensional data using PCA;
for *each PCA direction* **do**
| Compute M Laplacian eigenfunctions with the smallest eigenvalues;
end
Sort all eigenfunctions in terms of their corresponding eigenvalues
Pick the M eigenfunctions with the smallest eigenvalues
Threshold the eigenfunctions at 0 to binarize into hash codes

Algorithm 23.1: Spectral hashing.

One main challenge of SH, i.e., solving (23.17), is that it assumes a uniform distribution among the original data, which is not very realistic. Possible solutions are the use of *anchor graph hashing* [15], which alleviates the assumption of uniform distributions. Alternatively, *kernelized hashing* (KH) [16] could be employed, which generalizes the data-independent LSH method to handle linearly inseparable data. KH employs kernels to map the original feature vectors into a high-dimensional space and therefore make the linearly inseparable images easier to differentiate.

Similar to deep learning-based feature learning, also the hashing process can be learned unsupervised, e.g., by autoencoders, or supervised, using deep feedforward neural networks. This process can also be combined with feature learning to directly learn a compact binary image representation.

23.5 Exercises

1. A common metric to compare two global descriptors is the dot product of two ℓ_2-normalized descriptors.
   a. Show that this is equivalent to the cosine distance. What are the minimum/maximum values of this distance?
   b. Show that the cosine distance is directly proportional to the Euclidean distance and thus produces the same ranking.
2. NetVLAD [10] is a direct deep learning layer of VLAD encoding. Therefore, the vector to cluster association function Eq. (23.1) needs to be differentiable. How can this be achieved?
3. The triplet loss relies on one single negative sample that is negative to both the anchor sample and the other positive. Typically a mining step ensures that useful negative samples are chosen, e.g., the closest negative sample to the anchor sample is taken. Extend the original loss formulation (s. Eq. (23.8)) such that the closest negative sample to the anchor sample as well as the closest negative sample (different from the first negative one) to the positive sample is taken into account.

References

[1] Lew M.S., Sebe N., Djeraba C., Jain R., Content-based multimedia information retrieval: State of the art and challenges, ACM Transactions on Multimedia Computing, Communications, and Applications (TOMM) 2 (1) (2006) 1–19.

[2] Akgül C.B., Rubin D.L., Napel S., Beaulieu C.F., Greenspan H., Acar B., Content-based image retrieval in radiology: current status and future directions, Journal of Digital Imaging (ISSN 1618-727X) 24 (2) (Apr. 2011) 208–222, https://doi.org/10.1007/s10278-010-9290-9.

[3] Conjeti S., Learning to Hash for Large-Scale Medical Image Retrieval, PhD thesis, Technische Universität München, 2018.

[4] Sivic J., Zisserman A., Video Google: a text retrieval approach to object matching in videos, in: Ninth IEEE International Conference on Computer Vision (ICCV), vol. 2, Nice, Oct. 2003, pp. 1470–1477, https://doi.org/10.1109/ICCV.2003.1238663.

[5] Fei-Fei L., Perona P., A Bayesian hierarchical model for learning natural scene categories, in: 2005 IEEE Computer Society Conference on Computer Vision and Pattern Recognition (CVPR), vol. 2, IEEE, San Diego, ISBN 0-7695-2372-2, June 2005, pp. 524–531, https://doi.org/10.1109/CVPR.2005.16.

[6] Buzo A., Gray A., Gray R., Markel J., Speech coding based upon vector quantization, IEEE Transactions on Acoustics, Speech, and Signal Processing (ISSN 0096-3518) 28 (5) (Oct. 1980) 562–574, https://doi.org/10.1109/TASSP.1980.1163445.

[7] Murray N., Jégou H., Perronnin F., Zisserman A., Interferences in match kernels, IEEE Transactions on Pattern Analysis and Machine Intelligence (ISSN 0162-8828) 39 (9) (Oct. 2016) 1797–1810, https://doi.org/10.1109/TPAMI.2016.2615621, arXiv: 1611.08194.

[8] Jégou H., Perronnin F., Douze M., Sánchez J., Pérez P., Schmid C., Aggregating local image descriptors into compact codes, IEEE Transactions on Pattern Analysis and Machine Intelligence (ISSN 1939-3539) 34 (9) (Sept. 2012) 1704–1716, https://doi.org/10.1109/TPAMI.2011.235.

[9] Sánchez J., Perronnin F., Mensink T., Verbeek J., Image classification with the Fisher vector: theory and practice, International Journal of Computer Vision (ISSN 0920-5691) 105 (3) (June 2013) 222–245, https://doi.org/10.1007/s11263-013-0636-x.

[10] Arandjelović R., Gronat P., Torii A., Pajdla T., Sivic J., NetVLAD: CNN architecture for weakly supervised place recognition, IEEE Transactions on Pattern Analysis and Machine Intelligence 40 (6) (June 2018) 1437–1451, https://doi.org/10.1109/TPAMI.2017.2711011.

[11] Christlein V., Spranger L., Seuret M., Nicolaou A., Král P., Maier A., Deep Generalized Max Pooling, in: 2019 15th IAPR International Conference on Document Analysis and Recognition (ICDAR), Sept. 2019, pp. 1090–1096, https://doi.org/10.1109/ICDAR.2019.00177.

[12] Schroff F., Kalenichenko D., Philbin J., FaceNet: a unified embedding for face recognition and clustering, in: 2015 IEEE Conference on Computer Vision and Pattern Recognition (CVPR), June 2015, pp. 815–823, https://doi.org/10.1109/CVPR.2015.7298682.

[13] Li Z., Zhang X., Müller H., Zhang S., Large-scale retrieval for medical image analytics: A comprehensive review, Medical Image Analysis 43 (2018) 66–84.

[14] Weiss Y., Torralba A., Fergus R., Spectral hashing, in: NIPS, 2008, pp. 1753–1760.

[15] Liu W., Wang J., Kumar S., Chang S., Hashing with graphs, in: ICML, 2011.

[16] Liu W., Wang J., Ji R., Jiang Y.-G., Chang S.-F., Supervised hashing with kernels, in: 2012 IEEE Conference on Computer Vision and Pattern Recognition, June 2012, pp. 2074–2081, https://doi.org/10.1109/CVPR.2012.6247912.

PART

Evaluation in medical image analysis

CHAPTER

Assessment of image computing methods

24

Ipek Oguz[a], **Melissa Martin**[b], **and Russell T. Shinohara**[b]

[a]*Department of Computer Science, Vanderbilt University, Nashville, TN, United States*

[b]*Penn Statistics in Imaging and Visualization Center, Department of Biostatistics, Epidemiology, and Informatics, University of Pennsylvania, Philadelphia, PA, United States*

Learning points

- Task-specific algorithmic performance assessment
- Classification, regression, segmentation, registration
- Experimental design for learning algorithms
- Training, testing, and cross-validation
- Intra-rater and inter-rater variability

24.1 The fundamental methodological concept

Throughout this book, many different algorithms for classification, segmentation, and registration problems are presented. The algorithms presented here are far from an exhaustive list of the current state of the art, and many new methods are becoming available every day. As such, it is evident that there is a need for sound methods of assessing the performance of algorithms in quantitative and objective ways. This is useful both for users who need to choose among existing algorithms for a specific task and for algorithm developers who need to determine whether their new approach surpasses the state of the art. This chapter focuses on the question of how to compare a given set of methods for a medical image analysis task, for classification, segmentation, and registration problems.

24.2 Introduction

In this chapter, we review approaches for assessing the performance of medical image processing tools. In Section 24.3, we begin by introducing notation and describing how methods performing binary and multilabel classification tasks are assessed (Section 24.3.1). We then extend these approaches, which stem from epidemiology, to the case of continuous prediction and regression models in Section 24.3.2. In Section 24.4, we discuss study designs for the assessment of in-sample and out-of-sample

Medical Image Analysis. https://doi.org/10.1016/B978-0-12-813657-7.00039-X

predictive performance and the dangers of overfitting. Segmentation is a special case of classification in which spatial structure is important, and we consider metrics for segmentation performance in Section 24.5. In Section 24.6, we describe approaches for measuring registration performance. We then discuss inter- and intra-rater comparisons for situations in which no gold standard exists in Section 24.7.

24.3 Evaluation for classification tasks

The evaluation of classification and regression methods requires careful consideration of the goals of each analysis conducted. Although several other textbooks focus on the epidemiological aspects of these assessments, in this chapter we focus on an introduction to the most commonly used image analysis aspects of this problem. We begin by discussing binary classification performance assessments, and proceed to describe techniques for assessing regression tasks for classical scenarios. Throughout, we refer to the problem of diagnosis of a disease as a motivating example.

Briefly, we define some notation: suppose our goal is to predict the value of Y_i based on a set of features $\mathbf{X}_i$ measured for each subject $i = 1, \dots, n$. Further suppose that our goal is to assess the performance of an existing classifier ϕ; we postpone the concerns regarding the training of such a classifier to Section 24.4. In the remainder of this section, we focus on measures of the performance of ϕ based on discrepancies between $\phi(\mathbf{X}_i) \in \mathbb{R}$ and Y_i across subjects.

24.3.1 Evaluation for binary classification tasks

The case in which $Y_i \in \{0, 1\}$ is binary corresponds to the classical biostatistical scenario of diagnostic testing which has been studied extensively in the literature [1]. For convenience, and in accordance with this literature, we refer to the outcome $Y_i = 0$ as "non-diseased" and to $Y_i = 1$ as "diseased." Generally, continuous predictions $\phi(\mathbf{X}_i)$ of Y_i are thresholded into two categories corresponding to predictions of $Y_i = 0$ versus $Y_i = 1$ by determining for each subject whether $\phi(\mathbf{X}_i) \geq \tau$ or $\phi(\mathbf{X}_i) < \tau$. There are four possibilities for each subject which are listed in so-called *two-by-two* Table 24.1.

As the diagonal elements of this table indicate favorable performance of the classifier ϕ and the off-diagonal elements indicate unfavorable outcomes, the goal of classification is to maximize the numbers of true positive (TP) and true negative (TN) and minimize the numbers of false positive (FP) and false negative (FN). Some

Table 24.1 Classical epidemiological two-by-two table.

	$Y_i = 1$	$Y_i = 0$
$\phi(\mathbf{X}_i) < \tau$	True positive (TP)	False positive (FP)
$\phi(\mathbf{X}_i) \geq \tau$	False negative (FN)	True negative (TN)

classification methods yield binary values of $\phi(\mathbf{X}_i)$, and for these classifiers the reporting of these four pivotal quantities comprehensively describes the performance of the classifier.

Two key quantities that may be calculated from Table 24.1 are the classifier sensitivity and specificity. The sensitivity (also referred to as the recall or TP rate [TPR]) of the classifier denotes the proportion of subjects classified as diseased ($\phi(\mathbf{X}_i) < \tau$) out of those that truly have the disease ($Y_i = 1$). The specificity, on the other hand, is the proportion of those who do not have the disease ($Y_i = 0$) and are correctly classified as non-diseased ($\phi(\mathbf{X}_i) \geq \tau$). These two quantities are estimates of conditional probabilities of repeated application of the classifier in populations of subjects with $Y_i = 0$ and $Y_i = 1$ separately. This affords these two quantities the important distinction that they measure the performance of the classifier itself – and not that of the classifier's performance on a given population or sample of subjects. This stands in contrast to quantities such as the accuracy of a classifier ($\frac{TP+TN}{n}$, where n is the total number of samples), which depends on both the classifier characteristics and the population on which the classifier's performance is being assessed.

While an ideal classifier demonstrates high sensitivity and specificity, in practice there is often a trade-off between the two. At the extremes, the trivial examples of a classifier that always classifies subjects as non-diseased will yield high specificity and low sensitivity, whereas classifying all subjects as diseased yields perfect sensitivity and no specificity. For classifiers with continuous predictions (including the majority of those described in this textbook), the analyst can tune this trade-off to provide a binary classifier in the choice of τ, the selected threshold. For large τ, the classifier is highly sensitive and has lower specificity, and for small τ vice versa. This flexibility allows for the optimization of a classifier for a particular task, but makes the comparison between classifiers more complex. To address this, receiver operating characteristic (ROC) analysis studies the curve of test performance across thresholds τ in the space of sensitivity and 1-specificity. An example of such a curve is shown in Fig. 24.1 in solid green. Note that an ideal classifier will have a curve that sits squarely to the top left of the plotting area, and a chance classifier (given by a coin toss) corresponds to an ROC curve that lies on the diagonal (dashed) line.

The comparison of two classifiers through their ROC curves for a given dataset can be simple; if one sits higher than the other throughout the range of possible values of specificity, that classifier is clearly superior in performance. However, it is often the case that two classifiers have ambiguous relative performance according to that standard, with each providing superior sensitivity for different levels of specificity (see blue and green curves in Fig. 24.1). To address this, it is customary to report the area under the ROC curve (AUC), which is a single number that summarizes test performance across possible specificity values. It is also common to focus only on values of the specificity which are meaningful; in such cases, the partial AUC (pAUC) is the relative proportion of area under the curve for values of specificity between 1 and a pre-specified value (for example, see the gray-shaded area in Fig. 24.1). In cases where the relative performance of classifiers is more dependent on comparisons be-

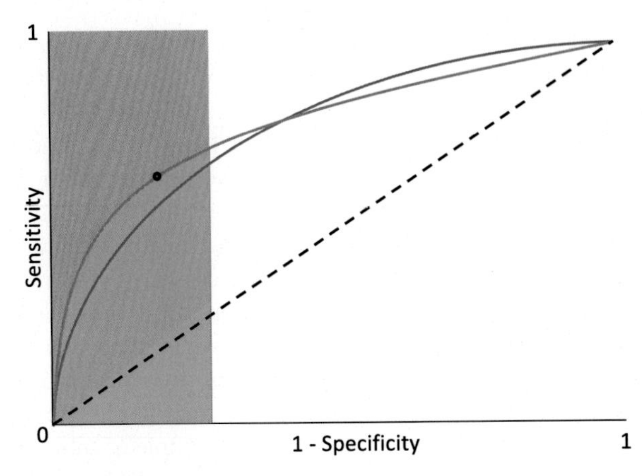

FIGURE 24.1 Two example ROC curves.

This figure shows two ROC curves in solid blue and green lines. Note that neither is superior across the range of specificity. The green area demonstrates the partial AUC within the gray area, and the black circle shows the classifier whose performance is closest to the top left (optimal) corner of the ROC space.

tween binarized classifiers, or in which a single binary classification is necessary for a single classifier, it is common to select the value of τ that corresponds to the maximal value of the sum of sensitivity and specificity, or that for which the ROC curve is closest to the top left corner of the space (see the black circle on the green curve in Fig. 24.1). The assessment of statistical comparison and assessment of uncertainty in ROC curves and AUCs is often conducted in validation datasets using specialized hypothesis tests such as DeLong's test, or using resampling techniques such as the bootstrap method.

Although the sensitivity and specificity of a classifier fully describe its performance, there are often cases in which the interpretation of a classification result is of primary interest. In such cases, the questions "what is the likelihood of disease given a positive test result" and "what is the likelihood of non-disease given a negative test result" are answered by estimating the positive predictive value (PPV, also referred to as precision) and negative predictive value (NPV). These quantities, which may be calculated from the sensitivity and specificity, depend strongly on the prevalence of the disease in the population, which can be estimated by $(TP + FN)/n$. As the trade-off between sensitivity and specificity explored through ROC analysis may be of less interest in cases where primary interest is in the test as applied to a specific population, graphical comparisons of precision–recall curves can provide for a complementary comprehensive exploration of classifier performance.

Similar ideas extend to the evaluation of multiclass classification tasks. For brevity, we omit the details here and refer the reader to the existing literature (e.g., [1]).

24.3.2 Evaluation for continuous prediction tasks

In cases where the outcome to be predicted is a continuous variable, the assessment of performance is slightly more complex. Although it is common to dichotomize (or, in the case of multiclass tasks, stratify) continuous outcomes, especially when there are clinically relevant cutoffs that are of interest, it is also often the goal to assess the performance of continuous predictions of continuous outcomes directly. In such cases, the framework of ROC analysis is not directly applicable and it is generally more common to assess performance using a loss function. More specifically, for a predictive model ϕ based on data $\mathbf{X}_i$ for an outcome $Y_i \in \mathbb{R}$, we define a loss function Ψ such that smaller values of $\Psi\{\phi(\mathbf{X}_i), Y_i\}$ denote more successful predictions, i.e., the goal is a smaller error. The most common example of a loss function is the squared error loss, $\Psi(a,b) = (a-b)^2$. To evaluate a loss function over a sample of data, we generally use a summary statistic of the subject-specific loss values (that is, across i) such as the mean or median.

A popular choice for evaluating continuous predictions is the mean square error (MSE) defined by $n^{-1}\sum_{i=1}^{n}\{\phi(\mathbf{X}_i) - Y_i\}^2$. This criterion has several advantages including mathematical convenience (for example, providing closed-form solutions for the least square regression problem) as well as interpretability. Indeed, it is easy to show that the population-level MSE can be decomposed into two terms: the squared bias $[\mathbb{E}\{\phi(\mathbf{X}_i) - Y_i\}]^2$ and the variance $\mathbb{E}\{\phi(\mathbf{X}_i) - Y_i\}^2$, where $\mathbb{E}$ denotes expectation. Thus, by optimizing the MSE, one is implicitly trading off bias with variance. In cases with some subjects for whom performance may be unusually poor (outliers), the MSE will be driven by these subjects more than others. This can be attributed to two aspects of the loss function: the operation of squaring the errors, which emphasizes large departures of $\phi(\mathbf{X}_i)$ from Y_i, and the averaging of each error across subjects i, which treats each error $\{\phi(\mathbf{X}_i) - Y_i\}^2$ equally, including those with especially large values. To mitigate this, it is common to use the median absolute error, $\text{median}\{|\phi(\mathbf{X}_i) - Y_i|; i = 1, \ldots, n\}$, which is both robust to outliers through the median operation and treats errors on a linear scale. Depending on the particular application under study, emphasizing worst case scenarios (via the MSE) or typical errors (via the median absolute error) may be appropriate. In some cases, reporting the correlation between $\phi(\mathbf{X}_i)$ and Y_i may also be appropriate; however, the particular choice of correlation metric is important, and the interpretation of correlated outcomes and predictions may not be sufficient. Indeed, if $\phi(\mathbf{X}_i)$ and Y_i are correlated this does not mean that $\phi(\mathbf{X}_i)$ is a useful prediction for Y_i, and large values of $\phi(\mathbf{X}_i)$ may have equal values to small values of Y_i due to scale-invariance.

A more complex situation for prediction involves the case of censoring in survival analyses. In particular, for prognostic studies it is often the case that some individuals live beyond the study's follow-up period (Fig. 24.2). For such subjects, we do not observe the outcome; rather, we simply know that the outcomes are larger than some value, say C_i. For example, if all subjects are followed 2 years after imaging, then $C_i = 2$ for all i. The data for each subject i then become the predictor $\mathbf{X_i}$, as well as the censoring indicator $1[Y_i \leq C_i]$ as well as $Z_i = \min(Y_i, C_i)$. This case, which is referred to as right-censoring in the statistical literature, is more difficult to address

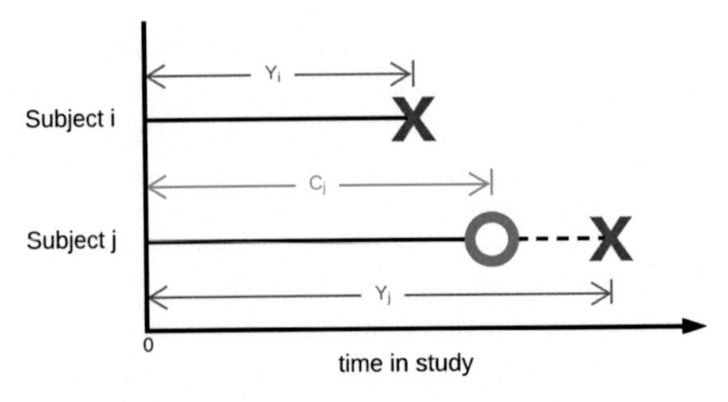

FIGURE 24.2 Survival analysis.

In prognostic studies it is often the case that some subjects (such as subject j in this example) live beyond the study's follow-up period ($Y_j > C_j$).

than the standard continuous outcomes situation. A simplistic option is to report the performance of ϕ for predicting Z_i, but this approach is flawed as it introduces dependence on the censoring mechanism (the distribution of C_i), which is a property of the study design as opposed to the relationship between $\mathbf{X_i}$ and Y_i. Thus, in such cases additional consideration to the goal of prediction and the metric to be used are required.

24.4 Learning and validation

A primary concern with the training and validation of classifiers and segmentation learners is referred to as overfitting. This occurs when the same data are used to estimate the learner (or model) ϕ and to benchmark the performance of $\phi(\mathbf{X}_i)$. As the complexity of ϕ increases (for example, using machine learning or deep learning models with large effective numbers of parameters), the learner has more flexibility to explain the relationship between Y_i and $\mathbf{X}_i$. While this can be an inherent strength, it can also be a weakness. Indeed, a sufficiently flexible classifier is able to explain all variation in Y_i based on $\mathbf{X}_i$ by "memorizing" the relationship between Y_i and $\mathbf{X}_i$, but in doing so may neglect to extract the important patterns in $\mathbf{X}_i$ that provide generalizable knowledge for prediction on future, unseen data. Thus, a trade-off is present between allowing infinite flexibility in learning and imposing lower-dimensional structure. Thus, in determining the performance of these classifiers our goal is to estimate the average loss expected in new subjects which were not used for training.

More formally, suppose that data $\mathbb{X}_i$ from subjects $i = 1, \ldots, n$ (the training dataset) are used to estimate the learner ϕ which results in an estimated $\hat{\phi}$. Depending on the level of flexibility of ϕ, $\hat{\phi}(\mathbb{X}_i)$ may be very close to Y_i; however, for too flexible a classifier, for another subject, say $i = n + 1$, $\hat{\phi}(\mathbb{X}_{n+1})$ may be very far from Y_{n+1}. Thus, as our goal is to assess performance of the fitted learner $\hat{\phi}$ in more general settings, we aim to estimate and compare $\mathbb{E}\Psi\{\hat{\phi}(\mathbf{X}_i), Y_i\}$, where $\mathbb{E}$ denotes the expectation operator for the population of subjects, not the sample in which $\hat{\phi}$ was learned. Note that for some loss functions, for example the median absolute error, the expectation operator $\mathbb{E}$ may be replaced by another operator that aggregates subject-level errors.

In assessing prediction error, there are two major criteria to consider: internal validity and generalizability. *Internal validity* involves the performance of the classifier on data that are similar to those used for training in every aspect. Indeed, these would be subjects imaged with the same scanner and protocol, and from the same population as the training dataset. Measures of internal validity can be assessed using a single sample of subjects if sufficiently large. To accomplish internal validation, two related techniques are most common: split-sample analyses and cross-validation. Split-sample experiments are as simple as dividing the subjects in the sample into two datasets which we refer to as the training ($\mathcal{T}$) and validation ($\mathcal{V}$) datasets. Generally, this splitting is conducted either at random or in a stratified fashion to ensure similarity between them. The learner is then fit using data from $\mathcal{T}$ to yield $\hat{\phi}$, and then applied to $\mathcal{V}$ to generate predictions $\{\hat{\phi}(X); X \in \mathcal{V}\}$. We then report the performance of $|\mathcal{V}|^{-1} \sum_{i \in \mathcal{V}} \Psi\{\hat{\phi}(\mathbf{X}_i), Y_i\}$.

While split-sample experiments are intuitive, they are susceptible to the issue that particular subjects being in either the training or validation sets could yield unusual observations of performance estimates; for example, if by chance outliers are present in $\mathcal{T}$, then the performance on $\mathcal{V}$ may be uncharacteristically poor. However, in such a case if the assignments of $\mathcal{T}$ and $\mathcal{V}$ were flipped, the performance may be much better. To deal with this issue, cross-validation may be employed. In its simplest form, so-called two-fold cross-validation, an assignment of $\mathcal{T}$ and $\mathcal{V}$ is constructed, $|\mathcal{V}|^{-1} \sum_{i \in \mathcal{V}} \Psi\{\hat{\phi}(\mathbf{X}_i), Y_i\}$ is calculated, and then the roles of $\mathcal{T}$ and $\mathcal{V}$ are reversed; $\mathcal{V}$ is used for training the model $\hat{\phi}^*$, and $\mathcal{T}$ is used for calculating $|\mathcal{T}|^{-1} \sum_{i \in \mathcal{T}} \Psi\{\hat{\phi}^*(\mathbf{X}_i), Y_i\}$. We then calculate the average of the two errors, i.e.,

$$\left[|\mathcal{V}|^{-1} \sum_{i \in \mathcal{V}} \Psi\{\hat{\phi}(\mathbf{X}_i), Y_i\} + |\mathcal{T}|^{-1} \sum_{i \in \mathcal{T}} \Psi\{\hat{\phi}^*(\mathbf{X}_i), Y_i\} \right] /2, \tag{24.1}$$

and report this as the cross-validated error.

While two-fold cross-validation is intuitive, it describes the performance of ϕ when fit using only half of the data – which may be much inferior to the performance of ϕ when fit on the whole dataset. To address this, K-fold cross-validation splits the dataset into K equally sized subsets $(\mathcal{S}_1, \ldots, \mathcal{S}_K)$ and repeatedly uses $K - 1$ of these subsets $(\mathcal{S}_1, \ldots, \mathcal{S}_{j-1}, \mathcal{S}_{j+1}, \ldots, \mathcal{S}_K)$ to train $\hat{\phi}^{(j)}$ and the remaining subset $(\mathcal{S}_j)$ to estimate $(K)/n \sum_{i \in \mathcal{S}_j} \Psi\{\hat{\phi}^{(j)}(\mathbf{X}_i), Y_i\}$. These are then averaged and this average error is reported. As K grows towards n, the computational complexity of conducting the cross-validation increases. On the other hand, the approximation to the performance of ϕ when trained on datasets of size n also increases; this can be especially important for smaller sample sizes. When $K = n$, the cross-validation

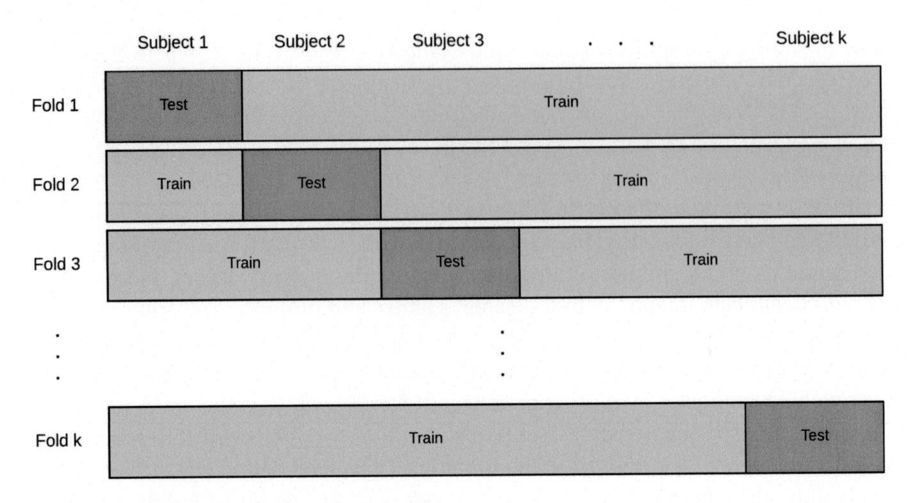

FIGURE 24.3 Leave-one-out cross-validation.

Leave-one-out cross-validation scheme. At the limit of the K-fold cross-validation, in leave-one-out cross-validation, each fold consists of a single subject X_i. The model is trained on all the other subjects and tested on X_i, and the error is averaged over the n folds.

scheme is known as leave-one-out cross-validation (Fig. 24.3), which is often considered to be optimal for assessing internal validity. However, since this requires training n models, the computational requirements can become prohibitive for large n.

While K-fold cross-validation can provide measures of out-of-sample performance for assessing internal validity, reporting uncertainty in these measures is often also key. To accomplish this, bootstrapping can be helpful. In particular, by repeatedly resampling training and validation sets, confidence intervals for performance measures may be estimated. The classical non-parametric bootstrap method accomplishes this, but at potentially great computational cost; more modern and computationally efficient bootstrapping procedures have also been proposed.

Beyond internal validity, the second important consideration is *generalizability*. This refers to the performance of the learner on data that come from different samples from it was trained on. An example of good generalizability is high performance on data acquired on a different imaging apparatus than the learner was trained on. To assess generalizability, there is usually an expectation that data from another study are available for analysis. Multicenter studies are often used for assessing generalizability; this can be conducted by training learners on data acquired at one study site and validating a learner on another study site. This idea has also been extended to leave-one-site-out cross-validation, which refers to the special case where K corresponds to the number of imaging sites, and the partitioning for cross-validation is by study site (Fig. 24.4). Multicenter studies are desirable for assessing generalizability as they also tend to include subjects with different demographics and disease severity at different study sites, and thus excellent performance on validation datasets implies

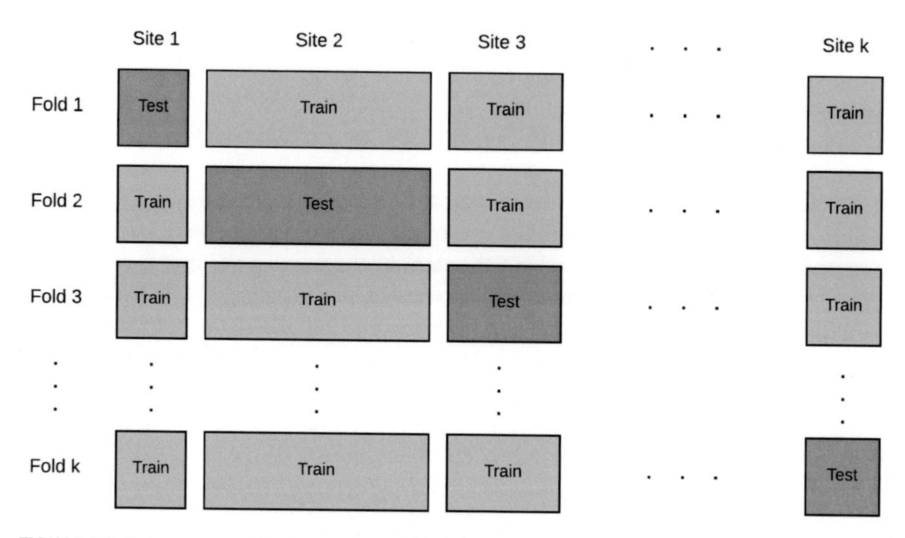

FIGURE 24.4 Leave-one-site-out cross-validation.

Leave-one-site-out cross-validation scheme. In multicenter imaging studies, generalizability of the trained model can be assessed by testing the model on data from a study site different than the study site where the training data were acquired. The leave-one-out cross-validation scheme can be applied to this scenario as a leave-one-site-out cross-validation.

generalizability to new image acquisition protocols as well as different populations of interest.

24.5 Evaluation for segmentation tasks

For segmentation tasks, the "true" segmentation is often impossible to obtain, since even highly trained experts can disagree with each other and/or with themselves on repeated tests. Nevertheless, for evaluation purposes, we need to define a baseline to compare against. This can be, for example, a manual segmentation or a combination of manual segmentations from different experts. Nevertheless, we refrain from referring to this baseline as "true" segmentation or "ground truth" and instead call it the independent gold standard or the reference standard. It is important that the baseline should be independent from the evaluated method. For example, using automated segmentation results as a starting point and manually correcting the errors would not be considered an independent standard, as it would be heavily biased towards agreement with the method being evaluated.

Given an independent gold standard segmentation X and a segmentation result Y that we are trying to evaluate, we can define TP, TN, FP, and FN similarly to their counterparts in binary classification (Section 24.3.1), since binary segmentation can be seen as a binary classification (foreground vs. background) at each voxel. Unlike generic classification tasks, segmentation also assigns spatial meaning to each classified sample; as such, we can think of these measures as the volume (or area in 2D images) of overlapping regions. For instance, we can define $\mathrm{TP} = |X \cap Y|$ and $\mathrm{TN} = |\bar{X} \cap \bar{Y}|$.

These numbers form the confusion matrix for the assessment. Since TP, TN, FP, and FN are directly correlated with the size of the target objects, a more standardized assessment can be achieved by taking the ratio of the overlap volume to the whole object volume, rather than absolute overlap volume. For instance, the TPR is defined as $\mathrm{TPR} = \frac{\mathrm{TP}}{|X|} = \frac{|X \cap Y|}{|X|}$. The TN, FP, and FN rates are similarly defined. Precision and accuracy are also often used in practice for evaluating segmentation performance.

A common way to summarize these measures into a single summary measure is to use the Dice–Sorensen coefficient (DSC), also referred to as the Dice score, the F1 score, or the Zijdenbos similarity index. The Dice score is defined as

$$\mathrm{DSC} = \frac{2\mathrm{TP}}{2\mathrm{TP} + \mathrm{FP} + \mathrm{FN}}. \tag{24.2}$$

Rearranging terms, this can be written more intuitively as the ratio of the overlap volume to the mean target volume:

$$\mathrm{DSC} = \frac{|X \cap Y|}{\frac{1}{2}(|X| + |Y|)} = \frac{2|X \cap Y|}{|X| + |Y|}. \tag{24.3}$$

Note that DSC has values in the range [0, 1]. When X and Y are non-overlapping, DSC is 0. For identical X and Y, DSC becomes 1. Within this range, higher values of DSC indicate closer agreement between the two segmentations X and Y.

Another common summary measure is the Jaccard coefficient (JC), which is defined as the ratio of the overlap to union:

$$\mathrm{JC} = \frac{|X \cap Y|}{|X \cup Y|}. \tag{24.4}$$

Like the DSC, the JC also has values in the range [0..1]. However, unlike DSC, the JC only counts the TPs once in both the numerator and denominator.

Finally, the volume similarity (VS) measure only considers the size of the segmentations, regardless of their overlap. It is defined as follows:

$$\mathrm{VS} = \frac{|X| - |Y|}{\frac{1}{2}(|X| + |Y|)}. \tag{24.5}$$

VS may be appropriate for situations where the segmentation is an intermediate step in a study of object size, such as overall lesion load in multiple sclerosis patients. However, note that two segmentations X and Y that are completely disjoint can still perform well on the VS metric if their (disjoint) sizes are similar.

We note that these volumetric overlap measures can be extended to multilabel segmentation tasks.

Volumetric overlap measures can sometimes be ineffective indicators of segmentation quality. For example, thin, elongated portions, such as the tail of the caudate nucleus of the brain, do not contribute much to the overall volume of the whole object. This means a segmentation method could largely mis-segment such thin features and still achieve a high Dice (or Jaccard) score. Similarly, a segmentation method could produce FP portions, but if these are thin enough, this may be largely unpenalized by the Dice score and other volumetric overlap measures.

Another example is very small segmentation targets, where the stability of measures such as the Dice score is compromised. Recall that the Dice score measures the overlap as a ratio of the target volume. If the target volume is very small, such as a small lesion, changing a single voxel in the segmentation can make a huge impact on the Dice score.

One potential solution to these shortcomings is using surface-based measures. Given an independent gold standard segmentation X and a segmentation result Y that we are trying to evaluate, we can represent X and Y as discretized surfaces X_{surf} and Y_{surf}, such as discrete curves in 2D or triangle meshes in 3D. Then, for each point p_i in X_{surf}, we can compute the distance d_i from p_i to the nearest point in Y_{surf}; d_i can be considered a measure of local error in the segmentation. Then, we can report summary statistics (such as mean and standard deviation) on all d_i along the entire surface. While the mean distance provides a measure of bulk error, the maximum error can provide insight into the lower bound of performance as well as identify the location(s) of poor performance. These distances can also be reported individually (e.g., in the form of a color overlay on a surface rendering) to provide insight into the spatial distribution of the segmentation error. Note that while for volumetric overlap measures such as Dice, a high value indicates a good segmentation, here, a low distance indicates a good segmentation.

Since the distance is computed between two discrete surfaces, the discretization of the surface can be a factor in the distance measurement. For instance, the mean distance from surface X to surface Y is typically not equal to the mean distance from surface Y to surface X (Fig. 24.5). It is thus good practice to consider both directions, or at least to report which direction has been used. A popular distance measure, the Hausdorff distance, computes the maximum of minimum distances from X to Y and the maximum of minimum distances from Y to X, and takes the maximum of these two, thus providing a lower bound on segmentation accuracy.

One important concern is how these segmentation surfaces are obtained. Some surface-based segmentation algorithms will directly produce a surface-based representation. Others will produce a volumetric segmentation (e.g., a binary image), which can be transformed into a surface representation using an algorithm like Marching Cubes [2]. Since the discrete representation of the surfaces directly affects the distance measurements, it is important to report how these surfaces were obtained. For example, the distance between two segmentations can be substantially different dependent on whether the surfaces were smoothed and how much. This is

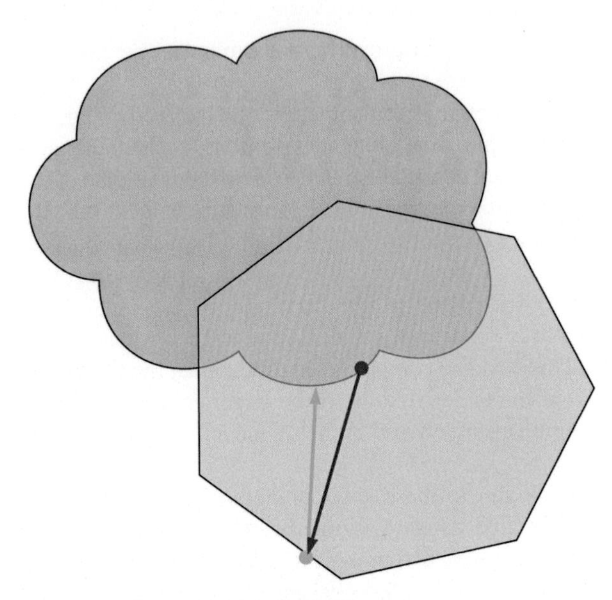

FIGURE 24.5 Surface-based segmentation evaluation.

Surface-based segmentation evaluation. Given two surfaces representing the independent gold standard segmentation and a segmentation result, the distance between the surfaces can be measured to assess segmentation accuracy. Note that distance computation is not necessarily symmetric. For example, the green point is the closest point on the green surface to the red point; however, the closest point to the green point on the red surface is not the red point.

especially important for smaller objects, as well as objects with thin features that may disappear with aggressive smoothing.

Another important concern is whether to use signed or unsigned distances. In the intuitive sense, distance does not have a sign and is always considered positive. This provides a measure of bulk error. However, sometimes it is helpful to differentiate between one surface being inside or outside another surface, rather than simply measuring the distance. For example, when measuring the tissue thickness, if the automated segmentation surface is inside the true surface, the thickness will be underestimated, whereas if the automated segmentation surface is outside the true surface, the thickness will be overestimated. The unsigned and signed distances can thus be thought of as analogous to error and bias, respectively. A good segmentation will have a low mean (or max) unsigned distance to the true surface, and the mean signed distance will be close to zero.

By this point, it should be clear that there is a large variety of measures that can be used to evaluate different aspects of segmentation performance. In some cases, such as in the context of a Grand Challenge comparing many algorithms, it may be helpful to use a summary score as a shorthand. However, no single measure can provide a comprehensive assessment of the data, and thus it is important for studies to report multiple measures to provide a more thorough evaluation of algorithm performance.

24.6 Evaluation of registration tasks

As discussed in Chapters 13 and 14, registration is a common task in medical image analysis. One of the most straightforward evaluations for registration success is through the same image similarity metrics that are being optimized in the registration process itself. For example, given two images I_{fixed} and I_{moving}, the registration may seek to find a transformation that minimizes a similarity metric such as the sum of squared differences between the images, i.e., $T_{optimal} = \operatorname{argmin}_T \int_D (I_{fixed} - T \circ I_{moving})^2$, where T is the transformation and D is the image domain. Then, the registration performance can be assessed based on the optimal value achieved by the optimizer, i.e., $\int_D (I_{fixed} - T_{optimal} \circ I_{moving})^2$. The same process can be used with different registration metrics, such as mutual information or normalized cross-correlation. In principle, these measurements can provide an indication of how well the optimizer has worked: for example, if the optimizer is stuck in a local minimum far from the globally optimal solution, a very poor value of the cost function can help diagnose this problem.

However, using the cost function as the sole measure of registration success is ill-advised, as this would inappropriately focus on just the voxel intensities and ignore the physical plausibility of the transformation itself. This can be abused with an overly simplistic model that seeks to simply shuffle voxel intensities in the image without regard to transformation plausibility [3]. In contrast, registration assessment should also incorporate measures of semantic success of the registration. These can include overlap measures (such as Dice score) of manually defined regions of interest in the fixed image and in the registered moving image. Similarly, distance between manually defined landmarks on the fixed and the registered moving image can be used as a way to assess registration performance (Fig. 24.6).

Surface-based registration or point correspondence methods are often even more difficult to evaluate directly, because of the difficulty of obtaining the ground truth alignment of surface models. Distance to manually defined landmarks can be used. However, these are typically sparse compared to densely sampled surface representations, and models can be overfitted to a small number of landmarks. It is therefore good practice to also consider indirect measures of performance by evaluating statistical shape models implied by the correspondence, such as the generalization, specificity, and compactness measures [4]. The generalization refers to the ability of the statistical shape model to represent unseen shapes, and can be estimated in leave-one-out or cross-validation experiments. The specificity of the shape model can be estimated by generating new random synthetic shapes from the implied shape distribution and computing the distance between these synthetic shapes and real shapes in the dataset. The compactness measure favors models that require fewer parameters

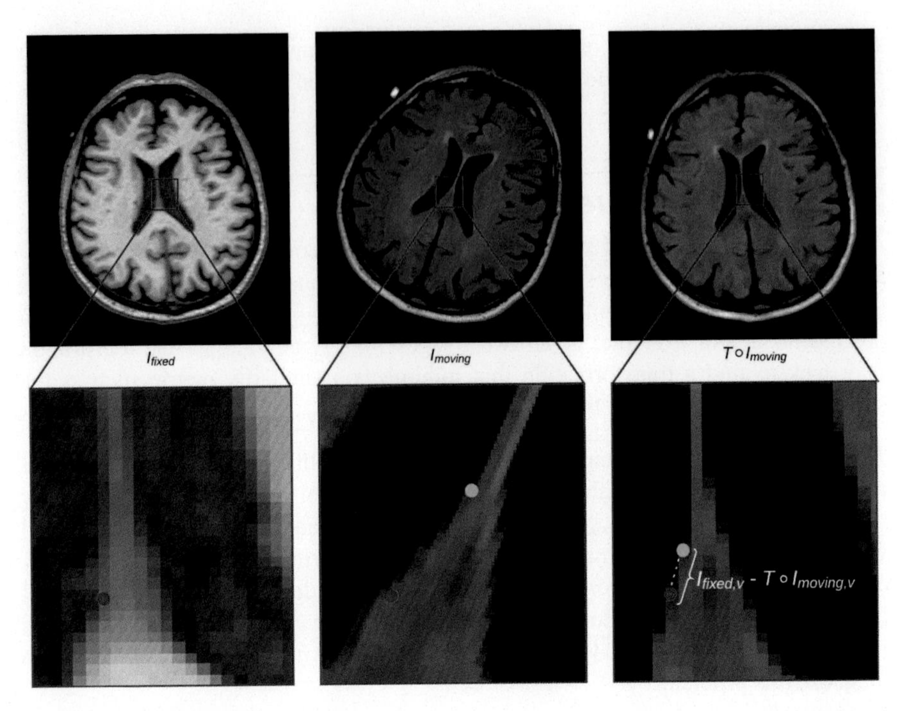

FIGURE 24.6 Registration evaluation.

This figure illustrates the use of landmarks for evaluating registration accuracy. The fixed image and moving image both have landmark annotations, illustrated by the red and green dots, respectively. After registration, the resulting transformation can be applied to both the moving image and to its associated landmarks. This should effectively move the green point closer to the red point than the initial configuration prior to the registration. The distance between the fixed landmarks and the registered moving landmarks can be used as a measure of registration performance.

to represent a shape. These measures of statistical shape model quality allow the assessment of underlying surface correspondence methods and can be complementary to accuracy measures, such as distance to a set of landmarks as discussed above.

24.7 Intra-rater and inter-rater comparisons

A common scenario for classification and segmentation tasks involves uncertainty in the assessment of the outcome. In such cases, oftentimes experts might disagree as to the outcomes; for example, two radiologists might provide different delineations of white matter lesions. To compare these assessments, it is important to extract information about what the raters agree on, and where they disagree. Note that,

Table 24.2 ICC and Kappa interpretation guidelines based on recommendations from Cicchetti et al. [5].

Value	Interpretation
<0.4	Poor
0.4–0.59	Fair
0.6–0.74	Good
0.75–1	Excellent

while somewhat related, this scenario differs from that described in Section 24.5 on segmentation performance assessment as in this case neither of the assessments represents the "ground truth" or the independent gold standard on its own.

Quantitative analyses of intra- and inter-rater assessments have a long history in statistics. The most common measures for assessing agreement between two continuous measurements, say $Y_i^{(1)}$ and $Y_i^{(2)}$ for subjects $i = 1, \ldots, n$, involve correlational analysis. The most common forms of correlation coefficients, Pearson ($\rho(Y_i^{(1)}, Y_i^{(2)})$) and Spearman correlations, both extract information about the common features across assessments. While Pearson's correlation coefficient ρ is susceptible to assumptions concerning linearity, Spearman's correlation coefficients are more robust. A quantity related to ρ that extends naturally to the case of $j = 1, \ldots, m$ raters is referred to as the intra-class correlation coefficient (ICC). ICC is easily interpretable in the context of assessments of the same quantity. Several versions of the ICC have been proposed, but the most common measure is easily described in the context of a mixed effect model,

$$Y_i^{(j)} = a_i + \epsilon_{ij}, \tag{24.6}$$

where a_i is a random effect and ϵ_{ij} is the residual with mean $\mu = \mathbb{E}(Y_i^{(j)})$. We define the ICC to be

$$\text{ICC} = \frac{\text{Var}\,(a_i)}{\text{Var}\,(a_i) + \text{Var}\,(\epsilon_{ij})}, \tag{24.7}$$

which can be interpreted as the proportion of variation explained by inter-subject differences as opposed to inter-rater differences. Thus, ICC values closer to 0 denote more pronounced inter-rater differences, and ICC values closer to 1 indicate better agreement between raters. There are several sets of guidelines as to which values of ICC denote various levels of agreement; those of Cicchetti et al. [5] have been most broadly adopted and are shown in Table 24.2.

Although correlational analyses provide insight into the degree of agreement across raters, they do not explore potential biases between assessments. A common bias that may exist between two assessments, say $Y_i^{(1)}$ and $Y_i^{(2)}$, is that larger values of predictions may be associated with greater (or lesser) departures between the measurements. For example, while the disagreement between two raters might be small for typically sized structures, very large or very small structures might bring out more discrepancy between the two raters' annotations. To address this, Bland–Altman analyses may be performed by calculating $(Y_i^{(1)} + Y_i^{(2)})/2$ and $Y_i^{(1)} - Y_i^{(2)}$ and comparing these quantities. The Bland–Altman plot (also known as a Tukey mean difference plot) includes these quantities on the horizontal and vertical axes, respectively. The plot also displays the limits of agreement, defined as the average difference between the quantities, plus or minus 1.96 times the standard deviation thereof. Measurements are referred to as agreeing well if all the data points fall within these limits of agreement.

For the case of categorical predictions, the ICC and Bland–Altman analyses do not apply directly. Instead, it is common practice to compare the degree to which two raters agree with that which would be expected by chance. More formally, for $Y_i^{(1)}, Y_i^{(2)}, \ldots, Y_i^{(m)}$, each having potential values $1, \ldots, l$, we denote the proportion of agreeing assessments by p and the proportion of assessments expected to agree by chance by p_0. We then define the kappa coefficient,

$$\kappa = \frac{p - p_0}{1 - p_0} = 1 - \frac{1 - p}{1 - p_0}. \tag{24.8}$$

As agreement increases, we note that κ approaches 1; as agreement decreases, κ declines to 0. Guidelines for clinical relevance of κ are similar to those used for ICC, and are provided in Table 24.2.

24.8 Recapitulation

This chapter provided an overview of quantitative assessment methods for medical image analysis algorithms. Specifically, assessments of classification, regression, segmentation, and registration tasks were discussed, as well as more general concerns for experimental design, dangers of overfitting, and creation of a "ground truth" against which the algorithm performance can be compared. An overarching theme has been that relying on a single measure for assessing performance is often too simplistic and can lead to undesired biases in assessment. Rather, a carefully selected set of measures should be used together for a more thorough performance assessment.

24.9 Exercises

Consider two concentric disks, as shown in Fig. 24.7. Disk A has a fixed radius $R = 100$, while disk B has an unknown radius r.

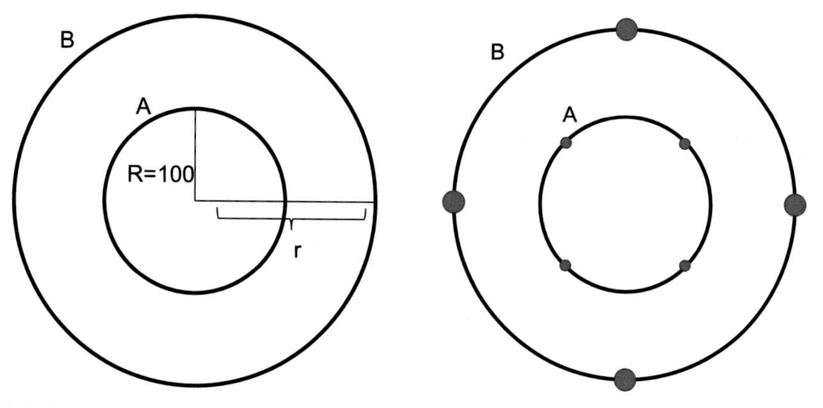

FIGURE 24.7 Exercises.

Left: Two concentric disks A (radius $R = 100$) and B (unknown radius r). Right: Sparse discretization with only four points per surface.

1. For values of r in the range $r \in [1, 1000]$, compute the Dice score between the two disks and plot it as a function of r.
   Does your answer change if the disks are represented in an image of low resolution?
2. For values of r in the range $r \in [1, 1000]$, compute the Hausdorff distance between the two surfaces and plot it as a function of r. Assume each surface is densely discretized, i.e., for each point p_A on A, there is a point p_B on B such that p_A, p_B, and the center of the two disks are collinear.
   Does your answer change if the surfaces are sparsely discretized by only four points equally spaced along the surface and the point sets for A and B are at a 45 degree offset from each other (Fig. 24.7, right panel)?
3. Given your answers to the above, consider which metric is more suitable in each of the following tasks:
   a. Evaluating the segmentation of a large object, such as the whole brain.
   b. Evaluating the segmentation of a small object, such as a blood vessel.
   c. Evaluating the segmentation of an object of unknown size, such as a lesion or tumor.
   d. Evaluating the segmentation of an object with highly convoluted boundary, such as the cortical surface of the human brain.

References

[1] Pepe M.S., The Statistical Evaluation of Medical Tests for Classification and Prediction, Oxford Statistical Science Series, Oxford University Press, ISBN 9780198565826, 2004.

[2] Lorensen W.E., Cline H.E., Marching cubes: a high resolution 3D surface construction algorithm, ACM SIGGRAPH Computer Graphics (ISSN 0097-8930) 21 (4) (Aug. 1987) 163–169, https://doi.org/10.1145/37402.37422.
[3] Rohlfing T., Image similarity and tissue overlaps as surrogates for image registration accuracy: widely used but unreliable, IEEE Transactions on Medical Imaging 31 (2) (Feb. 2012) 153–163.
[4] Styner M.A., Rajamani K.T., Nolte L.P., Zsemlye G., Székely G., Taylor C.J., Davies R.H., Evaluation of 3D correspondence methods for model building, Information Processing in Medical Imaging (IPMI) 18 (July 2003) 63–75.
[5] Cicchetti D.V., Guidelines, criteria, and rules of thumb for evaluating normed and standardized assessment instruments in psychology, Psychological Assessment 6 (4) (1994) 284.

Index

D

E

F

G

H

I

J

K

L

M

P

Q

R

S

T

Printed in the United States
by Baker & Taylor Publisher Services